Systems & Control: Foundations & Applications

Contents

Preface

In the analysis and synthesis of contemporary systems, engineers and scientists are frequently confronted with increasingly complex models that may simultaneously include components whose states evolve along continuous time (continuous dynamics) and discrete instants (discrete dynamics); components whose descriptions may exhibit hysteresis nonlinearities, time lags or transportation delays, lumped parameters, spatially distributed parameters, uncertainties in the parameters, and the like; and components that cannot be described by the usual classical equations (ordinary differential equations, difference equations, functional differential equations, partial differential equations, and Volterra integrodifferential equations), as in the case of discrete-event systems, logic commands, Petri nets, and the like. The qualitative analysis of systems of this type may require results for finite-dimensional systems as well as infinite-dimensional systems; continuous-time systems as well as discrete-time systems; continuous continuous-time systems as well as discontinuous continuous-time systems (DDS); and hybrid systems involving a mixture of continuous and discrete dynamics.

Presently, there are no books on stability theory that are suitable to serve as a single source for the analysis of system models of the type described above. Most existing engineering texts on stability theory address finite-dimensional systems described by ordinary differential equations, and discrete-time systems are frequently treated as analogous afterthoughts, or are relegated to books on sampled-data control systems. On the other hand, books on the stability theory of infinite-dimensional dynamical systems usually focus on specific classes of systems (determined, e.g., by functional differential equations, partial differential equations, and so forth). Finally, the literature on the stability theory of discontinuous dynamical systems (DDS) is presently scattered throughout journals and conference proceedings. Consequently, to become reasonably proficient in the stability analysis of contemporary dynamical systems of the type described above may require considerable investment of time. The present book aims to fill this void. To accomplish this, the book addresses four general areas: the representation and modeling of a variety of dynamical systems of the type described above; the presentation of the Lyapunov and Lagrange stability theory for dynamical systems defined on general metric spaces; the specialization of this stability theory to finite-dimensional dynamical systems; and the specialization of this stability theory to infinite-dimensional dynamical systems. Throughout the book, the applicability of the developed theory is demonstrated by means of numerous specific examples and applications to important classes of systems.

We first develop the Lyapunov and Lagrange stability results for general dynamical systems defined on metric spaces. Next, we present corresponding results for finite-dimensional dynamical systems and infinite-dimensional dynamical systems. Our presentation is very efficient, because in many cases the stability and boundedness results of finite-dimensional and infinite-dimensional dynamical systems are direct consequences of the corresponding stability and boundedness results of general dynamical systems defined on metric spaces.

In developing the subject at hand, we first present stability and boundedness results that are simultaneously applicable to discontinuous dynamical systems as well as continuous dynamical systems. (We refer to these in the following simply as "DDS results.") Because every discrete-time dynamical system can be associated with a DDS with identical stability and boundedness properties, the DDS results are also applicable to discrete-time dynamical systems. Accordingly, the DDS results constitute a unifying Lyapunov and Lagrange stability theory for continuous dynamical systems, discrete-time dynamical systems, and discontinuous dynamical systems. We further show that when the hypotheses of the *classical* Lyapunov stability and Lagrange stability results are satisfied, then the hypotheses of the corresponding DDS stability and boundedness results are also satisfied. This approach enables us to establish the classical Lyapunov and Lagrange stability results for continuous dynamical systems and discrete-time dynamical systems in an efficient manner. This also shows that the DDS results are, in general, less conservative than the corresponding classical Lyapunov and Lagrange stability results for continuous dynamical systems and discrete-time dynamical systems.

The book is suitable for a formal graduate course in stability theory of dynamical systems or for self-study by researchers and practitioners with an interest in systems theory in the following areas: all engineering disciplines, computer science, physics, chemistry, life sciences, and economics. It is assumed that the reader of this book has some background in linear algebra, analysis, and ordinary differential equations.

The authors are indebted to Tom Grasso, Birkhäuser's Computational Sciences and Engineering Editor, for the consideration, support, and professionalism that he rendered during the preparation and production of this book. The authors would also like to thank their families for their understanding during the writing of this book.

Summer 2007

Anthony N. Michel
Ling Hou
Derong Liu

Chapter 1

Introduction

In this book we present important results from the Lyapunov and Lagrange stability theory of dynamical systems. Our approach is sufficiently general to be applicable to finite- as well as infinite-dimensional dynamical systems whose motions may evolve along a continuum (continuous-time dynamical systems), discrete-time (discrete-time dynamical systems), and in some cases, a mixture of these (hybrid dynamical systems). In the case of continuous-time dynamical systems, we consider motions that are continuous with respect to time (continuous dynamical systems) and motions that allow discontinuities in time (discontinuous dynamical systems). The behavior of the dynamical systems that we consider may be described by various types of (differential) equations encountered in the physical sciences and the engineering disciplines, or they may defy descriptions by equations of this type. In the present chapter, we summarize the aims and scope of this book.

1.1 Dynamical Systems

A *dynamical system* is a four-tuple $\{T, X, A, S\}$ where T denotes *time set*, X is the *state-space* (a metric space with metric d), A is the *set of initial states*, and S denotes a *family of motions*. When $T = \mathbb{R}^+ = [0, \infty)$, we speak of a *continuous-time dynamical system*; and when $T = \mathbb{N} = \{0, 1, 2, 3, \dots\}$, we speak of a *discrete-time dynamical system*. For any *motion* $x(\cdot, x_0, t_0) \in S$, we have $x(t_0, x_0, t_0) = x_0 \in A \subset X$ and $x(t, x_0, t_0) \in X$ for all $t \in [t_0, t_1) \cap T$, $t_1 > t_0$, where t_1 may be finite or infinite. The set of motions S is obtained by varying (t_0, x_0) over $(T \times A)$. A dynamical system is said to be *autonomous*, if every $x(\cdot, x_0, t_0) \in S$ is defined on $T \cap [t_0, \infty)$ and if for each $x(\cdot, x_0, t_0) \in S$ and for each τ such that $t_0 + \tau \in T$, there exists a motion $x(\cdot, x_0, t_0 + \tau) \in S$ such that $x(t + \tau, x_0, t_0 + \tau) = x(t, x_0, t_0)$ for all t and τ satisfying $t + \tau \in T$.

A set $M \subset A$ is said to be *invariant* with respect to the set of motions S if $x_0 \in M$ implies that $x(t, x_0, t_0) \in M$ for all $t \geq t_0$, for all $t_0 \in T$, and for all $x(\cdot, x_0, t_0) \in S$. A point $p \in X$ is called an *equilibrium* for the dynamical system $\{T, X, A, S\}$ if the singleton $\{p\}$ is an invariant set with respect to the motions S. The term *stability* (more specifically, *Lyapunov stability*) usually refers to the qualitative behavior of motions relative to an invariant set (resp., an equilibrium) whereas the term *boundedness* (more specifically, *Lagrange stability*) refers to the (global) boundedness properties of the motions of a dynamical system. Of the many different types of Lyapunov stability that have been considered in the literature, perhaps the most important ones include *stability*, *uniform stability*, *asymptotic stability*, *uniform asymptotic stability*, *exponential stability*, *asymptotic stability in the large*, *uniform asymptotic stability in the large*, *exponential stability in the large*, *instability*, and *complete instability*. The most important Lagrange stability types include *boundedness*, *uniform boundedness*, and *uniform ultimate boundedness* of motions.

Classification of dynamical systems

When the state-space X is a finite-dimensional normed linear space, we speak of *finite-dimensional dynamical systems*, and otherwise, of *infinite-dimensional dynamical systems*. Also, when all motions of a continuous-time dynamical system are continuous with respect to time t, we speak of a *continuous dynamical system* and when one or more of the motions are not continuous with respect to t, we speak of a *discontinuous dynamical system* (DDS).

Continuous-time finite-dimensional dynamical systems may be determined, for example, by the solutions of ordinary differential equations and ordinary differential inequalities. These arise in a multitude of areas in science and engineering, including mechanics, circuit theory, power and energy systems, chemical processes, feedback control systems, certain classes of artificial neural networks, socioeconomic systems, and so forth. *Discrete-time finite-dimensional dynamical systems* may be determined, for example, by the solutions of ordinary difference equations and inequalities. These arise primarily in cases when digital computers or specialized digital hardware are an integral part of the system or when the system model is defined only at discrete points in time. Examples include digital control systems, digital filters, digital signal processing, digital integrated circuits, certain classes of artificial neural networks, and the like. In the case of both continuous-time and discrete-time finite-dimensional dynamical systems one frequently speaks of *lumped parameter systems*.

Infinite-dimensional dynamical systems, frequently viewed as *distributed parameter systems*, may be determined, for example, by the solutions of differential-difference equations (delay differential equations), functional differential equations (retarded and neutral types), Volterra integrodifferential equations, various classes of partial differential equations, and others. Also, continuous and discrete-time autonomous finite-dimensional and infinite-dimensional dynamical systems may be generated by linear and nonlinear semigroups. Infinite-dimensional dynamical systems are capable of incorporating effects that cannot be captured in finite-dimensional dynamical systems, including time lags and transportation delays, hysteresis effects, spatial distributions of system parameters, and so forth. Some specific examples of such systems include control systems with time delays, artificial neural network models endowed with time delays, multicore nuclear reactor models (represented by a class of Volterra integrodifferential equations), systems represented by the heat equation, systems represented by the wave equation, and many others.

There are many classes of dynamical systems whose motions cannot be determined by classical equations or inequalities of the type enumerated above. One of the most important of these is *discrete-event systems*. Examples of such systems include load balancing in manufacturing systems and in computer networks.

Discontinuous dynamical systems, both finite-dimensional and infinite-dimensional, arise in the modeling process of a variety of systems, including hybrid dynamical systems, discrete-event systems, switched systems, intelligent control systems, systems subjected to impulsive effects, and the like. In Figure 1.1.1, we depict in block diagram form a configuration that is applicable to many classes of such systems. There is a block that contains continuous-time dynamics, a block that contains phenomena which evolve at discrete points in time (discrete-time dynamics) or at discrete events, and a block that contains interface elements for the above two system components. The block that contains the continuous-time dynamics is usually characterized by one of the types of equations enumerated above and the block on the right in Figure 1.1.1 is usually characterized by difference equations, or it may involve other types of discrete characterizations, such as Petri nets, logic commands, various types of discrete-event systems, and the like. The block labeled Interface Elements may vary from the very simple to the very complicated. At the simplest level, this block may involve samplers and sample and hold elements. The sampling process may involve only one uniform rate, or it may be nonuniform (variable rate sampling), or there may be several different (uniform or nonuniform) sampling rates occurring simultaneously (multirate sampling). Perhaps the simplest specific example of the above class of systems is *digital control systems* where the continuous-time dynamics are described by ordinary differential equations, the discrete-time dynamics are characterized by ordinary difference equations, and the interface elements consist of sampling elements and sampling and hold elements.

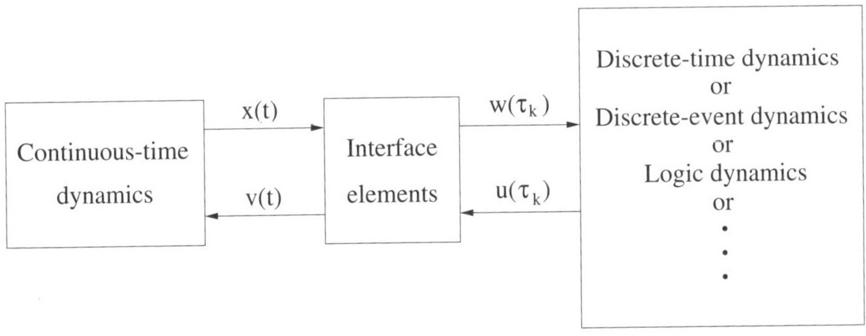

Figure 1.1.1: A discontinuous dynamical system configuration.

1.2 A Brief Perspective on the Development of Stability Theory

In his famous doctoral dissertation, Aleksandr Mikhailovich Lyapunov [45] developed the stability theory of dynamical systems determined by nonlinear time-varying ordinary differential equations. In doing so, he formulated his concepts of stability and instability and he developed two general methods for the stability analysis of an equilibrium: *Lyapunov's Direct Method*, also called *The Second Method of Lyapunov*, and *The Indirect Method of Lyapunov*, also called *The First Method*. The former involves the existence of scalar-valued auxiliary functions of the state space (called *Lyapunov functions*) to ascertain the stability properties of an equilibrium, whereas the latter seeks to deduce the stability properties of an equilibrium of a system described by a nonlinear differential equation from the stability properties of its linearization. In the process of discovering *The First Method*, Lyapunov established some important stability results for *linear systems* (involving the *Lyapunov Matrix Equation*). These results are equivalent to the independently discovered results by Routh (five years earlier) and Hurwitz (three years later).

Lyapunov did not use the concept of uniformity in his definitions of stability and asymptotic stability. Because his asymptotic stability theorem yields actually more than he was aware of (namely, uniform asymptotic stability) he was unable to establish necessary conditions (called *Converse Theorems* in the literature) for the *Second Method*. Once the issue of uniformity was settled by Malkin [46], progress on establishing *Converse Theorems* was made rapidly (Massera [47], [48]).

In the proofs of the various *Converse Theorems*, the Lyapunov functions are constructed in terms of the system solutions, and as such, these results can in general not be used to generate Lyapunov functions; they are, however, indispensable in establishing all kinds of general results. Thus, the principal disadvantage of the *Direct Method* is that there are no general rules for determining Lyapunov functions. In an attempt to overcome these difficulties, results which now comprise the *comparison theory* were discovered. In this approach, the stability properties of a given (complicated) system under study are deduced from the properties of a corresponding (simpler) system, called the *comparison system*. The system under study is related to the comparison system by means of a *stability preserving mapping*, which may be viewed as a generalization of the concept of the *Lyapunov function*. Some of the earliest comparison results are due to Müller [60] and Kamke [33], followed by the subsequent work reported in Wazewski [73], Matrosov [49], Bellman [8], Bailey [4], Lakshmikantham and Leela [37], Michel and Miller [53], Siljak [66], Grujic *et al.* [18], and others. In Michel *et al.* [57], a comparison theory for general dynamical systems is developed, using stability preserving mappings.

Of major importance in the further development of the *Direct Method* were results for autonomous dynamical systems determined by ordinary differential equations, due to Barbashin and Krasovskii [6] and LaSalle [38], [39], comprising the *Invariance Theory*. Among other issues, these results provide an effective means of estimating the *domain of attraction* of an asymptotically stable equilibrium, and more importantly, in the case of asymptotic stability, they require that the time derivative of a Lyapunov function along the motions of the system only be negative semidefinite, rather than negative definite.

One of the first important applications of the *Direct Method* was in the stability analysis of a class of nonlinear feedback control systems (regulator systems consisting of a linear part (described by linear, time-invariant ordinary differential equations) and a nonlinearity that is required to satisfy certain *sector conditions*). The formulation of this important class of systems constitutes the so called *absolute stability problem*. It was first posed and solved by Luré and Postnikov [44] who used a Lyapunov function consisting of a quadratic term in the states plus an integral term involving the system nonlinearity. An entirely different approach to the problem of absolute stability was developed by Popov [64]. His results are in terms of the frequency response of the linear part of the system and the sector conditions of the nonlinearity. Subsequently, Yacubovich [74] and Kalman [32] established a connection between the Luré type of results and the Popov type of results. A fairly complete account of the results concerning absolute stability is provided in the books by Aizerman and Gantmacher [1], Lefschetz [42], and Narendra and Taylor [61].

As mentioned earlier, there are many areas of applications of the Lyapunov stability theory, and to touch upon even a small fraction of these would be futile. However, we would like to point to a few of them, including applications to large-scale systems (see, e.g., Matrosov [49], Bailey [4], Michel and Miller [53], Siljak [66], and Grujic *et al.* [18]), robustness issues in stabilization of control systems (see, e.g., Zames [79], Michel and Wang [56], Wang and Michel [70], [71], Wang *et al.* [72], and Ye *et al.* [77]), adaptive control (see, e.g., Ioannou and Sun [31] and Åström and Wittenmark [3]), power systems (see, e.g., Pai [62]), and artificial neural networks (see, e.g., Michel and Liu [52]).

The results discussed thus far, pertaining to continuous finite-dimensional dynamical systems, are presented in numerous texts and monographs, including Hahn [20], LaSalle and Lefschetz [41], Krasovskii [35], Yoshizawa [78], Hale [23], Vidyasagar [68], Miller and Michel [59], and Khalil [34].

Lyapunov's stability theory for continuous finite-dimensional dynamical systems has been extended and generalized in every which way. Thus, the theory described above has been fully developed for discrete-time finite-dimensional dynamical systems determined by ordinary difference equations as well (see, e.g., LaSalle [40], Franklin and Powell [15], and Antsaklis and Michel [2]). The stability of infinite-dimensional dynamical systems determined by differential-difference equations are addressed, for example, in Bellman and Cooke [9], Halanay [22], and Hahn [21]; for functional differential equations they are treated, for example, in Krasovskii [35], Yoshizawa [78], and Hale [24]; for Volterra integrodifferential equations they are developed, for example, in Barbu and Grossman [7], Miller [58], Walter [69], Hale [25], and Lakshmikantham and Leela [37]; and for partial differential equations they are considered, for example, in Friedman [16], Hörmander [27], [28], and Garabedian [17]. In a more general approach, the stability analysis of infinite-dimensional dynamical systems is accomplished in the context of analyzing systems determined by differential equations and inclusions on Banach space (e.g., Krein [36], Lakshmikantham and Leela [37], and Daleckii and Krein [13]); linear and nonlinear semigroups (e.g., Hille and Phillips [26], Pazy [63], Crandall [11], and Crandall and Liggett [12]); and general dynamical systems (e.g., Hahn [21], Sell [65], Zubov [80], and Michel *et al.* [57]).

Much of the stability analysis of discontinuous dynamical systems has thus far been concerned with finite-dimensional dynamical systems (e.g., Ye *et al.* [75], Branicky [10], Michel [50], Michel and Hu [51], Liberzon and Morse [43], DeCarlo *et al.* [14], and Haddad *et al.* [19]). The stability analysis of infinite-dimensional DDS determined by functional differential equations, semigroups, and differential equations defined on Banach spaces is addressed in Sun *et al.* [67], Michel and Sun [54], and Michel *et al.* [55] and the stability analysis of general DDS defined on metric space is treated in Ye *et al.* [75], Michel [50], and Michel and Hu [51]. Some of the applications of these results include the stability analysis of systems with impulsive effects (see, e.g., Bainov and Simeonov [5], and Ye *et al.* [76]), digital control systems (see, e.g., Hou *et al.* [30]), pulse-width-modulated feedback control systems (see, e.g., Hou and Michel [29]), switched systems (see, e.g., DeCarlo *et al.* [14] and Ye *et al.* [75]), and digital control systems with delays (see, e.g., Sun *et al.* [67]).

1.3 Scope and Contents of the Book

Contemporary models of dynamical systems encountered by engineers and scientists may vary from being very simple to being very complicated. The motions (resp., states) of such systems may evolve along continuous time, discrete time, or a mixture, where parts of the motion evolve along continuous time and other parts evolve along discrete time. In the case when the states evolve along continuous time, the motions may be continuous at all points in time, or they may be discontinuous with respect to time. The behavior of some systems may adequately be captured by "lumped parameter" models, which means that such systems may be described by finite-dimensional dynamical systems determined by ordinary differential or difference equations. On the other hand, when systems exhibit, for example, hysteresis effects, or the effects of transportation delays or time lags, or the effects of spatially distributed parameters, then a finite-dimensional system description will no longer be adequate. In such cases, the behavior of the motions is captured by infinite-dimensional dynamical systems determined by the types of classical equations enumerated earlier. We need to hasten to add, however, that there are system descriptions for which the various classes of the classical equations enumerated earlier are inappropriate (e.g., discrete-event systems, systems characterized by Petri nets, and so forth). It is clear that a successful qualitative analysis of such systems may frequently require results for finite-dimensional systems as well as for infinite-dimensional systems; for continuous-time systems as well as for discrete-time systems; for continuous dynamical systems as well as for discontinuous dynamical systems; and for (hybrid) systems involving a mixture of both continuous-time and discrete-time system components. In the case when a system is not described by one of the traditional equations or inequalities, the qualitative analysis might have to take place, for example, in the setting of an abstract metric space, rather than a vector space.

Presently, there are no books on stability theory that are suitable to serve as a single source for the analysis of some of the system models enumerated above. Most of the engineering texts on stability theory are concerned with finite-dimensional continuous dynamical systems described by ordinary differential equations. The stability theory of finite-dimensional discrete-time dynamical systems described by difference equations is frequently addressed only *briefly* in books on sampled-data control systems, or as analogous afterthoughts in stability books dealing primarily with systems described by ordinary differential equations. As we have seen earlier, texts and monographs on the stability theory of infinite-dimensional dynamical systems usually focus on specific classes of systems (determined, e.g., by functional differential equations, partial differential equations, etc.). Finally, as noted previously, the literature concerning the stability of discontinuous dynamical systems is scattered throughout journal publications and conference proceedings. As a consequence, to become proficient in the stability analysis of contemporary dynamical systems of the type described above may require considerable investment of time. Therefore, there seems to be need for a book on stability theory that addresses continuous-time as well as discrete-time systems; continuous as well as discontinuous systems; finite-dimensional as well as infinite-dimensional systems; and systems involving descriptions by classical equations and inequalities as well as systems that cannot be described by such equations and inequalities. We aim to fill this void in the present book.

Finally, in addition to the objectives and goals stated above, we believe that the present book will serve as a guide to enable the reader to pursue study of further topics in greater depth, as needed.

Chapter Contents

The remainder of this book is organized in eight chapters.

In *Chapter 2* we introduce the concept of a dynamical system defined on a metric space (more formally than was done earlier), we give a classification of dynamical systems, and we present several important specific classes of finite- and infinite-dimensional dynamical systems determined by the various classical differential equations encountered in science and engineering. In a subsequent chapter (Chapter 5), we also present examples of dynamical systems that cannot be described by such equations.

The classes of dynamical systems that we consider include continuous-time and discrete-time finite-dimensional dynamical systems determined by ordinary differential equations and inequalities and ordinary difference equations and inequalities, respectively, and by infinite-dimensional dynamical systems described by differential-difference equations, functional differential equations, Volterra integrodifferential equations, certain classes of partial differential equations, and more generally, differential equations and inclusions defined on Banach spaces, and by linear and nonlinear semigroups. For the cases of continuous-time systems, in addition to continuous systems, we consider discontinuous dynamical systems as well.

In addition to the above, we also introduce the notion of a composite dynamical system, consisting of a mixture of different equations (defined for the same time set T). Also, in a subsequent chapter (Chapter 8) we consider a specific class of hybrid dynamical systems consisting of a mixture of equations defined on different time sets.

In *Chapter 3* we establish the Principal Lyapunov Stability and Boundedness Results, including Converse Theorems, for dynamical systems defined on metric spaces. By considering the most general setting first (dynamical systems defined on metric spaces), we are able to utilize some of the results of the present chapter in establishing in an efficient manner corresponding results presented in subsequent chapters for important classes of finite- and infinite-dimensional dynamical systems.

We first introduce the notions of an invariant set (resp., equilibrium) with respect to the motions of a dynamical system and we give the definitions of the various concepts of Lyapunov and Lagrange stability (including stability, uniform stability, local and global asymptotic stability, local and global uniform asymptotic stability, local and global exponential stability, instability, complete instability, uniform boundedness, and uniform ultimate boundedness).

Next, we establish the Principal Lyapunov and Lagrange Stability Results (sufficient conditions for the above stability, instability, and boundedness concepts) for discontinuous dynamical systems, continuous dynamical systems, and discrete-time dynamical systems, respectively. Because continuous dynamical systems constitute special cases of DDS, the stability, instability, and boundedness results for DDS are applicable to continuous dynamical systems as well. To prove the various Principal Lyapunov and Lagrange stability results for continuous dynamical systems, we show that when the hypotheses of any one of these results are satisfied, then the hypotheses of the corresponding DDS results are also satisfied; that is, the classical Lyapunov and Lagrange stability results for continuous dynamical systems reduce to the corresponding Lyapunov and Lagrange stability results that we established for DDS. This shows that the DDS results are more general than the corresponding classical Lyapunov and Lagrange stability results for continuous dynamical systems. Indeed, a specific example is presented of a continuous dynamical system with an equilibrium that can be shown to be uniformly asymptotically stable, using the uniform asymptotic stability result for DDS, and we prove that for the *same* example, there does *not* exist a Lyapunov function that satisfies the classical Lyapunov theorem for uniform asymptotic stability for continuous dynamical systems.

Next, we show that for every discrete-time dynamical system there exists an *associated* DDS with identical Lyapunov and Lagrange stability properties. Making use of such associated DDS, we prove, similarly as in the case of continuous dynamical systems, that the Lyapunov and Lagrange stability results for DDS are more general than the corresponding results for the classical Lyapunov and Lagrange stability results for discrete-time dynamical systems. We give an example of a discrete-time dynamical system with an equilibrium that can be shown to be uniformly asymptotically stable, by applying the uniform asymptotic stability result for DDS to the associated DDS, and we prove that for the *same original* discrete-time dynamical system there does *not* exist a Lyapunov function that satisfies the classical uniform asymptotic stability theorem for discrete-time dynamical systems.

In addition to proving that the classical Lyapunov and Lagrange stability results for continuous dynamical systems and discrete-time dynamical systems reduce to the corresponding DDS results, our approach described above establishes also a *unifying theory for DDS, continuous dynamical systems, and discrete-time dynamical systems.*

Next, under some additional mild conditions, we establish Converse Theorems (necessary conditions) for the above results for DDS, continuous dynamical systems, and discrete-time dynamical systems.

Finally, in an appendix section we present a comparison result involving maximal and minimal solutions of ordinary differential equations, which is required in some of the proofs of this chapter.

In *Chapter 4* we present important *specialized* Lyapunov and Lagrange stability results for dynamical systems defined on metric spaces. We first show that under some reasonable assumptions, in the case of autonomous dynamical systems, stability and asymptotic stability of an invariant set imply uniform stability and uniform asymptotic stability of an invariant set, respectively. Furthermore, we establish necessary and sufficient conditions for stability and asymptotic stability of an invariant set for autonomous dynamical systems. Next, for continuous and discrete-time autonomous dynamical systems, we present generalizations of LaSalle-type theorems that comprise the invariance theory for dynamical systems defined by semigroups in metric spaces. Also, for both continuous and discrete-time dynamical systems we present several results that make up a comparison theory for various Lyapunov and Lagrange stability types. In these results we deduce the qualitative properties of a complex dynamical system (the object of inquiry) from corresponding qualitative properties of a simpler and well-understood dynamical system (the comparison system). Finally, we present Lyapunov-like results that ensure the uniqueness of motions for continuous and discrete-time dynamical systems defined on metric spaces.

In *Chapter 5* we apply the results of Chapters 3 and 4 in the stability analysis of an important class of discrete-event systems with applications to a computer load-balancing problem and a manufacturing system.

In the preceding three chapters, we concern ourselves with the qualitative analysis of dynamical systems defined on metric spaces. In the next three chapters we address the Lyapunov and Lagrange stability of continuous-time and discrete-time finite-dimensional dynamical systems determined by ordinary differential equations and difference equations, respectively. For the case of continuous-time dynamical systems we consider continuous dynamical systems and discontinuous dynamical systems. In these three chapters our focus is on the qualitative analysis of equilibria (rather than general invariant sets). Throughout the next three chapters, we include numerous specific examples to demonstrate the applicability of the various results that are presented.

In *Chapter 6* we first present some preliminary material that is required throughout the next three chapters, including material on ordinary differential equations and ordinary difference equations; definition of the time-derivative of Lyapunov functions evaluated along the solutions of ordinary differential equations; evaluation of the first forward difference of Lyapunov functions along the solutions of difference equations; characterizations of Lyapunov functions, including quadratic forms; and a motivation and geometric interpretation for Lyapunov stability results for two-dimensional systems. Next, we present the Principal Lyapunov and Lagrange Stability Results (sufficient conditions) for continuous dynamical systems determined by ordinary differential equations; for discrete-time dynamical systems determined by difference equations; and for DDS determined by ordinary differential equations. In most cases, the proofs of these results are direct consequences of corresponding results that were presented in Chapter 3. Finally, we present converse theorems (necessary conditions) for the above Lyapunov and Lagrange stability results. In an appendix section we give some results concerning the continuous dependence of solutions of ordinary differential equations with respect to initial conditions.

In *Chapter 7* we continue our study of finite-dimensional dynamical systems with the presentation of some important specialized results for continuous and discrete-time systems. We first show that if for dynamical systems determined by autonomous and periodic ordinary differential equations, the equilibrium $x_e = 0$ is stable or asymptotically stable, then the equilibrium $x_e = 0$ is uniformly stable or uniformly asymptotically stable, respectively. Also, for such kind of dynamical systems, we present converse theorems for asymptotically stable systems. Next, for continuous and discrete-time dynamical systems determined by autonomous ordinary differential equations and ordinary difference equations, we establish LaSalle-type stability results that comprise the invariance theory for such systems. These results are direct consequences of corresponding results that were established in Chapter 3 for autonomous dynamical systems defined on metric spaces. For autonomous dynamical systems determined by ordinary differential equations, we next present two methods of determining estimates for the domain of attraction of an asymptotically stable equilibrium (including Zubov's Theorem). Next, we present the main Lyapunov stability and boundedness results for dynamical systems determined by linear homogeneous systems of ordinary differential equations (and difference equations), linear autonomous homogeneous ordinary differential equations (and difference equations), and linear periodic ordinary differential equations. Some of these results require explicit knowledge of state transition matrices whereas other results involve Lyapunov matrix equations. This is followed by a detailed study of the stability properties of the equilibrium $x_e = 0$ of dynamical systems determined by linear, second-order autonomous homogeneous systems of ordinary differential equations. Next, we investigate the qualitative properties of perturbed linear systems. In doing so, we develop Lyapunov's First Method (also called Lyapunov's Indirect Method) for continuous and discrete-time dynamical systems, and we study the existence of stable and unstable manifolds and the stability of periodic motions in continuous linear perturbed systems. Finally, similarly as in Chapter 4, we establish Lyapunov and Lagrange stability results for continuous and discrete-time dynamical systems that comprise a comparison theory for finite-dimensional dynamical systems.

In *Chapter 8* we apply the results presented in Chapters 6 and 7 in the analysis of several important classes of continuous, discontinuous, and discrete-time finite-dimensional dynamical systems. We first address the absolute stability problem of nonlinear regulator systems, by presenting Luré's result for direct control systems and Popov's result for indirect control systems. Next, we establish global and local Lyapunov stability results for Hopfield neural networks. This is followed by an investigation of an important class of hybrid systems, digital control systems. We consider system models with quantizers and without quantizers. Next, we present stability results for an important class of pulse-width-modulated (PWM) feedback control systems. Finally, we study the stability properties of systems with saturation nonlinearities with applications to digital filters.

In *Chapter 9* we address the Lyapunov and Lagrange stability of infinite-dimensional dynamical systems determined by differential equations defined on Banach spaces and semigroups. As in Chapters 6 through 8, we focus on the qualitative properties of equilibria and we consider continuous as well as discontinuous dynamical systems. Throughout this chapter, we present several specific examples to demonstrate the applicability of the presented results. These include systems determined by functional differential equations, Volterra integrodifferential equations, and partial differential equations. In addition, we apply the results of this chapter in the analysis of two important classes of infinite-dimensional dynamical systems: a point kinetics model of a multicore nuclear reactor (described by Volterra integrodifferential equations) and Cohen–Grossberg neural networks with time delays (described by differential-difference equations). As in Chapters 6 and 7, several of the results presented in this chapter are direct consequences of the results given in Chapters 3 and 4 for dynamical systems defined on metric spaces.

We first present the Principal Lyapunov and Lagrange Stability Results (sufficient conditions) for dynamical systems determined by general differential equations defined on Banach spaces. Most of these results are direct consequences of the corresponding results established in Chapter 3 for dynamical systems defined on metric spaces. We also present converse theorems (necessary conditions) for several of the above results. Most of these are also direct consequences of corresponding results given in Chapter 3 for dynamical systems defined on metric spaces. Next, we present LaSalle-type stability results that comprise the invariance theory for autonomous differential equations defined on Banach spaces. Essentially, these results are also direct consequences of corresponding results that are established in Chapter 4 for dynamical systems defined on metric spaces. This is followed by the presentation of several Lyapunov and Lagrange stability results that comprise a comparison theory for general differential equations defined on Banach spaces. Next, we present stability results for composite dynamical systems defined on Banach spaces that are described by a mixture of different types of differential equations. As mentioned earlier, we apply some of the results enumerated above in the analysis of a point kinetics model of a multicore nuclear reactor (described by Volterra integrodifferential equations). For the special case of functional differential equations, it is possible to improve on the Lyapunov stability results for general differential equations defined on Banach spaces by taking into account some of the specific properties of functional differential equations. We present improved Lyapunov stability results for dynamical systems determined by retarded functional differential equations. Some of these results include Razumikhin-type theorems. As pointed out earlier, we apply these results in the qualitative analysis of a class of artificial neural networks with time delays (described by differential-difference equations). Next, we establish Lyapunov and Lagrange stability results for discontinuous dynamical systems defined on Banach and Hilbert spaces. We consider DDS determined by differential equations defined on Banach spaces, and by DDS determined by linear and nonlinear semigroups.

Bibliography

[1] M. A. Aizerman and F. R. Gantmacher, *Absolute Stability of Regulator Systems*, San Francisco: Holden-Day, 1964.

[2] P. J. Antsaklis and A. N. Michel, *Linear Systems*, Boston: Birkhäuser, 2006.

[3] K. J. Åström and B. Wittenmark, *Adaptive Control*, 2nd Edition, New York: Addison-Wesley, 1995.

[4] F. N. Bailey, "The application of Lyapunov's Second Method to interconnected systems," *SIAM J. Control*, vol. 3, pp. 443–462, 1966.

[5] D. D. Bainov and P. S. Simeonov, *Systems with Impulse Effects: Stability Theory and Applications*, New York: Halsted, 1989.

[6] A. E. Barbashin and N. N. Krasovskii, "On the stability of motion in the large," *Dokl. Akad. Nauk.*, vol. 86, pp. 453-456, 1952.

[7] V. Barbu and S. I. Grossman, "Asymptotic behavior of linear integrodifferential systems," *Trans. Amer. Math. Soc.*, vol. 171, pp. 277–288, 1972.

[8] R. Bellman, "Vector Lyapunov functions," *SIAM J. Control*, vol. 1, pp. 32–34, 1962.

[9] R. Bellman and K. L. Cooke, *Differential-Difference Equations*, New York: Academic Press, 1963.

[10] M. S. Branicky, "Multiple Lyapunov functions and other analysis tools for switched and hybrid systems," *IEEE Trans. Autom. Control*, vol. 43, pp. 475–482, 1998.

[11] M. G. Crandall, "Semigroups of nonlinear transformations on general Banach spaces," *Contributions to Nonlinear Functional Analysis*, E. H. Zarantonello, Ed., New York: Academic Press, 1971.

[12] M. G. Crandall and T. M. Liggett, "Generation of semigroups of nonlinear transformations on general Banach spaces," *Amer. J. Math.*, vol. 93, pp. 265–298, 1971.

[13] J. L. Daleckii and S. G. Krein, *Stability of Solutions of Differential Equations in Banach Spaces*, Translations of Mathematical Monographs, vol. 43, Providence, RI: American Mathematical Society, 1974.

[14] R. DeCarlo, M. Branicky, S. Pettersson, and B. Lennartson, "Perspectives and results on the stability and stabilizability of hybrid systems," *Proc. IEEE*, vol. 88, pp. 1069–1082, 2000.

[15] G. F. Franklin and J. D. Powell, *Digital Control of Dynamical Systems*, Reading, MA: Addison-Wesley, 1980.

[16] A. Friedman, *Partial Differential Equations of Parabolic Type*, Englewood Cliffs, NJ: Prentice Hall, 1964.

[17] P. R. Garabedian, *Partial Differential Equations*, New York: Chelsea, 1986.

[18] L. T. Grujic, A. A. Martynyuk, and M. Ribbens-Pavella, *Large Scale Systems Stability Under Structural and Singular Perturbations*, Berlin: Springer-Verlag, 1987.

[19] W. M. Haddad, V. S. Chellaboina, and S. G. Nersesov, *Impulsive and Hybrid Dynamical Systems: Stability, Dissipativity and Control*, Princeton, NJ: Princeton University Press, 2006.

[20] W. Hahn, *Theorie und Anwendung der direkten Methode von Ljapunov*, Heidelberg: Springer-Verlag, 1959.

[21] W. Hahn, *Stability of Motion*, Berlin: Springer-Verlag, 1967.

[22] A. Halanay, *Differential Equations: Stability, Oscillations, Time Lags*, New York: Academic Press, 1966.

[23] J. K. Hale, *Ordinary Differential Equations*, New York: Wiley-Interscience, 1969.

[24] J. K. Hale, *Functional Differential Equations*, Berlin: Springer-Verlag, 1971.

[25] J. K. Hale, "Functional differential equations with infinite delays," *J. Math. Anal. Appl.,* vol. 48, pp. 276–283, 1974.

[26] E. Hille and R. S. Phillips, *Functional Analysis and Semigroups*, Amer. Math. Soc. Colloquium Publ. vol. 33, Providence, RI: American Mathematical Society, 1957.

[27] L. Hörmander, *Linear Partial Differential Equations*, Berlin: Springer-Verlag, 1963

[28] L. Hörmander, *The Analysis of Linear Partial Differential Operators*, vol. 1, 2, 3, 4, Berlin: Springer-Verlag, 1983–1985.

[29] L. Hou and A. N. Michel, "Stability analysis of pulse-width-modulated feedback systems," *Automatica*, vol. 37, pp. 1335–1349, 2001.

[30] L. Hou, A. N. Michel, and H. Ye, "Some qualitative properties of sampled-data control systems," *IEEE Trans. Autom. Control*, vol. 42, pp. 1721–1725, 1997.

[31] P. A. Ioannou and J. Sun, *Robust Adaptive Control*, Upper Saddle River, NJ: Prentice Hall, 1996.

[32] R. E. Kalman, "Lyapunov functions for the problem of Luré in automatic control," *Proc. Nat. Acad. Sci. USA*, vol. 49, pp. 201–205, 1963.

[33] E. Kamke, "Zur Theorie der Systeme gewöhnlicher Differentialgleichungen, II," *Acta Math.*, vol. 58, pp. 57–85, 1932.

[34] H. K. Khalil, *Nonlinear Systems*, New York: Macmillan, 1992.

[35] N. N. Krasovskii, *Stability of Motion*, Stanford, CA: Stanford University Press, 1963.

[36] S. G. Krein, *Linear Differential Equations in Banach Spaces*, Translation of Mathematical Monographs, vol. 29, Providence, RI: American Mathematical Society, 1970.

[37] V. Lakshmikantham and S. Leela, *Differential and Integral Inequalities*, vol. 1 and 2, New York: Academic Press, 1969.

[38] J. P. LaSalle, "The extent of asymptotic stability," *Proc. Nat. Acad. Sci.*, vol. 48, pp. 363–365, 1960.

[39] J. P. LaSalle, "Some extensions of Liapunov's Second Method," *IRE Trans. Circ. Theor.*, vol. 7, pp. 520–527, 1960.

[40] J. P. LaSalle, *The Stability and Control of Discrete Processes*, New York: Springer-Verlag, 1986.

[41] J. P. LaSalle and S. Lefschetz, *Stability of Lyapunov's Direct Method*, New York: Academic Press, 1961.

[42] S. Lefschetz, *Stability of Nonlinear Control Systems*, New York: Academic Press, 1965.

[43] D. Liberzon and A. S. Morse, "Basic problems in stability and design of switched systems," *IEEE Control Syst. Mag.*, vol. 19, pp. 59–70, 1999.

[44] A. I. Luré and V. N. Postnikov, "On the theory of stability of control systems," *Prikl. Mat. i. Mehk.*, vol. 8, pp. 3–13, 1944.

[45] A. M. Liapounoff, "Problème générale de la stabilité de mouvement," *Annales de la Faculté des Sciences de l'Université de Toulouse*, vol. 9, pp. 203–474, 1907. (Translation of a paper published in *Comm. Soc. Math.*, Kharkow, 1892, reprinted in *Ann. Math. Studies*, vol. 17, Princeton, NJ: Princeton, 1949. The French version was translated into English by A. T. Fuller, and was published in the *International Journal of Control*, vol. 55, pp. 531–773, 1992.)

[46] I. G. Malkin, "On the question of reversibility of Lyapunov's theorem on automatic stability," *Prikl. Mat. i Mehk.*, vol.18, pp. 129–138, 1954 (in Russian).

[47] J. L. Massera, "On Liapunoff's conditions of stability," *Ann. Mat.*, vol. 50, pp. 705–721, 1949.

[48] J. L. Massera, "Contributions to stability theory," *Ann. Mat.*, vol. 64, pp. 182–206, 1956.

[49] V. M. Matrosov, "The method of Lyapunov-vector functions in feedback systems," *Automat. Remote Control*, vol. 33, pp. 1458–1469, 1972.

[50] A. N. Michel, "Recent trends in the stability analysis of hybrid dynamical systems," *IEEE Trans. Circ. and Syst.–I: Fund. Theor. Appl.*, vol. 46, pp. 120–134, 1999.

[51] A. N. Michel and B. Hu, "Towards a stability theory of general hybrid dynamical systems," *Automatica*, vol. 35, pp. 371–384, 1999.

[52] A. N. Michel and D. Liu, *Qualitative Analysis and Synthesis of Recurrent Neural Networks*, New York: Marcel Dekker, 2002.

[53] A. N. Michel and R. K. Miller, *Qualitative Analysis of Large Scale Dynamical Systems*, New York: Academic Press, 1977.

[54] A. N. Michel and Y. Sun, "Stability of discontinuous Cauchy problems in Banach space," *Nonlinear Anal.*, vol. 65, pp. 1805–1832, 2006.

[55] A. N. Michel, Y. Sun, and A. P. Molchanov, "Stability analysis of discontinuous dynamical systems determined by semigroups," *IEEE Trans. Autom. Control*, vol. 50, pp. 1277–1290, 2005.

[56] A. N. Michel and K. Wang, "Robust stability: perturbed systems with perturbed equilibria," *Syst. and Control Lett.*, vol. 21, pp.155–162, 1993.

[57] A. N. Michel, K. Wang, and B. Hu, *Qualitative Theory of Dynamical Systems: The Role of Stability Preserving Mappings*, 2nd Edition, New York: Marcel Dekker, 2001.

[58] R. K. Miller, *Nonlinear Volterra Integral Equations*, Menlo Park, CA: W. A. Benjamin, 1971.

[59] R. K. Miller and A. N. Michel, *Ordinary Differential Equations*, New York: Academic Press, 1982.

[60] M. Müller, "Über das Fundamentaltheorem in der Theorie der gewöhnlichen Differentialgleichungen," *Mat. Zeit.*, vol. 26, pp. 615–645, 1926.

[61] K. S. Narendra and J. H. Taylor, *Frequency Domain Stability for Absolute Stability*, New York: Academic Press, 1973.

[62] M. A. Pai, *Power System Stability*, New York: North-Holland, 1981.

[63] A. Pazy, *Semigroups of Linear Operators and Applications to Partial Differential Equations*, New York: Springer-Verlag, 1983.

[64] V. M. Popov, "Absolute stability of nonlinear systems of automatic control," *Automation Remote Control*, vol. 22, pp. 857–875, 1961.

[65] G. R. Sell, *Lectures on Topological Dynamics and Differential Equations*, Princeton, NJ: Van Nostrand, 1969.

[66] D. D. Siljak, *Large-Scale Dynamical Systems: Stability and Structure*, New York: North Holland, 1978.

[67] Y. Sun, A. N. Michel, and G. Zhai, "Stability of discontinuous retarded functional differential equations with applications," *IEEE Trans. Autom. Control*, vol. 50, pp. 1090–1105, 2005.

[68] M. Vidyasagar, *Nonlinear Systems Analysis*, 2nd Edition, Englewood Cliffs, NJ: Prentice Hall, 1993.

[69] W. Walter, *Differential and Integral Inequalities*, Berlin: Springer-Verlag, 1970.

[70] K. Wang and A. N. Michel, "On sufficient conditions for stability of interval matrices," *Syst. Control Lett.*, vol. 20, pp. 345–351, 1993.

[71] K. Wang and A. N. Michel, "Qualitative analysis of dynamical systems determined by differential inequalities with applications to robust stability," *IEEE Trans. Circ. Syst.–I: Fund. Theor. Appl.*, vol. 41, pp. 377–386, 1994.

[72] K. Wang, A. N. Michel, and D. Liu, "Necessary and sufficient conditions for the Hurwitz and Schur stability of interval matrices," *IEEE Trans. Autom. Control*, vol. 39, pp. 1251–1255, 1996.

[73] T. Wazewski, "Systemés des equations et des inégalités différentielles ordinaires aux deuxiémes membres monotones et leurs applications," *Ann. Soc. Poln. Mat.*, vol. 23, pp. 112–166, 1950.

[74] V. A. Yacubovich, "Solution of certain matrix inequalities occurring in the theory of automatic control," *Dokl. Acad. Nauk. SSSR*, pp. 1304–1307, 1962.

[75] H. Ye, A. N. Michel, and L. Hou, "Stability theory for hybrid dynamical systems," *IEEE Trans. Autom. Control*, vol. 43, pp. 461–474, 1998.

[76] H. Ye, A. N. Michel, and L. Hou, "Stability analysis of systems with impulse effects," *IEEE Trans. Autom. Control*, vol. 43, pp. 1719–1923, 1998.

[77] H. Ye, A. N. Michel, and K. Wang, "Robust stability of nonlinear time-delay systems with applications to neural networks," *IEEE Trans. Circ. Syst.–I: Fund. Theor. Appl.*, vol. 43, pp. 532–543, 1996.

[78] T. Yoshizawa, *Stability Theory by Lyapunov's Second Method*, Tokyo: Math. Soc. Japan, 1966.

[79] G. Zames, "Input-output feedback stability and robustness, 1959–85," *IEEE Control Syst. Mag.*, vol. 16, pp. 61–66, 1996.

[80] V. I. Zubov, *Methods of A. M. Lyapunov and Their Applications*, Groningen, The Netherlands: P. Noordhoff, 1964.

Chapter 2

Dynamical Systems

Our main objective in the present chapter is to define a dynamical system and to present several important classes of dynamical systems. The chapter is organized into twelve parts.

In the first section we establish some of the notation that we require in this chapter, as well as in the subsequent chapters. Next, in the second section we present precise definitions for dynamical system and related concepts. We introduce finite-dimensional dynamical systems determined by ordinary differential equations in the third section, by differential inequalities in the fourth section, and by ordinary difference equations and inequalities in the fifth section. In the sixth section, we address infinite-dimensional dynamical systems determined by differential equations and inclusions defined on Banach spaces and in the seventh and eighth sections we consider special cases of infinite-dimensional dynamical systems determined by functional differential equations and Volterra integrodifferential equations, respectively. In the ninth section we discuss dynamical systems determined by semigroups defined on Banach and Hilbert spaces and in the tenth section we treat dynamical systems determined by specific classes of partial differential equations. Finally, we address composite dynamical systems in the eleventh section and discontinuous dynamical systems in the twelfth section.

The specific classes of dynamical systems that we consider in this chapter are very important. However, there are of course many more important classes of dynamical systems, not even alluded to in the present chapter. We address one such class of systems in Chapter 5, determined by discrete-event systems.

Much of the material presented in Sections 2.3–2.10 constitutes background material and concerns the well posedness (existence, uniqueness, continuation, and continuity with respect to initial conditions of solutions) of a great variety of equations (resp., systems). Even if practical, it still would distract from our objectives on hand if we were to present proofs for these results. Instead, we give detailed references where to find such proofs, and in some cases, we give hints (in the problem section) on how to prove some of these results. The above is in contrast with our presentations in the remainder of this book where we prove all results (except some, concerning additional background material).

2.1 Notation

Let Y, Z be arbitrary sets. Then $Y \cup Z, Y \cap Z, Y - Z$, and $Y \times Z$ denote the union, intersection, difference, and Cartesian product of Y and Z, respectively. If Y is a subset of Z, we write $Y \subset Z$ and if x is an element of Y, we write $x \in Y$. We denote a mapping f of Y into Z by $f\colon Y \to Z$ and we denote the set of all mappings from Y into Z by $\{Y \to Z\}$. Let $\emptyset$ denote the empty set.

Let $\mathbb{R}$ denote the set of real numbers, let $\mathbb{R}^+ = [0, \infty)$, let $\mathbb{N}$ denote the set of nonnegative integers (i.e., $\mathbb{N} = \{0, 1, 2, \dots\}$), and let $\mathbb{C}$ denote the set of complex numbers. Let $J \subset \mathbb{R}$ denote an interval (i.e., $J = [a, b), (a, b], [a, b]$, or $(a, b), b > a$, with $J = (-\infty, \infty) = \mathbb{R}$ allowed). If $Y_1, \dots, Y_n$ are n arbitrary sets, their Cartesian product is denoted by $Y_1 \times \cdots \times Y_n$, and if in particular $Y = Y_1 = \cdots = Y_n$ we write Y^n.

Let $\mathbb{R}^n$ denote real n-space. If $x \in \mathbb{R}^n, x^T = (x_1, \dots, x_n)$ denotes the transpose of x. Also, if $x, y \in \mathbb{R}^n$, then $x \leq y$ signifies $x_i \leq y_i, x < y$ signifies $x_i < y_i$, and $x > 0$ signifies $x_i > 0$ for all $i = 1, \dots, n$. We let $|\cdot|$ denote the Euclidean norm; that is, for $x = (x_1, \dots, x_n)^T \in \mathbb{R}^n, |x| = (x^T x)^{1/2} = \left(\sum_{i=1}^n x_i^2\right)^{1/2}$.

Let $A = [a_{ij}]_{n \times n}$ denote a real $n \times n$ matrix (i.e., $A \in \mathbb{R}^{n \times n}$) and let A^T denote the transpose of A. The matrix norm $|\cdot|$, induced by the Euclidean vector norm (defined on $\mathbb{R}^n$), is defined by

$$|A| = \inf \left\{\alpha \in \mathbb{R}^+ \colon \alpha |x| \geq |Ax|, x \in \mathbb{R}^n\right\} = \left[\lambda_M(A^T A)\right]^{1/2}$$

where $\lambda_M(A^T A)$ denotes the largest eigenvalue of $A^T A$ (recall that the eigenvalues of symmetric matrices are real). In the interests of clarity, we also use the notation $\|\cdot\|$ to distinguish the norm of a matrix (e.g., $\|A\|$) from the norm of a vector (e.g., $|x|$).

We let $L_p[G, U], 1 \leq p \leq \infty$, denote the usual Lebesgue space of all Lebesgue measurable functions with domain G and range U. The norm in $L_p[G, U]$ is usually denoted $\|\cdot\|_p$, or $\|\cdot\|_{L_p}$ if more explicit notation is required.

We let (X, d) be a metric space, where X denotes the underlying set and d denotes the metric. When the choice of the particular metric used is clear from context, we speak of a metric space X, rather than (X, d).

If Y and Z are metric spaces and if $f\colon Y \to Z$, and if f is continuous, we write $f \in C[Y, Z]$; that is, $C[Y, Z]$ denotes the set of all continuous mappings from Y to Z. We denote the inverse of a mapping f, if it exists, by f^{-1}.

A function $\psi \in C[[0, r_1], \mathbb{R}^+]$ (resp., $\psi \in C[\mathbb{R}^+, \mathbb{R}^+]$) is said to belong to *class* $\mathcal{K}$ (i.e., $\psi \in \mathcal{K}$) if $\psi(0) = 0$ and if ψ is strictly increasing on $[0, r_1]$ (resp., on $\mathbb{R}^+$). If $\psi\colon \mathbb{R}^+ \to \mathbb{R}^+$, if $\psi \in \mathcal{K}$, and if $\lim_{r \to \infty} \psi(r) = \infty$, then ψ is said to belong to *class* $\mathcal{K}_\infty$ (i.e., $\psi \in \mathcal{K}_\infty$).

For a function $f: \mathbb{R} \to \mathbb{R}$, we denote the upper right-hand, upper left-hand, lower right-hand, and lower left-hand Dini derivatives by D^+f, D^-f, D_+f, and D_-f, respectively. When we have a fixed Dini derivative of f in mind, we simply write Df, in place of the preceding notation.

2.2 Dynamical Systems

In characterizing the notion of dynamical system, we require the concepts of motion and family of motions.

Definition 2.2.1 Let (X, d) be a metric space, let $A \subset X$, and let $T \subset \mathbb{R}$. For any fixed $a \in A$, $t_0 \in T$, a mapping $p(\cdot, a, t_0): T_{a,t_0} \to X$ is called a *motion* if $p(t_0, a, t_0) = a$ where $T_{a,t_0} = [t_0, t_1) \cap T$, $t_1 > t_0$, and t_1 is finite or infinite. □

Definition 2.2.2 A subset S of the set

$$\bigcup_{(a,t_0) \in A \times T} \{T_{a,t_0} \to X\}$$

is called a *family of motions* if for every $p(\cdot, a, t_0) \in S$, we have $p(t_0, a, t_0) = a$. □

Definition 2.2.3 The four-tuple $\{T, X, A, S\}$ is called a *dynamical system.* □

In Definitions 2.2.1 and 2.2.2 we find it useful to think of X as *state space*, T as *time set*, t_0 as *initial time*, a as the *initial condition* of the motion $p(\cdot, a, t_0)$, and A as the *set of initial conditions.* Note that in our definition of motion, we allow in general more than one motion to initiate from a given pair of *initial data*, (a, t_0).

When in Definition 2.2.3, $T = J \subset \mathbb{R}^+$ (with $J = \mathbb{R}^+$ allowed), we speak of a *continuous-time dynamical system* and when $T = J \cap \mathbb{N}$ (with $J \cap \mathbb{N} = \mathbb{N}$ allowed) we speak of a *discrete-time dynamical system.* Also, when in Definition 2.2.3, X is a finite-dimensional vector space, we speak of a *finite-dimensional dynamical system*, and otherwise, of an *infinite-dimensional dynamical system.* Furthermore, if in a continuous-time dynamical system all motions (i.e., all elements of S) are continuous with respect to time t, we speak of a *continuous dynamical system.* If at least one motion of a continuous-time dynamical system is not continuous with respect to t, we speak of a *discontinuous dynamical system.*

When in Definition 2.2.3, T, X, and A are known from context, we frequently speak of a *dynamical system S*, or even of a *system S*, rather than a dynamical system $\{T, X, A, S\}$.

Definition 2.2.4 A dynamical system $\{T, X_1, A_1, S_1\}$ is called a *dynamical subsystem*, or simply, a *subsystem* of a dynamical system $\{T, X, A, S\}$ if $X_1 \subset X$, $A_1 \subset A$, and $S_1 \subset S$. □

Definition 2.2.5 A motion $p = p(\cdot, a, t_0)$ in a dynamical system $\{T, X, A, S\}$ is said to be *bounded* if there exist an $x_0 \in X$ and a $\beta > 0$ such that $d(p(t, a, t_0), x_0) < \beta$ for all $t \in T_{a,t_0}$. □

Definition 2.2.6 A motion $p^* = p^*(\cdot, a, t_0)$ defined on $[t_0, c) \cap T$ is called a *continuation* of another motion $p = p(\cdot, a, t_0)$ defined on $[t_0, b) \cap T$ if $p = p^*$ on $[t_0, b) \cap T, c > b$, and $[b, c) \cap T \neq \emptyset$. We say that p is *noncontinuable* if no continuation of p exists. Also, $p = p(\cdot, a, t_0)$ is said to be *continuable forward for all time* if there exists a continuation $p^* = p^*(\cdot, a, t_0)$ of p that is defined on $[t_0, \infty) \cap T$, where it is assumed that for any $\alpha > 0$, $[\alpha, \infty) \cap T \neq \emptyset$. □

In the remainder of this chapter, we present several important classes of dynamical systems. Most of this material serves as required background for the remainder of this book.

2.3 Ordinary Differential Equations

In this section we summarize some essential facts from the qualitative theory of ordinary differential equations that we require as background material and we show that the solutions of differential equations determine continuous, finite-dimensional dynamical systems.

A. Initial value problems

Let $D \subset \mathbb{R}^{n+1}$ be a domain (an open connected set), let $x = (x_1, \ldots, x_n)^T$ denote elements of $\mathbb{R}^n$, and let elements of D be denoted by (t, x). When x is a vector-valued function of t, let

$$\dot{x} = \frac{dx}{dt} = \left(\frac{dx_1}{dt}, \ldots, \frac{dx_n}{dt}\right)^T = (\dot{x}_1, \ldots, \dot{x}_n)^T.$$

For a given function $f_i \colon D \to \mathbb{R}$, $i = 1, \ldots, n$, let $f = (f_1, \ldots, f_n)^T$. Consider *systems of first-order ordinary differential equations* given by

$$\dot{x}_i = f_i(t, x_1, \ldots, x_n), \qquad i = 1, \ldots, n. \tag{E_i}$$

Equation (E_i) can be written more compactly as

$$\dot{x} = f(t, x). \tag{E}$$

A *solution* of (E) is an n vector-valued differentiable function φ defined on a real interval $J = (a, b)$ (we express this by $f \in C^1[J, \mathbb{R}^n]$) such that $(t, \varphi(t)) \in D$ for all $t \in J$ and such that

$$\dot{\varphi}(t) = f(t, \varphi(t))$$

for all $t \in J$. We also allow the cases when $J = [a, b)$, $J = (a, b]$, or $J = [a, b]$. When $J = [a, b]$, then $\dot{\varphi}(a)$ is interpreted as the right-side derivative and $\dot{\varphi}(b)$ is interpreted as the left-side derivative.

For $(t_0, x_0) \in D$, the *initial value problem* associated with (E) is given by

$$\dot{x} = f(t, x), \qquad x(t_0) = x_0. \tag{I_E}$$

An n vector-valued function φ is a *solution* of (I_E) if φ is a solution of (E) which is defined on $[t_0, b)$ and if $\varphi(t_0) = x_0$. To denote the dependence of the solutions of (I_E) on the initial data (t_0, x_0), we frequently write $\varphi(t, t_0, x_0)$. However, when the initial data are clear from context, we often write $\varphi(t)$ in place of $\varphi(t, t_0, x_0)$.

When $f \in C[D, \mathbb{R}^n]$, φ is a solution of (I_E) if and only if φ satisfies the integral equation

$$\varphi(t) = x_0 + \int_{t_0}^{t} f(s, \varphi(s))ds \tag{$\widetilde{E}$}$$

for $t \in [t_0, b)$. In $(\widetilde{E})$, we have used the notation

$$\int_{t_0}^{t} f(s, \varphi(s))ds = \left[\int_{t_0}^{t} f_1(s, \varphi(s))ds, \ldots, \int_{t_0}^{t} f_n(s, \varphi(s))ds\right]^T.$$

B. Existence, uniqueness, and continuation of solutions

The following examples demonstrate that we need to impose restrictions on the right-hand side of (E) to ensure the existence, uniqueness, and continuation of solutions of the initial value problem (I_E).

Example 2.3.1 For the scalar initial value problem

$$\dot{x} = g(x), \qquad x(0) = 0 \tag{2.3.1}$$

where $x \in \mathbb{R}$ and

$$g(x) = \begin{cases} 1, & x = 0 \\ 0, & x \neq 0 \end{cases}$$

there is no differentiable function φ that satisfies (2.3.1). Therefore, this initial value problem has *no solution* (in the sense defined above). □

Example 2.3.2 The initial value problem

$$\dot{x} = x^{1/3}, \qquad x(t_0) = 0$$

where $x \in \mathbb{R}$, has at least two solutions given by

$$\varphi_1(t) = \left[\frac{2}{3}(t - t_0)\right]^{3/2}$$

and $\varphi_2(t) = 0$ for $t \geq t_0$. □

Example 2.3.3 The scalar initial value problem

$$\dot{x} = ax, \qquad x(t_0) = x_0$$

where $x \in \mathbb{R}$, has a unique solution given by $\varphi(t) = e^{a(t-t_0)}x(t_0)$ for $t \geq t_0$. □

The following result, called the *Peano–Cauchy Existence Theorem*, provides a set of sufficient conditions for the existence of solutions of the initial value problem (I_E).

Theorem 2.3.1 Let $f \in C[D, \mathbb{R}^n]$. Then for any $(t_0, x_0) \in D$, the initial value problem (I_E) has a solution defined on $[t_0, t_0 + c)$ for some $c > 0$. □

The next result provides a set of sufficient conditions for the uniqueness of solutions of the initial value problem (I_E).

Theorem 2.3.2 Let $f \in C[D, \mathbb{R}^n]$. Assume that for every compact set $K \subset D$, f satisfies the *Lipschitz condition*

$$\left|f(t,x) - f(t,y)\right| \le L_K|x-y| \tag{2.3.2}$$

for all $(t,x), (t,y) \in K$ where L_K is a constant depending only on K. Then (I_E) has at most one solution on any interval $[t_0, t_0 + c), c > 0$. □

In the problem section we provide details for the proofs of Theorems 2.3.1 and 2.3.2. Alternatively, the reader may wish to refer, for example, to Miller and Michel [37] for proofs of these results.

Next, let φ be a solution of (E) on an interval J. By a *continuation* or *extension* of φ we mean an extension φ_0 of φ to a larger interval J_0 in such a way that the extension solves (E) on J_0. Then φ is said to be *continued* or *extended* to the larger interval J_0. When no such continuation is possible, then φ is said to be *noncontinuable*.

Example 2.3.4 The differential equation

$$\dot{x} = x^2$$

has a solution $\varphi(t) = 1/(1-t)$ defined on $J = (-1, 1)$. This solution is continuable to the left to $-\infty$ and is not continuable to the right. □

Example 2.3.5 The differential equation

$$\dot{x} = x^{1/3} \tag{2.3.3}$$

where $x \in \mathbb{R}$, has a solution $\psi(t) \equiv 0$ on $J = (-\infty, 0)$. This solution is continuable to the right in more than one way. For example, both $\psi_1(t) \equiv 0$ and $\psi_2(t) = (2t/3)^{3/2}$ are solutions of (2.3.3) for $t \ge 0$. □

Before stating the next result, we require the following concept.

Definition 2.3.1 A solution φ of (E) defined on the interval (a, b) is said to be *bounded* if there exists a $\beta > 0$ such that $|\varphi(t)| < \beta$ for all $t \in (a, b)$, where β may depend on φ. □

In the next result we provide a set of sufficient conditions for the continuability of solutions of (E).

Theorem 2.3.3 Let $f \in C[J \times \mathbb{R}^n, \mathbb{R}^n]$ where $J = (a, b)$ is a finite or an infinite interval. Assume that every solution of (E) is bounded. Then every solution of (E) can be continued to the entire interval $J = (a, b)$. □

In the problem section we give details for the proof of the above result. Alternatively, the reader may want to refer, for example, to Miller and Michel [37] for the proof of this result.

In Chapter 6 we establish sufficient conditions that ensure the boundedness of the solutions of (E), using the Lyapunov stability theory (refer to Example 6.2.9).

C. Dynamical systems determined by ordinary differential equations

On $\mathbb{R}^n$ we define the metric d, using the Euclidean norm $|\cdot|$, by

$$d(x, y) = |x - y| = \left[\sum_{i=1}^{n} (x_i - y_i)^2\right]^{1/2}$$

for all $x, y \in \mathbb{R}^n$. Let $A \subset \mathbb{R}^n$ be an open set, let $J \subset \mathbb{R}$ be a finite or an infinite open interval, and let $D = J \times A$. Assume that for (E) and (I_E) $f \in C[D, \mathbb{R}^n]$. In view of Theorem 2.3.1, (I_E) has at least one solution on $[t_0, t_0 + c)$ for some $c > 0$. Let S_{t_0,x_0} denote the set of all the solutions of (I_E) and let $S_E = \cup_{(t_0,x_0)\in D} S_{t_0,x_0}$. Then S_E constitutes the set of all the solutions of (E) that are defined on any half closed (resp., half open) interval $[a, b) \subset J$.

Let $T = J$ and $A \subset X = \mathbb{R}^n$. Then $\{T, X, A, S_E\}$ is a dynamical system in the sense of Definition 2.2.3. When $D = J \times A$ is understood from context, we refer to this dynamical system simply as S_E and we call S_E the *dynamical system determined by* (E).

We note in particular if $D = \mathbb{R}^+ \times \mathbb{R}^n$ and if for (E), $f \in C[D, \mathbb{R}^n]$, and if every motion in S_E is bounded, then in view of Theorem 2.3.3, every motion of S_E is continuable forward for all time (see Definition 2.2.6).

We conclude this subsection with the following important example.

Example 2.3.6 Let $A \in C[\mathbb{R}^+, \mathbb{R}^{n\times n}]$ and consider the *linear homogeneous ordinary differential equation*

$$\dot{x} = A(t)x. \tag{LH}$$

The existence and uniqueness of solutions of the initial value problems determined by (LH) are ensured by Theorems 2.3.1 and 2.3.2. In Chapter 6 (see Example 6.2.9) we show that all the motions of the dynamical systems S_{LH} determined by (LH) are continuable forward for all time (resp., all the solutions of (LH) can be continued to ∞). □

D. Two specific examples

In the following we consider two important special cases which we revisit several times.

Example 2.3.7 *Conservative dynamical systems*, encountered in classical mechanics, contain no energy-dissipating elements and are characterized by means of the *Hamiltonian function* $H(p,q)$, where $q^T = (q_1,\ldots,q_n)$ denotes *n generalized position coordinates* and $p^T = (p_1,\ldots,p_n)$ denotes *n generalized momentum coordinates.* We assume that $H(p,q)$ is of the form

$$H(p,q) = T(q,\dot{q}) + W(q) \tag{2.3.4}$$

where T denotes the *kinetic energy*, W denotes the *potential energy* of the system, and $\dot{q} = dq/dt$. These energy terms are determined from the path-independent line integrals

$$T(q,\dot{q}) = \int_0^{\dot{q}} p(q,\xi)^T d\xi = \int_0^{\dot{q}} \sum_{i=1}^{n} p_i(q,\xi)d\xi_i \tag{2.3.5}$$

$$W(q) = \int_0^{q} f(\eta)^T d\eta = \int_0^{q} \sum_{i=1}^{n} f_i(\eta)d\eta_i \tag{2.3.6}$$

where f_i, $i = 1,\ldots,n$, denote *generalized potential forces.*

Necessary and sufficient conditions for the path independence of the integral (2.3.5) are given by

$$\frac{\partial p_i}{\partial \dot{q}_j}(q,\dot{q}) = \frac{\partial p_j}{\partial \dot{q}_i}(q,\dot{q}), \qquad i,j = 1,\ldots,n. \tag{2.3.7}$$

A similar statement can be made for (2.3.6).

Conservative dynamical systems are now given by the system of $2n$ differential equations

$$\begin{cases} \dot{q}_i = \dfrac{\partial H}{\partial p_i}(p,q), & i = 1,\ldots,n \\ \dot{p}_i = -\dfrac{\partial H}{\partial q_i}(p,q), & i = 1,\ldots,n. \end{cases} \tag{2.3.8}$$

If we compute the derivative of $H(p,q)$ with respect to time t, evaluated along the solutions of (2.3.8) (given by $q_i(t), p_i(t), i = 1,\ldots,n$), we obtain

$$\begin{aligned} \frac{dH}{dt}(p(t),q(t)) &= \sum_{i=1}^{n} \frac{\partial H}{\partial p_i}(p,q)\dot{p}_i + \sum_{i=1}^{n} \frac{\partial H}{\partial q_i}(p,q)\dot{q}_i \\ &= -\sum_{i=1}^{n} \frac{\partial H}{\partial p_i}(p,q)\frac{\partial H}{\partial q_i}(p,q) + \sum_{i=1}^{n} \frac{\partial H}{\partial q_i}(p,q)\frac{\partial H}{\partial p_i}(p,q) \\ &\equiv 0. \end{aligned}$$

Thus, in a conservative dynamical system (2.3.8), the Hamiltonian (i.e., the total energy in the system) is constant along the solutions of (2.3.8).

Along with initial data $q_i(t_0), p_i(t_0), i = 1, \ldots, n$, the equations (2.3.8) determine an initial value problem. If the right-hand side of (2.3.8) is Lipschitz continuous, then according to Theorems 2.3.1 and 2.3.2, this initial value problem has unique solutions for all initial data that can be continued forward for all time. The set of the solutions of (2.3.8) generated by varying the initial data $(t_0, q(t_0), p(t_0))$ over $\mathbb{R} \times \mathbb{R}^{2n}$ determines a dynamical system in the sense of Definition 2.2.3. □

Example 2.3.8 (*Lagrange's Equation*) If the preceding dynamical system is modified to contain elements that dissipate energy, such as viscous friction elements in mechanical systems and resistors in electric circuits, we employ *Lagrange's equation* in describing such systems. For a system of *n degrees of freedom*, this equation is given by

$$\frac{d}{dt}\left(\frac{\partial L}{\partial \dot{q}_i}(q, \dot{q})\right) - \frac{\partial L}{\partial q_i}(q, \dot{q}) + \frac{\partial D}{\partial \dot{q}_i}(\dot{q}) = F_i, \qquad i = 1, \ldots, n \tag{2.3.9}$$

where $q^T = (q_1, \ldots, q_n)$ denotes the *generalized position vector*. The function $L(q, \dot{q})$ is called the *Lagrangian* and is defined as

$$L(q, \dot{q}) = T(q, \dot{q}) - W(q);$$

that is, it is the difference between the kinetic energy T (see (2.3.5)) and the potential energy W (see (2.3.6)).

The function $D(\dot{q})$ denotes *Rayleigh's dissipation function* which is assumed to be of the form

$$D(\dot{q}) = \frac{1}{2}\sum_{i=1}^{n}\sum_{j=1}^{n} \beta_{ij}\dot{q}_i\dot{q}_j$$

where $Q = [\beta_{ij}]$ is a symmetric, positive semidefinite matrix. The dissipation function D represents one-half the rate at which energy is dissipated as heat (produced by friction in mechanical systems and resistance in electric circuits).

The term $F_i, i = 1, \ldots, n$, in (2.3.9) denotes applied force and includes all external forces associated with the ith coordinate. The force F_i is defined to be positive when it acts to increase the value of q_i.

System (2.3.9) consists of n second-order ordinary differential equations that can be changed into a system of $2n$ first-order ordinary differential equations by letting $x_1 = q_1, x_2 = \dot{q}_1, \ldots, x_{2n-1} = q_n, x_{2n} = \dot{q}_n$. This system of equations, along with given initial data $x_i(t_0), i = 1, \ldots, 2n$, constitutes an initial value problem. If the functions L and D are sufficiently smooth, as in the preceding example, then for every set of initial data, the initial value problem has unique solutions that can be continued forward for all time. Furthermore, similarly as in the preceding example, this initial value problem determines a dynamical system. □

2.4 Ordinary Differential Inequalities

Let $J \subset \mathbb{R}$ be a finite or an infinite interval and let D denote a fixed Dini derivative. (E.g., if $\varphi \in C[J, \mathbb{R}^n]$, then $D\varphi$ denotes one of the four different Dini derivatives $D^+\varphi, D_+\varphi, D^-\varphi, D_-\varphi$.) Let $g \in C[J\times (\mathbb{R}^+)^n, \mathbb{R}^n]$ where $g(t,0)\geq 0$ for all $t\in J$. We consider differential inequalities given by

$$Dx \leq g(t,x). \qquad (EI)$$

We say that $\varphi \in C\big[[t_0, t_1), (\mathbb{R}^+)^n\big]$ is a *solution* of (EI) if $(D\varphi)(t) \leq g(t, \varphi(t))$ for all $t \in [t_0, t_1) \subset J$. Associated with (EI) we consider the *initial value problem*

$$Dx \leq g(t,x), \qquad x(t_0) = x_0 \qquad (I_{EI})$$

where $t_0 \in J$ and $x_0 \in \mathbb{R}^n_+ \cup \{0\}$ and where $\mathbb{R}_+ = (0, \infty)$. $\varphi \in C\big[[t_0, t_1), (\mathbb{R}^+)^n\big]$ is said to be a *solution* of (I_{EI}) if φ is a solution of (EI) and if $\varphi(t_0) = x_0$ (recall that $\mathbb{R}^+ = [0, \infty)$).

For $x_0 \in \mathbb{R}^n_+$, the existence of solutions of (I_{EI}) follows from the existence of the initial value problem

$$\dot{x} = g(t,x), \qquad x(t_0) = x_0$$

where $t_0 \in J$ and $x_0 \in \mathbb{R}^n_+$. Note that when $x_0 = 0$, then $\varphi(t) \equiv 0$ is a solution of (I_{EI}).

Let $T = J$, $A = \mathbb{R}^n_+ \cup \{0\} \subset X = (\mathbb{R}^+)^n$, and let X be equipped with the Euclidean metric. Let S_{t_0,x_0} denote the set of all solutions of (I_{EI}), and let

$$S_{EI} = \cup_{(t_0,x_0)\in J\times A} S_{t_0,x_0}.$$

Then S_{EI} is the set of all the solutions of (EI) with their initial values belonging to A. It now follows that $\{T, X, A, S_{EI}\}$ is a dynamical system. We refer to this system simply as *system* S_{EI}. We have occasion to use this system in subsequent chapters as a *comparison system*.

2.5 Difference Equations and Inequalities

The present section consists of two parts.

A. Difference equations

We now consider systems of *first-order difference equations* of the form

$$x(k+1) = f(k, x(k)) \qquad (D)$$

where $k \in \mathbb{N}$, $x(k) \in \mathbb{R}^n$, and $f\colon \mathbb{N} \times \mathbb{R}^n \to \mathbb{R}^n$.

Associated with (D) we have the *initial value problem*

$$x(k+1) = f(k, x(k)), \qquad x(k_0) = x_0 \qquad (I_D)$$

where $k_0 \in \mathbb{N}$, $x_0 \in \mathbb{R}^n$, and $k \in \mathbb{N}_{k_0} \triangleq [k_0, \infty) \cap \mathbb{N}$. We say that an n vector-valued function φ defined on $\mathbb{N}_{k_0}$ is a *solution* of (I_D) if $\varphi(k+1) = f(k, \varphi(k))$ and $\varphi(k_0) = x_0$ for all $k \in \mathbb{N}_{k_0}$. Any solution of (I_D) is also said to be a solution of (D).

Because f in (D) is a function, there are no difficulties that need to be addressed concerning the existence, uniqueness, and continuation of solutions of (I_D). Indeed, these issues follow readily from induction and the fact that the solutions of (I_D) are defined on $\mathbb{N}_{k_0}$.

Let $\varphi(\cdot, k_0, x_0)\colon \mathbb{N}_{k_0} \to \mathbb{R}^n$ denote the unique solution of (I_D) for $x(k_0) = x_0$ and let $S_D = \cup_{(k_0, x_0) \in \mathbb{N} \times \mathbb{R}^n} \{\varphi(\cdot, k_0, x_0)\}$. Then S_D is the set of all possible solutions of (D) defined on $\mathbb{N}_{k_0}$ for all $k_0 \in \mathbb{N}$.

Let $T = \mathbb{N}$ and $X = A = \mathbb{R}^n$ and let X be equipped with the Euclidean metric. Then $\{T, X, A, S_D\}$ is a discrete-time, finite-dimensional dynamical system (see Definition 2.2.3). Moreover, every motion of this dynamical system, which for short we denote by S_D, is continuable forward for all time.

Example 2.5.1 Important examples of dynamical systems determined by difference equations include *second-order sections* of *digital filters* in *direct form*, depicted in the block diagram of Figure 2.5.1.

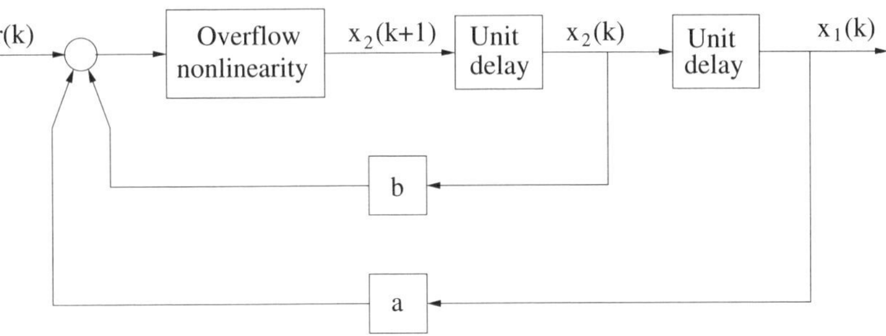

Figure 2.5.1: Digital filter in direct form.

In such filters, the type of overflow nonlinearity that is used depends on the type of arithmetic used. Frequently used overflow nonlinearities include the saturation function defined by

$$\text{sat}(\theta) = \begin{cases} 1, & \theta \geq 1 \\ \theta, & -1 < \theta < 1 \\ -1, & \theta \leq -1. \end{cases} \tag{2.5.1}$$

Letting r denote the external input to the filter, the equations that describe the filter are now given by

$$\begin{cases} x_1(k+1) = x_2(k) \\ x_2(k+1) = \text{sat}[ax_1(k) + bx_2(k) + r(k)]. \end{cases} \tag{2.5.2}$$

With $r(k)$ given for $k \in \mathbb{N}$, (2.5.2) possesses a unique solution $\varphi(k, k_0, x_0)$ for every set of initial data $(k_0, x_0) \in \mathbb{N} \times \mathbb{R}^n$ that exists for all $k \geq k_0$, where

$x_0 = [x_1(k_0), x_2(k_0)]^T$. The set of all solutions of (2.5.2) generated by varying (k_0, x_0) over $\mathbb{N} \times \mathbb{R}^n$, determines a dynamical system. □

B. Difference inequalities

We conclude the present section with a brief discussion of *systems of difference inequalities* given by

$$x(k+1) \leq g(k, x(k)) \tag{DI}$$

where $k \in \mathbb{N}$ and $g\colon \mathbb{N} \times (\mathbb{R}^+)^n \to (\mathbb{R}^+)^n$ with $g(k,0) \geq 0$ for all $k \in \mathbb{N}$. A function $\varphi\colon \mathbb{N}_{k_0} \to (\mathbb{R}^+)^n$ is a *solution* of (DI) if

$$\varphi(k+1) \leq g(k, \varphi(k))$$

for all $k \in \mathbb{N}_{k_0}$. In this case $\varphi(k_0)$ is an *initial value*. For *any* initial value $x_0 \in (\mathbb{R}^+)^n$, solutions of (DI) exist. For example, the solution of the initial value problem

$$x(k+1) = g(k, x(k)), \qquad x(k_0) = x_0$$

is such a solution of (DI) (refer to Part A above).

Let $T = \mathbb{N}$, $A = X = (\mathbb{R}^+)^n$ and let S_{DI} denote the set of all solutions of (DI) defined on $\mathbb{N}_{k_0}$ for any $k_0 \in \mathbb{N}$. Then $\{T, X, A, S_{DI}\}$ is a finite-dimensional, discrete-time dynamical system. We have occasion to make use of this system as a *comparison system* in subsequent chapters.

2.6 Differential Equations and Inclusions Defined on Banach Spaces

The present section consists of two parts.

A. Differential equations defined on Banach spaces

In order to put the presentations of the subsequent sections of this chapter into a clearer context, we briefly consider differential equations in Banach spaces. A general form of a system of first-order differential equations in a Banach space X is given by

$$\dot{x}(t) = F(t, x(t)) \tag{GE}$$

where $F\colon \mathbb{R}^+ \times C \to X$, $C \subset X$. Associated with (GE) we have the *initial value problem* given by

$$\dot{x}(t) = F(t, x(t)), \qquad x(t_0) = x_0 \tag{I_{GE}}$$

where $t_0 \in \mathbb{R}^+$, $t \geq t_0 \geq 0$, and $x_0 \in C \subset X$. Under appropriate assumptions, which ensure the existence of solutions of (GE), the initial value problem (I_{GE}) determines a continuous-time, infinite-dimensional dynamical system, denoted by S_{GE}, which consists of all the solutions $x(t, t_0, x_0)$ of (I_{GE}) with $x(t_0, x_0, t_0) = x_0$ for all $t_0 \in \mathbb{R}^+$ and $x_0 \in C$.

For the conditions of existence, uniqueness, and continuation of solutions of the initial value problem (I_{GE}), the reader may want to refer, for example, to Lakshmikantham and Leela [26] and Lasota and Yorke [27]. For example, if F is continuously differentiable, or at least locally Lipschitz continuous, then the theory of existence, uniqueness, and continuation of solutions of (I_{GE}) is essentially the same as for the finite-dimensional case we addressed in Section 2.3 when discussing ordinary differential equations (see, e.g., Dieudonné [11, Chapter 10, Section 4]). This is further demonstrated in Sections 2.7 and 2.8, where we concern ourselves with special classes of dynamical systems defined on Banach spaces, described by functional differential equations and Volterra integrodifferential equations, respectively. In general, however, issues concerning the well posedness of initial value problems (I_{GE}) can be quite complicated. For example, as shown in Godunov [15], if F in (GE) is only continuous, then (I_{GE}) may not have a solution. Throughout this book, we assume that (I_{GE}) and the associated dynamical systems are well posed.

Important classes of infinite-dimensional continuous-time dynamical systems are determined by partial differential equations. Such systems are addressed in Section 2.10. In the analysis of initial and boundary value problems determined by partial differential equations, semigroups play an important role. Semigroups, which are important in their own right in determining a great variety of dynamical systems, are treated in Section 2.9. We show how such systems may frequently be viewed as special cases of (GE) and (I_{GE}).

B. Differential inclusions defined on Banach spaces

In many applications (e.g., in certain classes of partial differential equations), the function F in (GE) may be discontinuous or even multivalued. This generality gives rise to *differential inclusions in Banach spaces*. One such form of systems of differential inclusions is briefly discussed in the following.

Let Ω be an open subset of a Banach space X, let 2^X denote the set of all subsets of X, let $\emptyset$ be the empty set, and let $F\colon \mathbb{R}^+ \times \Omega \to 2^X - \emptyset$ be a *set-valued mapping*. We consider systems of differential inclusions given by ([1], [34])

$$\dot{x}(t) \in F(t, x) \tag{GI}$$

where $t \in \mathbb{R}^+$, $x \in \Omega$, and $\dot{x}(t) = dx(t)/dt$. Associated with (GI), we have the *initial value problem*

$$\dot{x}(t) \in F(t, x), \qquad x(t_0) = x_0 \tag{I_{GI}}$$

where $t_0 \in \mathbb{R}^+$ and $x_0 \in \Omega$.

A differentiable function φ defined on an interval $[t_0, t_1)$ (t_1 may be infinite) is said to be a *solution of* (I_{GI}) if $\varphi(t_0) = x_0$ and if $\dot{\varphi}(t) \in F(t, \varphi(t))$ for all $t \in [t_0, t_1)$. We call any solution of (I_{GI}) a *solution of* (GI).

Now let

$$S_{GI} = \big\{\varphi(\cdot, t_0, x_0)\colon \varphi(\cdot, t_0, x_0) \text{ is a solution of } (I_{GI}) \text{ defined on } [t_0, t_1), t_1 > t_0, t_0 \in \mathbb{R}^+, x_0 \in \Omega\big\}.$$

Then S_{GI} is a dynamical system that we call the *dynamical system determined by* (GI).

In the following, we consider some specific cases.

Example 2.6.1 Let Ω be an open subset of $\mathbb{R}^n$ and let $f^m, f^M \in C[\mathbb{R}^+ \times \Omega, \mathbb{R}^n]$ where $f^m(t,x) \leq f^M(t,x)$ for all $(t,x) \in \mathbb{R}^+ \times \Omega$ where inequality of vectors is to be interpreted componentwise. Now consider systems of differential inequalities given by

$$f^m(t,x) \leq \dot{x} \leq f^M(t,x) \tag{IE}$$

where $\dot{x} = dx/dt$.

A function $\varphi \in C^1\big[[t_0, t_1), \Omega\big]$, where $t_0 \in \mathbb{R}^+$ and where t_1 may be finite or infinite, is said to be a solution of (IE) if for all $t \in [t_0, t_1)$,

$$f^m(t, \varphi(t)) \leq \dot{\varphi}(t) \leq f^M(t, \varphi(t)).$$

We refer to the set of all the solutions of (IE), denoted by S_{IE}, as the *dynamical system determined by* (IE).

The existence of the solutions of (IE) is guaranteed by the existence of the solutions of systems of ordinary differential equations. Thus, for any $f \in C[\Omega \times \mathbb{R}^+, \mathbb{R}^n]$ satisfying

$$f^m(t,x) \leq f(t,x) \leq f^M(t,x) \tag{IE}$$

for all $(t,x) \in \mathbb{R}^+ \times \Omega$, any solution of the equation

$$\dot{x} = f(t,x) \tag{E}$$

must also be a solution of (IE).

It is clear that S_{IE} is a specific example of a dynamical system determined by differential inclusions. □

Example 2.6.2 Consider systems described by the set of equations

$$\dot{x} = Ax + Bu \tag{2.6.1a}$$

where $x \in \mathbb{R}^n, u \in \mathbb{R}^m, A \in \mathbb{R}^{n \times n}, B \in \mathbb{R}^{n \times m}, \dot{x} = dx/dt$, and

$$u(t) = \big[g_1(c_1^T x(t-\tau)), \dots, g_m(c_m^T x(t-\tau))\big]^T \tag{2.6.1b}$$

where $\tau > 0$, $C = [c_1, \dots, c_m] \in \mathbb{R}^{m \times m}$, and $g_i \in C[\mathbb{R}, \mathbb{R}]$, $i = 1, \dots, n$, satisfy the *sector conditions*

$$\delta_i \sigma^2 \leq g_i(\sigma)\sigma \leq \Delta_i \sigma^2 \tag{2.6.1c}$$

where $\Delta_i \geq \delta_i \geq 0$, $i = 1, \dots, m$.

System (2.6.1) defines a feedback control system consisting of a linear plant and nonlinear controllers that take transportation delays into account. The sector conditions (2.6.1c) allow for deterministic uncertainties associated with the control actuators.

We refer to the set of all the solutions of system (2.6.1a)–(2.6.1c), denoted by $S_{(2.6.1)}$, as the dynamical system determined by (2.6.1). It is clear that $S_{(2.6.1)}$ is a specific example of a dynamical system determined by differential inclusions. □

We conclude by noting that the system (2.6.1a)–(2.6.1c) is a *differential-difference equation*. Such equations are special cases of *functional differential equations*, which we address next.

2.7 Functional Differential Equations

Let C_r denote the set $C\big[[-r,0],\mathbb{R}^n\big]$ with norm defined by

$$\|\varphi\| = \max\big\{|\varphi(t)|:\ -r \le t \le 0\big\}. \tag{2.7.1}$$

For a given function $x(\cdot)$ defined on $[-r,c)$, $c>0$, let x_t be the function determined by $x_t(s) = x(t+s)$ for $-r \le s \le 0$ and $t \in [0,c)$. A *retarded functional differential equation* with *delay r* is defined as

$$\dot{x}(t) = F(t, x_t) \tag{F}$$

where $F\colon \Omega \to \mathbb{R}^n$ and Ω is an open set in $\mathbb{R}\times C_r$. A differentiable function $p \in C\big[[t_0-r, t_0+c),\mathbb{R}^n\big]$, $c>0$, is a solution of equation (F) if $(t,p_t)\in\Omega$ for $t\in[t_0,t_0+c)$ and $\dot{p} = F(t,p_t)$ for $t\in[t_0,t_0+c)$.

At first glance it may appear that the functional differential equation (F) is not a special case of the general differential equation (GE) defined on a Banach space X (refer to Subsection 2.6A), because for the former, the range of the function F is in $\mathbb{R}^n$ (and not in C_r), and for the latter, the range of the function F is in $C \subset X$. However, it turns out that the functional differential equation (F) can be transformed into an equivalent equation which is a special case of (GE). To see this, we note that

$$\begin{aligned}
\dot{x}_t(s) &= \dot{x}(t+s), \quad -r \le s \le 0\\
&= \lim_{h\to 0^+}\frac{1}{h}[x(t+h+s) - x(t+s)], \quad -r \le s \le 0\\
&= \lim_{h\to 0^+}\frac{1}{h}[x_{t+h}(s) - x_t(s)]\\
&\stackrel{\triangle}{=} \frac{d}{dt}x_t(s).
\end{aligned}$$

Defining F_t by

$$F_t(t,x_t)(s) = F(t+s, x_{t+s}), \quad -r \le s \le 0,$$

it follows that the functional differential equation (F) can equivalently be expressed by the equation

$$\dot{x}_t = \frac{d}{dt}x_t = F_t(t,x_t) \tag{$\widetilde{F}$}$$

which is a special case of equation (GE) because the range of F_t is in C_r.

Example 2.7.1 *Linear retarded functional differential equations* have the form

$$\dot{x} = L(x_t) \tag{LF}$$

where L is a linear operator defined on C_r given by the Riemann–Stieltjes integral

$$L(\varphi) = \int_{-r}^{0} [dB(s)]\varphi(s) \tag{2.7.2}$$

where $B(s) = [b_{ij}(s)]$ is an $n \times n$ matrix whose entries are functions of bounded variation on $[-r, 0]$ (see, e.g., Yoshizawa [47]).

A special case of (LF) are *linear differential-difference equations* given by

$$\dot{x}(t) = A_1 x(t) + B_1 x(t - r) \tag{2.7.3}$$

where A_1 and B_1 are constant matrices (see, e.g., Bellman and Cooke [4]). □

Example 2.7.2 As a special case of the above example, we consider the *scalar equation*

$$\dot{x}(t) = \int_{-r}^{0} x(t + s) d\eta(s) \tag{2.7.4}$$

where η is a function of bounded variation on $[-r, 0]$ and the integral in (2.7.4) denotes a Riemann–Stieltjes integral. Defining $L\colon C_r \to \mathbb{R}$ by

$$L(\varphi) = \int_{-r}^{0} \varphi(s) d\eta(s),$$

we can rewrite (2.7.4) as

$$\dot{x}(t) = L(x_t).$$

If in particular, we consider the *scalar differential-difference equation*

$$\dot{x}(t) = ax(t) + bx(t - r), \tag{2.7.5}$$

where a, b are real constants and $t \in [0, c)$, and if we let

$$\eta(s) = \begin{cases} 0, & s = -r \\ b, & -r < s < 0 \\ a + b, & s = r \end{cases}$$

then we obtain in the present case

$$L(\varphi) = \int_{-r}^{0} \varphi(s) d\eta(s) = a\varphi(0) + b\varphi(-r).$$ □

We now associate with (F) the *initial value problem*

$$\dot{x}(t) = F(t, x_t), \qquad x_{t_0} = \psi \tag{I_F}$$

where $(t_0, \psi) \in \Omega \subset \mathbb{R} \times C_r$. A function $p \in C\big[[t_0 - r, t_0 + c), \mathbb{R}^n\big], c > 0$, is a *solution* of (I_F) if p is a solution of (F) and if $p_{t_0} = \psi$ (i.e., $p_{t_0}(s) = p(t_0+s) = \psi(s)$ for $-r \leq s < 0$).

If in (F) the function F is continuous, then $p \in C\big[[t_0 - r, t_0 + c), \mathbb{R}^n\big], c > 0$, is a solution of (I_F) if and only if

$$\begin{cases} p(t) = \psi(t - t_0), & t_0 - r \leq t \leq t_0 \\ p(t) = \psi(0) + \displaystyle\int_{t_0}^{t} F(s, p_s)ds, & t > t_0. \end{cases} \tag{2.7.6}$$

Alternatively, if we define an operator T on the function space $C\big[[t_0-r, t_0+c), \mathbb{R}^n\big]$ by

$$\begin{cases} (T)(t) = \psi(t - t_0), & t_0 - r \leq t \leq t_0 \\ (T)(t) = \psi(0) + \displaystyle\int_{t_0}^{t} F(x, p_s)ds, & t > t_0 \end{cases} \tag{2.7.7}$$

then p is a solution of (I_F) if and only if p is a *fixed point* of the operator T, that is, if and only if $Tp = p$. Note that when p satisfies (2.7.6), then the continuity of p implies the differentiability of p on $[t_0, c)$.

Similarly as in the case of ordinary differential equations (see Theorem 2.3.1), the following result provides a set of sufficient conditions for the existence of solutions of the initial value problem (I_F).

Theorem 2.7.1 Let Ω be an open set in $\mathbb{R} \times C_r$ and let $F \in C[\Omega, \mathbb{R}^n]$. Then for any $(t_0, \psi) \in \Omega$, (I_F) has a solution defined on $[t_0 - r, t_0 + c)$ for some $c > 0$. □

In the problem section we provide details for the proof of Theorem 2.7.1.

Similarly as in the case of ordinary differential equations (see Theorem 2.3.2), the next result provides a set of sufficient conditions for the uniqueness of solutions of the initial value problem (I_F).

Theorem 2.7.2 Let Ω be an open set in $\mathbb{R} \times C_r$ and assume that on every compact set $K \subset \Omega$, F satisfies the Lipschitz condition

$$\big|F(t, x) - F(t, y)\big| \leq L_K \|x - y\| \tag{2.7.8}$$

for all $(t, x), (t, y) \in K$, where L_K is a constant that depends only on K, $|\cdot|$ is a norm on $\mathbb{R}^n$, and $\|\cdot\|$ is the norm defined on C_r in (2.7.1). Then (I_F) has at most one solution on the interval $[t_0 - r, t_0 + c)$ for any $c > 0$. □

In the problem section we provide details for the proof of Theorem 2.7.2. Also, in Chapter 4, we prove a more general uniqueness result, applicable to differential equations defined on Banach spaces, in the context of the Lyapunov theory. Theorem 2.7.2 is a special case of that result (refer to Example 4.4.1).

Now let $p \in C\big[[t_0 - r, b), \mathbb{R}^n\big]$ be a solution of (F) where $b > t_0$. We say that p_0 is a *continuation* of p if there exists a $b_0 > b$ such that $p_0 \in C\big[[t_0 - r, b_0), \mathbb{R}^n\big]$ is a solution of (F) with the property that $p_0(t) = p(t)$ for $t \in [t_0 - r, b)$. A solution p of (F) is said to be *noncontinuable* if no such continuation exists.

Before giving a continuation result for (F), we recall that a mapping $F : X_1 \to X_2$, where X_1 and X_2 are metric spaces, is said to be *completely continuous* if F is continuous and if the closure of $F(B) = \{F(x) \colon x \in B\}$ is compact for every bounded closed set $B \subset X_1$.

Theorem 2.7.3 Let $\Omega = [t_0 - r, a) \times C_r$ where $a > t_0$ is finite or infinite. Assume that $F \colon \Omega \to \mathbb{R}^n$ is completely continuous and that every solution of (F) is bounded. Then every solution of (F) can be extended to the entire interval $[t_0 - r, a)$. □

In the problem section we provide details for the proof of Theorem 2.7.3. In Chapter 3 we present results that ensure the boundedness of the solutions of (F), using Lyapunov stability theory.

Now let $A \subset C_r$ be an open set, let $J \subset \mathbb{R}$ be a finite or an infinite interval, and let $\Omega = J \times A$. Assume that $F \in C[\Omega, \mathbb{R}^n]$. Then (I_F) has at least one solution defined on $[t_0 - r, t_0 + c)$ (see Theorem 2.7.1). Let $S_{t_0,\psi}$ denote the set of all the solutions of (I_F) and let $S_F = \cup_{(t_0,\psi)\in\Omega} S_{t_0,\psi}$. Then S_F is the set of the solutions of (F) that are defined on any half closed (resp., half open) interval $[a, b) \subset J$.

Next, let $T = J$ and let $A \subset X = C_r$ with the metric determined by the norm $\|\cdot\|$ given in (2.7.1). Then $\{T, X, A, S_F\}$ is a dynamical system in the sense of Definition 2.2.3. When T, X, and A are known from context, we refer to this dynamical system simply as S_F and we speak of the *dynamical system determined by* (F).

Finally, we note that if in particular $\Omega = \mathbb{R}^+ \times C_r$ and $F \colon \Omega \to \mathbb{R}^n$ is completely continuous and if every motion of S_F is bounded, then in view of Theorem 2.7.3, every motion of S_F is continuable forward for all time.

When F in equation (F) is a function of t, x_t, and $\dot{x}_t$ (rather than t and x_t), then the resulting equation is called a *neutral functional differential equation.* As in the case of retarded functional differential equations, such equations determine dynamical systems. We do not pursue systems of this type in this book.

2.8 Volterra Integrodifferential Equations

Volterra integrodifferential equations may be viewed as retarded functional differential equations with infinite delay; that is,

$$\dot{x}(t) = F(t, x_t) \tag{V}$$

where the interval $[-r, 0]$ is replaced by the interval $(-\infty, 0]$. This necessitates the use of a *fading memory space* X which consists of all measurable functions $\varphi \colon (-\infty, 0] \to \mathbb{R}^n$ with the property that φ is continuous on $-h \leq t \leq 0$ and that for every $\varphi \in X$, the function $\|\cdot\|$ defined by

$$\|\varphi\| = \sup\{|\varphi(t)| \colon -h \leq t \leq 0\} + \int_{-\infty}^{-h} p(t)|\varphi(t)|dt \tag{2.8.1}$$

is finite, where $p\colon (-\infty, -h) \to \mathbb{R}$ is a positive, continuously differentiable function such that $\dot{p}(t) \geq 0$ on $(-\infty, -h)$. It can easily be verified that this function is a norm on X.

More generally other choices of norms for X include

$$\|\varphi\| = \sup\big\{|\varphi(t)|\colon -h \leq t \leq 0\big\} + \left[\int_{-\infty}^{-h} p(t)|\varphi(t)|^q dt\right]^{1/q} \tag{2.8.2}$$

where $q \in [1, \infty)$. If in particular $q = 2$ and $h = 0$, then the norm (2.8.2) is induced by the inner product

$$\|\varphi\|^2 = \langle \varphi, \varphi \rangle = \langle \varphi(0), \varphi(0) \rangle + \int_{-\infty}^{0} p(t)\langle \varphi(t), \varphi(t) \rangle dt. \tag{2.8.3}$$

It can readily be shown that when X is equipped with (2.8.2), then $(X, \|\cdot\|)$ is a Banach space and when X is equipped with the inner product (2.8.3), then $(X, \langle\cdot,\cdot\rangle)$ is a Hilbert space.

Associated with (V) is the *initial value problem*

$$\dot{x}(t) = F(t, x_t), \qquad x_{t_0} = \psi \tag{I_V}$$

where $(t_0, \psi) \in \mathbb{R}^+ \times X$. A function $\varphi \in C[(-\infty, t_0 + c), \mathbb{R}^n]$, $c > 0$, is a *solution* of (I_V) if φ is a solution of (V) (i.e., $\dot{\varphi}(t) = F(t, \varphi_t)$ for $t \in [t_0, t_0 + c)$), and if $\varphi_{t_0} = \psi$ (i.e., $\varphi_{t_0}(s) = \varphi(t_0 + s) = \psi(s)$ for $-\infty < s \leq 0$).

We do not present results here concerning the existence, uniqueness, and continuation of solutions of (I_V). Instead, we refer the reader to Hale [20] for such results.

Let $T = \mathbb{R}^+$ and $A \subset X$, let $S_{t_0,\psi}$ denote the set of all the solutions of (I_V) and let $S_V = \cup_{(t_0,\psi)} S_{t_0,\psi}$. Then S_V denotes the set of all the solutions of (V) that are defined on any interval $[a, b) \subset \mathbb{R}^+$ and $\{T, X, A, S_V\}$ is a dynamical system. When the context is clear, we simply speak of the *dynamical system* S_V.

An important class of Volterra integrodifferential equations are *linear Volterra integrodifferential equations* of the form

$$\dot{x}(t) = Ax_t(0) + \int_{-\infty}^{0} K(s)x_t(s)ds \tag{LV}$$

which can equivalently be expressed as

$$\dot{x}(t) = Ax(t) + \int_{-\infty}^{t} K(s - t)x(s)ds \tag{2.8.4}$$

for $t \geq 0$, where $A \in \mathbb{R}^{n\times n}$ and $K = [k_{ij}]$ is a matrix-valued function with elements $k_{ij} \in L_1[(-\infty, 0), \mathbb{R}]$, $1 \leq i, j \leq n$.

Now let

$$X_p = \big\{\psi\colon (-\infty, 0] \to \mathbb{R}^n \text{ and } \psi\colon (-\infty, 0) \to \mathbb{R}^n \text{ belong to } L_p\big[(-\infty, 0), \mathbb{R}^n\big]\big\}$$

and let X_p be equipped with a norm given by

$$\|\psi\| = |\psi(0)| + \left[\int_{-\infty}^{0} |\psi(t)|^p dt\right]^{1/p} \tag{2.8.5}$$

where $p \in [1, \infty)$, and let

$$Y_p = \big\{\psi \in X_p \colon \dot{\psi} \in L_p\big[(-\infty, 0), \mathbb{R}^n\big] \text{ and } \psi(t) = \psi(0) + \int_0^t \dot{\psi}(s)ds \text{ for all } t \geq 0\big\}.$$

Associated with (LV) we have the *initial value problem*

$$\begin{cases} \dot{x}(t) = Ax_t(0) + \int_{-\infty}^{0} K(s)x_t(s)ds, & t \geq 0 \\ x(t) = \psi(t), & t \leq 0 \end{cases} \tag{I_{LV}}$$

where $\psi \in Y_p$.

In Barbu and Grossman [3], the following result is established for (I_{LV}).

Theorem 2.8.1 For any $\psi \in Y_p$, the initial value problem (I_{LV}) has a unique solution $x(t, \psi)$ that is defined on $(-\infty, \infty)$. □

For any $t_0 \in \mathbb{R}$, let $y(t, \psi, t_0) = x(t - t_0, \psi)$ where $x(t, \psi)$ denotes the unique solution of (I_{LV}). Let

$$S_{LV} = \big\{y = y(t, \psi, t_0) \colon t_0 \in \mathbb{R}, \psi \in Y_p\big\},$$

let $T = \mathbb{R}$, and let $A = Y_p \subset X_p = X$. Then $\{T, X, A, S_{LV}\}$ is a dynamical system, which for short, we simply refer to as *dynamical system* S_{LV}, or as the *dynamical system determined* by (LV).

In the following example we consider a simple model of the dynamics of a multicore nuclear reactor. We revisit this model in Chapter 9.

Example 2.8.1 [31] (*Point kinetics model of a multicore nuclear reactor*) We consider the point kinetics model of a multicore nuclear reactor with l cores described by the equations

$$\begin{cases} \Lambda_i \dot{p}_i(t) = \big[\rho_i(t) - \varepsilon_i - \beta_i\big]p_i(t) + \rho_i(t) + \sum_{k=1}^{6} \beta_{ki} c_{ki}(t) \\ \qquad + \sum_{j=1}^{l} \varepsilon_{ji}(P_{j0}/P_{i0}) \int_{-\infty}^{t} h_{ji}(t - s)p_j(s)ds \\ \dot{c}_{ki}(t) = \lambda_{ki}[p_i(t) - c_{ki}(t)], \qquad i = 1, \dots, l, \qquad k = 1, \dots, 6 \end{cases} \tag{2.8.6}$$

where $p_i : \mathbb{R} \to \mathbb{R}$ and $c_{ki} : \mathbb{R} \to \mathbb{R}$ represent the power in the ith core and the concentration of the kth precursor in the ith core, respectively. The constants $\Lambda_i, \varepsilon_i, \beta_{ki}, \varepsilon_{ji}$, P_{i0}, and λ_{ki} are all positive and

$$\beta_i = \sum_{k=1}^{6} \beta_{ki}.$$

The functions $h_{ji} \in L_1(\mathbb{R}^+, \mathbb{R})$. They determine the coupling between cores due to neutron migration from the jth to the ith core. The function ρ_i represents the reactivity of the ith core which we assume to have the form

$$\rho_i(t) = \int_{-\infty}^{t} w_i(t-s)p_i(s)ds \tag{2.8.7}$$

where $w_i \in L_1[\mathbb{R}^+, \mathbb{R}]$. The functions $p_i(t)$ and $c_{ki}(t)$ are assumed to be known, bounded, continuous functions defined on $-\infty < t < 0$.

In the present context, a physically realistic assumption is that $c_{ki}(t)e^{\lambda_{ki}t} \to 0$ as $t \to -\infty$. Under this assumption, we can solve for c_{ki} in terms of p_i to obtain

$$c_{ki}(t) = \int_{-\infty}^{t} \lambda_{ki} e^{-\lambda_{ki}(t-s)} p_i(s)ds. \tag{2.8.8}$$

Using (2.8.7) and (2.8.8) to eliminate ρ_i and c_{ki} from (2.8.6), we obtain l Volterra integrodifferential equations for $p_i(t)$, $i = 1, \ldots, l$. To express these equations in a more compact form, we let

$$\begin{aligned} F_i(t) &= \Lambda_i^{-1}\left[w_i(t) + \sum_{k=1}^{6} \beta_{ki}\lambda_{ki}e^{-\lambda_{ki}t} + \varepsilon_{ii}h_{ii}\right], \\ K_i &= \Lambda_i^{-1}[\varepsilon_i + \beta_i], \\ n_i(t) &= \Lambda_i^{-1} w_i(t), \quad \text{and} \\ G_{ij} &= \frac{\varepsilon_{ij}P_{j0}h_{ji}(t)}{\Lambda_i P_{i0}}. \end{aligned}$$

With $p_i(t)$ defined on $-\infty < t < \infty$, we have

$$\begin{aligned} \dot{p}_i(t) = -K_i p_i(t) &+ \int_{-\infty}^{t} F_i(t-s)p_i(s)ds + p_i(t)\int_{-\infty}^{t} n_i(t-s)p_i(s)ds \\ &+ \sum_{j=1, i\neq j}^{l} \int_{-\infty}^{t} G_{ij}(t-s)p_j(s)ds, \qquad i = 1, \ldots, l \end{aligned}$$

for $t \geq 0$ and $p_i(t) = \varphi_i(t)$ defined on $-\infty < t \leq 0$ where $\varphi_i \in Z_i$, the fading memory space of all absolutely continuous functions ψ_i defined on $(-\infty, 0]$ such that

$$\|\psi_i\|^2 = |\psi_i(0)|^2 + \int_{-\infty}^{0} |\psi_i(s)|^2 e^{L_i s} ds < \infty,$$

where $L_i > 0$ is a constant. We address the choice of L_i in Chapter 9, when studying the stability properties of (2.8.6). The set of all solutions of system (2.8.6), generated by varying φ_i over Z_i, $i = 1, \ldots, l$, determines a dynamical system. □

2.9 Semigroups

We now address linear and nonlinear semigroups that generate large classes of dynamical systems. Before addressing the subject on hand we need to introduce some additional notation.

A. Notation

Let X and Z denote Banach spaces and let $\|\cdot\|$ denote norms on such spaces. Also, Hilbert spaces are denoted X, Z, or H with inner product $\langle\cdot,\cdot\rangle$. In this case, the norm of $x \in H$ is given by $\|x\| = \langle x, x\rangle^{1/2}$.

Let A be a linear operator defined on a domain $D(A) \subset X$ with range in Z. We call A *closed* if its graph, $Gr(A) = \{(x, Ax) \in X \times Z \colon x \in D(A)\}$ is a closed subset of $X \times Z$ and we call A *bounded* if it maps bounded sets in X into bounded sets in Z, or equivalently, if it is continuous.

Subsequently, $I\colon X \to X$ denotes the identity transformation. Given a closed linear operator $A\colon D(A) \to X, D(A) \subset X$, we define the *resolvent set* of A, $\rho(A)$, as the set of all points λ in the complex plane such that the linear transformation $(A - \lambda I)$ has a bounded inverse, $(A - \lambda I)^{-1}\colon X \to X$. The complement of $\rho(A)$, denoted $\sigma(A)$, is called the *spectral set* or the *spectrum* of A.

Finally, given a bounded linear operator $A\colon D(A) \to Z, D(A) \subset X$, its norm is defined by

$$\|A\| = \sup\big\{\|Ax\|\colon \|x\| = 1\big\}.$$

B. C_0-semigroups

Consider a process whose evolution in time can be described by a linear differential equation

$$\dot{x}(t) = Ax(t), \qquad x(0) = x_0 \in D(A) \tag{I_L}$$

for $t \in \mathbb{R}^+$. Here $A\colon D(A) \to X$ is assumed to be a linear operator with domain $D(A)$ dense in X. Moreover, A is always assumed to be closed or else to have an extension $\overline{A}$ that is closed. By a *strong solution* $x(t)$ of (I_L) we mean a function $x\colon \mathbb{R}^+ \to D(A)$ such that $\dot{x}(t)$ exists and is continuous on $\mathbb{R}^+ \to X$ and such that (I_L) is true. The *abstract initial value problem* (I_L) is said to be *well posed* if for each $x_0 \in D(A)$, there is one and only one strong solution $x(t, x_0)$ of (I_L) defined on $0 \le t < \infty$ and if in addition $x(t, x_0)$ *depends continuously* on (t, x_0) in the sense that given any $N > 0$ there is an $M > 0$ such that $\|x(t, x_0)\| \le M$ when $0 \le t \le N$ and $\|x_0\| \le N$.

If (I_L) is well posed, then there is an operator T defined by $T(t)x_0 = x(t, x_0)$ which is (for each fixed t) a bounded linear mapping from $D(A)$ to X. We call $T(t)x_0 = x(t, x_0)$, $t \geq 0$, a *trajectory* of (I_L) for x_0. Because $T(t)$ is bounded, it has a continuous extension from $D(A)$ to the larger domain X. The trajectories $x(t, x_0) = T(t)x_0$ for $x_0 \in X$ but $x_0 \notin D(A)$ are called *generalized solutions* of (I_L). The resulting family of operators $\{T(t)\colon t \in \mathbb{R}^+\}$ is called a *C_0-semigroup* or a *linear semigroup*.

Independent of the above discussion, we now define C_0-semigroup.

Definition 2.9.1 [21], [23], [39] A one-parameter family of bounded linear operators $T(t)\colon X \to X, t \in \mathbb{R}^+$, is said to be a *$C_0$-semigroup*, or a *linear semigroup*, if

(i) $T(0) = I$ (I is the identity operator on X);

(ii) $T(t + s) = T(t)T(s)$ for any $t, s \in \mathbb{R}^+$; and

(iii) $\lim_{t\to 0^+} T(t)x = x$ for all $x \in X$. □

Evidently, every C_0-semigroup is generated by some abstract differential equation of the form (I_L).

Definition 2.9.2 Given any C_0-semigroup $T(t)$, its *infinitesimal generator* is the operator defined by

$$Ax = \lim_{t\to 0^+} \frac{T(t)x - x}{t}$$

where $D(A)$ consists of all $x \in X$ for which this limit exists. □

Theorem 2.9.1 [39] For a C_0-semigroup $T(t)$, there exist an $\omega \geq 0$ and an $M \geq 1$ such that

$$\|T(t)\| \leq Me^{\omega t}.$$ □

The next result provides necessary and sufficient conditions for a given linear operator A to be the infinitesimal generator of some C_0-semigroup.

Theorem 2.9.2 [21], [39] (*Hille–Yoshida–Phillips Theorem*) A linear operator A is the infinitesimal generator of a C_0-semigroup $T(t)$ satisfying $\|T(t)\| \leq Me^{\omega t}$, if and only if

(i) A is closed and $D(A)$ is dense in X;

(ii) the resolvent set $\rho(A)$ of A contains (ω, ∞); and

$$\left\|(A - \lambda I)^{-n}\right\| \leq \frac{M}{(\lambda - \omega)^n} \quad \text{for all } \lambda > \omega,$$

$n = 1, 2, \ldots$, where I denotes the identity operator on X. □

A *C_0-semigroup of contractions* is a C_0-semigroup $T(t)$ satisfying $\|T(t)\| \leq 1$ (i.e., in Theorem 2.9.1, $M = 1$ and $\omega = 0$). Such semigroups are of particular interest in Hilbert spaces.

Definition 2.9.3 A linear operator $A\colon D(A) \to H$, $D(A) \subset H$, on a Hilbert space H is said to be *dissipative* if $\text{Re}\langle Ax, x\rangle \leq 0$ for all $x \in D(A)$. □

For C_0-semigroups of contractions we have the following result.

Theorem 2.9.3 If A is the infinitesimal generator of a C_0-semigroup of contractions on a Hilbert space H, then A is dissipative and the range of $(A - \lambda I)$ is all of H for any $\lambda > 0$. Conversely, if A is dissipative and if the range of $(A - \lambda I)$ is H for at least one constant $\lambda_0 > 0$, then A is closed and A is the infinitesimal generator of a C_0-semigroup of contractions. □

The above result is useful in the study of parabolic partial differential equations (Section 2.10).

For linear semigroups with generator A one can deduce many important qualitative properties by determining the spectrum of A. Some of these are summarized in the following results (refer to Slemrod [42]).

Theorem 2.9.4 Given any two real numbers α and β with $\alpha < \beta$ there exists a C_0-semigroup $T(t)$ on a Hilbert space H such that $\text{Re}\lambda \leq \alpha$ for all $\lambda \in \sigma(A)$ and in addition $\|T(t)\| = e^{\beta t}$ for all $t \geq 0$. □

The next result applies to the following class of semigroups.

Definition 2.9.4 A C_0-semigroup $T(t)$ is called *differentiable* for $t > r$ if for each $x \in X$, $T(t)x$ is continuously differentiable on $r < t < \infty$. □

For example, a system of linear time-invariant functional differential equations with delay $[-r, 0]$ (as discussed in the last subsection of this section) determines a semigroup that is differentiable for $t > r$. Also, systems of parabolic partial differential equations (as discussed in the next section) normally generate semigroups that are differentiable for $t > 0$. In the finite-dimensional case (when $X = \mathbb{R}^n$), for linear semigroups the generator A must be an $n \times n$ matrix whose spectrum is the set of eigenvalues $\{\lambda\}$ of A. Such semigroups are differentiable as well for $t > 0$.

Following Slemrod [42] we have the following result.

Theorem 2.9.5 If $T(t)$ is a C_0-semigroup that is differentiable for $t > r$, if A is its generator, and if $\text{Re}\lambda \leq -\alpha_0$ for all $\lambda \in \sigma(A)$, then given any positive $\alpha < \alpha_0$, there is a constant $K(\alpha) > 0$ such that $\|T(t)\| \leq K(\alpha)e^{-\alpha t}$ for all $t > r$. □

We conclude by defining the *dynamical system determined by a C_0-semigroup* $T(t)$ as

$$S_{C_0} = \{p = p(\cdot, x_0, t_0)\colon p(t, x, t_0) \triangleq T(t - t_0)x,\ t_0 \in \mathbb{R}^+, t \geq t_0, x \in X\}.$$

We consider some specific examples of dynamical systems determined by C_0-semigroups in the last subsection of this section.

C. Nonlinear semigroups

A nonlinear semigroup is a generalization of the notion of C_0-semigroup. In arriving at this generalization, the linear initial value problem (I_L) is replaced by the nonlinear initial value problem

$$\dot{x}(t) = A(x(t)), \qquad x(0) = x_0 \tag{I_N}$$

where $A\colon D(A) \to X$ is a nonlinear mapping. As mentioned already in Section 2.6 (in connection with initial value problem (I_{GE})) if A is continuously differentiable (or at least locally Lipschitz continuous), then the theory of existence, uniqueness, and continuation of solutions of (I_N) is the same as in the finite-dimensional case (see Dieudonné [11, Chapter 10, Section 4]). If A is only continuous, then (I_N) needs not to have any solution at all (see Dieudonné [11, p. 287, Problem 5]). In general, one wishes to have a theory that includes nonlinear partial differential equations. This mandates that A be allowed to be only defined on a dense set $D(A)$ and to be discontinuous. For such functions A, the accretive property (defined later) generalizes the Lipschitz property.

Definition 2.9.5 [5], [8], [9], [15], [25], [27] Assume that C is a subset of a Banach space X. A family of one-parameter (nonlinear) operators $T(t)\colon C \to C$, $t \in \mathbb{R}^+$, is said to be a *nonlinear semigroup* defined on C if

(i) $T(0)x = x$ for $x \in C$;
(ii) $T(t+s)x = T(t)T(s)x$ for $t, s \in \mathbb{R}^+$, $x \in C$; and
(iii) $T(t)x$ is continuous in (t, x) on $\mathbb{R}^+ \times C$. □

A nonlinear semigroup $T(t)$ is called a *quasi-contractive* semigroup if there is a number $w \in \mathbb{R}$ such that

$$\big\|T(t)x - T(t)y\big\| \le e^{wt}\|x - y\| \tag{2.9.1}$$

for all $t \in \mathbb{R}^+$ and for all $x, y \in C$. If in (2.9.1) $w \le 0$, then $T(t)$ is called a *contraction semigroup*. Note that $C = X$ is allowed as a special case.

The mapping A in (I_N) is sometimes multivalued (i.e., a relation) and in general must be extended to be multivalued if it is to generate a quasi-contractive semigroup. Thus, we assume that $A(x)$, $x \in X$, is a subset of X and we identify A with its graph,

$$Gr(A) = \big\{(x, y)\colon x \in X \text{ and } y \in A(x)\big\} \subset X \times Y.$$

In this case the *domain* of A, written as $D(A)$, is the set of all $x \in X$ for which $A(x) \ne \emptyset$, the *range* of A is the set

$$Ra(A) = \cup\big\{A(x)\colon x \in D(A)\big\},$$

and the *inverse* of A at any point y is defined as the set

$$A^{-1}(y) = \big\{x \in X : y \in A(x)\big\}.$$

Let λ be a real or complex scalar. Then λA is defined by

$$(\lambda A)(x) = \{\lambda y \colon y \in A(x)\}$$

and $A + B$ is defined by

$$(A + B)(x) = A(x) + B(x) = \{y + z \colon y \in A(x), z \in B(x)\}.$$

Definition 2.9.6 A multivalued operator A is said to generate a nonlinear semigroup $T(t)$ on C if

$$T(t)x = \lim_{n \to \infty} \left(I - \frac{t}{n} A\right)^{-n} (x)$$

for all $x \in C$.

The *infinitesimal generator* A_s of a nonlinear semigroup $T(t)$ is defined by

$$A_s(x) = \lim_{t \to 0^+} \frac{T(t)x - x}{t}, \qquad x \in D(A_s)$$

for all x such that this limit exists. The operator A and the infinitesimal generator A_s are generally different operators. □

Definition 2.9.7 A multivalued operator A on X is said to be *w-accretive* if

$$\|(x_1 - \lambda y_1) - (x_2 - \lambda y_2)\| \geq (1 - \lambda w)\|x_1 - x_2\| \tag{2.9.2}$$

for all $\lambda \geq 0$ and for all $x_i \in D(A)$ and $y_i \in A(x_i)$, $i = 1, 2$. □

If, in particular, X is a Hilbert space, then (2.9.2) reduces to

$$\langle (wx_1 - y_1) - (wx_2 - y_2), x_1 - x_2 \rangle \geq 0. \tag{2.9.3}$$

The above property for the nonlinear case is analogous to $(A - wI)$ being dissipative in the linear symmetric case.

Theorem 2.9.6 Assume that A is w-accretive and that for each $\lambda \in (0, \lambda_0)$,

$$Ra(I - \lambda A) \supset C = \overline{D(A)}$$

where $\overline{D(A)}$ denotes the closure of $D(A)$ and $\lambda_0 > 0$ is a constant. Then A generates a quasi-contractive semigroup $T(t)$ on C with

$$\|T(t)x - T(t)y\| \leq e^{wt}\|x - y\|$$

for all $t \in \mathbb{R}^+$ and for all $x, y \in C$. □

In general, the trajectories $T(t)x$ determined by the semigroup in Theorem 2.9.6 are generalized solutions of (I_N) that need not be differentiable. Indeed, an example is discussed in Crandall and Liggett [9, Section 4], where $w = 0$, $\overline{D(A)} = X$, A generates a quasi-contraction $T(t)$ but the infinitesimal generator A_s has an empty domain. This means that not even one trajectory $T(t)x$ is differentiable at even one time t. If the graph of A is closed, then A is always an extension of the infinitesimal generator A_s. So whenever $x(t) = T(t)x$ has a derivative, then $\dot{x}(t)$ must be in $A(x(t))$.

The situation is more reasonable in the setting of a Hilbert space H. If A is w-accretive and closed (i.e., its graph is a closed subset of $H \times H$), then for any $x \in D(A)$ the set $A(x)$ is closed and convex. Thus, there is an element $A^0(x) \in A(x)$ such that $A^0(x)$ is the element of $A(x)$ closest to the origin. Given a trajectory $x(t) = T(t)x$, the right derivative

$$D^+x(t) = \lim_{h \to 0^+} \frac{x(t+h) - x(t)}{h}$$

must exist at all points $t \in \mathbb{R}^+$ and be continuous except possibly at a countably infinite set of points. The derivative $\dot{x}(t)$ exists and is equal to $D^+x(t)$ at all points where $D^+x(t)$ is continuous. Furthermore,

$$D^+x(t) = A^0(x(t))$$

for all $t \geq 0$. These results can be generalized to any space X that is uniformly convex. (Refer to Dunford and Schwarz [12, p. 74], for the definition of a uniformly convex space. In particular, any L_p space, $1 < p < \infty$, is a uniformly convex space.)

Definition 2.9.8 A trajectory $x(t) = T(t)x_0$ is called a *strong solution* of (I_N) if $x(t)$ is absolutely continuous on any bounded subset of $\mathbb{R}^+$ (so that $\dot{x}(t)$ exists almost everywhere) if $x(t) \in D(A)$ and if $\dot{x}(t) \in A(x(t))$ almost everywhere on $\mathbb{R}^+$. □

We also have

Definition 2.9.9 The initial value problem (I_N) is called *well posed* on C if there is a semigroup $T(t)$ such that for any $x_0 \in D(A)$, $T(t)x_0$ is a strong solution of (I_N), and if $\overline{D(A)} = C$. □

We summarize the above discussion in the following theorem.

Theorem 2.9.7 If X is a Hilbert space or a uniformly convex Banach space and if A is w-accretive and closed, then the initial value problem (I_N) is well posed on $C = \overline{D(A)}$ and $\dot{x}(t) = A^0(x(t))$ almost everywhere on $\mathbb{R}^+$. □

We conclude by defining the *dynamical system determined by a nonlinear semigroup* $T(t)$ as

$$S_N = \big\{p = p(\cdot, x, t_0) \colon p(t, x, t_0) \triangleq T(t - t_0)x, t_0 \in \mathbb{R}^+, t \geq t_0, x \in C\big\}.$$

We consider in the next subsection several specific examples of semigroups.

D. Examples of semigroups

We now consider several classes of important semigroups that arise in applications and we provide some related background material which we find useful in subsequent chapters.

Example 2.9.1 (*Ordinary differential equations*) Consider initial value problems described by a system of *autonomous* first-order ordinary differential equations given by

$$\dot{x} = g(x), \qquad x(0) = x_0 \tag{2.9.4}$$

where $g \colon \mathbb{R}^n \to \mathbb{R}^n$ and where it is assumed that g satisfies the Lipschitz condition

$$\big|g(x) - g(y)\big| \le L|x - y| \tag{2.9.5}$$

for all $x, y \in \mathbb{R}^n$. In this case g is w-accretive with $w = L$ and (2.9.5) implies that g is continuous on $\mathbb{R}^n$. This continuity implies that the graph of g is closed. By Theorem 2.9.7 there exist a semigroup $T(t)$ and a subset $D \subset \mathbb{R}^n$ such that D is dense in $\mathbb{R}^n$ and for any $x_0 \in D$, any solution $x(t) = T(t)x_0$ of (2.9.4) is absolutely continuous on any finite interval in $\mathbb{R}^+$. In the present case $D = \mathbb{R}^n$ and $T(t)$ is a quasi-contractive semigroup with

$$\big|T(t)x - T(t)y\big| \le e^{Lt}|x - y| \tag{2.9.6}$$

for all $x, y \in \mathbb{R}^n$ and $t \in \mathbb{R}^+$.

Now assume that in (2.9.4) $g(x) = Ax$ where $A \in \mathbb{R}^{n \times n}$; that is,

$$\dot{x} = Ax, \qquad x(0) = x_0. \tag{2.9.7}$$

In the present case (2.9.7) determines a differentiable C_0-semigroup with generator A. The spectrum of A, $\sigma(A)$, coincides with the set of all eigenvalues of A, $\{\lambda\}$. Now according to Theorem 2.9.5, if $\mathrm{Re}\lambda \le -\alpha_0$ for all $\lambda \in \sigma(A)$, where $\alpha_0 > 0$ is a constant, then given any positive $\alpha < \alpha_0$, there is a constant $K(\alpha) > 0$ such that

$$\big\|T(t)\big\| \le K(\alpha)e^{-\alpha t}, \qquad t \in \mathbb{R}^+. \tag{2.9.8}$$

□

Example 2.9.2 (*Functional differential equations*) Consider initial value problems described by a system of *autonomous* first-order functional differential equations

$$\begin{cases} \dot{x}(t) = F(x_t), & t > 0 \\ x(t) = \psi(t), & -r \le t \le 0 \end{cases} \tag{2.9.9}$$

where $F \colon C_r \to \mathbb{R}^n$. (For the notation used in this example, refer to Section 2.7.) Assume that F satisfies the Lipschitz condition

$$\big|F(\xi) - F(\eta)\big| \le K\|\xi - \eta\| \tag{2.9.10}$$

for all $\xi, \eta \in C_r$. Under these conditions, the initial value problem (2.9.9) has a unique solution for every initial condition ψ, denoted by $p(t, \psi)$ which is defined for all $t \in \mathbb{R}^+$ (refer to Section 2.7). In this case $T(t)\psi = p_t(\cdot, \psi)$, or equivalently, $(T(t)\psi)(s) = p(t + s, \psi)$ defines a quasi-contractive semigroup on C_r. Define $A \colon D(A) \to C_r$ by

$$A\psi = \dot{\psi}, \qquad D(A) = \big\{\psi \in C_r \colon \dot{\psi} \in C_r \text{ and } \dot{\psi}(0) = F(\psi)\big\}. \tag{2.9.11}$$

Then $D(A)$ is dense in C_r, A is the generator and also the infinitesimal generator of $T(t)$, and $T(t)$ is differentiable for $t > r$.

If in (2.9.9) $F = L$ is the linear mapping from C_r to $\mathbb{R}^n$ defined in (2.7.2), we have

$$\dot{x} = L(x_t) \tag{2.9.12}$$

where

$$L(\varphi) = \int_{-r}^{0} \big[dB(s)\big]\varphi(s). \tag{2.9.13}$$

In this case the semigroup $T(t)$ is a C_0-semigroup. The spectrum of its generator consists of all solutions of the equation

$$\det\left(\int_{-r}^{0} e^{\lambda s}B(s) - \lambda I\right) = 0. \tag{2.9.14}$$

If all solutions of (2.9.14) satisfy the relation Re$\lambda \leq -\gamma_0$ for some $\gamma_0 > 0$, then given any positive $\gamma < \gamma_0$, there is a constant $K(\gamma) > 0$ such that

$$\big\|T(t)\big\| \leq K(\gamma)e^{-\gamma t}, \qquad t \in \mathbb{R}^+ \tag{2.9.15}$$

(refer to Theorem 2.9.5). □

Example 2.9.3 (*Volterra integrodifferential equations*) We discuss the class of Volterra integrodifferential equations given in Section 2.8,

$$\begin{cases} \dot{x}(t) = Ax(t) + \displaystyle\int_{-\infty}^{t} K(s-t)x(s)ds, & t \geq 0 \\ x(t) = \varphi(t), & -\infty < t \leq 0 \end{cases} \tag{2.9.16}$$

where $A \in \mathbb{R}^{n\times n}$ and $K \in L_1\big[(-\infty,0),\mathbb{R}^{n\times n}\big]$; that is, K is an $n\times n$ matrix-valued function whose entries $k_{ij} \in L_1\big[(-\infty,0),\mathbb{R}\big]$. Let X_p, $1 \leq p < \infty$, be defined as in Section 2.8. Then

$$X_p \simeq L_p\big[(-\infty,0),\mathbb{R}^n\big] \times \mathbb{R}^n \tag{2.9.17}$$

where $\simeq$ denotes an isomorphic relation. To see this, note that for any $\varphi \in X_p$, $\varphi|_{(-\infty,0)} \in L_p[(-\infty,0),\mathbb{R}^n]$, $\varphi(0) \in \mathbb{R}^n$. Conversely, for any $\psi \in L_p[(-\infty,0),\mathbb{R}^n]$ and $Z \in \mathbb{R}^n$, there is a unique $\varphi \in X_p$ such that $\varphi|_{(-\infty,0)} = \psi$, and $\varphi(0) = Z$. In this case, if we denote $\varphi = (Z,\psi)$, the norm defined by (2.8.5) can now be written as

$$\|\varphi\| = \|(Z,\psi)\| = |Z| + \left[\int_{-\infty}^{0} |\psi(s)|^p ds\right]^{1/p}, \qquad 1 \leq p < \infty. \tag{2.9.18}$$

We now define an operator $\widetilde{A}$ by

$$\widetilde{A}(Z,\psi) = \left(AZ + \int_{-\infty}^{0} K(s)\psi(s)ds, \dot{\psi}\right) \tag{2.9.19}$$

on the domain

$$D(\widetilde{A}) = \big\{(Z,\psi)\colon \dot{\psi} \in L_p\big[(-\infty,0),\mathbb{R}^n\big] \text{ and } \psi(t) = Z + \int_0^t \dot{\psi}(s)ds \text{ for all } t \le 0\big\}. \tag{2.9.20}$$

Then $\widetilde{A}$ is an infinitesimal generator of a C_0-semigroup $T(t)$ on X_p. Furthermore, when $(Z,\psi) \in D(\widetilde{A})$, the equation

$$\big(x(t), x_t\big) = T(t)(Z,\psi) \tag{2.9.21}$$

determines a function $x(t)$ which is the unique solution of (2.9.16) (refer to Barbu and Grossman [3]).

If Re$\lambda > 0$, then $\lambda \in \sigma(\widetilde{A})$ if and only if

$$\det\left(A + \int_{-\infty}^0 e^{\lambda s}K(s)ds - \lambda I\right) = 0. \tag{2.9.22}$$

On the other hand, if Re$\lambda \le 0$, then λ is always in $\sigma(\widetilde{A})$. □

There are many other important classes of semigroups, including those that are determined by partial differential equations. We address some of these in the next section.

2.10 Partial Differential Equations

In our discussion of partial differential equations we require additional nomenclature.

A. Notation

A *vector index* or *exponent* is a vector $\alpha^T = (\alpha_1,\ldots,\alpha_n)$ whose components are nonnegative integers, $|\alpha| = \sum_{j=1}^n \alpha_j$, and for any $x \in \mathbb{R}^n$,

$$x^\alpha = (x_1,x_2,\ldots,x_n)^\alpha = x_1^{\alpha_1}\cdots x_n^{\alpha_n}.$$

Let $D_k = i(\partial/\partial x_k)$ for $k = 1,\ldots,n$, where $i = (-1)^{1/2}$ and let $D = (D_1, D_2, \ldots, D_n)$ so that

$$D^\alpha = D_1^{\alpha_1}\cdots D_n^{\alpha_n}. \tag{2.10.1}$$

In the sequel we let Ω be a domain in $\mathbb{R}^n$ (i.e., Ω is a connected set) with boundary $\partial\Omega$ and closure $\overline{\Omega}$. We assume that $\partial\Omega$ is of *class* C^k for suitable $k \ge 1$. By this we mean that for each $x \in \partial\Omega$, there is a ball B with center at x such that $\partial\Omega \cap B$ can be represented in the form

$$x_i = \varphi(x_1,\ldots,x_{i-1},x_{i+1},\ldots,x_n)$$

for some $i, i = 1,\ldots,n$, with φ continuously differentiable up to order k. This smoothness is easily seen to be true for the type of regions that normally occur in applications.

Also, let $C^l[\Omega, \mathbb{C}]$ denote the set of all complex-valued functions defined on Ω whose derivatives up to order l are continuous. For $u \in C^l[\Omega, \mathbb{C}]$, $l \in \mathbb{N}$, we define the norm

$$\|u\|_l = \left(\int_\Omega \sum_{|\alpha| \leq l} |D^\alpha u|^2 \right)^{1/2}. \tag{2.10.2}$$

Let

$$\widetilde{C}^l[\Omega, \mathbb{C}] = \{u \in C^l[\Omega, \mathbb{C}] \colon \|u\|_l < \infty\}$$

and let

$$C_0^l[\Omega, \mathbb{C}] = \{u \in C^l[\Omega, \mathbb{C}] \colon u = 0 \text{ in a neighborhood of } \partial\Omega\}.$$

We define $H^l[\Omega, \mathbb{C}]$ and $H_0^l[\Omega, \mathbb{C}]$ to be the completions in the norm $\|\cdot\|_l$ of the spaces $\widetilde{C}^l[\Omega, \mathbb{C}]$ and $C_0^l[\Omega, \mathbb{C}]$, respectively. In a similar manner, we can define the spaces $H^l[\Omega, \mathbb{R}]$ and $H_0^l[\Omega, \mathbb{R}]$. The spaces defined above are sometimes called *Sobolev spaces*. Their construction builds "zero boundary conditions" into, for example, $H_0^l[\Omega, \mathbb{R}]$.

Finally, we define $C^\infty[\Omega, \mathbb{C}] = \cap_{l \in \mathbb{N}} C^l[\Omega, \mathbb{C}]$ and we say that $u \in C^\infty[\overline{\Omega}, \mathbb{C}]$ if $D^\alpha u$ can be extended to be a continuous function on $\overline{\Omega}$ for any $\alpha \in \mathbb{N}^n$. We define $C^\infty[\Omega, \mathbb{R}]$ and $C^\infty[\overline{\Omega}, \mathbb{R}]$ in a similar manner. Occasionally, we say that u is a real-valued *smooth function on* Ω (on $\overline{\Omega}$) if $u \in C^\infty[\Omega, \mathbb{R}]$ (if $u \in C^\infty[\overline{\Omega}, \mathbb{R}]$). Complex-valued smooth functions on Ω (on $\overline{\Omega}$) are defined similarly.

B. Linear equations with constant coefficients

Given $r \times r$ complex constant square matrices A_α, $\alpha \in \mathbb{N}^n$, let

$$A(D) = \sum_{|\alpha| \leq m} A_\alpha D^\alpha,$$

and consider the initial value problem

$$\frac{\partial u}{\partial t}(t, x) = A(D)u(t, x), \qquad u(0, x) = \psi(x) \tag{I_P}$$

where $t \in \mathbb{R}^+$, $x \in \mathbb{R}^n$, $\psi \in L_2[\mathbb{R}^n, \mathbb{C}]$ are given, and $u \colon \mathbb{R}^+ \times \mathbb{R}^n \to \mathbb{C}^r$ is to be determined.

Proceeding intuitively for the moment, we apply L_2-Fourier transforms to (I_P) to obtain

$$\frac{\partial \widetilde{u}(t, \omega)}{\partial t} = A(\omega)\widetilde{u}(t, \omega), \qquad \widetilde{u}(0, \omega) = \widetilde{\psi}(\omega)$$

where $A(\omega) = \sum_{|\alpha| \leq m} A_\alpha \omega^\alpha$ for all $\omega \in \mathbb{R}^n$. In order to have a solution such that $u(t, x)$ and $(\partial u / \partial t)(t, x)$ are in L_2 over $x \in \mathbb{R}^n$, it is necessary that $A(\omega)\widetilde{u}(t, \omega)$ be in L_2 over $\omega \in \mathbb{R}^n$. This places some restrictions on $A(\omega)$. For the proof of the next result, refer to Krein [23, p. 163].

Theorem 2.10.1 The mapping $T(t)\psi = u(t,\cdot)$ defined by the solutions $u(t,x)$ of (I_P) determines a C_0-semigroup on $X = L_2[\mathbb{R}^n, \mathbb{C}]$ if and only if there exists a nonsingular matrix $S(\omega)$ and a constant $K > 0$ such that for all $\omega \in \mathbb{R}^n$, the following conditions are satisfied.

(i) $|S(\omega)| \leq K$ and $|S(\omega)^{-1}| \leq K$.

(ii) $S(\omega)A(\omega)S(\omega)^{-1} = [C_{ij}(\omega)]$ is upper triangular.

(iii) $\text{Re}C_{rr}(\omega) \leq \cdots \leq \text{Re}C_{11}(\omega) \leq K$.

(iv) $|C_{ik}(\omega)| \leq K(1 + |\text{Re}C_{ii}(\omega)|)$ for $k = i+1, \ldots, r$. □

Parabolic equations (i.e., equations for which $A(D)$ is strongly elliptic, defined later) satisfy these conditions whereas hyperbolic equations do not. We demonstrate this in the next examples.

Example 2.10.1 Consider a special case of (I_P) with $r = 1, m = n = 2$, given by

$$\begin{cases} \dfrac{\partial u}{\partial t} = \dfrac{\partial^2 u}{\partial x^2} + \dfrac{\partial^2 u}{\partial y^2} + a\dfrac{\partial u}{\partial x} + b\dfrac{\partial u}{\partial y} + cu \\ u(0,x) = \psi(x). \end{cases} \tag{2.10.3}$$

For $\omega = (\omega_1, \omega_2)^T \in \mathbb{R}^2$ we have

$$A(\omega) = -\omega_1^2 - \omega_2^2 + ia\omega_1 + ib\omega_2 + c = C_{11}(\omega).$$

Clearly, $\text{Re}A(\omega) = -\omega_1^2 - \omega_2^2 + c \leq c$ for all $\omega \in \mathbb{R}^2$. Therefore, all the hypotheses of Theorem 2.10.1 are satisfied and thus, (2.10.3) determines a C_0-semigroup on $X = L_2[\mathbb{R}^2, \mathbb{C}]$. □

Example 2.10.2 Consider the initial value problem determined by the *wave equation*

$$\begin{cases} \dfrac{\partial^2 u}{\partial t^2} = \dfrac{\partial^2 u}{\partial x^2} \\ u(0,x) = \psi(x). \end{cases} \tag{2.10.4}$$

The above equation can equivalently be expressed by

$$\frac{\partial u_1}{\partial t} = u_2, \qquad \frac{\partial u_2}{\partial t} = \frac{\partial^2 u_1}{\partial x^2}$$

with $u_1 = u$ and $u_2 = \partial u/\partial t$. Equation (2.10.4) is a specific case of (I_P) with $r = 2, m = 2, n = 1$, and

$$A(\omega) = \begin{bmatrix} 0 & 1 \\ -\omega^2 & 0 \end{bmatrix}.$$

The eigenvalues of $A(\omega)$ are given by $C_{11}(\omega) = i\omega$ and $C_{22}(\omega) = -i\omega$. In order that the hypotheses of Theorem 2.10.1 be satisfied, there must exist an $S(\omega)$ such that

$S(\omega)A(\omega)S(\omega)^{-1} = C(\omega)$ where $C(\omega)$ is upper triangular with diagonal elements $C_{11}(\omega)$ and $C_{22}(\omega)$. Then

$$A(\omega)S(\omega)^{-1} = S(\omega)^{-1}\begin{bmatrix} i\omega & C_{12}(\omega) \\ 0 & -i\omega \end{bmatrix}.$$

Let

$$S(\omega)^{-1} = \begin{bmatrix} x_1(\omega) & y_1(\omega) \\ x_2(\omega) & y_2(\omega) \end{bmatrix}.$$

A straightforward calculation yields

$$S(\omega)^{-1} = \begin{bmatrix} x_1(\omega) & y_1(\omega) \\ i\omega x_1(\omega) & C_{12}(\omega)x_1(\omega) - i\omega y_1(\omega) \end{bmatrix}$$

and

$$S(\omega) = \frac{1}{[C_{12}(\omega)x_1(\omega)^2 - 2i\omega x_1(\omega)y_1(\omega)]}\begin{bmatrix} C_{12}(\omega)x_1(\omega) - i\omega y_1(\omega) & -y_1(\omega) \\ -i\omega x_1(\omega) & x_1(\omega) \end{bmatrix}.$$

Because $\mathrm{Re}C_{11}(\omega) = 0$, condition (iv) in Theorem 2.10.1 implies that $|C_{12}(\omega)| \leq K$ and condition (i) of this theorem implies that all elements of $S(\omega)$ and $S(\omega)^{-1}$ are bounded by K. Thus,

$$|C_{12}(\omega)x_1(\omega) - i\omega y_1(\omega)| \leq K$$

and

$$|\omega||C_{12}(\omega)x_1(\omega) - 2i\omega y_1(\omega)|^{-1} \leq K$$

can be combined to yield

$$\begin{aligned} |\omega|/K &\leq |C_{12}(\omega)x_1(\omega) - i\omega y_1(\omega)| + |i\omega y_1(\omega)| \\ &\leq K + |i\omega y_1(\omega)| \\ &\leq 2K + |C_{12}(\omega)x_1(\omega)|. \end{aligned}$$

Using $|C_{12}(\omega)| \leq K$ and $|x_1(\omega)| \leq K$ for all $\omega \in \mathbb{R}$, we obtain

$$|\omega|/K \leq 2K + K^2$$

for all $\omega \in \mathbb{R}$. But this is impossible. Thus, no matrix $S(\omega)$ as asserted above exists. Therefore, *the solutions of* (2.10.4) *do not generate a* C_0*-semigroup.* □

C. Linear parabolic equations with smooth coefficients

In the following $\Omega \subset \mathbb{R}^n$ is assumed to be a bounded domain with smooth boundary $\partial\Omega$. We consider the differential operator of order $2m$ given by

$$A(t, x, D) = \sum_{|\alpha| \leq 2m} a_\alpha(t, x)D^\alpha \tag{2.10.5}$$

where $\alpha \in \mathbb{N}^n$, D^α is defined in (2.10.1) and the coefficients $a_\alpha(t,x)$ are complex-valued functions defined on $[0, T_0) \times \overline{\Omega}$ where $T_0 > 0$ is allowed to be infinite. The *principal part* of $A(t, x, D)$ is the operator given by

$$A'(t,x,D) = \sum_{|\alpha|=2m} a_\alpha(t,x) D^\alpha \tag{2.10.6}$$

and $A(t, x, D)$ is said to be *strongly elliptic* if there exists a constant $c > 0$ such that

$$\mathrm{Re} A'(t,x,\xi) \geq c|\xi|^{2m}$$

for all $t \in [0, T_0)$, $x \in \Omega$, and $\xi \in \mathbb{R}^n$.

In the following, we consider *linear, parabolic partial differential equations* with initial conditions and boundary conditions given by

$$\begin{cases} \dfrac{\partial u}{\partial t}(t,x) + A(t,x,D)u(t,x) = f(t,x) & \text{on } (0,T_0) \times \Omega \\ D^\alpha u(t,x) = 0, \quad |\alpha| < m & \text{on } (0,T_0) \times \partial\Omega \\ u(0,x) = u_0(x) & \text{on } \Omega \end{cases} \tag{I_{PP}}$$

where f and u_0 are complex-valued functions defined on $(0, T_0) \times \Omega$ and Ω, respectively.

Using the theory of Sobolev spaces, generalized functions (distributions), and differentiation in the distribution sense, the following result concerning the well posedness of (I_{PP}) (involving generalized solutions for (I_{PP})) has been established (see, e.g., Pazy [39] and Friedman [14]).

Theorem 2.10.2 For (I_{PP}), assume the following.

(i) $A(t, x, D)$ is strongly elliptic.

(ii) $f, a_\alpha \in C^\infty\big[[0, T_0] \times \overline{\Omega}, \mathbb{C}\big]$ for all $|\alpha| \leq 2m$.

(iii) $u_0 \in C^\infty[\overline{\Omega}, \mathbb{C}]$.

(iv) $\lim_{x \to \partial\Omega} D^\alpha u_0(x) = 0$ for all $|\alpha| < m$.

Then there exists a unique solution $u \in C^\infty\big[[0, T_0] \times \overline{\Omega}, \mathbb{C}\big]$. □

If the operator $A(t, x, D)$ and the functions f and u_0 are real-valued, then Theorem 2.10.2 is still true with the solution u of (I_{PP}) being real-valued.

Now let $T = [0, T_0]$ and $X = A = C^\infty[\overline{\Omega}, \mathbb{C}]$ and let S_{t_0,u_0} denote the set of the (unique) solutions of (I_{PP}), where in (I_{PP}), $u(0, x) = u_0(x)$ on Ω is replaced by $u(t_0, x) = u_0(x)$ on Ω with $t_0 \in [0, T_0)$. Let $S_{PP} = \cup_{(t_0,u_0) \in [0,T_0) \times A} S_{t_0,u_0}$. Then $\{T, X, A, S_{PP}\}$ is a dynamical system. When T, X, and A are known from context, we refer to this system simply as *dynamical system* S_{PP}.

Because $A(t, x, D)$ is in general time-varying, (I_{PP}) will in general not generate a semigroup. However, in the special case when $A(t, x, D) \equiv A(x, D)$, the following result has been established (refer, e.g., to Pazy [39]).

Theorem 2.10.3 In (I_{PP}), let

$$A(x, D) = \sum_{|\alpha| \leq 2m} a_\alpha(x) D^\alpha$$

be strongly elliptic on Ω and let $Au \triangleq A(x, D)u$ be defined on

$$D(A) = H^{2m}[\Omega, \mathbb{C}] \cap H_0^m[\Omega, \mathbb{C}].$$

Then A is the infinitesimal generator of a C_0-semigroup on $L_2[\Omega, \mathbb{C}]$. □

We conclude by pointing out that dynamical systems (as well as nonlinear semigroups) are determined by nonlinear partial differential equations as well. We do not pursue this topic in this book.

2.11 Composite Dynamical Systems

Problems that arise in science and technology are frequently described by a mixture of equations. For example, in control theory, feedback systems usually consist of an interconnection of several blocks, such as the plant, the sensors, the actuators, and the controller. Depending on the application, these components are characterized by different types of equations. For example, in the case of distributed parameter systems, the plant may be described by a partial differential equation, a functional differential equation, or by a Volterra integrodifferential equation, and the remaining blocks may be characterized by ordinary differential equations or ordinary difference equations. In particular, the description of digital controllers involves ordinary difference equations.

The above is an example of a large class of *composite systems*. Depending on the context, such systems are also referred to in the literature as *interconnected systems* and *decentralized systems* (e.g., [31]). When the motions of some of the system components evolve along different notions of time (continuous time $\mathbb{R}^+$ and discrete time $\mathbb{N}$) such systems are usually referred to as *hybrid systems* (e.g., [45], [46]).

In the present section, we confine our attention to interconnected (resp., composite) dynamical systems whose motion components all evolve along the *same* notion of time. In the next section, where we address discontinuous dynamical systems, and specific examples of hybrid dynamical systems, we relax this requirement. A metric space (X, d) is said to be *nontrivial* if X is neither empty nor a singleton, it is said to be *decomposable* if there are nontrivial metric spaces (X_1, d_1) and (X_2, d_2) such that $X = X_1 \times X_2$, and it is said to be *undecomposable* if it is not decomposable.

Now let (X, d), (X_i, d_i), $i = 1, \ldots, l$, be metric spaces. We assume that $X = X_1 \times \cdots \times X_l$ and that there are constants $c_1 > 0$ and $c_2 > 0$ such that

$$c_1 d(x, y) \leq \sum_{i=1}^{l} d_i(x_i, y_i) \leq c_2 d(x, y)$$

for all $x, y \in X$, where $x = [x_1, \dots, x_l]^T$, $y = [y_1, \dots, y_l]^T$, $x_i \in X_i$, and $y_i \in X_i$, $i = 1, \dots, l$. We can define the metric d on X in a variety of ways, including, for example,

$$d(x, y) = \sum_{i=1}^{l} d_i(x_i, y_i).$$

Definition 2.11.1 [34] A dynamical system $\{T, X, A, S\}$ is called a *composite dynamical system* if the metric space (X, d) can be decomposed as $X = X_1 \times \cdots \times X_l$, $l \geq 2$, where $X_1, \dots, X_l$ are nontrivial and undecomposable metric spaces with metrics $d_1, \dots, d_l$, respectively, and if there exist two metric spaces X_i and X_j, $i, j = 1, \dots, l$, $i \neq j$, that are not isometric. □

The following example may be viewed as a distributed control (in contrast to a boundary control) of a plant that is governed by the heat equation and a controller that is governed by a system of first-order ordinary differential equations. The variables for the controller and the plant are represented by $z_1 = z_1(t)$ and $z_2 = z_2(t, x)$, respectively.

Example 2.11.1 [31], [40] We consider the composite system described by the equations

$$\begin{cases} \dot{z}_1(t) = Az_1(t) + \int_\Omega f(x) z_2(t, x) dx, & t \in \mathbb{R}^+ \\ \dfrac{\partial z_2}{\partial t}(t, x) = \alpha \Delta z_2(t, x) + g(x) c^T z_1(t), & (t, x) \in \mathbb{R}^+ \times \Omega \\ z_2(t, x) = 0 & (t, x) \in \mathbb{R}^+ \times \partial\Omega \end{cases} \qquad (2.11.1)$$

where $z_1 \in \mathbb{R}^m$, $z_2 \in \mathbb{R}$, $A \in \mathbb{R}^{m \times m}$, $c \in \mathbb{R}^m$, f and $g \in L_2[\Omega, \mathbb{R}]$, $\alpha > 0$, Ω is a bounded domain in $\mathbb{R}^n$ with a smooth boundary $\partial\Omega$, and Δ denotes the Laplacian (i.e., $\Delta = \sum_{i=1}^{n} \partial^2 / \partial x_i^2$). The system of equations (2.11.1) may be viewed as a differential equation in the Banach space $X \triangleq \mathbb{R}^m \times H_0[\Omega, \mathbb{R}]$ where $H_0[\Omega, \mathbb{R}]$ is the completion of $C_0[\Omega, \mathbb{R}]$ with respect to the L_2-norm and $H_0[\Omega, \mathbb{R}] \subset L_2[\Omega, \mathbb{R}]$ (refer to Section 2.10). For every initial condition $z_0 = [z_{10}, z_{20}] \in \mathbb{R}^m \times H_0[\Omega, \mathbb{R}]$, there exists a unique solution $z(t, z_0)$ which depends continuously on z_0. For a proof of the well posedness of system (2.11.1), refer to [31].

The set of all solutions of (2.11.1) clearly determines a composite dynamical system. □

2.12 Discontinuous Dynamical Systems

All of the various types of dynamical systems that we have considered thus far include either discrete-time dynamical systems or continuous continuous-time dynamical systems (which we simply call continuous dynamical systems). In the present section we address discontinuous dynamical systems (continuous-time dynamical systems with motions that need not be continuous), which we abbreviate as DDS. Although the classes of DDS which we consider are very general, we have to put some restrictions on the types of discontinuities that we allow. To motivate the discussion of this section and to fix some of the ideas involved, we first consider an important specific example.

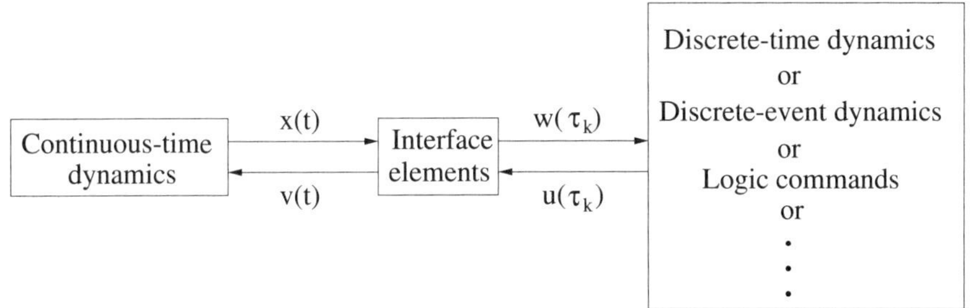

Figure 2.12.1: DDS configuration.

In Figure 2.12.1 we depict in block diagram form a configuration that is applicable to many classes of DDS, including hybrid systems and switched systems. There is a block that contains continuous-time dynamics, a block that contains phenomena which evolve at discrete points in time (discrete-time dynamics) or at discrete events, and a block that contains interface elements for the above system components. The block that contains the continuous-time dynamics is usually characterized by one or several types of the equations or inequalities defined on $\mathbb{R}^+$ enumerated in the previous sections (Sections 2.3, 2.4, and 2.6–2.10) whereas the block on the right in Figure 2.12.1 is usually characterized by difference equations or difference inequalities of the type addressed in Section 2.5 or it may contain other types of discrete characterizations involving, for example, Petri nets, logic commands, various types of discrete-event systems, and the like. The block labeled Interface Elements may vary from the very simple to the very complicated. At the simplest level, this block involves samplers and sample and hold elements. The sampling process may involve only one uniform rate, or it may be nonuniform (variable rate sampling), or there may be several different (uniform or nonuniform) sampling rates occurring simultaneously (multirate sampling).

Example 2.12.1 [29], [46] Perhaps the simplest specific example of the above class of systems are *sampled-data control systems* described by the equations

$$\begin{cases} \dot{x}(t) = A_k x(t) + B_k v(t), & \tau_k \le t < \tau_{k+1} \\ u(\tau_{k+1}) = C_k u(\tau_k) + D_k w(\tau_k), & \\ v(t) = u(\tau_k), & \tau_k \le t < \tau_{k+1} \\ w(\tau_k) = x(\tau_{k+1}^-), & \end{cases} \tag{2.12.1}$$

where $k \in \mathbb{N}, t \in \mathbb{R}^+$, $x(t) \in \mathbb{R}^n$, $u(\tau_k) \in \mathbb{R}^m$, $\{\tau_k\}$ denotes sampling instants, A_k, B_k, C_k, D_k are real matrices of appropriate dimensions, $v(\cdot)$ and $w(\cdot)$ are interface variables, and $x(\tau^-) = \lim_{\theta \to 0^+} x(\tau - \theta)$.

Now define $\widetilde{x}(t) = x(t), t \geq \tau_0$ and $\widetilde{u}(t) = v(t) = u(\tau_k), \tau_k \leq t < \tau_{k+1}, k \in \mathbb{N}$. Then $\widetilde{x}(t) = x(t^-)$ at $t = \tau_k$ and $\widetilde{u}(t^-) = u(\tau_k)$ at $t = \tau_{k+1}$ for all $k \in \mathbb{N}$. Let $y(t)^T = [\widetilde{x}(t)^T, \widetilde{u}(t)^T]$. Letting

$$F_k = \begin{bmatrix} A_k & B_k \\ 0 & 0 \end{bmatrix}, \qquad H_k = \begin{bmatrix} I & 0 \\ D_k & C_k \end{bmatrix}$$

where I denotes the $n \times n$ identity matrix, the system (2.12.1) can be described by the discontinuous ordinary differential equation

$$\begin{cases} \dot{y}(t) = F_k y(t), & \tau_k \leq t < \tau_{k+1} \\ y(t) = H_k y(t^-), & t = \tau_{k+1}, \quad k \in \mathbb{N}. \end{cases} \tag{2.12.2}$$

Next, for $k \in \mathbb{N}$, let $y_k(t, y_k, \tau_k), t \geq \tau_k$, denote the unique solution of the initial value problem

$$\begin{cases} \dot{y}(t) = F_k y(t), \\ y(\tau_k) = y_k. \end{cases} \tag{2.12.3}$$

Then clearly, for every $y_0 \in \mathbb{R}^{n+m}$, the unique solution of the DDS (2.12.2) is given by

$$y(t, y_0, \tau_0) = y_k(t, y_k, \tau_k), \qquad \tau_k \leq t < \tau_{k+1}, \ k \in \mathbb{N}.$$

Thus, the solutions of (2.12.2) are made up of an infinite sequence of solution segments determined by the solutions of (2.12.3), $k \in \mathbb{N}$, and these solutions may be discontinuous at the points of discontinuity given by $\{\tau_k\}, k = 1, 2, \ldots$. Finally, it is clear that the solutions of (2.12.2) determine a DDS. □

In Chapter 3 we develop a stability theory for general DDS, $\{\mathbb{R}^+, X, A, S\}$, defined on metric spaces, and in subsequent chapters, we specialize this theory for specific classes of finite-dimensional and infinite-dimensional dynamical systems determined by various equations and semigroups of the type described in the present chapter. In order to establish meaningful and reasonable results, it is necessary to impose some restrictions on the discontinuities of the motions $p \in S$, which of course should conform to assumptions that one needs to make in the modeling process of the DDS. Unless explicitly stated otherwise, we assume throughout this book that for a given discontinuous motion $p \in S$, the set of discontinuities is *unbounded* and *discrete* and is of the form

$$E_{1p} = \left\{\tau_1^p, \tau_2^p, \ldots : \tau_1^p < \tau_2^p < \cdots \right\}.$$

In the above expression, E_{1p} signifies the fact that in general, different motions may possess different sets of times at which discontinuities may occur. Because in most cases, the particular set E_{1p} in question is clear from context, we usually suppress the p-notation and simply write

$$E_1 = \left\{\tau_1, \tau_2, \ldots : \tau_1 < \tau_2 < \cdots \right\}.$$

In the remainder of this section we consider several important specific classes of DDS.

A. Ordinary differential equations

The sampled-data control system (2.12.1) which equivalently is represented by the discontinuous differential equation (2.12.2) is a special case of discontinuous ordinary differential equations of the form

$$\begin{cases} \dot{x}(t) = f_k(t, x(t)), & \tau_k \le t < \tau_{k+1}, \\ x(t) = g_k(x(t^-)), & t = \tau_{k+1},\ k \in \mathbb{N} \end{cases} \tag{SE}$$

where for each $k \in \mathbb{N}$, $f_k \in C[\mathbb{R}^+ \times \mathbb{R}^n, \mathbb{R}^n]$, $g_k \colon \mathbb{R}^n \to \mathbb{R}^n$, and $x(\tau^-)$ is given in Example 2.12.1.

Associated with (SE), we consider the family of initial value problems given by

$$\begin{cases} \dot{x}(t) = f_k(t, x(t)) \\ x(\tau_k) = x_k, \end{cases} \tag{SE_k}$$

$k \in \mathbb{N}$. We assume that for (τ_k, x_k), (SE_k) possesses a unique solution $x^{(k)}(t, x_k, \tau_k)$ which exists for all $t \in [\tau_k, \infty)$ (refer to Section 2.3 for conditions that ensure this). Then for every $(t_0, x_0) \in \mathbb{R}^+ \times \mathbb{R}^n$, $t_0 = \tau_0$, (SE) has a unique solution $x(t, x_0, t_0)$ that exists for all $t \in [t_0, \infty)$. This solution is made up of a sequence of continuous solution segments $x^{(k)}(t, x_k, \tau_k)$, defined over the intervals $[\tau_k, \tau_{k+1})$ with initial conditions (τ_k, x_k), $k \in \mathbb{N}$, where $x_{k+1} = x(\tau_{k+1}) = g_k(x(\tau_{k+1}^-))$, $k \in \mathbb{N}$ and the initial conditions $(\tau_0 = t_0, x_0)$ are given. At the points $\{\tau_{k+1}\}$, $k \in \mathbb{N}$, the solutions of (SE) have possible jumps (determined by $g_k(\cdot)$).

The set of all the solutions of (SE), S_{SE}, determines a DDS, $\{\mathbb{R}^+, X, A, S_{SE}\}$, where $X = A = \mathbb{R}^n$.

B. Functional differential equations ([43])

For the notation that we use in the present subsection, the reader should refer to Section 2.7.

We first consider a family of initial value problems described by *continuous* retarded functional differential equations (RFDEs) of the form

$$\begin{cases} \dot{x}(t) = F_k(t, x_t), \\ x_{\tau_k} = \varphi_k, \end{cases} \tag{SF_k}$$

$k \in \mathbb{N}$. For each $k \in \mathbb{N}$ we assume that $F_k \in C[\mathbb{R}^+ \times C_r, \mathbb{R}^n]$ and that (SF_k) is well posed so that for every $(\tau_k, \varphi_k) \in \mathbb{R}^+ \times C_r$, (SF_k) possesses a unique *continuous* solution $x^{(k)}(t, \varphi_k, \tau_k)$ that exists for all $t \in [\tau_k, \infty)$. (For conditions that ensure this, refer to Section 2.7.)

We now consider *discontinuous* RFDEs of the form

$$\begin{cases} \dot{x}(t) = F_k(t, x_t), & \tau_k \le t < \tau_{k+1} \\ x_{\tau_{k+1}} = G_k(x_{\tau_{k+1}^-}), & k \in \mathbb{N} \end{cases} \tag{SF}$$

where for each $k \in \mathbb{N}$, F_k is assumed to possess the identical properties given in (SF_k) and $G_k\colon C_r \to C_r$. Thus, at $t = \tau_{k+1}$, the mapping $G_k(\cdot)$ assigns to every state $x_{\tau_{k+1}^-}$ $(x_{\tau_{k+1}^-}(\theta) = x(\tau_{k+1}^- + \theta), -r \le \theta \le 0)$ unambiguously a state $x_{\tau_{k+1}}$ $(x_{\tau_{k+1}}(\theta) = x(\tau_{k+1} + \theta), -r \le \theta \le 0)$.

Under the above assumptions for (SF) and (SF_k), it is now clear that for every $(t_0, \varphi_0) \in \mathbb{R}^+ \times C_r, t_0 = \tau_0, (SF)$ has a unique solution $x(t, \varphi_0, t_0)$ that exists for all $t \in [t_0, \infty)$. This solution is made up of a sequence of continuous solution segments $x^{(k)}(t, \varphi_k, \tau_k)$ defined over the intervals $[\tau_k, \tau_{k+1}), k \in \mathbb{N}$, with initial conditions (τ_k, φ_k), where $\varphi_k = x_{\tau_k}$, $k = 1, 2, \ldots$ and where $(\tau_0 = t_0, \varphi_0)$ are given. At the points $\{\tau_{k+1}\}, k \in \mathbb{N}$, the solutions of (SF) have possible jumps (determined by $G_k(\cdot)$).

It is clear that (SF) determines an *infinite-dimensional* DDS, $\{T, X, A, S\}$, where $T = \mathbb{R}^+$, $X = A = C_r$, the metric on X is determined by the norm $\| \cdot \|$ defined on C_r (i.e., $d(\varphi, \eta) = \| \varphi - \eta \|$ for all $\varphi, \eta \in C_r$), and S denotes the set of all the solutions of (SF) corresponding to all possible initial conditions $(t_0, \varphi_0) \in \mathbb{R}^+ \times C_r$. In the interests of brevity, we refer to this DDS as "system (SF)" or as "(SF)".

C. Differential equations in Banach spaces ([32])

We first consider a family of *initial value Cauchy problems in Banach space* X of the form

$$\begin{cases} \dot{x}(t) = F_k(t, x(t)), & t \ge \tau_k, \\ x(\tau_k) = x_k \end{cases} \qquad (SG_k)$$

for $k \in \mathbb{N}$. For each $k \in \mathbb{N}$, we assume that $F_k\colon \mathbb{R}^+ \times X \to X$ and that $\dot{x} = dx/dt$. We assume that for every $(\tau_k, x_k) \in \mathbb{R}^+ \times X$, (SG_k) possesses a unique solution $x^{(k)}(t, x_k, \tau_k)$ that exists for all $t \in [\tau_k, \infty)$. We express this by saying that (SG_k) is *well posed*.

We now consider *discontinuous initial value problems in Banach space* X given by

$$\begin{cases} \dot{x}(t) = F_k(t, x(t)), & \tau_k \le t < \tau_{k+1} \\ x(\tau_{k+1}) = g_k(x(\tau_{k+1}^-)), & k \in \mathbb{N} \end{cases} \qquad (SG)$$

where for each $k \in \mathbb{N}$, F_k is assumed to possess the identical properties given in (SG_k) and where $g_k\colon X \to X$. Under these assumptions, it is clear that for every $(t_0, x_0) \in \mathbb{R}^+ \times X$, $t_0 = \tau_0$, (SG) has a unique solution $x(t, x_0, t_0)$ that exists for all $t \in [t_0, \infty)$. This solution is made up of a sequence of solution segments $x^{(k)}(t, x_k, \tau_k)$, defined over the intervals $[\tau_k, \tau_{k+1})$, $k \in \mathbb{N}$, with initial conditions (τ_k, x_k), where $x_k = x(\tau_k)$, $k = 1, 2, \ldots$, and where $(\tau_0 = t_0, x_0)$ are given. At the points $\{\tau_{k+1}\}, k \in \mathbb{N}$, the solutions of (SG) have possible jumps (determined by $g_k(\cdot)$).

Consistent with the characterization of a *discontinuous dynamical system* given in Section 2.2, it is clear from the above that system (SG) determines a DDS, $\{T, X, A, S\}$, where $T = \mathbb{R}^+$, $A = X$, the metric on X is determined by the norm $\| \cdot \|$ defined on X (i.e., $d(x, y) = \| x - y \|$ for all $x, y \in X$), and S denotes the set of all the solutions of (SG) corresponding to all possible initial conditions $(t_0, x_0) \in \mathbb{R}^+ \times X$.

In the interests of brevity, we refer to this DDS simply as "system (SG)", or simply as "(SG)".

D. Semigroups ([33])

We require a given collection of linear or nonlinear semigroups $\mathcal{T} = \{T_i(t)\}$ defined on a Banach space X, or on a set $C \subset X$, respectively; and a given collection of linear and continuous operators $\mathcal{H} = \{H_j\}(H_j\colon X \to X)$, or of nonlinear and continuous operators $(H_j\colon C \to C)$; and a given discrete and unbounded set $E = \{t_0 = \tau_0, \tau_1, \tau_2, \dots : \tau_0 < \tau_1 < \tau_2 \cdots\} \subset \mathbb{R}^+$. The number of elements in $\mathcal{T}$ and $\mathcal{H}$ may be *finite* or *infinite*.

We now consider dynamical systems whose *motions* $y(\cdot, y_0, t_0)$ with *initial time* $t_0 = \tau_0 \in \mathbb{R}^+$ and *initial state* $y(t_0) = y_0 \in X$ (resp., $y_0 \in C \subset X$) are given by

$$\begin{cases} y(t, y_0, t_0) = T_k(t - \tau_k)(y(\tau_k)), & \tau_k \le t < \tau_{k+1}, \\ y(t) = H_k(y(t^-)), & t = \tau_{k+1}, \qquad k \in \mathbb{N}. \end{cases} \tag{SH}$$

We define the DDS *determined by semigroups* as

$$\begin{aligned} S = \big\{y = y(\cdot, x, t_0)\colon & y(t, x, t_0) = T_k(t - \tau_k)(y(\tau_k)), \ \ \tau_k \le t < \tau_{k+1}, \\ & y(t) = H_k(y(t^-)), \ \ t = \tau_{k+1}, \ \ k \in \mathbb{N}, \ \ t_0 = \tau_0 \in \mathbb{R}^+, \\ & y(\tau_0) = x \in X, \ \ \text{resp., } x \in C \subset X\big\}. \end{aligned} \tag{2.12.4}$$

Note that every motion $y(\cdot, x, t_0)$ is unique, with $y(t_0, x, t_0) = x$, exists for all $t \ge t_0$, and is continuous with respect to t on $[t_0, \infty) - \{\tau_1, \tau_2, \dots\}$, and that at $t = \tau_k$, $k = 1, 2, \dots$, $y(\cdot, x, t_0)$ *may* be discontinuous. We call the set $E_1 = \{\tau_1, \tau_2, \dots\}$ the set of discontinuities for the motion $y(\cdot, x, t_0)$.

When in (2.12.4), $\mathcal{T}$ consists of C_0-semigroups, we speak of a DDS *determined by linear semigroups* and we denote this system by S_{DC_0}. Similarly, when in (2.12.4), $\mathcal{T}$ consists of nonlinear semigroups, we speak of a DDS *determined by nonlinear semigroups* and we denote this system by S_{DN}. When the types of the elements in $\mathcal{T}$ are not specified, we simply speak of a DDS *determined by semigroups* and we denote this system, as in (2.12.4), by S.

Finally, if in the case of S_{DC_0}, the elements in $\mathcal{H}$ are linear, we use in (SH) the notation $T_k(t - \tau_k)(y(\tau_k)) = T_k(t - \tau_k)y(\tau_k)$ and $H_k(y(t^-)) = H_k y(t^-)$.

Next, a few observations may be in order:

(a) For different initial conditions (x, t_0), resulting in different motions $y(\cdot, x, t_0)$, we allow the set of discontinuities $E_1 = \{\tau_1, \tau_2, \dots\}$, the set of semigroups $\{T_k\} \subset \mathcal{T}$, and the set of functions $\{H_k\} \subset \mathcal{H}$ to differ, and accordingly, the notation $E_1^{x,t_0} = \{\tau_1^{x,t_0}, \tau_2^{x,t_0}, \dots\}$, $\{T_k^{x,t_0}\}$ and $\{H_k^{x,t_0}\}$ might be more appropriate. However, because in all cases, all meaning is clear from context, we do not use such superscripts.

(b) S_{DC_0} and S_{DN} are very general classes of DDS and include large classes of finite-dimensional dynamical systems determined by ordinary differential equations and inequalities and large classes of infinite-dimensional dynamical systems determined by differential-difference equations, functional differential equations, Volterra integrodifferential equations, certain classes of partial differential equations, and the like.

(c) The dynamical system models S_{DC_0} and S_{DN} are very flexible and include as special cases, many of the DDS considered in the literature (e.g., [2], [10], [28], [29], [30], [46]), as well as general autonomous continuous dynamical systems: (i) If $T_k(t) = T(t)$ for all k ($\mathcal{T}$ has only one element) and if $H_k = I$ for all k, where I denotes the identity transformation, then S_{DC_0} reduces to an *autonomous, linear, continuous dynamical system* and S_{DN} reduces to an *autonomous, nonlinear, continuous dynamical system*. (ii) In the case of *dynamical systems subjected to impulse effects* (see, e.g., [2]), one would choose $T_k(t) = T(t)$ for all k whereas the impulse effects are captured by an infinite family of functions $\mathcal{H} = \{H_k\}$. (iii) In the case of *switched systems*, frequently only a *finite number* of systems that are being switched are required and so in this case one would choose a finite family of semigroups, $\mathcal{T} = \{T_i(t)\}$ (see, e.g., [10], [46]); and so forth. (iv) Perhaps it needs pointing out that even though systems S_{DN} and S_{DC_0} are *determined* by families of semigroups (and nonlinearities), by themselves they are *not* semigroups, inasmuch as in general, they are time-varying and do not satisfy the hypotheses (i)–(iii) in Definitions 2.9.1 and 2.9.5.

We conclude with a specific example involving partial differential equations.

Example 2.12.2 [33] (*DDS determined by the heat equation*) We let $\Omega \subset \mathbb{R}^n$ be a bounded domain with smooth boundary $\partial\Omega$ and we let $\Delta = \sum_{i=1}^n \partial^2/\partial x_i^2$ denote the *Laplacian*. Also, we let $X = H^2[\Omega, \mathbb{R}] \cap H_0^1[\Omega, \mathbb{R}]$ where $H_0^1[\Omega, \mathbb{R}]$ and $H^2[\Omega, \mathbb{R}]$ are *Sobolev spaces* (refer to Section 2.10). For any $\varphi \in X$, we define the H^1-norm by

$$\| \varphi \|_{H^1}^2 = \int_\Omega (| \nabla \varphi|^2 + |\varphi|^2) dx \tag{2.12.5}$$

where $\nabla \varphi^T = (\partial\varphi/\partial x_1, \ldots, \partial\varphi/\partial x_n)$.

We now consider DDS determined by the equations

$$\begin{cases} \dfrac{\partial u}{\partial t} = a_k \Delta u, & (t,x) \in [\tau_k, \tau_{k+1}) \times \Omega \\ u(t,\cdot) = g_k(u(t^-,\cdot)) \overset{\triangle}{=} \varphi_{k+1}(\cdot), & t = \tau_{k+1} \\ u(t_0, x) = \varphi_0(x), & x \in \Omega \\ u(t,x) = 0, & (t,x) \in [t_0, \infty) \times \partial\Omega, \quad k \in \mathbb{N} \end{cases} \tag{2.12.6}$$

where $\varphi_0 \in X$, $a_k > 0$, $k \in \mathbb{N}$, are constants, $\{g_k\}$ is a given family of mappings with $g_k \in C[X, X], k \in \mathbb{N}$, and $E = \{t_0 = \tau_0, \tau_1, \ldots : \tau_0 < \tau_1 < \tau_2 < \cdots\}$ is a given unbounded and discrete set. We assume that $g_k(0) = 0$ and there exists a constant $d_k > 0$ such that

$$\| g_k(\varphi) \|_{H^1} \leq d_k \| \varphi \|_{H^1} \tag{2.12.7}$$

for all $\varphi \in X, k \in \mathbb{N}$.

Associated with (2.12.6) we have a family of *initial and boundary value problems* determined by

$$\begin{cases} \dfrac{\partial u}{\partial t} = a_k \Delta u, & (t,x) \in [\tau_k, \infty) \times \Omega \\ u(\tau_k, x) = \varphi_k(x), & x \in \Omega \\ u(t,x) = 0, & (t,x) \in [\tau_k, \infty) \times \partial\Omega, \end{cases} \tag{2.12.8}$$

$k \in \mathbb{N}$. It has been shown (e.g., [39]) that for each $(\tau_k, \varphi_k) \in \mathbb{R}^+ \times X$, the initial and boundary value problem (2.12.8) has a unique solution $u_k = u_k(t,x), t \geq \tau_k, x \in \Omega$, such that $u_k(t,\cdot) \in X$ for each fixed $t \geq \tau_k$ and $u_k(t,\cdot) \in X$ is a continuously differentiable function from $[\tau_k, \infty)$ to X with respect to the H^1-norm given in (2.12.5).

It now follows that for every $\varphi_0 \in X$, (2.12.6) possesses a unique solution $u(t,\cdot)$ that exists for all $t \geq \tau_0 \geq 0$, given by

$$u(t,\cdot) = \begin{cases} u_k(t,\cdot), & \tau_k \leq t < \tau_{k+1} \\ g_k(u_k(t^-,\cdot)) \triangleq \varphi_{k+1}(\cdot), & t = \tau_{k+1},\ k \in \mathbb{N} \end{cases} \tag{2.12.9}$$

with $u(t_0, x) = \varphi_0(x)$. Notice that $u(t,\cdot)$ is continuous with respect to t on the set $[t_0, \infty) - \{\tau_1, \tau_2, \dots\}$, and that at $t = \tau_k, k = 1, 2, \dots, u(t,\cdot)$ *may* be discontinuous (depending on the properties of $g_k(\cdot)$).

For each $k \in \mathbb{N}$, (2.12.8) can be cast as an *initial value problem* in the space X with respect to the H^1-norm, letting $u_k(t,\cdot) = U_k(t)$,

$$\begin{cases} \dot{U}_k(t) = A_k U_k(t), & t \geq \tau_k \\ U_k(\tau_k) = \varphi_k \in X \end{cases} \tag{2.12.10}$$

where $A_k = a_k \sum_{i=1}^n \partial^2/\partial x_i^2$ and $U_k(t, \varphi_k), t \geq \tau_k$, denotes the solution of (2.12.10) with $U(\tau_k, \varphi_k) = \varphi_k$. It has been shown (see, e.g., [39]) that (2.12.10) determines a C_0-semigroup $T_k(t-\tau_k) : X \to X$, where for any $\varphi_k \in X, U_k(t, \varphi_k) = T(t-\tau_k)\varphi_k$.

Letting $u_k(t,\cdot) = T_k(t-\tau_k)u_k(\tau_k)$ in (2.12.9), system (2.12.6) can now be characterized as

$$\begin{cases} u(t,\cdot) = T_k(t-\tau_k)u_k(\tau_k,\cdot), & \tau \leq t < \tau_{k+1} \\ u(t,\cdot) = g_k(u_k(t^-,\cdot)), & t = \tau_{k+1},\ k \in \mathbb{N}. \end{cases} \tag{2.12.11}$$

Finally, it is clear that (2.12.6) (resp., (2.12.11)) determines a DDS which is a special case of the DDS (SH). □

2.13 Notes and References

Depending on the applications, different variants of dynamical systems have been employed (e.g., Hahn [18], Willems [44], and Zubov [48]). Our concept of dynamical system (Definition 2.2.3) was first used in [35], [36] and extensively further refined in [34] in the study of the role of stability-preserving mappings in stability analysis of dynamical systems. In the special case when X is a normed linear space and each motion $p(t, a, t_0)$ is assumed to be continuous with respect to a, t, and t_0, the definition of a dynamical system given in Definition 2.2.3 reduces to the definition of a dynamical system used in Hahn [18, pp. 166–167] (called a *family of motions* in [18]). When the motions satisfy additional requirements that we do not enumerate, Definition 2.2.3 reduces to the definition of a dynamical system, defined on metric space, used by Zubov [48, p. 199] (called a *general system* in [48]). The notion of a dynamical system employed in [44] is defined on normed linear space and involves variations to Definition 2.2.3 which we do not specify here.

In the problem section we provide hints on how to prove the results given in Section 2.3. For the complete proofs of these results (except Theorem 2.3.3) and for additional material on ordinary differential equations, refer to Miller and Michel [37]. Our treatment of the continuation of solutions (Theorem 2.3.3) is not conventional, but very efficient, inasmuch as it involves Lyapunov results developed in subsequent chapters.

Ordinary differential inequalities (and ordinary difference inequalities) play an important role in the qualitative analysis of dynamical systems (see, e.g., [26]) and are employed throughout this book.

Good sources on ordinary difference equations, with applications to control systems and signal processing include Franklin and Powell [13] and Oppenheim and Schafer [38], respectively.

For the complete proofs of Theorems 2.7.1–2.7.3, and additional material on functional differential equations, refer to Hale [19]. Hale is perhaps the first to treat Volterra integrodifferential equations as functional differential equations with infinite delay [20]. For a proof of Theorem 2.8.1, refer to Barbu and Grossman [3].

For the proofs of Theorems 2.9.1–2.9.4 and for additional material concerning C_o-semigroups, refer to Hille and Phillips [21], Krein [23] (Chapter 1), and Pazy [39]. For the proof of Theorem 2.9.5, refer to Slemrod [42]. For the proofs of Theorems 2.9.6 and 2.9.7 and for additional material concerning nonlinear semigroups and differential inclusions defined on Banach spaces, refer to Crandall [8], Crandall and Liggett [9], Brezis [5], Kurtz [25], Godunov [15], Lasota and Yorke [27], and Aubin and Cellina [1]. Our presentation in Section 2.9 on semigroups and differential inclusions defined on Banach spaces (see also Section 2.6) is in the spirit of the presentation given in Michel and Miller [31] (Chapter 5), and Michel *et al.* [34].

For the proofs of Theorems 2.10.1–2.10.3, and additional material concerning partial differential equations, refer to Krein [23], Friedman [14], and Pazy [39]. Additional sources on partial differential equations include Hörmander [22] and Krylov [24]. Our presentation on partial differential equations in Section 2.10 is in the spirit of Michel and Miller [31, Chapter 5] and Michel *et al.* [34, Chapter 2].

Our presentation on composite dynamical systems in Section 2.11 is primarily based on material from Michel and Miller [31], Michel *et al.* [34, Chapter 6], and Rasmussen and Michel [40], and Section 2.12 on discontinuous dynamical systems relies primarily on material from Michel [29], Michel and Hu [30], Michel and Sun [32], Michel *et al.* [33], Sun *et al.* [43], and Ye *et al.* [46]. Finally, for a general formulation of a hybrid dynamical system defined on a metric space (involving a notion of generalized time), refer to Ye *et al.* [45] with subsequent developments given in Ye *et al.* [46], Sun *et al.* [43], Michel *et al.* [33], Michel and Sun [32], Michel and Hu [30], and Michel [29].

2.14 Problems

Problem 2.14.1 Consider a class of scalar *nth-order ordinary differential equations* given by

$$y^{(n)} = g(t, \dot{y}, \ldots, y^{(n-1)}) \tag{E_n}$$

where $t \in J \subset \mathbb{R}$, J is a finite or an infinite interval, $y \in \mathbb{R}$, $\dot{y} = y^{(1)} = dy/dt, \ldots,$ $y^{(n)} = d^n y/dt^n$, and $g \in C[J \times \mathbb{R}^n, \mathbb{R}]$. Initial value problems associated with (E_n) are given by

$$\begin{cases} y^{(n)} = g(t, \dot{y}, \ldots, y^{(n-1)}) \\ y(t_0) = y_0, \quad \dot{y}(t_0) = y_1, \quad \ldots, \quad y^{(n-1)}(t_0) = y_{n-1} \end{cases} \tag{I_{E_n}}$$

where $t_0 \in J$ and $y_0, y_1, \ldots, y_{n-1} \in \mathbb{R}$.

Show that (E_n) determines a dynamical system (in the sense of Definition 2.2.3) that we denote by S_{E_n}.

Hint: Show that (E_n) (and (I_{E_n})) can equivalently be represented by a system of n first-order ordinary differential equations. □

Problem 2.14.2 Consider a *class of nth-order ordinary scalar difference equations* given by

$$y(k) = g(k, y(k-1), \ldots, y(k-n)) \tag{D_n}$$

where $k \in \mathbb{N}_n \triangleq [n, \infty) \cap \mathbb{N}$, $n \in \mathbb{N}$, $y \colon \mathbb{N} \to \mathbb{R}$, and $g \colon \mathbb{N}_n \times \mathbb{R}^n \to \mathbb{R}$. Associated with (D_n), consider *initial value problems* given by

$$\begin{cases} y(k) = g(k, y(k-1), \ldots, y(k-n)) \\ y(0) = y_0, \qquad y(1) = y_1, \ldots, y(n-1) = y_{n-1} \end{cases} \tag{I_{D_n}}$$

where $y_0, y_1, \ldots, y_{n-1} \in \mathbb{R}$.

Show that (D_n) determines a dynamical system (in the sense of Definition 2.2.3) which we denote by S_{D_N}.

Hint: Show that (D_n) (and (I_{D_n})) can equivalently be represented by a system of n first-order ordinary difference equations. □

Problem 2.14.3 Let D denote a fixed Dini derivative and let $g \in C[J \times (\mathbb{R}^+)^n, \mathbb{R}^n]$ where $g(t, 0) \geq 0$ for all $t \in J$. Consider differential inequalities given by

$$Dx \geq g(t, x) \tag{2.14.1}$$

and define a *solution* of (2.14.1) as a function $\varphi \in C\big[[t_0, t_1), (\mathbb{R}^+)^n\big]$ that satisfies $(D\varphi)(t) \geq g(t, \varphi(t))$ for all $t \in [t_0, t_1) \subset J$. Associated with (2.14.1), we consider *initial value problems* given by

$$Dx \geq g(t, x), \qquad x(t_0) = x_0 \tag{2.14.2}$$

where $t_0 \in J$ and $x_0 \in (0,\infty)^n \cup \{0\}$. We say that $\varphi \in C\big[[t_0,t_1),(\mathbb{R}^+)^n\big]$ is a solution of (2.14.2) if $\varphi(t_0) = x_0$.

Show that (2.14.1) determines a dynamical system that we denote by $S_{(2.14.1)}$. □

Problem 2.14.4 Consider *ordinary difference inequalities* given by

$$x(k+1) \geq g(k, x(k)) \tag{2.14.3}$$

where $k \in \mathbb{N}$ and $g\colon \mathbb{N} \times (\mathbb{R}^+)^n \to (\mathbb{R}^+)^n$ with $g(k,0) \geq 0$ for all $k \in \mathbb{N}$. A function $\varphi\colon \mathbb{N}_{k_0} \to (\mathbb{R}^+)^n$ is a *solution* of (2.14.3) if

$$\varphi(k+1) \geq g(k, \varphi(k))$$

for all $k \in \mathbb{N}_{k_0}$. In this case $\varphi(k_0)$ is an *initial value.*

Show that (2.14.3) determines a dynamical system that we denote by $S_{(2.14.3)}$. □

Problem 2.14.5 (a) In Figure 2.14.1, M_1 and M_2 denote point masses, K_1, K_2, K denote spring constants, and x_1, x_2 denote displacements of the masses M_1 and M_2, respectively. Use the Hamiltonian formulation of dynamical systems described in Example 2.3.7 to derive a system of first-order ordinary differential equations that characterize this system. Verify your answer by using Newton's second law of motion to derive the same system of equations. By specifying $x_1(t_0), \dot{x}_1(t_0), x_2(t_0)$, and $\dot{x}_2(t_0)$, the above yields an initial value problem.

(b) Show that the above mechanical system determines a dynamical system in the sense of Definition 2.2.3. □

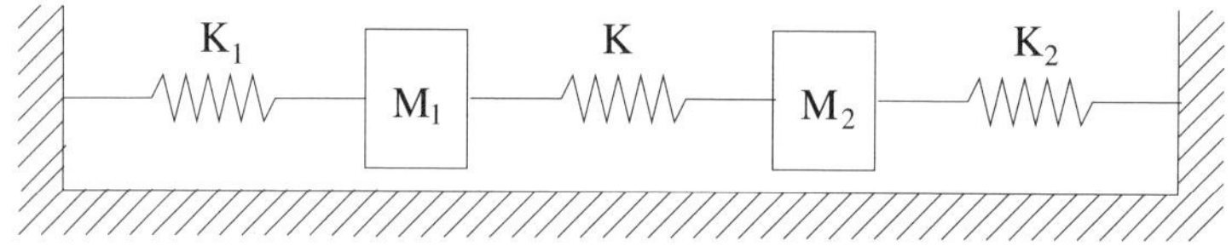

Figure 2.14.1: Example of a conservative dynamical system.

Problem 2.14.6 (a) In Figure 2.14.2, K_1, K_2, K, M_1, and M_2 are the same as in Figure 2.14.1 and B_1, B_2, and B denote viscous damping coefficients. Use the Lagrange formulation of dynamical systems described in Example 2.3.8 to derive two second-order ordinary differential equations that characterize this system. Transform these equations into a system of first-order ordinary differential equations. Verify your answer by using Newton's second law of motion to derive the same system of equations. By specifying $x_1(t_0), \dot{x}_1(t_0), x_2(t_0)$, and $\dot{x}_2(t_0)$, the above yields an initial value problem.

(b) Show that the above mechanical system determines a dynamical system in the sense of Definition 2.2.3. □

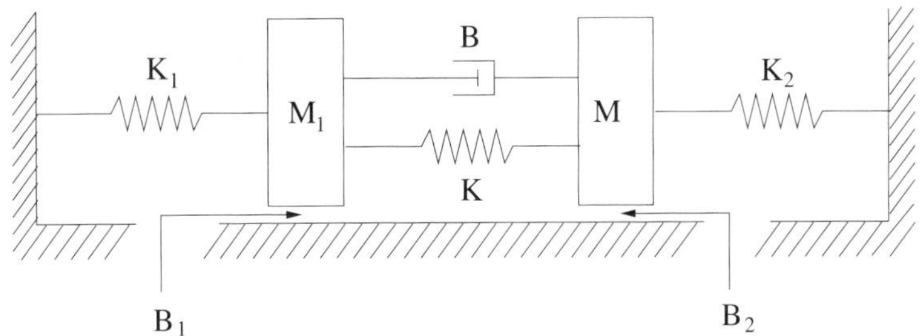

Figure 2.14.2: Example of a mechanical system with energy dissipation.

Problem 2.14.7 The following result, called the *Ascoli–Arzela* Lemma, is required in the proof of Problem 2.14.8 given below.

Let D be a closed and bounded subset of $\mathbb{R}^n$ and let $\{f_m\}$ be a sequence of functions in $C[D, \mathbb{R}^n]$. If $\{f_m\}$ is *equicontinuous* and *uniformly bounded* on D, then there is a subsequence $\{f_{mk}\}$ and a function $f \in C[D, \mathbb{R}^n]$ such that $\{f_{mk}\}$ converges to f uniformly on D. Recall that $\{f_m\}$ is equicontinuous on D if for every $\varepsilon > 0$ there is a $\delta > 0$ (independent of x, y, and m) such that

$$|f_m(x) - f_m(y)| < \varepsilon \text{ whenever } |x - y| < \delta$$

for all $x, y \in D$ and for all m. Recall also that $\{f_m\}$ is uniformly bounded if there is a constant $M > 0$ such that $|f_m(x)| \leq M$ for all $x \in D$ and for all m.

Hint: To prove the Ascoli–Arzela Lemma, let $\{r_k\}, k \in \mathbb{N}$, be a dense subset of D. Determine a subsequence $\{f_{km}\}$ and a function f defined on $\{r_k\}$ such that $f_{km}(r_k) \to f(r_k)$ as $m \to \infty$ for all $k \in \mathbb{N}$. Next, prove that the subsequence $\{f_{mm}\}$ converges to $\{f\}$ on $\{r_k\}, k \in \mathbb{N}$, uniformly as $m \to \infty$. Conclude, by extending the domain of f from $\{r_k\}$ to D.

For a complete statement of the proof outlined above, refer to Miller and Michel [37]. □

Problem 2.14.8 Prove Theorem 2.3.1.

Hint: First, show that for every $\varepsilon > 0$ there exists a piecewise linear function $\varphi_\varepsilon \colon J \to \mathbb{R}^n$ such that $\varphi_\varepsilon(t_0) = x_0$, $(t_0, x_0) \in D$, and $|\dot{\varphi}_\varepsilon(t) - f(t, \varphi_\varepsilon(t))| < \varepsilon$ for all $t \in [t_0, t_0 + c]$ (where $\dot{\varphi}_\varepsilon$ is defined) for some $c > 0$ and $(t, \varphi_\varepsilon(t)) \in D$ for all $t \in [t_0, t_0 + c]$. (φ_ε is called an *ε-approximate solution of* (I_E).)

Next, let φ_m be an ε-approximate solution of (I_E) with $\varepsilon_m = 1/m$. Show that the sequence $\{\varphi_m\}$ is uniformly bounded and equicontinuous.

Finally, apply the Ascoli–Arzela Lemma to show that there is a subsequence $\{\varphi_{mk}\}$ of $\{\varphi_m\}$ given above and a $\varphi \in C\big[[t_0, t_0 + c], \mathbb{R}^n\big]$ such that $\{\varphi_{mk}\}$ converges to φ uniformly on $[t_0, t_0 + c]$, and such that φ satisfies

$$\varphi(t) = x_0 + \int_{t_0}^{t} f(s, \varphi(s))ds$$

for $t \in [t_0, t_0 + c]$. Therefore, φ is a solution of (I_E).

For a complete statement of the proof outlined above, refer to Miller and Michel [37]. □

Problem 2.14.9 The following result, called the *Gronwall Inequality* is required in the proof of Problem 2.14.10 given below.

Let $r, k \in C\big[[a,b], \mathbb{R}^+\big]$ and let $\delta \geq 0$ such that

$$r(t) \leq \delta + \int_a^t k(s)r(s)ds, \qquad a \leq t \leq b. \tag{2.14.4}$$

Then

$$r(t) \leq \delta \exp\left[\int_a^t k(s)ds\right], \qquad a \leq t \leq b. \tag{2.14.5}$$

Hint: For $\delta > 0$, integrate both sides of

$$\frac{k(s)r(s)}{\delta + \int_a^s k(\eta)r(\eta)d\eta} \leq k(s)$$

from a to t. Use inequality (2.14.4) to conclude the result when $\delta \neq 0$. When $\delta = 0$, consider a positive sequence $\{\delta_n\}$ such that $\delta_n \to 0$ as $n \to \infty$ and apply it to (2.14.5).

For a complete statement of the proof outlined above, refer to Miller and Michel [37]. □

Problem 2.14.10 Prove Theorem 2.3.2

Hint: Apply the Gronwall inequality given above in Problem 2.14.9.

For a complete statement of the proof, refer to Miller and Michel [37]. □

Problem 2.14.11 The following result is required in the proof of Problem 2.14.12 given below.

Let $D \subset \mathbb{R} \times \mathbb{R}^n$ be a domain. Let $f \in C[D, \mathbb{R}^n]$ with f bounded on D and let φ be a solution of (E) on the interval (a, b). Show that

(a) The two limits $\lim_{t\to a^+} \varphi(t) = \varphi(a^+)$ and $\lim_{t\to b^-} \varphi(t) = \varphi(b^-)$ exist.

(b) If $(a, \varphi(a^+)) \in D$ (resp., $(b, \varphi(b^-)) \in D$), then the solution φ can be continued to the left past the point $t = a$ (resp., to the right past the point $t = b$).

(A complete statement of the proof of the above result can be found in Miller and Michel [37].) □

Problem 2.14.12 Prove Theorem 2.3.3.

Hint: Use the result given in Problem 2.14.11. □

Problem 2.14.13 Prove Theorem 2.7.1.

Hint: To prove this result, use *Schauder's Fixed Point Theorem*: A continuous mapping of a compact convex set in a Banach space X into itself has at least one fixed point. Let T be the operator defined by (2.7.7). Find a compact convex set

$X \subset C\big[[t_0 - r, t_0 + c], \mathbb{R}^n\big]$ for some $c > 0$ such that $T(X) \subset X$. Now apply Schauder's Fixed Point Theorem. A possible choice of X is given by

$$X = \big\{x \in C\big[[-r+t_0, t_0+c], \mathbb{R}^n\big] : x_{t_0} = \psi, \| x_t - \psi \| \le d \text{ for all } t \in [t_0, t_0+c]\big\},$$

where $0 < c \le d/M$, $d > 0$ sufficiently small, with $M \ge |f(t, \varphi)|$ for all (t, φ) in a fixed neighborhood of (t_0, ψ) in Ω.

For the complete proof of Theorem 2.7.1 outlined above, refer to Hale [19]. □

Problem 2.14.14 Prove Theorem 2.7.2.

Hint: Let $x(t)$ and $y(t)$ be two solutions of (I_F). Then

$$x(t) - y(t) = \int_{t_0}^{t} [f(s, x_s) - f(s, y_s)]\, ds, \qquad t \ge t_0, \;\; x_{t_0} - y_{t_0} = 0.$$

Using the above, show that there exists a $c_0 > 0$ such that $x(t) = y(t)$ for all $t \in [t_0 - r, t_0 + c_0]$. To complete the proof, repeat the above for successive intervals of length c_0.

For the complete proof of Theorem 2.7.2 outlined above, refer to Hale [19]. □

Problem 2.14.15 The following result is required in the proof of Problem 2.14.16 given below.

Let Ω be an open set in $\mathbb{R} \times C_r$ and let $F : \Omega \to \mathbb{R}^n$ be completely continuous. Assume that $p \in C\big[[t_0 - r, b), \mathbb{R}^n\big]$ is a noncontinuable solution of (F). Show that for any bounded closed set U in $\mathbb{R} \times C_r, U \subset \Omega$, there exists a $t_U \in (t_0, b)$ such that $(t, p_t) \notin U$ for every $t \in [t_U, b]$.

Hint: The case $b = \infty$ is clear. Suppose that b is finite. The case $r = 0$ reduces to an ordinary differential equation. So assume that $r > 0$. Now prove the assertion by contradiction, assuming that $b < \infty$ and $r > 0$. □

Problem 2.14.16 Prove Theorem 2.7.3.

Hint: Apply the result given in Problem 2.14.15. For the complete proof, refer to Hale [19]. □

Problem 2.14.17 Prove Theorem 2.8.1.

Hint: Using the theory of C_0-semigroups, refer to Example 2.9.3 for a choice of the infinitesimal generator for the C_0-semigroup (refer to [3]). □

Problem 2.14.18 Consider the initial value problem

$$\begin{cases} \dot{x} = A(t)x \\ x(t_0) = x_0 \end{cases} \qquad (LH)$$

where $A \in C[\mathbb{R}^+, \mathbb{R}^{n \times n}]$.

(a) Show that the set of solutions obtained for (LH) by varying (t_0, x_0) over $(\mathbb{R}^+, \mathbb{R}^n)$ determines a dynamical system in the sense of Definition 2.2.3.

(b) Show that in general, (LH) does not determine a C_0-semigroup.

(c) Show that when $A(t) \equiv A$, (LH) determines a C_0-semigroup. □

Problem 2.14.19 Prove the assertion made in Example 2.9.1 that the initial value problem (2.9.4) determines a quasi-contractive semigroup. □

Problem 2.14.20 Consider the initial value problem for the *heat equation*

$$\begin{cases} \dfrac{\partial u}{\partial t} = a^2 \Delta u, x \in \mathbb{R}^n, & t \in \mathbb{R}^+ \\ u(0,x) = \varphi(x), & x \in \mathbb{R}^n \end{cases} \tag{2.14.6}$$

where $a > 0$, $\Delta = \sum_{i=1}^{n} \partial^2/\partial x_i^2$, and $\varphi \in C[\mathbb{R}^n, \mathbb{R}]$ is bounded.

(a) Verify that the unique solutions of (2.14.6) are given by *Poisson's formula*,

$$u(t,x) = \frac{1}{(2a\sqrt{\pi t})^n} \int_{\mathbb{R}^n} e^{-|x-y|^2/(4a^2 t)} \varphi(y) dy.$$

(b) Show that the operators $T(t), t \in \mathbb{R}^+$, determined by $u(t, \cdot) = T(t)\varphi$, determine a C_0-semigroup. □

Problem 2.14.21 Consider the initial value problem for the *one-dimensional wave equation*

$$\begin{cases} \dfrac{\partial^2 u}{\partial t^2} = c^2 \dfrac{\partial^2 u}{\partial x^2}, & x \in \mathbb{R}, \ t \in \mathbb{R}^+ \\ u(0,x) = \varphi(x), \dfrac{\partial u}{\partial t}(0,x) = \psi(x), & x \in \mathbb{R} \end{cases} \tag{2.14.7}$$

where $c > 0$, $\varphi \in C^2[\mathbb{R}, \mathbb{R}]$, and $\psi \in C^1[\mathbb{R}, \mathbb{R}]$.

(a) Verify that the unique solution of (2.14.7) is given by *d'Alembert's formula*

$$u(t,x) = \frac{1}{2}\left[\varphi(x-ct) + \varphi(x+ct)\right] + \frac{1}{2c}\int_{x-ct}^{x+ct} \psi(\eta) d\eta.$$

(b) Let $\psi \equiv 0$. For $\varphi \in C^2[\mathbb{R}, \mathbb{R}]$, define the operators $T(t)$, $t \in \mathbb{R}^+$, by $T(t)\varphi = u(t, \cdot)$. Show that $T(t), t \in \mathbb{R}^+$, do not satisfy the semigroup property (specifically, they do not satisfy the property $T(t)T(s) = T(t+s), t, s \in \mathbb{R}^+$).

(c) Now let $u(t, \varphi, t_0)$ denote the solutions of

$$\begin{cases} \dfrac{\partial^2 u}{\partial t^2} = c^2 \dfrac{\partial^2 u}{\partial x^2}, & x \in \mathbb{R}, \ t \geq t_0 \\ u(t_0,x) = \varphi(x), \ \dfrac{\partial u}{\partial t}(t_0,x) = 0, & x \in \mathbb{R} \end{cases} \tag{2.14.8}$$

where $t_0 \in \mathbb{R}^+$ and $\varphi \in C^2[\mathbb{R}, \mathbb{R}]$. Show that for all $\varphi \in C^2[\mathbb{R}, \mathbb{R}]$, the resulting solutions $u(t, \varphi, t_0)$ form a dynamical system in the sense of Definition 2.2.3 with $T = \mathbb{R}^+, X = A = C^2[\mathbb{R}, \mathbb{R}]$ where we assume that X is equipped with some norm (e.g., $\| \varphi \| = \max_{x \in \mathbb{R}} |\varphi(x)|$). □

Problem 2.14.22 We now consider a specific class of *multirate digital feedback control systems*. The *plant* is described by

$$\begin{cases} \dot{x}(t) = Ax(t) + B_1 u_{1c}(t) + B_2 u_{2c}(t) \\ y(t) = \begin{bmatrix} y_1(t) \\ y_2(t) \end{bmatrix} = \begin{bmatrix} D_1 x(t) \\ D_2 x(t) \end{bmatrix} \end{cases} \tag{2.14.9}$$

where $x \in \mathbb{R}^n$, $A \in \mathbb{R}^{n\times n}$, $B_1 \in \mathbb{R}^{n\times n_1}$, $B_2 \in \mathbb{R}^{n\times n_2}$, $D_1 \in \mathbb{R}^{m_1\times n}$, $D_2 \in \mathbb{R}^{m_2\times n}$, $u_{1c} \in \mathbb{R}^{n_1}$, $u_{2c} \in \mathbb{R}^{n_2}$, and

$$\begin{cases} u_{1c}(t) = u_1(k), & kT_b \leq t < (k+1)T_b, & k \in \mathbb{N}, \\ u_{2c}(t) = u_2(2k), & 2kT_b \leq t < 2(k+1)T_b, & k \in \mathbb{N}. \end{cases} \tag{2.14.10}$$

In (2.14.10), $T_b > 0$ is the *basic sampling period* whereas $u_1(k)$ and $u_2(2k)$ are specified by output feedback equations of the form

$$\begin{aligned} u_1(k+1) &= F_1 u_1(k) + K_1 y_1(kT_b) \\ &= F_1 u_1(k) + K_1 D_1 x(kT_b), \qquad k \in \mathbb{N} \\ u_2(2(k+1)) &= F_2 u_2(2k) + K_2 y_2(2kT_b) \\ &= F_2 u_2(2k) + K_2 D_2 x(2kT_b), \qquad k \in \mathbb{N} \end{aligned} \tag{2.14.11}$$

where K_1, K_2, F_1, and F_2 are matrices of appropriate dimensions. The system inputs $u_{1c}(t)$ and $u_{2c}(t)$ are realized by *multirate zero-order hold elements*.

Similarly as in Example 2.12.1, show that the above hybrid system can equivalently be represented by a system of discontinuous ordinary differential equations that generate a discontinuous dynamical system. □

Bibliography

[1] J. P. Aubin and A. Cellina, *Differential Inclusions*, New York: Springer-Verlag, 1984.

[2] D. D. Bainov and P. S. Simeonov, *Systems with Impulse Effects: Stability Theory and Applications*, New York: Halsted Press, 1989.

[3] V. Barbu and S. I. Grossman, "Asymptotic behavior of linear integrodifferential systems," *Trans. Amer. Math. Soc.*, vol. 171, pp. 277–288, 1972.

[4] R. Bellman and K. L. Cooke, *Differential-Difference Equations*, New York: Academic Press, 1963.

[5] H. Brezis, *Operateurs Maximaux Monotones*, Amsterdam: North-Holland, 1973.

[6] R. W. Brockett, "Hybrid models for motion control systems," *Essays on Control Perspectives in the Theory and Its Applications*, H. L. Trentelman and J. C. Willems, Eds., Boston: Birkhäuser, pp. 29–53, 1993.

[7] B. Brogliato, S. I. Niculescu, and P. Orhant, "On the control of finite-dimensional systems with unilateral constraints," *IEEE Trans. Autom. Control*, vol. 42, pp. 200–215, 1997.

[8] M. G. Crandall, "Semigroups of nonlinear transformations on general Banach spaces," *Contributions to Nonlinear Functional Analysis*, E. H. Zarantonello, Ed., New York: Academic Press, 1971.

[9] M. G. Crandall and T. M. Liggett, "Generation of semigroups of nonlinear transformations on general Banach spaces," *Amer. J. Math.*, vol. 93, pp. 265–298, 1971.

[10] R. DeCarlo, S. Branicky, S. Pettersson, and B. Lennartson, "Perspectives and results on the stability and stabilizability of hybrid systems," *Proc. IEEE*, vol. 88, pp. 1069–1082, 2000.

[11] J. Dieudonné, *Foundations of Modern Analysis*, New York: Academic Press, 1960.

[12] N. Dunford and J. T. Schwarz, *Linear Operators*, Part I, New York: Wiley, 1958.

[13] G. F. Franklin and J. D. Powell, *Digital Control of Dynamical Systems*, Reading, MA: Addison-Wesley, 1980.

[14] A. Friedman, *Partial Differential Equations of the Parabolic Type*, Englewood Cliffs, NJ: Prentice Hall, 1964.

[15] A. N. Godunov, "Peano's theorem in Banach spaces," *Functional Anal. Appl.*, vol. 9, pp. 53–56, 1975.

[16] A. Gollu and P. P. Varaiya, "Hybrid dynamical systems," *Proc. 28th IEEE Conf. on Decision and Control*, Tampa, FL, Dec. 1989, pp. 2708–2712.

[17] R. Grossman, A. Nerode, A. Ravn, and H. Rischel, Eds., *Hybrid Systems*, New York: Springer-Verlag, 1993.

[18] W. Hahn, *Stability of Motion*, Berlin: Springer-Verlag, 1967.

[19] J. K. Hale, *Functional Differential Equations*, Berlin: Springer-Verlag, 1971.

[20] J. K. Hale, "Functional differential equations with infinite delays," *J. Math. Anal. Appl.*, vol. 48, pp. 276–283, 1974.

[21] E. Hille and R. S. Phillips, *Functional Analysis and Semigroups*, Amer. Math. Soc. Colloquium Publ., vol. 33, Providence, RI: American Mathematical Society, 1957.

[22] L. Hörmander, *Linear Partial Differential Equations*, Berlin: Springer-Verlag, 1963.

[23] S. G. Krein, *Linear Differential Equations in Banach Spaces*, Translation of Mathematical Monographs, vol. 29, Providence, RI: American Mathematical Society, 1970.

[24] N. V. Krylov, *Nonlinear Elliptic and Parabolic Equations of the Second Order*, Boston: D. Reidel, 1987.

[25] T. Kurtz, "Convergence of sequences of semigroups of nonlinear equations with applications to gas kinetics," *Trans. Amer. Math. Soc.*, vol. 186, pp. 259–272, 1973.

[26] V. Lakshmikantham and S. Leela, *Differential and Integral Inequalities*, vol. I, II, New York: Academic Press, 1969.

[27] A. Lasota and J. A. Yorke, "The generic property of existence of solutions of differential equations in Banach space," *J. Differential Equations*, vol. 13, pp. 1–12, 1973.

[28] D. Liberzon and A. S. Morse, "Basic problems in stability and design of switched systems," *IEEE Control Syst. Mag.*, vol. 19, pp. 59–70, 1999.

[29] A. N. Michel, "Recent trends in the stability analysis of hybrid dynamical systems," *IEEE Trans. Circ. and Syst.*, vol. 46, pp. 120–134, 1999.

[30] A. N. Michel and B. Hu, "Towards a stability theory of general hybrid dynamical systems," *Automatica*, vol. 35, pp. 371–384, 1999.

[31] A. N. Michel, and R. K. Miller, *Qualitative Analysis of Large Scale Dynamical Systems*, New York: Academic Press, 1977.

[32] A. N. Michel and Y. Sun, "Stability of discontinuous Cauchy problems in Banach space," *Nonlinear Anal.*, vol. 65, pp. 1805–1832, 2006.

[33] A. N. Michel, Y. Sun, and A. P. Molchanov, "Stability analysis of discontinuous dynamical systems determined by semigroups," *IEEE Trans. Autom. Control*, vol. 50, pp. 1277–1290, 2005.

[34] A. N. Michel, K. Wang, and B. Hu, *Qualitative Theory of Dynamical Systems – The Role of Stability Preserving Mappings*, Second Edition, New York: Marcel Dekker, 2001.

[35] A. N. Michel, K. Wang, and K. M. Passino, "Qualitative equivalence of dynamical systems with applications to discrete event systems," *Proc. of the 31st IEEE Conf. on Decision and Control*, Tucson, AZ, Dec. 1992, pp. 731–736.

[36] A. N. Michel, K. Wang and K. M. Passino, "Stability preserving mappings and qualitative equivalence of dynamical systems, Part I," *Avtomatika i Telemekhanika*, vol. 10, pp. 3–12, 1994.

[37] R. K. Miller and A. N. Michel, *Ordinary Differential Equations*, New York: Academic Press, 1982.

[38] A. V. Oppenheim and R. W. Schafer, *Digital Signal Processing*, Englewood Cliffs, NJ: Prentice Hall, 1975.

[39] A. Pazy, *Semigroups of Linear Operators and Applications to Partial Differential Equations*, New York: Springer-Verlag, 1963.

[40] R. D. Rasmussen and A. N. Michel, "Stability of interconnected dynamical systems described on Banach spaces," *IEEE Trans. Autom. Control*, vol. 21, pp. 464–471, 1976.

[41] C. Rui, I. Kolmanovsky, and N. H. McClamroch, "Hybrid control for stabilization of a class of cascade nonlinear systems," *Proc. American Control Conf.*, Albuquerque, NM, June 1997, pp. 2800–2804.

[42] M. Slemrod, "Asymptotic behavior of C_0-semigroup as determined by the spectrum of the generator," *Indiana Univ. Math. J.*, vol. 25, pp. 783–791, 1976.

[43] Y. Sun, A. N. Michel, and G. Zhai, " Stability of discontinuous retarded functional differential equations with applications," *IEEE Trans. Autom. Control*, vol. 50, pp. 1090–1105, 2005.

[44] J. C. Willems, "Paradigms and puzzles in the theory of dynamical systems," *IEEE Trans. Autom. Control*, vol. 36, pp. 259–294, 1991.

[45] H. Ye, A. N. Michel, and P. J. Antsaklis "A general model for the qualitative analysis of hybrid dynamical systems," *Proc. of the 34th IEEE Conf. on Decision and Control*, New Orleans, LA, Dec. 1995, pp. 1473–1477.

[46] H. Ye, A. N. Michel, and L. Hou, "Stability theory for hybrid dynamical systems," *IEEE Trans. Autom. Control*, vol. 43, pp. 461–474, 1998.

[47] T. Yoshizawa, *Stability Theory by Lyapunov's Second Method*, Math. Soc., Japan, Tokyo, 1966.

[48] V. I. Zubov, *Methods of A.M. Lyapunov and Their Applications*, Groningen, The Netherlands: P. Noordhoff, 1964.

Chapter 3

Fundamental Theory: The Principal Stability and Boundedness Results on Metric Spaces

In this chapter we present the Principal Lyapunov and Lagrange Stability Results, including Converse Theorems for *continuous dynamical systems*, *discrete-time dynamical systems*, and *discontinuous dynamical systems* (DDS) defined on metric spaces. The results of this chapter constitute the *fundamental theory* for the entire book because most of the *general results* that we develop in the subsequent chapters concerning *finite-dimensional systems* (described on finite-dimensional linear spaces) and *infinite-dimensional systems* (defined on Banach and Hilbert spaces) can be deduced as consequences of the results of the present chapter. Most of the *specific applications* to the fundamental theory that we consider therefore are deferred to the later chapters where we address finite-dimensional and infinite-dimensional systems. However, after addressing in the next chapter additional (more specialized) stability and boundedness results for dynamical systems defined on metric spaces, we present applications to the results of this chapter in Chapter 5 in the analysis of a class of discrete-event systems (with applications to a manufacturing system and a computer load-balancing problem) that determine dynamical systems defined on metric spaces.

The conventional approach in proving the various Principal Lyapunov and Lagrange Stability Results for continuous, discrete-time, and discontinuous dynamical systems is to show that when a dynamical system satisfies a certain set of hypotheses, then the system possesses a certain type of stability or boundedness property. For the reasons discussed below we do not pursue this approach (however, we ask the reader to do so in the exercise section).

In establishing the results presented in this chapter, we first prove the Principal Lyapunov and Lagrange Stability Results for discontinuous dynamical systems, using basic principles and definitions. Next, to establish the Principal Lyapunov and Lagrange Stability Results for continuous dynamical systems, we show that whenever the hypotheses of a given stability or boundedness result for continuous dynamical systems are satisfied, then the hypotheses of the corresponding stability or boundedness result for DDS are also satisfied (using the fact that continuous dynamical systems may be viewed as special cases of DDS). This shows that the classical Lyapunov and Lagrange Stability Results for continuous dynamical systems reduce to the corresponding results for DDS (i.e., the classical Principal Lyapunov and Lagrange Stability Results for continuous dynamical systems are more conservative than the corresponding Lyapunov and Lagrange Stability Results for DDS). Indeed, we present a result for a continuous dynamical system whose equilibrium can be shown to be uniformly asymptotically stable, using the uniform asymptotic stability theorem for DDS, and we show that for the *same* example, no Lyapunov function exists that satisfies the classical uniform asymptotic stability theorem for continuous dynamical systems.

Next, we show that for every discrete-time dynamical system there exists an *associated DDS* with identical stability and boundedness properties. Making use of such associated DDSs, we show that when the hypotheses of a given classical Lyapunov or Lagrange stability result for discrete-time dynamical systems are satisfied, then the hypotheses of the corresponding Lyapunov and Lagrange stability result for DDS are satisfied. This shows that the classical Lyapunov and Lagrange stability results for discrete-time dynamical systems reduce to the corresponding results for DDS (i.e., the classical Principal Lyapunov and Lagrange stability results for discrete-time dynamical systems are more conservative than the corresponding Lyapunov and Lagrange stability results for DDS). We present a specific example of a discrete-time dynamical system whose equilibrium can be shown to be uniformly asymptotically stable using the uniform asymptotic stability theorem for DDS, and we show that for the *same* example, no Lyapunov function exists that satisfies the classical Lyapunov theorem for uniform asymptotic stability for discrete-time dynamical systems.

In addition to giving us a great deal of insight, the approach that we employ in proving the various stability and boundedness results culminates in a *unifying qualitative theory* for the analysis of continuous, discrete-time, and discontinuous dynamical systems. Furthermore, our approach in proving the results presented herein is more efficient than the conventional approach alluded to earlier.

This chapter is organized into eight sections. In the first section we address the qualitative characterization of invariant sets of dynamical systems. In the next three sections we present and prove the principal Lyapunov stability results and the Lagrange stability results (boundedness of motions) for discontinuous dynamical systems, continuous dynamical systems, and discrete-time dynamical systems, respectively. This is followed by three sections where we address converse theorems for DDS, continuous dynamical systems, and discrete-time dynamical systems. Finally, in Section 3.8 we present some required background material concerning ordinary differential equations.

Before proceeding with our subject on hand, we would like to remind the reader that our definition of dynamical system (Definition 2.2.3) does in general not require that time be reversible in the motions (in contrast to many dynamical systems determined, e.g., by various types of differential equations), nor are the motions required to be unique with respect to initial conditions. For such general systems, when required, we make an assumption that is akin to the semigroup property, but is more general, which essentially requires that for a dynamical system S, any partial motion is also a motion of S, and any composition of two motions is also a motion of S (refer to Assumption 3.5.1). Of course when in a dynamical system the semigroup property holds, then Assumption 3.5.1 is automatically implied.

3.1 Some Qualitative Characterizations of Dynamical Systems

Most, but not all the qualitative aspects of dynamical systems that we address concern qualitative characterizations of invariant sets. In the present section, we first introduce the notion of an invariant set of a dynamical system. Next, we present various concepts of stability of invariant sets, instability of sets, and boundedness of motions.

A. Invariant sets

In the following, we utilize the notation introduced in Chapter 2.

Definition 3.1.1 Let $\{T, X, A, S\}$ be a dynamical system. A set $M \subset A$ is said to be *invariant with respect to* S, or for short, (S, M) *is invariant*, if $a \in M$ implies that $p(t, a, t_0) \in M$ for all $t \in T_{a,t_0}$, all $t_0 \in T$, and all $p(\cdot, a, t_0) \in S$. □

Recall that $T_{a,t_0} = [t_0, t_1) \cap T$, $t_1 > t_0$, which means that in the above definition, *evolution in time is forward.*

In a broader context, the evolution in time is allowed to be forward as well as backward. In such cases, a distinction is made between *positive invariant set* (forward in time) and *negative invariant set* (backward in time) (see, e.g., [11]).

We note that the union of invariant sets is also an invariant set.

Example 3.1.1 (*Conservative dynamical systems*) Recall the *Hamiltonian system* given in Example 2.3.7, described by the equations

$$\begin{cases} \dot{q}_i = \dfrac{\partial H}{\partial p_i}(p, q), & i = 1, \dots, n, \\ \dot{p}_i = -\dfrac{\partial H}{\partial q_i}(p, q), & i = 1, \dots, n. \end{cases} \tag{3.1.1}$$

The solutions of (3.1.1) determine a continuous dynamical system with $T = \mathbb{R}$ and $X = A = \mathbb{R}^{2n}$. For any $c \in \mathbb{R}$ such that the set

$$M_c = \{(p, q)^T \in \mathbb{R}^{2n} \colon H(p, q) = c\}$$

is nonempty, M_c is an invariant set. This follows, because

$$\frac{dH}{dt}(p(t), q(t)) \equiv 0$$

for all solutions $(p(t), q(t))^T$ of (3.1.1), as shown in Example 2.3.7. □

Example 3.1.2 (*Heat equation*) Let $\Omega \subset \mathbb{R}^n$ be a domain with smooth boundary $\partial\Omega$ and consider the *initial value and boundary value problem* given by the equations (see also Chapter 2)

$$\begin{cases} \dfrac{\partial u}{\partial t} = a^2 \Delta u, & x \in \Omega,\ t \geq t_0 \\ u(t_0, x) = \varphi(x), & x \in \Omega \\ u(t, x) = 0, & x \in \partial\Omega,\ t > t_0 \end{cases} \tag{3.1.2}$$

where $t_0 \geq 0$, $a > 0$, $\Delta = \sum_{i=1}^{n} \partial^2/\partial x_i^2$, and $\varphi \in C[\Omega, \mathbb{R}]$ with $\lim_{x\to\partial\Omega} \varphi(x) = 0$. Let $T = \mathbb{R}^+$,

$$X = A = \{\varphi \in C[\Omega, \mathbb{R}] : \lim_{x\to\partial\Omega} \varphi(x) = 0\}$$

and let X and A be equipped with the norm

$$\|\varphi\| = \max_{x\in\Omega} |\varphi(x)|.$$

It has been shown that for every $\varphi \in A$, (3.1.2) possesses a unique solution $u(t, x)$ that exists for all $t \geq t_0$. It follows that $\{T, X, A, S_{(3.1.2)}\}$ is a dynamical system where the set of motions is determined by the solutions of (3.1.2).

Let $M \subset A = X$ denote the set given by

$$M = \{\varphi \in C^2[\Omega, \mathbb{R}] \cap X : \Delta\varphi(x) = 0 \text{ for all } x \in \Omega\}.$$

Then M is invariant with respect to $S_{(3.1.2)}$. Indeed, for any $\varphi \in M$, $u(t, x) \equiv \varphi(x)$ is a solution of (3.1.2). By the uniqueness of each solution, it follows that M is invariant with respect to $S_{(3.1.2)}$. □

The most important special case of invariant sets is an equilibrium.

Definition 3.1.2 We call $x_0 \in A$ an *equilibrium* (or an *equilibrium point*) of a dynamical system $\{T, X, A, S\}$ if the set $\{x_0\} \subset A$ is invariant with respect to S (i.e., $(S, \{x_0\})$ is invariant). □

In the following, we enumerate several specific examples of equilibria.

Example 3.1.3 (*Ordinary differential equations*) Let $f \in C[\mathbb{R}^+ \times \Omega, \mathbb{R}^n]$ where $\Omega \subset \mathbb{R}^n$ is a domain, assume that $x_e \in \Omega$ satisfies $f(t, x_e) = 0$ for all $t \in \mathbb{R}^+$, and that the system of first-order ordinary differential equations

$$\dot{x} = f(t, x) \tag{E}$$

has a unique solution for the initial condition $x(t_0) = x_e$. As pointed out in Chapter 2, this is true if, for example, f satisfies a Lipschitz condition. Then x_e is an *equilibrium* of the dynamical system S_E determined by the solutions of (E). In this case we also say that x_e is an equilibrium of (E). □

Example 3.1.4 (*Ordinary difference equations*) Let $f\colon \mathbb{N}\times\mathbb{R}^n \to \mathbb{R}^n$ and assume that there exists an $x_e \in \mathbb{R}^n$ such that $f(k, x_e) = x_e$ for all $k \in \mathbb{N}$. Then x_e is an *equilibrium* of the dynamical system S_D determined by the solutions of the system of ordinary difference equations given by

$$x(k+1) = f(k, x(k)), \tag{D}$$

$k \in \mathbb{N}$. □

Example 3.1.5 (*Heat equation*) In Example 3.1.2, each $\varphi \in M$ is an *equilibrium* of (3.1.2). □

Example 3.1.6 (*Ordinary differential equations in a Banach space*) Let X be a Banach space, let $C \subset X$, and let $F\colon \mathbb{R}^+ \times C \to X$. Assume that $F(t, x_e) = 0$ for all $t \in \mathbb{R}^+$ and that

$$\dot{x}(t) = F(t, x(t)), \qquad x(t_0) = x_e \tag{3.1.3}$$

has a unique solution for any $t_0 \in \mathbb{R}^+$. Then x_e *is an equilibrium of the dynamical system determined by the solutions of*

$$\dot{x}(t) = F(t, x(t)). \tag{3.1.4}$$

In this case we also say that x_e *is an equilibrium of* (3.1.4). □

Example 3.1.7 (*Semigroups*) Let $T(t)$, $t \in \mathbb{R}^+$, be a linear or nonlinear semigroup (see Chapter 2) defined on a subset C of a Banach space X. If there exists an $x_e \in C$ such that $T(t)x_e = x_e$ for all $t \in \mathbb{R}^+$, then x_e *is an equilibrium of the dynamical system determined by the semigroup* $T(t)$. In this case we also say that x_e *is an equilibrium of the semigroup* $T(t)$. □

We conclude this subsection by introducing several additional concepts that we require.

Definition 3.1.3 A dynamical system $\{T, X, A, S\}$ is said to satisfy the *uniqueness property* if for any $(a, t_0) \in A \times T$ there exists a unique noncontinuable motion $p(\cdot, a, t_0) \in S$ (refer to Definition 2.2.6 for the definition of noncontinuable motion). □

When a dynamical system is determined by equations of the type considered in Chapter 2, the uniqueness property of a dynamical system is equivalent to the uniqueness of solutions of initial value problems determined by such equations.

Definition 3.1.4 For each motion $p(\cdot, a, t_0) \in S$ in a dynamical system $\{T, X, A, S\}$, the set

$$C(p) = \{x \in X\colon x = p(t, a, t_0) \text{ for some } t \geq t_0 \text{ and } t \in T\}$$

is called a *trajectory*. □

In the literature where evolution in time is allowed to be forward and backward, a distinction is made between *positive semitrajectory* (forward in time) and a *negative semitrajectory* (backward in time) (refer, e.g., to [11]).

Definition 3.1.5 A motion $p(\cdot, a, t_0) \in S$ in a dynamical system $\{T, X, A, S\}$ is said to be *periodic* if there exists a constant $\omega > 0$ such that $t + \omega \in T$ for each $t \in T$, and such that $p(t + \omega, a, t_0) = p(t, a, t_0)$ for all $t \in T$, $t \geq t_0$, and $t \in T$. We call ω a *period* of the periodic motion $p(\cdot, a, t_0)$. □

If a dynamical system S satisfies the uniqueness property, then clearly any trajectory of a motion in S is an invariant set. In particular, the trajectory of a periodic motion, which is usually a closed curve in a metric space X, is an invariant set. Furthermore, the union of a family of trajectories is an invariant set of S.

B. Qualitative characterizations: Stability and boundedness

Let $\{T, X, A, S\}$ be a dynamical system and assume that $M \subset A$ is an invariant set of S, or for short, that (S, M) is invariant. In the definitions that follow, the phrase "(S, M) is said to be ..." is understood to mean "the set M that is invariant with respect to system S is said to be ...". Thus, in Definition 3.1.6 given below, "(S, M) is said to be stable ..." should be read as "the set M that is invariant with respect to system S is said to be stable ...".

Let (X, d) be a metric space. We recall that the distance between a point $a \in X$ and a set $M \subset X$ is defined as

$$d(a, M) = \inf_{x \in M} d(a, x).$$

Finally, the reader should make reference to Definition 2.2.1 for the meaning of the set T_{a,t_0}.

Definition 3.1.6 (S, M) is said to be *stable* if for every $\varepsilon > 0$ and every $t_0 \in T$, there exists a $\delta = \delta(\varepsilon, t_0) > 0$ such that $d(p(t, a, t_0), M) < \varepsilon$ for all $t \in T_{a,t_0}$ and for all $p(\cdot, a, t_0) \in S$, whenever $d(a, M) < \delta$. (S, M) is said to be *uniformly stable* if it is stable and if in the above, δ is independent of t_0 (i.e., $\delta = \delta(\varepsilon)$). □

In the following definitions, we address asymptotic properties of invariant sets with respect to dynamical systems. Throughout this book, whenever we deal with asymptotic properties, we assume that for any $(a, t_0) \in A \times T$, $T_{a,t_0} = [t_0, \infty) \cap T$ and that $T \cap [\alpha, \infty) \neq \emptyset$ for any $\alpha > 0$.

Definition 3.1.7 (S, M) is *attractive* if there exists an $\eta = \eta(t_0) > 0$ such that $\lim_{t \to \infty} d(p(t, a, t_0), M) = 0$ for all $p(\cdot, a, t_0) \in S$ whenever $d(a, M) < \eta$. □

We call the set of all $a \in A$ such that $\lim_{t \to \infty} d(p(t, a, t_0), M) = 0$ for all $p(\cdot, a, t_0) \in S$ the *domain of attraction of* (S, M) *at time* t_0.

Definition 3.1.8 (S, M) is *asymptotically stable* if it is stable and attractive. □

Definition 3.1.9 (S, M) is *uniformly asymptotically stable* if

(i) it is uniformly stable; and

(ii) for every $\varepsilon > 0$ and every $t_0 \in T$, there exist a $\delta > 0$, independent of t_0 and ε, and a $\tau = \tau(\varepsilon) > 0$, independent of t_0, such that $d(p(t, a, t_0), M) < \varepsilon$ for all $t \in T_{a,t_0+\tau}$ and for all $p(\cdot, a, t_0) \in S$ whenever $d(a, M) < \delta$. □

When condition (ii) in the above definition is satisfied, we say that (S, M) is *uniformly attractive.*

Definition 3.1.10 (S, M) is *exponentially stable* if there exists an $\alpha > 0$, and for every $\varepsilon > 0$ and every $t_0 \in T$, there exists a $\delta = \delta(\varepsilon) > 0$ such that

$$d(p(t, a, t_0), M) < \varepsilon e^{-\alpha(t-t_0)}$$

for all $t \in T_{a,t_0}$ and for all $p(\cdot, a, t_0) \in S$ whenever $d(a, M) < \delta$. □

Note that the exponential stability of (S, M) implies the uniform asymptotic stability of (S, M).

Definition 3.1.11 (S, M) is *unstable* if it is not *stable.* □

The preceding definitions concern *local* characterizations. In the remaining definitions we address *global* characterizations.

Definition 3.1.12 A motion $p(\cdot, a, t_0) \in S$ is *bounded* if there exists a $\beta > 0$ such that $d(p(t, a, t_0), a) < \beta$ for all $t \in T_{a,t_0}$. □

Definition 3.1.13 A dynamical system S is *uniformly bounded* if for every $\alpha > 0$ and for every $t_0 \in T$ there exists a $\beta = \beta(\alpha) > 0$ (independent of t_0) such that if $d(a, x_0) < \alpha$, then for $p(\cdot, a, t_0) \in S$, $d(p(t, a, t_0), x_0) < \beta$ for all $t \in T_{a,t_0}$, where x_0 is a fixed point in X. □

Definition 3.1.14 A uniformly bounded dynamical system S is *uniformly ultimately bounded* if there exists a $B > 0$ and if corresponding to any $\alpha > 0$ and $t_0 \in T$, there exists a $\tau = \tau(\alpha) > 0$ (independent of t_0) such that for all $p(\cdot, a, t_0) \in S$, $d(p(t, a, t_0), x_0) < B$ for all $t \in T_{a,t_0+\tau}$ whenever $d(a, x_0) < \alpha$, where x_0 is a fixed point in X. □

In the above two definitions, the constants β and B may in general depend on the choice of $x_0 \in X$. However, the definitions themselves are independent of the choice of x_0. More generally, we may replace $x_0 \in X$ in these definitions by any fixed bounded set in X.

Definition 3.1.15 (S, M) is *asymptotically stable in the large* if

(i) it is stable; and

(ii) for every $p(\cdot, a, t_0) \in S$ and for all $(t_0, a) \in T \times A$,

$$\lim_{t\to\infty} d(p(t, a, t_0), M) = 0.$$ □

When Definition 3.1.15 is satisfied, the domain of attraction of (S, M) is the entire set A.

Definition 3.1.16 (S, M) is *uniformly asymptotically stable in the large* if

(i) it is uniformly stable;

(ii) S is uniformly bounded; and

(iii) for every $\alpha > 0$, for every $\varepsilon > 0$, and for every $t_0 \in T$, there exists a $\tau = \tau(\varepsilon, \alpha) > 0$ (independent of t_0), such that if $d(a, M) < \alpha$, then for all $p(\cdot, a, t_0) \in S$, $d(p(t, a, t_0), M) < \varepsilon$ for all $t \in T_{a,t_0+\tau}$. □

When condition (iii) in the above definition is satisfied, we say that (S, M) is *globally uniformly attractive*.

Definition 3.1.17 (S, M) is *exponentially stable in the large* if there exist an $\alpha > 0$, a $\gamma > 0$, and for every $\beta > 0$, there exists a $k(\beta) > 0$ such that

$$d(p(t, a, t_0), M) \leq k(\beta)[d(a, M)]^{\gamma} e^{-\alpha(t-t_0)}$$

for all $p(\cdot, a, t_0) \in S$ and for all $t \in T_{a,t_0}$ whenever $d(a, M) < \beta$. □

The preceding notions of stability, instability, asymptotic stability, and exponential stability are referred to in the literature as stability concepts *in the sense of Lyapunov* whereas the preceding concepts involving the *boundedness of motions of dynamical systems*, are referred to in the literature as *Lagrange stability*.

We now consider a few specific examples.

Example 3.1.8 (*Linear ordinary differential equations with constant coefficients*) For the system of linear ordinary differential equations

$$\dot{x} = Ax, \tag{3.1.5}$$

where $x \in \mathbb{R}^n$ and $A \in \mathbb{R}^{n \times n}$, the point $x_e = 0$ is an equilibrium. For the initial conditions $x(t_0) = x_0$, the solution of (3.1.5) is given by

$$\varphi(t, x_0, t_0) = e^{A(t-t_0)} x_0$$

where

$$e^{At} = I + \sum_{j=1}^{\infty} \frac{t^j}{j!} A^j$$

(see, e.g., [1]). Letting $P \in \mathbb{R}^{n \times n}$ denote a nonsingular matrix, we obtain

$$e^{At} = P^{-1} e^{(PAP^{-1})t} P.$$

If we choose P so that PAP^{-1} is in Jordan canonical form, we can see readily that the following statements are true (see, e.g., [1]).

(a) The equilibrium $x_e = 0$ of (3.1.5) is *stable* if and only if all eigenvalues of A have nonpositive real parts, and every eigenvalue with zero real part has an associated Jordan block of order one.

(b) When $x_e = 0$ of (3.1.5) is stable, it is also *uniformly stable.*

(c) When $x_e = 0$ is stable, the dynamical system determined by (3.1.5) is *uniformly bounded.*

(d) The equilibrium $x_e = 0$ of (3.1.5) is *attractive* if and only if all eigenvalues of A have negative real parts.

(e) When $x_e = 0$ of (3.1.5) is attractive, it is also *uniformly attractive.*

(f) The equilibrium $x_e = 0$ of (3.1.5) is *uniformly asymptotically stable*, in fact, *uniformly asymptotically stable in the large* if and only if all eigenvalues of A have negative real parts. In this case, the dynamical system determined by (3.1.5) is *uniformly ultimately bounded.*

(g) The equilibrium $x_e = 0$ of (3.1.5) is *exponentially stable*, in fact, *exponentially stable in the large* if and only if all eigenvalues of A have negative real parts.

(h) When the conditions given in (a) are not satisfied, the equilibrium $x_e = 0$ of (3.1.5) is *unstable.* □

Example 3.1.9 (*Linear ordinary difference equations with constant coefficients*) For the system of linear ordinary difference equations

$$x(k+1) = Ax(k), \tag{3.1.6}$$

where $k \in \mathbb{N}$, $x(k) \in \mathbb{R}^n$ and $A \in \mathbb{R}^{n\times n}$, the point $x_e = 0$ is an equilibrium. For the initial conditions $x(k_0) = x_0$, the solutions of (3.1.6) are given by

$$\varphi(k, x_0, k_0) = A^{(k-k_0)}x_0.$$

Similarly as in Example 3.1.8, we can transform the system (3.1.6) so that the matrix A is in Jordan canonical form to come to the following conclusions (see, e.g., [1]).

(a) The equilibrium $x_e = 0$ of (3.1.6) is *stable*, in fact *uniformly stable*, if and only if all eigenvalues of A have magnitude less than or equal to one and every eigenvalue of A with magnitude equal to one has an associated *Jordan block of order one*. In this case, the dynamical system determined by (3.1.6) is *uniformly bounded.*

(b) The equilibrium $x_e = 0$ of (3.1.6) is *uniformly asymptotically stable in the large* (in fact, *exponentially stable in the large*) if and only if all eigenvalues of A have magnitude less than one. In this case, the dynamical system determined by (3.1.6) is *uniformly ultimately bounded.*

(c) When the conditions of (a) are not satisfied, the equilibrium $x_e = 0$ of (3.1.6) is *unstable.* □

Example 3.1.10 (*Heat equation*) Consider the *initial-value problem* given by

$$\begin{cases} \dfrac{\partial u}{\partial t} = a^2 \Delta u, & x \in \mathbb{R}^n, \quad t \geq t_0 \geq 0 \\ u(t_0, x) = \varphi(x), & x \in \mathbb{R}^n \end{cases} \tag{3.1.7}$$

where $a > 0$, $t \in \mathbb{R}^+$, $\Delta = \sum_{i=1}^n \partial^2/\partial x_i^2$, and $\varphi \in C[\mathbb{R}^n, \mathbb{R}]$ is bounded. Let $T = \mathbb{R}^+$ and let $X = A$ be the set of real-valued and bounded functions defined on $\mathbb{R}^n$ with norm given by

$$\|\varphi\| = \max\big\{|\varphi(x)| \colon x \in \mathbb{R}^n\big\}.$$

We let $S_{(3.1.7)}$ denote the dynamical system determined by (3.1.7). For any constant c, $\varphi_e(x) = c$ ($x \in \mathbb{R}^n$) is an equilibrium for $S_{(3.1.7)}$.

For any initial condition φ, the solution of the heat equation (3.1.7) is given by *Poisson's formula* (see Chapter 2)

$$u(t, \varphi, t_0) = \frac{1}{\big[2a\sqrt{\pi(t - t_0)}\,\big]^n} \int_{\mathbb{R}^n} e^{-|x-y|^2/[4a^2(t-t_0)]} \varphi(y)dy.$$

From this it is easily verified that

$$\|u(t, \varphi, t_0) - c\| \leq \|\varphi - c\| \tag{3.1.8}$$

for any $\varphi \in X$ and for all $t \geq t_0$. Therefore, $\varphi_e \equiv c$ is *uniformly stable*. However, the equilibrium $\varphi_e \equiv c$ is *not attractive*, because for $\varphi \equiv c + \varepsilon$, $u(t, \varphi, t_0) = c + \varepsilon$ for any $\varepsilon \in \mathbb{R}$. Therefore, the equilibrium $\varphi_e \equiv c$ is *not asymptotically stable* and *not uniformly asymptotically stable*.

Next, let $M \subset X$ be the set of all constant functions. Then $(S_{(3.1.7)}, M)$ is *uniformly asymptotically stable in the large*. To show this, note that for any $\varphi \in X$, $\lim_{t\to\infty} u(t, \varphi, t_0) = \widetilde{u}(x)$ exists and satisfies $\Delta\widetilde{u} \equiv 0$; that is, $\widetilde{u}$ is a harmonic function. Furthermore, $\widetilde{u}$ is bounded because φ is bounded, by (3.1.8). By Liouville's Theorem, any bounded harmonic function on $\mathbb{R}^n$ must be constant. Therefore, $\widetilde{u} \in M$. Hence, conditions (ii) and (iii) of Definition 3.1.16 are satisfied. The uniform stability of $(S_{(3.1.7)}, M)$ follows from the uniform stability of $(S_{(3.1.7)}, \{c\})$ for each $c \in \mathbb{R}$. Therefore, $(S_{(3.1.7)}, M)$ is uniformly asymptotically stable in the large.

Finally, inequality (3.1.8) implies that S is *uniformly bounded*. However, $S_{(3.1.7)}$ is *not uniformly ultimately bounded*. □

Example 3.1.11 Consider the scalar *differential-difference equation*

$$\dot{x}(t) = x(t - 1), \qquad t \geq t_0 \geq 1. \tag{3.1.9}$$

Let $T = [1, \infty)$ and let $X = A = C\big[[-1, 0], \mathbb{R}\big]$, with the norm given by

$$\|\varphi\| = \max\big\{|\varphi(t)| \colon -1 \leq t \leq 0\big\}.$$

Associated with (3.1.9) we have the *initial-value problem*

$$\begin{cases} \dot{x}(t) = x(t-1), & t \geq t_0 \\ x(t) = \varphi(t - t_0), & t \in [t_0 - 1, t_0] \end{cases} \tag{3.1.10}$$

which has a unique solution $x(t, \varphi, t_0)$ for each $t_0 \in T$ and each $\varphi \in X = A$.

Let $S_{(3.1.10)}$ be the dynamical system determined by (3.1.10). Then $\varphi_e = 0$ is an equilibrium of $S_{(3.1.10)}$. In the following, we show that $\varphi_e = 0$ is *unstable*.

For $\varphi(t) \equiv \varepsilon$, $t \in [-1, 0]$, we have that

$$x(t, \varphi, 1) = \varepsilon + \sum_{j=1}^{N} \frac{(t-j)^j}{j!}, \qquad N \leq t \leq (N+1), \tag{3.1.11}$$

$N = 0, 1, 2, \ldots$. It follows from (3.1.11) that

$$x(N, \varphi, 1) \geq \varepsilon(1 + N - 1) = N\varepsilon.$$

Therefore, for *arbitrarily small* $\varepsilon > 0$, when $N \geq 1/\varepsilon$, we have $x(N, \varphi, 1) \geq 1$ for $\varphi(t) \equiv \varepsilon$, $t \in [-1, 0]$. Hence, $\varphi_e = 0$ is unstable. □

Before proceeding any further, it should be pointed out that the notions of stability (Definition 3.1.6) and attractivity (Definition 3.1.7) are *independent* concepts. This is demonstrated by considering the specific example

$$\begin{cases} \dot{x}_1 = \dfrac{x_1^2(x_2 - x_1) + x_2^5}{(x_1^2 + x_2^2)[1 + (x_1^2 + x_2^2)^2]}, \\[2ex] \dot{x}_2 = \dfrac{x_2^2(x_2 - 2x_1)}{(x_1^2 + x_2^2)[1 + (x_1^2 + x_2^2)^2]}. \end{cases} \tag{3.1.12}$$

The origin $x_e = 0 \in \mathbb{R}^2$ is an equilibrium of (3.1.12). It is shown in [2, pp. 191–194], that the equilibrium $x_e = 0$ is *attractive and unstable*.

In Definition 3.1.11, we defined *instability of a set* M *that is invariant* with respect to a dynamical system S. It turns out that we require a more general concept, namely, *instability of any set* $M \subset A$ with respect to a dynamical system S.

Definition 3.1.18 Let $\{T, X, A, S\}$ be a dynamical system and let $M \subset A$. The set M is *unstable* with respect to S if for every $\delta > 0$, there exists a $p(\cdot, a, t_0) \in S$ with t_0 independent of δ, and a $t_1 \in T_{a,t_0}$ such that $d(a, M) < \delta$ and $d(p(t_1, a, t_0), M) \geq \varepsilon_0$ for some $\varepsilon_0 > 0$ which is independent of the δ. □

Note that when (S, M) is invariant, then Definitions 3.1.11 and 3.1.18 coincide.

A severe case of instability is the concept of *complete instability*. To introduce this concept, we require the following property of a set M.

Definition 3.1.19 Let $\{T, X, A, S\}$ be a dynamical system and let $M \subset A$. The set M is said to be *proper with respect to* S if for every $\delta > 0$, there exists a $p(\cdot, a, t_0) \in S$ with $T_{a,t_0} \neq \emptyset$ and $0 < d(a, M) < \delta$. □

Definition 3.1.20 Let $\{T, X, A, S\}$ be a dynamical system and let $M \subset A$. The set M is said to be *completely unstable* with respect to S if for every subsystem $\widetilde{S}$ of S such that M is proper with respect to $\widetilde{S}$, M is unstable with respect to $\widetilde{S}$. □

We conclude the present section with an example.

Example 3.1.12 Consider the scalar differential inequality with initial conditions given by

$$\begin{cases} \dot{y}(t) \geq cy(t), & t \geq t_0, \\ y(t_0) = y_0, & t_0 \in \mathbb{R}^+, \ y_0 \in \mathbb{R}^+. \end{cases} \tag{3.1.13}$$

Let $T = \mathbb{R}$, $X = A = \mathbb{R}^+$ and let $S_{(3.1.13)}$ denote the set of all solutions of (3.1.13). Then $\{T, X, A, S_{(3.1.13)}\}$ is a dynamical system.

We show that if $c > 0$, the set $M = \{0\}$ is *completely unstable* with respect to $S_{(3.1.13)}$.

First we note that (3.1.13) implies that $y(t) \geq y_0 e^{c(t-t_0)}$. For any subsystem $\widetilde{S} \subset S_{(3.1.13)}$ such that the set $\{0\}$ is proper with respect to $\widetilde{S}$, we can prove that $\{0\}$ is unstable with respect to $\widetilde{S}$. To see this, let $\varepsilon_0 = 1$. Because $\{0\}$ is proper with respect to $\widetilde{S}$, then for any $\delta > 0$ there exists a $y(\cdot, y_0, t_0) \in \widetilde{S}$ such that $0 < y_0 < \min\{1, \delta\}$ and such that $y(t, y_0, t_0) \geq y_0 e^{c(t-t_0)}$. Let $t_1 \geq t_0 + (1/c)\ln(1/y_0) > t_0$. Then $y(t_1, t_0, y_0) \geq y_0 e^{c(t_1-t_0)} \geq 1 = c_0$. By Definition 3.1.18, $\{0\}$ is unstable with respect to $\widetilde{S}$.

It now follows from Definition 3.1.20 that $\{0\}$ is completely unstable with respect to $S_{(3.1.13)}$. □

3.2 The Principal Lyapunov and Lagrange Stability Results for Discontinuous Dynamical Systems

Before proceeding with our task on hand, we recall from the preceding chapter that we assume throughout, that for every motion $p \in S$ in a DDS, $\{\mathbb{R}^+, X, A, S\}$, the set of times at which discontinuities *may* occur is unbounded and discrete and is of the form

$$E_p = \{\tau_1^p, \tau_2^p, \ldots : \tau_1^p < \tau_2^p < \cdots\}.$$

In the above expression, E_p signifies the fact that, in general, different motions may possess different sets of times at which discontinuities may occur. Because in most cases, the particular set E_p in question is clear from context, we usually suppress the p-notation and simply write

$$E = \{\tau_1, \tau_2, \ldots : \tau_1 < \tau_2 < \cdots\}.$$

A. Local stability results

In the results that follow, we require the notion of a neighborhood of a set.

Definition 3.2.1 U is called a *neighborhood of a set* M if U contains an open neighborhood of $\overline{M}$, the closure of M. □

Theorem 3.2.1 Let $\{\mathbb{R}^+, X, A, S\}$ be a dynamical system and let $M \subset A$ be closed. Assume that there exist a function $V: X \times \mathbb{R}^+ \to \mathbb{R}^+$ and two functions $\varphi_1, \varphi_2 \in \mathcal{K}$ defined on $\mathbb{R}^+$ such that

$$\varphi_1(d(x, M)) \leq V(x, t) \leq \varphi_2(d(x, M)) \tag{3.2.1}$$

for all $x \in X$ and $t \in \mathbb{R}^+$. (We recall that functions of *class* $\mathcal{K}$ are defined earlier in Section 2.1.)

Assume that for any motion $p(\cdot, a, t_0) \in S$, $V(p(t, a, t_0), t)$ is continuous everywhere on $\mathbb{R}^+_{\tau_0} = \{t \in \mathbb{R}^+ : t \geq \tau_0\}$ except on an unbounded and discrete subset $E = \{\tau_1, \tau_2, \ldots : \tau_1 < \tau_2 < \cdots\}$ of $\mathbb{R}^+_{\tau_0}$. Also, assume that there exists a neighborhood U of M such that for all $a \in U$ and for all $p(\cdot, a, \tau_0) \in S$, $V(p(\tau_n, a, \tau_0), \tau_n)$ is nonincreasing for $n \in \mathbb{N} = \{0, 1, 2, \ldots\}$. Furthermore, assume that there exists a function $f \in C[\mathbb{R}^+, \mathbb{R}^+]$, independent of $p \in S$, such that $f(0) = 0$ and such that

$$V(p(t, a, \tau_0), t) \leq f(V(p(\tau_n, a, \tau_0), \tau_n)) \tag{3.2.2}$$

for all $t \in (\tau_n, \tau_{n+1})$, $n \in \mathbb{N}$.

Then, (S, M) is *invariant* and *uniformly stable*.

Proof. We first prove that (S, M) is invariant. If $a \in M$, then $V(p(\tau_0, a, \tau_0), \tau_0) = 0$ because $d(a, M) = 0$ and $V(p(\tau_0, a, \tau_0), \tau_0) = V(a, \tau_0) \leq \varphi_2(d(a, M)) = 0$. Therefore, we know that $V(p(\tau_n, a, \tau_0), \tau_n) = 0$ for all $n \in \mathbb{N}$ because $V(p(\tau_n, a, \tau_0), \tau_n)$ is nonincreasing. Furthermore $V(p(t, a, \tau_0), t) = 0$ for all $t \in (\tau_n, \tau_{n+1})$, $n \in \mathbb{N}$, because $V(p(t, a, \tau_0), t) \leq f(V(p(\tau_n, a, \tau_0), \tau_n)) = 0$. It is then implied that $p(t, a, \tau_0) \in M$ for all $t \geq \tau_0$. Therefore (S, M) is invariant by definition.

Because f is continuous and $f(0) = 0$, then for any $\varepsilon > 0$ there exists a $\delta = \delta(\varepsilon) > 0$ such that $f(r) < \varphi_1(\varepsilon)$ as long as $0 \leq r < \delta$. We can assume that $\delta < \varphi_1(\varepsilon)$. Thus for any motion $p(\cdot, a, \tau_0) \in S$, as long as the initial condition $d(a, M) < \varphi_2^{-1}(\delta)$ is satisfied, then

$$V(p(\tau_n, a, \tau_0), \tau_n) \leq V(p(\tau_0, a, \tau_0), \tau_0) \leq \varphi_2(d(a, M)) < \delta < \varphi_1(\varepsilon)$$

for $n = 1, 2, \ldots$. Furthermore, for any $t \in (\tau_n, \tau_{n+1})$ we can conclude that

$$V(p(t, a, \tau_0), t) \leq f(V(p(\tau_n, a, \tau_0), \tau_n)) < \varphi_1(\varepsilon).$$

Thus, we have shown that $V(p(t, a, \tau_0), t) < \varphi_1(\varepsilon)$ is true for all $t \in \mathbb{R}^+_{\tau_0}$. In view of (3.2.1), we have

$$d(p(t, a, \tau_0), M) \leq \varphi_1^{-1}(V(p(t, a, \tau_0), t)) < \varphi_1^{-1}(\varphi_1(\varepsilon)) = \varepsilon.$$

Therefore, by definition, (S, M) is uniformly stable. □

Theorem 3.2.2 If in addition to the assumptions given in Theorem 3.2.1 there exists a function $\varphi_3 \in \mathcal{K}$ defined on $\mathbb{R}^+$ such that for all $a \in U$, for all $p(\cdot, a, \tau_0) \in S$, and for all $n \in \mathbb{N}$,

$$DV(p(\tau_n, a, \tau_0), \tau_n) \leq -\varphi_3(d(p(\tau_n, a, \tau_0), M)) \tag{3.2.3}$$

where

$$DV(p(\tau_n, a, \tau_0), \tau_n) \triangleq \frac{1}{\tau_{n+1} - \tau_n}\big[V(p(\tau_{n+1}, a, \tau_0), \tau_{n+1}) - V(p(\tau_n, a, \tau_0), \tau_n)\big] \tag{3.2.4}$$

then (S, M) is *uniformly asymptotically stable.*

Proof. For any $a \in U$ and for any $p(\cdot, a, \tau_0) \in S$, letting $z_n = V(p(\tau_n, a, \tau_0), \tau_n)$, $n \in \mathbb{N}$, we obtain from the assumptions of the theorem that

$$z_{n+1} - z_n \leq -(\tau_{n+1} - \tau_n)\big(\varphi_3 \circ \varphi_2^{-1}\big)(z_n)$$

for all $n \in \mathbb{N}$. If we denote $\varphi = \varphi_3 \circ \varphi_2^{-1}$, then $\varphi \in \mathcal{K}$ and the above inequality becomes

$$z_{n+1} - z_n \leq -(\tau_{n+1} - \tau_n)\varphi(z_n).$$

Inasmuch as $\{z_n\}$ is nonincreasing and $\varphi \in \mathcal{K}$, it follows that

$$z_{k+1} - z_k \leq -\varphi(z_k)(\tau_{k+1} - \tau_k) \leq -\varphi(z_n)(\tau_{k+1} - \tau_k)$$

for all $k \leq n$. We thus obtain that

$$z_{n+1} - z_0 \leq -(\tau_{n+1} - \tau_0)\varphi(z_n),$$

which in turn yields

$$\varphi(z_n) \leq \frac{z_0 - z_{n+1}}{\tau_{n+1} - \tau_0} \leq \frac{z_0}{\tau_{n+1} - \tau_0}, \tag{3.2.5}$$

for all $n \in \mathbb{N}$.

Now consider a fixed $\delta > 0$. For any given $\varepsilon > 0$, we can choose a $\gamma > 0$ such that

$$\max\left\{\varphi_1^{-1}\left(\varphi^{-1}\left(\frac{\varphi_2(\delta)}{\gamma}\right)\right), \varphi_1^{-1}\left(f\left(\varphi^{-1}\left(\frac{\varphi_2(\delta)}{\gamma}\right)\right)\right)\right\} < \varepsilon \tag{3.2.6}$$

because $\varphi_1, \varphi_2, \varphi \in \mathcal{K}$ and $f(0) = 0$. For any $a \in A$ with $d(a, M) < \delta$ and any $\tau_0 \in \mathbb{R}^+$, we are now able to show that $d(p(t, a, \tau_0), M) < \varepsilon$ whenever $t \geq \tau_0 + \gamma$. This is because for any $t \geq \tau_0 + \gamma$, t must belong to some interval $[\tau_n, \tau_{n+1})$ for some $n \in \mathbb{N}$, that is, $t \in [\tau_n, \tau_{n+1})$. Therefore we know that $\tau_{n+1} - \tau_0 > \gamma$. It follows from (3.2.5) that

$$\varphi(z_n) \leq \frac{z_0}{\gamma} = \frac{V(a, \tau_0)}{\gamma} \leq \frac{\varphi_2(\delta)}{\gamma},$$

which implies that

$$V(p(\tau_n, a, \tau_0), \tau_n) = z_n \leq \varphi^{-1}\left(\frac{\varphi_2(\delta)}{\gamma}\right), \tag{3.2.7}$$

and

$$V(p(t, a, \tau_0), t) \leq f\left(\varphi^{-1}\left(\frac{\varphi_2(\delta)}{\gamma}\right)\right) \tag{3.2.8}$$

if $t \in (\tau_n, \tau_{n+1})$. In the case when $t = \tau_n$, it follows from (3.2.7) that

$$d(p(\tau_n, a, \tau_0), M) < \varphi_1^{-1}(V(p(\tau_n, a, \tau_0), \tau_n)) < \varepsilon,$$

noticing that (3.2.6) holds. In the case when $t \in (\tau_n, \tau_{n+1})$, we can conclude from (3.2.8) that

$$d(p(t, a, \tau_0), M) < \varphi_1^{-1}(V(p(t, a, \tau_0), t)) < \varepsilon.$$

This proves that (S, M) is uniformly asymptotically stable. □

Theorem 3.2.3 Let $\{\mathbb{R}^+, X, A, S\}$ be a dynamical system and let $M \subset A$ be closed. Assume that there exist a function $V\colon X \times \mathbb{R}^+ \to \mathbb{R}^+$ and four positive constants c_1, c_2, c_3, and b such that

$$c_1[d(x, M)]^b \leq V(x, t) \leq c_2[d(x, M)]^b \tag{3.2.9}$$

for all $x \in X$ and $t \in \mathbb{R}^+$.

Assume that there exists a neighborhood U of M such that for all $a \in U$ and for all $p(\cdot, a, \tau_0) \in S$, $V(p(t, a, \tau_0), t)$ is continuous everywhere on $\mathbb{R}^+_{\tau_0}$ except on an unbounded and discrete subset $E = \{\tau_1, \tau_2, \ldots : \tau_1 < \tau_2 < \cdots\}$ of $\mathbb{R}^+_{\tau_0}$. Furthermore, assume that there exists a function $f \in C[\mathbb{R}^+, \mathbb{R}^+]$ such that

$$V(p(t, a, \tau_0), t) \leq f(V(p(\tau_n, a, \tau_0), \tau_n)) \tag{3.2.10}$$

for $t \in (\tau_n, \tau_{n+1})$, $n \in \mathbb{N}$, and that for some positive constant q, f satisfies

$$f(r) = o(r^q) \quad \text{as } r \to 0^+ \tag{3.2.11}$$

(i.e., $\lim_{r \to 0^+} f(r)/r^q = 0$). Assume that for all $n \in \mathbb{N}$,

$$DV(p(\tau_n, a, \tau_0), \tau_n) \leq -c_3[d(p(\tau_n, a, \tau_0), M)]^b \tag{3.2.12}$$

for all $a \in U$ and all $p(\cdot, a, \tau_0) \in S$, where $DV(p(\tau_n, a, \tau_0), \tau_n)$ is given in (3.2.4).

Then (S, M) is *exponentially stable*.

Proof. It follows from Theorem 3.2.1 that under the present hypotheses, M is an invariant set of S. For any $a \in U$ and $p(\cdot, a, \tau_0) \in S$, let $z_n = V(p(\tau_n, a, \tau_0), \tau_n)$, $n \in \mathbb{N}$, and $z(t) = V(p(t, a, \tau_0), t)$. We obtain from (3.2.9) and (3.2.12) that

$$\frac{z_{n+1} - z_n}{\tau_{n+1} - \tau_n} \leq -\frac{c_3}{c_2} z_n,$$

which yields

$$z_{n+1} \leq [1 - \eta(\tau_{n+1} - \tau_n)]z_n,$$

where $\eta = c_3/c_2$. If $1 - \eta(\tau_{n+1} - \tau_n) \leq 0$ is true for some n, then $z_k = 0$ and $z(t) \leq f(z_k) = 0$ for all $t \in (\tau_k, \tau_{k+1})$ and all $k > n$. Thus, $d(p(t, a, \tau_0), M) = 0$ for all $t > \tau_{n+1}$. In the following we assume that $1 - \eta(\tau_{n+1} - \tau_n) > 0$ for all $n \geq 0$.

Because $e^{-\eta r} \geq 1 - \eta r$, it follows that

$$z_{n+1} \leq e^{-\eta(\tau_{n+1} - \tau_n)} z_n.$$

Hence,

$$z_{n+1} \leq e^{-\eta(\tau_{n+1} - \tau_0)} z_0$$

is true for all $n \geq 0$. It now follows from (3.2.9) that

$$d(p(\tau_n, a, \tau_0), M) \leq \left(\frac{z_0}{c_1}\right)^{1/b} e^{-(\eta/b)(\tau_n - \tau_0)} \leq \left(\frac{c_2}{c_1}\right)^{1/b} d(a, M) e^{-(\eta/b)(\tau_n - \tau_0)}. \tag{3.2.13}$$

In the last step, we have made use of the fact that

$$z_0 = V(p(\tau_0, a, \tau_0), \tau_0) \leq c_2[d(a, M)]^b.$$

Inasmuch as $f(r) = o(r^q)$ as $r \to 0^+$, it is easily seen that $f(r)/r^q \in C[\mathbb{R}^+, \mathbb{R}^+]$. Let

$$\lambda_{d(a,M)} = \sup_{r \in (0,\ c_2(d(a,M))^b]} \frac{f(r)}{r^q}.$$

Then $f(r) \leq \lambda_{d(a,M)} r^q$ for all $r \in [0,\ c_2(d(a, M))^b]$. It follows from (3.2.10) that for all $t \in (\tau_n, \tau_{n+1})$, it is true that

$$\begin{aligned} z(t) &\leq f(z_n) \\ &\leq \lambda_{d(a,M)} z_n^q \\ &\leq \lambda_{d(a,M)} e^{-\eta q(\tau_n - \tau_0)} z_0^q \\ &= \lambda_{d(a,M)} e^{\eta q(t - \tau_n)} e^{-\eta q(t - \tau_0)} z_0^q \\ &\leq \lambda_{d(a,M)} e^q e^{-\eta q(t - \tau_0)} z_0^q. \end{aligned}$$

The last inequality follows because $t - \tau_n \leq \tau_{n+1} - \tau_n \leq 1/\eta$. Thus,

$$d(p(t, a, \tau_0), M) \leq \left(\frac{z(t)}{c_1}\right)^{1/b} \leq \left[\frac{\lambda_{d(a,M)} e^q c_2^q}{c_1}\right]^{1/b} [d(a, M)]^q e^{-(\eta q/b)}(t - \tau_0). \tag{3.2.14}$$

For any $\varepsilon > 0$ there exists a $\delta > 0$ such that

$$\varepsilon \geq \min\left\{\left(\frac{c_2}{c_1}\right)^{1/b} d(a, M),\ \left[\frac{\lambda_{d(a,M)} e^q c_2^q}{c_1}\right]^{1/b} [d(a, M)]^q\right\}$$

for any $a \in U$ whenever $d(a, M) < \delta$. Letting

$$\alpha = \min\left\{\frac{\eta}{b}, \frac{\eta q}{b}\right\},$$

we have, in view of (3.2.13) and (3.2.14), that

$$d(p(t, a, \tau_0), M) \leq \varepsilon e^{-\alpha(t-\tau_0)}$$

for all $p(\cdot, a, \tau_0) \in S$ and $t \in \mathbb{R}^+_{\tau_0}$, whenever $d(a, M) < \delta$. Therefore (S, M) is exponentially stable. This concludes the proof of the theorem. □

B. Global stability and boundedness results

Next, we address global results.

Theorem 3.2.4 Let $\{\mathbb{R}^+, X, A, S\}$ be a dynamical system, let $M \subset A$, and assume that M is bounded. Assume that there exist a function $V\colon X \times \mathbb{R}^+ \to \mathbb{R}^+$ and two strictly increasing functions $\varphi_1, \varphi_2 \in C[\mathbb{R}^+, \mathbb{R}^+]$ with $\lim_{r\to\infty} \varphi_i(r) = \infty$, $i = 1, 2$, such that

$$\varphi_1(d(x, M)) \leq V(x, t) \leq \varphi_2(d(x, M)) \tag{3.2.15}$$

for all $x \in X$ and for all $t \in \mathbb{R}^+$ whenever $d(x, M) \geq \Omega$, where Ω is a positive constant.

Assume that for every $p(\cdot, a, \tau_0) \in S$, $V(p(t, a, \tau_0), t)$ is continuous everywhere on $\mathbb{R}^+_{\tau_0}$ except on an unbounded and discrete subset $E = \{\tau_1, \tau_2, \ldots : \tau_1 < \tau_2 < \cdots\}$ of $\mathbb{R}^+_{\tau_0}$. Also, assume that for all $p(\cdot, a, \tau_0) \in S$,

$$V(p(\tau_{n+1}, a, \tau_0), \tau_{n+1}) \leq V(p(\tau_n, a, \tau_0), \tau_n) \tag{3.2.16}$$

for all τ_n whenever $d(p(\tau_n, a, \tau_0), M) \geq \Omega$.

Furthermore, assume that there exists a function $f \in C[\mathbb{R}^+, \mathbb{R}^+]$, independent of $p(\cdot, a, t_0) \in S$, such that for all $n \in \mathbb{N}$ and all $p(\cdot, a, t_0) \in S$

$$V(p(t, a, \tau_0), t) \leq f(V(p(\tau_n, a, \tau_0), \tau_n)) \tag{3.2.17}$$

for all $t \in (\tau_n, \tau_{n+1})$ whenever $d(p(t, a, \tau_0), M) \geq \Omega$.

Furthermore, assume that there exists a constant $\Gamma > 0$ such that

$$d(p(\tau_{n+1}, a, \tau_0), M) \leq \Gamma$$

whenever $d(p(\tau_n, a, \tau_0), M) \leq \Omega$ for all $p(\cdot, a, \tau_0) \in S$.

Then, S is *uniformly bounded.*

Proof. For any $\alpha > 0, \tau_0 \in \mathbb{R}^+$, $a \in A$ such that $d(a, M) < \alpha$, and $p(\cdot, a, \tau_0) \in S$, let $z_n = V(p(\tau_n, a, \tau_0), \tau_n)$ and let $z(t) = V(p(t, a, \tau_0), t)$. If $d(a, M) \geq \Omega$, it follows from (3.2.15) and (3.2.16) that

$$\varphi_1(d(p(\tau_n, a, \tau_0), M)) \leq z_n \leq z_0 \leq \varphi_2(\alpha).$$

Thus $d(p(\tau_n, a, \tau_0), M) \le (\varphi_1^{-1} \circ \varphi_2)(\alpha)$ for as long as $d(p(\tau_k, a, \tau_0), M) \ge \Omega$, for all $k < n$.

If $d(p(\tau_n, a, \tau_0), M)$ starts at a value less than Ω or if it reaches a value less than Ω for some n_0 (i.e., if $d(p(\tau_{n_0}, a, \tau_0), M) \le \Omega$), then $d(p(\tau_{n_0+1}, a, \tau_0), M) \le \Gamma$, by assumption. We can now replace α in the foregoing argument by Γ and obtain that $d(p(\tau_n, a, \tau_0), M) \le (\varphi_1^{-1} \circ \varphi_2)(\Gamma)$ for as long as $d(p(\tau_k, a, \tau_0), M) \le \Omega$, for all k such that $n_0 < k < n$.

By induction, we conclude that

$$d(p(\tau_n, a, \tau_0), M) \le \beta_1(\alpha) \triangleq \max\left\{\Gamma, (\varphi_1^{-1} \circ \varphi_2)(\Gamma), (\varphi_1^{-1} \circ \varphi_2)(\alpha)\right\}$$

for all $n \in \mathbb{N}$.

Because $f \in C[\mathbb{R}^+, \mathbb{R}^+]$, there exists a $\beta_2 = \beta_2(\alpha)$ such that $f(r) \le \beta_2$ whenever $r \in [0, \varphi_2(\beta_1(\alpha))]$. For any $t \in (\tau_n, \tau_{n+1})$, we have that $z(t) \le f(z_n) \le \beta_2$.

If we let

$$\beta(\alpha) = \max\left\{\beta_1(\alpha), \varphi_1^{-1}(\beta_2(\alpha))\right\}, \tag{3.2.18}$$

then it is easily seen that $d(p(t, a, \tau_0), M) \le \beta(\alpha)$ for all $t \in \mathbb{R}^+_{\tau_0}$ and $a \in A$ whenever $d(a, M) < \alpha$. Because M is bounded, S is uniformly bounded. The proof is completed. □

Theorem 3.2.5 If in addition to the assumptions in Theorem 3.2.4 there exists a function $\varphi_3 \in \mathcal{K}$ defined on $\mathbb{R}^+$ such that for all $p(\cdot, a, \tau_0) \in S$

$$DV(p(\tau_n, a, \tau_0), \tau_n) \le -\varphi_3(d(p(\tau_n, a, \tau_0), M)) \tag{3.2.19}$$

for all τ_n whenever $d(p(\tau_n, a, \tau_0), M) \ge \Omega$, where DV in (3.2.19) is defined in (3.2.4).

Then S is *uniformly ultimately bounded.*

Proof. Let $B = \beta(\Omega)$, where $\beta(\cdot)$ is given in (3.2.18). We show that corresponding to any $\alpha > 0$ and $\tau_0 \in \mathbb{R}^+$, there exists a $\tau = \tau(\alpha) > 0$ such that $d(p(t, a, \tau_0), M) \le B$ for all $t > \tau_0 + \tau$ and $p(\cdot, a, \tau_0) \in S$ whenever $d(a, M) < \alpha$.

If $d(p(\tau_k, a, \tau_0), M) \ge \Omega$ for all $k < n$, we obtain, using the same argument as that for (3.2.5), that

$$\varphi_3(d(p(\tau_n, a, \tau_0), M)) \le \frac{z_0 - z_n}{\tau_{n+1} - \tau_0} \le \frac{z_0}{\tau_{n+1} - \tau_0} \le \frac{\varphi_2(\alpha)}{\tau_{n+1} - \tau_0}. \tag{3.2.20}$$

Let $\tau = \varphi_2(\alpha)/\varphi_3(\Omega)$. For any $t > \tau_0 + \tau$, there exists an $n \in \mathbb{N}$ such that $t \in [\tau_n, \tau_{n+1})$. Thus $\tau_{n+1} - \tau_0 > \tau$. There must exist a $k_0 \le n$ such that $d(p(\tau_{k_0}, a, \tau_0), M) < \Omega$. Otherwise, in view of (3.2.19), $d(p(\tau_n, a, \tau_0), M) < \varphi_3^{-1}(\varphi_2(\alpha)/\tau) < \varphi_3^{-1}(\varphi_3(\Omega)) = \Omega$. We have arrived at a contradiction. Therefore, $d(p(\tau_{k_0}, a, \tau_0), M) < \Omega$ for some $k_0 \le n$. By the same argument as that in the proof of Theorem 3.2.4, we know that $d(p(t, a, \tau_0), M) \le B$. Hence, we have shown that S is uniformly ultimately bounded. □

Theorem 3.2.6 Let $\{\mathbb{R}^+, X, A, S\}$ be a dynamical system. Let $M \subset A$ be bounded and closed. Assume that there exist a function $V\colon X \times \mathbb{R}^+ \to \mathbb{R}^+$ and two functions $\varphi_1, \varphi_2 \in \mathcal{K}_\infty$ such that

$$\varphi_1(d(x, M)) \le V(x, t) \le \varphi_2(d(x, M)) \tag{3.2.21}$$

for all $x \in X$ and $t \in \mathbb{R}^+$.

Assume that for any $p(\cdot, a, \tau_0) \in S$, $V(p(t, a, \tau_0), t)$ is continuous everywhere on $\mathbb{R}^+_{\tau_0}$ except on an unbounded and discrete subset $E = \{\tau_1, \tau_2, \dots : \tau_1 < \tau_2 < \cdots\}$ of $\mathbb{R}^+_{\tau_0}$. Furthermore, assume that there exists a function $f \in C[\mathbb{R}^+, \mathbb{R}^+]$ with $f(0) = 0$ such that for any $p(\cdot, a, \tau_0) \in S$,

$$V(p(t, a, \tau_0), t) \le f(V(p(\tau_n, a, \tau_0), \tau_n)) \tag{3.2.22}$$

for $t \in (\tau_n, \tau_{n+1})$, $n \in \mathbb{N}$.

Assume that there exists a function $\varphi_3 \in \mathcal{K}$ defined on $\mathbb{R}^+$ such that for any $p(\cdot, a, \tau_0) \in S$,

$$DV(p(\tau_n, a, \tau_0), \tau_n) \le -\varphi_3(d(p(\tau_n, a, \tau_0), M)) \tag{3.2.23}$$

$n \in \mathbb{N}$, where DV in (3.2.23) is defined in (3.2.4).

Then, (S, M) is *uniformly asymptotically stable in the large*. (Recall that functions of class $\mathcal{K}_\infty$ are defined in Section 2.1.)

Proof. It follows from Theorem 3.2.1 that under the present hypotheses, M is an invariant set of S and (S, M) is uniformly stable. We need to show that conditions (ii) and (iii) in Definition 3.1.16 are also satisfied.

Consider arbitrary $\alpha > 0$, $\varepsilon > 0$, $\tau_0 \in \mathbb{R}^+$, and $a \in A$ such that $d(a, M) < \alpha$. Letting $z_n = V(p(\tau_n, a, \tau_0), \tau_n)$ and $z(t) = V(p(t, a, \tau_0), t)$, we obtain from the assumptions of the theorem that $\{z_n\}$ is nonincreasing and that

$$z(t) \le \max\left\{\varphi_2(\alpha), \max_{r \in [0, \varphi_2(\alpha)]} f(r)\right\}$$

whenever $d(a, M) < \alpha$. Thus S is uniformly bounded.

Let $\varphi = \varphi_3 \circ \varphi_2^{-1}$. Using the same argument as that in the proof of Theorem 3.2.2, we obtain that

$$\varphi(z_n) \le \frac{z_0 - z_n}{\tau_{n+1} - \tau_0} \le \frac{z_0}{\tau_{n+1} - \tau_0}.$$

Let $\gamma_1 = \gamma_1(\varepsilon, \alpha) = \varphi_2(\alpha)/\varphi(\varphi_1(\varepsilon)) > 0$ and choose a $\delta > 0$ such that $\max_{r \in [0,\delta]} f(r) < \varphi_1(\varepsilon)$. Let $\gamma_2 = \varphi_2(\alpha)/\varphi(\delta)$ and $\gamma = \max\{\gamma_1, \gamma_2\}$. For any $a \in A$ with $d(a, M) < \alpha$ and any $\tau_0 \in \mathbb{R}^+$, we are now able to show that $d(p(t, a, \tau_0), M) < \varepsilon$ whenever $t \ge \tau_0 + \gamma$. The above statement is true because for any $t \ge \tau_0 + \gamma$, t must belong to some interval $[\tau_n, \tau_{n+1})$ for some $n \in \mathbb{N}$; that is, $t \in [\tau_n, \tau_{n+1})$. Therefore we know that $\tau_{n+1} - \tau_0 > \gamma$ and that

$$\varphi(z_n) \le \frac{z_0}{\gamma} < \frac{\varphi_2(\alpha)}{\gamma},$$

which implies that

$$V(p(\tau_n, a, \tau_0), \tau_n) = z_n < \varphi^{-1}\left(\frac{\varphi_2(\alpha)}{\gamma}\right) \le \min\{\varphi_1(\varepsilon), \delta\}.$$

We thus have $d(p(\tau_n, a, \tau_0), M) < \varepsilon$ and $V(p(t, a, \tau_0), t) \le f(z_n) \le \varphi_1(\varepsilon)$ for all $t \in (\tau_n, \tau_{n+1})$, and hence, $d(p(t, a, \tau_0), M) < \varepsilon$. This proves that (S, M) is uniformly asymptotically stable in the large. □

Theorem 3.2.7 Let $\{\mathbb{R}^+, X, A, S\}$ be a dynamical system. Let $M \subset A$ be bounded and closed. Assume that there exist a function $V: X \times \mathbb{R}^+ \to \mathbb{R}^+$ and four positive constants c_1, c_2, c_3, and b such that

$$c_1[d(x, M)]^b \le V(x, t) \le c_2[d(x, M)]^b \tag{3.2.24}$$

for all $x \in X$ and $t \in \mathbb{R}^+$.

Assume that for every $p(\cdot, a, \tau_0) \in S$, $V(p(t, a, \tau_0), t)$ is continuous everywhere on $\mathbb{R}^+_{\tau_0}$ except on an unbounded subset $E = \{\tau_1, \tau_2, \ldots : \tau_1 < \tau_2 < \cdots\}$ of $\mathbb{R}^+_{\tau_0}$. Furthermore, assume that there exists a function $f \in C[\mathbb{R}^+, \mathbb{R}^+]$ with $f(0) = 0$ such that

$$V(p(t, a, \tau_0), t) \le f(V(p(\tau_n, a, \tau_0), \tau_n)) \tag{3.2.25}$$

for $t \in (\tau_n, \tau_{n+1})$, $n \in \mathbb{N}$, and such that for some positive constant q, f satisfies

$$f(r) = \mathcal{O}(r^q) \quad \text{as } r \to 0^+. \tag{3.2.26}$$

Assume that

$$DV(p(\tau_n, a, \tau_0), \tau_n) \le -c_3[d(p(\tau_n, a, \tau_0), M)]^b \tag{3.2.27}$$

for all $p(\cdot, a, \tau_0) \in S$ and all $a \in A$ where DV in (3.2.27) is defined in (3.2.4).

Then (S, M) is *exponentially stable in the large.*

Proof. It follows from Theorem 3.2.1 that under the present hypotheses, M is an invariant set of S.

For any $\beta > 0$ and any a such that $d(a, M) < \beta$, using the same argument as that in the proof of Theorem 3.2.3, we obtain that

$$d(p(t, a, \tau_0), M) \le \left(\frac{z(t)}{c_1}\right)^{1/b} \le \left[\frac{\lambda_{d(a,M)} e^q c_2^q}{c_1}\right]^{1/b} [d(a, M)]^q e^{-\eta q(t-\tau_0)/b}$$

for all $t \in \mathbb{R}^+_{\tau_0}$, where $\eta = c_3/c_2$ and $\lambda_{d(a,M)}$ is chosen such that $f(r) \le \lambda_{d(a,M)} r^q$ for all $r \in [0,\ c_2(d(a, M))^b]$. Let

$$\mu = \min\left\{\frac{\eta}{b},\ \frac{\eta q}{b}\right\},$$
$$\lambda = \lambda_\beta,$$
$$\gamma = \min\{1,\ q\},$$

and

$$k(\beta) = \max\left\{\left(\frac{c_2}{c_1}\right)^{1/b}\beta^{1-\gamma}, \left(\frac{\lambda e^q c_2^q}{c_1}\right)^{1/b}\beta^{q-\gamma}\right\}.$$

Then

$$d(p(t,a,\tau_0),M) \le k(\beta)[d(a,M)]^\gamma e^{-\mu(t-\tau_0)}$$

for all $p(\cdot,a,\tau_0) \in S$ and $t \in \mathbb{R}^+_{\tau_0}$. Therefore (S,M) is exponentially stable in the large. This concludes the proof of the theorem. □

Remark 3.2.1 The hypotheses of Theorem 3.2.1 can be relaxed by requiring only that $V(p(\tau'_n,a,\tau_0),\tau'_n)$ is nonincreasing for $n \in \mathbb{N}$ and that

$$V(p(t,a,\tau_0),t) \le f(V(p(\tau'_n,a,\tau_0),\tau'_n))$$

for all $t \in (\tau'_n, \tau'_{n+1})$, $n \in \mathbb{N}$, where $E' = \{\tau'_1, \tau'_2, \dots,\}$ is a strictly increasing unbounded subsequence of the set $E = \{\tau_1, \tau_2, \dots\}$. In the same spirit, we can replace in Theorem 3.2.2 inequality (3.2.3) by

$$DV(p(\tau'_n,a,\tau_0),\tau'_n) \le \varphi_3(d(p(\tau'_n,a,\tau_0),M))$$

for all $n \in \mathbb{N}$, where $DV(p(\tau'_n,a,\tau_0),\tau'_n)$ is defined as in (3.2.4) and $\tau'_n \in E'$. Furthermore, the hypotheses in Theorems 3.2.3–3.2.7 can be altered in a similar manner. These assertions follow easily from the proofs of Theorems 3.2.1–3.2.7. □

C. Instability results

Thus far, we have concerned ourselves with stability and boundedness results. We now address instability.

Theorem 3.2.8 Let $\{\mathbb{R}^+, X, A, S\}$ be a dynamical system and let $M \subset A$ be a closed set. Assume that there exist a function $V\colon X \times \mathbb{R}^+ \to \mathbb{R}$ and a $\tau_0 \in \mathbb{R}^+$ that satisfy the following conditions.

(i) There exists a function $\varphi \in \mathcal{K}$ defined on $\mathbb{R}^+$ such that

$$V(x,t) \le \varphi(d(x,M)) \tag{3.2.28}$$

for all $x \in X$ and $t \in \mathbb{R}^+$.

(ii) In every neighborhood of M there is a point x such that $V(x,\tau_0) > 0$ and there exists a motion $p(\cdot,x,\tau_0) \in S$.

(iii) For any $a \in A$ such that $V(a,\tau_0) > 0$ and any $p(\cdot,a,\tau_0) \in S$, $V(p(t,a,\tau_0),t)$ is continuous everywhere on $\mathbb{R}^+_{\tau_0}$ except on an unbounded and discrete subset $E = \{\tau_1, \tau_2, \dots : \tau_1 < \tau_2 < \cdots\}$ of $\mathbb{R}^+_{\tau_0}$. Assume that there exists a function $\psi \in \mathcal{K}$ defined on $\mathbb{R}^+$ such that

$$DV(p(\tau_n,a,\tau_0),\tau_n) \ge \psi(|V(p(\tau_n,a,\tau_0),\tau_n)|) \tag{3.2.29}$$

for all $n \in \mathbb{N}$, where $DV(p(\tau_n,a,\tau_0),\tau_n)$ is given in (3.2.4).

Then M is *unstable* with respect to S.

Proof. By assumption, for every $\delta > 0$ there exists an $a \in A$ such that $d(a, M) < \delta$ and $V(a, \tau_0) > 0$. Let $z_n = V(p(\tau_n, a, \tau_0), \tau_n)$. Then $z_0 = V(a, \tau_0) > 0$. From assumption (iii) it follows that $\{z_n\}$ is increasing and

$$\begin{aligned} z_n &\geq z_{n-1} + (\tau_n - \tau_{n-1})\psi(z_{n-1}) \\ &\geq z_0 + (\tau_n - \tau_0)\psi(z_0) \\ &> (\tau_n - \tau_0)\psi(V(a, \tau_0)). \end{aligned}$$

Hence, as τ_n goes to ∞, $d(p(\tau_n, a, \tau_0), M) \geq \varphi^{-1}(V(p(\tau_n, a, \tau_0), \tau_n))$ can become arbitrarily large. Therefore, (S, M) is unstable. □

Theorem 3.2.9 In addition to the assumptions given in Theorem 3.2.8, assume that $V(x, \tau_0) > 0$ for all $x \notin M$. Then M is *completely unstable* with respect to S.

Proof. Because $V(a, \tau_0)$ is positive for every $a \notin M$ and every $\tau_0 \in \mathbb{R}^+$, the argument in the proof of Theorem 3.2.8 applies for all $a \notin M$; that is, along every motion $p(\cdot, a, \tau_0) \in S$, $d(p(\tau_n, a, \tau_0), M)$ tends to ∞ as n goes to ∞. We conclude that (S, M) is completely unstable. □

We conclude the present section with an important observation.

Remark 3.2.2 It is emphasized that because continuous dynamical systems constitute special cases of DDS, all the results of the present section are applicable to continuous dynamical systems as well. □

3.3 The Principal Lyapunov and Lagrange Stability Results for Continuous Dynamical Systems

In the present section we establish the Principal Lyapunov Stability and Boundedness Results for continuous dynamical systems. We show that these results are a direct consequence of the results of the preceding section (i.e., we show that when the hypotheses of the results of the present section for continuous dynamical systems are satisfied, then the hypotheses of the corresponding results of the preceding section for DDS are also satisfied). In this way, we establish a unifying link between the stability results of DDS and continuous dynamical systems. More important, we show that the results of the present section, which constitute the Principal Lyapunov and Lagrange Stability Results for continuous dynamical systems, are in general more conservative than the corresponding results for DDS. We include in this section a specific example that reinforces these assertions.

A. Local stability results

We first consider local results.

Theorem 3.3.1 Let $\{\mathbb{R}^+, X, A, S\}$ be a continuous dynamical system and let $M \subset A$ be closed. Assume that there exist a function $V: X \times \mathbb{R}^+ \to \mathbb{R}^+$ and two functions $\varphi_1, \varphi_2 \in \mathcal{K}$ defined on $\mathbb{R}^+$ such that

$$\varphi_1(d(x, M)) \leq V(x, t) \leq \varphi_2(d(x, M)) \tag{3.3.1}$$

for all $x \in X$ and $t \in \mathbb{R}^+$. Assume that there exists a neighborhood U of M such that for all $a \in U$ and for all $p(\cdot, a, \tau_0) \in S$, $V(p(t, a, \tau_0), t)$ is continuous and nonincreasing for all $t \in \mathbb{R}^+_{\tau_0}$. Then (S, M) is *invariant* and *uniformly stable*.

Proof. For any $p(\cdot, a, \tau_0) \in S$, let $E = \{\tau_1, \tau_2, \dots : \tau_1 < \tau_2 < \cdots\}$ be an arbitrary unbounded subset of $\mathbb{R}^+_{\tau_0}$. Let $f \in C[\mathbb{R}^+, \mathbb{R}^+]$ be the identity function; that is, $f(r) = r$.

By assumption, for any $a \in U$ and $p(\cdot, a, \tau_0) \in S$, $V(p(t, a, \tau_0), t)$ is continuous on $\mathbb{R}^+_{\tau_0}$ and $V(p(\tau_n, a, \tau_0), \tau_n)$ is nonincreasing for $n \in \mathbb{N}$. Furthermore,

$$V(p(t, a, \tau_0), t) \leq V(p(\tau_n, a, \tau_0), \tau_n) = f(V(p(\tau_n, a, \tau_0), \tau_n))$$

for all $t \in (\tau_n, \tau_{n+1})$, $n \in \mathbb{N}$. Hence, all the hypotheses of Theorem 3.2.1 are satisfied and thus, (S, M) is *invariant* and *uniformly stable*. □

Theorem 3.3.2 If in addition to the assumptions given in Theorem 3.3.1 there exists a function $\varphi_3 \in \mathcal{K}$ defined on $\mathbb{R}^+$ such that for all $a \in U$ and for all $p(\cdot, a, \tau_0) \in S$ the upper right-hand Dini derivative $D^+V(p(t, a, \tau_0), t)$ satisfies

$$D^+V(p(t, a, \tau_0), t) \leq -\varphi_3(d(p(t, a, \tau_0), M)) \tag{3.3.2}$$

for all $t \in \mathbb{R}^+_{\tau_0}$, then (S, M) is *uniformly asymptotically stable*.

Proof. For any $a \in U$ and any $p(\cdot, a, \tau_0) \in S$, choose $\widetilde{E} = \{s_1, s_2, \dots\}$ recursively in the following manner. For $n \in \mathbb{N}$, let $s_0 = \tau_0$ and $s_{n+1} = s_n + \min\{1, \alpha_n\}$, where

$$\alpha_n = \sup\left\{\tau \colon V(p(t, a, \tau_0), t) \geq \frac{1}{2} V(p(s_n, a, \tau_0), s_n) \text{ for all } t \in (s_n, s_n + \tau)\right\};$$

that is, $V(p(t, a, \tau_0)) \geq V(p(s_n, a, \tau_0))/2$ for all $t \in (s_n, s_{n+1})$.

If $\widetilde{E}$ is unbounded then simply let $\tau_n = s_n$, $n \in \mathbb{N}$. The set $E = \{\tau_1, \tau_2, \dots\}$ is clearly unbounded and discrete. It follows from the assumptions of the theorem and from the choice of τ_n that we have for any $t \in (\tau_n, \tau_{n+1})$,

$$\begin{aligned} d(p(t, a, \tau_0), M) &\geq (\varphi_2^{-1} \circ V)(p(t, a, \tau_0), t) \\ &\geq \left(\varphi_2^{-1} \circ \frac{1}{2} V\right)(p(\tau_n, a, \tau_0), \tau_n) \\ &\geq \left(\varphi_2^{-1} \circ \frac{1}{2} \varphi_1\right)(d(p(\tau_n, a, \tau_0), M)). \end{aligned}$$

Now refer to the Appendix, Section 3.8. Letting

$$g(t, V(p(t, a, \tau_0), t)) = -\varphi_3(d(p(t, a, \tau_0), M)), \quad t_0 = \tau_n, x_0 = V(p(\tau_n, a, \tau_0), \tau_n),$$

the (maximal) solution of (I_E) is given by

$$\varphi_M(\tau_{n+1}) = V(p(\tau_n, a, \tau_0), \tau_n) - \int_{\tau_n}^{\tau_{n+1}} \varphi_3(d(p(t, a, \tau_0), M))dt.$$

It now follows from the *Comparison Theorem* (Theorem 3.8.1) that

$$\begin{aligned} V(p(\tau_{n+1}, a, \tau_0), \tau_{n+1}) &- V(p(\tau_n, a, \tau_0), \tau_n) \\ &\leq - \int_{\tau_n}^{\tau_{n+1}} \varphi_3(d(p(t, a, \tau_0), M))dt \\ &\leq - \int_{\tau_n}^{\tau_{n+1}} \Big(\varphi_3 \circ \varphi_2^{-1} \circ \frac{1}{2}\varphi_1\Big)(d(p(\tau_n, a, \tau_0), M))dt \\ &= - (\tau_{n+1} - \tau_n)\Big(\varphi_3 \circ \varphi_2^{-1} \circ \frac{1}{2}\varphi_1\Big)(d(p(\tau_n, a, \tau_0), M)). \end{aligned}$$

It follows readily from the above inequality that for all $n \in \mathbb{N}$

$$DV(p(\tau_n, a, \tau_0), \tau_n) \leq -\Big(\varphi_3 \circ \varphi_2^{-1} \circ \frac{1}{2}\varphi_1\Big)(d(p(\tau_n, a, \tau_0), M)), \tag{3.3.3}$$

where DV is defined in (3.2.4).

Next, we consider the case when $\widetilde{E}$ is bounded; that is, $\sup\{s_n : n \in \mathbb{N}\} = L < \infty$. Because s_n is strictly increasing, it must be true that $L = \lim_{n\to\infty} s_n$. Therefore there exists an $n_0 \in \mathbb{N}$ such that $s_n \in (L-1, L)$ for all $n \geq n_0$. Furthermore, it follows from the continuity of $V(p(t, a, \tau_0), t)$ that

$$V(p(s_{n+1}, a, \tau_0), s_{n+1}) = \frac{1}{2} V(p(s_n, a, \tau_0), s_n),$$

which yields $V(p(L, a, \tau_0), L) = \lim_{n\to\infty} V(p(s_n, a, \tau_0), s_n) = 0$. Let $\tau_n = s_n$, if $n \leq n_0$, and $\tau_n = s_{n_0} + (n - n_0)$ if $n > n_0$. The set $E = \{\tau_1, \tau_2, \dots\}$ is clearly unbounded and discrete. Similarly as shown above, (3.3.3) holds for any $n < n_0$. For all $n > n_0$, we have

$$V(p(\tau_n, a, \tau_0), \tau_n) \leq V(p(L, a, \tau_0), L) = 0.$$

Therefore (3.3.3) is also satisfied. When $n = n_0$, we have $\tau_{n_0+1} = \tau_{n_0} + 1 > L$, $V(p(\tau_{n_0+1}, a, \tau_0), \tau_{n_0+1}) \leq V(p(L, a, \tau_0), L) = 0$, and

$$DV(p(\tau_{n_0}, a, \tau_0), \tau_{n_0}) = -V(p(\tau_{n_0}, a, \tau_0), \tau_{n_0}) \leq -\varphi_1(d(p(\tau_{n_0}, a, \tau_0), M)). \tag{3.3.4}$$

If we let $\widetilde{\varphi}_3$ defined on $\mathbb{R}^+$ be given by

$$\widetilde{\varphi}_3(r) = \min\big\{\varphi_1(r), \big(\varphi_3 \circ \varphi_2^{-1} \circ \frac{1}{2}\varphi_1\big)(r)\big\},$$

then $\widetilde{\varphi}_3 \in \mathcal{K}$. In view of (3.3.3) and (3.3.4), we have shown that

$$DV(p(\tau_n, a, \tau_0), \tau_n) \leq -\widetilde{\varphi}_3(d(p(\tau_n, a, \tau_0), M))$$

for all $n \in \mathbb{N}$.

Combining with Theorem 3.3.1, we have shown that the hypotheses of Theorem 3.2.2 are satisfied. Therefore (S, M) is *uniformly asymptotically stable.* □

Theorem 3.3.3 Let $\{\mathbb{R}^+, X, A, S\}$ be a continuous dynamical system and let $M \subset A$ be closed. Assume that there exist a function $V: X \times \mathbb{R}^+ \to \mathbb{R}^+$ and four positive constants c_1, c_2, c_3, and b such that

$$c_1[d(x, M)]^b \le V(x, t) \le c_2[d(x, M)]^b \tag{3.3.5}$$

for all $x \in X$ and $t \in \mathbb{R}^+$.

Assume that there exists a neighborhood U of M such that for all $a \in U$ and for all $p(\cdot, a, \tau_0) \in S$, $V(p(t, a, \tau_0), t)$ is continuous and the upper right-hand Dini derivative $D^+V(p(t, a, \tau_0), t)$ satisfies

$$D^+V(p(t, a, \tau_0), t) \le -c_3[d(p(t, a, \tau_0), M)]^b \tag{3.3.6}$$

for all $t \in \mathbb{R}^+_{\tau_0}$.

Then (S, M) is *exponentially stable.*

Proof. Let $\widetilde{c}_3 = \min\{c_1, c_1c_3/(2c_2)\}$ and let $\varphi_1, \varphi_2, \varphi_3$, and $\widetilde{\varphi}_3 \in \mathcal{K}$ defined on $\mathbb{R}^+$ be given by $\varphi_k(r) = c_k r^b$, $k = 1, 2, 3$, and $\widetilde{\varphi}_3(r) = \widetilde{c}_3 r^b$. Let $f \in C[\mathbb{R}^+, \mathbb{R}^+]$ be the identity function. It follows from the proof of Theorem 3.3.2 that (3.2.9), (3.2.10), and (3.2.12), are all satisfied. In addition, (3.2.11) is clearly satisfied with any $q \in (0, 1)$. Therefore, the hypotheses of Theorem 3.2.3 are satisfied and thus, (S, M) is *exponentially stable.* □

B. Global stability and boundedness results

Next, we address global results.

Theorem 3.3.4 Let $\{\mathbb{R}^+, X, A, S\}$ be a continuous dynamical system. Let $M \subset A$ be bounded. Assume that there exist a function $V: X \times \mathbb{R}^+ \to \mathbb{R}^+$ and two strictly increasing functions $\varphi_1, \varphi_2 \in C[\mathbb{R}^+, \mathbb{R}^+]$ with $\lim_{r\to\infty} \varphi_i(r) = \infty$, $i = 1, 2$, such that

$$\varphi_1(d(x, M)) \le V(x, t) \le \varphi_2(d(x, M)) \tag{3.3.7}$$

for all $x \in X$ and $t \in \mathbb{R}^+$ whenever $d(x, M) \ge \Omega$, where Ω is a positive constant.

Also, assume that $V(p(t, a, \tau_0), t)$ is continuous and nonincreasing for all $t \in \mathbb{R}^+$ and $p(\cdot, a, \tau_0) \in S$ whenever $d(p(t, a, \tau_0), M) \ge \Omega$.

Then S is *uniformly bounded.*

Proof. Let $\widetilde{\Omega} = \Omega + 1$. For any $a \in A$ and $p(\cdot, a, \tau_0) \in S$, choose $E = \{\tau_1, \tau_2, \dots\}$ recursively in the following manner. For $n \in \mathbb{N}$ let $\tau_{n+1} = \tau_n + \min\{1, \alpha_n\}$, where

$$\alpha_n \triangleq \begin{cases} \sup\{\tau: d(p(t, a, \tau_0), M) > \Omega \text{ for all } t \in (\tau_n, \tau_n + \tau)\}, \\ \qquad\qquad \text{if } d(p(\tau_n, a, \tau_0), M) \ge \widetilde{\Omega}; \\ \sup\{\tau: d(p(t, a, \tau_0), M) < \Omega + 2 \text{ for all } t \in (\tau_n, \tau_n + \tau)\}, \\ \qquad\qquad \text{if } d(p(\tau_n, a, \tau_0), M) < \widetilde{\Omega}. \end{cases}$$

We first show that E is unbounded. Suppose that $\sup_{n\in\mathbb{N}}\{\tau_n\} = L < \infty$. Because $\{\tau_n\}$ is strictly increasing, it must be true that $L = \lim_{n\to\infty} \tau_n$. Therefore there exists an $n_0 \in \mathbb{N}$ such that $\alpha_n < 1$ for all $n \geq n_0$. It follows from the definition of α_n and the continuity of $V(p(t, a, \tau_0), t)$ that if $d(p(\tau_n, a, \tau_0), M) < \widetilde{\Omega}$ then

$$d(p(\tau_{n+1}, a, \tau_0), M) = d(p(\tau_n + \alpha_n, a, \tau_0), M) = \Omega + 2 > \widetilde{\Omega}$$

and if $d(p(\tau_n, a, \tau_0), M) \geq \widetilde{\Omega}$ then

$$d(p(\tau_{n+1}, a, \tau_0), M) = d(p(\tau_n + \alpha_n, a, \tau_0), M) = \Omega < \widetilde{\Omega}.$$

Therefore, $\lim_{n\to\infty} p(\tau_n, a, \tau_0)$ does not exist. On the other hand, $p(t, a, \tau_0)$ is continuous and thus, $\lim_{n\to\infty} p(\tau_n, a, \tau_0)$ must exist. This is a contradiction. Therefore E is unbounded. Clearly E is also discrete.

For any $n \in \mathbb{N}$, if $d(p(\tau_n, a, \tau_0), M) \geq \widetilde{\Omega}$, it follows from the choice of τ_{n+1} that $d(p(t, a, \tau_0), M) \geq \Omega$ for all $t \in (\tau_n, \tau_{n+1}]$. Thus, by the assumption that $V(p(t, a, \tau_0), t)$ is nonincreasing whenever $d(p(t, a, \tau_0), M) \geq \Omega$, we have

$$V(p(\tau_{n+1}, a, \tau_0), \tau_{n+1}) \leq V(p(\tau_n, a, \tau_0), \tau_n)$$

and

$$V(p(t, a, \tau_0), t) \leq V(p(\tau_n, a, \tau_0), \tau_n)$$

for all $t \in (\tau_n, \tau_{n+1})$ whenever $d(p(\tau_n, a, \tau_0), M) \geq \widetilde{\Omega}$. Thus (3.2.16) and (3.2.17) are satisfied with $f \in C[\mathbb{R}^+, \mathbb{R}^+]$ being the identity function.

If $d(p(\tau_n, a, \tau_0), M) < \widetilde{\Omega}$, then $d(p(t, a, \tau_0), M) \leq \Omega + 2 \triangleq \Gamma$ is true for all $t \in (\tau_n, \tau_{n+1})$ by the choice of τ_{n+1}.

Hence, all the hypotheses of Theorem 3.2.4 are satisfied and thus, S is *uniformly bounded.* □

Theorem 3.3.5 If in addition to the assumptions given in Theorem 3.3.4 there exists a function $\varphi_3 \in \mathcal{K}$ defined on $\mathbb{R}^+$ such that for all $p(\cdot, a, \tau_0) \in S$ the upper right-hand Dini derivative $D^+V(p(t, a, \tau_0), t)$ satisfies

$$D^+V(p(t, a, \tau_0), t) \leq -\varphi_3(d(p(t, a, \tau_0), M)) \tag{3.3.8}$$

for all $t \in \mathbb{R}^+_{\tau_0}$ whenever $d(p(t, a, \tau_0), M) > \Omega$, then S is *uniformly ultimately bounded.*

Proof. Let $\widetilde{\Omega} = \Omega + 1$. For any $a \in A$ and $p(\cdot, a, \tau_0) \in S$, choose $E = \{\tau_1, \tau_2, \dots\}$ recursively in the following manner. For $n \in \mathbb{N}$ let $\tau_{n+1} = \tau_n + \min\{1, \alpha_n\}$, where

$$\alpha_n \triangleq \begin{cases} \sup\big\{\tau\colon d(p(t, a, \tau_0), M) > \Omega \text{ and} \\ \qquad V(p(t, a, \tau_0), t) \geq \frac{1}{2} V(p(\tau_n, a, \tau_0), \tau_n) \\ \qquad \text{for all } t \in (\tau_n, \tau_n + \tau)\big\}, \quad \text{if } d(p(\tau_n, a, \tau_0), M) \geq \widetilde{\Omega}; \\ \sup\big\{\tau\colon d(p(t, a, \tau_0), M) < \Omega + 2 \text{ for all } t \in (\tau_n, \tau_n + \tau)\big\}, \\ \qquad\qquad \text{if } d(p(\tau_n, a, \tau_0), M) < \widetilde{\Omega}. \end{cases}$$

We need to show that E is unbounded. Suppose that $\sup_{n\in\mathbb{N}}\{\tau_n\} = L < \infty$. Because τ_n is strictly increasing, it must be true that $L = \lim_{n\to\infty} \tau_n$. Therefore there exists an $n_0 \in \mathbb{N}$ such that $\alpha_n < 1$ for all $n \geq n_0$. It follows from the choice of α_n and the continuity of $p(t, a, \tau_0)$ and $V(p(t, a, \tau_0), t)$ that if $d(p(\tau_n, a, \tau_0), M) < \widetilde{\Omega}$ then

$$d(p(\tau_{n+1}, a, \tau_0), M) = d(p(\tau_n + \alpha_n, a, \tau_0), M) = \Omega + 2 > \widetilde{\Omega},$$

and if $d(p(\tau_n, a, \tau_0), M) \geq \widetilde{\Omega}$ then either

$$d(p(\tau_{n+1}, a, \tau_0), M) = d(p(\tau_n + \alpha_n, a, \tau_0), M) = \Omega < \widetilde{\Omega},$$

or

$$V(p(\tau_{n+1}, a, \tau_0), \tau_{n+1}) = \frac{1}{2} V(p(\tau_n, a, \tau_0), \tau_n).$$

Therefore, either $\lim_{n \to \infty} p(\tau_n, a, \tau_0)$ or $\lim_{n \to \infty} V(p(\tau_n, a, \tau_0), \tau_n)$ does not exist. On the other hand, both $p(t, a, \tau_0)$ and $V(p(t, a, \tau_0), t)$ are continuous and their limit as $\{\tau_n\}$ approaches L must exist. This is a contradiction. Therefore E is unbounded. Clearly E is also discrete.

For any $n \in \mathbb{N}$, if $d(p(\tau_n, a, \tau_0), M) \geq \widetilde{\Omega}$, it follows from the choice of τ_{n+1} that $d(p(t, a, \tau_0), M) \geq \Omega$ and $V(p(t, a, \tau_0), t) \geq 0.5V(p(\tau_n, a, \tau_0), \tau_n)$ for all $t \in (\tau_n, \tau_{n+1}]$. In view of (3.3.7) we have that

$$\begin{aligned} d(p(t, a, \tau_0), M) &\geq \left(\varphi_2^{-1} \circ V\right)(p(t, a, \tau_0), t) \\ &\geq \left(\varphi_2^{-1} \circ \frac{1}{2} V\right)(p(\tau_n, a, \tau_0), \tau_n) \\ &\geq \left(\varphi_2^{-1} \circ \frac{1}{2} \varphi_1\right)(d(p(\tau_n, a, \tau_0), M)) \end{aligned}$$

for all $t \in (\tau_n, \tau_{n+1}]$. As in the proof of Theorem 3.3.2, it follows from the *Comparison Theorem* (see Theorem 3.8.1 in the Appendix of this chapter) and (3.3.8) that

$$\begin{aligned} &V(p(\tau_{n+1}, a, \tau_0), \tau_{n+1}) - V(p(\tau_n, a, \tau_0), \tau_n) \\ &\quad \leq - \int_{\tau_n}^{\tau_{n+1}} \varphi_3(d(p(t, a, \tau_0), M)) dt \\ &\quad \leq - \int_{\tau_n}^{\tau_{n+1}} \left(\varphi_3 \circ \varphi_2^{-1} \circ \frac{1}{2} \varphi_1\right)(d(p(\tau_n, a, \tau_0), M)) dt \\ &\quad = - (\tau_{n+1} - \tau_n) \left(\varphi_3 \circ \varphi_2^{-1} \circ \frac{1}{2} \varphi_1\right)(d(p(\tau_n, a, \tau_0), M)). \end{aligned}$$

Let $\widetilde{\varphi}_3 \triangleq \varphi_3 \circ \varphi_2^{-1} \circ \frac{1}{2} \varphi_1$. It follows readily from the above inequality that for all $n \in \mathbb{N}$

$$DV(p(\tau_n, a, \tau_0), \tau_n) \leq -\widetilde{\varphi}_3(d(p(\tau_n, a, \tau_0), M))$$

whenever $d(p(\tau_n, a, \tau_0), M) \geq \widetilde{\Omega}$. Combining with Theorem 3.3.4, we have shown that the hypotheses of Theorem 3.2.5 are satisfied. Therefore S is *uniformly ultimately bounded*. □

Theorem 3.3.6 Let $\{\mathbb{R}^+, X, A, S\}$ be a continuous dynamical system. Let $M \subset A$ be bounded and closed. Assume that there exist a function $V: X \times \mathbb{R}^+ \to \mathbb{R}^+$ and two functions $\varphi_1, \varphi_2 \in \mathcal{K}_\infty$ such that

$$\varphi_1(d(x, M)) \leq V(x, t) \leq \varphi_2(d(x, M)) \tag{3.3.9}$$

for all $x \in X$ and $t \in \mathbb{R}^+$.

Assume that for all $p(\cdot, a, \tau_0) \in S$ and $t \in \mathbb{R}^+_{\tau_0}$, $V(p(t, a, \tau_0), t)$ is continuous. Furthermore, assume that there exists a function $\varphi_3 \in \mathcal{K}$ defined on $\mathbb{R}^+$ such that for all $a \in A$ and all $p(\cdot, a, \tau_0) \in S$, the upper right-hand Dini derivative $D^+V(p(t, a, \tau_0), t)$ satisfies

$$D^+V(p(t, a, \tau_0), t) \leq -\varphi_3(d(p(t, a, \tau_0), M)) \tag{3.3.10}$$

for all $t \in \mathbb{R}^+_{\tau_0}$.

Then (S, M) is *uniformly asymptotically stable in the large.*

Proof. For any $a \in A$ and $p(\cdot, a, \tau_0) \in S$, choose $E = \{\tau_1, \tau_2, \ldots : \tau_1 < \tau_2 < \cdots\}$ in the same manner as in the proof of Theorem 3.3.2. Let $f \in C[\mathbb{R}^+, \mathbb{R}^+]$ be the identity function; that is, $f(r) = r$.

It follows from (3.3.10) and the *Comparison Theorem* (Theorem 3.8.1 in the Appendix of this chapter) that

$$V(p(t, a, \tau_0), t) - V(p(\tau_n, a, \tau_0), \tau_n) \leq -\int_{\tau_n}^{t} \varphi_3(d(p(s, a, \tau_0), M))ds \leq 0,$$

and thus,

$$V(p(t, a, \tau_0), t) \leq V(p(\tau_n, a, \tau_0), \tau_n) = f(V(p(\tau_n, a, \tau_0), \tau_n))$$

for all $t \in (\tau_n, \tau_{n+1})$, $n \in \mathbb{N}$.

Similarly as in the proof of Theorem 3.3.2, we can show that

$$DV(p(\tau_n, a, \tau_0), \tau_n) \leq -\widetilde{\varphi}_3(d(p(\tau_n, a, \tau_0), M)),$$

for all $n \in \mathbb{N}$, where $\widetilde{\varphi}_3 \in \mathcal{K}$ is given by $\widetilde{\varphi}_3(r) = \min\{\varphi_1(r), (\varphi_3 \circ \varphi_2^{-1} \circ \frac{1}{2}\varphi_1)(r)\}$.

Hence, we have shown that the hypotheses of Theorem 3.2.6 are satisfied. Therefore (S, M) is *uniformly asymptotically stable in the large.* □

Theorem 3.3.7 Let $\{\mathbb{R}^+, X, A, S\}$ be a continuous dynamical system and let $M \subset A$ be closed and bounded. Assume that there exist a function $V: X \times \mathbb{R}^+ \to \mathbb{R}^+$ and four positive constants c_1, c_2, c_3, and b such that

$$c_1[d(x, M)]^b \leq V(x, t) \leq c_2[d(x, M)]^b \tag{3.3.11}$$

for all $x \in X$ and $t \in \mathbb{R}^+$.

Assume that for all $p(\cdot, a, \tau_0) \in S$ and $t \in \mathbb{R}^+_{\tau_0}$, $V(p(t, a, \tau_0), t)$ is continuous. Furthermore, assume that for all $a \in A$ and for all $p(\cdot, a, \tau_0) \in S$, the upper right-hand Dini derivative $D^+V(p(t, a, \tau_0), t)$ satisfies

$$D^+V(p(t, a, \tau_0), t) \leq -c_3[d(p(t, a, \tau_0), M)]^b \tag{3.3.12}$$

for all $t \in \mathbb{R}^+_{\tau_0}$.

Then (S, M) is *exponentially stable in the large.*

Proof. Let $\widetilde{c}_3 = \min\{c_1, c_1c_3/(2c_2)\}$ and let $\varphi_1, \varphi_2, \varphi_3$, and $\widetilde{\varphi}_3 \in \mathcal{K}$ defined on $\mathbb{R}^+$ be given by $\varphi_k(r) = c_k r^b$, $k = 1, 2, 3$, and $\widetilde{\varphi}_3(r) = \widetilde{c}_3 r^b$. Let $f \in C[\mathbb{R}^+, \mathbb{R}^+]$ be the identity function. It follows from the proof of Theorem 3.3.3 that (3.2.24), (3.2.25), and (3.2.27) are all satisfied. In addition, (3.2.26) is clearly satisfied for any $q \in (0, 1)$. Therefore, the hypotheses of Theorem 3.2.7 are satisfied and thus, (S, M) is *exponentially stable in the large*. □

C. Instability results

Next, we consider instability results of a set M with respect to S.

Theorem 3.3.8 (*Lyapunov's First Instability Theorem*) Let $\{\mathbb{R}^+, X, A, S\}$ be a dynamical system and let $M \subset A$ be closed, where A is assumed to be a neighborhood of M. Assume that every motion $p(\cdot, a, \tau_0) \in S$ is a continuous function of t on $\mathbb{R}^+_{\tau_0}$ and assume that there exist a function $V\colon X \times \mathbb{R}^+ \to \mathbb{R}$ and a $t_0 \in \mathbb{R}^+$ that satisfy the following conditions.

(i) There exists a function $\varphi \in \mathcal{K}$ defined on $\mathbb{R}^+$ such that

$$V(x,t) \le \varphi(d(x, M)) \tag{3.3.13}$$

for all $x \in X$ and $t \in \mathbb{R}^+$.

(ii) In every neighborhood of M, there is a point x such that $V(x, t_0) > 0$ and there exists a motion $p(\cdot, x, \tau_0) \in S$.

(iii) There exists a function $\psi \in \mathcal{K}$ defined on $\mathbb{R}^+$ such that

$$D^+V(p(t, a, t_0), t) \ge \psi(|V(p(t, a, t_0), t)|) \tag{3.3.14}$$

for all $p(\cdot, a, t_0) \in S$ and for all $t \in \mathbb{R}^+_{t_0}$, where D^+ denotes the upper right-hand Dini derivative with respect to t.

Then M is *unstable* with respect to S.

Proof. Note that assumptions (i) and (ii) are identical to those of Theorem 3.2.8. We now show that assumption (iii) reduces to assumption (iii) of Theorem 3.2.8.

For any $a \in A$ and $p(\cdot, a, t_0) \in S$, choose arbitrarily an unbounded and discrete subset $E = \{t_1, t_2, \dots : t_1 < t_2 < \cdots\}$ of $\mathbb{R}^+_{t_0}$.

It follows from assumption (iii) that $V(p(t, a, t_0), t)$ is nondecreasing. Therefore for any $a \in A$ such that $V(a, t_0) > 0$ and any $p(\cdot, a, t_0) \in S$, we have $V(p(t, a, t_0), t) > 0$ for all $t > t_0$. By the *Comparison Theorem* (Theorem 3.8.1 in the Appendix) we obtain

$$\begin{aligned}
V(p(t_{n+1}, a, t_0), t_{n+1}) &- V(p(t_n, a, t_0), t_n) \\
&\ge \int_{t_n}^{t_{n+1}} \psi(|V(p(t, a, t_0), t)|)dt \\
&\ge \int_{t_n}^{t_{n+1}} \psi(|V(p(t_n, a, t_0), t_n)|)dt \\
&= (t_{n+1} - t_n)\psi(V(p(t_n, a, t_0), t_n)).
\end{aligned}$$

Hence, inequality (3.2.29) is satisfied.

Therefore, all the hypotheses of Theorem 3.2.8 are satisfied and thus, M is *unstable* with respect to S. □

Theorem 3.3.9 In addition to the assumptions given in Theorem 3.3.8, assume that $V(x, t_0) > 0$ for all $x \notin M$. Then M is *completely unstable* with respect to S.

Proof. Note that combining with Theorem 3.3.8, the present assumptions reduce to those of Theorem 3.2.9. Therefore we conclude that M is completely unstable with respect to S. □

In our next result we require the following notion.

Definition 3.3.1 Let $\{T, X, A, S\}$ be a dynamical system and let $Y \subset X$. We denote by $S|_Y$ the family of motions of S restricted to Y. Thus, $\widetilde{p}(\cdot, a, \tau_0) \in S|_Y$ with domain $\widetilde{T}_{a,\tau_0}$ if and only if $a \in A \cap Y$ and there exists a $p(\cdot, a, \tau_0) \in S$ such that $p(t, a, \tau_0) = \widetilde{p}(t, a, \tau_0)$ whenever $p(t, a, \tau_0) \in Y$, and $\widetilde{T}_{a,\tau_0}$ is the subset of T_{a,τ_0} which consists of all t such that $p(t, a, \tau_0) \in Y$. We call $S|_Y$ the *restriction of system S on Y*. □

Theorem 3.3.10 (*Lyapunov's Second Instability Theorem*) Let $\{\mathbb{R}^+, X, A, S\}$ be a dynamical system and let $M \subset A$ be closed, where A is assumed to be a neighborhood of M. Assume that every motion $p(\cdot, a, \tau_0) \in S$ is a continuous function of t on $\mathbb{R}^+_{\tau_0}$, and that there exist a $\tau_0 \in T$ and a function $V \in C[M_\varepsilon \times (\tau_0, \infty), \mathbb{R}]$, where $M_\varepsilon = \{x \in X\colon d(x, M) < \varepsilon\}$, $\varepsilon > 0$, such that the following conditions are satisfied.

(i) V is bounded on $M_\varepsilon \times [\tau_0, \infty)$.

(ii) For all $p(\cdot, a, \tau_0) \in S|_{M_\varepsilon}$ and $t \in \mathbb{R}^+_{\tau_0}$,

$$DV(p(t, a, \tau_0), t) \geq \lambda V(p(t, a, \tau_0), t) \tag{3.3.15}$$

where $\lambda > 0$ is a constant and D denotes a fixed Dini derivative with respect to t.

(iii) In every neighborhood of M, there exists an x such that $V(x, t_1) > 0$ for a fixed $t_1 \geq \tau_0$.

Then M is *unstable* with respect to S.

Proof. By contradiction. If M is invariant and stable with respect to S, then for the $\varepsilon > 0$ and $t_1 \in \mathbb{R}^+$, there exists a $\delta = \delta(\varepsilon, t_1) > 0$ such that $d(p(t, a, t_1), M) < \varepsilon$ for all $p(\cdot, a, t_1) \in S$ and for all $t \in \mathbb{R}^+_{t_1}$ whenever $d(a, M) < \delta$. Because A is a neighborhood of M, there exists by condition (iii) an $x_1 \in \{a \in A\colon d(a, M) < \delta\}$

such that $V(x_1, t_1) > 0$. Let $y(t) = V(p(t, x_1, t_1), t)$. Then $y(t_1) = V(x_1, t_1) > 0$. By condition (ii),

$$Dy(t) \geq \lambda y(t) \quad \text{for } t \geq t_1.$$

Let $z(t) = y(t)e^{-\lambda t}$. Then

$$Dz(t) = e^{-\lambda t} Dy(t) - \lambda y(t) e^{-\lambda t} \geq 0.$$

Therefore, $z(t)$ is nondecreasing. For any $t \geq t_1$ we have $z(t) \geq z(t_1)$ and thus, $y(t) \geq y(t_1)e^{\lambda(t-t_1)}$. Because $y(t_1) > 0$, we have $\lim_{t \to +\infty} y(t) = \infty$. This contradicts condition (i) and completes the proof. □

D. An example

The scalar differential equation

$$\dot{x} = \begin{cases} (\ln 2)x, & \text{if } t \in [t_0 + 2k, t_0 + 2k + 1), \\ -(\ln 4)x, & \text{if } t \in [t_0 + 2k + 1, t_0 + 2(k+1)), \end{cases} \tag{3.3.16}$$

where $k \in \mathbb{N}, x \in \mathbb{R}$, and $t_0 \in \mathbb{R}^+$, determines a dynamical system $\{\mathbb{R}^+, X, A, S\}$ with $X = A = \mathbb{R}$ and with $p(\cdot, a, t_0) \in S$ determined by the solutions of (3.3.16) (obtained by integrating (3.3.16)),

$$p(t, a, t_0) = \begin{cases} \dfrac{a}{2^k} e^{(\ln 2)(t - t_0 - 2k)}, & \text{if } t \in [t_0 + 2k, t_0 + 2k + 1], \\ \dfrac{a}{2^{k-1}} e^{-(\ln 4)(t - t_0 - 2k - 1)}, & \text{if } t \in [t_0 + 2k + 1, t_0 + 2(k+1)], \end{cases} \tag{3.3.17}$$

for each pair $(a, t_0) \in \mathbb{R} \times \mathbb{R}^+$ and for all $k \in \mathbb{N}$ and $t \geq t_0$. The plot of a typical motion for this system is given in Figure 3.3.1. Note that for every $(a, t_0) \in \mathbb{R} \times \mathbb{R}^+$, there exists a *unique* $p(\cdot, a, t_0) \in S$ that is *defined* and *continuous* for $t \geq t_0$ and that $M = \{0\}$ is invariant with respect to S.

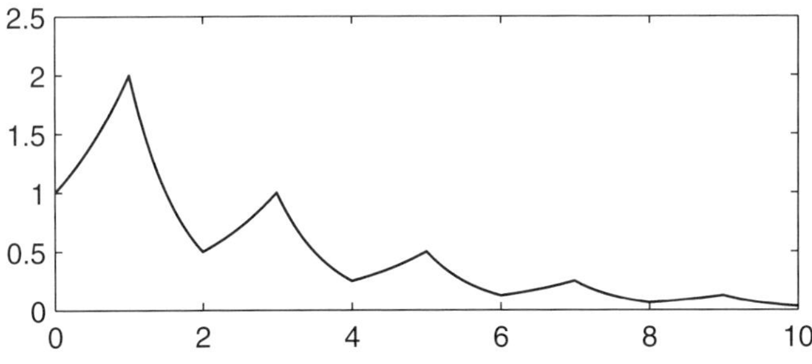

Figure 3.3.1: Plot of the motion, $p(t, 1, 0) \in S$.

The block diagram of system (3.3.16) is depicted in Figure 3.3.2. This system can be viewed as a switched system with switching occurring every unit of time since initial time t_0.

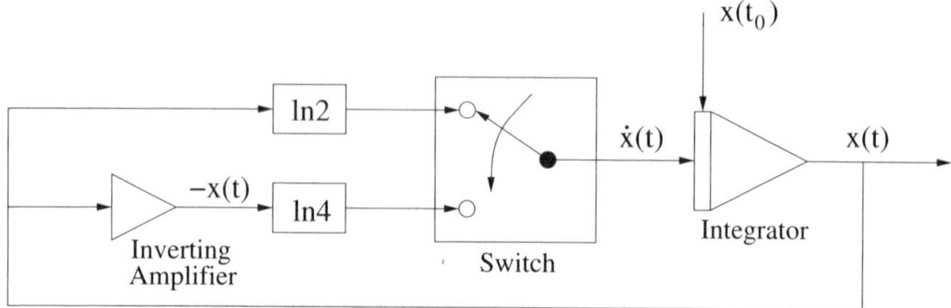

Figure 3.3.2: Block diagram for system (3.3.16).

In the following we show that

(a) there exists a function $V\colon \mathbb{R}\times\mathbb{R}^+ \to \mathbb{R}^+$ that satisfies Theorem 3.2.2 and therefore, $(S,\{0\})$ is uniformly asymptotically stable; and

(b) there does not exist a Lyapunov function $V\colon \mathbb{R}\times\mathbb{R}^+ \to \mathbb{R}^+$ that satisfies the hypotheses of Theorem 3.3.2 and therefore, Theorem 3.3.2 cannot be used to prove that $(S,\{0\})$ is uniformly asymptotically stable.

Proof. (a) Let $V\colon \mathbb{R} \to \mathbb{R}^+$ be chosen as $V(x) = |x|$ for all $x \in \mathbb{R}$. For any $p(\cdot, a, t_0)$, choose the set $E = \{t_1, t_2, \dots : t_k = t_0 + 2k, k = 1, 2, \dots\}$. By (3.3.17), $V(p(t_k, a, t_0)) = |a/2^k|$, and $V(p(t, a, t_0)) \le 2V(p(t_k, a, t_0))$ for all $t \in [t_k, t_{k+1}]$, $k \in \mathbb{N}$. Therefore all hypotheses of Theorem 3.2.2 are satisfied and hence, $(S,\{0\})$ is uniformly asymptotically stable.

(b) For purposes of contradiction, assume that there exist a Lyapunov function $V\colon \mathbb{R}\times\mathbb{R}^+ \to \mathbb{R}^+$ and two functions $\varphi_1, \varphi_2 \in \mathcal{K}$ defined on $\mathbb{R}^+$ such that

$$\varphi_1(|x|) \le V(x,t) \le \varphi_2(|x|) \tag{3.3.18}$$

for all $(x,t) \in \mathbb{R}\times\mathbb{R}^+$, and there exists a neighborhood U of 0 such that for all $a \in U$ and for all $p(\cdot, a, t_0) \in S$, $V(p(t, a, t_0), t)$ is nonincreasing for all $t \ge t_0$, $t \in \mathbb{R}^+$. Without loss of generality, we assume that $1 \in U$.

By (3.3.17), $p(t_0 + 1, a, t_0) = 2a$ for any $(a, t_0) \in \mathbb{R}\times\mathbb{R}^+$. In particular, for any $n \in \mathbb{N}$, because

$$\begin{aligned} p\Big(2, \frac{1}{2^n}, 1\Big) &= \frac{1}{2^{n-1}}, \\ p\Big(3, \frac{1}{2^{n-1}}, 2\Big) &= \frac{1}{2^{n-2}}, \\ &\vdots \\ p\Big(n, \frac{1}{2}, n-1\Big) &= 1, \end{aligned}$$

and because $V(p(t, a, t_0), t)$ is nonincreasing for all $p(\cdot, a, t_0) \in S$, we have that

$$\begin{aligned}
&V\Big(\frac{1}{2^n}, 1\Big) \geq V\Big(p\Big(2, \frac{1}{2^n}, 1\Big), 2\Big) = V\Big(\frac{1}{2^{n-1}}, 2\Big)\\
&\quad \text{(along the motion } p\Big(\cdot, \frac{1}{2^n}, 1\Big));\\
&V\Big(\frac{1}{2^{n-1}}, 2\Big) \geq V\Big(p\Big(3, \frac{1}{2^{n-1}}, 2\Big), 3\Big) = V\Big(\frac{1}{2^{n-2}}, 3\Big)\\
&\quad \text{(along the motion } p\Big(\cdot, \frac{1}{2^{n-1}}, 2\Big));\\
&\quad \vdots\\
&V\Big(\frac{1}{2}, n-1\Big) \geq V\Big(p\Big(n, \frac{1}{2}, n-1\Big), n\Big) = V(1, n)\\
&\quad \text{(along the motion } p\Big(\cdot, \frac{1}{2}, n-1\Big)).
\end{aligned}$$

Therefore,

$$V\Big(\frac{1}{2^n}, 1\Big) \geq V(1, n).$$

On the other hand,

$$\varphi_1\Big(\frac{1}{2^n}\Big) \leq V\Big(\frac{1}{2^n}, 1\Big) \leq \varphi_2\Big(\frac{1}{2^n}\Big) \quad \text{and} \quad \varphi_1(1) \leq V(1, n) \leq \varphi_2(1).$$

Thus,

$$\varphi_2\Big(\frac{1}{2^n}\Big) \geq \varphi_1(1)$$

is true for all $n \in \mathbb{N}$, which implies that

$$\varphi_2(0) = \lim_{n\to\infty} \varphi_2\Big(\frac{1}{2^n}\Big) \geq \varphi_1(1) > 0.$$

However, by the assumption $\varphi_2 \in \mathcal{K}$, we know that $\varphi_2(0) = 0$. We have arrived at a contradiction. Therefore, there does not exist a Lyapunov function that satisfies the hypotheses of the classical Lyapunov Theorem for uniform asymptotic stability for continuous dynamical systems, Theorem 3.3.2. □

3.4 The Principal Lyapunov and Lagrange Stability Results for Discrete-Time Dynamical Systems

In this section we present the Principal Lyapunov Stability and Boundedness Results for discrete-time dynamical systems. As in the case of continuous dynamical systems, we show that these results are a direct consequence of the corresponding stability and boundedness results for DDS given in Section 3.2. To accomplish this, we first embed the class of discrete-time systems considered herein into a class of DDS with equivalent stability properties. Using this class of DDS, we then show that when the hypotheses of the stability and boundedness results for the discrete-time systems are satisfied, then the hypotheses of the corresponding DDS results given in Section 3.2 are also satisfied. This shows that the results of the present section, which constitute the Principal Lyapunov and Lagrange Stability Results for discrete-time dynamical systems, are in general more conservative than the corresponding results for DDS. We include in this section a specific example that reinforces this assertion. Furthermore, by establishing a link between the stability and boundedness results for DDS and for discrete-time dynamical systems, we have completed a unifying stability theory for continuous dynamical systems, discrete-time dynamical systems, and discontinuous dynamical systems.

Every discrete-time dynamical system, $\{\mathbb{N}, X, A, S\}$, can be associated with a DDS, $\{\mathbb{R}^+, X, A, \widetilde{S}\}$, where

$$\widetilde{S} = \big\{\widetilde{p}(t, a, \tau_0 = n_0) \colon \widetilde{p}(t, a, \tau_0) = p(n, a, n_0) \text{ for } t \in [n, n+1), \\ n \geq n_0, n, n_0 \in \mathbb{N}\big\}.$$

For such *associated systems*, the following result follows directly from definitions.

Lemma 3.4.1 The discrete-time dynamical system, $\{\mathbb{N}, X, A, S\}$, and the *associated* DDS, $\{\mathbb{R}^+, X, A, \widetilde{S}\}$, have identical stability properties. □

A. Local stability results

We first present local results.

Theorem 3.4.1 Let $\{\mathbb{N}, X, A, S\}$ be a discrete-time dynamical system and let $M \subset A$ be closed. Assume that there exist a function $V \colon X \times \mathbb{N} \to \mathbb{R}^+$ and two functions $\varphi_1, \varphi_2 \in \mathcal{K}$ defined on $\mathbb{R}^+$ such that

$$\varphi_1(d(x, M)) \leq V(x, n) \leq \varphi_2(d(x, M)) \tag{3.4.1}$$

for all $x \in X$ and $n \in \mathbb{N}$. Assume that there exists a neighborhood U of M such that for all $a \in U$ and for all $p(\cdot, a, n_0) \in S$, $V(p(n, a, n_0), n)$ is nonincreasing for all $n \in \mathbb{N}_{n_0}$ (i.e., $n \geq n_0$, $n, n_0 \in \mathbb{N}$). Then (S, M) is *invariant* and *uniformly stable*.

Proof. First, let $\{\mathbb{R}^+, X, A, \widetilde{S}\}$ be the *associated* DDS and let $\widetilde{V} \colon X \times \mathbb{R}^+ \to \mathbb{R}^+$ be defined as $\widetilde{V}(x, t) = V(x, n)$ for all $x \in X$ and $t \in [n, n+1)$, $n \in \mathbb{N}$. It follows directly from (3.4.1) that

$$\varphi_1(d(x, M)) \leq \widetilde{V}(x, t) \leq \varphi_2(d(x, M))$$

for all $x \in X$ and $t \in \mathbb{R}^+$.

For any $a \in U$ and $p(\cdot, a, n_0) \in S$, the associated motion $\widetilde{p}(t, a, \tau_0 = n_0)$ is continuous everywhere on $\mathbb{R}^+_{\tau_0}$ except possibly on $E = \{\tau_1 = n_0 + 1, \tau_2 = n_0 + 2, \dots\}$.

E is clearly unbounded and discrete. Let $f \in C[\mathbb{R}^+, \mathbb{R}^+]$ be the identity function. It follows from the assumptions that $\widetilde{V}(\widetilde{p}(\tau_n, a, n_0), \tau_n)$ is nonincreasing and

$$\widetilde{V}(\widetilde{p}(t, a, \tau_0), t) = \widetilde{V}(\widetilde{p}(\tau_n, a, \tau_0), n) = f(\widetilde{V}(\widetilde{p}(\tau_n, a, \tau_0), n)),$$

for all $t \in (\tau_n, \tau_{n+1})$, $n \in \mathbb{N}$.

Hence the associated DDS, $\{\mathbb{R}^+, X, A, \widetilde{S}\}$, and the set M satisfy the hypotheses of Theorem 3.2.1 and thus, (S, M) is *invariant* and *uniformly stable*. □

Theorem 3.4.2 If in addition to the assumptions given in Theorem 3.4.1 there exists a function $\varphi_3 \in \mathcal{K}$ defined on $\mathbb{R}^+$ such that for all $a \in U$ and for all $p(\cdot, a, n_0) \in S$

$$V(p(n+1, a, n_0), n+1) - V(p(n, a, n_0), n) \leq -\varphi_3(d(p(n, a, n_0), M)) \quad (3.4.2)$$

for all $n \in \mathbb{N}_{n_0}$, then (S, M) is *uniformly asymptotically stable*.

Proof. Let $\{\mathbb{R}^+, X, A, \widetilde{S}\}$ be the associated DDS and let $\widetilde{V}\colon X \times \mathbb{R}^+ \to \mathbb{R}^+$ be defined as $\widetilde{V}(x, t) = V(x, n)$ for all $x \in X$ and $t \in [n, n+1)$, $n \in \mathbb{N}$.

For any $a \in U$ and $p(\cdot, a, n_0) \in S$, the associated motion $\widetilde{p}(t, a, \tau_0 = n_0)$ is continuous everywhere on $\mathbb{R}^+_{\tau_0}$ except possibly on $E = \{\tau_1 = n_0 + 1, \tau_2 = n_0 + 2, \dots\}$. E is clearly unbounded and discrete. Noting that $\tau_n = n_0 + n$ and $\tau_{n+1} - \tau_n = 1$, along the motion $\widetilde{p}(t, a, n_0)$ we have that

$$\begin{aligned} D\widetilde{V}&(\widetilde{p}(\tau_n, a, n_0), \tau_n) \\ &= \frac{1}{\tau_{n+1} - \tau_n}\Big(\widetilde{V}(\widetilde{p}(\tau_{n+1}, a, n_0), \tau_{n+1}) - \widetilde{V}(\widetilde{p}(\tau_n, a, n_0), \tau_n)\Big) \\ &= V(p(n_0 + n + 1, a, n_0), n_0 + n + 1) - V(p(n_0 + n, a, n_0), n_0 + n) \\ &\leq -\varphi_3(d(\widetilde{p}(\tau_n, a, n_0), M)) \end{aligned}$$

for all $n \in \mathbb{N}$.

In view of the proof of Theorem 3.4.1, we have shown that the associated DDS, $\{\mathbb{R}^+, X, A, \widetilde{S}\}$, and the set M satisfy the hypotheses of Theorem 3.2.2. Therefore (S, M) is *uniformly asymptotically stable*. □

Theorem 3.4.3 Let $\{\mathbb{N}, X, A, S\}$ be a discrete-time dynamical system and let $M \subset A$ be closed. Assume that there exist a function $V\colon X \times \mathbb{N} \to \mathbb{R}^+$ and four positive constants c_1, c_2, c_3, and b such that

$$c_1[d(x, M)]^b \leq V(x, n) \leq c_2[d(x, M)]^b \quad (3.4.3)$$

for all $x \in X$ and $n \in \mathbb{N}$.

Assume that there exists a neighborhood U of M such that for all $a \in U$, for all $p(\cdot, a, n_0) \in S$ and for all $n \in \mathbb{N}_{n_0}$,

$$V(p(n+1, a, n_0), n+1) - V(p(n, a, n_0), n) \leq -c_3[d(p(n, a, n_0), M)]^b. \quad (3.4.4)$$

Then (S, M) is *exponentially stable*.

Proof. Let $\varphi_1, \varphi_2, \varphi_3 \in \mathcal{K}$ defined on $\mathbb{R}^+$ be given by $\varphi_k(r) = c_k r^b, k = 1, 2, 3$. Let $f \in C[\mathbb{R}^+, \mathbb{R}^+]$ be the identity function. It follows from the proof of Theorem 3.4.2 that (3.2.9), (3.2.10), and (3.2.12) are all satisfied. In addition, (3.2.11) is clearly satisfied with $q \in (0, 1)$. Therefore, the associated DDS, $\{\mathbb{R}^+, X, A, \widetilde{S}\}$, and the set M satisfy the hypotheses of Theorem 3.2.3 and thus, (S, M) is exponentially stable. □

B. Global stability and boundedness results

Next, we address global results.

Theorem 3.4.4 Let $\{\mathbb{N}, X, A, S\}$ be a dynamical system and let $M \subset A$ be bounded. Assume that there exist a function $V: X \times \mathbb{N} \to \mathbb{R}^+$ and two strictly increasing functions $\varphi_1, \varphi_2 \in C[\mathbb{R}^+, \mathbb{R}^+]$ with $\lim_{r\to\infty} \varphi_i(r) = \infty, i = 1, 2$, such that

$$\varphi_1(d(x, M)) \le V(x, n) \le \varphi_2(d(x, M)) \tag{3.4.5}$$

for all $x \in X$ and $n \in \mathbb{N}$ whenever $d(x, M) \ge \Omega$, where Ω is a positive constant.

Also, assume that $V(p(n, a, n_0), n)$ is nonincreasing for all $p(\cdot, a, n_0) \in S$ and for all $n \in \mathbb{N}_{n_0}$ whenever $d(p(n, a, n_0), M) \ge \Omega$. Assume that there exists a constant $\Gamma > 0$ such that $d(p(n+1, a, n_0), M) \le \Gamma$ whenever $d(p(n, a, n_0), M) \le \Omega$.

Then S is *uniformly bounded.*

Proof. First, let $\{\mathbb{R}^+, X, A, \widetilde{S}\}$ be the associated DDS and let $\widetilde{V}: X \times \mathbb{R}^+ \to \mathbb{R}^+$ be defined as $\widetilde{V}(x, t) = V(x, n)$ for all $x \in X$ and $t \in [n, n+1), n \in \mathbb{N}$. It follows directly from (3.4.5) that

$$\varphi_1(d(x, M)) \le \widetilde{V}(x, t) \le \varphi_2(d(x, M))$$

for all $x \in X$ and $t \in \mathbb{R}^+$ whenever $d(x, M) \ge \Omega$.

For any $a \in A$ and $p(\cdot, a, n_0) \in S$, the associated motion $\widetilde{p}(t, a, \tau_0 = n_0)$ is continuous everywhere on $\mathbb{R}^+_{\tau_0}$ except possibly on $E = \{\tau_1 = n_0 + 1, \tau_2 = n_0 + 2, \dots\}$. E is clearly unbounded and discrete. Let $f \in C[\mathbb{R}^+, \mathbb{R}^+]$ be the identity function. It follows from the assumptions that $\widetilde{V}(\widetilde{p}(\tau_n, a, n_0), \tau_n)$ is nonincreasing whenever $d(\widetilde{p}(\tau_n, a, \tau_0), M) \ge \Omega$ and

$$\widetilde{V}(\widetilde{p}(t, a, \tau_0), t) = \widetilde{V}(\widetilde{p}(\tau_n, a, \tau_0), n) = f(\widetilde{V}(\widetilde{p}(\tau_n, a, \tau_0), n)),$$

for $t \in (\tau_n, \tau_{n+1}), n \in \mathbb{N}$, whenever $d(\widetilde{p}(t, a, \tau_0), M) \ge \Omega$.

It is easily seen that $d(\widetilde{p}(\tau_{n+1}, a, \tau_0), M) \le \Gamma$ whenever $d(\widetilde{p}(\tau_n, a, \tau_0), M) \le \Omega$.

Hence the associated DDS, $\{\mathbb{R}^+, X, A, \widetilde{S}\}$, and the set M satisfy the hypotheses of Theorem 3.2.4 and thus, S is *uniformly bounded.* □

Theorem 3.4.5 If in addition to the assumptions given in Theorem 3.4.4 there exists a function $\varphi_3 \in \mathcal{K}$ defined on $\mathbb{R}^+$ such that for all $p(\cdot, a, n_0) \in S$

$$V(p(n+1, a, n_0), n+1) - V(p(n, a, n_0), n) \le -\varphi_3(d(p(n, a, n_0), M)) \tag{3.4.6}$$

for all $n \in \mathbb{N}_{n_0}$ whenever $d(p(n, a, n_0), M) \ge \Omega$, then S is *uniformly ultimately bounded.*

Proof. Let $\{\mathbb{R}^+, X, A, \widetilde{S}\}$ be the associated DDS and let $\widetilde{V}: X \times \mathbb{R}^+ \to \mathbb{R}^+$ be defined as $\widetilde{V}(x,t) = V(x,n)$ for all $x \in X$ and $t \in [n, n+1), n \in \mathbb{N}$.

For any $a \in A$ and $p(\cdot, a, n_0) \in S$, the associated motion $\widetilde{p}(t, a, \tau_0 = n_0)$ is continuous everywhere on $\mathbb{R}^+_{\tau_0}$ except possibly on $E = \{\tau_1 = n_0 + 1, \tau_2 = n_0 + 2, \dots\}$. E is clearly unbounded and discrete. Noting that $\tau_n = n_0 + n$ and $\tau_{n+1} - \tau_n = 1$, along the motion $\widetilde{p}(t, a, n_0)$ we have that

$$\begin{aligned}
&D\widetilde{V}(\widetilde{p}(\tau_n, a, n_0), \tau_n)\\
&\quad= \frac{1}{\tau_{n+1} - \tau_n}\Big(\widetilde{V}(\widetilde{p}(\tau_{n+1}, a, n_0), \tau_{n+1}) - \widetilde{V}(\widetilde{p}(\tau_n, a, n_0), \tau_n)\Big)\\
&\quad= V(p(n_0 + n + 1, a, n_0), n_0 + n + 1) - V(p(n_0 + n, a, n_0), n_0 + n)\\
&\quad\le -\varphi_3(d(\widetilde{p}(\tau_n, a, n_0), M))
\end{aligned}$$

for all $n \in \mathbb{N}$ whenever $d(\widetilde{p}(\tau_n, a, n_0), M) \ge \Omega$.

In view of the proof of Theorem 3.4.4, we have shown that the associated DDS, $\{\mathbb{R}^+, X, A, \widetilde{S}\}$, and the set M satisfy the hypotheses of Theorem 3.2.5. Therefore S is *uniformly ultimately bounded*. □

Theorem 3.4.6 Let $\{\mathbb{N}, X, A, S\}$ be a dynamical system and let $M \subset A$ be closed and bounded. Assume that there exist a function $V: X \times \mathbb{N} \to \mathbb{R}^+$ and two functions $\varphi_1, \varphi_2 \in \mathcal{K}_\infty$ such that

$$\varphi_1(d(x, M)) \le V(x, n) \le \varphi_2(d(x, M)) \tag{3.4.7}$$

for all $x \in X$ and $n \in \mathbb{N}$.

Assume that there exists a function $\varphi_3 \in \mathcal{K}$ defined on $\mathbb{R}^+$ such that for all $a \in A$ and for all $p(\cdot, a, n_0) \in S$,

$$V(p(n+1, a, n_0), n+1) - V(p(n, a, n_0), n) \le -\varphi_3(d(p(n, a, n_0), M)) \tag{3.4.8}$$

for all $n \in \mathbb{N}_{n_0}$.

Then (S, M) is *uniformly asymptotically stable in the large*.

Proof. Let $\{\mathbb{R}^+, X, A, \widetilde{S}\}$ be the associated DDS and let $\widetilde{V}: X \times \mathbb{R}^+ \to \mathbb{R}^+$ be defined as $\widetilde{V}(x,t) = V(x,n)$ for all $x \in X$ and $t \in [n, n+1), n \in \mathbb{N}$.

For any $a \in A$ and $p(\cdot, a, n_0) \in S$, the associated motion $\widetilde{p}(t, a, \tau_0 = n_0)$ is continuous everywhere on $\mathbb{R}^+_{\tau_0}$ except possibly on $E = \{\tau_1 = n_0 + 1, \tau_2 = n_0 + 2, \dots\}$. E is clearly unbounded and discrete. Let $f \in C[\mathbb{R}^+, \mathbb{R}^+]$ be the identity function; that is, $f(r) = r$. Similarly as in the proof of Theorem 3.4.5, we can show that the associated motions and the function $\widetilde{V}$ satisfy (3.2.21)–(3.2.23).

Thus, we have shown that the associated DDS $\{\mathbb{R}^+, X, A, \widetilde{S}\}$ and the set M satisfy the hypotheses of Theorem 3.2.6. Therefore (S, M) is *uniformly asymptotically stable in the large*. □

Theorem 3.4.7 Let $\{\mathbb{N}, X, A, S\}$ be a dynamical system and let $M \subset A$ be closed and bounded. Assume that there exist a function $V: X \times \mathbb{N} \to \mathbb{R}^+$ and four positive constants c_1, c_2, c_3, and b such that

$$c_1[d(x, M)]^b \leq V(x, n) \leq c_2[d(x, M)]^b \tag{3.4.9}$$

for all $x \in X$ and $n \in \mathbb{N}$.

Assume that for all $a \in A$ and for all $p(\cdot, a, n_0) \in S$,

$$V(p(n+1, a, n_0), n+1) - V(p(n, a, n_0), n) \leq -c_3[d(p(n, a, n_0), M)]^b \tag{3.4.10}$$

for all $n \in \mathbb{N}_{n_0}$.

Then (S, M) is *exponentially stable in the large.*

Proof. The proof proceeds similarly as that in the local exponential stability case. See the proof of Theorem 3.4.3. □

C. Instability results

We now address instability results of a set M with respect to S.

Theorem 3.4.8 (*Lyapunov's First Instability Theorem*) Let $\{\mathbb{N}, X, A, S\}$ be a dynamical system and let $M \subset A$ be closed, where A is assumed to be a neighborhood of M. Assume that there exist a function $V: X \times \mathbb{N} \to \mathbb{R}$ and a $k_0 \in \mathbb{N}$ that satisfy the following conditions.

(i) There exists a function $\psi \in \mathcal{K}$ defined on $\mathbb{R}^+$ such that

$$V(x, k) \leq \psi(d(x, M))$$

for all $(x, k) \in X \times \mathbb{N}$.

(ii) There exists a function $\varphi \in \mathcal{K}$ defined on $\mathbb{R}^+$ such that

$$V(p(k+1, a, k_0), k+1) - V(p(k, a, k_0), k) \geq \varphi(|V(p(k, a, k_0), k)|)$$

for all $p(\cdot, a, k_0) \in S$ and all $k \in \mathbb{N}_{k_0}$.

(iii) In every neighborhood of M there is a point x such that $V(x, k_0) > 0$ and there exists a motion $p(\cdot, x, k_0) \in S$.

Then M is *unstable* with respect to S.

Proof. Let $\{\mathbb{R}^+, X, A, \widetilde{S}\}$ be the associated DDS and let $\widetilde{V}: X \times \mathbb{R}^+ \to \mathbb{R}$ be defined as $\widetilde{V}(x, t) = V(x, n)$ for all $x \in X$ and $t \in [n, n+1), n \in \mathbb{N}$.

For any $a \in A$ and $p(\cdot, a, n_0) \in S$, the associated motion $\widetilde{p}(t, a, \tau_0 = n_0)$ is continuous everywhere on $\mathbb{R}^+_{\tau_0}$ except possibly on $E = \{\tau_1 = n_0+1, \tau_2 = n_0+2, \dots\}$. E is clearly unbounded and discrete. Along the motion $\widetilde{p}(\cdot, a, n_0)$ we have

$$D\widetilde{V}(\widetilde{p}(\tau_n, a, n_0), \tau_n) = V(p(\tau_{n+1}, a, n_0), \tau_{n+1}) - V(p(\tau_n, a, n_0), \tau_n).$$

It is easily seen that the associated DDS, $\{\mathbb{R}^+, X, A, \widetilde{S}\}$, and the set M satisfy the hypotheses of Theorem 3.2.8 and thus, M is unstable with respect to S. □

Theorem 3.4.9 In addition to the assumptions given in Theorem 3.4.8, assume that $V(x, k_0) > 0$ for all $x \notin M$. Then M is *completely unstable* with respect to S.

Proof. Note that by combining with Theorem 3.4.8, the present assumptions reduce to those of Theorem 3.2.9. Therefore, we conclude that M is completely unstable with respect to S. □

Theorem 3.4.10 (*Lyapunov's Second Instability Theorem*) Let $\{\mathbb{N}, X, A, S\}$ be a dynamical system and let $M \subset A$ be closed, where A is assumed to be a neighborhood of M. Assume that for any $(a, k_0) \in A \times \mathbb{N}$ and every $p(\cdot, a, k_0) \in S$, there exist a $k_0 \in \mathbb{N}$ and a function $V\colon M_\varepsilon \times \mathbb{N}_{k_0} \to \mathbb{R}$, where $M_\varepsilon = \{x \in X\colon d(x, M) < \varepsilon\}$, $\varepsilon > 0$, such that the following conditions are satisfied.

(i) V is bounded on $M_\varepsilon \times \mathbb{N}_{k_0}$.

(ii) For all $p(\cdot, a, k_0) \in S|_{M_\varepsilon}$ and $k \in \mathbb{N}_{k_0}$,

$$V(p(k+1, a, k_0), k+1) \geq \lambda V(p(k, a, k_0), k)$$

where $\lambda > 1$ is a constant.

(iii) In every neighborhood of M, there exists an x such that $V(x, k_1) > 0$ and there exists a motion $p(\cdot, x, k_1) \in S$ for a fixed $k_1 \geq k_0$.

Then M is *unstable* with respect to S.

Proof. By contradiction. If M is invariant and stable with respect to S, then for any $\varepsilon > 0$ and $k_1 \in \mathbb{R}^+$, there exists a $\delta = \delta(\varepsilon, k_1) > 0$ such that $d(p(k, a, k_1), M) < \varepsilon$ for all $p(\cdot, a, k_1) \in S$ and $k \in \mathbb{N}_{k_1}$ whenever $d(a, M) < \delta$. Because A is a neighborhood of M, it follows from condition (iii) that there exists an $x_1 \in \{a \in A\colon d(a, M) < \delta\}$ such that $V(x_1, k_1) > 0$. By condition (ii),

$$\begin{aligned} V(p(k+1, a, k_1), k+1) &\geq \lambda V(p(k, a, k_1), k) \\ &\geq \cdots \\ &\geq \lambda^{(k+1-k_1)} V(p(k_1, a, k_1), k_1) \end{aligned}$$

for all $k \geq k_1$. Because $V(p(k_1, a, k_1), k_1) = V(x_1, k_1) > 0$ and $\lambda > 1$, we have $\lim_{k \to +\infty} V(p(k+1, a, k_1), k+1) = \infty$. This contradicts condition (i) and completes the proof. □

D. An example

The scalar difference equation

$$x(n+1) = \begin{cases} 2x(n) & \text{if } n = n_0 + 2k, \\ x(n)/4 & \text{if } n = n_0 + 2k + 1, \end{cases} \tag{3.4.11}$$

with $x(n_0) = a$, $k \in \mathbb{N}$, where $a \in \mathbb{R}$ and $n_0 \in \mathbb{N}$, determines a dynamical system $\{\mathbb{N}, X, A, S\}$ with $X = A = \mathbb{R}$ and with $p(\cdot, a, n_0) \in S$ determined by the solutions of (3.4.11),

$$p(n, a, n_0) = \begin{cases} \dfrac{a}{2^k} & \text{if } n = n_0 + 2k, \\ \dfrac{a}{2^{k-1}} & \text{if } n = n_0 + 2k + 1, \end{cases} \tag{3.4.12}$$

$n \in \mathbb{N}$, for each pair $(a, k_0) \in \mathbb{R} \times \mathbb{N}$ and for all $n \geq k_0$. The plot of a typical motion for this system is given in Figure 3.4.1. Note that for each $(a, n_0) \in \mathbb{R} \times \mathbb{N}$, there exists a *unique* $p(\cdot, a, n_0) \in S$ that is defined for $n \geq n_0$. Clearly, $M = \{0\}$ is an invariant set with respect to S.

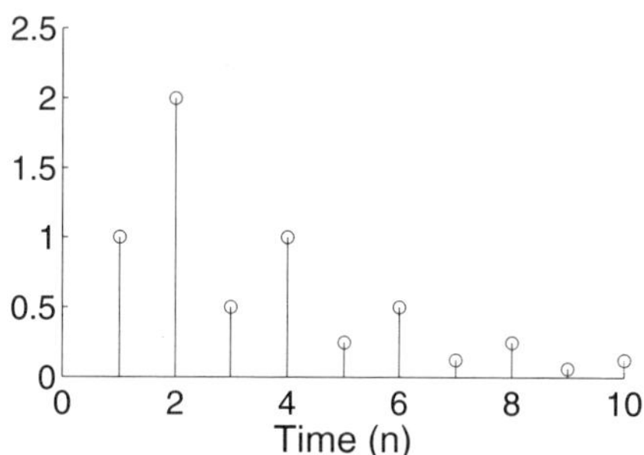

Figure 3.4.1: Plot of the motion $p(n, 1, 1) \in S$

We show in the following that

(a) for the associated DDS $\{\mathbb{R}^+, X, A, \widetilde{S}\}$ there exists a function $\widetilde{V}\colon \mathbb{R} \times \mathbb{N} \to \mathbb{R}^+$ that satisfies Theorem 3.2.2 and therefore, $(S, \{0\})$ is uniformly asymptotically stable; and

(b) for the discrete-time dynamical system $\{\mathbb{N}, X, A, S\}$, there does not exist a function $V\colon \mathbb{R} \times \mathbb{N} \to \mathbb{R}^+$ that satisfies the hypotheses of Theorem 3.4.2 and therefore, Theorem 3.4.2 cannot be used to prove that $(S, \{0\})$ is uniformly asymptotically stable.

Proof. (a) Let $\widetilde{V}\colon \mathbb{R} \to \mathbb{R}^+$ be given by $\widetilde{V}(x) = |x|$ for all $x \in \mathbb{R}$. For any $p(\cdot, a, n_0) \in S$, the associated motion is given by

$$\widetilde{p}(t, a, n_0) = \begin{cases} \dfrac{a}{2^k} & \text{if } t \in [n_0 + 2k, n_0 + 2k + 1), \\ \dfrac{a}{2^{k-1}} & \text{if } t \in [n_0 + 2k + 1, n_0 + 2(k + 1)), \end{cases}$$

$k \in \mathbb{N}$. Choose $E = \{\tau_1, \tau_2, \ldots\}$ with $\tau_k = n_0 + k$, and $E' = \{\tau_1', \tau_2', \ldots\}$ with $\tau_k' = n_0 + 2k$ (refer to Remark 3.2.1), $k \in \mathbb{N}$. By (3.4.12), $\widetilde{V}(p(\tau_k', a, \tau_0)) = |a/2^k|$, and $\widetilde{V}(p(t, a, t_0)) \leq 2\widetilde{V}(p(\tau_k', a, \tau_0))$ for all $t \in [\tau_k', \tau_{k+1}')$, $k \in \mathbb{N}$. Therefore all the conditions of Theorem 3.2.2 (and Remark 3.2.1) are satisfied and hence, $(\widetilde{S}, \{0\})$

is uniformly asymptotically stable. By Lemma 3.4.1 it now follows that $(S, \{0\})$ is uniformly asymptotically stable.

(b) For purposes of contradiction, assume that there exist a Lyapunov function $V\colon \mathbb{R}\times\mathbb{N}\to\mathbb{R}^+$ and two functions $\varphi_1, \varphi_2 \in \mathcal{K}$ defined on $\mathbb{R}^+$ such that

$$\varphi_1(|x|) \le V(x,n) \le \varphi_2(|x|) \tag{3.4.13}$$

for all $(x,n)\in\mathbb{R}\times\mathbb{N}$, and there exists a neighborhood U of 0 such that for all $a\in U$ and for all $p(\cdot,a,n_0)\in S$, $V(p(n,a,n_0),n)$ is nonincreasing for all $n\ge n_0$, $n\in\mathbb{N}$. Without loss of generality, we assume that $1\in U$.

By (3.4.12), $p(n_0+1,a,n_0)=2a$ for any $(a,n_0)\in\mathbb{R}\times\mathbb{N}$. In particular, for any $n\in\mathbb{N}$, because

$$p\Big(2,\frac{1}{2^n},1\Big)=\frac{1}{2^{n-1}},\qquad p\Big(3,\frac{1}{2^{n-1}},2\Big)=\frac{1}{2^{n-2}},\ldots,\ p\Big(n,\frac{1}{2},n-1\Big)=1,$$

and because $V(p(n,a,n_0),n)$ is nonincreasing for all $p(\cdot,a,n_0)\in S$, we have that

$$\begin{aligned}
&V\Big(\frac{1}{2^n},1\Big)\ge V\Big(p\Big(2,\frac{1}{2^n},1\Big),2\Big)=V\Big(\frac{1}{2^{n-1}},2\Big)\\
&\qquad\text{(along the motion } p\Big(\cdot,\frac{1}{2^n},1\Big));\\
&V\Big(\frac{1}{2^{n-1}},2\Big)\ge V\Big(p\Big(3,\frac{1}{2^{n-1}},2\Big),3\Big)=V\Big(\frac{1}{2^{n-2}},3\Big)\\
&\qquad\text{(along the motion } p\Big(\cdot,\frac{1}{2^{n-1}},2\Big));\\
&\qquad\vdots\\
&V\Big(\frac{1}{2},n-1\Big)\ge V\Big(p\Big(n,\frac{1}{2},n-1\Big),n\Big)=V(1,n)\\
&\qquad\text{(along the motion } p\Big(\cdot,\frac{1}{2},n-1\Big)).
\end{aligned}$$

Therefore,

$$V\Big(\frac{1}{2^n},1\Big)\ge V(1,n).$$

On the other hand,

$$\varphi_1\Big(\frac{1}{2^n}\Big)\le V\Big(\frac{1}{2^n},1\Big)\le\varphi_2\Big(\frac{1}{2^n}\Big)\quad\text{and}\quad\varphi_1(1)\le V(1,n)\le\varphi_2(1).$$

Thus,

$$\varphi_2\Big(\frac{1}{2^n}\Big)\ge\varphi_1(1)$$

is true for all $n\in\mathbb{N}$, which implies that

$$\varphi_2(0)=\lim_{n\to\infty}\varphi_2\Big(\frac{1}{2^n}\Big)\ge\varphi_1(1)>0.$$

However, by the assumption $\varphi_2 \in \mathcal{K}$, we know that $\varphi_2(0) = 0$. We have arrived at a contradiction. Therefore, there does not exist a Lyapunov function that satisfies the hypotheses of the classical Lyapunov Theorem for uniform asymptotic stability for discrete-time dynamical systems, Theorem 3.4.2. □

3.5 Converse Theorems for Discontinuous Dynamical Systems

The results of the previous three sections constitute sufficient conditions for various types of stability, instability, and boundedness for discontinuous dynamical systems, continuous dynamical systems, and discrete-time dynamical systems, respectively. It turns out that under some additional mild assumptions, these results constitute necessary conditions as well. Such results are referred to as *converse theorems* in the literature. The proofs of these results do not provide us with the means of constructing Lyapunov functions $V(p(\cdot, a, \tau_0), \cdot)$ in a systematic manner in applications. Nevertheless, converse theorems occupy an important place in the general development of the qualitative theory of dynamical systems. We address only converse theorems concerning local results.

A. Local results

In our first result, we require the following hypothesis.

Assumption 3.5.1 Let $\{\mathbb{R}^+, X, A, S\}$ be a DDS and assume that

(i) for any $p(\cdot, a, t_0) \in S$, there exists a $\widetilde{p}(\cdot, a_1, t_1) \in S$ with $a_1 = p(t_1, a, t_0)$ and $t_1 > t_0$ such that $\widetilde{p}(\cdot, a_1, t_1) = p(\cdot, a, t_0)$ for all $t \geq t_1$; and

(ii) for any two motions $p_i(\cdot, a_i, t_i) \in S$, $i = 1, 2$, $t_2 > t_1$, if $a_2 = p_1(t_2, a_1, t_1)$, then there exists a $\hat{p}(\cdot, a_1, t_1) \in S$ such that $\hat{p}(t, a_1, t_1) = p_1(t, a_1, t_0)$ for $t \in [t_1, t_2)$ and $\hat{p}(t, a_1, t_1) = p_2(t, a_2, t_2)$ for $t \geq t_2$. □

In part (i) of Assumption 3.5.1, $\widetilde{p}(\cdot, a_1, t_1)$ may be viewed as a *partial motion* of the motion $p(\cdot, a, t_0)$, and in part (ii), $\hat{p}(\cdot, a, t_1)$ may be viewed as a *composition* of $p_1(\cdot, a_1, t_1)$ and $p_2(\cdot, a_2, t_2)$. With this convention, Assumption 3.5.1 states that

(a) any partial motion is a motion in S; and

(b) any composition of two motions is a motion in S.

We require the above assumption in all converse theorems for dynamical systems defined on metric spaces. The reason for this is that in Definitions 2.2.1 and 2.2.2, the motions are defined for initial conditions and *forward* in time, and in general, time is not required to be reversible. (This is in contrast to many dynamical systems determined, e.g., by various types of differential equations, addressed in subsequent chapters.) We note, however, that when in a dynamical system the semigroup property holds, then Assumption 3.5.1 is automatically implied.

Theorem 3.5.1 Let $\{\mathbb{R}^+, X, A, S\}$ be a DDS and let $M \subset A$ be a closed invariant set, where A is assumed to be a neighborhood of M. Suppose that S satisfies Assumption 3.5.1 and that (S, M) is *uniformly stable*. Then there exist neighborhoods A_1 and X_1 of M such that $A_1 \subset X_1 \subset A$ and a mapping $V\colon X_1 \times \mathbb{R}^+ \to \mathbb{R}^+$ that satisfies the following conditions.

(i) There exist $\psi_1, \psi_2 \in \mathcal{K}$ such that

$$\psi_1(d(x, M)) \leq V(x, t) \leq \psi_2(d(x, M)) \tag{3.5.1}$$

for all $(x, t) \in X_1 \times \mathbb{R}^+$.

(ii) For every $p(\cdot, a, \tau_0) \in S$ with $a \in A_1$, $V(p(t, a, \tau_0), t)$ is nonincreasing for all $t \in \mathbb{R}^+_{\tau_0}$.

Proof. If (S, M) is uniformly stable, then in view of Lemma 3.10.3 (refer to Section 3.10, Problem 3.10.15), there exists a function $\varphi \in \mathcal{K}$ defined on $[0, h_0]$ for some $h_0 > 0$ such that

$$d(p(t, a, \tau_0), M) \leq \varphi(d(a, M)) \tag{3.5.2}$$

for all $p(\cdot, a, \tau_0) \in S$, for all $t \in \mathbb{R}^+_{\tau_0}$ and for all $\tau_0 \in \mathbb{R}^+$ whenever $d(a, M) < h_0$.

A is a neighborhood of M, therefore it follows that $X_1 = \{x \in A\colon d(x, M) < h_0\}$ is also a neighborhood of M. We now define $V\colon X_1 \times \mathbb{R}^+ \to \mathbb{R}^+$ by

$$V(x, t) = \sup\big\{d(p(t', x, t), M)\colon p(\cdot, x, t) \in S,\ t' \in \mathbb{R}^+_t\big\}.$$

Then for all $x \in X$, $t \in \mathbb{R}^+$, we have that

$$V(x, t) \geq d(p(t, x, t), M) = d(x, M)$$

and in view of (3.5.2) we have that $V(x, t) \leq \varphi(d(x, M))$. Therefore, V satisfies condition (i) of this theorem.

Next, let $A_1 = \{a \in X_1\colon d(a, M) < \varphi^{-1}(h_0)\}$ if $\varphi(h_0) > h_0$ and $A_1 = X_1$ otherwise. We now prove that for any $p_0(\cdot, a, \tau_0) \in S$ with $a \in X_1$, we have that $v(t) = V(p_0(t, a, \tau_0), t)$ is nonincreasing for all $t \in \mathbb{R}^+_{\tau_0}$.

Let $t_1, t_2 \in \mathbb{R}^+_{\tau_0}$ and $t_1 < t_2$. Let $a_i = p_0(t_i, a, \tau_0)$, $i = 1, 2$. Then,

$$v(t_i) = \sup\big\{d(p(t', a_i, t_i), M)\colon p(\cdot, a_i, t_i) \in S, t' \in \mathbb{R}^+_{t_i}\big\}, \qquad i = 1, 2.$$

To prove that $v(t_2) \leq v(t_1)$, it suffices to show that for every $p_2(\cdot, a_2, t_2) \in S$ and for every $t' \in \mathbb{R}^+_{t_2}$, there exists a $p_1(\cdot, a_1, t_1) \in S$ such that $p_2(t', a_2, t_2) = p_1(t', a_1, t_1)$.

By (i) in Assumption 3.5.1 there exists a $\widetilde{p}_0(\cdot, a_1, t_1) \in S$ such that

$$\widetilde{p}_0(t, a_1, t_1) = p_0(t, a, t_0)$$

for all $t \in \mathbb{R}^+_{t_1}$. By (ii) in Assumption 3.5.1, for $\widetilde{p}_0(\cdot, a_1, t_1)$ and $p_2(\cdot, a_2, t_2)$, where $a_2 = p_0(t_2, a, t_0) = \widetilde{p}_0(t_2, a_1, t_1)$, there exists a $p_1(\cdot, a_1, t_1) \in S$ such that

$$p_2(t', a_2, t_2) = p_1(t', a_1, t_1)$$

because $t' \in \mathbb{R}^+_{t_2}$. Therefore, $v(t_2) \leq v(t_1)$.

This concludes the proof of the theorem. □

In the proofs of the remaining results of the present section, we require the following additional assumption.

Assumption 3.5.2 Let $\{\mathbb{R}^+, X, A, S\}$ be a DDS. We assume that every motion $p(\cdot, a, \tau_0) \in S$ is continuous everywhere on $\mathbb{R}^+_{\tau_0}$, except possibly on an unbounded and discrete set $E = \{\tau_1, \tau_2, \dots : \tau_1 < \tau_2 < \cdots\}$ (recall that in general E depends on $p(\cdot, a, \tau_0) \in S$), and that $l_E \triangleq \inf_{k\in\{1,2,\dots\}, p\in S}\{\tau_{k+1} - \tau_k\} > 0$, and that $L_E \triangleq \sup_{k\in\{1,2,\dots\}, p\in S}\{\tau_{k+1} - \tau_k\} < \infty$. □

In the proof of the converse theorem for uniform asymptotic stability, we require a preliminary result.

Definition 3.5.1 A continuous function $\sigma\colon [s_1, \infty) \to \mathbb{R}^+$ is said to belong to *class* $\mathcal{L}$ if σ is strictly decreasing on $[s_1, \infty)$ and if $\lim_{s\to\infty} \sigma(s) = 0$ where $s_1 \in \mathbb{R}^+$. □

Lemma 3.5.1 Let $\beta \in \mathcal{L}$ be defined on $\mathbb{R}^+$. Then there exists a function $\alpha \in \mathcal{K}$ defined on $\mathbb{R}^+$ such that for any discrete subset $\{\tau_0, \tau_1, \dots\} \subset \mathbb{R}^+$ satisfying $\inf\{\tau_{k+1} - \tau_k\colon k = 1, 2, \dots\} > 0$, it is true that $\sum_{i=0}^{\infty} \alpha(\beta(\tau_i - \tau_0)) < \infty$.

Proof. We define $\eta \in C[(0,\infty), (0,\infty)]$ as

$$\eta(t) = \begin{cases} \beta(t)/t, & t \in (0,1), \\ \beta(t), & t \in [1, \infty). \end{cases}$$

Clearly, $\eta(t)$ is strictly decreasing for all $t > 0$, $\lim_{t\to 0^+} \eta(t) = +\infty$, and $\eta(t) \geq \beta(t)$ for all $t > 0$. Furthermore, η is invertible, and $\eta^{-1} \in C[(0,\infty), (0,\infty)]$ is strictly decreasing, and $\eta^{-1}(\beta(\tau)) \geq \eta^{-1}(\eta(\tau)) = \tau$ for all $\tau > 0$.

We now define $\alpha(0) = 0$ and

$$\alpha(t) = e^{-\eta^{-1}(t)}, \qquad t > 0.$$

Then $\alpha \in \mathcal{K}$, and

$$\alpha(\beta(\tau)) = e^{-\eta^{-1}(\beta(\tau))} \leq e^{-\tau}.$$

If we denote $l_m = \inf\{\tau_{j+1} - \tau_j\colon j = 1, 2 \dots\}$, we know that $\tau_j - \tau_0 \geq (j-1)l_m$. Hence it is true that

$$\begin{aligned} \sum_{j=0}^{\infty} \alpha(\beta(\tau_j - \tau_0)) \leq \sum_{j=0}^{\infty} e^{-(\tau_j - \tau_0)} &\leq 1 + \sum_{j=1}^{\infty} e^{-(j-1)l_m} \\ &= 1 + \frac{1}{1 - e^{-l_m}} < +\infty. \end{aligned}$$

This completes the proof. □

We are now in a position to prove the following result.

Theorem 3.5.2 Let $\{\mathbb{R}^+, X, A, S\}$ be a DDS and let $M \subset A$ be a closed invariant set, where A is assumed to be a neighborhood of M. Assume that S satisfies Assumptions 3.5.1 and 3.5.2, and furthermore, assume that for every $(a, \tau_0) \in A \times \mathbb{R}^+$, there exists a *unique* motion $p(\cdot, a, \tau_0) \in S$. Let (S, M) be *uniformly asymptotically stable*. Then there exist neighborhoods A_1, X_1 of M such that $A_1 \subset X_1 \subset A$, and a mapping $V: X_1 \times \mathbb{R}^+ \to \mathbb{R}^+$ that satisfies the following conditions.

(i) There exist functions $\psi_1, \psi_2 \in \mathcal{K}$ (defined on $\mathbb{R}^+$) such that

$$\psi_1(d(x, M)) \leq V(x, t) \leq \psi_2(d(x, M)) \tag{3.5.3}$$

for all $(x, t) \in X_1 \times \mathbb{R}^+$;

(ii) There exists a function $\psi_3 \in \mathcal{K}$, defined on $\mathbb{R}^+$, such that for all $p(\cdot, a, \tau_0) \in S$, we have

$$DV(p(\tau_k, a, \tau_0), \tau_k) \leq -\psi_3(d(p(\tau_k, a, \tau_0), M)) \tag{3.5.4}$$

where $a \in A_1$, $k \in \mathbb{N}$ and where DV is defined in (3.2.4).

(iii) There exists a function $f \in C[\mathbb{R}^+, \mathbb{R}^+]$ such that $f(0) = 0$ and such that

$$V(p(t, a, \tau_0), t) \leq f(V(p(\tau_k, a, \tau_0), \tau_k)) \tag{3.5.5}$$

for every $p(\cdot, a, \tau_0) \in S$ and all $t \in (\tau_k, \tau_{k+1})$, $a \in A_1$ and $\tau_0 \in \mathbb{R}^+$.

Proof. Inasmuch as (S, M) is uniformly asymptotically stable, we know by Theorem 3.5.1 that there exist some neighborhoods $\widetilde{A}_1$ and $\widetilde{X}_1$ of M such that $\widetilde{A}_1 \subset \widetilde{X}_1 \subset A$, and a mapping $\widetilde{V}: \widetilde{X}_1 \times \mathbb{R}^+ \to \mathbb{R}^+$ that satisfies the following conditions.

(a) There exist two functions $\widetilde{\varphi}_1, \widetilde{\varphi}_2 \in \mathcal{K}$ such that

$$\widetilde{\varphi}_1(d(x, M)) \leq \widetilde{V}(x, t) \leq \widetilde{\varphi}_2(d(x, M))$$

for all $(x, t) \in \widetilde{X}_1 \times \mathbb{R}^+$.

(b) For every $p(\cdot, a, \tau_0) \in S$ with $a \in \widetilde{A}_1$, $\widetilde{V}(p(t, a, \tau_0), t)$ is nonincreasing for all $t \geq \tau_0$.

From (a) and (b) above, we conclude that for any $t \in [\tau_k, \tau_{k+1})$, it is true that

$$\begin{aligned}\widetilde{\varphi}_1(d(p(t, a, \tau_0), M)) \leq \widetilde{V}(p(t, a, \tau_0), t) &\leq \widetilde{V}(p(\tau_k, a, \tau_0), \tau_k) \\ &\leq \widetilde{\varphi}_2(d(p(\tau_k, a, \tau_0), M))\end{aligned}$$

which implies that

$$d(p(t, a, \tau_0), M) \leq (\widetilde{\varphi}_1^{-1} \circ \widetilde{\varphi}_2)(d(p(\tau_k, a, \tau_0), M)) \tag{3.5.6}$$

for all $t \in [\tau_k, \tau_{k+1})$ and $k \in \mathbb{N}$.

By Lemma 3.10.5 (see Problem 3.10.17 in Section 3.10), there exist a function $\varphi \in \mathcal{K}$ defined on $[0, h_0]$ for some $h_0 > 0$, and a function $\sigma \in \mathcal{L}$, defined on $\mathbb{R}^+$, such that for all $t \geq \tau_0$

$$d(p(t, a, \tau_0), M) < \varphi(d(a, M))\sigma(t - \tau_0) \tag{3.5.7}$$

for all $p(\cdot, a, \tau_0) \in S$ whenever $d(a, M) < h_0$. Let $X_1 = \{x \in \widetilde{A}_1 : d(x, M) < h_0\}$ and $A_1 = \{a \in X_1 : d(a, M) < \varphi^{-1}(h_0)\}$ if $\varphi^{-1}(h_0) \leq h_0$ and $A_1 = X_1$ otherwise.

We now define the Lyapunov function $V(x, \tau_0)$ for $(x, \tau_0) \in X_1 \times \mathbb{R}^+$. Because for any $(x, \tau_0) \in X_1 \times \mathbb{R}^+$, there exists a unique motion $p(\cdot, x, \tau_0)$ that is continuous everywhere on $\mathbb{R}^+_{\tau_0}$ except on $E = \{\tau_1, \tau_2 \ldots : \tau_1 < \tau_2 < \cdots\}$, we define

$$V(x, \tau_0) = \sum_{j=0}^{\infty} u\big(d(p(\tau_j, x, \tau_0), M)\big) \tag{3.5.8}$$

where $u \in \mathcal{K}$, defined on $\mathbb{R}^+$, is specified later in such a manner that the above summation converges. Obviously,

$$V(x, \tau_0) \geq u\big(d(p(\tau_0, x, \tau_0), M)\big) = u\big(d(x, M)\big).$$

Hence, if we define $\psi_1 = u$, then $V(x, \tau_0) \geq \psi_1(d(x, M))$ for all $(x, \tau_0) \in X_1 \times \mathbb{R}^+$.

Consider $p(\cdot, x, \tau_0) \in S$ and the corresponding set $E = \{\tau_1, \tau_2, \ldots\}$. If we denote $\widetilde{x} = p(\tau_k, a, \tau_0)$, and $\widetilde{\tau}_0 = \tau_k$ for some $k \geq 1$, we know there exists a unique motion $\widetilde{p}(\cdot, \widetilde{x}, \widetilde{\tau}_0) \in S$ that is continuous everywhere on $t \geq \widetilde{\tau}_0$ except on $\{\widetilde{\tau}_1, \widetilde{\tau}_2, \ldots\}$. By the definition of V given in (3.5.8), we know that

$$V(\widetilde{x}, \widetilde{\tau}_0) = \sum_{j=0}^{\infty} u\big(d(\widetilde{p}(\widetilde{\tau}_j, \widetilde{x}, \widetilde{\tau}_0), M)\big).$$

By the uniqueness property and Assumption 3.5.1(i), we know that $\widetilde{\tau}_j = \tau_{k+j}$, and

$$\widetilde{p}(\widetilde{\tau}_j, \widetilde{x}, \widetilde{\tau}_0) = p(\tau_{k+j}, p(\tau_k, a, \tau_0), \tau_k) = p(\tau_{k+j}, a, \tau_0).$$

Therefore, it is clear that

$$V(p(\tau_k, a, \tau_0), \tau_k) = \sum_{j=k}^{\infty} u\big(d(p(\tau_j, a, \tau_0), M)\big). \tag{3.5.9}$$

Similarly, for any $t \in (\tau_k, \tau_{k+1}), k \in \mathbb{N}$, $V(p(t, a, \tau_0), t)$ is defined as

$$V(p(t, a, \tau_0), t) = u\big(d(p(t, a, \tau_0), M)\big) + \sum_{j=k+1}^{\infty} u\big(d(p(\tau_j, a, \tau_0), M)\big). \tag{3.5.10}$$

It follows that

$$\begin{aligned}
&DV(p(\tau_k, a, \tau_0), \tau_k) \\
&\quad = \frac{1}{\tau_{k+1} - \tau_k}\big[V(p(\tau_{k+1}, a, \tau_0), \tau_{k+1}) - V(p(\tau_k, a, \tau_0), \tau_k)\big] \\
&\quad = \frac{1}{\tau_{k+1} - \tau_k}\left[\sum_{j=k+1}^{\infty} u\big(d(p(\tau_j, a, \tau_0), M)\big) - \sum_{j=k}^{\infty} u\big(d(p(\tau_j, a, \tau_0), M)\big)\right] \\
&\quad = -\frac{1}{\tau_{k+1} - \tau_k} u\big(d(p(\tau_k, a, \tau_0), M)\big)
\end{aligned}$$

for $k \in \mathbb{N}$. Because $\tau_{k+1} - \tau_k \leq L_E$ by Assumption 3.5.2, it follows that

$$DV(p(\tau_k, a, \tau_0), \tau_k) \leq -u\big(d(p(\tau_k, a, \tau_0), M)\big)/L_E = -\psi_3(d(p(\tau_k, a, \tau_0), M))$$

where we have defined $\psi_3 = u/L_E$.

We now show how to choose $u \in \mathcal{K}$ so that the infinite summation in (3.5.8) converges. It follows from (3.5.7) that for any $(x, \tau_0) \in X_1 \times \mathbb{R}^+$, we have

$$\begin{aligned} u\big(d(p(t, x, \tau_0), M)\big) &< u\big(\varphi(d(x, M))\sigma(t - \tau_0)\big) \\ &\leq \big[u\big(\varphi(d(x, M))\sigma(0)\big)\big]^{1/2}\big[u\big(\varphi(h_0)\sigma(t - \tau_0)\big)\big]^{1/2}. \end{aligned} \tag{3.5.11}$$

Let $\beta(\tau) = \varphi(h_0)\sigma(\tau)$. Then $\beta \in \mathcal{L}$. Hence, by Lemma 3.5.1, there exists a function $\alpha \in \mathcal{K}$ defined on $\mathbb{R}^+$ such that $\sum_{i=0}^{\infty} \alpha\big(\beta(\tau_i - \tau_0)\big) < \infty$. If we define $u(r) = [\alpha(r)]^2$, then it follows that

$$\big[u\big(\varphi(h_0)\sigma(t - \tau_0)\big)\big]^{1/2} = \alpha\big(\varphi(h_0)\sigma(t - \tau_0)\big) = \alpha\big(\beta(t - \tau_0)\big). \tag{3.5.12}$$

Hence, we conclude from (3.5.8)–(3.5.10) that

$$\begin{aligned} V(x, \tau_0) &= \sum_{j=0}^{\infty} u\big(d(p(\tau_j, x, \tau_0), M)\big) \\ &< \sum_{j=0}^{\infty} \big[u\big(\varphi(d(x, M))\sigma(0)\big)\big]^{1/2}\big[u\big(\varphi(h_0)\sigma(\tau_j - \tau_0)\big)\big]^{1/2} \\ &= \big[u\big(\varphi(d(x, M))\sigma(0)\big)\big]^{1/2} \sum_{j=0}^{\infty} \alpha\big(\beta(\tau_j - \tau_0)\big) \\ &< \big[u\big(\varphi(d(x, M))\sigma(0)\big)\big]^{1/2}\big[1 + 1/(1 - e^{-l_E})\big], \end{aligned}$$

where l_E is the lower bound given in Assumption 3.5.2. If we define $\psi_2 \in \mathcal{K}$ by

$$\psi_2(r) = [u(\varphi(\sigma(0)r))]^{1/2}[1 + 1/(1 - e^{-l_E})],$$

then it follows that $V(x, \tau_0) \leq \psi_2(d(x, M))$. Thus we have proved conditions (i) and (ii) of the theorem.

To prove condition (iii) of the theorem, let $t \in (\tau_k, \tau_{k+1})$. We have already shown that

$$V(p(t, a, \tau_0), t) \leq \psi_2(d(p(t, a, \tau_0), M)).$$

Furthermore, because $a \in A_1 \subset \widetilde{A}_1$, (3.5.6) is satisfied. Hence, we know that

$$V(p(t, a, \tau_0), t) \leq \Big(\psi_2 \circ \widetilde{\varphi}_1^{-1} \circ \widetilde{\varphi}_2\Big)(d(p(\tau_k, a, \tau_0), M)). \tag{3.5.13}$$

On the other hand, we have also shown that

$$V(p(\tau_k, a, \tau_0), \tau_k) \geq \psi_1(d(p(\tau_k, a, \tau_0), M)),$$

which implies that

$$(\psi_1^{-1} \circ V)(p(\tau_k, a, \tau_0), \tau_k) \geq d(p(\tau_k, a, \tau_0), M). \tag{3.5.14}$$

Combining (3.5.13) and (3.5.14), we obtain that

$$V(p(t, a, \tau_0), t) \leq \left(\psi_2 \circ \widetilde{\varphi}_1^{-1} \circ \widetilde{\varphi}_2 \circ \psi_1^{-1}\right)(V(p(\tau_k, a, \tau_0), \tau_k))$$

for all $t \in (\tau_k, \tau_{k+1})$, $k \in \mathbb{N}$, and all $(a, \tau_0) \in A_1 \times \mathbb{R}^+$. If we define $f \in C[\mathbb{R}^+, \mathbb{R}^+]$ as $f = \psi_2 \circ \widetilde{\varphi}_1^{-1} \circ \widetilde{\varphi}_2 \circ \psi_1^{-1}$, then $f(0) = 0$ and

$$V(p(t, a, \tau_0), t) \leq f(V(p(\tau_k, a, \tau_0), \tau_k)).$$

This concludes the proof of the theorem. □

The hypotheses in the next result are not exactly symmetric with the corresponding hypotheses given in Theorem 3.2.3. Nevertheless, they do provide a set of necessary conditions for exponential stability.

Theorem 3.5.3 Let $\{\mathbb{R}^+, X, A, S\}$ be a DDS and let $M \subset A$ be a closed invariant set, where A is a neighborhood of M. Suppose that system S satisfies Assumptions 3.5.1 and 3.5.2 and that for every $(a, \tau_0) \in A \times \mathbb{R}^+$, there exists a *unique* motion $p(\cdot, a, \tau_0) \in S$. Let (S, M) be *exponentially stable*. Then there exist neighborhoods A_1 and X_1 of M such that $A_1 \subset X_1 \subset A$, and a mapping $V: X_1 \times \mathbb{R}^+ \to \mathbb{R}^+$ that satisfies the following conditions.

(i) There exist functions $\psi_1, \psi_2 \in \mathcal{K}$, defined on $\mathbb{R}^+$, such that

$$\psi_1(d(x, M)) \leq V(x, t) \leq \psi_2(d(x, M)) \tag{3.5.15}$$

for all $(x, t) \in X_1 \times \mathbb{R}^+$.

(ii) There exists a constant $c > 0$ such that for every $p(\cdot, a, \tau_0) \in S$,

$$DV(p(\tau_k, a, \tau_0), \tau_k) \leq -cV(p(\tau_k, a, \tau_0), \tau_k) \tag{3.5.16}$$

for $k \in \mathbb{N}$, where $a \in A_1$ and where DV is defined in (3.2.4).

(iii) There exists a function $f \in C[\mathbb{R}^+, \mathbb{R}^+]$ with $f(0) = 0$ and

$$f(r) = \mathcal{O}(r^q) \quad \text{as } r \to 0^+ \tag{3.5.17}$$

for some constant $q > 0$ such that

$$V(p(t, a, \tau_0), t) \leq f(V(p(\tau_k, a, \tau_0), \tau_k)) \tag{3.5.18}$$

for every $p(\cdot, a, \tau_0) \in S$, $t \in (\tau_k, \tau_{k+1})$, $k \in \mathbb{N}$, $a \in A_1$, and $\tau_0 \in \mathbb{R}^+$.

Proof. By Lemma 3.10.6 (see Problem 3.10.18, Section 3.10), there exist a function $\varphi \in \mathcal{K}$ defined on $[0, h_0]$ for some $h_0 > 0$, and an $\alpha > 0$ such that for all $t \in \mathbb{R}^+_{\tau_0}$

$$d(p(t, a, \tau_0), M) \leq \varphi(d(a, M))e^{-\alpha(t-\tau_0)} \tag{3.5.19}$$

for all $p(\cdot, a, \tau_0) \in S$ whenever $d(a, M) < h_0$. Let $X_1 = \{x \in A\colon d(x, M) < h_0\}$ and $A_1 = \{a \in X_1\colon d(a, M) < \varphi^{-1}(h_0)\}$ if $\varphi^{-1}(h_0) \leq h_0$ and $A_1 = X_1$ otherwise.

For $(x, \tau_0) \in X_1 \times \mathbb{R}^+$, there exists a unique motion $p(\cdot, x, \tau_0) \in S$. We define

$$V(x, \tau_0) = \sup_{t' \geq \tau_0} \{d(p(t', x, \tau_0), M)e^{\alpha(t'-\tau_0)}\}. \tag{3.5.20}$$

Now for $(a, \tau_0) \in A_1 \times \mathbb{R}^+$ and $p(t, a, \tau_0), t \in \mathbb{R}^+_{\tau_0}$, it must be true by Assumption 3.5.1 that the unique motion $p(t', p(t, a, \tau_0), t) = p(t', a, \tau_0)$ for all $t' \in \mathbb{R}^+_t$. Thus,

$$\begin{aligned} V(p(t, a, \tau_0), t) &= \sup_{t' \geq t} \{d(p(t', p(t, a, \tau_0), t), M)e^{\alpha(t'-t)}\} \\ &= \sup_{t' \geq t} \{d(p(t', a, \tau_0), M)e^{\alpha(t'-t)}\}. \end{aligned} \tag{3.5.21}$$

For $k \in \mathbb{N}$, we have

$$\begin{aligned} V(p(\tau_{k+1}, a, \tau_0), \tau_{k+1}) &= \sup_{t' \geq \tau_{k+1}} \{d(p(t', a, \tau_0), M)e^{\alpha(t'-\tau_k)}e^{-\alpha(\tau_{k+1}-\tau_k)}\} \\ &\leq \sup_{t' \geq \tau_{k+1}} \{d(p(t', a, \tau_0), M)e^{\alpha(t'-\tau_k)}\}e^{-\alpha l_E} \\ &\leq \sup_{t' \geq \tau_k} \{d(p(t', a, \tau_0), M)e^{\alpha(t'-\tau_k)}\}e^{-\alpha l_E} \\ &= e^{-\alpha l_E} V(p(\tau_k, a, \tau_0), \tau_k), \end{aligned}$$

where l_E is the lower limit given in Assumption 3.5.2. Letting $c = (1 - e^{-\alpha l_E})/L_E$, where L_E is the upper limit given in Assumption 3.5.2, we obtain

$$\begin{aligned} DV(p(\tau_k, a, \tau_0), \tau_k) &= \frac{1}{\tau_{k+1} - \tau_k}[V(p(\tau_{k+1}, a, \tau_0), \tau_{k+1}) - V(p(\tau_k, a, \tau_0), \tau_k)] \\ &\leq -\frac{1}{L_E}(1 - e^{-\alpha L_E})V(p(\tau_k, a, \tau_0), \tau_k) \\ &= -cV(p(\tau_k, a, \tau_0), \tau_k). \end{aligned}$$

Also, (3.5.19)–(3.5.21) imply that $d(x, M) \leq V(x, \tau_0) \leq \varphi(d(x, M))$ for all $(x, t) \in X_1 \times \mathbb{R}^+$. By (3.5.21), for every $t \in (\tau_k, \tau_{k+1})$ we have that

$$\begin{aligned} V(p(t, a, \tau_0), t) &= \sup_{t' \geq t}\{d(p(t', a, \tau_0), M)e^{\alpha(t'-\tau_k)}e^{-\alpha(t-\tau_k)}\} \\ &\leq \sup_{t' \geq t}\{d(p(t', a, \tau_0), M)e^{\alpha(t'-\tau_k)}\} \\ &\leq \sup_{t' \geq \tau_k}\{d(p(t', a, \tau_0), M)e^{\alpha(t'-\tau_k)}\} \\ &= V(p(\tau_k, a, \tau_0), \tau_k). \end{aligned}$$

The proof is completed by letting $f(r) = r$ and $q = 1/2$. □

We conclude by noting that converse theorems for DDSs for *uniform boundedness, uniform ultimate boundedness, uniform asymptotic stability in the large, exponential stability in the large*, and *instability* can also be established, using the methodology employed in the preceding results.

B. Refinements: Continuity of Lyapunov functions

The converse theorems presented in this section involve Lyapunov functions that need not necessarily be continuous. In the present subsection, we show that under some additional very mild assumptions, the Lyapunov functions for the converse theorems are continuous with respect to initial conditions.

In the proof of Theorem 3.5.2, the Lyapunov function V is constructed based on the unique motion that starts at $(x, \tau_0) \in A \times \mathbb{R}^+$. In the following, we show that under some additional very mild assumptions (Assumption 3.5.3) the function V given in the converse Theorem 3.5.2 is continuous (i.e., $V(x_{0m}, \tau_{0m})$ approaches $V(x_0, \tau_0)$ as $m \to \infty$ if $x_{0m} \to x_0$ and $\tau_{0m} \to \tau_0$ as $m \to \infty$). We then define continuous dependence on the initial conditions for motions of DDSs and show that Assumption 3.5.3 is satisfied when the motions are continuous with respect to initial conditions.

Assumption 3.5.3 Let $\{\mathbb{R}^+, X, A, S\}$ be a DDS and let $\{x_{0m}\} \subset A$, $\{\tau_{0m}\} \subset \mathbb{R}^+$, $x_{0m} \to x_0 \in A$, and $\tau_{0m} \to \tau_0$ as $m \to \infty$. The motion starting at (x_{0m}, τ_{0m}) is denoted by $p_m(t, x_{0m}, \tau_{0m})$ with the discontinuity set

$$E_{(x_{0m}, \tau_{0m})} = \{\tau_{1m}, \tau_{2m}, \dots : \tau_{0m} < \tau_{1m} < \tau_{2m} < \cdots\},$$

$m \in \mathbb{N}$. Assume that

(a) $\tau_{km} \to \tau_k$ as $m \to \infty$, for all $k \in \mathbb{N}$; and

(b) $p_m(\tau_{km}, x_{0m}, \tau_{0m}) \to x_k = p(\tau_k, x_0, \tau_0)$ as $m \to \infty$ for all $k \in \mathbb{N}$. □

We first strengthen Lemma 3.5.1 as follows.

Lemma 3.5.2 Let $\beta \in \mathcal{L}$ be defined on $\mathbb{R}^+$. Then there exists a function $\alpha \in \mathcal{K}$ defined on $\mathbb{R}^+$ such that for any discrete subset $\{r_0, r_1, \dots\} \subset \mathbb{R}^+$ satisfying $l_E = \inf\{r_{n+1} - r_n : n = 1, 2, \dots\} > 0$, it is true that

$$\sum_{i=0}^{\infty} \alpha(\beta(r_i - r_0)) < +\infty,$$

and

$$\sum_{i=k}^{\infty} \alpha(\beta(r_i - r_0)) < \frac{\exp(-(k-1)l_E)}{1 - \exp(-l_E)},$$

for all $k \geq 1$.

Proof. Let η and $\alpha \in C[(0,\infty),(0,\infty)]$ be the same as in the proof of Lemma 3.5.1. Then

$$\begin{aligned}\sum_{j=k}^{\infty}\alpha(\beta(\tau_j-\tau_0)) &\leq \sum_{j=k}^{\infty}\exp(-(\tau_j-\tau_0))\\ &\leq \sum_{j=k}^{\infty}\exp(-(j-1)l_E)\\ &= \frac{\exp\left(-(k-1)l_E\right)}{1-\exp(-l_E)},\end{aligned}$$

and

$$\sum_{j=0}^{\infty}\alpha(\beta(\tau_j-\tau_0)) < +\infty,$$

as shown in Lemma 3.5.1. The proof is completed. □

We are now in a position to present our first result.

Theorem 3.5.4 If in addition to the assumptions given in Theorem 3.5.2, the motions in S also satisfy Assumption 3.5.3, then the Lyapunov function in the Converse Theorem 3.5.2 is continuous with respect to initial conditions.

Proof. It follows from the proof of Theorem 3.5.2 that there exist a function $\varphi \in \mathcal{K}$ defined on $[0,h_0]$ for some $h_0 > 0$, and a function $\sigma \in \mathcal{L}$ defined on $\mathbb{R}^+$, such that for all $t \geq \tau_0$

$$d(p(t,a,\tau_0),M) < \varphi(d(a,M))\sigma(t-\tau_0) \tag{3.5.22}$$

for all $p(\cdot,a,\tau_0)\in S$ whenever $d(a,M) < h_0$. Let $X_1 = \{x \in A\colon d(x,M) < h_0\}$, and $A_1 = \{a \in X_1\colon d(a,M) < \varphi^{-1}(h_0)\}$ if $\varphi^{-1}(h_0) \leq h_0$ and $A_1 = X_1$ otherwise.

Let $\beta(\tau) = \varphi(h_0)\sigma(\tau)$, $\alpha \in \mathcal{K}$ be defined on $\mathbb{R}^+$ such that Lemma 3.5.2 is true, and $u(r) = [\alpha(r)]^2$. For any $(x,\tau_0) \in X_1 \times \mathbb{R}^+$, the Lyapunov function $V(x,\tau_0)$ is defined as

$$V(x,\tau_0) = \sum_{j=0}^{\infty} u\big(d(p(\tau_j,x,\tau_0),M)\big). \tag{3.5.23}$$

It follows from (3.5.22) that for any $(x,\tau_0)\in X_1\times\mathbb{R}^+$, we have

$$\begin{aligned}u\big(d(p(t,x,\tau_0),M)\big) &< u\big(\varphi(d(x,M))\sigma(t-\tau_0)\big)\\ &\leq \big[u\big(\varphi(d(x,M))\sigma(0)\big)\big]^{1/2}\big[u\big(\varphi(h_0)\sigma(t-\tau_0)\big)\big]^{1/2}.\end{aligned} \tag{3.5.24}$$

From the choice of u, we have

$$\big[u\big(\varphi(h_0)\sigma(t-\tau_0)\big)\big]^{1/2} = \alpha\big(\varphi(h_0)\sigma(t-\tau_0)\big) = \alpha\big(\beta(t-\tau_0)\big). \tag{3.5.25}$$

We now show that V is continuous with respect to initial conditions. Suppose $x_{0m} \to x_0$ and $\tau_{0m} \to \tau_0$ as $m \to \infty$. We denote $p_m(\tau_{km}, x_{0m}, \tau_{0m})$ by x_{km}. Then

$$\begin{aligned} V(x_{0m}, \tau_{0m}) &= \sum_{i=0}^{\infty} u(d(p_m(\tau_{im}, x_{0m}, \tau_{0m}), M)) \\ &= \sum_{i=0}^{\infty} u(d(x_{im}, M)). \end{aligned} \tag{3.5.26}$$

We show that $V(x_{0m}, \tau_{0m})$ approaches $V(x_0, \tau_0) = \sum_{i=0}^{\infty} u(d(x_i, M))$ as $m \to \infty$.

It follows from (3.5.24), (3.5.25), and Lemma 3.5.2 that

$$\begin{aligned} \sum_{i=k}^{\infty} u(d(p(\tau_i, x_0, \tau_0), M)) &< \sum_{i=k}^{\infty} [u(\varphi(d(x_0, M))\sigma(0))]^{1/2} [u(\varphi(h_0)\sigma(\tau_i - \tau_0))]^{1/2} \\ &\leq [u(\varphi(h_0)\sigma(0))]^{1/2} \sum_{i=k}^{\infty} \alpha(\beta(\tau_i - \tau_0)) \\ &< [u(\varphi(h_0)\sigma(0))]^{1/2} \frac{\exp(-(k-1)l_E)}{1 - \exp(-l_E)}. \end{aligned}$$

For every $\varepsilon > 0$, in view of the above inequality, there exists an $n_0 > 0$ such that

$$\sum_{i=n_0}^{\infty} u(d(x_i, M)) < \varepsilon/4 \tag{3.5.27}$$

for all $x_0 \in A_1$. Similarly,

$$\sum_{i=n_0}^{\infty} u(d(x_{im}, M)) < \varepsilon/4 \tag{3.5.28}$$

for all $x_{0m} \in A_1$.

On the other hand, for every $k \leq n_0$, there exists a $\delta_k > 0$ such that

$$|u(r) - u(d(x_k, M))| < \frac{\varepsilon}{2n_0}$$

whenever $|r - d(x_k, M)| < \delta_k$ (because $u(\cdot)$ is continuous everywhere on $\mathbb{R}^+$). Because $x_{km} \to x_k$ as $m \to \infty$, there exists for each $k \leq n_0$ an $m_k > 0$ such that $d(x_{km}, x_k) < \delta_k$ is true for all $m \geq m_k$. Now let $m_\varepsilon = \max_{k \leq n_0}\{m_k\}$. For every $m > m_\varepsilon$ we have $|d(x_k, M) - d(x_{km}, M)| \leq d(x_k, x_{km}) < \delta_k$ and thus

$$\begin{aligned} \left| \sum_{k=0}^{n_0-1} u(d(x_k, M)) - \sum_{k=0}^{n_0-1} u(d(x_{km}, M)) \right| & \\ &\leq \sum_{k=0}^{n_0-1} |u(d(x_k, M)) - u(d(x_{km}, M))| \\ &< \frac{\varepsilon}{2}. \end{aligned}$$

Therefore we have shown that

$$\begin{aligned}|V(x_0,\tau_0)-V(x_{0m},\tau_{0m})| &= \left|\sum_{k=0}^{\infty}u(d(x_k,M))-\sum_{k=0}^{\infty}u(d(x_{km},M))\right| \\ &\le \left|\sum_{k=0}^{n_0-1}u(d(x_k,M))-\sum_{k=0}^{n_0-1}u(d(x_{km},M))\right| \\ &\quad+\sum_{k=n_0}^{\infty}u(d(x_k,M))+\sum_{k=n_0}^{\infty}u(d(x_{km},M)) \\ &< \varepsilon. \end{aligned} \tag{3.5.29}$$

Therefore, we conclude that V is continuous with respect to initial conditions (x_0, τ_0). □

The following concept of continuous dependence on initial conditions for DDS is motivated by a corresponding term for ordinary differential equations (see, e.g., [11]), and is used as a sufficient condition for Assumption 3.5.3.

Definition 3.5.2 Suppose $\{x_{0m}\} \subset A \subset X, \{\tau_{0m}\} \subset \mathbb{R}^+$, $x_{0m} \to x_0 \in A$ and $\tau_{0m} \to \tau_0$ as $m \to \infty$. Assume that the motions are given by

$$p(t,x_0,\tau_0)=p^{(k)}(t,x_k,\tau_k), \qquad t\in[\tau_k,\tau_{k+1}),$$

and

$$p_m(t,x_{0m},\tau_{0m})=p_m^{(k)}(t,x_{km},\tau_{km}), \qquad t\in[\tau_{km},\tau_{(k+1)m}),$$

$k \in \mathbb{N}$, where $p^{(k)}(t, x_k, \tau_k)$ and $p_m^{(k)}(t, x_{km}, \tau_{km})$ are continuous for all $t \in \mathbb{R}^+$ with

$$p^{(k)}(\tau_k,x_k,\tau_k)=p(\tau_k,x_0,\tau_0)=x_k$$

and

$$p_m^{(k)}(\tau_{km},x_{km},\tau_{km})=p_m(\tau_{km},x_{0m},\tau_{0m})=x_{km}.$$

The motions in S are said to be *continuous with respect to the initial conditions* (x_0, τ_0) if

(a) $\tau_{km} \to \tau_k$ as $m \to \infty$, for all $k \in \mathbb{N}$; and

(b) for every compact set $K \subset \mathbb{R}^+$ and every $\varepsilon > 0$ there exists an $L = L(K, \varepsilon) > 0$ such that for all $t \in K$ and $k \in \mathbb{N}$ such that $K \cap [\tau_k, \tau_{k+1}) \neq \emptyset$,

$$d(p_m^{(k)}(t,x_{km},\tau_{km}),p^{(k)}(t,x_k,\tau_k))<\varepsilon$$

whenever $m > L$. □

An example of the set of continuous functions $p^{(k)}(t, x_k, \tau_k)$ is

$$p^{(k)}(t,x_k,\tau_k)=\begin{cases} x_k, & t<\tau_k,\\ p(t,x_0,\tau_0), & t\in[\tau_k,\tau_{k+1}),\\ p(\tau_{k+1}^-,x_0,\tau_0), & t\ge\tau_{k+1}.\end{cases}$$

Another example of $p^{(k)}(t, x_k, \tau_k)$ is given in Example 3.5.1, following the next result.

Theorem 3.5.5 If in addition to the assumptions given in Theorem 3.5.2, the motions in S are continuous with respect to initial conditions, then the Lyapunov function given in (3.5.23) is continuous with respect to initial conditions (x_0, τ_0).

Proof. We show that under the present hypotheses, Assumption 3.5.3 is satisfied and hence V is continuous with respect to initial conditions by Theorem 3.5.4.

Suppose $x_{0m} \to x_0$ and $\tau_{0m} \to \tau_0$ as $m \to \infty$. Assumption 3.5.3(a) is the same as Definition 3.5.2(a). We only need to show Assumption 3.5.3(b) is satisfied; that is, $x_{km} \to x_k$ as $m \to \infty$ for all $k \in \mathbb{N}$.

For a fixed $k > 0$, $k \in \mathbb{N}$, let $K = [\tau_k - l_E/2, \tau_k + l_E/2]$. For every $\varepsilon > 0$ there exists an $L = L(K, \varepsilon/2) > 0$ such that for all $t \in K$

$$d(p_m^{(k)}(t, x_{km}, \tau_{km}), p^{(k)}(t, x_k, \tau_k)) < \varepsilon/2 \tag{3.5.30}$$

whenever $m > L$. Because $p^{(k)}(t, x_k, \tau_k)$ is continuous on $\mathbb{R}^+$, there exists a $\delta > 0$ such that $d(p^{(k)}(t', x_k, \tau_k), p^{(k)}(\tau_k, x_k, \tau_k)) < \varepsilon/2$ whenever $|t' - \tau_k| < \delta$. Because $\tau_{km} \to \tau_k$ as $m \to \infty$, there exists an $L_1 > 0$ such that $\tau_{km} \in K$ and $|\tau_{km} - \tau_k| < \delta$ for all $m > L_1$. Therefore, when $m > \max\{L, L_1\}$, we have by (3.5.30)

$$d(p_m^{(k)}(\tau_{km}, x_{km}, \tau_{km}), p^{(k)}(\tau_{km}, x_k, \tau_k)) < \varepsilon/2,$$

and by the continuity of $p^{(k)}(t, x_k, \tau_k)$

$$d(p^{(k)}(\tau_{km}, x_k, \tau_k), p^{(k)}(\tau_k, x_k, \tau_k)) < \varepsilon/2.$$

By the triangle inequality we have

$$\begin{aligned} d(p_m^{(k)}&(\tau_{km}, x_{km}, \tau_{km}), p^{(k)}(\tau_k, x_k, \tau_k)) \\ &\leq d(p_m^{(k)}(\tau_{km}, x_{km}, \tau_{km}), p^{(k)}(\tau_{km}, x_k, \tau_k)) \\ &\quad + d(p^{(k)}(\tau_{km}, x_k, \tau_k), p^{(k)}(\tau_k, x_k, \tau_k)) \\ &< \varepsilon. \end{aligned}$$

This shows that $x_{km} = p_m^{(k)}(\tau_{km}, x_{km}, \tau_{km}) \to x_k$ as $m \to \infty$. This completes the proof. □

We conclude the present subsection by considering a specific example to demonstrate that the assumptions concerning the continuous dependence of the solutions (motions) on initial data, is a realistic assumption.

Example 3.5.1 Consider systems with impulse effects, which are described by equations of the form

$$\begin{cases} \dfrac{dx}{dt} = f(x, t), & t \neq t_k, \\ x(t_k) = g(x(t_k^-)), \end{cases} \tag{3.5.31}$$

where $x \in \mathbb{R}^n$ denotes the state, $g \in C[\mathbb{R}^n, \mathbb{R}^n]$, and $f \in C[\mathbb{R}^n \times \mathbb{R}^+, \mathbb{R}^n]$ satisfies a Lipschitz condition with respect to x that guarantees the existence and uniqueness of solutions of system (3.5.31) for given initial conditions. The set $E = \{t_1, t_2, \dots : t_1 < t_2 < \cdots\} \subset \mathbb{R}^+$ denotes the set of times when jumps occur. Assume that E is fixed in the interest of simplicity.

A function $\varphi\colon [t_0, \infty) \to \mathbb{R}^n$ is said to be a *solution* of the system with impulse effects (3.5.31) if (i) $\varphi(t)$ is left continuous on $[t_0, \infty)$ for some $t_0 \geq 0$; (ii) $\varphi(t)$ is differentiable and $(d\varphi/dt)(t) = f(\varphi(t), t)$ everywhere on (t_0, ∞) except on an unbounded subset $E \cap \{t\colon t > t_0\}$; and (iii) for any $t = t_k \in E \cap \{t\colon t > t_0\}$,

$$\varphi(t^+) = \lim_{t' \to t, t' > t} \varphi(t') = g(\varphi(t^-)).$$

Suppose $\tau_0 \in [t_{k_0}, t_{k_0+1})$ for some $k_0 \in \mathbb{N}$. The motion $p(t, x_0, \tau_0)$ is given by

$$p(t, x_0, \tau_0) = \begin{cases} p_{(d)}(t, x_k, t_k), & t \in [t_k, t_{k+1}), \quad k > k_0 \\ g(p_{(d)}(t_{k+1}^-, x_k, t_k)), & t = t_{k+1} \end{cases}$$

and $p(t, x_0, \tau_0) = p_{(d)}(t, x_0, \tau_0)$, $t \in [\tau_0, t_{k_0+1})$, where $x_k = p(t_k, x_0, \tau_0)$, and where $p_{(d)}(t, x_k, t_k)$, $t \in \mathbb{R}^+$ is the solution of the following ordinary differential equation

$$\frac{dx}{dt} = f(x, t), \qquad x(t_k) = x_k. \tag{3.5.32}$$

Suppose $x_{0m} \to x_0$ and $\tau_{0m} \to \tau_0$ as $m \to \infty$. Without loss of generality, we may assume that $\tau_0 < t_1 \in E$. By the assumption that E is fixed it follows that for sufficiently large m, the discontinuity set is $\{\tau_{km} = t_k\}$, for all $k > 0$. From the continuous dependence on initial conditions of ordinary differential equations, we know that $\{p_{(d)}(t, x_{0m}, \tau_{0m})\} \to p_{(d)}(t, x_0, \tau_0)$ for t in any compact set of $\mathbb{R}^+$ as $m \to \infty$.

Because $g(\cdot)$ is continuous, we have

$$x_{1m} = g(p_{(d)}(t_1^-, x_{0m}, \tau_{0m})) \to x_1 = g(p_{(d)}(t_1^-, x_0, \tau_0)) \text{ as } m \to \infty.$$

In turn, we have $p_{(d)}(t, x_{1m}, t_1) \to p_{(d)}(t, x_1, t_1)$ for t in any compact set of $\mathbb{R}^+$ as $m \to \infty$ and thus,

$$x_{2m} = g(p_{(d)}(t_2^-, x_{1m}, t_1)) \to x_2 = g(p_{(d)}(t_2^-, x_1, t_1)) \text{ as } m \to \infty.$$

By induction, we can show that $x_{km} \to x_k$ as $m \to \infty$ for all $k > 0$. Therefore we have shown that the motions of (3.5.31) are continuous with respect to initial conditions. □

3.6 Converse Theorems for Continuous Dynamical Systems

We address only local converse theorems.

A. Local results

Our first result, concerning uniform stability, is identical to the converse theorem for uniform stability for DDS.

Theorem 3.6.1 Let $\{\mathbb{R}^+, X, A, S\}$ be a continuous dynamical system and let $M \subset A$ be a closed invariant set, where A is assumed to be a neighborhood of M. Suppose that S satisfies Assumption 3.5.1. Assume that (S, M) is *uniformly stable*. Then there exist neighborhoods A_1 and X_1 of M such that $A_1 \subset X_1 \subset A$ and a mapping $V\colon X_1 \times \mathbb{R}^+ \to \mathbb{R}^+$ that satisfies the following conditions.

(i) There exist functions $\psi_1, \psi_2 \in \mathcal{K}$ such that

$$\psi_1(d(x, M)) \leq V(x, t) \leq \psi_2(d(x, M))$$

for all $(x, t) \in X_1 \times \mathbb{R}^+$.

(ii) For every $p(\cdot, a, \tau_0) \in S$ with $a \in A_1$, $V(p(t, a, \tau_0), t)$ is nonincreasing for all $t \in \mathbb{R}^+_{\tau_0}$.

Proof. The proof is identical to the proof of Theorem 3.5.1 and is not repeated here. □

Before proceeding further, it might be instructive to comment on the hypotheses of the next two results, the converse theorems for uniform asymptotic stability and exponential stability. In such results, for the case of continuous dynamical systems (see, e.g., Hahn [2], Miller and Michel [11], and Yoshizawa [14]), it is usually assumed that the motions are *unique forward in time*, unique backward in time, and that they satisfy the semigroup property; that is, for any $p(\cdot, a, t_0) \in S$ and $t_0 \leq t_1 \leq t$, $p(t, p(t_1, a, t_0), t_1) = p(t, a, t_0)$. The latter property ensures that Assumption 3.5.1 concerning partial motions is satisfied.

In contrast, as in the case of DDS, we require in the present section in the converse theorems for uniform asymptotic stability and exponential stability for continuous dynamical systems the weaker assumptions that the motions of a dynamical system are unique forward in time and that they satisfy Assumption 3.5.1 concerning partial motions.

We note in passing that for discrete-time dynamical systems determined by difference equations, the motions are in general not unique backward in time, unless the right-hand side of the difference equation is a bijective function which is only rarely the case.

Examples of dynamical systems whose motions are not unique forward in time, nor backward in time, and that do not satisfy Assumption 3.5.1 concerning partial motions include the examples given in Subsections 3.3D and 3.4D. To see this, we consider in particular the example given in Subsection 3.3D. Examining Figure 3.6.1, where we depict two solutions with initial conditions $(t_0 = 0, x(0) = 1)$ and $(t_0 = 1, x(1) = 2)$, we see that the motions are *unique with respect to initial conditions:* for each initial condition there exists one and only one motion that exists for all $t \geq t_0$. However, because these motions intersect at different time instants, the motions of

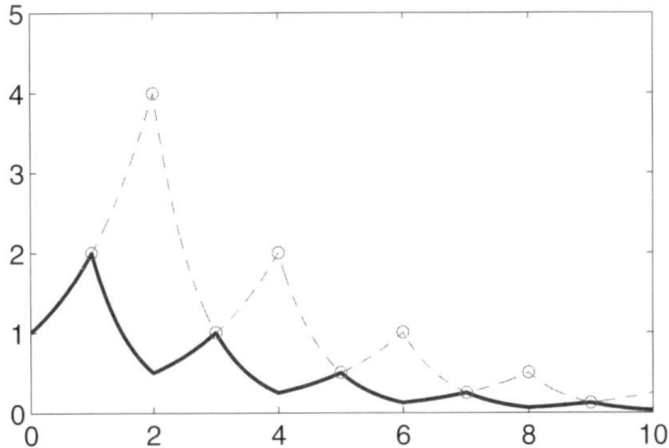

Figure 3.6.1: Two motions that intersect.

this dynamical system are not unique forward in time, nor are they unique backward in time.

Moreover, because the composition of some parts of these motions do not result in a partial motion, Assumption 3.5.1 is also not satisfied in the present example.

In the proof of our next result, we require the following preliminary result.

Lemma 3.6.1 Let $\beta \in \mathcal{L}$ be defined on $\mathbb{R}^+$. Then there exists a function $\alpha \in \mathcal{K}$ defined on $\mathbb{R}^+$ such that

$$\int_0^\infty \alpha(\beta(\tau))d\tau \leq 1.$$

Proof. We define $\eta \in C[(0,\infty),(0,\infty)]$ by

$$\eta(t) = \begin{cases} \beta(t)/t, & t \in (0,1), \\ \beta(t), & t \in [1,\infty). \end{cases}$$

By construction, $\eta(t)$ is strictly decreasing for all $t > 0$, $\lim_{t\to 0^+} \eta(t) = +\infty$, and $\eta(t) \geq \beta(t)$ for all $t > 0$. Furthermore, η^{-1} exists and is strictly decreasing, and $\eta^{-1}(\beta(t)) \geq \eta^{-1}(\eta(t)) = t$ for all $t > 0$.

We now define $\alpha(0) = 0$ and $\alpha(t) = e^{-\eta^{-1}(t)}$ for all $t > 0$. Then α is a class $\mathcal{K}$ function, $\alpha(\beta(t)) = e^{-\eta^{-1}(\beta(t))} \leq e^{-t}$, and

$$\int_0^\infty \alpha(\beta(\tau))d\tau \leq \int_0^\infty e^{-\tau}d\tau \leq 1.$$

□

Theorem 3.6.2 Let $\{\mathbb{R}^+, X, A, S\}$ be a continuous dynamical system and let $M \subset A$ be a closed invariant set, where A is assumed to be a neighborhood of M. Assume that S satisfies Assumption 3.5.1 and that for every $(a,\tau_0) \in A \times \mathbb{R}^+$, there exists a *unique* motion $p(\cdot, a, \tau_0) \in S$ that is defined and continuous for all $t \in \mathbb{R}^+_{\tau_0}$. Let

(S, M) be *uniformly asymptotically stable*. Then there exist neighborhoods A_1 and X_1 of M such that $A_1 \subset X_1 \subset A$, and a mapping $V\colon X_1 \times \mathbb{R}^+ \to \mathbb{R}^+$ that satisfies the following conditions.

(i) There exist two functions $\psi_1, \psi_2 \in \mathcal{K}$ such that

$$\psi_1(d(x, M)) \leq V(x, t) \leq \psi_2(d(x, M))$$

for all $(x, t) \in X_1 \times \mathbb{R}^+$.

(ii) There exists a function $\psi_3 \in \mathcal{K}$ such that for all $p(\cdot, a, \tau_0) \in S$ and for all $t \in [\tau_0, \infty)$, we have

$$D^+V(p(t, a, \tau_0), t) \leq -\psi_3(d(p(t, a, \tau_0), M))$$

whenever $a \in A_1$.

Proof. By Lemma 3.10.5 (see Problem 3.10.17, Section 3.10), there exist a function $\varphi \in \mathcal{K}$ defined on $[0, h_0]$ for some $h_0 > 0$, and a function $\sigma \in \mathcal{L}$ defined on $\mathbb{R}^+$, such that for all $t \geq \tau_0$

$$d(p(t, a, \tau_0), M) < \varphi(d(a, M))\sigma(t - \tau_0) \tag{3.6.1}$$

for all $p(\cdot, a, \tau_0) \in S$ whenever $d(a, M) < h_0$. Let $X_1 = \{x \in A\colon d(x, M) < h_0\}$ and let

$$A_1 = \begin{cases} \{a \in X_1\colon d(a, M) < \varphi^{-1}(h_0)\} & \text{if } \varphi(h_0) > h_0, \\ X_1 & \text{otherwise.} \end{cases}$$

We define

$$Z(x, t) = \int_t^\infty u\big(d(p(\tau, x, t), M)\big)d\tau \tag{3.6.2}$$

where $u \in \mathcal{K}$ is to be determined later and is such that the integral converges for all $(x, t) \in X_1 \times \mathbb{R}^+$. For $p(\cdot, a, \tau_0) \in S$, $p(\tau, p(t, a, \tau_0), t) = p(\tau, a, \tau_0)$ because of Assumption 3.5.1 and the uniqueness of the motion $p(\cdot, a, \tau_0)$. Therefore, the integrand in the right-hand side of (3.6.2) is independent of t for $x = p(t, a, \tau_0)$ where $a \in A_1$. Because $u(d(p(\tau, a, \tau_0), M))$ is a continuous function of τ, it follows that $Z(p(t, a, \tau_0), t)$ is differentiable with respect to t and that

$$\frac{d}{dt}Z(p(t, a, \tau_0), t) = -u\big(d(p(t, x, \tau_0), M)\big) \tag{3.6.3}$$

for all $(a, \tau_0) \in A_1 \times \mathbb{R}^+$ and $t \geq \tau_0$.

To determine how to choose $u \in \mathcal{K}$ so that the integral in (3.6.2) converges for all $(x, t) \in X_1 \times \mathbb{R}^+$, we use (3.6.1). For $x \in X_1$, $t \in \mathbb{R}^+$, and $\tau \geq t$, we have

$$d(p(\tau, x, t), M) \leq \varphi(d(x, M))\sigma(\tau - t).$$

Because $\varphi(d(x, M)) \leq \varphi(h_0)$ for $x \in X_1$ and because $\sigma(\tau - t) \leq \sigma(0)$, we have that

$$u\big(\varphi(d(x, M))\sigma(\tau - t)\big) \leq \big[u\big(\varphi(d(x, M))\sigma(0)\big)\big]^{1/2}\big[u\big(\varphi(h_0)\sigma(\tau - t)\big)\big]^{1/2}$$

for $x \in X_1$ and $\tau \geq t \geq 0$. Therefore,

$$\begin{aligned} Z(x,t) &\leq \big[u\big(\varphi(d(x,M))\sigma(0)\big)\big]^{1/2} \int_t^\infty \big[u\big(\varphi(h_0)\sigma(\tau - t)\big)\big]^{1/2} d\tau \\ &= \big[u\big(\varphi(d(x,M))\sigma(0)\big)\big]^{1/2} \int_0^\infty \big[u\big(\varphi(h_0)\sigma(\tau)\big)\big]^{1/2} d\tau. \end{aligned}$$

In applying Lemma 3.6.1, we choose $\beta(\tau) = \varphi(h_0)\sigma(\tau)$ and $u(r) = \big[\alpha(r)\big]^2$. Then

$$Z(x,t) \leq \big[u\big(\varphi(d(x,M))\sigma(0)\big)\big]^{1/2} = \alpha\big(\varphi(d(x,M))\sigma(0)\big). \tag{3.6.4}$$

For $(x,t) \in X_1 \times \mathbb{R}^+$, we now define

$$W(x,t) = \sup_{t' \geq t} \big\{d(p(t',x,t),M)\big\}.$$

Then

$$d(x,M) = d(p(t,x,t),M) \leq W(x,t) \leq \varphi(d(x,M))\sigma(0). \tag{3.6.5}$$

Let $V(x,t) = Z(x,t) + W(x,t)$. In the proof of Theorem 3.5.1 we have shown that $W(p(t,a,\tau_0),t)$ is nonincreasing for all $t \geq \tau_0$ (i.e., $D^+W(p(t,a,\tau_0),t) \leq 0$). Therefore, (3.6.3) implies that

$$D^+V(p(t,a,\tau_0),t) \leq -u\big(d(p(t,a,\tau_0),M)\big)$$

for all $a \in A_1$ and $t \geq \tau_0$; that is, V satisfies condition (ii) of the theorem.

To show that V satisfies condition (i), we note that

$$d(x,M) \leq V(x,t) \leq \alpha\big(\varphi(d(x,M))\sigma(0)\big) + \varphi(d(x,M))\sigma(0).$$

for all $x \in X_1$ and $t \in \mathbb{R}^+$, where we have used (3.6.4) and (3.6.5). This concludes the proof of the theorem. □

The hypotheses in our next result are not precisely symmetric with the corresponding assumptions in Theorem 3.3.3 for exponential stability of (S,M). Nevertheless, they do constitute necessary conditions for exponential stability of (S,M).

Theorem 3.6.3 Let $\{\mathbb{R}^+, X, A, S\}$ be a continuous dynamical system and let $M \subset A$ be a closed invariant set, where A is assumed to be a neighborhood of M. Assume that S satisfies Assumption 3.5.1, and furthermore, assume that for every $(a,\tau_0) \in A \times \mathbb{R}^+$, there exists a unique continuous motion $p(\cdot,a,\tau_0) \in S$ that is defined and continuous for all $t \in [\tau_0,\infty)$. Let (S,M) be *exponentially stable*. Then there exist neighborhoods A_1 and X_1 of M such that $A_1 \subset X_1 \subset A$, and a mapping $V\colon X_1 \times \mathbb{R}^+ \to \mathbb{R}^+$ that satisfies the following conditions.

(i) There exist two functions $\psi_1, \psi_2 \in \mathcal{K}$ such that

$$\psi_1(d(x,M)) \leq V(x,t) \leq \psi_2(d(x,M))$$

for all $(x,t) \in X_1 \times \mathbb{R}^+$.

(ii) There exists a constant $c > 0$ such that for every $p(\cdot, a, \tau_0) \in S$ and for all $t \in [\tau_0, \infty)$,
$$D^+V(p(t, a, \tau_0), t) \leq -cV(p(t, a, \tau_0), t)$$
where $a \in A_1$.

Proof. By Lemma 3.10.6 (see Problem 3.10.18, Section 3.10), there exist a function $\varphi \in \mathcal{K}$, defined on $[0, h_0]$ for some $h_0 > 0$, and a constant $\alpha > 0$ such that
$$d(p(t, a, \tau_0), M) < \varphi(d(a, M))e^{-\alpha(t-\tau_0)} \tag{3.6.6}$$
for all $p(\cdot, a, \tau_0) \in S$ and all $t \geq \tau_0$ whenever $d(a, M) < h_0$.

Let $X_1 = \{x \in A\colon d(x, M) < h_0\}$ and let
$$V(x, t) = \sup_{t' \geq t} \{d(p(t', x, t), M)e^{\alpha(t'-t)}\} \tag{3.6.7}$$
for all $(x, t) \in X_1 \times \mathbb{R}^+$. Let $A_1 = \{a \in X_1 : d(a, M) < \varphi^{-1}(h_0)\}$ if $\varphi(h_0) > h_0$ and $A_1 = X_1$ otherwise. Then for $a \in A_1$ and $\tau_0 \in \mathbb{R}^+$, we have by Assumption 3.5.1 and the uniqueness of the motions that
$$\begin{aligned} V(p(t, a, \tau_0), t) &= \sup_{t' \geq t} \{d(p(t', p(t, a, \tau_0), t), M)e^{\alpha(t'-t)}\} \\ &= \sup_{t' \geq t} \{d(p(t', a, \tau_0), M)e^{\alpha(t'-t)}\}. \end{aligned}$$
Therefore, for $\Delta t > 0$, we have
$$\begin{aligned} V(p(t + \Delta t, a, \tau_0), t + \Delta t) &= \sup_{t' \geq t+\Delta t} \{d(p(t', p(t, a, \tau_0), t), M)e^{\alpha(t'-t-\Delta t)}\} \\ &= \sup_{t' \geq t+\Delta t} \{d(p(t', a, \tau_0), M)e^{\alpha(t'-t)}\}e^{-\alpha\Delta t} \\ &\leq V(p(t, a, \tau_0), t)e^{-\alpha\Delta t}. \end{aligned}$$
The above inequality yields
$$D^+V(p(t, a, \tau_0), t) \leq \lim_{\Delta t \to 0^+} V(p(t, a, \tau_0), t)\frac{e^{-\alpha\Delta t} - 1}{\Delta t} = -\alpha V(p(t, a, \tau_0), t).$$
Finally, (3.6.6) and (3.6.7) imply that
$$d(x, M) \leq V(x, t) \leq \varphi(d(x, M))$$
for all $(x, t) \in X_1 \times \mathbb{R}^+$. This concludes the proof of the theorem. □

We conclude by noting that converse theorems for continuous dynamical systems for *uniform boundedness*, *uniform ultimate boundedness*, *uniform asymptotic stability in the large*, *exponential stability in the large*, and *instability* can also be established, using the methodology employed in the preceding results.

B. Refinements: Continuity of Lyapunov functions

In this subsection, we first define the notion of continuity with respect to initial conditions for continuous dynamical systems. We then show that the Lyapunov functions in the converse theorems (Theorems 3.6.2 and 3.6.3) are continuous.

Definition 3.6.1 Let $\{\mathbb{R}^+, X, A, S\}$ be a dynamical system. Let $\{a_m\} \subset A \subset X$, $\{t_m\} \subset \mathbb{R}^+$, $a_m \to a \in A$, and $t_m \to t_0$ as $m \to \infty$, let $\{p_m\} = \{p_m(\cdot, a_m, t_m)\}$ be a sequence of noncontinuable motions defined on intervals $J_m = [t_m, c_m)$, and let $p = p(\cdot, a, t_0)$ be a noncontinuable motion defined on an interval $[t_0, c_0)$. We say that the motion p is *continuous with respect to initial conditions* if there is a subsequence $\{m_j\} \subset \{m\}$ such that

(i) $\lim_{j\to\infty} \inf[0, c_{m_j} - t_{m_j}) \supset J_0 = [0, c_0 - t_0)$; and

(ii) $p_{m_j}(t + t_{m_j}, a_{m_j}, t_{m_j}) \to p(t + t_0, a, t_0)$ uniformly on compact subsets of J_0 as $j \to \infty$.

If in particular, the motion p is unique, then it is required that the *entire* sequence $\{p_m(t + t_m, a_m, t_m)\}$ tends to $p(t + t_0, a, t_0)$ uniformly in t on compact subsets of J_0. □

For the motivation of Definition 3.6.1, the reader may want to refer to Theorem 6.8.1 in the appendix section of Chapter 6 (Section 6.8), where conditions for the continuous dependence of the solutions of ordinary differential equations on initial conditions and parameters are presented (as required background material).

Theorem 3.6.4 Let $\{\mathbb{R}^+, X, A, S\}$ be a continuous dynamical system and let $M \subset A$ be a closed invariant set, where A is assumed to be a neighborhood of M. Assume that S satisfies Assumption 3.5.1 and the motions in S are *continuous with respect to initial conditions*, and furthermore, assume that for every $(a, \tau_0) \in A \times \mathbb{R}^+$, there exists a *unique* motion $p(\cdot, a, \tau_0) \in S$ that is defined and continuous for all $t \in \mathbb{R}^+_{\tau_0}$. Let (S, M) be *uniformly asymptotically stable*. Then there exist neighborhoods A_1 and X_1 of M such that $A_1 \subset X_1 \subset A$, and a function $V \in C[X_1 \times \mathbb{R}^+, \mathbb{R}^+]$ that satisfies the conditions of Theorem 3.6.2.

Proof. Let $V(x,t) = W(x,t) + Z(x,t)$, where $Z(x,t)$ and $W(x,t)$ are the same as in the proof of Theorem 3.6.2; that is,

$$W(x,t) = \sup_{t' \geq t} \{d(p(t', x, t), M)\},$$

$$Z(x,t) = \int_t^\infty u\big(d(p(\tau, x, t), M)\big) d\tau,$$

for all $(x,t) \in X_1 \times \mathbb{R}^+$, where X_1 is given in the proof of Theorem 3.6.2. We show in the following that both $W(\cdot,\cdot)$ and $Z(\cdot,\cdot)$ are continuous in (x,t), and hence, $V(x,t)$ is continuous.

Let $\varepsilon > 0$ be arbitrary. Suppose that $\{(x_m, t_m)\} \subset X_1 \times \mathbb{R}^+$, $m = 1, 2, \ldots$, and $(x_m, t_m) \to (x_0, t_0)$ as $m \to \infty$. There exists an $L_1 > 0$ such that $\sigma(s) < \varepsilon/\varphi(h_0)$ for all $s > L_1$, where $\sigma(\cdot) \in \mathcal{L}$ is given in (3.6.1). Then $d(p(\tau + t, x, t), M) \leq \varphi(h_0)\sigma(\tau) < \varepsilon$ for all $\tau > L_1$. Thus,

$$\begin{aligned}
&|W(x_m, t_m) - W(x_0, t_0)| \\
&\quad = \sup_{\tau \geq 0} \{d(p(\tau + t_m, x_m, t_m), M)\} - \sup_{\tau \geq 0} \{d(p(\tau + t_0, x_0, t_0), M)\} \\
&\quad \leq \sup_{0 \leq \tau \leq L_1} \{|d(p(\tau + t_m, x_m, t_m), M) - d(p(\tau + t_0, x_0, t_0), M)|\} + 2\varepsilon.
\end{aligned}$$

Because the motions in S are continuous with respect to initial conditions, $p(t, x_m, t_m)$ converges to $p(t, x_0, t_0)$ uniformly on the compact set $[0, L_1]$; that is, there exists an $m_0 \in \mathbb{N}$ such that $\sup_{0 \leq \tau \leq L_1} |d(p(\tau, x_m, t_m), M) - d(p(\tau, x_0, t_0), M)| < \varepsilon$ for all $m > m_0$. The continuity of $W(\cdot, \cdot)$ now follows immediately.

Similarly, there exists an $L_2 > 0$ such that

$$\int_{t+L_2}^{\infty} u(d(p(\tau, x, t), M))d\tau \leq \alpha(\varphi(h_0)\sigma(0)) \int_{L_2}^{\infty} \alpha(\varphi(h_0))\sigma(\tau)d\tau < \varepsilon,$$

for all $(x, t) \in X_1 \times \mathbb{R}^+$, where u and α are given in the proof of Theorem 3.6.2. Hence,

$$\begin{aligned}
&|Z(x_m, t_m) - Z(x_0, t_0)| \\
&= \int_{t_m}^{\infty} u(d(p(\tau, x_m, t_m), M))d\tau - \int_{t_0}^{\infty} u(d(p(\tau, x_0, t_0), M))d\tau \\
&\leq \int_0^{L_2} \Big| u(d(p(\tau + t_m, x_m, t_m), M)) - u(d(p(\tau + t_0, x_0, t_0), M)) \Big| d\tau \\
&\quad + \int_{t_m+L_2}^{\infty} u(d(p(\tau, x_m, t_m), M))d\tau + \int_{t_0+L_2}^{\infty} u(d(p(\tau, x_0, t_0), M))d\tau \\
&\leq \int_0^{L_2} \Big| u(d(p(\tau + t_m, x_m, t_m), M)) - u(d(p(\tau + t_0, x_0, t_0), M)) \Big| d\tau + 2\varepsilon.
\end{aligned}$$

The term

$$\int_0^{L_2} \big| u(d(p(\tau + t_m, x_m, t_m), M)) - u(d(p(\tau + t_0, x_0, t_0), M)) \big| d\tau$$

becomes arbitrarily small for sufficiently large m because $p(t, x_m, t_m)$ converges to $p(t, x_0, t_0)$ uniformly on the compact set $[0, L_2]$. Therefore we have shown that $Z(x, t)$ is continuous in (x, t). □

In our next result, the Lyapunov function is constructed slightly differently from that in the proof of Theorem 3.6.3 to ensure the continuity of the Lyapunov function.

Theorem 3.6.5 Let $\{\mathbb{R}^+, X, A, S\}$ be a continuous dynamical system and let $M \subset A$ be a closed invariant set, where A is assumed to be a neighborhood of M. Assume that S satisfies Assumption 3.5.1 and the motions in S are *continuous with respect to initial conditions*, and furthermore, assume that for every $(a, \tau_0) \in A \times \mathbb{R}^+$, there exists a *unique* motion $p(\cdot, a, \tau_0) \in S$ that is defined and continuous for all $t \in \mathbb{R}^+_{\tau_0}$. Let (S, M) be *exponentially stable*. Then there exist neighborhoods A_1 and X_1 of M such that $A_1 \subset X_1 \subset A$, and a function $V \in C[X_1 \times \mathbb{R}^+, \mathbb{R}^+]$ that satisfies the conditions of Theorem 3.6.3.

Proof. By Lemma 3.10.6 (refer to Problem 3.10.18), there exist a function $\varphi \in \mathcal{K}$, defined on $[0, h_0]$ for some $h_0 > 0$, and a constant $\alpha > 0$ such that

$$d(p(t, a, \tau_0), M) < \varphi(d(a, M))e^{-\alpha(t-\tau_0)} \tag{3.6.8}$$

for all $p(\cdot, a, \tau_0) \in S$ and all $t \geq \tau_0$ whenever $d(a, M) < h_0$.

Let $X_1 = \{x \in A: d(x, M) < h_0\}$ and let

$$V(x,t) = \sup_{t' \geq t} \{d(p(t', x, t), M)e^{\lambda\alpha(t'-t)}\} \tag{3.6.9}$$

for all $(x, t) \in X_1 \times \mathbb{R}^+$, where $0 < \lambda < 1$ is a constant. Let

$$A_1 = \begin{cases} \{a \in X_1: d(a, M) < \varphi^{-1}(h_0)\} & \text{if } \varphi(h_0) > h_0, \\ X_1 & \text{otherwise.} \end{cases}$$

It can be shown in the manner as in the proof of Theorem 3.6.3 that the V function satisfies the conditions in Theorem 3.6.3 with the constant $c = \lambda\alpha$.

To establish the continuity of V, we let $\varepsilon > 0$ be arbitrary, $\{(x_m, t_m)\} \subset X_1 \times \mathbb{R}^+$, $m = 1, 2, \ldots$, and $(x_m, t_m) \to (x_0, t_0)$ as $m \to \infty$. It follows from (3.6.8) that

$$d(p(\tau + t, x, t), M)e^{\lambda\alpha\tau} \leq \varphi(h_0)e^{-(1-\lambda)\alpha\tau} < \varepsilon$$

for all $\tau > L \triangleq \ln(\varepsilon/\varphi(h_0))/[(1-\lambda)\alpha]$. Thus,

$$\begin{aligned} &|V(x_m, t_m) - V(x_0, t_0)| \\ &\quad = \sup_{\tau \geq 0} \{d(p(\tau + t_m, x_m, t_m), M)e^{\lambda\alpha\tau}\} - \sup_{\tau \geq 0} \{d(p(\tau + t_0, x_0, t_0), M)e^{\lambda\alpha\tau}\} \\ &\quad \leq \sup_{0 \leq \tau \leq L} \{|d(p(\tau + t_m, x_m, t_m), M) - d(p(\tau + t_0, x_0, t_0), M)|e^{\lambda\alpha\tau}\} + 2\varepsilon. \end{aligned}$$

Because the motions in S are continuous with respect to initial conditions, $p(t, x_m, t_m)$ converges to $p(t, x_0, t_0)$ uniformly on the compact set $[0, L]$ as $m \to \infty$; that is, there exists an $m_0 \in \mathbb{N}$ such that

$$\sup_{0 \leq \tau \leq L} |d(p(\tau + t_m, x_m, t_m), M) - d(p(\tau + t_0, x_0, t_0), M)|e^{\lambda\alpha\tau} < \varepsilon$$

for all $m > m_0$. Therefore V is continuous in $X_1 \times \mathbb{R}^+$. □

3.7 Converse Theorems for Discrete-Time Dynamical Systems

In the present section we address local converse theorems for discrete-time systems.

Our first result, concerning uniform stability, is similar to the converse theorems for uniform stability for DDS and continuous dynamical systems.

Theorem 3.7.1 Let $\{\mathbb{N}, X, A, S\}$ be a discrete-time dynamical system and let $M \subset A$ be a closed invariant set, where A is assumed to be a neighborhood of M. Suppose that S satisfies Assumption 3.5.1 (modified in the obvious way for discrete-time systems). Let (S, M) be *uniformly stable*. Then there exist neighborhoods A_1 and

X_1 of M such that $A_1 \subset X_1 \subset A$ and a mapping $V\colon X_1 \times \mathbb{N} \to \mathbb{R}^+$ that satisfies the following conditions.

(i) There exist functions $\psi_1, \psi_2 \in \mathcal{K}$ such that

$$\psi_1(d(x, M)) \le V(x, n) \le \psi_2(d(x, M)) \tag{3.7.1}$$

for all $(x, n) \in X_1 \times \mathbb{N}$.

(ii) For every $p(\cdot, a, n_0) \in S$ with $a \in A_1$, $V(p(n, a, n_0), n)$ is nonincreasing for all $n \in \mathbb{N}_{n_0}$ (i.e., $n \ge n_0, n, n_0 \in \mathbb{N}$).

Proof. The proof is similar to the proof of Theorem 3.5.1 and is not repeated here. □

Theorem 3.7.2 Let $\{\mathbb{N}, X, A, S\}$ be a discrete-time dynamical system and let $M \subset A$ be a closed invariant set, where A is assumed to be a neighborhood of M. Assume that S satisfies Assumption 3.5.1 and that for every $(a, n_0) \in A \times \mathbb{N}$ there exists a *unique* motion $p(\cdot, a, n_0) \in S$ that is defined for all $n \in \mathbb{N}_{n_0}$ (i.e., $n \ge n_0, n, n_0 \in \mathbb{N}$). Let (S, M) be *uniformly asymptotically stable*. Then there exist neighborhoods A_1 and X_1 of M such that $A_1 \subset X_1 \subset A$ and a mapping $V\colon X_1 \times \mathbb{N} \to \mathbb{R}^+$ that satisfies the following conditions.

(i) There exist functions $\psi_1, \psi_2 \in \mathcal{K}$ such that

$$\psi_1(d(x, M)) \le V(x, n) \le \psi_2(d(x, M)) \tag{3.7.2}$$

for all $(x, n) \in X_1 \times \mathbb{N}$.

(ii) There exists a function $\psi_3 \in \mathcal{K}$ such that for all $p(\cdot, a, n_0) \in S$ and for all $n \in \mathbb{N}_{n_0}$, we have

$$DV(p(n, a, n_0), n) \le -\psi_3(d(p(n, a, n_0), M)) \tag{3.7.3}$$

where $a \in A_1$, and

$$DV(p(n, a, n_0), n) = V(p(n+1, a, n_0), n+1) - V(p(n, a, n_0), n). \tag{3.7.4}$$

Proof. By Lemma 3.10.5 (refer to Problem 3.10.17, Section 3.10), there exist a function $\varphi \in \mathcal{K}$ defined on $[0, h_0]$ for some $h_0 > 0$, and a function $\sigma \in \mathcal{L}$ defined on $\mathbb{R}^+$, such that for all $n \in \mathbb{N}_{n_0}^+$,

$$d(p(n, a, n_0), M) < \varphi(d(a, M))\sigma(n - n_0) \tag{3.7.5}$$

for all $p(\cdot, a, n_0) \in S$ whenever $d(a, M) < h_0$. Let $X_1 = \{x \in A\colon d(x, M) < h_0\}$ and let

$$A_1 = \begin{cases} \{a \in X_1\colon d(a, M) < \varphi^{-1}(h_0)\} & \text{if } \varphi(h_0) > h_0, \\ X_1 & \text{otherwise.} \end{cases}$$

We define

$$V(x, n_0) = \sum_{j=n_0}^{\infty} u(d(p(j, x, n_0), M)) \tag{3.7.6}$$

where $u \in \mathcal{K}$ is to be determined later and is such that the summation converges for all $(x, n_0) \in X_1 \times \mathbb{N}$. For $p(\cdot, a, n_0) \in S$, $p(n, p(n_1, a, n_0), n_1) = p(n, a, n_0)$ because of Assumption 3.5.1 and the uniqueness of the motion $p(\cdot, a, n_0)$. Therefore, the summation in the right-hand side of (3.7.6) is independent of n_0 for $x = p(n, a, n_0)$ where $a \in A_1$.

To determine how to choose $u \in \mathcal{K}$ so that the summation in (3.7.6) converges for all $(x, n_0) \in X_1 \times \mathbb{N}$, we apply Lemma 3.5.1. It follows from (3.7.5) that for any $(x, n_0) \in X_1 \times \mathbb{N}$, we have

$$\begin{aligned} u(d(p(n, x, n_0), M)) &< u(\varphi(d(x, M))\sigma(n - n_0)) \\ &\leq [u(\varphi(d(x, M))\sigma(0))]^{1/2}[u(\varphi(h_0)\sigma(n - n_0))]^{1/2}. \end{aligned} \tag{3.7.7}$$

Let $\beta(\tau) = \varphi(h_0)\sigma(\tau)$. Then $\beta \in \mathcal{L}$. Hence, by Lemma 3.5.1, there exists a function $\alpha \in \mathcal{K}$ defined on $\mathbb{R}^+$ such that

$$\sum_{i=0}^{\infty} \alpha(\beta((n_0 + i) - n_0)) = \sum_{j=n_0}^{\infty} \alpha(\beta(j - n_0)) < \infty.$$

If we define $u(r) = [\alpha(r)]^2$, then it follows that

$$[u(\varphi(h_0)\sigma(n - n_0))]^{1/2} = [\alpha(\varphi(h_0)\sigma(n - n_0))]^{1/2} = \alpha(\beta(n - n_0)).$$

Hence, we conclude that

$$\begin{aligned} V(x, n_0) &= \sum_{j=n_0}^{\infty} u(d(p(j, x, n_0), M)) \\ &< \sum_{j=n_0}^{\infty} [u(\varphi(d(x, M))\sigma(0))]^{1/2}[u(\varphi(h_0)\sigma(j - n_0))]^{1/2} \\ &= [u(\varphi(d(x, M))\sigma(0))]^{1/2} \sum_{j=n_0}^{\infty} \alpha(\beta(j - n_0)) \\ &< [u(\varphi(d(x, M))\sigma(0))]^{1/2}[1 + 1/(1 - e^{-1})]. \end{aligned}$$

If we define $\psi_2 \in \mathcal{K}$ by

$$\psi_2(r) = [u(\varphi(r)\sigma(0))]^{1/2}[1 + 1/(1 - e^{-1})],$$

then it follows that $d(x, M) \leq V(x, n_0) \leq \psi_2(d(x, M))$. Thus we have proved condition (i) of the theorem.

For any $p(\cdot, a, n_0) \in S$ and any $n > n_0$, it follows from the uniqueness assumption that

$$V(p(n, a, n_0), n) = \sum_{j=n}^{\infty} u(d(p(j, p(n, a, n_0), n), M)) = \sum_{j=n}^{\infty} u(d(p(j, x, n_0), M)).$$

Along each motion $p(\cdot, a, n_0) \in S$, we have that

$$\begin{aligned} DV(p(n,a,n_0),n) &= \sum_{j=n+1}^{\infty} u(d(p(j,a,n_0),M)) - \sum_{j=n}^{\infty} u(d(p(j,a,n_0),M)) \\ &= -\, u(d(p(n,x,n_0),M)) \end{aligned}$$

for all $(a, n_0) \in A_1 \times \mathbb{N}$ and $n \geq n_0$; that is, V satisfies condition (ii) of the theorem. This concludes the proof of the theorem. □

The hypotheses in our next result are not exactly symmetric with the corresponding assumptions in Theorem 3.4.3. However, they do provide a set of necessary conditions for exponential stability of (S, M).

Theorem 3.7.3 Let $\{\mathbb{N}, X, A, S\}$ be a discrete-time dynamical system and let $M \subset A$ be a closed invariant set, where A is assumed to be a neighborhood of M. Assume that S satisfies Assumption 3.5.1 and that for every $(a, n_0) \in A \times \mathbb{N}$ there exists a unique motion $p(\cdot, a, n_0) \in S$ that is defined for all $n \in \mathbb{N}_{n_0}$. Let (S, M) be *exponentially stable*. Then there exist neighborhoods A_1 and X_1 of M such that $A_1 \subset X_1 \subset A$ and a mapping $V\colon X_1 \times \mathbb{N} \to \mathbb{R}^+$ that satisfies the following conditions.

(i) There exist functions $\psi_1, \psi_2 \in \mathcal{K}$ such that

$$\psi_1(d(x,M)) \leq V(x,n) \leq \psi_2(d(x,M)) \tag{3.7.8}$$

for all $(x, n) \in X_1 \times \mathbb{N}$.

(ii) There exists a constant $c > 0$ such that for all $p(\cdot, a, n_0) \in S$ and for all $n \in \mathbb{N}_{n_0}$, we have

$$DV(p(n,a,n_0),n) \leq -cV(p(n,a,n_0),n) \tag{3.7.9}$$

where $a \in A_1$, $c > 0$ is a constant and $DV(p(n, a, n_0), n)$ is defined in (3.7.4).

Proof. By Lemma 3.10.6 (refer to Problem 3.10.18, Section 3.10), there exist a function $\varphi \in \mathcal{K}$, defined on $[0, h_0]$ for some $h_0 > 0$, and a constant $\alpha > 0$ such that

$$d(p(n,a,n_0),M) < \varphi(d(a,M))e^{-\alpha(n-n_0)} \tag{3.7.10}$$

for all $p(n, a, n_0) \in S$ and $n \geq n_0$ whenever $d(a, M) < h_0$.

Let $X_1 = \{x \in A\colon d(x, M) < h_0\}$ and let

$$V(x,n_0) = \sup_{n' \geq n_0} \{d(p(n',x,n_0),M)e^{\alpha(n'-n_0)}\} \tag{3.7.11}$$

for all $(x, n_0) \in X_1 \times \mathbb{N}$. Let $A_1 = \{a \in X_1 : d(a, M) < \varphi^{-1}(h_0)\}$ if $\varphi(h_0) > h_0$ and $A_1 = X_1$ otherwise. Then for $a \in A_1$ and $n_0 \in \mathbb{N}$, we have by Assumption 3.5.1 and the uniqueness of the motions that

$$\begin{aligned} V(p(n,a,n_0),n) &= \sup_{n' \geq n} \{d(p(n',p(n,a,n_0),n),M)e^{\alpha(n'-n)}\} \\ &= \sup_{n' \geq n} \{d(p(n',a,n_0),M)e^{\alpha(n'-n)}\}. \end{aligned}$$

Therefore, we have

$$
\begin{aligned}
V(p(n+1,a,n_0),n+1) &= \sup_{n'\geq n+1}\left\{d(p(n',a,n_0),M)e^{\alpha(n'-(n+1))}\right\}\\
&= \sup_{n'\geq n+1}\left\{d(p(n',a,n_0),M)e^{\alpha(n'-n)}\right\}e^{-\alpha}\\
&\leq \sup_{n'\geq n}\left\{d(p(n',a,n_0),M)e^{\alpha(n'-n)}\right\}e^{-\alpha}\\
&= V(p(n,a,n_0),n)e^{-\alpha}. \qquad (3.7.12)
\end{aligned}
$$

Equation (3.7.12) yields

$$DV(p(n,a,n_0),n) \leq -(1-e^{-\alpha})V(p(n,a,n_0),n).$$

Finally, (3.7.10) and (3.7.11) imply that

$$d(x,M) \leq V(x,n) \leq \varphi(d(x,M))$$

for all $(x,n)\in X_1\times\mathbb{N}$. This concludes the proof of the theorem. □

We conclude by noting that converse theorems for discrete-time dynamical systems for *uniform boundedness*, *uniform ultimate boundedness*, *uniform asymptotic stability in the large*, *exponential stability in the large*, and *instability* can also be established, using the methodology employed in the preceding results.

3.8 Appendix: Some Background Material on Differential Equations

In this section we present a result that is required in the proof of some of the results of Section 3.3, relating the maximal solution of (I_E),

$$\dot{x} = g(t,x), \qquad x(t_0) = x_0 \qquad (I_E)$$

to the solutions of (EI),

$$Dx \leq g(t,x) \qquad (EI)$$

where $g\in C[\mathbb{R}^+\times\mathbb{R}^l,\mathbb{R}^l]$, D denotes a Dini derivative, and inequality of vectors is to be interpreted componentwise. In the proof of the main result of this section, we require several preliminary results that we state and prove first.

Definition 3.8.1 Let Ω be a connected set in $\mathbb{R}^l$. A function $g\colon \mathbb{R}^+\times\Omega\to\mathbb{R}^l$ is said to be *quasi-monotone nondecreasing* if for each component g_j of g, $j=1,\dots,l$, the inequality $g_j(t,y)\leq g_j(t,z)$ is true whenever $y,z\in\Omega$ and $y_i\leq z_i$ for all $i\neq j$, $i,j=1,\dots,l$ and $y_j=z_j$. □

We note that when g is a scalar-valued function, it is automatically quasi-monotone nondecreasing.

Definition 3.8.2 A solution φ_M of the initial value problem (I_E) is called a *maximal solution* on $[t_0, t_0 + c)$ if for any solution φ defined on $[t_0, t_0 + c)$ it is true that $\varphi_M(t) \geq \varphi(t)$ for all $t \in [t_0, t_0 + c)$, where $c > 0$ and the inequality (for vectors) is understood to be componentwise. □

A *minimal solution* of (I_E) on some interval is defined similarly. By definition, a maximal (resp., minimal) solution of (I_E), if it exists, must be *unique*.

Lemma 3.8.1 Let $g \in C[\mathbb{R}^+ \times \Omega, \mathbb{R}^l]$, let $v, w \in C\big[[t_0, t_0 + c), \mathbb{R}^l\big]$, $t_0 \in \mathbb{R}^+$, and $c > 0$, and assume that the following conditions are true.

(i) g is quasi-monotone nondecreasing.

(ii) $v(t_0) < w(t_0)$.

(iii) $D_- v(t) \leq g(t, v(t))$ and $D_- w(t) > g(t, w(t))$ for $t \in (t_0, t_0 + c)$, where D_- denotes the lower-left Dini derivative.

Then it is true that

$$v(t) < w(t) \tag{3.8.1}$$

for $t \in [t_0, t_0 + c)$.

Proof. Let $u(t) = w(t) - v(t)$. Then condition (ii) reads as $u(t_0) > 0$. Suppose that the assertion (3.8.1) is not true. Then the set

$$F = \bigcup_{i=1}^{l} \Big\{ t \in [t_0, t_0 + c) \colon u_i(t) \leq 0 \Big\} \neq \emptyset.$$

Let $t_1 = \inf F$. Because $u_i(t_0) > 0$, $1 \leq i \leq l$, and $u \in C\big[[t_0, t_0 + c), \mathbb{R}^l\big]$, it is clear that $t_1 > t_0$. The set F is closed, and thus $t_1 \in F$. We now show that there exists a $j \in \{1, 2, \ldots, l\}$ such that

$$u_j(t_1) = 0. \tag{3.8.2}$$

If (3.8.2) is not true (i.e., if $u_i(t_1) < 0$ for all $1 \leq i \leq l$), then $u_i(t) < 0$ in a sufficiently small neighborhood to the left of t_1 by the continuity of u. This contradicts the definition of t_1, and therefore (3.8.2) holds. Moreover, $t_1 = \inf F$ implies that

$$u_i(t_1) \geq 0, \qquad i \neq j \tag{3.8.3}$$

and

$$D_- u_j(t_1) \leq 0. \tag{3.8.4}$$

Combining (3.8.4) and condition (iii), we obtain

$$g_j(t_1, w(t_1)) < g_j(t_1, v(t_1)). \tag{3.8.5}$$

On the other hand, (3.8.2), (3.8.3), and the quasi-monotonicity condition imply that

$$g_j(t_1, w(t_1)) > g_j(t_1, v(t_1))$$

which contradicts (3.8.5). This concludes the proof. □

The above result makes possible the proof of the next result.

Lemma 3.8.2 Let $v, f \in C\big[[t_0, t_0 + c), \mathbb{R}^l\big]$. If for a fixed Dini derivative D it is true that $Dv(t) \leq f(t)$ for $t \in [t_0, t_0 + c)$, $c > 0$, then

$$D_- v(t) \leq f(t) \tag{3.8.6}$$

for $t \in (t_0, t_0 + c)$.

Proof. Because $D_- v(t) \leq D^- v(t)$ and $D_+ v(t) \leq D^+ v(t)$, we only need to prove the lemma for $D = D_+$. Let

$$u(t) = v(t) - \int_{t_0}^{t} f(s)ds.$$

Then $D_+ u(t) = D_+ v(t) - f(t) \leq 0$ for $t \in [t_0, t_0 + c)$. We want to prove that $u(t)$ is nonincreasing on $[t_0, t_0 + c)$, which is equivalent to proving that $m(t) = -u(-t)$ is nonincreasing on $(-t_0 - c, -t_0]$. Note that $D_- m(t) = D^+ u(-t) \leq 0$. We apply Lemma 3.8.1 to show that for any $t_1, t_2 \in (-t_0 - c, -t_0]$, $t_1 < t_2$, $m(t_1) \geq m(t_2)$. Let $w(t) = m(t_1) + \varepsilon(t - t_1 + \varepsilon)$ for $t \in [t_1, -t_0]$ where $\varepsilon > 0$. Then $D_- w(t) = \varepsilon > 0$, $D_- m(t) \leq 0$, and $m(t_1) < w(t_1) + \varepsilon^2$. By Lemma 3.8.1, $m(t) < w(t)$ for all $t \in [t_1, -t_0]$. In particular, $m(t_2) < w(t_2) = m(t_1) + \varepsilon(t_2 - t_1 + \varepsilon)$. Because $\varepsilon > 0$ is arbitrary, we obtain $m(t_2) \leq m(t_1)$ by letting $\varepsilon \to 0$.

We have proved that $u(t)$ is nondecreasing on $[t_0, t_0 + c)$. Therefore, $D_- u(t) \leq 0$ and $D_- v(t) = D_- u(t) + f(t) \leq f(t)$ for $t \in (t_0, t_0 + c)$. □

We require one more preliminary result.

Lemma 3.8.3 Let $g \in C[\mathbb{R}^+ \times \Omega, \mathbb{R}^l]$ and assume that g is quasi-monotone nondecreasing. Then for each $(t_0, x_0) \in \mathbb{R}^+ \times \mathbb{R}^l$, there exists a $c > 0$ such that the maximal solution of (I_E) exists on $[t_0, t_0 + c]$.

Proof. Let

$$D_{a,b} = D_{a,b}(t_0, x_0) = \big\{(t, x) \in \mathbb{R}^+ \times \Omega\colon t_0 \leq t \leq t_0 + a, \;\; |x - x_0| \leq b\big\}.$$

Inasmuch as g is continuous, we may assume that $|g(t, x)| \leq M$ for all $(t, x) \in D_{a,b}$. According to Theorem 2.3.1 and Problem 2.14.8, we may choose $c_1 = \min\{a, b/M\}$ such that (I_E) has a solution defined on $[t_0, t_0 + c_1]$.

Now consider the differential equation with the initial condition given by

$$\dot{y} = g(t, y) + \varepsilon, \qquad y(t_0) = x_0 + \varepsilon \tag{3.8.7}$$

where $0 \le \varepsilon < b/2$. We note that for

$$D'_{a,(b/2)} = D'_{a,(b/2)}(t_0, x_0+\varepsilon) = \left\{(t,y) \in \mathbb{R}^+ \times \Omega \colon t_0 \le t \le t_0 + a,\ |x - x_0| \le \frac{b}{2}\right\}$$

we have $|g(t,y) + \varepsilon| \le M + b/2$ for all $(t,y) \in D'_{a,(b/2)}$. Therefore, (3.8.7) has a solution $y(t,\varepsilon)$ defined on $[t_0, t_0 + c]$, where

$$c = \min\left\{a, \frac{b/2}{M + b/2}\right\} = \min\left\{a, \frac{b}{2M + b}\right\}.$$

For $0 < \varepsilon_2 < \varepsilon_1 \le \varepsilon$, Lemma 3.8.1 implies that $y(t,0) < y(t,\varepsilon_2) < y(t,\varepsilon_1)$ for $t \in [t_0, t_0 + c]$. Therefore, $\lim_{\varepsilon \to 0^+} y(t,\varepsilon) = y^*(t)$ exists and the convergence is uniform for $t \in [t_0, t_0 + c]$. Hence $y^*(t)$ is a solution of (I_E), because

$$\begin{aligned} y^*(t) &= \lim_{\varepsilon \to 0^+} y(t,\varepsilon) \\ &= \lim_{\varepsilon \to 0^+} \left(x_0 + \varepsilon + \int_{t_0}^{t} (g(s, y(s,\varepsilon)) + \varepsilon) ds\right) \\ &= x_0 + \int_{t_0}^{t} g(s, y^*(s)) ds. \end{aligned}$$

Because $y(t,\varepsilon) > y(t,0)$ for $t \in [t_0, t_0 + c]$ we obtain $y^*(t) \ge y(t,0)$ for $t \in [t_0, t_0 + c]$. Because $y(t,0)$ is any solution of (I_E) defined on $[t_0, t_0 + c]$, it follows that y^* is a maximal solution of (I_E). □

The above result concerns the *local existence* of a maximal solution for (I_E). The existence of a *noncontinuable* maximal solution follows by invoking Zorn's lemma (see, e.g., [11]).

We now state and prove the main result of this section.

Theorem 3.8.1 *(Comparison Theorem)* Assume that $g \in C[\mathbb{R}^+ \times \Omega, \mathbb{R}^l]$ is quasi-monotone nondecreasing. Let $x(t)$ be a solution of (EI) defined on $[t_0, t_0 + c]$, $c > 0$, and let $r(t)$, defined on $[t_0, t_0 + c]$, be the maximal solution of (I_E), where $r(t_0) = x(t_0) = x_0$. Then

$$x(t) \le r(t)$$

for all $t \in [t_0, t_0 + c)$.

Proof. Let

$$F = \{t' \in [t_0, t_0 + c) \colon x(t) \le r(t) \text{ for all } t \in [t_0, t']\}.$$

Then $t_0 \in F$, because $x(t_0) = r(t_0)$. It suffices to show that $\sup F = t_0 + c$. If this is not true (i.e., if $\sup F = t_1 < t_0 + c$), then $r(t_1) \ge x(t_1)$. We consider

$$\dot{y} = g(t,y) + \varepsilon, \qquad y(t_1) = r(t_1) + \varepsilon \tag{3.8.8}$$

where $\varepsilon > 0$ is sufficiently small. By the proof of Lemma 3.8.3, there exists a $c_1 > 0$ such that solutions $y(t, \varepsilon)$ of (3.8.8) are defined on $[t_1, t_1 + c_1]$, and $\lim_{\varepsilon \to 0^+} y(t, \varepsilon)$ exists as the maximal solution of

$$\dot{y} = g(t, y), \qquad y(t_1) = r(t_1)$$

for $t \in [t_1, t_1 + c_1]$. By the uniqueness of the maximal solution, $r(t) = \lim_{\varepsilon \to 0^+} y(t, \varepsilon)$ for $t \in [t_1, t_1 + c_1]$. We assume without loss of generality that $t_1 + c_1 < t_0 + c$ (for otherwise, we choose c_1 smaller).

Now $D_- y(t, \varepsilon) = \dot{y}(t, \varepsilon) > g(t, y(t, \varepsilon))$ together with Lemma 3.8.2 implies that $D_- x(t) \leq g(t, x(t))$ for all $t \in (t_1, t_1 + c_1)$. In addition, $y(t_1, \varepsilon) = r(t_1) + \varepsilon > r(t_1) \geq x(t_1)$. By Lemma 3.8.1, $y(t, \varepsilon) > x(t)$ for all $t \in [t_1, t_1 + c_1)$. Letting $\varepsilon \to 0^+$, we see that $r(t) \geq x(t)$ for all $t \in [t_1, t_1 + c_1)$. Therefore, for any $t' \in (t_1, t_1 + c_1)$, we have that $t' \in F$. But this contradicts the fact that $t_1 = \sup F$. This concludes the proof of the theorem. □

In the scalar case ($l = 1$), we can forgo the quasi-monotone condition.

3.9 Notes and References

The material given in Section 3.1 is standard fare in the qualitative analysis of dynamical systems (see, e.g., Zubov [15], Hahn [2], and Michel *et al.* [10]).

The stability and boundedness results for discontinuous dynamical systems presented in Section 3.2 were first reported in Ye [12] and Ye *et al.* [13], with subsequent developments given in Hou [3], Hu [5], Michel [8], and Michel and Hu [9].

The Principal Lyapunov Stability Results given in Sections 3.3 and 3.4 are included in several texts (see, e.g., Hahn [2], Zubov [15], and Michel *et al.* [10]). In [2] and [15], these results are proved using the basic stability and boundedness definitions and fundamental methods of analysis, whereas in [10], these results are established by invoking a comparison theory, making use of stability preserving mappings. Our approach of proving these results by using the stability and boundedness results for DDS established in Section 3.2 (and thus, establishing a unifying stability theory for discontinuous, continuous, and discrete-time dynamical systems) is novel and new (refer to Hou and Michel [4]).

The converse theorems for DDS presented in Section 3.5 were originally established in [12] and [13] with subsequent refinements and developments given in Hou [3], Hu [5], Michel [8], and Michel and Hu [9].

The converse theorems for uniform stability, Theorem 3.6.1 and Theorem 3.7.1, are in the spirit of results given in Zubov [15]. The converse theorems for uniform asymptotic stability, Theorem 3.6.2 and Theorem 3.7.2 and their proofs, including Lemma 3.6.1, are adaptations of material given in Hahn [2] and the converse theorems for exponential stability, Theorem 3.6.3 and Theorem 3.7.3, are based on a result in Massera [7].

References for the background material on differential equations given in Section 3.8 include Lakshmikantham and Leela [6], Miller and Michel [11], and Michel *et al.* [10].

3.10 Problems

Problem 3.10.1 Consider the autonomous system of first-order ordinary differential equations

$$\dot{x} = f(x) \tag{3.10.1}$$

where $f \in C[\mathbb{R}^n, \mathbb{R}^n]$. Assume that there exists a function $V \in C^1[\mathbb{R}^n, \mathbb{R}]$ such that

$$\dot{V}_{(3.10.1)}(x) \triangleq [\nabla V(x)]^T f(x) = 0$$

for all $x \in \mathbb{R}^n$, where $\nabla V(x) = \left[\partial V/\partial x_1, \ldots, \partial V/\partial x_n\right]^T$. Let

$$B_\lambda = \{x \in \mathbb{R}^n : V(x) = \lambda\} \tag{3.10.2}$$

$$C_\lambda = \{x \in \mathbb{R}^n : V(x) \leq \lambda\} \tag{3.10.3}$$

and

$$D_\lambda = \{x \in \mathbb{R}^n : V(x) \geq \lambda\}. \tag{3.10.4}$$

It should be noted that each of these sets may consist of several disjoint component sets. Prove that the sets B_λ, C_λ, and D_λ are invariant with respect to (3.10.1). Prove that each disjoint component set of B_λ, C_λ, and D_λ is invariant with respect to (3.10.1). □

Problem 3.10.2 Consider the autonomous system of first-order difference equations

$$x(k+1) = f(x(k)) \tag{3.10.5}$$

where $k \in \mathbb{N} = \{0, 1, 2, \ldots\}$ and $f : \mathbb{R}^n \to \mathbb{R}^n$. Assume that there exists a function $V : \mathbb{R}^n \to \mathbb{R}$ such that

$$DV_{(3.10.5)}(x) \triangleq V(f(x)) - V(x) = 0$$

for all $x \in \mathbb{R}^n$. Let B_λ, C_λ, and D_λ be defined by (3.10.2), (3.10.3), and (3.10.4), respectively. Prove that B_λ, C_λ, and D_λ are invariant with respect to (3.10.5). Prove that each disjoint component set of B_λ, C_λ, and D_λ is invariant with respect to (3.10.5). □

Problem 3.10.3 For a dynamical system $\{\mathbb{R}^+, X, A, S\}$ assume that there exists a function $V \in C[X, \mathbb{R}]$ such that

$$D^+V_{(S)}(x) \triangleq \overline{\lim_{\Delta t \to 0^+}} \sup_{\substack{p(\cdot, x, t_0) \in S \\ p(t_0, x, t_0) = x}} \frac{1}{\Delta t}\{V(p(t+\Delta t, x, t_0)) - V(p(t, x, t_0))\} \leq 0$$

for all $x \in X$. Let C_λ be defined by (3.10.3). Prove that C_λ is an invariant set with respect to S. □

In Sections 3.3 and 3.4 we proved the Principal Lyapunov and Lagrange stability results for continuous dynamical systems and discrete-time dynamical systems using in most cases the corresponding results for DDS. In the following, we ask the reader to prove these results, using definitions of stability and boundedness (as was done in proving the stability and boundedness results for DDS in Section 3.2).

Problem 3.10.4 Prove Theorems 3.3.1 and 3.4.1 by using the definition of *uniform stability* (given in Definition 3.1.6). □

Problem 3.10.5 Prove Theorems 3.3.2 and 3.4.2 by using the definition of *uniform asymptotic stability* (given in Definition 3.1.9). □

Problem 3.10.6 Prove Theorems 3.3.3 and 3.4.3 by using the definition of *exponential stability* (given in Definition 3.1.10). □

Problem 3.10.7 Prove Theorems 3.3.4 and 3.4.4 by using the definition of *uniform boundedness* (given in Definition 3.1.13). □

Problem 3.10.8 Prove Theorems 3.3.5 and 3.4.5 by using the definition of *uniform ultimate boundedness* (given in Definition 3.1.14). □

Problem 3.10.9 Prove Theorems 3.3.6 and 3.4.6 by using the definition of *uniform asymptotic stability in the large* (given in Definition 3.1.16). □

Problem 3.10.10 Prove Theorems 3.3.7 and 3.4.7 by using the definition of *exponential stability in the large* (given in Definition 3.1.17). □

Problem 3.10.11 Prove Theorems 3.3.8 and 3.3.10 and Theorems 3.4.8 and 3.4.10 by using the definition of *instability* (given in Definition 3.1.18). □

Problem 3.10.12 Prove Theorems 3.3.9 and 3.4.9 by using the definition of *complete instability* (given in Definition 3.1.20). □

For most of the boundedness and stability concepts that we introduced in Section 3.1 there are *equivalent* definitions which frequently make the proofs of the stability and boundedness results easier and more systematic. These definitions involve certain comparison functions whose properties we ask the reader to explore in the next two problems.

Problem 3.10.13 Prove the following results.

Lemma 3.10.1 [2] A continuous function $\sigma\colon [s_1, \infty) \to \mathbb{R}^+$ is said to belong to *class* $\mathcal{L}$ if σ is strictly decreasing on $[s_1, \infty)$ and if $\lim_{s\to\infty} \sigma(s) = 0$ where $s_1 \in \mathbb{R}^+$ (refer to Definition 3.5.1). Show that the functions of class $\mathcal{K}$, class $\mathcal{K}_\infty$, and class $\mathcal{L}$ possess the following properties.

(i) If $\varphi_1, \varphi_2 \in \mathcal{K}$, then $\varphi_1 \circ \varphi_2 \in \mathcal{K}$, where $(\varphi_1 \circ \varphi_2)(r) = \varphi_1(\varphi_2(r))$.

(ii) If $\varphi \in \mathcal{K}$ and $\sigma \in \mathcal{L}$, then $\varphi \circ \sigma \in \mathcal{L}$.

(iii) If $\varphi \in \mathcal{K}$, then φ^{-1} exists and $\varphi^{-1} \in \mathcal{K}$; also, if $\varphi \in \mathcal{K}_\infty$, then $\varphi^{-1} \in \mathcal{K}_\infty$.

(iv) If $\varphi \in \mathcal{K}$ and if φ is defined on $[0, k^2]$, then there exist $\varphi_1, \varphi_2 \in \mathcal{K}$, defined on $[0, k]$, $k > 0$, such that

$$\varphi(r_1 r_2) \leq \varphi_1(r_1)\varphi_2(r_2)$$

for all $r_1, r_2 \in [0, k]$.

Hint: In (iv), choose $\varphi_1(r) = \varphi_2(r) = \sqrt{\varphi(kr)}$. □

Problem 3.10.14 Prove the following results.

Lemma 3.10.2 [2] A real-valued function $l = l(r, s)$ is said to belong to the *class* $\mathcal{KL}$ if

(i) it is defined for $0 \leq r \leq r_1$ (resp., $0 \leq r < \infty$) and for $0 \leq s_0 \leq s < \infty$; and

(ii) for each fixed s it belongs to class $\mathcal{K}$ with respect to r and it is monotone decreasing to zero as s increases (it need not be strictly monotone decreasing).

Let $l \in \mathcal{KL}$. Show that there exist functions $\varphi \in \mathcal{K}$ and $\sigma \in \mathcal{L}$ such that

$$l(r, s) \leq \varphi(r)\sigma(s)$$

for the following two cases.

(a) $l(r, s)$ is bounded with respect to r (i.e., $l(r, s) \leq l_0(s)$).

(b) $0 \leq r < \infty$ and $l(r, s)/l(r_0, s)$ is monotone decreasing for all $r \geq r_0$ as s increases.

Hint: In (a), assume without loss of generality that $l_0 \in \mathcal{L}$ and choose

$$\varphi(r) = \sqrt{l(r, s_0)}, \qquad \sigma(s) = \sqrt{l_0(s)}.$$

In (b) we have

$$l(r, s) < l(r, s_0) l_0(r_0, s)/l(r_0, s_0).$$ □

In the next four problems we ask the reader to establish several equivalent stability definitions phrased in terms of comparison functions discussed above.

Problem 3.10.15 Prove the following results.

Lemma 3.10.3 [2] Show that (S, M) is *stable* if and only if for each $t_0 \in T$ there exists a function $\varphi \in \mathcal{K}$ defined on $[0, r_0]$, $r_0 > 0$, such that

$$d(p(t, a, t_0), M) \leq \varphi(d, (a, M))$$

for all $p(\cdot, a, t_0) \in S$ and for all $t \in T_{a,t_0}$ whenever $d(a, M) < r_0$, where φ may depend on t_0.

Prove that (S, M) is *uniformly stable* if and only if in the above, φ is independent of t_0. □

Problem 3.10.16 Prove the following results.

Lemma 3.10.4 [2] Show that (S, M) is *attractive* if and only if for each $t_0 \in T$ there exists an $\eta = \eta(t_0) > 0$ such that for each $p(\cdot, a, t_0) \in S$, there exists a function $\sigma \in \mathcal{L}$ (where σ may depend on t_0 and $p(\cdot, a, t_0)$) such that if $d(a, M) < \eta$, then $d(p(t, a, t_0), M) < \sigma(t - t_0)$ for all $t \in T_{a,t_0}$. (The class of functions $\mathcal{L}$ is defined in Problem 3.10.13.)

Show that (S, M) is *uniformly attractive* if and only if the above is true for η independent of t_0 and for σ independent of t_0 and of $p(\cdot, a, t_0)$. □

Problem 3.10.17 Prove the following results.

Lemma 3.10.5 [2] Prove that (S, M) is *asymptotically stable* if and only if for each $t_0 \in T$ there exists a function $\varphi \in \mathcal{K}$ on $[0, r_0]$, $r_0 > 0$, such that for each $p(\cdot, a, t_0) \in S$, there exists a function $\sigma \in \mathcal{L}$ such that if $d(a, M) < r_0$, then

$$d(p(t, a, t_0), M) \leq \varphi(d(a, M))\sigma(t - t_0)$$

for all $t \in T_{a,t_0}$.

Prove that (S, M) is *uniformly asymptotically stable* if and only if the above is true for φ independent of t_0 and for σ independent of t_0 and of $p(\cdot, a, t_0)$.

Prove that (S, M) is *uniformly asymptotically stable in the large* if and only if the above is true for φ independent of t_0 and for σ independent of t_0 and of $p(\cdot, a, t_0)$, and furthermore, $\varphi \in \mathcal{K}_\infty$.

Hint: Use the results of Problem 3.10.14 □

Problem 3.10.18 Prove the following results.

Lemma 3.10.6 [2] Show that (S, M) is *exponentially stable* (resp., *exponentially stable in the large*) if and only if (S, M) is uniformly asymptotically stable (resp., uniformly asymptotically stable in the large) *and* in the statement for uniform asymptotic stability in Problem 3.10.17 $\sigma(s) = e^{-\alpha s}$ with $\alpha > 0$. □

In the next six problems we ask the reader to prove several of the stability results of Sections 3.3 and 3.4 by utilizing the equivalent definitions for stability established above.

Problem 3.10.19 Prove Theorem 3.3.1 by utilizing the equivalent definition of *uniform stability* given in Problem 3.10.15. □

Problem 3.10.20 Prove Theorem 3.3.2 and Theorem 3.3.6 by utilizing the equivalent definitions of *uniform asymptotic stability* and *uniform asymptotic stability in the large* given in Problem 3.10.17. □

Problem 3.10.21 Prove Theorem 3.3.3 and Theorem 3.3.7 by utilizing the equivalent definitions of *exponential stability* and *exponential stability in the large* given in Problem 3.10.18. □

Problem 3.10.22 Prove Theorem 3.4.1 by utilizing the equivalent definition of *uniform stability* given in Problem 3.10.15. □

Problem 3.10.23 Prove Theorem 3.4.2 and Theorem 3.4.6 by utilizing the equivalent definitions of *uniform asymptotic stability* and *uniform asymptotic stability in the large* given in Problem 3.10.17. □

Problem 3.10.24 Prove Theorem 3.4.3 and Theorem 3.4.7 by utilizing the equivalent definitions of *exponential stability* and *exponential stability in the large* given in Problem 3.10.18. □

Problem 3.10.25 Let S be the dynamical system determined by the scalar differential equation

$$\dot{y} = -\psi(y), \qquad y \in \mathbb{R}^+$$

where $\psi \in \mathcal{K}$. Prove that $(S, \{0\})$ is *uniformly asymptotically stable*.
Hint [2]: Let G denote a primitive function of $-1/\psi$. Then

$$y(t) = G^{-1}(t - t_0 + G(y_0))$$

where G^{-1} denotes the inverse of G. If the function $-1/\psi$ is integrable near 0, then there exists a finite number t_1 such that $y(t) = 0$ for all $t \geq t_1$. If the function $-1/\psi$ is not integrable near 0, then $G^{-1} \in \mathcal{L}$ (see Problem 3.10.13 for the definition of class $\mathcal{L}$). In a similar manner as in Problem 3.10.13(iv), conclude that for $s_1 \geq c_1$, $s_2 \geq c_2$ and $\sigma \in \mathcal{L}$, there exist $\sigma_1, \sigma_2 \in \mathcal{L}$ such that

$$\sigma(s_1 + s_2) \leq \sigma_1(s_1)\sigma_2(s_2).$$

Next, apply the above inequality to $G^{-1} \in \mathcal{L}$ with $s_1 = t - t_0$ and $s_2 = G(y_0)$ to obtain

$$y(t) \leq \sigma_1(t - t_0)\sigma_2(G(y_0)).$$

This inequality implies the uniform attractivity of $(S, \{0\})$ (refer to Definition 3.1.7 and Problem 3.10.16). The uniform stability of $(S, \{0\})$ follows from the uniform stability of $(S_1, \{0\})$ where $S \subset S_1$ and S_1 is the dynamical system determined by

$$\dot{y} \leq 0, \qquad y \in \mathbb{R}^+.$$ □

Problem 3.10.26 Prove that if in Problem 3.10.25 $\psi \in \mathcal{K}_\infty$, then $(S, \{0\})$ is *uniformly asymptotically stable in the large*. □

Problem 3.10.27 Let $y \in C\big[[t_0, \infty), \mathbb{R}\big]$ and assume that $y(t_0) > 0$ and that

$$Dy(t) \geq \psi(y(t))$$

for all $t \geq t_0$ where D is a fixed Dini derivative and $\psi \in \mathcal{K}$ is defined on $\mathbb{R}^+$. Show that $\lim_{t\to\infty} y(t) = \infty$.
Hint: Apply Theorem 3.8.1 to show that $y(t) \geq r(t)$ where $r(t)$ is the minimal solution of

$$\dot{y} = \psi(y)$$

with the initial condition $r(t_0) = y(t_0)$. Problem 3.10.27 can now be reduced to proving that $\lim_{t\to\infty} r(t) = \infty$. □

Problem 3.10.28 Let S be the dynamical system determined by the scalar difference equation

$$y_{k+1} \leq y_k - \psi(y_k), \qquad y_k \in \mathbb{R}^+, \qquad \psi \in \mathcal{K},$$

for all $k \in T_{a,k_0}$. Prove that $(S, \{0\})$ is *uniformly asymptotically stable.*
Hint: $(S, \{0\})$ is uniformly stable because every motion of S is a decreasing non-negative sequence. To show that $(S, \{0\})$ is uniformly attractive, first establish that $y_{k+1} - y_{k_0} \leq -(k - k_0 + 1)\psi(y_k)$ and conclude that

$$\psi(y_k) \leq \frac{y_{k_0} - y_{k+1}}{k - k_0 + 1} \leq \frac{y_{k_0}}{k - k_0 + 1}. \tag{3.10.6}$$

Choose $\delta > 0$ in such a manner that ψ^{-1} exists on $[0, \delta]$, and for every $\varepsilon > 0$, $k_0 \in \mathbb{N}$, choose $\tau = \delta/\psi(\varepsilon)$. Then for $k \geq k_0 + \tau$, we have $k - k_0 + 1 \geq \tau + 1 > \tau = \delta/\psi(\varepsilon)$. By (3.10.6), we have that

$$|y_k| = y_k = \psi^{-1}\Big(\frac{y_{k_0}}{k - k_0 + 1}\Big) < \psi^{-1}\Big(\frac{\delta}{k - k_0 + 1}\Big) < \varepsilon$$

for all $k \geq k_0 + \tau$ whenever $|y_{k_0}| = y_{k_0} < \delta$. Therefore, $(S, \{0\})$ is uniformly attractive (see Definition 3.1.9). □

Problem 3.10.29 Prove that if in Problem 3.10.28 $\psi \in \mathcal{K}_\infty$, then $(S, \{0\})$ is *uniformly asymptotically stable in the large.* □

Problem 3.10.30 Prove the assertions made in Examples 3.1.8 and 3.1.9. □

Bibliography

[1] P. J. Antsaklis and A. N. Michel, *Linear Systems*, Boston: Birkhäuser, 2006.

[2] W. Hahn, *Stability of Motion*, Berlin: Springer-Verlag, 1967.

[3] L. Hou, *Qualitative Analysis of Discontinuous Deterministic and Stochastic Dynamical Systems*, Ph.D. Dissertation, University of Notre Dame, Notre Dame, IN, 2000.

[4] L. Hou and A. N. Michel, "Unifying theory for stability of continuous, discontinuous and discrete-time dynamical systems," *Nonlinear Anal. Hybrid Syst.*, vol. 1, no. 2, pp.154-172, 2007.

[5] B. Hu, *Qualitative Analysis of Hybrid Dynamical Systems*, Ph.D. Dissertation, University of Notre Dame, Notre Dame, IN, 1999.

[6] V. Lakshmikantham and S. Leela, *Differential and Integral Inequalities*, vol. I and vol. II, New York: Academic Press, 1969.

[7] J. L. Massera, "Contributions to stability theory," *Ann. Math.*, vol. 64, pp. 182–206, 1956.

[8] A. N. Michel, "Recent trends in the stability analysis of hybrid dynamical systems," *IEEE Trans. Circ. Syst. I: Fund. Theor. Appl.*, vol. 46, pp. 120–134, 1999.

[9] A. N. Michel and B. Hu, "Towards a stability theory of general hybrid dynamical systems," *Automatica*, vol. 35, pp. 371–384, 1999.

[10] A. N. Michel, K. Wang, and B. Hu, *Qualitative Theory of Dynamical Systems-The Role of Stability Preserving Mappings*, 2nd Edition, New York: Marcel Dekker, 2001.

[11] R. K. Miller and A. N. Michel, *Ordinary Differential Equations*, New York: Academic Press, 1982.

[12] H. Ye, *Stability Analysis of Two Classes of Dynamical Systems: General Hybrid Dynamical Systems and Neural Networks with Delays*, Ph.D. Dissertation, University of Notre Dame, Notre Dame, IN, 1996.

[13] H. Ye, A. N. Michel, and L. Hou, "Stability theory for hybrid dynamical systems," *IEEE Trans. Autom. Control*, vol. 43, pp. 461–474, 1998.

[14] T. Yoshizawa, *Stability Theory by Lyapunov's Second Method*, Tokyo: Math. Soc. of Japan, 1966.

[15] V. I. Zubov, *Methods of A. M. Lyapunov and their Applications*, Groningen, The Netherlands: P. Noordhoff, 1964.

Chapter 4

Fundamental Theory: Specialized Stability and Boundedness Results on Metric Spaces

In this chapter we present a number of important specialized stability and boundedness results for dynamical systems defined on metric spaces. These include results for *autonomous dynamical systems* (in Section 4.1), results that comprise the *Invariance Theory* (in Section 4.2), some results that go under the heading of *Comparison Theory* (in Section 4.3), and a result that addresses the *uniqueness of motions* in dynamical systems (in Section 4.4).

Before proceeding with our subject on hand, we would like to remind the reader once more that our definition of dynamical system (Definition 2.2.3) does in general not require that time be reversible in the motions (in contrast to many dynamical systems determined, e.g., by various types of differential equations), nor are the motions required to be unique with respect to initial conditions. For such general systems, when required, we make an assumption that is akin to the semigroup property, but is more general, which essentially requires that for a dynamical system S, any partial motion is also a motion of S, and any composition of two motions is also a motion of S (refer to Assumption 3.5.1). Of course when in a dynamical system the semigroup property holds, then Assumption 3.5.1 is automatically implied.

4.1 Autonomous Dynamical Systems

In the present section we show that under reasonable assumptions, in the case of autonomous dynamical systems, the stability and asymptotic stability of an invariant set $M \subset A$, with respect to S, imply the uniform stability and uniform asymptotic stability of (S, M), respectively. We also establish necessary and sufficient conditions for stability and asymptotic stability of (S, M) for autonomous systems in this section.

Definition 4.1.1 A dynamical system $\{T, X, A, S\}$ is said to be an *autonomous dynamical system* if

(i) every $p(\cdot, a, t_0) \in S$ is defined on $T_{a,t_0} = T \cap [t_0, \infty)$; and

(ii) for each $p(\cdot, a, t_0) \in S$ and for each τ such that $t_0 + \tau \in T$, there exists a motion $p(\cdot, a, t_0 + \tau) \in S$ such that $p(t + \tau, a, t_0 + \tau) = p(t, a, t_0)$ for all $t \in T_{a,t_0}$ and all τ satisfying $t + \tau \in T$. □

Examples of *autonomous* dynamical systems include *linear* and *nonlinear semigroups*. These systems have motions that are *unique* and *continuous* with respect to initial conditions.

In general we do not require that autonomous dynamical systems satisfy the uniqueness property. For example, autonomous dynamical systems determined by *differential inclusions defined on Banach spaces* (refer to Chapter 2) do not satisfy the uniqueness property.

In the next two results we assume that the motions of the dynamical systems are continuous with respect to initial conditions in the sense of Definition 3.6.1.

Theorem 4.1.1 Let $\{\mathbb{R}^+, X, A, S\}$ be an autonomous dynamical system that satisfies Assumption 3.5.1 and for which the motions are continuous with respect to initial conditions. Let $M \subset A$ be a closed and invariant set. If (S, M) is *stable*, then (S, M) is *uniformly stable*.

Proof. Because $\{\mathbb{R}^+, X, A, S\}$ is autonomous, for each $p(\cdot, a, t_0) \in S$, there exists a motion $p(\cdot, a, 0) \in S$ such that $p(t, a, t_0) = p(t - t_0, a, 0)$ for all $t \geq t_0$.

Because (S, M) is *stable*, for every $\varepsilon > 0$ there exists a $\delta = \delta(\varepsilon, 0) > 0$ such that $d(p(t, a, 0), M) < \varepsilon$ for all $t \geq 0$ and all $p(\cdot, a, 0) \in S$ whenever $d(a, M) < \delta$. Therefore, $d(p(t, a, t_0), M) = d(p(t - t_0, a, 0), M) < \varepsilon$ for all $t \geq t_0$. Note that δ is independent of t_0. We have shown that (S, M) is uniformly stable. □

An analogous result and proof of Theorem 4.1.1 for discrete-time dynamical systems can be established by making obvious modifications.

Theorem 4.1.2 Let $\{\mathbb{R}^+, X, A, S\}$ be an autonomous dynamical system for which the motions are continuous with respect to initial conditions and that satisfies Assumption 3.5.1. Let $M \subset A$ be a closed and invariant set and assume that A is compact. If (S, M) is *asymptotically stable*, then (S, M) is *uniformly asymptotically stable*.

Proof. The uniform stability of (S, M) follows from Theorem 4.1.1. We only need to show the uniform attractivity of (S, M); that is, we need to show that there exists a $\delta > 0$, and for every $\varepsilon > 0$ and every $t_0 \in \mathbb{R}^+$, there exists a

$\tau = \tau(\varepsilon) > 0$, independent of t_0, such that $d(p(t, a, t_0), M) < \varepsilon$ for all $t \geq t_0 + \tau$ and for all $p(\cdot, a, t_0) \in S$ whenever $d(a, M) < \delta$. Because $\{\mathbb{R}^+, X, A, S\}$ is autonomous, for each $p(\cdot, a, t_0) \in S$, there exists a motion $p(\cdot, a, 0) \in S$ such that $p(t, a, t_0) = p(t - t_0, a, 0)$ for all $t \geq t_0$. Therefore, it is equivalent to show that there exists a $\delta > 0$, and for every $\varepsilon > 0$, there exists a $\tau = \tau(\varepsilon) > 0$, such that $d(p(t, a, 0), M) < \varepsilon$ for all $t \geq \tau$ and for all $p(\cdot, a, 0) \in S$ whenever $d(a, M) < \delta$.

Assume that (S, M) is not uniformly attractive. In view of the above statement, for every $\delta > 0$, there exists an $\varepsilon > 0$, two sequences $\{a_m : m \in \mathbb{N}\} \subset X$ and $\{t_m : m \in \mathbb{N}\} \subset \mathbb{R}^+$ with $\lim_{m\to\infty} t_m = \infty$ such that $d(a_m, M) < \delta$ and $d(p(t_m, a_m, 0), M) > \varepsilon$ for all $m \in \mathbb{N}$.

Because A is compact, $\{a_m : m \in \mathbb{N}\}$ has a convergent subsequence. Without loss of generality, we may assume that $a_m \to a^* \in A$.

The uniform stability of (S, M) implies that there exists a $\delta^* > 0$ such that $d(p(t, a, t_0), M) < \varepsilon/2$ whenever $d(a, M) < \delta^*$. The attractivity of (S, M) implies that there exists a $\tau > 0$ such that $d(p(t, a^*, 0), M) < \delta^*$ for all $t \geq \tau$. In particular, $d(p(\tau, a^*, 0), M) < \delta^*$. Therefore,

$$d(p(t, a^*, 0), M) = d(p(t, p(\tau, a^*, 0), \tau), M) < \varepsilon/2$$

for all $t \geq \tau$. On the other hand, by continuity with respect to initial conditions, $p(\tau, a_m, 0) \to p(\tau, a^*, 0)$. Together with $\lim_{m\to\infty} t_m = \infty$, there exists an m' such that $t_{m'} > \tau$ and $d(p(\tau, a_{m'}, 0), M) < \delta^*$. Thus,

$$d(p(t_{m'}, a_{m'}, 0), M) = d(p(t_{m'}, p(\tau, a_{m'}, 0), \tau), M) < \varepsilon/2.$$

We have arrived at a contradiction. Therefore, (S, M) is *uniformly asymptotically stable.* □

Similarly as in Theorem 4.1.2, we can also show that when $\{\mathbb{R}^+, X, A, S\}$ is an autonomous dynamical system with motions that are continuous with respect to initial conditions and that satisfies Assumption 3.5.1, and if $M \subset A$ is invariant and A is compact, then if (S, M) is *asymptotically stable in the large*, then (S, M) is *uniformly asymptotically stable in the large*. Also, we can establish an analogous result of Theorem 4.1.2 for discrete-time dynamical systems by making appropriate modifications. In proving converse theorems for the uniform asymptotic stability of invariant sets $M \subset A$ for dynamical systems $\{T, X, A, S\}$, we require in Theorem 3.6.2 ($T = \mathbb{R}^+$) and Theorem 3.7.2 ($T = \mathbb{N}$) that the systems satisfy the uniqueness property of the motions. In the next two results we remove this restriction for autonomous dynamical systems. In doing so, we are able to establish *necessary and sufficient conditions* for *stability* and *asymptotic stability* of invariant sets. In these results, the Lyapunov functions V are independent of t (i.e., $V(x, t) \equiv V(x)$).

Theorem 4.1.3 Let $\{T, X, A, S\}$ be an autonomous dynamical system with $T = \mathbb{R}^+$ or $T = \mathbb{N}$, and let $M \subset A$ be a closed invariant set, where A is assumed to be a neighborhood of M. Assume that S satisfies Assumption 3.5.1. Then (S, M) is *stable* if and only if there exist neighborhoods A_1 and X_1 of M such that $A_1 \subset X_1 \subset A$

and a mapping $V\colon X_1 \to \mathbb{R}^+$ that satisfies the following conditions.

(i) There exist $\psi_1, \psi_2 \in \mathcal{K}$ such that

$$\psi_1(d(x, M)) \le V(x) \le \psi_2(d(x, M))$$

for all $x \in X_1$.

(ii) For every $p(\cdot, a, t_0) \in S$ with $a \in A_1$, $V(p(t, a, t_0))$ is nonincreasing for all $t \in T \cap [t_0, \infty)$.

Proof. (*Sufficiency*) We define S_{A_1} by

$$S_{A_1} = \{p(\cdot, a, t_0) \in S\colon a \in A_1, t_0 \in T\}. \tag{4.1.1}$$

Then $\{T, X, A_1, S_1\}$ is a dynamical system. It follows from Theorem 3.3.1 ($T = \mathbb{R}^+$) or Theorem 3.4.1 ($T = \mathbb{N}$) that (S_{A_1}, M) is stable. Because A_1 is a neighborhood of M, it is straightforward to verify that the stability of (S_{A_1}, M) implies the stability of (S, M).

(*Necessity*) We apply Theorem 3.5.1 (resp., Theorem 3.6.1). It suffices to show that in the proof of that theorem, $V(x, t)$ is independent of t when S is an autonomous system.

Recall that in the proof of Theorem 3.5.1,

$$V(x, t) = \sup\{d(p(t', x, t), M)\colon p(\cdot, x, t) \in S, t' \in T_{x,t}\}.$$

We prove that for any $t_1, t_2 \in T$, $V(x, t_1) = V(x, t_2)$. Let $t_2 = t_1 + \tau$. We note that for either $T = \mathbb{R}^+$ or $T = \mathbb{N}$, $t' \in T \cap [t_1, \infty)$ implies that $t' + \tau \in T \cap [t_2, \infty)$. For every motion $p(\cdot, x, t_1) \in S$, there exists a motion $p(\cdot, x, t_2) \in S$ such that $p(t', x, t_1) = p(t' + \tau, x, t_2)$ for all $t' \in T \cap [t_1, \infty)$. Therefore, by (4.1.1) it follows that $V(x, t_1) \le V(x, t_2)$. In a similar manner as above, it also follows that $V(x, t_2) \le V(x, t_1)$. Therefore, we have $V(x, t_1) = V(x, t_2)$ for any $t_1, t_2 \in T$, which means that $V(x, t)$ is independent of t. □

In the final result of this section, we establish necessary and sufficient conditions for the asymptotic stability of invariant sets for autonomous dynamical systems.

Theorem 4.1.4 Let $\{T, X, A, S\}$ be an autonomous dynamical system with $T = \mathbb{R}^+$ or $T = \mathbb{N}$, and let $M \subset A$ be a closed invariant set, where A is assumed to be a neighborhood of M. Assume that S satisfies Assumption 3.5.1 and that all motions in S are continuous with respect to initial conditions. Then (S, M) is *asymptotically stable* if and only if there exist neighborhoods A_1 and X_1 of M such that $A_1 \subset X_1 \subset A$ and a mapping $V\colon X_1 \to \mathbb{R}^+$ that satisfies the following conditions.

(i) There exist $\psi_1, \psi_2 \in \mathcal{K}$ such that

$$\psi_1(d(x, M)) \le V(x) \le \psi_2(d(x, M))$$

for all $x \in X_1$.

(ii) For every $p(\cdot, a, t_0) \in S$ with $a \in A_1$, $V(p(t, a, t_0))$ is nonincreasing for all $t \in T \cap [t_0, \infty)$ and $\lim_{t\to\infty} V(p(t, a, t_0)) = 0$.

Proof. The necessity and sufficiency of stability follow from Theorem 4.1.3. Therefore, we only need to address the attractivity of (S, M).

(*Sufficiency*) We choose $\eta > 0$ such that $\{a \in A\colon d(a, M) < \eta\} \subset A_1$. Then, whenever $d(a, M) < \eta$, we have

$$\lim_{t\to\infty} \psi_1(d(p(t, a, t_0), M)) \leq \lim_{t\to\infty} V(p(t, a, t_0)) = 0$$

which implies that $\lim_{t\to\infty} d(p(t, a, t_0), M) = 0$. This implies the attractivity of (S, M).

(*Necessity*) If (S, M) is attractive, then there exists an $\eta > 0$ such that

$$\lim_{t\to\infty} d(p(t, a, t_0), M) = 0$$

for all $p(\cdot, a, t_0) \in S$ whenever $d(a, M) < \eta$. Therefore, for every $p(\cdot, a, t_0) \in S$ with $a \in \{x \in A\colon d(a, M) < \eta\}$, we have that

$$\lim_{t\to\infty} V(p(t, a, t_0)) \leq \lim_{t\to\infty} \psi_2(d(p(t, a, t_0), M)) = 0. \qquad \square$$

4.2 Invariance Theory

In the present section we establish sufficient conditions for the asymptotic stability of invariant sets for dynamical systems determined by semigroups defined on metric spaces. These sufficient conditions may be easier to satisfy than the corresponding results given in Sections 3.3 and 3.4. We first need to generalize the notion of a semigroup defined on Banach spaces, presented in Chapter 2. In the following definition, we allow $T = \mathbb{R}^+$, for a *continuous semigroup on metric space* and $T = \mathbb{N}$, for a *discrete-time semigroup on metric space.*

Definition 4.2.1 Let X be a metric space. A family of mappings $G(t)\colon X \to X$, $t \in T$ ($T = \mathbb{R}^+$ or $T = \mathbb{N}$), is said to be a *semigroup* defined on X if

(i) $G(0)x = x$ for all $x \in X$;

(ii) $G(t + s)x = G(t)G(s)x$ for all $t, s \in T$ and $x \in X$; and

(iii) $G(t)x$ is continuous in $x \in X$ for each $t \in T$. $\square$

When $T = \mathbb{R}^+$ and the metric space is a subset of a Banach space, then the above definition coincides with Definition 2.9.5.

As was shown in Chapter 2, semigroups determine dynamical systems, in fact autonomous dynamical systems. We denote a dynamical system determined by a semigroup (as defined above) by S_G.

We require the following concept.

Definition 4.2.2 Let $\{T, X, A, S\}$ be a dynamical system where $T = \mathbb{R}^+$ or $T = \mathbb{N}$. For each motion $p(\cdot, a, t_0) \in S$ which is defined on $T \cap [t_0, \infty)$, the set

$$\omega(p) = \Big\{x \in X : x = \lim_{n\to\infty} p(t_n, a, t_0) \quad \text{where } \{t_n\} \subset T_{a,t_0} \text{ is any increasing sequence such that } \lim_{n\to\infty} t_n = \infty\Big\}$$

is called the *ω-limit set* of the motion $p(\cdot, a, t_0)$. □

It can be shown that

$$\omega(p) = \bigcap_{t \in T\cap[t_0,\infty)} \overline{\{p(t', a, t_0) : t' \in T \cap [t_0, \infty)\}} \tag{4.2.1}$$

where $\overline{B}$ denotes the closure in X of the set B.

In the subsequent results of the present section, we require the following preliminary result concerning limit sets.

Lemma 4.2.1 Let $\{T, X, A, S_G\}$ be a dynamical system determined by semigroup G where $T = \mathbb{R}^+$ or $T = \mathbb{N}$ and G is defined on the metric space $X = A$. For a motion $p(\cdot, a, t_0) \in S_G$, assume that the *trajectory*

$$\gamma^+(p) = \big\{p(t, a, t_0) \in X : t \in T \cap [t_0, \infty)\big\} \subset X_0$$

where X_0 is a compact subset of X. Then the *ω-limit set* $\omega(p)$ is nonempty, compact, and invariant with respect to S_G. Furthermore, $p(t, a, t_0) \to \omega(p)$ as $t \to \infty$.

Proof. By (4.2.1), $\omega(p)$ is closed and $\omega(p) \subset X_0$. Because any closed subset of a compact set is compact (see [3]), it follows that $\omega(p)$ is compact. Furthermore, because $\gamma^+(p) \subset X_0$ and any compact set is sequentially compact (see [3]), it follows that $\omega(p)$ is nonempty.

For any $y \in \omega(p)$, there exists by definition an increasing sequence $\{t_n > t_0\} \subset T$ with $\lim_{n\to\infty} t_n = \infty$ and $t_n \in T \cap [t_0, \infty)$ such that $\lim_{n\to\infty} p(t_n, a, t_0) = y$. Let $u_n(t) = p(t_n + t, a, t_0)$ for all $t \in T$ and $a \in X$. Then $u_n(t) \in X_0$ for all $t \in T$. The compactness of X_0 implies that the sequence of functions $\{u_n\}$ is uniformly bounded on $t \in T$ for $n \in \mathbb{N}$.

When $T = \mathbb{N}$, there exists for each $k \in \mathbb{N}$ a subsequence $\{u_{n_{j,k}}\}$ such that $u_{n_{j,k}}(k) \to u(k)$ as $j \to \infty$. Therefore, for the (diagonalized) subsequence $\{u_{n_{j,j}}\}$, we have that $u_{n_{j,j}}(k) \to u(k)$ as $j \to \infty$ for all $k \in \mathbb{N}$.

For the case $T = \mathbb{R}^+$, we have

$$\begin{aligned} u_n(t) &= p(t_n + t, a, t_0) \\ &= G(t_n + t - t_0)a \\ &= G(t)G(t_n - t_0)a \\ &= G(t)p(t_n, a, t_0). \end{aligned} \tag{4.2.2}$$

Because $p(t_n, a, t_0) \in X_0$ and X_0 is compact, and because $G(t)x$ is uniformly continuous for $(t, x) \in [0, k] \times X_0$, where $k \in \mathbb{N}$, it follows that $u_n(t)$ is equicontinuous for $t \in [0, k]$ (refer to Problem 2.14.7). By the Ascoli–Arzela Lemma, there exists for each fixed $k \in \mathbb{N}$ a subsequence $\{u_{n_{j,k}}\}$ that converges uniformly on $[0, k]$ as $j \to \infty$ (refer to Problem 2.14.7). We have shown that $\{u_{n_{j,j}}\}$ converges to a continuous function, say u, on $\mathbb{R}^+$.

Summarizing, when $T = \mathbb{N}$, there exists a subsequence $\{u_{n_j}\}$ that converges on T to a function $u\colon T \to X$, and when $T = \mathbb{R}^+$, there exists a subsequence $\{u_{n_j}\}$ that converges to $u \in C[T, X]$ on T.

Clearly, for each $t \in T$, $u(t) \in \omega(p)$ because $p(t_{n_j} + t, a, t_0) = u_{n_j}(t) \to u(t)$ as $j \to \infty$.

Now for each $t \in T$,

$$G(t)y = \lim_{n\to\infty} G(t)p(t_n, a, t_0) = \lim_{n\to\infty} u_n(t) = u(t) \tag{4.2.3}$$

where we have used (4.2.2). This implies that $G(t)y = u(t) \in \omega(p)$ for all $t \in T$. Because $y \in \omega(p)$ was arbitrarily chosen at the outset, we have proved that $\omega(p)$ is invariant with respect to S_G.

To complete the proof, we need to show that $p(t, a, t_0) \to \omega(p)$ as $t \to \infty$. If this is not the case, then there is an $\varepsilon > 0$ and an increasing sequence $\{t_m\}$ with $\lim_{m\to\infty} t_m = \infty$ such that $d(p(t_m, a, t_0), \omega(p)) \geq \varepsilon$ for all $m \in \mathbb{N}$. Because $p(t_m, a, t_0) \in X_0$ and because X_0 is compact, there exists a subsequence $\{t_{m_j}\}$ such that $\lim_{j\to\infty} t_{m_j} = \infty$ and such that $\{p(t_{m_j}, a, t_0)\}$ converges to a point, say y_0, in X_0 (see [3]). By definition, $y_0 \in \omega(p)$. On the other hand, $d(p(t_{m_j}, a, t_0), \omega(p)) \geq \varepsilon$ implies that $d(y_0, \omega(p)) \geq \varepsilon$. We have thus arrived at a contradiction. This proves that $p(t, a, t_0) \to \omega(p)$ as $t \to \infty$. This completes the proof of the theorem. □

For a continuous function V we now define the *derivative of V with respect to S_G* when $T = \mathbb{R}^+$ and the *difference of V with respect to S_G* when $T = \mathbb{N}$ in the following manner.

Definition 4.2.3 Let G be a continuous or discrete semigroup on a metric space X and let S_G denote the dynamical system determined by G. For each $V \in C[X_1, \mathbb{R}]$, $X_1 \subset X$, we define a function $D_{(G)}V\colon X_1 \to \mathbb{R}$ in the following manner.

For $T = \mathbb{R}^+$,

$$D_{(G)}V(x) = \overline{\lim_{t\to 0^+}} \left(\frac{1}{t}\right)[V(G(t)x) - V(x)] \tag{4.2.4}$$

and for $T = \mathbb{N}$,

$$D_{(G)}V(x) = V(G(1)x) - V(x). \tag{4.2.5}$$

□

We also require the next preliminary result.

Lemma 4.2.2 Let G be a semigroup defined on a metric space X and let X_1 be a compact subset of X. Let $V \in C[X_1, \mathbb{R}]$ and assume that $D_{(G)}V(x) \leq 0$ for all $x \in X_1$. Then for any $p(\cdot, a, t_0) \in S_G$ such that the trajectory for $p(\cdot, a, t_0)$ is a subset of X_1 (i.e., $\gamma^+(p) \subset X_1$), the following relation holds.

$$\omega(p) \subset Z \triangleq \Big\{x \in X_1 \colon D_{(G)}V(x) = 0\Big\}. \tag{4.2.6}$$

Proof. We first show that $v(t) \triangleq V(p(t, a, t_0))$ is nonincreasing for $t \in T$. This follows immediately because for $T = \mathbb{R}^+$, we have $D^+v(t) = D_{(G)}V(p(t, a, t_0)) \leq 0$ and because for $T = \mathbb{N}$, we have $v(t+1) - v(t) = D_{(G)}V(p(t, a, t_0)) \leq 0$.

Because any continuous function on a compact set is bounded, V is bounded on X_1 and in particular, the nonincreasing function $v(t) = V(p(t, a, t_0))$ is bounded. This implies that $\lim_{t\to\infty} V(p(t, a, t_0)) = v_0 \in \mathbb{R}$ exists.

To prove that for any $y \in \omega(p)$ it is true that $y \in Z$ (i.e., $D_{(G)}V(y) = 0$), it suffices to show that $V(G(t)y)$ is independent of $t \in T$. Indeed, there exists an increasing sequence $\{t_n\} \subset T \cap [t_0, \infty)$ such that

$$V(G(t)y) = \lim_{n\to\infty} V(G(t)p(t_n, a, t_0)) = \lim_{n\to\infty} V(p(t_n + t, a, t_0)) = v_0.$$

This completes the proof. □

We are now in a position to state and prove the main *invariance results* for dynamical systems determined by semigroups on metric spaces.

Theorem 4.2.1 Let G be a continuous semigroup ($T = \mathbb{R}^+$) or a discrete semigroup ($T = \mathbb{N}$) defined on a metric space X, let S_G be a dynamical system determined by G, and let X_1 be a compact subset of X. Assume that there exists a $V \in C[X_1, \mathbb{R}]$ such that $D_{(G)}V(x) \leq 0$ for all $x \in X_1$ (where $D_{(G)}V$ is defined by (4.2.4) when $T = \mathbb{R}^+$ and by (4.2.5) when $T = \mathbb{N}$). Then for any $p(\cdot, a, t_0) \in S_G$ such that the trajectory $\gamma^+(p) \subset X_1$, $p(t, a, t_0) \to M$ as $t \to \infty$, where M is the largest invariant set in Z with respect to S_G and Z is defined in (4.2.6).

Proof. By Lemma 4.2.1, $p(t, a, t_0) \to \omega(p)$ as $t \to \infty$ and $\omega(p)$ is invariant with respect to S_G. By Lemma 4.2.2, $\omega(p) \subset Z$. Inasmuch as M is the largest invariant set in Z, we have $\omega(p) \subset M$. Therefore, $p(t, a, t_0) \to M$ as $t \to \infty$. □

Corollary 4.2.1 In addition to the assumptions in Theorem 4.2.1, suppose that the largest invariant set $M \subset Z$ is the minimal set determined by the function V on a neighborhood X_0 of M, where $X_0 \subset X_1$, and M is given by

$$M = \big\{x \in X_0 \colon V(x) \leq V(y) \text{ for all } y \in X_0\big\}. \tag{4.2.7}$$

Then M is *asymptotically stable* with respect to S_G.

Proof. It is clear that $V(x)$ is a constant for all $x \in M$. We denote this constant by $V(M)$. Now let $V_1(x) = V(x) - V(M)$. Then by the assumptions there exist $\psi_1, \psi_2 \in \mathcal{K}$ such that

$$\psi_1(d(x, M)) \leq V_1(x) \leq \psi_2(d(x, M))$$

for all $x \in X_0$. In fact, we may choose ψ_1 and ψ_2 as

$$\psi_1(r) = \min\{V_1(y): r \leq d(y, M), y \in X_0\},$$
$$\psi_2(r) = \max\{V_1(y): d(y, M) \leq r\},$$

which are defined on $[0, r_0], r_0 > 0$, where we assume that $\{y: d(y, M) \leq r_0\} \subset X_0$.

It now follows from Theorem 3.3.1 (resp., Theorem 3.4.1) that (S_G, M) is uniformly stable and from Theorem 4.2.1 that (S_G, M) is attractive. Therefore, (S_G, M) is asymptotically stable. □

In the last result of the present section, we require the following concept.

Definition 4.2.4 A metric space X is said to be *locally compact* if any bounded closed subset of X is compact. □

Corollary 4.2.2 Let G be a continuous semigroup ($T = \mathbb{R}^+$) or a discrete semigroup ($T = \mathbb{N}$) defined on a metric space X, and let S_G be the dynamical system determined by G. Assume that X is locally compact and that there exists a $V \in C[X, R]$ that satisfies the following conditions.

(i) $D_{(G)}V(x) \leq 0$ for all $x \in X$, where $D_{(G)}V$ is defined in (4.2.4) when $T = \mathbb{R}^+$ and by (4.2.5) when $T = \mathbb{N}$.

(ii) The largest invariant set M in the set $Z = \{x \in X: D_{(G)}V(x) = 0\}$ is bounded and is the minimal set in X determined by V; that is,

$$M = \{x \in X: V(x) \leq V(y) \text{ for all } y \in X\}.$$

(iii) $V(x) \to \infty$ as $d(x, M) \to \infty$.

Then M is *uniformly asymptotically stable in the large* with respect to S_G; that is, (S_G, M) is *uniformly asymptotically stable in the large*.

Proof. In a similar manner as in the proof of Corollary 4.2.1, for

$$V_1(x) = V(x) - V(M)$$

there exist $\psi_1, \psi_2 \in \mathcal{K}_\infty$ such that

$$\psi_1(d(x, M)) \leq V_1(x) \leq \psi_2(d(x, M))$$

for all $x \in X$, where we need to use hypothesis (iii) to conclude that $\psi_1, \psi_2 \in \mathcal{K}_\infty$.

By Corollary 4.2.1, (S_G, M) is uniformly stable. By Theorem 3.3.4, (S_G, M) is uniformly bounded. We now apply Theorem 4.2.1 to prove that (S_G, M) is globally attractive. For any $\alpha > 0$, it follows from the uniform boundedness of (S_G, M) that there exists a $\beta = \beta(\alpha) > 0$ such that if $d(a, M) \leq \alpha$, then for all $p(\cdot, a, t_0) \in S_G$, $d(p(t, a, t_0), M) \leq \beta$ for all $t \in T \cap [t_0, \infty)$. Choose $X_1 = \{x \in X: d(x, M) \geq \beta\}$ in applying Theorem 4.2.1. It now follows from Theorem 4.2.1 that $p(t, a, t_0) \to M$ as $t \to \infty$ whenever $d(a, M) < \alpha$.

We have proved that (S_G, M) is uniformly asymptotically stable in the large. □

We note that in contrast to the results of Section 3.2, where we require that along the motions $p(\cdot, a, t_0)$ of a dynamical system we have $D^+V(p(t,a,t_0),t) \le -\varphi_3(d(p(t,a,t_0),M)$, where $\varphi_3 \in \mathcal{K}$ (see Theorem 3.3.2), we can relax the corresponding condition in the results of the present section by requiring that $D_{(G)}V(x) \le 0$. The significance of this becomes more apparent in applications of these results, presented in subsequent chapters. Identical statements can be made for discrete-time dynamical systems as well.

4.3 Comparison Theory

In this section we present results that make it possible to deduce the qualitative properties of a dynamical system, *the object of inquiry*, from the properties of another dynamical system, *the comparison system*. This type of analysis is generally referred to as *comparison theory*. It is usually used to simplify the analysis of complex systems, which otherwise might be intractable.

We address both continuous dynamical systems and discrete-time dynamical systems.

A. Continuous dynamical systems

We begin by considering a system of ordinary differential equations given by

$$\dot{x} = g(t,x) \tag{E}$$

where $g \in C[\mathbb{R}^+ \times (\mathbb{R}^+)^l, \mathbb{R}^l]$, and an associated system of ordinary differential inequalities given by

$$Dx \le g(t,x) \tag{EI}$$

where D denotes a Dini derivative. We assume that $g(t,x_e) \equiv 0$ if $x_e = 0$, so that $x_e = 0$ is an equilibrium for (E). We first identify under what conditions one can deduce the qualitative properties of the dynamical system S_{EI} (determined by (EI)) from the qualitative properties of the dynamical system S_E (determined by (E)). Next we use these results in establishing a comparison theory that enables us to deduce the qualitative properties of an invariant set with respect to a dynamical system S (more specifically, a dynamical system $\{\mathbb{R}^+, X, A, S\}$) from the corresponding qualitative properties of the invariant set $\{0\} \subset \mathbb{R}^l$ with respect to the dynamical system S_E determined by the differential equation (E).

Theorem 4.3.1 Assume that $g \in C[\mathbb{R}^+ \times (\mathbb{R}^+)^l, \mathbb{R}^l]$ is quasi-monotone nondecreasing and that $g(t,0) = 0$ for all $t \in \mathbb{R}^+$. Then the following statements are true.

(a) If $x_e = 0$ is an *equilibrium* of S_E, then $x_e = 0$ is also an equilibrium of S_{EI}.

(b) The *stability*, *uniform stability*, *asymptotic stability*, *uniform asymptotic stability*, *exponential stability*, *uniform asymptotic stability in the large*, and *exponential stability in the large* of $(S_E, \{0\})$ imply the same corresponding types of stability of $(S_{EI}, \{0\})$.

(c) The *uniform boundedness* and *uniform ultimate boundedness* of S_E imply the same corresponding types of boundedness of S_{EI}.

Proof. It follows from Theorem 3.8.1 that for any motion $x(\cdot, x_0, t_0) \in S_{EI}$, where $(t_0, x_0) \in \mathbb{R}^+ \times (\mathbb{R}^+)^l$ we have that

$$x(t, x_0, t_0) \leq r(t, x_0, t_0) \tag{4.3.1}$$

for all $t \geq t_0$, where $r(\cdot, x_0, t_0)$ denotes the maximal solution of (E) (and therefore, $r(\cdot, x_0, t_0) \in S_E$), and inequality is to be interpreted componentwise. In addition, we also have that

$$x(t, x_0, t_0) \geq 0 \tag{4.3.2}$$

for all $t \geq t_0$ by the way S_{EI} is defined (i.e., $g \in C[\mathbb{R}^+ \times (\mathbb{R}^+)^l, \mathbb{R}^l]$).

All conclusions of the theorem follow now from (4.3.1) and (4.3.2) and from Definitions 3.1.2, and 3.1.6–3.1.17. □

We now state and prove the main result of this subsection.

Theorem 4.3.2 Let $\{\mathbb{R}^+, X, A, S\}$ be a dynamical system and let $M \subset A$. Assume that there exists a function $V\colon X \times \mathbb{R}^+ \to (\mathbb{R}^+)^l$ that satisfies the following conditions.

(i) There exists a function $g \in C[\mathbb{R}^+ \times (\mathbb{R}^+)^l, \mathbb{R}^l]$ that is quasi-monotone nondecreasing such that $g(t, 0) = 0$ for all $t \in \mathbb{R}^+$ and such that

$$D[V(p(t, a, t_0), t)] \leq g(t, V(p(t, a, t_0), t))$$

for all $p(\cdot, a, t_0) \in S$ and $t \in \mathbb{R}^+_{t_0}$, where D denotes a fixed Dini derivative with respect to t.

(ii) There exist $\psi_1, \psi_2 \in \mathcal{K}$ defined on $\mathbb{R}^+$ such that

$$\psi_1(d(x, M)) \leq |V(x, t)| \leq \psi_2(d(x, M))$$

for all $(x, t) \in X \times \mathbb{R}^+$, where $d(\cdot, \cdot)$ denotes the metric on X and $|\cdot|$ is the Euclidean norm on $\mathbb{R}^l$.

If M is closed, the following statements are true.

(a) The *invariance* of $(S_E, \{0\})$ implies the invariance of (S, M).

(b) The *stability*, *asymptotic stability*, *uniform stability*, and *uniform asymptotic stability* of $(S_E, \{0\})$ imply the same corresponding types of stability of (S, M).

(c) If in hypothesis (ii), $\psi_1(r) = ar^b, a > 0, b > 0$, then the *exponential stability* of $(S_E, \{0\})$ implies the exponential stability of (S, M).

(d) If M is bounded and if in hypothesis (ii), $\psi_1, \psi_2 \in \mathcal{K}_\infty$, then the *uniform asymptotic stability in the large* of $(S_E, \{0\})$ implies the uniform asymptotic stability in the large of (S, M).

(e) If in (c) and in hypothesis (ii), $\psi_i(r) = a_i r^b$, $a_i > 0$, $b > 0$, $i = 1, 2$, then the *exponential stability in the large* of $(S_E, \{0\})$ implies the exponential stability in the large of (S, M).

If M is bounded, but not necessarily closed, the following statement is true.

(f) If in (ii), $\psi_1, \psi_2 \in \mathcal{K}_\infty$, then the *uniform boundedness* and *uniform ultimate boundedness* of S_E imply the same corresponding types of boundedness of S.

Proof. For any $a \in A$, $t_0 \in \mathbb{R}^+$, $p(\cdot, a, t_0) \in S$, it follows from (i) that

$$V(p(t, a, t_0), t) = \widetilde{p}(t, V(a, t_0), t_0)$$

is a motion in S_{EI}.

(a) It follows from Theorem 4.3.1 that $x_e = 0$ is an equilibrium of S_{EI}. For any $a \in M$, $t_0 \in \mathbb{R}^+$, $p(\cdot, a, t_0) \in S$, it follows from (ii) that

$$|V(p(t_0, a, t_0), t_0)| \leq \psi_2(d(a, M)) = 0.$$

It follows from the invariance of $(S_{EI}, \{0\})$ that $V(p(t, a, t_0), t) = 0$ for all $t \in \mathbb{R}^+_{t_0}$. Thus $d(p(t, a, t_0), M) \leq \psi^{-1}(|V(p(t, a, t_0), t)|) = 0$ for all $t \in R^+_{t_0}$. Because M is closed, $p(t, a, t_0) \in M$, which implies the invariance of (S, M).

(b) Assume that $(S_E, \{0\})$ is stable. Then $(S_{EI}, \{0\})$ is stable by Theorem 4.3.1. For every $\varepsilon > 0$ and every $t_0 \in \mathbb{R}^+$, there exists a $\delta = \delta(\varepsilon, t_0) > 0$ such that $|\widetilde{p}(t, \widetilde{a}, t_0)| < \varepsilon$ for all $t \in \mathbb{R}^+_{t_0}$ and for all $\widetilde{p}(\cdot, \widetilde{a}, t_0) \in S_{EI}$, whenever $|\widetilde{a}| < \delta$. It follows from (ii) that whenever $d(a, M) < \psi_2^{-1}(\delta)$, $|V(a, t_0)| < \delta$. Hence, $d(p(t, a, t_0), M) \leq \psi_1^{-1}(|V(p(t, a, t_0), t)|) = \psi_1^{-1}(|\widetilde{p}(t, V(a, t_0), t_0)|) < \psi_1^{-1}(\varepsilon)$ for all $t \in \mathbb{R}^+_{t_0}$ and for all $p(\cdot, a, t_0) \in S$ whenever $d(a, M) < \psi_2^{-1}(\delta)$. Therefore, (S, M) is stable. Similarly, we can show that the asymptotic stability, uniform stability, and uniform asymptotic stability of $(S_E, \{0\})$ imply the same corresponding types of stability of (S, M).

(c) Assume that $(S_E, \{0\})$ is exponentially stable. Then $(S_{EI}, \{0\})$ is exponentially stable by Theorem 4.3.1. There exists $\alpha > 0$, and for every $\varepsilon > 0$ and every $t_0 \in \mathbb{R}^+$, there exists a $\delta = \delta(\varepsilon) > 0$ such that $|\widetilde{p}(t, \widetilde{a}, t_0)| < \varepsilon e^{-\alpha(t-t_0)}$ for all $t \in \mathbb{R}^+_{t_0}$ and for all $\widetilde{p}(\cdot, \widetilde{a}, t_0) \in S_{EI}$ whenever $|\widetilde{a}| < \delta$. It follows from (ii) that whenever $d(a, M) < \psi_2^{-1}(\delta)$, $|V(a, t_0)| < \delta$. Hence,

$$\begin{aligned} d(p(t, a, t_0), M) &\leq \psi_1^{-1}(|V(p(t, a, t_0), t)|) \\ &= \psi_1^{-1}(|\widetilde{p}(t, V(a, t_0), t_0)|) \\ &< \psi_1^{-1}\left(\varepsilon e^{-\alpha(t-t_0)}\right) \\ &= (\varepsilon/a)^{1/b} e^{-(\alpha/b)(t-t_0)} \end{aligned}$$

for all $t \in \mathbb{R}^+_{t_0}$ and for all $p(\cdot, a, t_0) \in S$ whenever $d(a, M) < \psi_2^{-1}(\delta)$. Therefore, (S, M) is exponentially stable.

(d) We have already shown in part (b) that (S, M) is uniformly stable. Because S_E is uniformly bounded by assumption, S_{EI} is uniformly bounded by Theorem 4.3.1. Then for every $\alpha > 0$ and for every $t_0 \in \mathbb{R}^+$ there exists a $\beta = \beta(\alpha) > 0$ such that

$|\widetilde{p}(t,\widetilde{a},t_0)-x_0|<\beta$ for all $t\in\mathbb{R}^+_{t_0}$ and $\widetilde{p}(\cdot,\widetilde{a},t_0)\in S_{EI}$ if $|\widetilde{a}|<\alpha$, where x_0 is a fixed point in $(\mathbb{R}^+)^l$. Hence, for all $t\in\mathbb{R}^+_{t_0}$ and for all $p(\cdot,a,t_0)\in S$

$$\begin{aligned} d(p(t,a,t_0),M) &\le \psi_1^{-1}(|V(p(t,a,t_0),t)|) \\ &= \psi_1^{-1}(|\widetilde{p}(\cdot,V(a,t_0),t_0)|) \\ &< \psi_1^{-1}(\beta+|x_0|). \end{aligned}$$

Inasmuch as M is bounded, we conclude that S is uniformly bounded. Lastly, we show that (S,M) is globally uniformly attractive. Because $(S_{EI},\{0\})$ is uniformly asymptotically stable in the large by Theorem 4.3.1, for every $\alpha>0,\varepsilon>0$, and for every $t_0\in\mathbb{R}^+$, there exists a $\tau=\tau(\varepsilon,\alpha)>0$ such that if $|\widetilde{a}|<\varphi_2(\alpha)$, then for all $\widetilde{p}(\cdot,\widetilde{a},t_0)\in S_{EI}$, $|\widetilde{p}(t,\widetilde{a},t_0)|<\varphi_1(\varepsilon)$ for all $t\in\mathbb{R}^+_{t_0+\tau}$. Hence, it follows from (ii) that whenever $d(a,M)<\alpha$, $V(a,t)\le\varphi_2(\alpha)$ and hence it is true that $|V(p(t,a,t_0),t)|=|\widetilde{p}(t,V(a,t_0),t_0)|<\varphi_1(\varepsilon)$ for all $t\in\mathbb{R}^+_{t_0+\tau}$. Therefore, $d(p(t,a,t_0),M)\le\varphi_1^{-1}(|V(p(t,a,t_0),t)|)<\varepsilon$ for all $t\in\mathbb{R}^+_{t_0+\tau}$. It now follows that (S,M) is uniformly asymptotically stable in the large.

(e) Under the assumption, $(S_{EI},\{0\})$ is exponentially stable in the large by Theorem 4.3.1. Then there exist an $\alpha>0$, a $\gamma>0$, and for every $\beta>0$, there exists a $k(\beta)>0$ such that $|\widetilde{p}(t,\widetilde{a},t_0)|<k(\beta)|\widetilde{a}|^\gamma e^{-\alpha(t-t_0)}$ for all $\widetilde{p}(\cdot,\widetilde{a},t_0)\in S_{EI}$ and $t\in\mathbb{R}^+_{t_0}$ whenever $|\widetilde{a}|<a_2\beta^b$. Hence, it follows from (ii) that whenever $d(a,M)<\beta$, $V(a,t)<a_2\beta^b$ and hence,

$$|V(p(t,a,t_0),t)|=|\widetilde{p}(t,V(a,t_0),t_0)|<k(\beta)|V(a,t_0)|^\gamma e^{-\alpha(t-t_0)}.$$

Then

$$\begin{aligned} d(p(t,a,t_0),M) &\le \Big[|V(p(t,a,t_0),t)|/a_1\Big]^{1/b} \\ &< \Big[k(\beta)|V(a,t_0)|^\gamma e^{-\alpha(t-t_0)}\Big]^{1/b} \\ &< [k(\beta)a_2^\gamma]^{1/b}[d(a,M)]^\gamma e^{-(\alpha/b)(t-t_0)}. \end{aligned}$$

Let $k_1(\beta)=[k(\beta)a_2^\gamma]^{1/b}$, and $\alpha_1=\alpha/b$. Then

$$d(p(t,a,t_0),M)<k_1(\beta)[d(a,M)]^\gamma e^{-\alpha_1(t-t_0)}$$

for all $p(\cdot,a,t_0)\in S$ and for all $t\in\mathbb{R}^+_{t_0}$ whenever $d(a,M)<\beta$.

(f) The uniform boundedness of S is shown in (d). Note that in this part of the proof, M is only required to be bounded. The uniform ultimate boundedness can be shown similarly. □

For obvious reasons, we call the function V in Theorem 4.3.2 a *vector Lyapunov function.*

If in equation (E), $g\in[\mathbb{R}^+\times\mathbb{R}^l,\mathbb{R}^l]$ and if in inequality (EI) we restrict the domain of g to $\mathbb{R}^+\times(\mathbb{R}^+)^l$, then the statements of Theorem 4.3.2 are still true. Specifically, if $\widetilde{S}_E$ denotes the dynamical system determined by (E) for $g\in[\mathbb{R}^+\times\mathbb{R}^l,\mathbb{R}^l]$

and if S_E denotes the dynamical system determined by (E) with the domain of g restricted to $\mathbb{R}^+ \times (\mathbb{R}^+)^l$, then S_E is a subsystem of $\widetilde{S}_E$. Therefore, if we replace S_E by $\widetilde{S}_E$ in the statements of Theorem 4.3.2, the conclusions of this theorem are still true.

We conclude the present subsection with a specific example.

Example 4.3.1 We choose in particular

$$g(t,x) = Bx$$

where $B \in \mathbb{R}^{l\times l}$. Then g is quasi-monotone nondecreasing if and only if all the off-diagonal elements of $B = [b_{ij}]$ are nonnegative. In view of Theorem 4.3.2 and the results given in Example 3.1.8, we have the following results:

Let $\{\mathbb{R}^+, X, A, S\}$ be a dynamical system and let $M \subset A$ be closed. Assume that there exists a continuous function $V\colon X \times \mathbb{R}^+ \to (\mathbb{R}^+)^l$ that satisfies the following conditions:

(i) For all $p(\cdot, a, t_0) \in S$ and all $t \in \mathbb{R}^+_{t_0}$,

$$DV(p(t,a,t_0),t) \le BV(p(t,a,t_0),t)$$

where the off-diagonal elements of $B \in \mathbb{R}^{l\times l}$ are nonnegative and D is a fixed Dini derivative.

(ii) There exist $\psi_1, \psi_2 \in \mathcal{K}$ such that

$$\psi_1(d(x,M)) \le |V(x,t)| \le \psi_2(d(x,M))$$

for all $x \in X$ and $t \in \mathbb{R}^+$, where d is the metric defined on X and $|\cdot|$ denotes the Euclidean norm on $\mathbb{R}^l$.

Then the following statements are true.

(a) If the eigenvalues of B have nonpositive real parts and every eigenvalue of B with zero real part has an associated Jordan block of order one, then (S, M) is *invariant* and *uniformly stable*;

(b) If all eigenvalues of B have negative real parts, then (S, M) is *uniformly asymptotically stable*. In addition, if in hypothesis (ii) above, $\psi_1, \psi_2 \in \mathcal{K}_\infty$ and M is bounded, then (S, M) is *uniformly asymptotically stable in the large*.

(c) If in part (b), $\psi_i(r) = a_i r^b$, $a_i > 0$, $b > 0$, $i = 1, 2$, then (S, M) is *exponentially stable in the large*.

Finally, recalling that a matrix $H \in \mathbb{R}^{l\times l}$ is called an *M-matrix* if all the off-diagonal elements of H are nonpositive and if all the eigenvalues of H have positive real parts, we can rephrase condition (b) given above by stating that $-B$ is an M-matrix, in place of "all eigenvalues of B have negative real parts." For the properties of M-matrices, refer, for example, to [4] and to Definition 7.7.1. □

B. Discrete-time dynamical systems

Next, we consider a system of difference equations given by

$$x(k+1) = h(k, x(k)), \tag{D}$$

where $h\colon \mathbb{N} \times (\mathbb{R}^+)^l \to \mathbb{R}^l$, and the associated system of difference inequalities given by

$$x(k+1) \leq h(k, x(k)), \tag{DI}$$

where for all $k \in \mathbb{N}$, $x(k) \in (\mathbb{R}^+)^l$. We denote the dynamical systems determined by (D) and (DI) by S_D and S_{DI}, respectively.

Definition 4.3.1 A function $g\colon \mathbb{N} \times \Omega \to \mathbb{R}^l$ is said to be *monotone nondecreasing* if $g(k, x) \leq g(k, y)$ for all $x \leq y$, $x, y \in \Omega$ and all $k \in \mathbb{N}$, where $\Omega \subset \mathbb{R}^l$ is a subset of $\mathbb{R}^l$ and where inequality of vectors is to be interpreted componentwise. □

Lemma 4.3.1 Assume that $h\colon \mathbb{N} \times (\mathbb{R}^+)^l \to \mathbb{R}^l$ is monotone nondecreasing and that $h(k, 0) = 0$ for all $k \in \mathbb{N}$. Then the following statements are true.

(a) If $x_e = 0$ is an *equilibrium* of S_D, then $x_e = 0$ is also an equilibrium of S_{DI}.

(b) The *stability*, *uniform stability*, *asymptotic stability*, *uniform asymptotic stability*, *exponential stability*, *uniform asymptotic stability in the large*, and *exponential stability in the large* of $(S_D, \{0\})$ imply the same corresponding types of stability of $(S_{DI}, \{0\})$.

(c) The *uniform boundedness* and *uniform ultimate boundedness* of S_D imply the same corresponding types of boundedness of S_{DI}.

Proof. For any motion $x(\cdot, x_0, n_0) \in S_{DI}$ and any motion $r(\cdot, x_0, n_0) \in S_D$, where $(n_0, x_0) \in \mathbb{N} \times (\mathbb{R}^+)^l$ we have that

$$\begin{aligned}
x(n_0+1, x_0, n_0) &\leq h(n_0, x_0) \\
&= r(n_0+1, x_0, n_0) \\
x(n_0+2, x_0, n_0) &\leq h(n_0+1, x(n_0+1, x_0, n_0)) \\
&\leq h(n_0+1, r(n_0+1, x_0, n_0)) \\
&= r(n_0+2, x_0, n_0) \\
&\vdots \\
x(n+1, x_0, n_0) &\leq h(n, x(n, x_0, n_0)) \\
&\leq h(n, r(n, x_0, n_0)) \\
&= r(n+1, x_0, n_0)
\end{aligned} \tag{4.3.3}$$

for all $n \geq n_0$, and inequality is to be interpreted componentwise. In addition, we also have that

$$x(n, x_0, n_0) \geq 0 \tag{4.3.4}$$

for all $n \geq n_0$ by the way S_{DI} is defined (i.e., $h\colon \mathbb{N} \times (\mathbb{R}^+)^l \to \mathbb{R}^l$).

All conclusions of the theorem follow now from (4.3.3) and (4.3.4) and from Definitions 3.1.2, and 3.1.6–3.1.17. □

We now present the main result of this subsection.

Theorem 4.3.3 Let $\{\mathbb{N}, X, A, S\}$ be a dynamical system and let $M \subset A$. Assume that there exists a function $V\colon X \times \mathbb{N} \to (\mathbb{R}^+)^l$ that satisfies the following conditions.

(i) There exists a function $h\colon \mathbb{N} \times (\mathbb{R}^+)^l \to \mathbb{R}^l$ which is monotone nondecreasing such that $h(k, 0) = 0$ for all $k \in \mathbb{N}$, and

$$V(p(k+1, a, k_0), k+1) \leq h(k, V(p(k, a, k_0), k))$$

for all $p(\cdot, a, k_0) \in S$ and $k \in \mathbb{N}_{k_0}$.

(ii) There exist $\psi_1, \psi_2 \in \mathcal{K}$ defined on $\mathbb{R}^+$ such that

$$\psi_1(d(x, M)) \leq |V(x, k)| \leq \psi_2(d(x, M))$$

for all $(x, k) \in X \times \mathbb{N}$, where $d(\cdot, \cdot)$ denotes the metric on X and $|\cdot|$ is the Euclidean norm on $\mathbb{R}^l$.

If M is closed, then the following statements are true.

(a) The *invariance* of $(S_D, \{0\})$ implies the invariance of (S, M).

(b) The *stability, asymptotic stability, uniform stability*, and *uniform asymptotic stability* of $(S_D, \{0\})$ imply the same corresponding types of stability of (S, M).

(c) If in hypothesis (ii), $\psi_1(r) = ar^b, a > 0, b > 0$, then the *exponential stability* of $(S_D, \{0\})$ implies the exponential stability of (S, M).

(d) If M is bounded and if in hypothesis (ii), $\psi_1, \psi_2 \in \mathcal{K}_\infty$, then the *uniform asymptotic stability in the large* of $(S_D, \{0\})$ implies the uniform asymptotic stability in the large of (S, M).

(e) If in (c), $\psi_i(r) = a_i r^b$, $a_i > 0$, $b > 0$, $i = 1, 2$, and M is bounded, then the *exponential stability in the large* of $(S_D, \{0\})$ implies the exponential stability in the large of (S, M).

If M is bounded, but not necessarily closed, the following statement is true.

(f) If in (ii), $\psi_1, \psi_2 \in \mathcal{K}_\infty$, then the *uniform boundedness* and *uniform ultimate boundedness* of S_D imply the same corresponding types of boundedness of S.

Proof. For any $a \in A$, $k_0 \in \mathbb{N}$, $p(\cdot, a, k_0) \in S$, it follows from (i) that

$$V(p(k, a, k_0), k) = \widetilde{p}(k, V(a, k_0), k_0)$$

is a motion in S_{DI}. The rest of the proof is similar to the proof of Theorem 4.3.2 and is not repeated here. □

If in equation (D), $h\colon \mathbb{N} \times \mathbb{R}^l \to \mathbb{R}^l$, and if in inequality (DI), we restrict the domain of h to $\mathbb{N} \times (\mathbb{R}^+)^l$, then the statements of Theorem 4.3.3 are still true, for the same reasons as given immediately after Theorem 4.3.2.

We conclude the present subsection with a specific example.

Example 4.3.2 We choose in particular

$$h(k, x) = Bx$$

where $B = [b_{ij}] \in \mathbb{R}^{l \times l}$. Then h is monotone nondecreasing if and only if $b_{ij} \geq 0$ for all $i, j = 1, \ldots, l$. In view of Theorem 4.3.3 and the results given in Example 3.1.9, we have the following results.

Let $\{\mathbb{N}, X, A, S\}$ be a dynamical system and let $M \subset A$ be closed. Assume that there exists a continuous function $V\colon X \times \mathbb{N} \to (\mathbb{R}^+)^l$ that satisfies the following conditions.

(i) For all $p(\cdot, a, k_0) \in S$ and all $k \in \mathbb{N}_{k_0}$,

$$V(p(k+1, a, k_0), k+1) \leq BV(p(k, a, k_0), k)$$

where $B = [b_{ij}] \in \mathbb{R}^{l \times l}$ with $b_{ij} \geq 0$ for all $i, j = 1, \ldots, l$.

(ii) There exist $\psi_1, \psi_2 \in \mathcal{K}$ defined on $\mathbb{R}^+$ such that

$$\psi_1(d(x, M)) \leq |V(x, k)| \leq \psi_2(d(x, M))$$

for all $x \in X$ and $k \in \mathbb{N}$, where d is the metric defined on X and $|\cdot|$ denotes the Euclidean norm on $\mathbb{R}^l$.

Then the following statements are true.

(a) If the eigenvalues of B have magnitude less than or equal to one and every eigenvalue of B with magnitude equal to one has an associated Jordan block of order one, then (S, M) is *invariant* and *uniformly stable*.

(b) If all eigenvalues of B have magnitude less than one, then (S, M) is *uniformly asymptotically stable*. In addition, if in hypothesis (ii) above, $\psi_1, \psi_2 \in \mathcal{K}_\infty$ and M is bounded, then (S, M) is *uniformly asymptotically stable in the large*.

(c) If in part (b), $\psi_i(r) = a_i r^b$, $a_i > 0$, $b > 0$, $i = 1, 2$, then (S, M) is *exponentially stable in the large*. □

4.4 Uniqueness of Motions

In several results that we have encountered thus far and which we will encounter, the dynamical systems are endowed with the uniqueness of motions property (refer to Definition 3.1.3). This property is especially prevalent in applications. In the present section we establish a Lyapunov-type result which ensures that a dynamical system possesses the uniqueness of motions property.

In the following, we let $T = \mathbb{R}^+$ or $T = \mathbb{N}$.

Theorem 4.4.1 Let $\{T, X, A, S\}$ be a dynamical system and assume that there exists a function $V\colon X \times X \times T \to \mathbb{R}^+$ that satisfies the following conditions.

(i) $V(x, y, t) = 0$ for all $t \in T$ if $x = y$.

(ii) $V(x, y, t) > 0$ for all $t \in T$ if $x \neq y$.

(iii) For any $p_i(\cdot, a, t_0) \in S, i = 1, 2, V(p_1(t, a, t_0), p_2(t, a, t_0), t)$ is nonincreasing in t.

Then S satisfies the uniqueness of motions property.

Proof. Let $p_i(\cdot, a, t_0) \in S$, $i = 1, 2$, and let $q(t) = V(p_1(t, a, t_0), p_2(t, a, t_0), t)$, for all $t \in T_{a,t_0}$. Then $q(t_0) = 0$ by (i). By (iii), $q(t)$ is nonincreasing. Therefore $q(t) = 0$ for all $t \in T_{a,t_0}$. Finally, by (ii), $p_1(t, a, t_0) = p_2(t, a, t_0)$ for all $t \in T_{a,t_0}$. We have proved that S satisfies the uniqueness property. □

We demonstrate the applicability of Theorem 4.4.1 by means of the following example.

Example 4.4.1 We consider dynamical systems determined by first-order differential equations in a Banach space X with norm $\|\cdot\|$, given by

$$\dot{x}(t) = F(t, x(t)) \tag{F}$$

where $t \in \mathbb{R}^+$, $F\colon \mathbb{R}^+ \times \mathrm{C} \to X$, and $x(t) \in \mathrm{C} \subset X$.

Associated with (F) is the initial value problem given by

$$\dot{x}(t) = F(t, x(t)), \qquad x(t_0) = x_0 \tag{I_F}$$

where $t_0 \in \mathbb{R}^+$, $t \geq t_0$, and $x_0 \in \mathrm{C} \subset X$. The following result yields sufficient conditions for the uniqueness of the solutions of the initial value problem (I_F).

Theorem 4.4.2 For (F), assume that on every compact set $K \subset \mathbb{R}^+ \times \mathrm{C}, F(\cdot, \cdot)$ satisfies the *Lipschitz condition*

$$\|F(t, x) - F(t, y)\| \leq L_K \|x - y\|$$

for all $(t, x), (t, y) \in K$, where L_K is a constant that depends only on the choice of K. Then for every $(t_0, x_0) \in \mathbb{R}^+ \times \mathrm{C}$, (F) has at most one solution $x(t)$ defined on $[t_0, t_0 + c)$ for some $c > 0$, that satisfies $x(t_0) = x_0$.

Proof. It suffices to show that (F) has at most one solution on $[t_0, b]$ that satisfies $x(t_0) = x_0$ where b is any finite number greater than t_0.

Let $x(t)$ and $y(t)$ be two solutions of (F) that are defined on $[t_0, b]$. By the continuity of $x(t)$ and $y(t)$, the set

$$K = \{(t, \varphi) \in [t_0, b] \times \mathrm{C}\colon \varphi = x(t) \text{ or } \varphi = y(t) \text{ for some } t \in [t_0, b]\}$$

is compact. Let $L = L_K$ be the Lipschitz constant for $F(\cdot, \cdot)$ corresponding to K, and let D^+ denote the upper-right Dini derivative in t. Choose $V(x, y, t) = \|x - y\| e^{-Lt}$,

$t \geq 0$. Then for $t \in [t_0, b]$,

$$\begin{aligned}
&D^+V(x(t), y(t), t)\\
&= \overline{\lim_{h\to 0^+}} \frac{1}{h}\Big[e^{-L(t+h)}\|x(t+h) - y(t+h)\| - e^{-Lt}\|x(t) - y(t)\|\Big]\\
&= \overline{\lim_{h\to 0^+}} \frac{1}{h}\Big[\big(e^{-L(t+h)} - e^{-Lt}\big)\|x(t) - y(t)\|\\
&\quad + e^{-L(t+h)}\big(\|x(t+h) - y(t+h)\| - \|x(t) - y(t)\|\big)\Big]\\
&\leq -e^{-Lt}L\|x(t) - y(t)\| + e^{-Lt}D^+\|x(t) - y(t)\|\\
&\leq -e^{-Lt}L\|x(t) - y(t)\| + e^{-Lt}\|\dot{x}(t) - \dot{y}(t)\|\\
&= e^{-Lt}\big[- L\|x(t) - y(t)\| + \|F(t, x(t)) - F(t, y(t))\| \big]\\
&\leq e^{-Lt}\big[- L\|x(t) - y(t)\| + L\|x(t) - y(t)\| \big]\\
&= 0.
\end{aligned}$$

Therefore, condition (iii) of Theorem 4.4.1 is satisfied. Conditions (i) and (ii) of Theorem 4.4.1 are clearly also satisfied. Therefore, $x(t) = y(t)$ for $t \in [t_0, t_0 + c)$ for some $c > 0$. □

4.5 Notes and References

The necessary and sufficient conditions for stability and asymptotic stability for autonomous dynamical systems given in Section 4.1, Theorems 4.1.3 and 4.1.4, are based on results presented in Zubov [8].

The invariance theory for continuous-time dynamical systems determined by semigroups defined on metric spaces, given in Section 4.2, is based on work reported in Hale [1], and the results for the discrete-time case were first reported in Michel *et al.* [5].

The results for the Comparison Theory presented in Section 4.3 are based on material presented in Lakshmikantham and Leela [2] and Miller and Michel [6] concerning Theorems 4.3.1 and 4.3.2, whereas Lemma 4.3.1 and Theorem 4.3.3 are based on material presented in Michel *et al.* [5].

The uniqueness result given in Section 4.4, Theorem 4.4.1, is motivated by existing results for dynamical systems determined by functional differential equations (Yoshizawa [7]) and differential equations in Banach space (Lakshmikantham and Leela [2]).

4.6 Problems

In Sections 3.3 and 3.4 we proved several stability and boundedness results for continuous dynamical systems and discrete-time dynamical systems making use of corresponding results for DDS. In Problems 3.10.4–3.10.12, we asked the reader to prove the results of Sections 3.3 and 3.4 by invoking the basic definitions for the various types of stability and boundedness. In Problems 3.10.19–3.10.24, we asked the reader to prove some of the results of Sections 3.3 and 3.4 by using the equivalent definitions for various stability and boundedness concepts (involving comparison functions), established in Problems 3.10.15–3.10.18. In the next four problems we ask the reader to prove some of the results of Sections 3.3 and 3.4 yet another way: by invoking the comparison theory established in Section 4.3.

Problem 4.6.1 Prove Theorems 3.3.4 and 3.4.4 by using the comparison theorems, Theorems 4.3.2(f) and 4.3.3(f), respectively.
Hint: Let $l = 1$. Let $y(t) = V(p(t, a, t_0), t)$ for the case when $t \in T = \mathbb{R}^+$ and $y_k = V(p(k, a, k_0), k)$ when $k \in T = \mathbb{N}$. Choose $g(t, y) \equiv 0$ in applying Theorem 4.3.2 for the case $T = \mathbb{R}^+$ and $h(k, y) \equiv 0$ in applying Theorem 4.3.3. □

Problem 4.6.2 Prove Theorems 3.3.6 and 3.4.6 by using the comparison theorems, Theorem 4.3.2(d) and 4.3.3(d), respectively.
Hint: Let $l = 1$. For $T = \mathbb{R}^+$, let $y(t) = V(p(t, a, t_0), t)$ and from (3.3.9) and (3.3.10), obtain for all $t \in T_{a,t_0}$

$$Dy(t) \leq -\psi(y(t)) \tag{4.6.1}$$

where $\psi = \varphi_3 \circ \varphi_2^{-1} \in \mathcal{K}$. In applying Theorem 4.3.2, let $g(t, y) = -\psi(y)$. In Problem 3.10.25 we ask the reader to prove that the equilibrium $y_e = 0$ is a uniformly asymptotically stable equilibrium of the dynamical system $S_E = S_{(4.6.2)}$ determined by the scalar differential equation

$$\dot{y} = -\psi(y), \qquad y \in \mathbb{R}^+ \tag{4.6.2}$$

where $\psi \in \mathcal{K}$.

Next, we note that $\psi \in \mathcal{K}_\infty$ if $\varphi_2, \varphi_3 \in \mathcal{K}_\infty$. In Problem 3.10.26 we ask the reader to prove that the equilibrium $y_e = 0$ of (4.6.2) is uniformly asymptotically stable in the large when $\psi \in \mathcal{K}_\infty$. It now follows from Theorem 4.3.2 that (S, M) is also uniformly asymptotically stable in the large.

The reader can show that for $T = \mathbb{N}$, the proof follows along similar lines, using Theorem 4.3.3 and Problems 3.10.28 and 3.10.29. □

Problem 4.6.3 Prove Theorems 3.3.7 and 3.4.7 by using the comparison theorems, Theorems 4.3.2(e) and 4.3.3(e), respectively.
Hint: In the hint given for Problem 4.6.2 we let $\varphi_i(r) = c_i r^b$, $c_i > 0$, $b > 0$, $r \geq 0$, $i = 1, 2, 3$. For $T = \mathbb{R}^+$, we have that $\psi(r) = (\varphi_3 \circ \varphi_2^{-1})(r) = ar$, where $a = c_3/c_2 > 0$. System $S_E = S_{(4.6.2)}$ is now determined by

$$\dot{y} = -ay, \qquad y \in \mathbb{R}^+,$$

so that $y(t) = y_0 e^{-a(t-t_0)}$, $t \geq t_0$. It is clear that in this case $(S_E, \{0\})$ is exponentially stable in the large. It now follows from Theorem 4.3.2 that (S, M) is exponentially stable in the large.

The reader can show that for $T = \mathbb{N}$, the proof follows along similar lines, using Theorem 4.3.3. □

Problem 4.6.4 Prove Theorems 3.3.5 and 3.4.5, using the comparison theorems, Theorem 4.3.2(f) and 4.3.3(f), respectively.

Hint: For both $T = \mathbb{R}^+$ and $T = \mathbb{N}$, if (S, M) is uniformly asymptotically stable in the large and if M is bounded, then S is uniformly ultimately bounded. This can be verified from Definitions 3.1.14 and 3.1.16, replacing $x_0 \in X$ in Definition 3.1.14 by a bounded set M. □

Problem 4.6.5 Consider the initial and boundary value problem for a parabolic partial differential equation given by

$$\begin{cases} \dfrac{\partial u}{\partial t}(t,x) = \dfrac{\partial^2 u}{\partial x^2}(t,x) + F(t,x,u), & x \in [\lambda_1(t), \lambda_2(t)], \quad t \in [a,b] \\ u(a,x) = g(x), & x \in [\lambda_1(a), \lambda_2(a)] \\ u(t, \lambda_i(t)) = h_i(t), & t \in [a,b], \ i = 1,2, \end{cases} \tag{4.6.3}$$

where $\lambda_1^0 \le \lambda_1(t) \le \lambda_2(t) \le \lambda_2^0$ for all $t \in [a,b]$, $F \in C\big[[a,b] \times [\lambda_1^0, \lambda_2^0] \times \mathbb{R}, \mathbb{R}\big]$ $\lambda_1, \lambda_2, h_1, h_2 \in C\big[[a,b], \mathbb{R}\big]$, $g \in C\big[[\lambda_1^0, \lambda_2^0], \mathbb{R}\big]$ and $g(\lambda_i(a)) = h_i(a)$, $i = 1,2$.

Assume that there exists a constant $K > 0$ such that

$$F(t,x,u_1) - F(t,x,u_2) \le K(u_1 - u_2)$$

for all $u_1 > u_2$ and for all $(t,x) \in [a,b] \times [\lambda_1^0, \lambda_2^0]$.

By applying Theorem 4.4.1, show that there exists at most one solution of system (4.6.3).

Hint: For any $v_1, v_2 \in X = C[\mathbb{R}, \mathbb{R}]$ choose

$$V(t, v_1, v_2) = e^{-2Kt} \int_{\lambda_1(t)}^{\lambda_2(t)} |v_1(x) - v_2(x)|^2 dx.$$

For any two solutions of (4.6.3), $u_i = u_i(t,x)$, $i = 1,2$, using the fact that

$$u_1(t, \lambda_1(t)) = u_2(t, \lambda_2(t))$$

for all $t \in [t_0, b]$, show that

$$D^+V(t, u_1(t,x), u_2(t,x)) \le -2e^{-2Kt} \int_{\lambda_1(t)}^{\lambda_2(t)} \left[\frac{\partial u_1}{\partial x}(t,x) - \frac{\partial u_2}{\partial x}(t,x)\right]^2 dx \le 0.$$

To complete the proof, show that the hypotheses of Theorem 4.4.1 are satisfied. □

Problem 4.6.6 Prove the following results.

Theorem 4.6.1 [5] (*Comparison Theorem*) Let $\{T, X_1, A_1, S_1\}$ and $\{T, X_2, A_2, S_2\}$ be two dynamical systems and let $M_1 \subset A_1 \subset X_1$ and $M_2 \subset A_2 \subset X_2$. Assume there exists a function $V\colon X_1 \times T \to X_2$ that satisfies the following hypotheses.

(i) $\mathcal{V}(S_1) \subset S_2$, where $\mathcal{V}(S_1)$ is defined as

$$\mathcal{V}(S_1) \triangleq \{q(\cdot, b, t_0)\colon q(t,b,t_0) = V(p(t,a,t_0),t), p(\cdot,a,t_0) \in S_1, \ t \in T, \\ \text{with } b = V(a,t_0) \text{ and } T_{b,t_0} = T_{a,t_0}, a \in A_1, t_0 \in T\}.$$

(ii) M_1 and M_2 satisfy the relation

$$M_2 \supset \{x_2 \in X_2 \colon x_2 = V(x_1, t') \text{ for some } x_1 \in M_1 \text{ and } t' \in T\},$$

and A_1 and A_2 satisfy the relation

$$A_2 \supset \{x_2 \in X_2 \colon x_2 = V(x_1, t') \text{ for some } x_1 \in A_1 \text{ and } t' \in T\}.$$

(iii) There exist $\psi_1, \psi_2 \in \mathcal{K}$ defined on $\mathbb{R}^+$, such that

$$\psi_1(d_1(x, M_1)) \leq d_2(V(x,t), M_2) \leq \psi_2(d_1(x, M_1)) \tag{4.6.4}$$

for all $x \in X_1$ and $t \in T$, where d_1 and d_2 are the metrics on X_1 and X_2, respectively.

If M_1 is closed, then the following statements are true.

(a) The invariance of (S_2, M_2) implies the *invariance* of (S_1, M_1).

(b) The stability, uniform stability, asymptotic stability, and uniform asymptotic stability of (S_2, M_2) imply the *stability*, *uniform stability*, *asymptotic stability*, and *uniform asymptotic stability* of (S_1, M_1), respectively.

(c) If in (4.6.4), $\psi_1(r) = \mu r^\nu$, $\mu > 0$, $\nu > 0$, then the exponential stability of (S_2, M_2) implies the *exponential stability* of (S_1, M_1).

(d) If in (4.6.4), $\psi_1, \psi_2 \in \mathcal{K}_\infty$, then the asymptotic stability in the large of (S_2, M_2) implies the *asymptotic stability in the large* of (S_1, M_1).

If M_1 and M_2 are bounded, but not necessarily closed, and if in (4.6.4), $\psi_1, \psi_2 \in \mathcal{K}_\infty$, then the following statement is true.

(e) The uniform boundedness and the uniform ultimate boundedness of S_2 imply the *uniform boundedness* and the *uniform ultimate boundedness* of S_1, respectively.

If M_1 and M_2 are bounded and closed, and if in (4.6.4), $\psi_1, \psi_2 \in \mathcal{K}_\infty$, then the following statement is true.

(f) The uniform asymptotic stability in the large of (S_2, M_2) implies the *uniform asymptotic stability in the large* of (S_1, M_1).

(g) If in addition, we have in (4.6.4) that $\psi_i(r) = \mu_i r^\nu$, $\mu_i > 0$, $\nu > 0$, $i = 1, 2$, then the exponential stability in the large of (S_2, M_2) implies the *exponential stability in the large* of (S_1, M_1).

Hint: In each case, use the definitions of the various stability and boundedness concepts to establish the indicated relationships. (The complete proof of this theorem is given in [5, Section 3.3]). □

In the next results we employ the continuous-time dynamical system S_{EI} determined by the differential inequality (EI) and discrete-time dynamical system S_{DI} determined by the difference inequality (DI), as comparison systems (refer to Subsections 4.3A and 4.3B).

Problem 4.6.7 Prove the following results.

Proposition 4.6.1 Let $\{T, X, A, S\}$ be a dynamical system and let $M \subset A \subset X$. Let $T = \mathbb{R}^+$ or $\mathbb{N}$. Assume that there exists a function $V: X \times T \to (\mathbb{R}^+)^l$ that satisfies the following conditions.

(i) When $T = \mathbb{R}^+$, there exists a function $g \in C[\mathbb{R}^+ \times (\mathbb{R}^+)^l, \mathbb{R}^l]$ such that $g(t, 0) = 0$ for all $t \in \mathbb{R}^+$, and such that

$$D[V(p(t, a, t_0), t)] \le g(t, V(p(t, a, t_0), t)) \tag{4.6.5}$$

for all $p(\cdot, a, t_0) \in S$, $t \in T_{a,t_0}$.

When $T = \mathbb{N}$, there exists a function $h: \mathbb{N} \times (\mathbb{R}^+)^l \to \mathbb{R}^l$ such that $h(k, 0) = 0$ for all $k \in \mathbb{N}$, and such that

$$V(p(k+1, a, k_0), k+1) \le h(k, V(p(k, a, k_0), k)) \tag{4.6.6}$$

for all $p(\cdot, a, k_0) \in S$, $k \in T_{a,k_0}$.

(ii) There exist functions $\psi_1, \psi_2 \in \mathcal{K}$ defined on $\mathbb{R}^+$ such that when $T = \mathbb{R}^+$,

$$\psi_1(d(x, M)) \le |V(x, t)| \le \psi_2(d(x, M)) \tag{4.6.7}$$

and when $T = \mathbb{N}$,

$$\psi_1(d(x, M)) \le |V(x, k)| \le \psi_2(d(x, M)) \tag{4.6.8}$$

for all $x \in X$ and $t \in \mathbb{R}^+$ (resp., $k \in \mathbb{N}$), where d denotes the metric defined on X and $|\cdot|$ denotes the Euclidean norm on $\mathbb{R}^l$.

If M is closed, then the following statements are true.

(a) The invariance of $(S_{EI}, \{0\})$ (resp., $(S_{DI}, \{0\})$), implies the *invariance* of (S, M).

(b) The stability, uniform stability, asymptotic stability, and uniform asymptotic stability of $(S_{EI}, \{0\})$ (resp., $(S_{DI}, \{0\})$), imply the corresponding types of *stability* of (S, M), respectively.

(c) If in (4.6.7) (resp., in (4.6.8)), $\psi_1(r) = \mu r^\nu$, $\mu > 0, \nu > 0$, then the exponential stability of $(S_{EI}, \{0\})$ (resp., $(S_{DI}, \{0\})$), implies the *exponential stability* of (S, M).

(d) If in (4.6.7) (resp., in (4.6.8)), $\psi_1, \psi_2 \in \mathcal{K}_\infty$, then the asymptotic stability in the large of $(S_{EI}, \{0\})$ (resp., $(S_{DI}, \{0\})$), implies the *asymptotic stability in the large* of (S, M).

If M is bounded (but not necessarily closed), and if in (4.6.7) (resp., in (4.6.8)), $\psi_1, \psi_2 \in \mathcal{K}_\infty$, then the following statement is true.

(e) The uniform boundedness and the uniform ultimate boundedness of S_{EI} (resp., S_{DI}), imply the *uniform boundedness* and the *uniform ultimate boundedness* of S, respectively.

If M is bounded and closed, and if in (4.6.7) (resp., in (4.6.8)), $\psi_1, \psi_2 \in \mathcal{K}_\infty$, then the following statements are true.

(f) The uniform asymptotic stability in the large of $(S_{EI}, \{0\})$ (resp., $(S_{DI}, \{0\})$), implies the *uniform asymptotic stability in the large* of (S, M).

(g) If in addition to the conditions of part (f), we have in (4.6.7) (resp., in (4.6.8)), that $\psi_i(r) = \mu_i r^\nu$, $\mu_i > 0$, $\nu > 0$, $i = 1, 2$, then the exponential stability in the large of $(S_{EI}, \{0\})$ (resp., $(S_{DI}, \{0\})$), implies the *exponential stability in the large* of (S, M).

Hint: In the notation of Theorem 4.6.1, let $X = X_1$, $A = A_1$, and $S = S_1$. Let $\mathbb{R}^l = X_2 = A_2$ and $S_{EI} = S_2$ (resp., $S_{DI} = S_2$). Let $M = M_1$, $\{0\} = M_2$, and note that $\mathcal{V}(S_1) \subset S_{EI}$ (resp., $\mathcal{V}(S_1) \subset S_{DI}$). All statements of the proposition are now a direct consequence of Theorem 4.6.1. □

In proving Theorems 4.3.2 and 4.3.3, we invoked the basic stability and boundedness definitions introduced in Section 3.1. In the next two problems we ask the reader to use Proposition 4.6.1 to prove these results.

Problem 4.6.8 Prove Theorem 4.3.2 using Proposition 4.6.1 and Theorem 4.3.1. □

Problem 4.6.9 Prove Theorem 4.3.3 using Proposition 4.6.1 and Lemma 4.3.1. □

Problem 4.6.10 Prove relation (4.2.1). □

Bibliography

[1] J. K. Hale, "Dynamical systems and stability," *J. Math. Anal. Appl.*, vol. 26, pp. 39–59, 1969.

[2] V. Lakshmikantham and S. Leela, *Differential and Integral Inequalities*, vol. I and vol. II, New York: Academic Press, 1969.

[3] A. N. Michel and C. J. Herget, *Algebra and Analysis for Engineers and Scientists*, Boston: Birkhäuser, 2007.

[4] A. N. Michel and R. K. Miller, *Qualitative Analysis of Large Scale Dynamical Systems*, New York: Academic Press, 1977.

[5] A. N. Michel, K. Wang, and B. Hu, *Qualitative Theory of Dynamical Systems—The Role of Stability Preserving Mappings*, 2nd Edition, New York: Marcel Dekker, 2001.

[6] R. K. Miller and A. N. Michel, *Ordinary Differential Equations*, New York: Academic Press, 1982.

[7] T. Yoshizawa, *Stability Theory by Liapunov's Second Method*, Tokyo: The Mathematical Society of Japan, 1966.

[8] V. I. Zubov, *Methods of A. M. Lyapunov and Their Applications*, Groningen, The Netherlands: P. Noordhoff, 1964.

Chapter 5

Applications to a Class of Discrete-Event Systems

In this chapter we apply the stability theory of dynamical systems defined on metric spaces in the analysis of an important class of discrete-event systems. We first give a description of the types of discrete-event systems that we consider and we then show that these discrete-event systems determine dynamical systems (Section 5.1). Next, we establish necessary conditions for the uniform stability, uniform asymptotic stability, and exponential stability of invariant sets with respect to the class of discrete-event systems considered herein (Section 5.2). We then apply these results in the analysis of two specific examples, a manufacturing system (Section 5.3) and a computer network (Section 5.4).

5.1 A Class of Discrete-Event Systems

Discrete-event systems (DES) are systems whose evolution in time is characterized by the occurrence of events at possibly irregular time intervals. For example, "logical" DES constitute a class of nonlinear discrete-time systems whose behavior can generally not be described by conventional nonlinear discrete-time systems defined on $\mathbb{R}^n$. Examples of logical DES models include the standard automata-theoretic models (e.g., the Moore and Mealy machines). A large class of the logical DES in turn, can be represented by *Petri Nets*.

We consider DES described by

$$G = (X, \mathcal{E}, f_e, g, \mathcal{E}_v) \tag{5.1.1}$$

where (X, d) is a metric space which denotes the *set of states* (the metric d is specified as needed), $\mathcal{E}$ is the *set of events*,

$$f_e \colon X \to X \tag{5.1.2}$$

for $e \in \mathcal{E}$ are operators,

$$g: X \to P(\mathcal{E}) - \{\emptyset\} \tag{5.1.3}$$

is the *enable function* and $\mathcal{E}_v \subset \mathcal{E}^{\mathbb{N}}$ is the *set of valid event trajectories*. Presently, for an arbitrary set Z, $Z^{\mathbb{N}}$ denotes the set of all sequences $\{z_k\}_{k\in\mathbb{N}}$, where $z_k \in Z$ for $k \in \mathbb{N}$ and $P(Z)$ denotes the power set of Z. We require that $f_e(x)$ be defined only when $e \in g(x)$. The inclusion of $P(\mathcal{E}) - \{\emptyset\}$ in the co-domain of g ensures that there will always exist some event that can occur. If for some physical system, it is possible that at some state no events occur, we model this by appending a *null event* e_0. When this occurs, the state remains the same while time advances. We call G defined in the above manner, a *discrete-event system.*

We associate "time" indices with states $x^k \in X$ and corresponding *enabled events* $e_k \in \mathcal{E}$ at time $k \in \mathbb{N}$ if $e_k \in g(x^k)$. Thus, if at state $x^k \in X$, event $e_k \in \mathcal{E}$ *occurs* at time $k \in \mathbb{N}$, then the next state is given by $x^{k+1} = f_{e_k}(x^k)$. Any sequence $\{x^k\} \in X^{\mathbb{N}}$ such that for all k, $x^{k+1} = f_{e_k}(x^k)$, where $e_k \in g(x^k)$, is a *state trajectory*. The set of all *event trajectories*, $\mathcal{E}^g \subset \mathcal{E}^{\mathbb{N}}$, is composed of sequences $\{e_k\} \in \mathcal{E}^{\mathbb{N}}$ having the property that there exists a state trajectory $\{x^k\} \in X^{\mathbb{N}}$ where for all k, $e_k \in g(x^k)$. Hence, *to each event trajectory, which specifies the order of the application of the operators* f_e, *there corresponds a unique state trajectory* (but, in general, not vice versa). We define the *set of valid event trajectories* $\mathcal{E}_v \subset \mathcal{E}^g \subset \mathcal{E}^{\mathbb{N}}$ as those event trajectories that are *physically possible* in the DES G. We let $\mathcal{E}_v(x^0) \subset \mathcal{E}_v$ denote the set of all event trajectories in $\mathcal{E}_v$ that initiate at $x^0 \in X$. We also utilize a *set of allowed event trajectories*, $\mathcal{E}_a \subset \mathcal{E}_v$, and correspondingly, $\mathcal{E}_a(x^0)$. All such event trajectories must be of infinite length. If one is concerned with the analysis of systems with finite length trajectories, this can be modeled by a null event as discussed above.

Next, for fixed $k \in \mathbb{N}$, let E_k denote an event sequence of k events that have occurred ($E_0 = \emptyset$ is the empty sequence). If $E_k = e_0, e_1, \dots, e_{k-1}$, let $E_k E \in \mathcal{E}_v(x^0)$ denote the *concatenation* of E_k and $E = e_k e_{k+1}, \dots$, i.e.,

$$E_k E = e_0, e_1, \dots, e_{k-1}, e_k, e_{k+1}, \dots.$$

We let $x(x^0, E_k, k)$ denote the *state reached at time* k from $x^0 \in X$ by application of an event sequence E_k such that $E_k E \in \mathcal{E}_v(x^0)$. By definition, $x(x^0, \emptyset, 0) = x^0$ for all $x^0 \in X$. We call $x(x^0, E_k, \cdot)$ a *DES motion*. Presently, we assume that for all $x^0 \in X$, if $E_k E \in \mathcal{E}_v(x^0)$ and $E_{k'} E' \in \mathcal{E}_v(x(x^0, E_k, k))$, then $E_k E_{k'} E' \in \mathcal{E}_v(x^0)$. Consequently, for all $x^0 \in X$, we have

$$x(x(x^0, E_k, k), E_{k'}, k') = x(x^0, E_k E_{k'}, k + k') \quad \text{for all } k, k' \in \mathbb{N}.$$

We now define $S_{G,\mathcal{E}_v}$ by

$$\begin{aligned} S_{G,\mathcal{E}_v} = \{p(\cdot, x^0, k_0) \colon p(k, x^0, k_0) = x(x^0, E_{k-k_0}, k - k_0), k \geq k_0, \\ k, k_0 \in \mathbb{N}, x^0 \in X, E_{k-k_0} E \in \mathcal{E}_v(x^0)\}. \end{aligned} \tag{5.1.4}$$

Let $T = \mathbb{N}$ and $A = X$. Then $\{T, X, A, S_{G,\mathcal{E}_v}\}$ is a dynamical system in the sense of Definition 2.2.3. Indeed, it is an *autonomous dynamical system* (refer to

Definition 4.1.1). We call $\{T, X, A, S_{G,\mathcal{E}_v}\}$ the *dynamical system determined by the discrete-event system* G. In the interests of brevity, we refer to this henceforth as a dynamical system $\{X, S_{G,\mathcal{E}_v}\}$. We define $S_{G,\mathcal{E}_a} \subset S_{G,\mathcal{E}_v}$ and $\{X, S_{G,\mathcal{E}_a}\}$ similarly. We note that (5.1.4) implies that $S_{G,\mathcal{E}_v}$ satisfies Assumption 3.5.1. In general, however, $S_{G,\mathcal{E}_a}$ does not satisfy Assumption 3.5.1.

5.2 Stability Analysis of Discrete-Event Systems

Because discrete-event systems of the type discussed above determine dynamical systems, the concepts of invariant sets and various types of stability of invariant sets arise in a natural manner. When $(S_{G,\mathcal{E}_v}, M)$ is invariant, stable, or asymptotically stable, we say that M is *invariant, stable*, or *asymptotically stable with respect to* $\mathcal{E}_v$, respectively. The invariance, stability, or asymptotic stability with respect to $\mathcal{E}_a$ are defined similarly.

Theorem 5.2.1 Let $\{X, S_{G,\mathcal{E}_v}\}$ be a discrete-event system and let $M \subset X$ be closed. Then M is *invariant* and *stable* with respect to $\mathcal{E}_v$ if and only if there exist neighborhoods of M, given by $B_i = \{x \in X : d(x, M) < r_i\}$, $i = 1, 2$, where $0 < r_2 \leq r_1$, and a mapping $V : B_1 \to \mathbb{R}^+$ that satisfies the following conditions.

(i) There exist $\psi_1, \psi_2 \in \mathcal{K}$ such that

$$\psi_1(d(x, M)) \leq V(x) \leq \psi_2(d(x, M))$$

for all $x \in B_1$.

(ii) $V(x(x^0, E_k, k))$ is a nonincreasing function for $k \in \mathbb{N}$ for all E_k such that $E_k E \in \mathcal{E}_v(x^0)$ whenever $x^0 \in B_2$.

Proof. Because $S_{G,\mathcal{E}_v}$ is an autonomous system that satisfies Assumption 3.5.1, the theorem is an immediate consequence of Theorem 4.1.3. The choices of B_1 and B_2 are given as X_1 and A_1 in Theorem 4.1.3. □

Theorem 5.2.2 Let $\{X, S_{G,\mathcal{E}_v}\}$ be a discrete-event system and let $M \subset X$ be closed. Then M is *invariant* and *asymptotically stable* with respect to $\mathcal{E}_v$ if and only if there exist neighborhoods of M given by $B_i = \{x \in M : d(x, M) < r_i\}$, $i = 1, 2$, where $0 < r_2 \leq r_1$, and a mapping $V : B_1 \to \mathbb{R}^+$ that satisfies conditions (i) and (ii) of Theorem 5.2.1, and furthermore, $\lim_{k\to\infty} V(x(x^0, E_k, k)) = 0$ for all E_k such that $E_k E \in \mathcal{E}_v(x^0)$ whenever $x^0 \in B_2$.

Proof. The proof of this theorem is a direct consequence of Theorem 4.1.4. □

When considering the stability or asymptotic stability of an invariant set M with respect to $\mathcal{E}_a$, if we replace $\mathcal{E}_v$ by $\mathcal{E}_a$ everywhere in the statements of Theorems 5.2.1 and 5.2.2, then the "if" parts (i.e., the sufficient conditions) remain true; however, the "only if" parts of these results (i.e., the necessary conditions) in general do not hold because we do not require that $S_{G,\mathcal{E}_a}$ satisfy Assumption 3.5.1.

5.3 Analysis of a Manufacturing System

In Figure 5.3.1 we depict a manufacturing system that processes batches of N different types of jobs according to a priority scheme. Presently, we use the term "job" in a very general sense, and the completion of a job may mean, for example, the processing of a batch of 10 *parts*, the processing of a batch of 6.53 *tasks*, and the like. There are N *producers* P_i, $i = 1, \ldots, N$, of different types of jobs. The producers P_i place batches of their jobs in their respective *buffers* B_i, $i = 1, \ldots, N$. The buffers B_i have *safe capacity limits* $b_i > 0$, $i = 1, \ldots, N$. Let x_i, $i = 1, \ldots, N$, denote the number of jobs in buffer B_i. Let x_i for $i = N+1, \ldots, 2N$ denote the number of P_{i-N} type jobs in the machine. The machine can safely process less than or equal to $M > 0$ jobs of any type at any time. As the machine finishes processing batches of P_i type jobs, they are placed in their respective *output bins* (P_i-bins). The producers P_i can only place batches of jobs in their buffers B_i if $x_i < b_i$. Also, there is a priority scheme whereby batches of P_i type jobs are only allowed to enter the machine when $x_j = 0$ for all j such that $j < i \leq N$, that is, only when there are no jobs in any buffers to the left of buffer B_i.

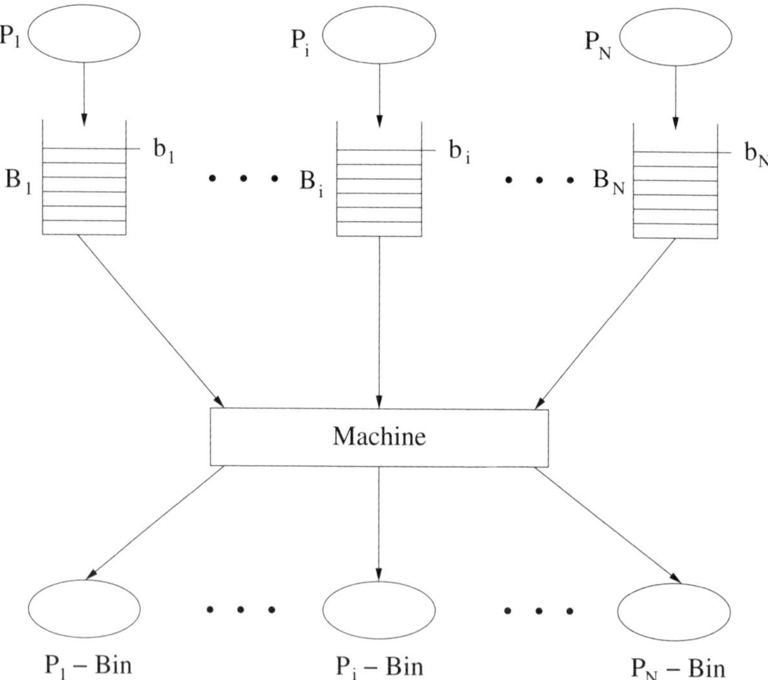

Figure 5.3.1: Manufacturing system.

We now specify the DES model G for the present manufacturing system. To this end we let $X = \mathbb{R}^{2N}$ and $x^k \in X$, where $(x^k)^T = (x_1, x_2, \ldots, x_{2N})^k$ denotes the state at time k. Let the set of events $\mathcal{E}$ be composed of the events e_{P_i}, $i = 1, \ldots, N$

(representing producer P_i placing a batch of α_{P_i} jobs into buffer B_i), the events e_{a_i}, $i = 1, \ldots, N$ (representing a batch of α_{a_i} P_i jobs from buffer B_i arriving at the machine for processing), and the events e_{d_i}, $i = 1, \ldots, N$ (representing a batch of α_{d_i} P_i jobs departing from the machine after they are processed and placed into their respective output bins). When using the term e_{P_i} type of event, e_{a_i} type of event, or e_{d_i} type of event, we mean an event e_{P_i}, e_{a_i}, or e_{d_i} for any α_{P_i}, α_{a_i}, or α_{d_i}, respectively. It is assumed that all jobs are *infinitely divisible*, so that, for example, a batch of $5\frac{1}{3}$ jobs can be placed into buffer B_i, 8.563 of these jobs can be placed into the machine for processing, then 3.14 of these could be processed. We note, however, that the results described in the following can be modified to be applicable for *discrete jobs* as well. Now let $\gamma \in (0, 1]$ denote a fixed *parameter*. According to the restrictions imposed in the preceding discussion, the *enable function* g and the *event operators* f_e for $e \in g(x^k)$, are now defined as follows.

(i) If $x_i < b_i$ for some $i = 1, \ldots, N$, then $e_{P_i} \in g(x^k)$ and

$$f_{e_{P_i}}(x^k)^T = (x_1, \ldots, x_i + \alpha_{P_i}, \ldots, x_N, x_{N+1}, \ldots, x_{2N}),$$

where $\alpha_{P_i} \le |x_i - b_i|$.

(ii) If $\sum_{j=N+1}^{2N} x_j < M$, and for some $i = 1, \ldots, N$, $x_i > 0$, and $x_l = 0$ for all l, $l < i \le N$, then $e_{a_i} \in g(x^k)$ and

$$f_{e_{a_i}}(x^k)^T = (x_1, \ldots, x_i - \alpha_{a_i}, \ldots, x_N, x_{N+1}, \ldots, x_{N+i} + \alpha_{a_i}, \ldots, x_{2N}),$$

where $\gamma x_i \le \alpha_{a_i} \le \min\left\{x_i, \left|\sum_{j=N+1}^{2N} x_j - M\right|\right\}$.

(iii) If $x_i > 0$ for any i, $i = 1, \ldots, N$, then $e_{d_i} \in g(x^k)$ and

$$f_{e_{d_i}}(x^k)^T = (x_1, \ldots, x_N, x_{N+1}, \ldots, x_{N+i} - \alpha_{d_i}, \ldots, x_{2N}),$$

where $\gamma x_{N+i} \le \alpha_{d_i} \le x_{N+i}$.

In case (i), every time that an event e_{P_i} occurs, some amount of jobs arrives at the buffers but the producers will never overfill the buffers.

In case (ii), the e_{a_i} are enabled only when the machine is not too full and the ith buffer has appropriate priority. The amount of jobs that can arrive at the machine is limited by the number available in the buffers and by how many the machine can process at once. We require that $\gamma x_i \le \alpha_{a_i}$ so that nonnegligible batches of jobs arrive when they are allowed.

In case (iii), the constraints on α_{d_i} ensure that the number of jobs that can depart the machine is limited by the number of jobs in the machine and that nonnegligible amounts of jobs depart from the machine.

We let $\mathcal{E}_v = \mathcal{E}^g$; that is, the set of all event trajectories is defined by g and f_e for $e \in g(x^k)$. The manufacturing system operates in a standard asynchronous fashion.

Now let

$$M = \left\{x \in X \colon x_i \le b_i, i = 1, \ldots, N, \text{ and } \sum_{j=N+1}^{2N} x_j \le M\right\} \tag{5.3.1}$$

which represents all states for which the manufacturing system is in a *safe operating mode*. It is easy to see that M is *invariant* by letting $x^k \in M$ and by showing that no matter which event occurs, it will always be true that the next state $x^{k+1} \in M$.

In the following, we study the *stability properties* of the manufacturing system formulated above. Our results show that under conditions when our manufacturing system starts in an unsafe mode (too many jobs in a buffer, or in the machine, or in both), it will eventually return to a safe operating condition.

To simplify our notation, we let $(x^k)^T = (x_1, \dots, x_{2N})$, $(x^{k+1})^T = (x'_1, \dots, x'_{2N})$, $\widetilde{x} = (\widetilde{x}_1, \dots, \widetilde{x}_{2N})^T$, and $\widetilde{x}' = (\widetilde{x}'_1, \dots, \widetilde{x}'_{2N})^T$, suppressing the "$k$" notation as indicated. For this manufacturing system example we take

$$d(x, M) = \inf\left\{ \sum_{j=1}^{2N} |x_j - \widetilde{x}_j| : \widetilde{x} \in M \right\}. \tag{5.3.2}$$

Proposition 5.3.1 For the manufacturing system, the closed invariant set M defined in (5.3.1) is *stable* with respect to $\mathcal{E}_v$. □

Proof. We choose $V(x^k) = d(x^k, M)$. We show that the function $V(x^k)$ satisfies hypotheses (i) and (ii) of Theorem 5.2.1 for all $x^k \notin M$.

Hypothesis (i) follows directly from the choice of $V(x^k)$. To verify that $V(x^k)$ satisfies hypothesis (ii) we show that $V(x^k) \geq V(x^{k+1})$ for all $x^k \notin M$, no matter what event $e \in g(x^k)$ occurs, causing $x^{k+1} = f_e(x^k)$, whenever it lies on an event trajectory in $\mathcal{E}_v$.

(a) For $x^k \notin M$, if e_{P_i} occurs for some i, $i = 1, \dots, N$, we need to show that

$$\inf\left\{ \sum_{j=1}^{2N} |x_j - \widetilde{x}_j| : \widetilde{x} \in M \right\} \geq \inf\left\{ \sum_{j=1, j\neq i}^{2N} |x_j - \widetilde{x}'_j| + |x_i + \alpha_{P_i} - \widetilde{x}'_i| : \widetilde{x}' \in M \right\}. \tag{5.3.3}$$

It suffices to show that for all $\widetilde{x} \in M$ at which the inf is achieved on the left of (5.3.3), there exists $\widetilde{x}' \in M$ such that

$$\sum_{j=1}^{2N} |x_j - \widetilde{x}_j| \geq \sum_{j=1, j\neq i}^{2N} |x_j - \widetilde{x}'_j| + |x_i + \alpha_{P_i} - \widetilde{x}'_i|. \tag{5.3.4}$$

If we choose $\widetilde{x}'_l = \widetilde{x}_l$ for all $l \neq i$, then it suffices to show that for all $\widetilde{x}_i$, $0 \leq \widetilde{x}_i \leq b_i$, at which the inf on the left side of (5.3.3) is achieved, there exists $\widetilde{x}'_i$, $0 \leq \widetilde{x}'_i \leq b_i$, such that

$$|x_i - \widetilde{x}_i| \geq |x_i + \alpha_{P_i} - \widetilde{x}'_i| \tag{5.3.5}$$

where $\alpha_{P_i} \leq |x_i - b_i|$. Choosing $\widetilde{x}'_i = x_i + \alpha_{P_i}$, so that $0 \leq \widetilde{x}'_i \leq b_i$, results in $\widetilde{x}' \in M$, and relation (5.3.5) is satisfied.

(b) For $x^k \notin M$, if e_{a_i} occurs for some i, $i = 1, \dots, N$, then following the above approach, it suffices to show that for all $\widetilde{x} \in M$ at which the inf is achieved, there exists $\widetilde{x}' \in M$ such that

$$\sum_{j=1}^{2N} |x_j - \widetilde{x}_j| \geq \sum_{j=1, j\neq i, N+i}^{2N} |x_j - \widetilde{x}'_j| + |x_i - \alpha_{a_i} - \widetilde{x}'_i| + |x_{N+i} + \alpha_{a_i} - \widetilde{x}'_{N+i}|. \tag{5.3.6}$$

Choose $\widetilde{x}_l' = \widetilde{x}_l$ for all $l \neq i, N+i$. It suffices to show that for all $\widetilde{x}_i$ and $\widetilde{x}_{N+i}$, there exist $\widetilde{x}_i', \widetilde{x}_{N+i}'$, such that

$$|x_i - \widetilde{x}_i| \geq |x_i - \alpha_{a_i} - \widetilde{x}_i'| \tag{5.3.7}$$

and

$$|x_{N+i} - \widetilde{x}_{N+i}| \geq |x_{N+i} + \alpha_{a_i} - \widetilde{x}_{N+i}'|. \tag{5.3.8}$$

For (5.3.7), if $x_i \leq b_i$, then the inf is achieved so that $|x_i - \widetilde{x}_i| = |x_i - \alpha_{a_i} - \widetilde{x}_i'| = 0$, whereas if $x_i > b_i$, the inf is achieved at $\widetilde{x}_i = b_i$. Therefore, $|x_i - b_i| \geq |x_i - \alpha_{a_i} - \widetilde{x}_i'|$, because either $\widetilde{x}_i' = b_i$ or $\widetilde{x}_i' = x_i - \alpha_{a_i}$. The case for (5.3.8) is similar to case (a) above.

The proof for when e_{d_i} occurs is similar to the case for (5.3.8). □

We note that for the above manufacturing system, the closed set M specified in (5.3.1) is not asymptotically stable in the large with respect to $\mathcal{E}_v$. (We ask the reader to prove this assertion in the problem section. Alternatively, the reader may wish to refer to [13, Proposition 2] for the proof.)

In the following, we identify a hypothesis that ensures asymptotic stability in the large for the above manufacturing system. To this end, we let $\mathcal{E}_a \subset \mathcal{E}_v$ denote the set of event trajectories such that each type of event e_{P_i}, e_{a_i} and e_{d_i}, $i = 1, \ldots, N$, occurs *infinitely often* on each event trajectory $E \in \mathcal{E}_a$. If we assume for the manufacturing system that only events which lie on event trajectories in $\mathcal{E}_a$ occur, then it is always the case that eventually each type of event (e_{P_i}, e_{a_i} and e_{d_i}, $i = 1, \ldots, N$) will occur.

Proposition 5.3.2 For the manufacturing system, the closed invariant set M given in (5.3.1) is *asymptotically stable in the large* with respect to $\mathcal{E}_a$ where $\mathcal{E}_a \subset \mathcal{E}_v$ is as defined above. □

Proof. By Proposition 5.3.1, M is stable with respect to $\mathcal{E}_a$. To establish asymptotic stability, we show that $V(x^k) \to 0$ for all E_k such that $E_k E \in \mathcal{E}_a(x^0)$ as $k \to \infty$ for all $x^k \notin M$.

Because $\alpha_{a_i} \geq \gamma x_i$ and $\alpha_{d_i} \geq \gamma x_{N+i}$, where $\gamma \in (0, 1]$, if e_{a_i} and e_{d_i}, $i = 1, \ldots, N$, occur infinitely often (as the restrictions on $\mathcal{E}_a$ guarantee), x_i and x_{N+i} will converge so that $V(x^k) \to 0$ as $k \to \infty$ (of course it could happen that $V(x^k) = 0$ for some finite k). Therefore, if the manufacturing system starts out in an unsafe operating mode, it will eventually enter a safe operating mode. □

5.4 Load Balancing in a Computer Network

We consider a computer network specified by a digraph (C, A) where $C = \{1, \ldots, N\}$ represents a set of computers labeled by $i \in C$ and $A \subset C \times C$ specifies the set of connections; that is, if $(i, j) \in A$, then computer i is connected to computer j. We require that every computer is connected to the network so that if $i \in C$, then there exists a pair $(i, j) \in A$ or a pair $(j, i) \in A$ for some $j \in C$. Also, we assume that if $(i, j) \in A$, then $(j, i) \in A$ and furthermore, if $(i, j) \in A$, then $i \neq j$. We assume that for each computer there is a buffer that holds tasks (the load), and we assume that each task (load) can be executed by any of the computers in the network. We let the load of computer $i \in C$ be denoted $x_i \geq 0$ and we assume that each connection in the network, $(i, j) \in A$, allows computer i to pass a portion of its load to computer j. We also assume that in the case of every connection (i, j), computer i is able to sense the size of the load of computer j, and furthermore, when $(i, j) \notin A$, then computer i cannot pass a load directly to computer j, nor can computer i sense the load of computer j.

We assume that the initial load distribution in the computer network is uneven and we wish to establish rules (resp., an algorithm) under which a more even load distribution in the computer network is realized. We assume that no tasks are performed by any of the computers during the load-balancing process.

In the literature, distinctions are made between *continuous loads* (also called fluid loads) and *discrete loads*. In the former case, it is assumed that the computer load is infinitely divisible whereas in the case of the latter, a load is a multiple of a uniformly sized block that is not divisible. In the following, we consider only continuous loads.

We next specify the discrete-event system model G for the computer balancing problem described above. To this end, we let $X = \mathbb{R}^N$ denote the state space, and we let $x^k = (x_1, \dots, x_N)^T$ and $x^{k+1} = (x'_1, \dots, x'_N)^T$ denote the state at times k and $k+1$, respectively. Let $e^{ij}_{\alpha_k}$ denote the event that a load of amount α_k is passed from computer i to computer j at time k. If the state is x^k, then for some $(i, j) \in A$, $e^{ij}_{\alpha_k}$ occurs to produce the next state x^{k+1}. Let $\mathcal{E} = \{e^{ij}_{\alpha} : (i, j) \in A, \alpha \in \mathbb{R}_+\}$ denote the infinite set of events. (Note that e^{ij}_0 are valid events.) In the following, "an event of type e^{ij}_{α}" means the passing of a load of the amount $\alpha > 0$ from computer i to computer j.

We now specify the enable function g and the event operator f_e for $e \in g(x^k)$. We choose the parameter $\gamma \in (0, 1/2]$.

(i) If for any $(i, j) \in A$, $x_i > x_j$, then $e^{ij}_{\alpha} \in g(x^k)$ and $f_e(x^k) = x^{k+1}$ where $e = e^{ij}_{\alpha}$, $x'_i = x_i - \alpha$, $x'_j = x_j + \alpha$, $x'_l = x_l$ for all $l \neq i, l \neq j$, and $\gamma |x_i - x_j| \leq \alpha \leq (1/2)|x_i - x_j|$.

(ii) If for any $(i, j) \in A$, $x_i = x_j$, then $e^{ij}_0 \in g(x^k)$ and $f_e(x^k) = x^k$ where $e = e^{ij}_0$.

Let $\mathcal{E}_v = \mathcal{E}^g$ and let $\mathcal{E}_a \subset \mathcal{E}_v$ denote the set of event trajectories such that events of each type e^{ij}_{α} occur infinitely often on each $E \in \mathcal{E}_a$. This assumption ensures that each pair of connected computers will continually try to balance the load between them.

In order to ensure load balancing, we make use of the set

$$M = \{x \in X : x_i = x_j \text{ for all } (i, j) \in A\}, \tag{5.4.1}$$

which represents *perfect load balancing*. It is easy to show that M is invariant by letting $x^k \in M$ and then showing that no matter which event occurs, $x^{k+1} \in M$.

Proposition 5.4.1 For the computer network load-balancing problem, the following is true.

(a) M is *stable* with respect to $\mathcal{E}_v$.

(b) M is *not asymptotically stable* with respect to $\mathcal{E}_v$.

(c) M is *asymptotically stable in the large* with respect to $\mathcal{E}_a \subset \mathcal{E}_v$. □

In proving Proposition 5.4.1, one proceeds similarly as was done in the proof of Propositions 5.3.1 and 5.3.2 for the manufacturing system. In the proof of Proposition 5.4.1, we find it convenient to choose as the distance function

$$d(x, M) = \inf\{ \max\{|x_1 - \widetilde{x}_1|, \ldots, |x_N - \widetilde{x}_N|\} : \widetilde{x} \in M\}$$

where $x = (x_1, \ldots, x_N)^T$ and $\widetilde{x} = (\widetilde{x}_1, \ldots, \widetilde{x}_N)^T$, and as the Lyapunov function

$$V(x) = d(x, M),$$

and applying the results of Section 5.2. We leave the details of these proofs as an exercise for the reader.

5.5 Notes and References

The applications to discrete-event systems presented in this chapter are based on [8], [9], and [13]. For additional background material on discrete-event systems, refer, for example, to [1], [3], and [15].

The manufacturing system considered in Section 5.3 is a generalization of systems used in the study of a simple "mutual exclusion problem" in the computer science literature (see, e.g., [7] and [10]) and is similar to several applications addressed in the DES literature (e.g., [6] and [15]).

The results of Section 5.3 have been extended in [11] and [12] to be applicable to "discrete" jobs.

Usage of the set $\mathcal{E}_a$ in Proposition 5.3.2 imposes what is called in the computer science literature a "fairness constraint" (see, e.g., [5]). One can guarantee that this constraint is met using a mechanism for sequencing access to the machine. Such fairness constraints have also been used in the study of temporal logic (e.g., [3] and [7]) and the mutual exclusion problem in the computer science literature (e.g., [14]).

The load-balancing problem presented in Section 5.4 was motivated by a similar problem studied in [1]. Various other forms of this problem have also been addressed in the DES literature (e.g., [2]) and in the computer science literature (e.g., [1], [2], and [4]), and in the numerous references cited in these sources.

5.6 Problems

Problem 5.6.1 For the manufacturing system discussed in Section 5.3, let M denote the closed invariant set defined in (5.3.1). Prove that M is not asymptotically stable in the large with respect to $\mathcal{E}_v$, where $\mathcal{E}_v$ is the same as in Proposition 5.3.1.

Hint: Let $x_i > b_i$ for all $i = 1, \ldots, N$, where the b_i are as defined in (5.3.1). Choose $x_{N+i} > 0$ for some i so that e_{d_i} occurs, to process P_i type jobs and to put them into the

P_i-bin. For each successive time $\alpha_{d_i} = \gamma x_{N+i}$, it can happen that $E = e_{d_i} e_{d_i} e_{d_i} \cdots$ (a constant string) and $E \in \mathcal{E}_v$. This shows that M is not asymptotically stable in the large with respect to $\mathcal{E}_v$. □

Problem 5.6.2 The matrix equations that describe the dynamical behavior of a *Petri net* P are given by

$$M_{k+1} = M_k + A^T u_k \tag{5.6.1}$$

where $k \in \mathbb{N}$, $M_k \in \mathbb{N}^m$, A is an $n \times m$ matrix of integers (the *incidence matrix*), and $u_k \in \{0,1\}^n$ denotes a *firing vector* (refer, e.g., to [8], [9], and [10] for background material on Petri nets). A Petri net P is said to be *uniformly bounded* (resp., *uniformly ultimately bounded*) if the motions of the dynamical system determined by (5.6.1) are uniformly bounded (resp., uniformly ultimately bounded). Prove that the following statements are true.

(a) A Petri net P is *uniformly bounded* if there exists an m-vector $\varphi > 0$ such that $A\varphi \leq 0$ (inequality of vectors is understood to be componentwise).

(b) A Petri net P is *uniformly ultimately bounded* if there exist an m-vector $\varphi > 0$ and an n-vector $b > 0$ such that $A\varphi \leq -b$.

Hint: Apply Theorems 3.4.4 and 3.4.5 with the choice of $M = \{0\}$ and $V(x) = x^T \varphi$ for $x = (n_1, \ldots, n_m) \in \mathbb{N}^m$. □

Problem 5.6.3 Prove Proposition 5.4.1, using the hints given in Section 5.4. □

Bibliography

[1] D. P. Bertsekas and J. N. Tsitsiklis, *Parallel and Distributed Computation: Numerical Methods*, Englewood Cliffs, NJ: Prentice Hall, 1989.

[2] R. K. Boel and J. H. van Schuppen, "Distributed routing for load balancing," *Proc. IEEE*, vol. 77, pp. 210–221, 1989.

[3] E. M. Clark, M. C. Browne, E. A. Emerson, and A. P. Sistla, "Using temporal logic for automatic verification of finite state systems," in *Logic and Models of Concurrent Systems*, K. R. Apt, ed., New York: Springer-Verlag, 1985, pp. 3–25.

[4] G. Cybenko, "Dynamic load balancing for distributed memory multiprocessors," *Technical Report 87-1*, Department of Computer Science, Tufts University, Medford, MA, 1987.

[5] N. Francez, *Fairness*, New York: Springer-Verlag, 1986.

[6] J. F. Knight and K. M. Passino, "Decidability for a temporal logic used in discrete event system analysis," *Int. J. Control*, vol. 52, pp. 1489–1506, 1990.

[7] Z. Manna and A. Pnueli, "Verification of concurrent programs: A temporal proof system," *Report No. STAN-CS-83-967*, Department of Computer Science, Stanford University, 1983.

[8] A. N. Michel, K. Wang, and K. M. Passino, "Qualitative equivalence of dynamical systems with applications to discrete event systems," *Proc. of the 31st IEEE Conference on Decision and Control*, Tuscon, AZ, Dec. 1992, pp. 731–736.

[9] A. N. Michel, K. Wang, and K. M. Passino, "Stability preserving mappings and qualitative equivalence of dynamical systems-Part I," *Avtomatika i Telemekhanika*, vol. 10, pp. 3–12, 1994.

[10] T. Murata, "Petri nets: Properties, analysis, and applications," *Proc. IEEE*, vol. 77, pp. 541–580, 1989.

[11] K. M. Passino, A. N. Michel, and P. J. Antsaklis, "Stability analysis of discrete event systems," *Proc. of the 28th Annual Allerton Conference on Communication, Control, and Computing*, University of Illinois at Urbana-Champaign, Oct. 1990, pp. 487–496.

[12] K. M. Passino, A. N. Michel, and P. J. Antsaklis, "Lyapunov stability of discrete event systems," *Proc. of the 1991 American Control Conference*, Boston, MA, June 1991, pp. 2911–2916.

[13] K. M. Passino, A. N. Michel, and P. J. Antsaklis, "Lyapunov stability of a class of discrete event systems," *IEEE Trans. Autom. Control*, vol. 39, pp. 269–279, 1994.

[14] M. Raynal, *Algorithms for Mutual Exclusion*, Cambridge, MA: MIT Press, 1986.

[15] J. G. Thistle and W. M. Wonham, "Control problems in a temporal logic framework," *Int. J. Control*, vol. 44, pp. 943–976, 1986.

Chapter 6

Finite-Dimensional Dynamical Systems

In the present chapter we apply the results of Chapter 3 (for the stability of invariant sets and the boundedness of motions of general dynamical systems defined on metric spaces) in the analysis of continuous finite-dimensional dynamical systems determined by differential equations (E), discrete-time finite-dimensional dynamical systems determined by difference equations (D), and finite-dimensional discontinuous dynamical systems. When considering various stability types, our focus is on invariant sets that are equilibria.

This chapter consists of eight parts. In the first section we introduce some important preliminary results which we require throughout the chapter. In the second, third, and fourth sections we present the principal Lyapunov stability and boundedness results for continuous dynamical systems, discrete-time dynamical systems, and discontinuous dynamical systems, respectively. Throughout these sections we consider specific examples to demonstrate applications of the various results. In the fifth, sixth, and seventh sections we establish converse theorems for the results of the second, third, and fourth sections, respectively. In the eighth section we provide some background material concerning the continuous dependence of the solutions of ordinary differential equations on initial conditions.

The results of the present chapter are general and fundamental in nature. In the next chapter, where we continue the qualitative analysis of finite-dimensional dynamical systems, we concentrate on important specialized results.

6.1 Preliminaries

In this section we present preliminary material which we require throughout this chapter. The present section is organized into seven parts. In the first three subsections we recall facts concerning finite-dimensional dynamical systems determined by ordinary differential equations, ordinary difference equations, and discontinuous dynamical systems, respectively. In the fourth subsection we rephrase the various definitions of stability and boundedness given in Chapter 3 for dynamical systems defined on metric spaces, for the case of finite-dimensional dynamical systems (defined on $\mathbb{R}^n$). In the fifth subsection we introduce several characterizations of Lyapunov functions and in the sixth subsection we discuss an important special class of Lyapunov functions, quadratic forms. In the final subsection we present some geometric interpretations and motivation for Lyapunov stability results (for two-dimensional autonomous systems).

Before proceeding, a comment concerning the notation that we employ in the remainder of this book to designate Lyapunov functions is in order. When addressing *general dynamical systems*, $\{T, X, A, S\}$, defined on metric spaces, we used uppercase letters in Chapters 3 and 4 to denote such functions (V-functions). In keeping with the notation most frequently used in texts on *specific classes* of *finite-dimensional* and *infinite-dimensional dynamical systems*, we use lowercase letters to denote Lyapunov functions when dealing with such systems.

A. Dynamical systems determined by ordinary differential equations

We consider systems of first-order ordinary differential equations of the form

$$\dot{x} = f(t, x) \tag{E}$$

where $t \in \mathbb{R}^+$, $x \in \Omega \subset \mathbb{R}^n$, $\dot{x} = dx/dt$, and $f\colon \mathbb{R}^+ \times \Omega \to \mathbb{R}^n$. We assume that Ω is an open connected set and that $0 \in \Omega$. We always assume that for each $(t_0, x_0) \in \mathbb{R}^+ \times \Omega$, (E) possesses at least one solution (refer to Theorem 2.3.1), we sometimes require that these solutions be unique (refer to Theorem 2.3.2) and we usually (but not always) require that the solutions, denoted by $\varphi(t, t_0, x_0)$, exist for all $t \in [t_0, \infty)$. Recall that $\varphi(t_0, t_0, x_0) = x_0$. Under the assumption that $f \in C[\mathbb{R}^+ \times \Omega, \mathbb{R}^n]$, the solutions $\varphi(t, t_0, x_0)$ of (E) are continuous with respect to initial conditions (t_0, x_0) (refer to the appendix section, Section 6.8). Recall that (E) determines a dynamical system $\{T, X, A, S_E\}$ where $T = \mathbb{R}^+$, $X = \mathbb{R}^n$, $A \subset X$, and S_E denotes the set of motions determined by (E). We usually denote this system simply by S_E (refer to Subsection 2.3C).

In the present chapter we study primarily the stability properties of invariant sets $M \subset \mathbb{R}^n$ for the special case when $M = \{x_e\} \subset \mathbb{R}^n$, and the boundedness of S_E. Recall that in this case we say that x_e is an *equilibrium* (resp., *equilibrium point*) of (E).

In Example 3.1.3 we noted that if $f(t, x_e) = 0$ for all $t \in \mathbb{R}^+$, then $\{x_e\} = M$ is an invariant set with respect to S_E (i.e., (S_E, M) is invariant). Furthermore, it is easily shown that when (E) possesses a unique solution for every $(t_0, x_0) \in \mathbb{R}^+ \times \Omega$, then $(S_E, \{x_e\})$ is invariant if and only if $f(t, x_e) = 0$ for all $t \in \mathbb{R}^+$.

To simplify our language throughout this chapter, we say that "x_e is an equilibrium of (E)", rather than "the set $\{x_e\}$ is invariant with respect to S_E" (or, "$(S_E, \{x_e\})$

is invariant"); "the equilibrium x_e of (E) is asymptotically stable", rather than "the invariant set $\{x_e\}$ of (E) is asymptotically stable" (or, "$(S_E, \{x_e\})$ is asymptotically stable"); and so forth.

Without loss of generality, we may assume that a given equilibrium x_e of (E) is located at the origin (i.e., $x_e = 0$). To see this, suppose that $x_e \neq 0$ is an equilibrium of (E). Let $y = x - x_e$ and $\bar{f}(t, y) = f(t, y + x_e)$. Then (E) can be written as

$$\dot{y} = \bar{f}(t, y),$$

with $\bar{f}(t, 0) = 0$ for all $t \in \mathbb{R}^+$.

As in Chapter 3, we employ Lyapunov functions $v \in C[B(r) \times \mathbb{R}^+, \mathbb{R}]$ where $B(r) \subset \Omega, B(r) = \{x \in \mathbb{R}^n : |x| < r\}$ with $r > 0$. For such functions, we define the *upper-right derivative of v with respect to t along the solutions of (E)* $(\varphi(\cdot, t, x))$ by

$$v'_{(E)}(x, t) = \overline{\lim_{\Delta t \to 0^+}} \sup_{\varphi(t,t,x)=x} \frac{1}{\Delta t}\big[v(\varphi(t + \Delta t, t, x), t + \Delta t) - v(x, t)\big] \quad (6.1.1)$$

which is defined on $B(r) \times \mathbb{R}^+$. When for each $(t_0, x_0) \in \mathbb{R}^+ \times B(r)$, (E) has a unique solution, (6.1.1) reduces to

$$v'_{(E)}(x, t) = \overline{\lim_{\Delta t \to 0^+}} \frac{1}{\Delta t}\big[v(\varphi(t + \Delta t, t, x), t + \Delta t) - v(x, t)\big]. \quad (6.1.2)$$

If in addition, v satisfies a local Lipschitz condition in x, that is, for each $T > 0$ there exists a constant $L > 0$ such that

$$|v(x, t) - v(y, t)| \leq L|x - y|$$

for all $x, y \in B(r)$ and all $t \in [0, T]$, then (6.1.2) can equivalently be expressed as

$$v'_{(E)}(x, t) = \overline{\lim_{\Delta t \to 0^+}} \frac{1}{\Delta t}\big[v(x + (\Delta t) f(t, x), t + \Delta t) - v(x, t)\big]. \quad (6.1.3)$$

(We ask the reader to verify (6.1.3).) Finally, if in addition to the above assumptions, we have $v \in C^1[B(r) \times \mathbb{R}^+, \mathbb{R}]$, then (6.1.3) assumes the equivalent form

$$v'_{(E)}(x, t) = \sum_{i=1}^{n} \frac{\partial v}{\partial x_i}(x, t) f_i(t, x) + \frac{\partial v}{\partial t}(x, t) \quad (6.1.4)$$

where $f(t, x) = [f_1(t, x), \ldots, f_n(t, x)]^T$ is given in (E).

B. Dynamical systems determined by ordinary difference equations

We now consider systems of first-order ordinary difference equations of the form

$$x(k + 1) = f(k, x(k)) \quad (D)$$

where $k \in \mathbb{N}$, $x(k) \in \Omega \subset \mathbb{R}^n$, and $f\colon \mathbb{N} \times \Omega \to \Omega$. We assume that Ω is an open connected set and that $0 \in \Omega$. For each $(k_0, x_0) \in \mathbb{N} \times \Omega$ there exists a unique solution $\varphi(k, k_0, x_0)$ with $\varphi(k_0, k_0, x_0) = x(k_0) = x_0$. We recall that (D) determines a dynamical system $\{T, X, A, S_D\}$ where $T = \mathbb{N}$, $X = \mathbb{R}^n$, $A \subset X$, and S_D denotes the set of motions determined by (D). We usually denote this system simply by S_D (refer to Section 2.5).

As in the case of dynamical systems determined by (E), we concentrate primarily in studying the qualitative properties of an equilibrium x_e of D (i.e., in studying the stability properties of an invariant set $M = \{x_e\}$). It is easily shown that a point $x_e \in \Omega$ is an equilibrium of (D) if and only if

$$x_e = f(k, x_e)$$

for all $k \in \mathbb{N}$. As in the case of ordinary differential equations, we may assume without loss of generality that the equilibrium x_e of (D) is located at the origin $(x_e = 0)$.

Finally, we let $\varphi(k, k_0, x_0)$ denote any solution of (D) with initial conditions $\varphi(k_0, k_0, x_0) = x(k_0) = x_0$. For a function $v \in C[\Omega \times \mathbb{N}, \mathbb{R}]$, we define the *first forward difference of v with respect to k along the solutions of* (D) by

$$\begin{aligned} \Delta_{(D)} v(x, k) &= v(\varphi(k+1, k, x), k+1) - v(\varphi(k, k, x), k) \\ &= v(f(k, x), k+1) - v(x, k). \end{aligned} \tag{6.1.5}$$

C. Discontinuous dynamical systems (DDS)

In the present chapter we address finite-dimensional discontinuous dynamical systems (finite-dimensional DDS), $\{T, X, A, S\}$, where $T = \mathbb{R}^+, \mathbb{R}^n = X \supset A$, and the motions of S are determined by the solutions $\varphi(\cdot, t_0, x_0)$ of discontinuous ordinary differential equations of the type specified later. As in Chapter 3 we assume that the set of times at which discontinuities may occur is unbounded and discrete and is of the form

$$E_\varphi = \{\tau_1^\varphi, \tau_2^\varphi, \dots : \tau_1^\varphi < \tau_2^\varphi < \cdots\}.$$

In the above expression, E_φ signifies that in general, different motions may possess different sets of times at which discontinuities may occur. Usually, the particular set E_φ in question is clear from context and accordingly, we suppress the φ-notation and simply write

$$E = \{\tau_1, \tau_2, \dots : \tau_1 < \tau_2 < \cdots\}.$$

We find it sometimes useful to express the motions (solutions) of finite-dimensional DDS by

$$\varphi(t, t_0, x_0) = x^{(k)}(t, \tau_k, x_k), \qquad \tau_k \le t < \tau_{k+1}, \qquad k \in \mathbb{N},$$

where $t_0 = \tau_0$ and x_0 are given initial conditions. Throughout, we assume that S contains the *trivial solution* $\varphi(t, t_0, 0) = 0$ for all $t \ge t_0$ so that $(S, \{0\})$ is invariant; that is, $x_e = 0$ is an *equilibrium* for the finite-dimensional DDS.

The most general specific class of finite-dimensional DDS that we consider is generated by differential equations of the form (refer to Subsection 2.12A).

$$\begin{cases} \dot{x}(t) = f_k(t, x(t)), & \tau_k \le t < \tau_{k+1} \\ x(t) = g_k(x(t^-)), & t = \tau_{k+1},\ k \in \mathbb{N}, \end{cases} \qquad (SE)$$

where for each $k \in \mathbb{N}$, $f_k \in C[\mathbb{R}^+ \times \mathbb{R}^n, \mathbb{R}^n]$, $g_k\colon \mathbb{R}^n \to \mathbb{R}^n$, and $x(t^-) = \lim_{t' \to t, t' < t} x(t')$ denotes the left limit of $x(t')$ at $t' = t$.

As in Subsection 2.12A, associated with (SE), we consider the family of *initial value problems* given by

$$\begin{cases} \dot{x}(t) = f_k(t, x(t)) \\ x(\tau_k) = x_k, \end{cases} \qquad (SE_k)$$

$k \in \mathbb{N}$. We assume that for (τ_k, x_k), (SE_k) possesses a unique solution $x^{(k)}(t, \tau_k, x_k)$ that exists for all $t \in [\tau_k, \infty)$ (refer to Section 2.3 for conditions that ensure this). Then for every $(t_0, x_0) \in \mathbb{R}^+ \times \mathbb{R}^n$, $t_0 = \tau_0$, (SE) has a unique solution $\varphi(t, t_0, x_0)$ that exists for all $t \in [t_0, \infty)$. This solution is made up of a sequence of continuous solution segments $x^{(k)}(t, \tau_k, x_k)$ defined over the intervals $[\tau_k, \tau_{k+1})$, $k \in \mathbb{N}$, with initial conditions (τ_k, x_k), where $x_k = x(\tau_k) = g_{k-1}(x(\tau_k^-))$, $k = 1, 2, \dots$, and the initial conditions $(\tau_0 = t_0, x_0)$ are given. At the points $\{\tau_{k+1}\}$, $k \in \mathbb{N}$, the solutions of (SE) may have possible jumps, or the four derivatives $D^+\varphi, D_+\varphi, D^-\varphi$, and $D_-\varphi$ may not be equal, or φ may be continuous.

We assume that for each $k \in \mathbb{N}$, $f_k(t, 0) = 0$ for all $t \ge \tau_k$. Then $x_e = 0$ will be an equilibrium for (SE_k) and (SE).

D. Qualitative characterizations: Stability and boundedness

At this point it might be instructive to rephrase the various stability and boundedness concepts given in Definitions 3.1.6–3.1.20 for the case of finite-dimensional dynamical systems. We consider here only systems determined by ordinary differential equations. The various stability and boundedness definitions for discrete-time systems determined by difference equations involve obvious modifications.

Let $X = \mathbb{R}^n, M = \{0\}, T = \mathbb{R}^+$, and $d(x, y) = |x - y|$ for all $x, y \in \mathbb{R}^n$ where $|\cdot|$ denotes any one of the equivalent norms on $\mathbb{R}^n$. Also, note that for any $x \in \mathbb{R}^n$, $d(x, 0) = |x|$. From Definitions 3.1.6–3.1.20 we now have the following characterizations of the equilibrium $x_e = 0$ of (E) and the solutions of (E).

Definition 6.1.1 (a) The equilibrium $x_e = 0$ of (E) is *stable* if for every $\varepsilon > 0$ and any $t_0 \in \mathbb{R}^+$ there exists a $\delta(\varepsilon, t_0) > 0$ such that for all solutions of (E),

$$|\varphi(t, t_0, x_0)| < \varepsilon \qquad \text{for all } t \ge t_0 \tag{6.1.6}$$

whenever

$$|x_0| < \delta(\varepsilon, t_0). \tag{6.1.7}$$

If in (6.1.7) $\delta(\varepsilon, t_0)$ is independent of t_0 (i.e., $\delta(\varepsilon, t_0) = \delta(\varepsilon)$), then the equilibrium $x_e = 0$ of (E) is said to be *uniformly stable*. (Note that in this definition the solutions $\varphi(t, t_0, x_0)$ exist over $[t_0, t_1)$ where t_1 may be finite or infinite.)

(b) The equilibrium $x_e = 0$ of (E) is *asymptotically stable* if

(1) it is stable; and

(2) for every $t_0 \geq 0$ there exists an $\eta(t_0) > 0$ such that $\lim_{t\to\infty} \varphi(t, t_0, x_0) = 0$ for all solutions of (E) whenever $|x| < \eta(t_0)$.

When (2) is true, we say that the equilibrium $x_e = 0$ of (E) is *attractive*. Also, the set of all $x_0 \in \mathbb{R}^n$ such that $\varphi(t, t_0, x_0) \to 0$ as $t \to 0$ for some $t_0 \geq 0$ is called the *domain of attraction* of the equilibrium $x_e = 0$ of (E) (at t_0).

(c) The equilibrium $x_e = 0$ of (E) is *uniformly asymptotically stable* if

(1) it is uniformly stable; and

(2) for every $\varepsilon > 0$ and every $t_0 \in \mathbb{R}^+$, there exist a $\delta_0 > 0$, independent of t_0 and ε, and a $T(\varepsilon) > 0$, independent of t_0, such that for all solutions of (E)

$$|\varphi(t, t_0, x_0)| < \varepsilon \qquad \text{for all } t \geq t_0 + T(\varepsilon)$$

whenever $|x_0| < \delta_0$.

When (2) is true, we say that the equilibrium $x_e = 0$ of (E) is *uniformly attractive*. Note that condition (2) is often paraphrased by saying that there exists a $\delta_0 > 0$ such that

$$\lim_{t\to\infty} \varphi(t + t_0, t_0, x_0) = 0$$

uniformly in (t_0, x_0) for $t_0 \geq 0$ and for $|x_0| \leq \delta_0$.

(d) The equilibrium $x_e = 0$ of (E) is *exponentially stable* if there exists an $\alpha > 0$ and for every $\varepsilon > 0$ and every $t_0 \geq 0$, there exists a $\delta(\varepsilon) > 0$ such that for all solutions of (E)

$$|\varphi(t, t_0, x_0)| \leq \varepsilon e^{-\alpha(t-t_0)} \qquad \text{for all } t \geq t_0$$

wherever $|x_0| < \delta(\varepsilon)$.

As in Chapter 3, we note that the exponential stability of the equilibrium $x_e = 0$ of (E) implies its uniform asymptotic stability.

(e) A solution $\varphi(t, t_0, x_0)$ of (E) is *bounded* if there exists a $\beta > 0$ such that $|\varphi(t, t_0, x_0)| < \beta$ for all $t \geq t_0$, where β may depend on each solution. System (E) is said to possess *Lagrange stability* if for each $t_0 \geq 0$ and $x_0 \in \mathbb{R}^n$, the solution $\varphi(t, t_0, x_0)$ is bounded.

(f) The solutions of (E) are *uniformly bounded* if for any $\alpha > 0$ and every $t_0 \in \mathbb{R}^+$, there exists a $\beta = \beta(\alpha) > 0$ (independent of t_0) such that if $|x_0| < \alpha$, then $|\varphi(t, t_0, x_0)| < \beta$ for all $t \geq t_0$. (To arrive at this definition, we choose in Definition 3.1.13, without loss of generality, that $x_0 = 0$.)

(g) The solutions of (E) are *uniformly ultimately bounded* (with bound B) if there exists a $B > 0$ and if corresponding to any $\alpha > 0$ and for every $t_0 \in \mathbb{R}^+$, there exists a $T = T(\alpha) > 0$ (independent of t_0) such that $|x_0| < \alpha$ implies that $|\varphi(t, t_0, x_0)| < B$ for all $t \geq t_0 + T(\alpha)$.

(h) The equilibrium $x_e = 0$ of (E) is *asymptotically stable in the large* if it is stable and if every solution of (E) tends to zero as $t \to \infty$. In this case, the domain of attraction of the equilibrium $x_e = 0$ of (E) is all of $\mathbb{R}^n$ and $x_e = 0$ is the *only* equilibrium of (E).

(i) The equilibrium $x_e = 0$ of (E) is *uniformly asymptotically stable in the large* if

(1) it is uniformly stable;

(2) the solutions of (E) are uniformly bounded; and

(3) for any $\alpha > 0$, any $\varepsilon > 0$ and every $t_0 \in \mathbb{R}^+$, there exists a $T(\varepsilon, \alpha) > 0$, independent of t_0, such that if $|x_0| < \alpha$, then for all solutions of (E), we have $|\varphi(t, t_0, x_0)| < \varepsilon$ for all $t \geq t_0 + T(\varepsilon, \alpha)$.

When (3) is true, we say that the equilibrium $x_e = 0$ of (E) is *globally uniformly attractive* (resp., *uniformly attractive in the large*).

(j) The equilibrium $x_e = 0$ of (E) is *exponentially stable in the large* if there exist an $\alpha > 0$ and a $\gamma > 0$, and for any $\beta > 0$, there exists a $k(\beta) > 0$ such that for all solutions of (E),

$$|\varphi(t, t_0, x_0)| \leq k(\beta)|x_0|^\gamma e^{-\alpha(t-t_0)} \qquad \text{for all } t \geq t_0$$

whenever $|x_0| < \beta$.

(k) The equilibrium $x_e = 0$ of (E) is *unstable* if it is not stable. In this case, there exist a $t_0 \geq 0$ and a sequence $x_{0m} \to 0$ of initial points and a sequence $\{t_m \geq 0\}$ such that $|\varphi(t_0 + t_m, t_0, x_{0m})| \geq \varepsilon$ for all m. □

E. Some characterizations of Lyapunov functions

We now address several important properties that Lyapunov functions may possess. We first consider the case $w\colon B(r) \to \mathbb{R}$ (resp., $w\colon \Omega \to \mathbb{R}$) where $B(r) \subset \Omega \subset \mathbb{R}^n$, $B(r) = \{x \in \mathbb{R}^n\colon |x| < r\}$ for some $r > 0$, Ω is an open connected set, and $0 \in \Omega$.

Definition 6.1.2 A function $w \in C[B(r), \mathbb{R}]$ (resp., $w \in C[\Omega, \mathbb{R}]$) is said to be *positive definite* if

(i) $w(0) = 0$; and

(ii) $w(x) > 0$ for all $x \neq 0$. □

Definition 6.1.3 A function $w \in C[B(r), \mathbb{R}]$ (resp., $w \in C[\Omega, \mathbb{R}]$) is said to be *negative definite* if $-w$ is positive definite. □

Definition 6.1.4 A function $w \in C[\mathbb{R}^n, \mathbb{R}]$ is said to be *radially unbounded* if

(i) $w(0) = 0$;

(ii) $w(x) > 0$ for all $x \in (\mathbb{R}^n - \{0\})$; and

(iii) $w(x) \to \infty$ as $|x| \to \infty$. □

Definition 6.1.5 A function $w \in C[B(r), \mathbb{R}]$ (resp., $w \in C[\Omega, \mathbb{R}]$) is said to be *indefinite* if

(i) $w(0) = 0$; and

(ii) in every neighborhood of the origin $x = 0$, w assumes negative and positive values. □

Definition 6.1.6 A function $w \in C[B(r), \mathbb{R}]$ (resp., $w \in C[\Omega, \mathbb{R}]$) is said to be *positive semidefinite* if

(i) $w(0) = 0$; and

(ii) $w(x) \geq 0$ for all $x \in B(r)$ (resp., $x \in \Omega$). □

Definition 6.1.7 A function $w \in C[B(r), \mathbb{R}]$ (resp., $w \in C[\Omega, \mathbb{R}]$) is said to be *negative semidefinite* if $-w$ is positive semidefinite. □

Next, we consider the case $v \in C[B(r) \times \mathbb{R}^+, \mathbb{R}]$ (resp., $v \in C[\Omega \times \mathbb{R}^+, \mathbb{R}]$).

Definition 6.1.8 A function $v \in C[B(r) \times \mathbb{R}^+, \mathbb{R}]$ (resp., $v \in C[\Omega \times \mathbb{R}^+, \mathbb{R}]$) is said to be *positive definite* if there exists a positive definite function $w \in C[B(r), \mathbb{R}]$ (resp., $w \in C[\Omega, \mathbb{R}]$) such that

(i) $v(0, t) = 0$ for all $t \geq 0$; and

(ii) $v(x, t) \geq w(x)$ for all $t \geq 0$ and all $x \in B(r)$ (resp., $x \in \Omega$). □

Definition 6.1.9 A function $v \in C[B(r) \times \mathbb{R}^+, \mathbb{R}]$ (resp., $v \in C[\Omega \times \mathbb{R}^+, \mathbb{R}]$) is said to be *negative definite* if $-v$ is positive definite. □

Definition 6.1.10 A function $v \in C[\mathbb{R}^n \times \mathbb{R}^+, \mathbb{R}]$ is said to be *radially unbounded* if there exists a radially unbounded function $w \in C[\mathbb{R}^n, \mathbb{R}]$ such that

(i) $v(0, t) = 0$ for all $t \geq 0$; and

(ii) $v(x, t) \geq w(x)$ for all $t \geq 0$ and all $x \in \mathbb{R}^n$. □

Definition 6.1.11 A function $v \in C[B(r) \times \mathbb{R}^+, \mathbb{R}]$ (resp., $v \in C[\Omega \times \mathbb{R}^+, \mathbb{R}]$) is said to be *decrescent* if there exists a positive definite function $w \in C[B(r), \mathbb{R}]$ (resp., $w \in C[\Omega, \mathbb{R}]$) such that

$$|v(x, t)| \leq w(x)$$

for all $(x, t) \in B(r) \times \mathbb{R}^+$ (resp., $(x, t) \in \Omega \times \mathbb{R}^+$). □

Definition 6.1.12 A function $v \in C[B(r) \times \mathbb{R}^+, \mathbb{R}]$ (resp., $v \in C[\Omega \times \mathbb{R}^+, \mathbb{R}]$) is said to be *positive semidefinite* if

(i) $v(0, t) = 0$ for all $t \in \mathbb{R}^+$; and

(ii) $v(x, t) \geq 0$ for all $t \in \mathbb{R}^+$ and all $x \in B(r)$ (resp., $x \in \Omega$). □

Definition 6.1.13 A function $v \in C[B(r) \times \mathbb{R}^+, \mathbb{R}]$ (resp., $v \in C[\Omega \times \mathbb{R}^+, \mathbb{R}]$) is said to be *negative semidefinite* if $-v$ is positive semidefinite. □

Some of the preceding characterizations of v-functions (and w-functions) can be rephrased in equivalent and very useful ways. In doing so, we use comparison functions of *class* $\mathcal{K}$ and *class* $\mathcal{K}_\infty$.

Theorem 6.1.1 A function $v \in C[B(r) \times \mathbb{R}^+, \mathbb{R}]$ (resp., $v \in C[\Omega \times \mathbb{R}^+, \mathbb{R}]$) is *positive definite* if and only if

(i) $v(0,t) = 0$ for all $t \in \mathbb{R}^+$; and

(ii) there exists a function $\psi \in \mathcal{K}$ defined on $[0, r]$ (resp., on $\mathbb{R}^+$) such that

$$v(x,t) \geq \psi(|x|)$$

for all $t \in \mathbb{R}^+$ and all $x \in B(r)$ (resp., $x \in \Omega$).

Proof. If $v(x,t)$ is positive definite, then there exists a positive definite function $w(x)$ such that $v(x,t) \geq w(x)$ for $t \in \mathbb{R}^+$ and $|x| \leq r$. Define

$$\psi_0(s) = \inf\{w(x) \colon s \leq |x| \leq r\}$$

for $0 < s \leq r$. Clearly ψ_0 is a positive and nondecreasing function such that $\psi_0(|x|) \leq w(x)$ on $0 < |x| \leq r$. Because ψ_0 is continuous, it is Riemann integrable. Define the function ψ by $\psi(0) = 0$ and

$$\psi(u) = u^{-1} \int_0^u (s/r)\psi_0(s)ds, \qquad 0 < u \leq r.$$

Clearly $0 < \psi(u) \leq \psi_0(u) \leq w(x) \leq v(x,t)$ if $t \geq 0$ and $|x| = u$. Moreover, ψ is continuous and increasing (i.e., $\psi \in \mathcal{K}$, by construction).

Conversely, assume that (i) and (ii) are true and define $w(x) = \psi(|x|)$. It now follows readily from Definition 6.1.8 that $v(x,t)$ is positive definite. □

Theorem 6.1.2 A function $v \in C[\mathbb{R}^n \times \mathbb{R}^+, \mathbb{R}]$ is *radially unbounded* if and only if

(i) $v(0,t) = 0$ for all $t \in \mathbb{R}^+$; and

(ii) there exists a function $\psi \in \mathcal{K}_\infty$ such that

$$v(x,t) \geq \psi(|x|)$$

for all $(x,t) \in \mathbb{R}^n \times \mathbb{R}^+$. □

We ask the reader to prove Theorem 6.1.2 in the problem section.

Theorem 6.1.3 A function $v \in C[B(r) \times \mathbb{R}^+, \mathbb{R}]$ (resp., $v \in C[\Omega \times \mathbb{R}^+, \mathbb{R}]$) is *decrescent* if and only if there exists a function $\psi \in \mathcal{K}$ defined on $[0, r]$ (resp., on $\mathbb{R}^+$) such that

$$|v(x,t)| \leq \psi(|x|)$$

for all $(x,t) \in B(r) \times \mathbb{R}^+$ (resp., $(x,t) \in \Omega \times \mathbb{R}^+$). □

We ask the reader to prove Theorem 6.1.3 in the problem section.

Note that when $w \in C[B(r), \mathbb{R}]$ (resp., $w \in C[\Omega, \mathbb{R}]$) is positive or negative definite, then it is also decrescent for in this case we can always find $\psi_1, \psi_2 \in \mathcal{K}$ such that

$$\psi_1(|x|) \leq |w(x)| \leq \psi_2(|x|)$$

for all $x \in B(r)$ for some $r > 0$. On the other hand, when $v \in C[B(r) \times \mathbb{R}^+, \mathbb{R}]$ (resp., $v \in C[\Omega \times \mathbb{R}^+, \mathbb{R}]$), care must be taken in establishing whether v is decrescent.

For the case of discrete-time dynamical systems determined by difference equations (D), we employ functions $v \in C[B(r) \times \mathbb{N}, \mathbb{R}]$ (resp., $v \in C[\Omega \times \mathbb{N}, \mathbb{R}]$). We define such functions as being *positive definite*, *negative definite*, *positive semidefinite*, *negative semidefinite*, *decrescent*, and *radially unbounded* by modifying Definitions 6.1.2–6.1.13 (and Theorems 6.1.1–6.1.3) in an obvious way.

Example 6.1.1 (a) For $v \in C[\mathbb{R}^2 \times \mathbb{R}^+, \mathbb{R}]$ given by $v(x,t) = (1+\cos^2 t)x_1^2 + 2x_2^2$, we have

$$\psi_1(|x|) \triangleq x^T x \leq V(x,t) \leq 2x^T x \triangleq \psi_2(|x|)$$

for all $x \in \mathbb{R}^2$ and $t \in \mathbb{R}^+$, where $\psi_1, \psi_2 \in \mathcal{K}_\infty$. Therefore, v is positive definite, decrescent, and radially unbounded.

(b) For $v \in C[\mathbb{R}^2 \times \mathbb{R}^+, \mathbb{R}]$ given by $v(x,t) = (x_1^2 + x_2^2)\cos^2 t$, we have

$$0 \leq v(x,t) \leq x^T x \triangleq \psi(|x|)$$

for all $x \in \mathbb{R}^2$ and $t \in \mathbb{R}^+$, where $\psi \in \mathcal{K}$. Thus, v is positive semidefinite and decrescent.

(c) For $v \in C[\mathbb{R}^2 \times \mathbb{R}^+, \mathbb{R}]$ given by $v(x,t) = (1+t)(x_1^2 + x_2^2)$, we have

$$\psi(|x|) \triangleq x^T x \leq v(x,t)$$

for all $x \in \mathbb{R}^2$ and $t \in \mathbb{R}^+$, where $\psi \in \mathcal{K}_\infty$. Thus, v is positive definite and radially unbounded. It is not decrescent.

(d) For $v \in C[\mathbb{R}^2 \times \mathbb{R}^+, \mathbb{R}]$ given by $v(x,t) = x_1^2/(1+t) + x_2^2$, we have

$$v(x,t) \leq x^T x \triangleq \psi(|x|)$$

for all $x \in \mathbb{R}^2$ and $t \in \mathbb{R}^+$, where $\psi \in \mathcal{K}_\infty$. Hence, v is decrescent and positive semidefinite. It is not positive definite.

(e) The function $v \in C[\mathbb{R}^2 \times \mathbb{R}^+, \mathbb{R}]$ given by $v(x,t) = (x_2 - x_1)^2(1+t)$ is positive semidefinite. It is not positive definite nor decrescent. □

F. Quadratic forms

We now consider an important class of Lyapunov functions, *quadratic forms*, given by

$$v(x) = x^T Bx = \sum_{i,k=1}^{n} b_{ik} x_i x_k \tag{6.1.8}$$

where $x \in \mathbb{R}^n$ and $B = [b_{ij}] \in \mathbb{R}^{n\times n}$ is assumed to be symmetric (i.e., $B = B^T$). Recall that in this case B is diagonalizable and all of its eigenvalues are real. For a proof of the next results, the reader should consult any text on linear algebra and matrices (e.g., Michel and Herget [16]).

Theorem 6.1.4 Let v be the quadratic form defined in (6.1.8). Then

(i) v is positive definite (and radially unbounded) if and only if all principal minors of B are positive, that is, if and only if

$$\det \begin{bmatrix} b_{11} & \cdots & b_{1k} \\ \cdot & & \cdot \\ \cdot & & \cdot \\ \cdot & & \cdot \\ b_{k1} & \cdots & b_{kk} \end{bmatrix} > 0, \qquad k = 1, \dots, n.$$

These inequalities are called the *Sylvester inequalities*.

(ii) v is negative definite if and only if

$$(-1)^k \det \begin{bmatrix} b_{11} & \cdots & b_{1k} \\ \cdot & & \cdot \\ \cdot & & \cdot \\ \cdot & & \cdot \\ b_{k1} & \cdots & b_{kk} \end{bmatrix} > 0, \qquad k = 1, \dots, n.$$

(iii) v is *definite* (i.e., either positive definite or negative definite) if and only if all eigenvalues are nonzero and have the same sign.

(iv) v is *semidefinite* (i.e., either positive semidefinite or negative semidefinite) if and only if the nonzero eigenvalues of B have the same sign.

(v) If $\lambda_1, \dots, \lambda_n$ denote all the eigenvalues of B (not necessarily distinct), if $\lambda_m = \min_{1\le i\le n} \lambda_i$, if $\lambda_M = \max_{1\le i\le n} \lambda_i$, and if $|\cdot|$ denotes the Euclidean norm ($|x| = (x^T x)^{1/2}$), then

$$\lambda_m |x|^2 \le v(x) \le \lambda_M |x|^2 \qquad \text{for all} \quad x \in \mathbb{R}^n.$$

(vi) v is indefinite if and only if B possesses both positive and negative eigenvalues. □

The purpose of the next example is to point out some of the geometric properties of quadratic forms.

Example 6.1.2 Let B be a real symmetric 2×2 matrix and let

$$v(x) = x^T Bx.$$

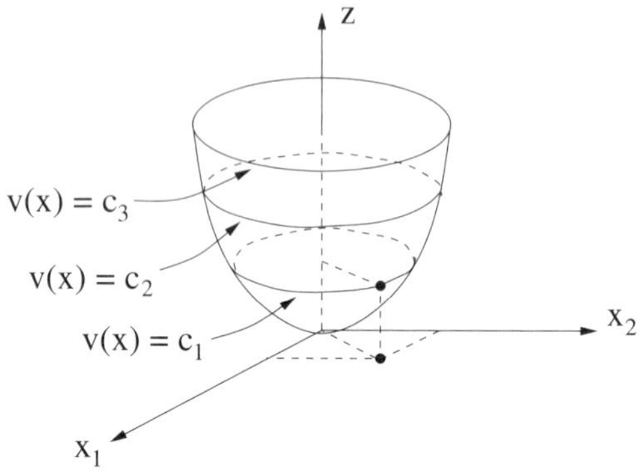

Figure 6.1.1: Cup-shaped surface of (6.1.9).

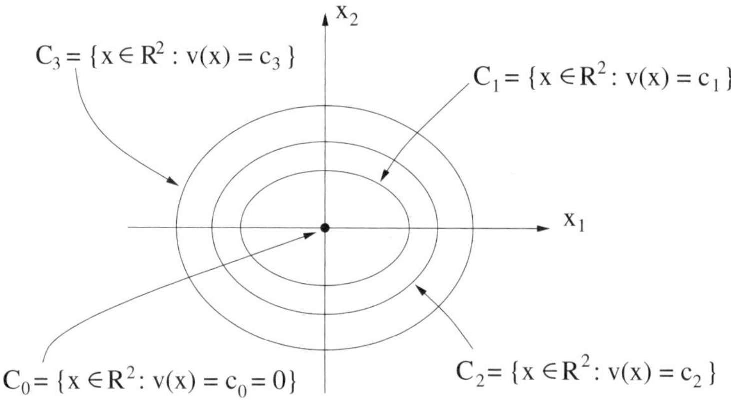

Figure 6.1.2: Level curves.

Assume that both eigenvalues of B are positive so that v is positive definite and radially unbounded. In $\mathbb{R}^3$, let us now consider the surface determined by

$$z = v(x) = x^T Bx. \tag{6.1.9}$$

Equation (6.1.9) describes a cup-shaped surface as depicted in Figure 6.1.1. Note that corresponding to every point on this cup-shaped surface there exists one and only one point in the x_1–x_2 plane. Note also that the loci defined by

$$C_i = \{x \in \mathbb{R}^2 \colon v(x) = c_i \geq 0\} \;\; (c_i = \text{constant})$$

determine closed curves in the x_1–x_2 plane as shown in Figure 6.1.2. We call these curves *level curves*. Note that $C_0 = \{0\}$ corresponds to the case when $z = c_0 = 0$. Note also that this function v can be used to cover the entire $\mathbb{R}^2$ plane with closed curves by selecting for z all values in $\mathbb{R}^+$.

In the case when $v(x) = x^T Bx$ is a positive definite quadratic form with $x \in \mathbb{R}^n$, the preceding comments are still true; however, in this case, the closed curves C_i

must be replaced by closed hypersurfaces in $\mathbb{R}^n$ and a simple geometric visualization as in Figures 6.1.1 and 6.1.2 is no longer possible. □

G. Lyapunov stability results: Geometric interpretation

Before stating and proving the principal Lyapunov stability and boundedness results, it might be instructive to give a geometric interpretation of some of these results. To this end, we consider systems of equations

$$\begin{cases} \dot{x}_1 = f_1(x_1, x_2) \\ \dot{x}_2 = f_2(x_1, x_2) \end{cases} \tag{6.1.10}$$

where $x = (x_1, x_2)^T \in \mathbb{R}^2$ and $f_i\colon \mathbb{R}^2 \to \mathbb{R}$, $i = 1, 2$. We assume that f_1 and f_2 are such that for every (t_0, x_0), $t_0 \geq 0$, (6.1.10) has a unique solution $\varphi(t, t_0, x_0)$ with $\varphi(t_0, t_0, x_0) = x_0$. We also assume that $x_e = (x_1, x_2)^T = (0, 0)^T$ is the only equilibrium in $B(h)$ for some $h > 0$.

Now let v be a positive definite function, and to simplify our discussion, assume that v is continuously differentiable with nonvanishing gradient $\nabla v(x)^T = ((\partial v/\partial x_1)(x_1, x_2), (\partial v/\partial x_2)(x_1, x_2))$ on $0 < |x| \leq h$. It can be shown that similarly as in the case of quadric forms, the equation

$$v(x) = c \qquad (c \geq 0)$$

defines for sufficiently small constants $c > 0$ a family of closed curves C_i which cover the neighborhood $B(h)$ as shown in Figure 6.1.3. Note that the origin $x = 0$ is located in the interior of each such curve and in fact $C_0 = \{0\}$.

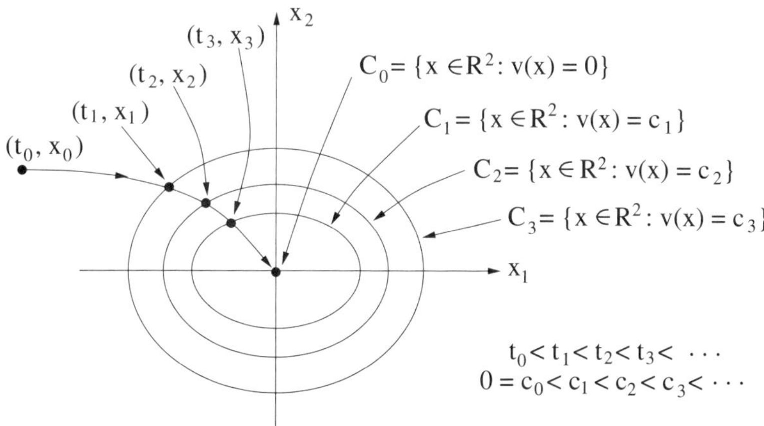

Figure 6.1.3: Family of closed curves C_i.

Next, suppose that all the trajectories of (6.1.10) originating from points on the circular disk $|x| \leq r_1 < h$ cross the curves $v(x) = c$ from the exterior towards the interior as we proceed along these trajectories in the direction of increasing values of t. Then we can conclude that these trajectories approach the origin as t increases; that is, the equilibrium $x_e = 0$ is in this case asymptotically stable.

Now in terms of the given v-function, we have the following geometric interpretation: for a given solution $\varphi(t, t_0, x_0)$ to cross the curve $v(x) = r$, $r = v(x_0)$, the angle between the outward normal vector $\nabla v(x_0)$ and the derivative of $\varphi(t, t_0, x_0)$ at $t = t_0$ must be greater than $\pi/2$; that is,

$$v'_{(6.1.10)}(x_0) = \nabla v(x_0)^T f(x_0) < 0$$

where $f(x_0) = (f_1(x_0), f_2(x_0))^T$. For this to happen at all points, we must have that $v'_{(6.1.10)}(x) < 0$ for $0 < |x| \le r_1$.

The same result can be arrived at using an *analytic* point of view: the function

$$V(t) \triangleq v(\varphi(t, t_0, x_0))$$

will decrease monotonically as t increases. But this will imply that the derivative $v'(\varphi(t, t_0, x_0))$ along the solution $\varphi(t, t_0, x_0)$ of (6.1.10) must be negative definite in $B(r)$ for $r > 0$ sufficiently small.

Next, assume that (6.1.10) has only one equilibrium (located at the origin $x_e = 0$) and that v is positive definite and radially unbounded. It can be shown that similarly as in the case of quadratic forms, the equation $v(x) = c$, $c \in \mathbb{R}^+$, can in this case be used to cover all of $\mathbb{R}^2$ by closed curves of the type depicted in Figure 6.1.3. Now if for arbitrary initial data (t_0, x_0), the corresponding solution of (6.1.10), $\varphi(t, t_0, x_0)$, behaves as discussed above, then it follows that the time derivative of v along this solution, $v'(\varphi(t, t_0, x_0))$, will be negative definite in $\mathbb{R}^2$.

The preceding discussion was given for arbitrary solutions of (6.1.10). Accordingly, we can make the following conjectures.

1. If there exists a positive definite function v such that $v'_{(6.1.10)}$ is negative definite, then the equilibrium $x_e = 0$ of (6.1.10) is *asymptotically stable*.

2. If there exists a positive definite and radially unbounded function v such that $v'_{(6.1.10)}$ is negative definite for all $x \in \mathbb{R}^2$, then the equilibrium $x_e = 0$ of (6.1.10) is *asymptotically stable in the large*.

Continuing the preceding discussion by making reference to Figure 6.1.4, let us assume that we can find for (6.1.10) a continuously differentiable function $v: \mathbb{R}^2 \to \mathbb{R}$ that is indefinite and which has the properties discussed in the following. Because v is indefinite, there exist in every neighborhood of the origin points for which $v > 0$, $v < 0$, and $v(0) = 0$. Confining our attention to $B(k)$ where $k > 0$ is sufficiently small, we let $D = \{x \in B(k) \colon v(x) < 0\}$, which may consist of several subdomains. The boundary of D, ∂D, consists of points in $\partial B(k)$ and points determined by $v(x) = 0$. Let H denote a subdomain of D having the property that $0 \in \partial H$. Assume that in the interior of H, v is bounded. Suppose that $v'_{(6.1.10)}(x)$ is negative definite in D and that $\varphi(t, t_0, x_0)$ is a solution of (6.1.10) that originates somewhere on the boundary of H ($x_0 \in \partial H$) with $v(x_0) = 0$. Then this solution will penetrate the boundary of H at points where $v = 0$ as t increases and it can never again reach a point where $v = 0$. In fact, as t increases, this solution will penetrate the set of points determined by $|x| = k$ (because by assumption, $v'_{(6.1.10)} < 0$ along this trajectory

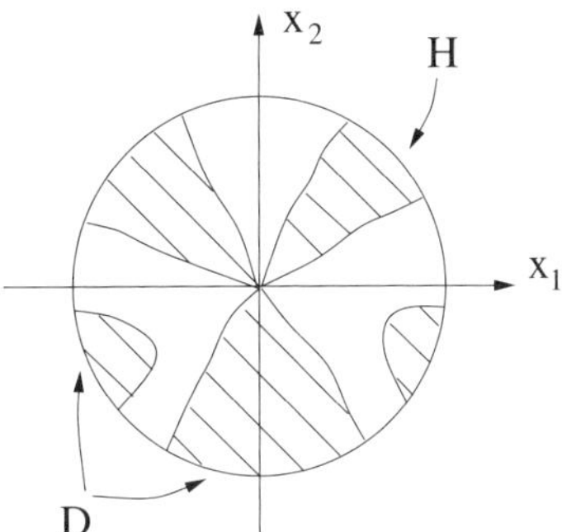

Figure 6.1.4: Domains D and H.

and $v < 0$ in H). But this shows that the equilibrium $x_e = 0$ of (6.1.10) is unstable. Accordingly, we can make the following conjecture.

3. Let a function $v\colon \mathbb{R}^2 \to \mathbb{R}$ be given that is continuously differentiable and which has the following properties.

(i) There exist points x arbitrarily close to the origin such that $v(x) < 0$; they form the domain D that is bounded by the set of points determined by $v = 0$ and the circle $|x| = k$. D may consist of several subdomains. Let H denote a subdomain of D having the property that $0 \in \partial H$.

(ii) In the interior of H, v is bounded.

(iii) In the interior of D, $v'_{(6.1.10)}$ is negative.

Then the equilibrium $x_e = 0$ of (6.1.10) is *unstable*.

In the present chapter, we state and prove results that include the foregoing conjectures as special cases.

6.2 The Principal Stability and Boundedness Results for Ordinary Differential Equations

In the present section we address stability and boundedness properties of continuous finite-dimensional dynamical systems determined by ordinary differential equations (E).

A. Stability

In our first two results we concern ourselves with the stability and uniform stability of the equilibrium $x_e = 0$ of (E).

Theorem 6.2.1 Assume that for some $r > 0$ and $B(r) \subset \Omega$, there exists a positive definite function $v \in C[B(r) \times \mathbb{R}^+, \mathbb{R}]$ such that $v'_{(E)}$ is negative semidefinite. Then the following statements are true.

a) The equilibrium $x_e = 0$ of (E) is *stable*.

b) If in addition, v is decrescent, then $x_e = 0$ of (E) is *uniformly stable*.

Proof. (a) According to Definition 6.1.1(a), we fix $\varepsilon > 0$ and $t_0 \geq 0$ and we seek a $\delta > 0$ such that

$$\big|\varphi(t, t_0, x_0)\big| < \varepsilon \qquad \text{for all } t \geq t_0$$

whenever $|x_0| < \delta$. Without loss of generality, we can assume that $\varepsilon < r$. Because $v(x,t)$ is positive definite, then by Theorem 6.1.1 there is a function $\psi \in \mathcal{K}$ such that $v(x,t) \geq \psi(|x|)$ for all $0 \leq |x| < r, t \geq 0$. Pick $\delta > 0$ so small that $v(x_0, t_0) < \psi(\varepsilon)$ if $|x_0| \leq \delta$. Because $v'_{(E)}(x,t) \leq 0$, then $v(\varphi(t, t_0, x_0), t)$ is monotone nonincreasing and $v(\varphi(t, t_0, x_0), t) < \psi(\varepsilon)$ for all $t \geq t_0$. Thus, $|\varphi(t, t_0, x_0)|$ cannot reach the value ε, because this would imply that $v(\varphi(t, t_0, x_0), t) \geq \psi(|\varphi(t, t_0, x_0)|) = \psi(\varepsilon)$.

(b) Because $v(x,t)$ is positive definite and decrescent, it follows from Theorems 6.1.1 and 6.1.3 that there exist two functions $\psi_1, \psi_2 \in \mathcal{K}$ such that

$$\psi_1(|x|) \leq v(x,t) \leq \psi_2(|x|)$$

for all $(x,t) \in B(r) \times \mathbb{R}^+$.

Let $\varphi(t, t_0, x_0)$ be a solution of (E) with initial condition $\varphi(t_0) = x_0$. Then we have

$$D^+ v\big(\varphi(t, t_0, x_0), t\big) \leq v'_{(E)}\big(\varphi(t, t_0, x_0), t\big)$$

for all $(t_0, x_0) \in \mathbb{R}^+ \times B(r)$ and all $t \in \mathbb{R}^+_{t_0}$ such that $\varphi(t, t_0, x_0) \in B(r)$, where D^+ denotes the upper-right Dini derivative with respect to t. Inasmuch as $v'_{(E)}$ is negative semidefinite, we conclude that $v(\varphi(t, t_0, x_0), t)$ is nonincreasing for all $t \in \mathbb{R}^+_{t_0}$. Statement (b) follows now directly from Theorem 3.3.1. □

Example 6.2.1 (*Simple pendulum*) Consider the simple pendulum described by the equations

$$\begin{cases} \dot{x}_1 = x_2 \\ \dot{x}_2 = -k \sin x_1 \end{cases} \tag{6.2.1}$$

where $k > 0$ is a constant. System (6.2.1) clearly has an equilibrium $x_e = 0$. The total energy for (6.2.1) is the sum of the kinetic energy and potential energy, given by

$$v(x) = \frac{1}{2}x_2^2 + k\int_0^{x_1} \sin\eta d\eta = \frac{1}{2}x_2^2 + k(1 - \cos x_1).$$

This function is continuously differentiable, $v(0) = 0$, and v is positive definite in $\Omega = \{(x_1, x_2)^T \in \mathbb{R}^2 \colon |x_1| < c < 2\pi\}$. Also, v is decrescent, because it does not depend on t.

Along the solutions of (6.2.1) we have

$$v'_{(6.2.1)}(x) = (k \sin x_1)\dot{x}_1 + x_2\dot{x}_2 = (k \sin x_1)x_2 + x_2(-k \sin x_1) = 0.$$

In accordance with Theorem 6.2.1(b), the equilibrium $x_e = 0$ of (6.2.1) is uniformly stable.

Note that because $v'_{(6.2.1)}(x) = 0$, the total energy for system (6.2.1) is constant for a given set of initial conditions for all $t \geq 0$. □

The major shortcoming of the results that comprise the Principal Lyapunov Stability Results (called the *Direct Method of Lyapunov*) is that there are no specific rules which tell us how to choose a v-function in a particular problem. The preceding example suggests that a good choice for a v-function is the total energy of a system. Another widely used class of v-functions consists of quadratic forms (refer to Subsection 6.1F).

Example 6.2.2 Consider the second-order system

$$\ddot{x} + \dot{x} + e^{-t}x = 0. \tag{6.2.2}$$

Letting $x = x_1$, $\dot{x} = x_2$, we can express (6.2.2) equivalently by

$$\begin{cases} \dot{x}_1 = x_2 \\ \dot{x}_2 = -x_2 - e^{-t}x_1. \end{cases} \tag{6.2.3}$$

This system has an equilibrium at the origin $x_e = (x_1, x_2)^T = (0, 0)^T = 0$. Let us choose the positive definite function

$$v(x_1, x_2) = x_1^2 + x_2^2.$$

Along the solutions of (6.2.3), we have

$$v'_{(6.2.3)}(x_1, x_2, t) = 2x_1x_2(1 - e^{-t}) - 2x_2^2.$$

The above choice of v-function does not satisfy the hypotheses of Theorem 6.2.1. Thus, we can reach no conclusion. Therefore, let us choose another v-function,

$$v(x_1, x_2, t) = x_1^2 + e^t x_2^2.$$

In this case we obtain

$$v'_{(6.2.3)}(x_1, x_2, t) = -e^t x_2^2.$$

This v-function is positive definite and $v'_{(6.2.3)}$ is negative semidefinite. Therefore, Theorem 6.2.1(a) is applicable and we can conclude that the equilibrium $x_e = 0$ is stable. However, because v is not decrescent, Theorem 6.2.1(b) is not applicable and we cannot conclude from this choice of v-function that the equilibrium $x_e = 0$ is uniformly stable. □

Example 6.2.3 (*Hamiltonian system*) Consider the conservative dynamical system with n degrees of freedom, which we discussed in Chapter 2 (Example 2.3.7) and which is given by

$$\begin{cases} \dot{q}_i = \dfrac{\partial H}{\partial p_i}(p, q), & i = 1, \dots, n \\ \dot{p}_i = -\dfrac{\partial H}{\partial q_i}(p, q), & i = 1, \dots, n \end{cases} \tag{6.2.4}$$

where $q^T = (q_1, \ldots, q_n)$ denotes the generalized position vector, $p^T = (p_1, \ldots, p_n)$ the generalized momentum vector, $H(p,q) = T(p)+W(q)$ the Hamiltonian, $T(p)$ the kinetic energy, and $W(q)$ the potential energy. The positions of the equilibrium points of (6.2.4) correspond to the points in $\mathbb{R}^{2n}$ where the partial derivatives of H vanish. In the following, we assume that $(p^T, q^T) = (0^T, 0^T)$ is an isolated equilibrium of (6.2.4), and without loss of generality we also assume that $H(0,0) = 0$. Furthermore, we assume that H is smooth and that $T(p)$ and $W(q)$ are of the form

$$T(p) = T_2(p) + T_3(p) + \cdots$$

and

$$W(q) = W_k(q) + W_{k+1}(q) + \cdots, \qquad k \geq 2.$$

Here $T_j(p)$ denotes the terms in p of order j and $W_j(q)$ denotes the terms in q of order j. The kinetic energy $T(p)$ is always assumed to be positive definite with respect to p. If the potential energy has an isolated minimum at $q = 0$, then W is positive definite with respect to q. We choose as a v-function

$$v(p,q) = H(p,q) = T(p) + W(q)$$

which is positive definite. Inasmuch as

$$v'_{(6.2.4)}(p,q) = \frac{dH}{dt}(p,q) = 0,$$

Theorem 6.2.1(a) is applicable and we conclude that the equilibrium at the origin is *stable*. Because v is independent of t, it is also decrescent, and so Theorem 6.2.1(b) is also applicable. Therefore, the equilibrium at the origin is also *uniformly stable*. Note that Example 6.2.1 (the simple pendulum) is a special case of the present example. □

B. Asymptotic stability

In the next two results we address the asymptotic stability of the equilibrium $x_e = 0$ of (E).

Theorem 6.2.2 For (E) we assume that there exists a positive definite and decrescent function $v \in C[B(r) \times \mathbb{R}^+, \mathbb{R}]$ such that $v'_{(E)}$ is negative definite. Then the equilibrium $x_e = 0$ of (E) is *uniformly asymptotically stable.*

Proof. Because $v'_{(E)}$ is negative definite, there exists a function $\psi_3 \in \mathcal{K}$ such that

$$v'_{(E)}(x,t) \leq -\psi_3(|x|)$$

for all $(x,t) \in B(r) \times \mathbb{R}^+$.

Let $\varphi(t, t_0, x_0)$ be a solution of (E) with initial condition $\varphi(t_0) = x_0$. Then we have

$$D^+ v(\varphi(t,t_0,x_0),t) \leq v'_{(E)}(\varphi(t,t_0,x_0),t) \leq -\psi_3(|\varphi(t,t_0,x_0)|)$$

for all $(x_0, t_0) \in B(r) \times \mathbb{R}^+$ and all $t \in \mathbb{R}^+_{t_0}$ such that $\varphi(t, t_0, x_0) \in B(r)$, where D^+ denotes the upper-right Dini derivative with respect to t. The rest of the proof follows directly from Theorem 3.3.2. □

Theorem 6.2.3 With $\Omega = \mathbb{R}^n$, assume that there exists a positive definite, decrescent, and radially unbounded function $v \in C[\mathbb{R}^n \times \mathbb{R}^+, \mathbb{R}]$ such that $v'_{(E)}$ is negative definite (on $\mathbb{R}^n \times \mathbb{R}^+$) (i.e., there exist $\psi_1, \psi_2 \in \mathcal{K}_\infty$ and $\psi_3 \in \mathcal{K}$ such that

$$\psi_1(|x|) \leq v(x,t) \leq \psi_2(|x|)$$

and

$$v'_{(E)}(x,t) \leq -\psi_3(|x|)$$

for all $(x,t) \in \mathbb{R}^n \times \mathbb{R}^+$). Then the equilibrium $x_e = 0$ of (E) is *uniformly asymptotically stable in the large.*

Proof. This result is a direct consequence of Theorem 3.3.6. □

Example 6.2.4 Consider the system

$$\begin{cases} \dot{x}_1 = (x_1 - c_2 x_2)(x_1^2 + x_2^2 - 1) \\ \dot{x}_2 = (c_1 x_1 + x_2)(x_1^2 + x_2^2 - 1) \end{cases} \tag{6.2.5}$$

which has an equilibrium at the origin $x_e = (x_1, x_2)^T = (0,0)^T = 0$. We choose as a v-function

$$v(x) = c_1 x_1^2 + c_2 x_2^2$$

and obtain

$$v'_{(6.2.5)}(x) = 2(c_1 x_1^2 + c_2 x_2^2)(x_1^2 + x_2^2 - 1).$$

If $c_1 > 0$ and $c_2 > 0$, then v is positive definite and radially unbounded and $v'_{(6.2.5)}$ is negative definite in the domain $x_1^2 + x_2^2 < 1$. Therefore, Theorem 6.2.2 is applicable and we conclude that the equilibrium $x_e = 0$ is uniformly asymptotically stable. On the other hand, Theorem 6.2.3 is not applicable and we cannot conclude that the equilibrium $x_e = 0$ is uniformly asymptotically stable in the large. □

Example 6.2.5 Consider the system

$$\begin{cases} \dot{x}_1 = x_2 + c x_1(x_1^2 + x_2^2) \\ \dot{x}_2 = -x_1 + c x_2(x_1^2 + x_2^2) \end{cases} \tag{6.2.6}$$

where c is a real constant. For this system, $x_e = 0$ is the only equilibrium. We choose as a v-function

$$v(x) = x_1^2 + x_2^2$$

and we obtain

$$v'_{(6.2.6)}(x) = 2c(x_1^2 + x_2^2)^2.$$

If $c = 0$, Theorem 6.2.1(b) is applicable and we conclude that the equilibrium $x_e = 0$ of (6.2.6) is *uniformly stable*. If $c < 0$, then Theorem 6.2.3 is applicable and we conclude that the equilibrium $x_e = 0$ of (6.2.6) is *uniformly asymptotically stable in the large.* □

C. Exponential stability

In the next two results we address the exponential stability of the equilibrium $x_e = 0$ of (E).

Theorem 6.2.4 Assume that there exist a function $v \in C[B(r) \times \mathbb{R}^+, \mathbb{R}]$ and four positive constants c_1, c_2, c_3, and b such that

$$c_1|x|^b \leq v(x,t) \leq c_2|x|^b$$

and

$$v'_{(E)}(x,t) \leq -c_3|x|^b$$

for all $(x,t) \in B(r) \times \mathbb{R}^+$. Then the equilibrium $x_e = 0$ of (E) is *exponentially stable*.

Proof. This result is a direct consequence of Theorem 3.3.3. □

Theorem 6.2.5 With $\Omega = \mathbb{R}^n$, assume that there exist a function $v \in C[\mathbb{R}^n \times \mathbb{R}^+, \mathbb{R}]$ and four positive constants c_1, c_2, c_3, and b such that

$$c_1|x|^b \leq v(x,t) \leq c_2|x|^b$$

and

$$v'_{(E)}(x,t) \leq -c_3|x|^b$$

for all $(x,t) \in \mathbb{R}^n \times \mathbb{R}^+$. Then the equilibrium $x_e = 0$ of (E) is *exponentially stable in the large*.

Proof. This result is a direct consequence of Theorem 3.3.7. □

Example 6.2.6 Consider the system

$$\begin{cases} \dot{x}_1 = -a(t)x_1 - bx_2 \\ \dot{x}_2 = bx_1 - c(t)x_2 \end{cases} \tag{6.2.7}$$

where b is a real constant and a and c are real and continuous functions defined for $t \geq 0$ satisfying $a(t) \geq \delta > 0$ and $c(t) \geq \delta > 0$ for all $t \geq 0$. We assume that $x_e = 0$ is the only equilibrium for (6.2.7).

Choosing

$$v(x) = \frac{1}{2}(x_1^2 + x_2^2),$$

we obtain

$$v'_{(6.2.7)}(x,t) = -a(t)x_1^2 - c(t)x_1^2 \leq -\delta(x_1^2 + x_2^2)$$

for all $t \geq 0$, $x \in \mathbb{R}^2$. All the hypotheses of Theorem 6.2.5 are satisfied and we conclude that the equilibrium $x_e = 0$ of (6.2.7) is *exponentially stable in the large*. □

Example 6.2.7 Consider the system

$$\begin{cases} \dot{x}_1 = x_2 - x_1(x_1^2 + x_2^2) \\ \dot{x}_2 = -x_1 - x_2(x_1^2 + x_2^2). \end{cases} \tag{6.2.8}$$

Clearly, $x_e = 0$ is an equilibrium for (6.2.8). Now choose

$$v(x) = x_1^2 + x_2^2$$

which is positive definite, radially unbounded, and decrescent. Along the solutions of (6.2.8), we have

$$v'_{(6.2.8)}(x) = -2(x_1^2 + x_2^2)^2$$

which is negative definite on $\mathbb{R}^2$. By Theorem 6.2.3, the equilibrium $x_e = 0$ of (6.2.8) is *uniformly asymptotically stable in the large*. However, the hypotheses of Theorem 6.2.5 are not satisfied and we cannot conclude that the equilibrium $x_e = 0$ of (6.2.8) is exponentially stable in the large. In fact, in Problem 6.10.10 we ask the reader to show that this equilibrium is *not* exponentially stable. □

D. Boundedness of solutions

In the next two results we concern ourselves with the uniform boundedness and the uniform ultimate boundedness of the solutions of (E).

Theorem 6.2.6 With $\Omega = \mathbb{R}^n$, assume that there exists a function $v \in C[\mathbb{R}^n \times \mathbb{R}^+, \mathbb{R}]$ that satisfies the following conditions.

(i) There exist two functions $\psi_1, \psi_2 \in C[\mathbb{R}^+, \mathbb{R}^+]$ that are strictly increasing with $\lim_{r\to\infty} \psi_i(r) = \infty$, $i = 1, 2$, and a constant $M > 0$, such that

$$\psi_1(|x|) \leq v(x,t) \leq \psi_2(|x|)$$

for all $|x| \geq M$ and $t \in \mathbb{R}^+$.

(ii) For all $|x| \geq M$ and $t \in \mathbb{R}^+$,

$$v'_{(E)}(x,t) \leq 0.$$

Then the solutions of (E) are *uniformly bounded*.

Proof. This result is a direct consequence of Theorem 3.3.4. □

Theorem 6.2.7 In addition to all hypotheses of Theorem 6.2.6, assume that there exists a function $\psi_3 \in \mathcal{K}$ such that

$$v'_{(E)}(x,t) \leq -\psi_3(|x|)$$

for all $|x| \geq M$ and $t \in \mathbb{R}^+$. Then the solutions of (E) are *uniformly ultimately bounded*.

Proof. This result is a direct consequence of Theorem 3.3.5. □

Example 6.2.8 Consider the system

$$\begin{cases} \dot{x} = -x - y \\ \dot{y} = -y - f(y) + x \end{cases} \tag{6.2.9}$$

where $f(y) = y(y^2 - 6)$. System (6.2.9) has equilibrium points located at $x = y = 0$, $x = -y = 2$, and $x = -y = -2$.

Choosing for a v-function

$$v(x, y) = \frac{1}{2}(x^2 + y^2)$$

we obtain

$$v'_{(6.2.9)}(x, y) = -x^2 - y^2(y^2 - 5) \leq -x^2 - \left(y^2 - \frac{5}{2}\right)^2 + \frac{25}{4}.$$

The v-function is positive definite and radially unbounded and $v'_{(6.2.9)}$ is negative for all (x, y) such that $x^2 + y^2 > R^2$, where, for example, $R = 10$ is acceptable. All the hypotheses of Theorem 6.2.6 are satisfied and we conclude that the solutions of (6.2.9) are *uniformly bounded.* Furthermore, all the hypotheses of Theorem 6.2.7 are satisfied and we conclude that the solutions of (6.2.9) are in fact *uniformly ultimately bounded.* □

Returning now to Theorem 2.3.3, we recall the following result concerning the *continuation* of solutions of (E). Let $f \in C[J \times \mathbb{R}^n, \mathbb{R}^n]$ where $J = (a, b)$ is a finite or infinite interval. Assume that every solution of (E) is bounded. Then every solution of (E) can be continued to the entire interval J.

The above result is in a certain sense wanting, because it assumes that all the solutions of (E) are bounded. In the next example, we provide reasonable conditions under which this assumption is satisfied.

Example 6.2.9 With $\Omega = \mathbb{R}^n$, assume for (E) that $f \in C[\mathbb{R}^+ \times \mathbb{R}^n, \mathbb{R}^n]$ and that $|f(t, x)| \leq \lambda(t)\psi(|x|)$ for all $t \in \mathbb{R}^+$ and all $|x| \geq M > 0$, where $\lambda \in C[\mathbb{R}^+, \mathbb{R}^+]$ has the property that $\int_0^\infty \lambda(t)dt < \infty$ and $\psi \in [[M, \infty), (0, \infty)]$ has the property that $\int_M^\infty (1/\psi(r))\, dr = \infty$. Then S_E, the set of all the solutions of (E), is *uniformly bounded.* To prove this, we choose the v-function

$$v(x, t) = -\int_0^t \lambda(s)ds + \int_M^{|x|} \frac{du}{\psi(u)}$$

and we apply Theorem 6.2.6. Condition (i) of the theorem is clearly satisfied. To show that condition (ii) is also satisfied, we note that

$$v'_{(E)}(x, t) \leq -\lambda(t) + \frac{1}{\psi(|x|)} \cdot \frac{|x|\,|f(t, x)|}{|x|} \leq 0$$

for all $t \in \mathbb{R}^+$ and $|x| \geq M$. □

For the case when $\int_0^\infty \lambda(t)dt = \infty$ along with Example 6.2.9, we have the following result.

Corollary 6.2.1 With $\Omega = \mathbb{R}^n$, assume for (E) that $f \in C[\mathbb{R}^+ \times \mathbb{R}^n, \mathbb{R}^n]$ and that $|f(t,x)| \le \lambda(t)\psi(|x|)$ for all $t \in \mathbb{R}^+$ and all $|x| \ge M$, where $\lambda \in C[\mathbb{R}^+, \mathbb{R}^+]$, $\psi \in C[[M,\infty),(0,\infty)]$, and $\int_M^\infty (1/\psi(r))dr = \infty$. Then every solution of (E) is continuable forward for all time.

Proof. It has been shown in Example 6.2.9 that the solutions of (E) are bounded when $\int_0^\infty \lambda(t)dt < \infty$. Therefore, it follows from Theorem 2.3.3 that every solution is continuable forward for all time. In the following we assume that $\int_0^\infty \lambda(t)dt = \infty$. We first show that for any $T > 0$, every solution $\varphi(\cdot, t_0, x_0)$ of (E) is bounded on (t_0, T). For otherwise there exist a $T > 0$ and a solution $\varphi(t, t_0, x_0)$ which is unbounded on (t_0, T). Let $\alpha_T = \int_0^T \lambda(s)ds$. Because $\int_M^\infty (1/\psi(r))dr = \infty$, there exist $b > a > M$ such that $\int_a^b (1/\psi(r))dr > 2\alpha_T$. Furthermore, there must exist t_1 and t_2, $t_0 \le t_1 \le t_2 \le T$, such that $|\varphi(t_1)| = a$, $|\varphi(t_2)| = b$, and $|x(t)| \ge M$ for all $t \in [t_1, t_2]$.

Let

$$v(\varphi(t), t) = -\int_0^t \lambda(s)ds + \int_M^{|\varphi(t)|} \frac{du}{\psi(u)}, \qquad t \in [t_1, t_2].$$

Then similarly as done in Example 6.2.9 we can show that $v'_{(E)}(\varphi(t), t) \le 0$ for all $t \in [t_1, t_2]$. Hence, $v(\varphi(t_2), t_2) \le v(\varphi(t_1), t_1)$.

On the other hand,

$$\begin{aligned} v(\varphi(t_2), t_2) &= -\int_0^{t_2} \lambda(s)ds + \int_M^{|\varphi(t_2)|} \frac{du}{\psi(u)} \\ &= -\int_0^{t_1} \lambda(s)ds - \int_{t_1}^{t_2} \lambda(s)ds + \int_M^{|\varphi(t_1)|} \frac{du}{\psi(u)} + \int_{|\varphi(t_1)|}^{|\varphi(t_2)|} \frac{du}{\psi(u)} \\ &> -\alpha_T + 2\alpha_T + v(\varphi(t_1), t_1) \\ &= \alpha_T + v(\varphi(t_1), t_1). \end{aligned}$$

We have arrived at a contradiction. Therefore, for any $T > 0$, every solution $\varphi(\cdot, t_0, x_0)$ of (E) is bounded on (t_0, T). It now follows from Theorem 2.3.3 that every solution is continuable to T. Because T is arbitrary, every solution is continuable forward for all time. □

Corollary 6.2.1 is readily applied to linear homogeneous systems

$$\dot{x} = A(t)x \qquad (LH)$$

where $A \in C[\mathbb{R}^+, \mathbb{R}^{n\times n}]$. In the present case we have $|f(t,x)| = |A(t)x| \le \lambda(t)\psi(|x|)$ where $\lambda(t) = \|A(t)\|$ and $\psi(|x|) = |x|$. It is readily verified that all the conditions of Corollary 6.2.1 are satisfied. Therefore, every solution of (LH) is continuable forward for all time.

E. Instability

We now present three instability results for (E).

Theorem 6.2.8 (*Lyapunov's First Instability Theorem*) The equilibrium $x_e = 0$ of (E) is *unstable* if there exist a $t_0 \in \mathbb{R}^+$ and a decrescent function $v \in C[B(r) \times \mathbb{R}^+, \mathbb{R}]$ for some $r > 0$ such that $v'_{(E)}$ is positive definite (negative definite) and if in every neighborhood of the origin there are points x such that $v(x, t_0) > 0$ $(v(x, t_0) < 0)$. Furthermore, if v is positive definite (negative definite), then the equilibrium $x_e = 0$ of (E) is *completely unstable* (refer to Definition 3.1.20).

Proof. This result is a direct consequence of Theorems 3.3.8 and 3.3.9. □

Example 6.2.10 If in Example 6.2.5, we have $c > 0$, then $v(x) = x_1^2 + x_2^2$ and $v'_{(6.2.6)}(x) = 2c(x_1^2 + x_2^2)^2$. We can conclude from Theorem 6.2.8 that the equilibrium $x_e = 0$ of (6.2.6) is *unstable*, in fact, *completely unstable.* □

Example 6.2.11 We now consider

$$\begin{cases} \dot{x}_1 = c_1 x_1 + x_1 x_2 \\ \dot{x}_2 = -c_2 x_2 + x_1^2 \end{cases} \tag{6.2.10}$$

where $c_1 > 0$ and $c_2 > 0$ are constants. We choose as a v-function

$$v(x) = x_1^2 - x_2^2$$

to obtain

$$v'_{(6.2.10)}(x) = 2(c_1 x_1^2 + c_2 x_2^2).$$

Because v is indefinite and $v'_{(6.2.10)}$ is positive definite, Theorem 6.2.8 applies and we conclude that the equilibrium $x_e = 0$ of (6.2.10) is *unstable*. □

Example 6.2.12 (*Hamiltonian system*) We now return to the conservative system considered in Example 6.2.3. In the present case we assume that $W(0) = 0$ is an isolated maximum. This is ensured by assuming that W_k is a negative definite homogeneous polynomial of degree k, where k is an even integer. Now recall that we assumed in Example 6.2.3 that T_2 is positive definite. We now choose as a v-function

$$v(p, q) = p^T q = \sum_{i=1}^{n} p_i q_i.$$

Along the solutions of (6.2.4) we now have

$$\begin{aligned} v'_{(6.2.4)}(p, q) &= \sum_{i=1}^{n} p_i \frac{\partial T_2}{\partial p_i} + \sum_{i=1}^{n} p_i \frac{\partial T_3}{\partial p_i} + \cdots - \sum_{i=1}^{n} q_i \frac{\partial W_k}{\partial q_i} - \sum_{i=1}^{n} q_i \frac{\partial W_{k+1}}{\partial q_i} - \cdots \\ &= 2T_2(p) + 3T_3(p) + \cdots - kW_k(q) - (k+1)W_{k+1}(q) - \cdots. \end{aligned}$$

In a sufficiently small neighborhood of the origin, the sign of $v'_{(6.2.4)}$ is determined by the sign of the term $2T_2(p) - kW_k(q)$, and thus, $v'_{(6.2.4)}$ is positive definite. Because v is indefinite, Theorem 6.2.8 applies. We conclude that the equilibrium $(p^T, q^T) = (0^T, 0^T) = 0$ is *unstable*. □

Theorem 6.2.9 (*Lyapunov's Second Instability Theorem*) Assume that for system (E) there exists a bounded function $v \in C[B(\varepsilon) \times [t_0, \infty), \mathbb{R}]$ where $\varepsilon > 0$ and $B(\varepsilon) \subset \Omega$, having the following properties.

(i) For all $(x, t) \in B(\varepsilon) \times [t_0, \infty)$,

$$v'_{(E)}(x, t) \geq \lambda v(x, t)$$

where $\lambda > 0$ is a constant.

(ii) In every neighborhood of the origin, there exists an x such that $v(x, t_1) > 0$ for a fixed $t_1 \geq t_0$.

Then the equilibrium $x_e = 0$ of (E) is *unstable*.

Proof. This result is a direct consequence of Theorem 3.3.10. □

Example 6.2.13 Consider the system

$$\begin{cases} \dot{x}_1 = x_1 + x_2 + x_1 x_2^4 \\ \dot{x}_2 = x_1 + x_2 - x_1^2 x_2. \end{cases} \tag{6.2.11}$$

This system clearly has an equilibrium at the origin. We choose the v-function $v(x) = \frac{1}{2}(x_1^2 - x_2^2)$ and compute

$$v'_{(6.2.11)}(x) = 2v(x) + x_1^2 x_2^4 + x_1^2 x_2^2 \geq 2v(x).$$

All the hypotheses of Theorem 6.2.9 are satisfied. Therefore, the equilibrium $x_e = 0$ of (6.2.11) is *unstable*. □

Theorem 6.2.10 (*Chetaev's Instability Theorem*) Assume that for system (E) there exist a function $v \in C[B(r) \times \mathbb{R}^+, \mathbb{R}]$ for some $r > 0$, where $B(r) \subset \Omega$, a $t_0 \in \mathbb{R}^+$, and an $h > 0, h < r$, such that the following conditions are satisfied.

(i) There exists a component D of the set $\{(x, t) \in B(r) \times \mathbb{R}^+ \colon v(x, t) < 0, |x| < h\}$ such that for every neighborhood of the origin there exists an x in this neighborhood such that $(x, t_0) \in D$.

(ii) v is bounded from below on D.

(iii) $v'_{(E)}(x, t) \leq -\psi(|v(x, t)|)$ for all $(x, t) \in D$, where $\psi \in \mathcal{K}$.

Then the equilibrium $x_e = 0$ of (E) is *unstable*.

Proof. Let $M > 0$ be a number such that $-M \leq v(x, t)$ on D. Given any $r_1 > 0$ choose $(x_0, t_0) \in B(r_1) \times \mathbb{R}^+ \cap D$. Then the solution $\varphi_0(t, t_0, x_0)$ must leave $B(h)$ in finite time. Indeed, $|\varphi_0(t)|$ must become equal to h in finite time. To see this, assume the contrary. Let $v(t) = v(\varphi_0(t), t)$. Because $v(t_0) < 0$ and $v'_{(E)}(x, t) \leq -\psi(|v(x, t)|)$, we have $v(t) \leq v(t_0) < 0$ for all $t \geq 0$. Thus

$$v(t) \leq v(t_0) - \int_{t_0}^{t} \psi(|v(t_0)|) ds \to -\infty$$

as $t \to \infty$. This contradicts the bound $v(t) \geq -M$. Hence there is a $t^* > t_0$ such that $(\varphi_0(t^*), t^*) \in \partial D$. But $v(t^*) < 0$, so the only part of ∂D that $(\varphi_0(t^*), t^*)$ can penetrate is that part where $|\varphi_0(t)| = h$. Because this can happen for arbitrarily small $|x_0|$, the instability of $x_e = 0$ is proved. □

For the case of *autonomous systems*,

$$\dot{x} = f(x) \tag{A}$$

where $x \in \Omega$, $f \in C[\Omega, \mathbb{R}^n]$, Ω is an open connected subset of $\mathbb{R}^n$ that contains the origin and $f(0) = 0$, we have the following simpler version of Theorem 6.2.10.

Corollary 6.2.2 Assume that for system (A) there exists a function $v \in C[B(r), \mathbb{R}]$, $r > 0$, $B(r) \subset \Omega$, that satisfies the following conditions.

(i) The open set $\{x \in B(h)\colon v(x) < 0\}$ for some $h > 0, h < r$, contains a component D for which $0 \in \partial D$.

(ii) $v'_{(A)}(x) < 0$ for all $x \in D$, $x \neq 0$.

Then the equilibrium $x_e = 0$ of (A) is *unstable*. □

Example 6.2.14 Consider the system

$$\begin{cases} \dot{x}_1 = x_1 + x_2 \\ \dot{x}_2 = x_1 - x_2 + x_1 x_2 \end{cases} \tag{6.2.12}$$

which clearly has an equilibrium at the origin $x_e = 0$. Choose

$$v(x) = -x_1 x_2$$

to obtain

$$v'_{(6.2.12)}(x) = -x_1^2 - x_2^2 - x_1^2 x_2.$$

Let

$$D = \{x \in \mathbb{R}^2\colon x_1 > 0,\ x_2 > 0 \text{ and } x_1^2 + x_2^2 < 1\}.$$

Then for all $x \in D$, $v(x) < 0$, and $v'_{(6.2.12)}(x) < 2v(x)$. All the hypotheses of Theorem 6.2.10 (and Corollary 6.2.2) are satisfied. Therefore, the equilibrium $x_e = 0$ of (6.2.12) is *unstable*. □

Example 6.2.15 Once more, we return to the conservative system considered in Examples 6.2.3 and 6.2.12. To complete the stability analysis of this system, we assume that $W(0) = 0$ is not a local minimum of the potential energy. Then there are points q arbitrarily close to the origin such that $W(q) < 0$. Because $H(0, q) = W(q)$, there are points $(p^T, q^T)^T$ arbitrarily close to the origin where $H(p, q) < 0$ for all p sufficiently near the origin. Therefore, there are points $(p^T, q^T)^T$ arbitrarily close to the origin such that $p^T q > 0$ and $-H(p, q) > 0$, simultaneously. Let U be some neighborhood of the origin and let

$$U_1 = \{(p^T, q^T)^T \in U\colon p^T q > 0 \text{ and } -H(p, q) > 0\}.$$

Now choose

$$v(p, q) = H(p, q) p^T q.$$

Using the fact that $(dH/dt)[p(t), q(t)] = 0$ along the solutions of (6.2.4) (refer to Example 6.2.3), we obtain

$$v'_{(6.2.4)}(p, q) = -H(p, q)[-2T_2(p) - 3T_3(p) - \cdots + kW_k(q) + \cdots]. \quad (6.2.13)$$

Now choose $U = B(r)$ with $r > 0$ sufficiently small so that $T(p) > 0$ within $U - \{0\}$. Because in U_1, $H(p, q) = T(p) + W(q) < 0$ and $T(p) > 0$, it must be true that $W(q) < 0$ within U_1. Therefore, for U sufficiently small, the term in brackets in (6.2.13) is negative within U_1 and $v'_{(6.2.4)}$ is negative within U_1. The origin is a boundary point of U_1, thus there exists a component D of U_1 such that the origin is on the boundary of D. Because any component of U_1 is a component of the set $\{(p^T, q^T)^T \in U : v(p, q) < 0\}$, because $v'_{(6.2.4)}$ is negative on D, and because v is bounded on any compact set containing D, it follows from Theorem 6.2.10 (or Corollary 6.2.2) that the equilibrium $(p^T, q^T)^T = 0$ of system (6.2.4) is *unstable*. □

We conclude this section by observing that frequently the results of the present section yield more than just stability (resp., instability and boundedness) information. For example, suppose that for system (A) there exist a continuously differentiable function $v: \mathbb{R}^n \to \mathbb{R}$ and three positive constants c_1, c_2, c_3 such that

$$c_1|x|^2 \le v(x) \le c_2|x|^2, \qquad v'_{(A)}(x) \le -c_3|x|^2 \quad (6.2.14)$$

for all $x \in \mathbb{R}^n$. Then in accordance with Theorem 6.2.5, the equilibrium $x_e = 0$ of system (A) is exponentially stable in the large. However, we know more: evaluating (6.2.14) along the solution $\varphi(t, t_0, x_0)$ we obtain

$$\begin{cases} c_1|\varphi(t, t_0, x_0)|^2 \le v(\varphi(t, t_0, x_0)) \le c_2|\varphi(t, t_0, x_0)|^2 \\ v'_{(A)}(\varphi(t, t_0, x_0)) \le -c_3|\varphi(t, t_0, x_0)|^2 \end{cases} \quad (6.2.15)$$

for all $t \ge t_0, x_0 \in \mathbb{R}^n$. It is now an easy matter to obtain from (6.2.15) the very useful estimate

$$|\varphi(t, t_0, x_0)| \le \sqrt{c_2/c_1}\, |x_0| e^{-[c_3/(2c_2)](t-t_0)}$$

for all $t \ge t_0$ and $x_0 \in \mathbb{R}^n$.

We present applications of the results of this section to specific important classes of dynamical systems determined by ordinary differential equations in Chapter 8.

6.3 The Principal Stability and Boundedness Results for Ordinary Difference Equations

In the present section we address stability and boundedness properties of discrete-time finite-dimensional dynamical systems determined by ordinary difference equations (D). As indicated in Subsection 6.1B, we assume without loss of generality that $x_e = 0$ is an equilibrium of (D). Also, $\Delta_{(D)}v(x,k)$ denotes the first forward difference of $v(x,k)$ along the solutions of (D) (refer to (6.1.5)).

A. Local stability results

In our first result we concern ourselves with the stability, uniform stability, uniform asymptotic stability, and exponential stability of the equilibrium $x_e = 0$ of (D).

Theorem 6.3.1 In the following, assume that $v \in C[\Omega \times \mathbb{N}, \mathbb{R}]$ and that v is positive definite.

(a) If $\Delta_{(D)}v$ is negative semidefinite, then the equilibrium $x_e = 0$ of (D) is *stable*.

(b) If v is decrescent and $\Delta_{(D)}v$ is negative semidefinite, then the equilibrium $x_e = 0$ of (D) is *uniformly stable*.

(c) If v is decrescent and $\Delta_{(D)}v$ is negative definite, then the equilibrium $x_e = 0$ of (D) is *uniformly asymptotically stable*.

(d) If there exist four positive constants c_1, c_2, c_3, and b such that

$$c_1|x|^b \leq v(x,k) \leq c_2|x|^b$$

and

$$\Delta_{(D)}v(x,k) \leq -c_3|x|^b$$

for all $(x,k) \in \Omega \times \mathbb{N}$, then the equilibrium $x_e = 0$ of (D) is *exponentially stable*.

Proof. The proof of statement (a) follows along similar lines as the proof of statement (a) of Theorem 6.2.1.

Statements (b), (c), and (d) are direct consequences of Theorems 3.4.1, 3.4.2, and 3.4.3, respectively. □

Example 6.3.1 Consider the linear system

$$\begin{cases} x_1(k+1) = x_1(k) + 0.5x_2(k) \\ x_2(k+1) = 0.5x_2(k). \end{cases} \tag{6.3.1}$$

The origin $x_e = 0$ is clearly an equilibrium of (6.3.1). Choose the v-function

$$v(x) = |x_1 + x_2| + |x_2|.$$

Along the solutions of (6.3.1) we have

$$\begin{aligned} \Delta_{(6.3.1)}v(x(k)) &= |x_1(k+1) + x_2(k+1)| + |x_2(k+1)| \\ &\quad - |x_1(k) + x_2(k)| - |x_2(k)| \\ &= |x_1(k) + 0.5x_2(k) + 0.5x_2(k)| + 0.5|x_2(k)| \\ &\quad - |x_1(k) + x_2(k)| - |x_2(k)| \\ &= -0.5|x_2(k)|, \end{aligned}$$

which is negative semidefinite. The function v is positive definite and decrescent. All conditions of Theorem 6.3.1(b) are satisfied. Therefore, the equilibrium $x_e = 0$ of (6.3.1) is *uniformly stable*. □

Example 6.3.2 Consider the linear system given by

$$x(k+1) = \left(1 - \frac{1}{2^{k+1}}\right) x(k). \tag{6.3.2}$$

The equilibrium $x_e = 0$ of (6.3.2) is uniformly stable. This can be shown by choosing $v(x) = |x|$. Then it is clear that $\Delta_{(6.3.2)}v$ is negative semidefinite. Therefore the trivial solution is *uniformly stable*. Furthermore, every motion tends to 0 as $k \to \infty$ (i.e., the trivial solution is *attractive*). Thus the trivial solution is *asymptotically stable*.

On the other hand the equilibrium $x_e = 0$ of (6.3.2) is *not uniformly asymptotically stable*. For any $\delta > 0$ and any $\tau \in \mathbb{N}$, there exists a $k_0 \in \mathbb{N}$ such that $x(k_0 + \tau) > 0.5\delta$, when $x(k_0) = 0.8\delta$. Condition (2) of Definition 6.1.1(c) is not satisfied. This is the result of the fact that the motions decrease very slowly as $k \to \infty$.

Exponential stability implies uniform asymptotic stability; thus the equilibrium $x_e = 0$ of (6.3.2) cannot be exponentially stable, either. □

Example 6.3.3 Consider the nonlinear autonomous system given by

$$x(k+1) = x(k) - x(k)^3 = x(k)\big(1 - x(k)^2\big). \tag{6.3.3}$$

This system clearly has an equilibrium $x_e = 0$. Let the v-function be given by

$$v(x) = |x|.$$

Along the solutions of (6.3.3) we have

$$\Delta_{(6.3.3)}v(x(k)) = |x(k+1)| - |x(k)| = -|x(k)|^3.$$

All the conditions of Theorem 6.3.1(c) are satisfied. Therefore, the equilibrium $x_e = 0$ of (6.3.3) is *uniformly asymptotically stable*.

For the v-function chosen above, there do not exist positive constants c_1, c_2, c_3, and b that satisfy the conditions of Theorem 6.3.1(d). As a matter of fact, because of the slow decreasing rate of $|x(k)|$, the equilibrium $x_e = 0$ of (6.3.3) is *not exponentially stable*. This can be shown by contradiction. Suppose that there exist an $\alpha > 0$ and a $\delta > 0$ (let ε in Definition 6.1.1(d) be 1) such that

$$|x(k)| < e^{-\alpha(k-k_0)} \tag{6.3.4}$$

for all $k \in \mathbb{N}$ whenever $|x(0)| < \delta$. Let m be sufficiently large such that

$$1 - e^{-2m\alpha} \geq e^{-\alpha/2}.$$

Let $x(0) = e^{-m\alpha}$. We then have $1 - x(0)^2 \geq e^{-\alpha/2}$. Because $x(k)$ is positive and $\{x(k)\}$ is decreasing, it is true that $1 - x(k)^2 \geq e^{-\alpha/2}$ for all $k \in \mathbb{N}$. Thus,

$$x(k+1) = x(k)(1 - x(k)^2) \geq x(k)e^{-\alpha/2},$$

which in turn yields

$$x(k) \geq x(0)e^{-k\alpha/2} = e^{-(m+k/2)\alpha}$$

for all $k \in \mathbb{N}$. Let $k = 4m$. Then

$$x(4m) \geq e^{-3m\alpha}.$$

On the other hand, it follows from (6.3.4) that

$$x(4m) < e^{-4m\alpha},$$

which is a contradiction. Therefore, the equilibrium $x_e = 0$ of (6.3.3) is not exponentially stable. □

Example 6.3.4 Consider the system

$$\begin{cases} x_1(k+1) = x_1(k)^2 + x_2(k)^2 \\ x_2(k+1) = x_1(k)x_2(k). \end{cases} \tag{6.3.5}$$

The origin $x_e = 0$ is clearly an equilibrium of (6.3.5). Choose the v-function

$$v(x) = |x_1| + |x_2| = |x|_1.$$

Along the solutions of (6.3.5) we have

$$\begin{aligned} \Delta_{(6.3.5)}v(x(k)) &= x_1(k)^2 + x_2(k)^2 + |x_1(k)x_2(k)| - |x_1(k)| - |x_2(k)| \\ &\leq x_1(k)^2 + 2|x_1(k)|\,|x_2(k)| + x_2(k)^2 - |x_1(k)| - |x_2(k)| \\ &= (|x_1(k)| + |x_2(k)|)^2 - (|x_1(k)| + |x_2(k)|) \\ &= |x(k)|_1^2 - |x(k)|_1 \\ &= (|x(k)|_1 - 1)|x(k)|_1. \end{aligned}$$

For any $|x|_1 < c < 1$, we have $\Delta_{(6.3.5)}v(x(k)) \leq (c-1)|x(k)|_1$. In accordance with Theorem 6.3.1(d), the equilibrium $x_e = 0$ of system (6.3.5) is *exponentially stable*. The *domain of attraction* of the equilibrium $x_e = 0$ is given by

$$\{x \in \mathbb{R}^2 \colon |x|_1 = |x_1| + |x_2| < c,\ 0 < c < 1\}.$$ □

B. Global stability results

In our second result we concern ourselves with the global uniform asymptotic stability and the global exponential stability of the equilibrium $x_e = 0$ of (D).

Theorem 6.3.2 (a) With $\Omega = \mathbb{R}^n$, assume that there exists a positive definite, decrescent, and radially unbounded function $v \in C[\mathbb{R}^n \times \mathbb{N}, \mathbb{R}]$ such that $\Delta_{(D)}v$ is negative definite; that is, there exist $\psi_1, \psi_2 \in \mathcal{K}_\infty$ and $\psi_3 \in \mathcal{K}$, such that

$$\psi_1(|x|) \leq v(x,k) \leq \psi_2(|x|) \tag{6.3.6}$$

and

$$\Delta_{(D)} v(x,k) \leq -\psi_3(|x|) \tag{6.3.7}$$

for all $(x,k) \in \mathbb{R}^n \times \mathbb{N}$. Then the equilibrium $x_e = 0$ of (D) is *uniformly asymptotically stable in the large.*

(b) In part (a), assume that inequalities (6.3.6) and (6.3.7) are of the form

$$c_1|x|^b \leq v(x,k) \leq c_2|x|^b$$

and

$$\Delta_{(D)} v(x,k) \leq -c_3|x|^b$$

for all $(x,k) \in \mathbb{R}^n \times \mathbb{N}$, where c_1, c_2, c_3, and b are positive constants. Then the equilibrium $x_e = 0$ of (D) is *exponentially stable in the large.*

Proof. (a) This result is a direct consequence of Theorem 3.4.6.

(b) This result is a direct consequence of Theorem 3.4.7. □

Example 6.3.5 Consider the system

$$\begin{cases} x_1(k+1) = x_1(k) - cx_1(k)\text{sat}\big(x_1(k)^2 + x_2(k)^2\big) \\ x_2(k+1) = x_2(k) - cx_2(k)\text{sat}\big(x_1(k)^2 + x_2(k)^2\big) \end{cases} \tag{6.3.8}$$

where sat$(\cdot)$ is the *saturation function* given by

$$\text{sat}(r) = \begin{cases} 1, & r > 1 \\ r, & -1 \leq r \leq 1 \\ -1, & r < -1. \end{cases} \tag{6.3.9}$$

The origin $x_e = 0$ is clearly an equilibrium of (6.3.8). Choose the v-function as

$$v(x) = x_1^2 + x_2^2.$$

Along the solutions of (6.3.8) we have

$$\begin{aligned} \Delta_{(6.3.8)} v(x(k)) &= x_1(k+1)^2 + x_2(k+1)^2 - x_1(k)^2 - x_2(k)^2 \\ &= \big(x_1(k)^2 + x_2(k)^2\big)\big(1 - c\,\text{sat}\big(x_1(k)^2 + x_2(k)^2\big)\big)^2 \\ &\quad - x_1(k)^2 - x_2(k)^2 \\ &= -c\big(x_1(k)^2 + x_2(k)^2\big)\text{sat}\big(x_1(k)^2 + x_2(k)^2\big) \\ &\quad \times \big(2 - c\,\text{sat}\big(x_1^2(k) + x_2(k)^2\big)\big). \end{aligned}$$

If $0 < c < 2$, Theorem 6.3.2 applies and we conclude that the equilibrium $x_e = 0$ of (6.3.8) is *uniformly asymptotically stable in the large.* Similarly as was done in Example 6.3.3, we can show that $x_e = 0$ is *not exponentially stable in the large.* □

Example 6.3.6 Consider the system

$$\begin{cases} x_1(k+1) = 0.5x_2(k) + cx_1(k)\text{sat}\big(x_1(k)^2 + x_2(k)^2\big) \\ x_2(k+1) = -0.5x_1(k) + cx_2(k)\text{sat}\big(x_1(k)^2 + x_2(k)^2\big) \end{cases} \tag{6.3.10}$$

where sat$(\cdot)$ is given by (6.3.9). The origin $x_e = 0$ is clearly an equilibrium of (6.3.10). Choose the v-function as

$$v(x) = x_1^2 + x_2^2.$$

Along the solutions of (6.3.10) we have

$$\begin{aligned} \Delta_{(6.3.10)}v(x(k)) &= x_1(k+1)^2 + x_2(k+1)^2 - x_1(k)^2 - x_2(k)^2 \\ &= -\big(x_1(k)^2 + x_2(k)^2\big)\big(0.75 - c^2\left[\text{sat}\big(x_1(k)^2 + x_2(k)^2\big)\right]^2\big). \end{aligned}$$

If $c < \sqrt{0.75}$, Theorem 6.3.2(a) applies and we conclude that the equilibrium $x_e = 0$ of (6.3.10) is *uniformly asymptotically stable in the large*. In fact, Theorem 6.3.2(b) also applies. Hence, the equilibrium $x_e = 0$ is also *exponentially stable in the large*. □

Example 6.3.7 Consider the linear time-varying system given by

$$\begin{cases} x_1(k+1) = \dfrac{1}{(k+2)}x_1(k) - \dfrac{2}{(k+2)}x_2(k) \\ x_2(k+1) = \dfrac{2}{(k+2)}x_1(k) + \dfrac{1}{(k+2)}x_2(k) \end{cases} \tag{6.3.11}$$

where $k \geq 1$. This system clearly has an equilibrium at the origin $x_e = (x_1, x_2)^T = 0$.
We choose as a v-function

$$v(x_1, x_2) = x_1^2 + x_2^2.$$

Along the solutions of (6.3.11) we have

$$\begin{aligned} \Delta_{(6.3.11)}v(x_1(k), x_2(k)) &= \frac{5}{(k+2)^2}(x_1(k)^2 + x_2(k)^2) - (x_1(k)^2 + x_2(k)^2) \\ &= \left(\frac{5}{(k+2)^2} - 1\right)(x_1(k)^2 + x_2(k)^2) \\ &\leq -\frac{4}{9}\big(x_1(k)^2 + x_2(k)^2\big), \qquad (x_1, x_2)^T \in \mathbb{R}^2. \end{aligned}$$

All the conditions of Theorem 6.3.2(b) are satisfied. Accordingly, the equilibrium $x_e = 0$ of system (6.3.11) is *exponentially stable in the large*. □

In the next result we address the uniform boundedness and the uniform ultimate boundedness of solutions of (D).

Theorem 6.3.3 With $\Omega = \mathbb{R}^n$, assume that there exists a function $v \in C[\mathbb{R}^n \times \mathbb{N}, \mathbb{R}]$ that satisfies the following conditions.

(i) There exist two functions $\psi_1, \psi_2 \in C[\mathbb{R}^+, \mathbb{R}^+]$ that are strictly increasing with $\lim_{r\to\infty} \psi_i(r) = \infty$, $i = 1, 2$, and a constant $M > 0$, such that

$$\psi_1(|x|) \leq v(x, k) \leq \psi_2(|x|)$$

for all $|x| \geq M$ and $k \in \mathbb{N}$.

(ii) For all $|x| \geq M$ and $k \in \mathbb{N}$,

$$\Delta_{(D)} v(x, k) \leq 0.$$

Then the solutions of (D) are *uniformly bounded*.

If in addition to (i) and (ii), there exists a function $\psi_3 \in \mathcal{K}_\infty$ such that

$$\Delta_{(D)} v(x, k) \leq -\psi_3(|x|)$$

for all $|x| \geq M$ and $k \in \mathbb{N}$, then the solutions of (D) are *uniformly ultimately bounded*.

Proof. This theorem is a direct consequence of Theorems 3.4.4 and 3.4.5. □

Example 6.3.8 Consider the system

$$\begin{cases} x_1(k+1) = ax_1(k) + f(x_2(k)) + 0.5M \\ x_2(k+1) = ax_2(k) + g(x_1(k)) + 0.5M \end{cases} \tag{6.3.12}$$

where $|a| < 1$, $|f(\eta)| \leq c|\eta|$ and $|g(\eta)| \leq c|\eta|$ for all $\eta \in \mathbb{R}$, and $M \in \mathbb{R}^+$.

Choose

$$v(x) = |x_1| + |x_2| = |x|_1.$$

Along the solutions of (6.3.12) we have for all $(x_1, x_2) \in \mathbb{R}^2$,

$$\begin{aligned} \Delta_{(6.3.12)} v(x(k)) &= \left|ax_1(k) + f(x_2(k)) + \frac{1}{2}M\right| \\ &\quad + \left|ax_2(k) + g(x_1(k)) + \frac{1}{2}M\right| - |x_1(k)| - |x_2(k)| \\ &\leq |a|\ |x_1(k)| + c|x_2(k)| + |a|\ |x_2(k)| + c|x_1(k)| + M \\ &\quad - |x_1(k)| - |x_2(k)| \\ &= (|a| + c - 1)|x_1(k)| + (|a| + c - 1)|x_2(k)| + M \\ &= (|a| + c - 1)|x(k)|_1 + M. \end{aligned}$$

Assume that a and c are such that $|a| + c - 1 < 0$. Then $\Delta_{(6.3.12)} v(x(k)) < 0$ for all $|x|_1 > M/(1 - |a| - c)$. All conditions of Theorem 6.3.3 are satisfied. Therefore, the solutions of (6.3.12) are *uniformly bounded*. □

C. Instability results

In the next results we concern ourselves with the instability of the equilibrium $x_e = 0$ of (D).

Theorem 6.3.4 (*Lyapunov's First Instability Theorem*) The equilibrium $x_e = 0$ of (D) is *unstable* if there exist a $k_0 \in \mathbb{N}$ and a decrescent function $v \in C[B(r) \times \mathbb{N}, \mathbb{R}]$ where $r > 0$, $B(r) \subset \Omega$, such that $\Delta_{(D)}v$ is positive definite (negative definite) and if in every neighborhood of the origin there are points x such that $v(x, k_0) > 0$ $(v(x, k_0) < 0)$. If in addition to the above assumptions, v is positive definite (negative definite), then the equilibrium $x_e = 0$ of (D) is *completely unstable* (refer to Definition 3.1.20).

Proof. By definition, the function v is decrescent implies there exists a function $\psi \in \mathcal{K}$ defined on $[0, r]$ such that

$$\left|v(x,k)\right| \leq \psi(|x|)$$

for all $(x, k) \in B(r) \times \mathbb{N}$.

Under the assumptions of the present theorem, v (or $-v$) satisfies the hypotheses of Theorem 3.4.8 for instability and the hypotheses of Theorem 3.4.9 for complete instability. The proof is completed. □

Theorem 6.3.5 (*Lyapunov's Second Instability Theorem*) Assume that for system (D) there exists a bounded function $v \in C[B(r) \times [k_0, \infty), \mathbb{R}]$, where $r > 0$, $B(r) \subset \Omega$, and $k_0 \in \mathbb{N}$, such that the following conditions are satisfied.

(i) For all $(x, k) \in B(r) \times [k_0, \infty)$,

$$\Delta_{(D)}v(x,k) \geq \lambda v(x,k),$$

where $\lambda > 0$ is a constant.

(ii) In every neighborhood of the origin, there exists an x such that $v(x, k_1) > 0$ for a fixed $k_1 \geq k_0$, $k_1 \in \mathbb{N}$.

Then the equilibrium $x_e = 0$ of (D) is *unstable*.

Proof. This result is a direct consequence of Theorem 3.4.10. □

Example 6.3.9 Consider the system

$$\begin{cases} x_1(k+1) = a^2|x_1(k)| + x_2(k)f(x_2(k)) \\ x_2(k+1) = a^2|x_2(k)| + x_1(k)g(x_1(k)) \end{cases} \tag{6.3.13}$$

where $a^2 > 1$, $f \in C[\mathbb{R}, \mathbb{R}]$, $g \in C[\mathbb{R}, \mathbb{R}]$, and $\eta f(\eta) \geq 0$ and $\eta g(\eta) \geq 0$ for all $\eta \in \mathbb{R}$. The origin $(x_1, x_2)^T = (0, 0)^T = 0$ is clearly an equilibrium of (6.3.13).

We choose as a v-function

$$v(x) = |x_1| + |x_2| = |x|_1.$$

Along the solutions of (6.3.13) we have for all $x \in \mathbb{R}^2$,

$$\begin{aligned}\Delta_{(6.3.13)}v(x(k)) = {} & \left|a^2|x_1(k)| + x_2(k)f(x_2(k))\right| + \left|a^2|x_2(k)| + x_1(k)g(x_1(k))\right| \\ & - |x_1(k)| - |x_2(k)|\end{aligned}$$

$$\begin{aligned}&\geq a^2|x_1(k)| + a^2|x_2(k)| - |x_1(k)| - |x_2(k)|\\&= a^2|x(k)|_1 - |x(k)|_1\\&= (a^2-1)|x(k)|_1.\end{aligned}$$

Because by assumption $a^2 > 1$, $\Delta_{(6.3.13)}v$ is positive definite. All the hypotheses of Theorem 6.3.4 are satisfied and we conclude that the equilibrium $x_e = 0$ of (6.3.13) is *unstable*, in fact, *completely unstable*. □

Example 6.3.10 Consider the system

$$\begin{cases} x_1(k+1) = a^2|x_1(k)| + x_2(k)f(x_2(k)) \\ x_2(k+1) = a^2 x_2(k) \end{cases} \tag{6.3.14}$$

where $a^2 > 1$, $f \in C[\mathbb{R}, \mathbb{R}]$, and $\eta f(\eta) \geq 0$ for all $\eta \in \mathbb{R}$. Choose as a v-function

$$v(x_1, x_2) = |x_1| - |x_2|.$$

Along the solutions of (6.3.14) we have

$$\begin{aligned}\Delta_{(6.3.14)}v(x_1,x_2) &= |a^2|x_1(k)| + x_2(k)f(x_2(k))| - |a^2x_2(k)| - |x_1(k)| + |x_2(k)|\\&\geq a^2|x_1(k)| - a^2|x_2(k)| - |x_1(k)| + |x_2(k)|\\&= (a^2-1)|x_1(k)| - (a^2-1)|x_2(k)|\\&= (a^2-1)(|x_1(k)| - |x_2(k)|)\\&= \lambda v(x_1,x_2)\end{aligned}$$

where $\lambda = a^2 - 1 > 0$ because $a^2 > 1$. In every neighborhood of the origin there are points $\bar{x}$ in $\mathbb{R}^2$ such that $v(\bar{x}) > 0$. Thus, all the hypotheses of Theorem 6.3.5 are satisfied. Therefore, the equilibrium $x_e = 0$ of (6.3.14) is *unstable*. □

We present applications to important specific classes of finite-dimensional dynamical systems determined by ordinary difference equations in Chapter 8.

6.4 The Principal Stability and Boundedness Results for Discontinuous Dynamical Systems

In this section we fist present local stability results, global stability and boundedness results, and instability results for finite-dimensional DDS (refer to Subsection 6.1C). We address applications of these results in the last subsection of this section and further, in Chapter 8. As noted in Subsection 6.1C, we assume that for the dynamical systems in question, the origin $x_e = 0 \in \mathbb{R}^n$ is an equilibrium.

A. Local stability results

We first address local results.

Theorem 6.4.1 Let $\{\mathbb{R}^+, \mathbb{R}^n, A, S\}$ be a finite-dimensional discontinuous dynamical system (for short, a finite-dimensional DDS S) and assume that $x_e = 0$ is an equilibrium. Assume that there exist a function $v\colon \mathbb{R}^n \times \mathbb{R}^+ \to \mathbb{R}^+$ and two functions $\psi_1, \psi_2 \in \mathcal{K}$ defined on $\mathbb{R}^+$ such that

$$\psi_1(|x|) \le v(x,t) \le \psi_2(|x|) \tag{6.4.1}$$

for all $x \in \mathbb{R}^n$ and $t \in \mathbb{R}^+$.

Assume that for any $\varphi(\cdot, t_0, x_0) \in S$ $(t_0 = \tau_0)$, $v(\varphi(t,t_0,x_0),t)$ is continuous everywhere on $\mathbb{R}^+_{t_0} = \{t \in \mathbb{R}^+ : t \ge t_0\}$ except on an unbounded subset $E = \{\tau_1, \tau_2, \dots : \tau_1 < \tau_2 < \cdots\}$ of $\mathbb{R}^+_{t_0}$. Also assume that there exists a neighborhood U of the origin $x_e = 0$ such that for all $x_0 \in U$ and $\varphi(\cdot, t_0, x_0) \in S$, $v(\varphi(\tau_k, t_0, x_0), \tau_k)$ is nonincreasing for $k \in \mathbb{N}$. Furthermore, assume that there exists a function $f \in C[\mathbb{R}^+, \mathbb{R}^+]$, independent of $\varphi \in S$, such that $f(0) = 0$ and such that

$$v(\varphi(t,t_0,x_0),t) \le f(v(\varphi(\tau_k,t_0,x_0),\tau_k)) \tag{6.4.2}$$

for all $t \in (\tau_k, \tau_{k+1})$, $k \in \mathbb{N}$.

Then the equilibrium $x_e = 0$ of the DDS S is *uniformly stable*.

Proof. This result is a direct consequence of Theorem 3.2.1. □

Theorem 6.4.2 If in addition to the assumptions given in Theorem 6.4.1, there exists a function $\psi_3 \in \mathcal{K}$ defined on $\mathbb{R}^+$ such that for all $x_0 \in U$, $\varphi(\cdot, t_0, x_0) \in S$, and $k \in \mathbb{N}$,

$$Dv(\varphi(\tau_k,t_0,x_0),\tau_k) \le -\psi_3(|\varphi(\tau_k,t_0,x_0)|) \tag{6.4.3}$$

where

$$Dv(\varphi(\tau_k,t_0,x_0),\tau_k) \triangleq \frac{1}{\tau_{k+1}-\tau_k}\big[v(\varphi(\tau_{k+1},t_0,x_0),\tau_{k+1}) - v(\varphi(\tau_k,t_0,x_0),\tau_k)\big], \tag{6.4.4}$$

then the equilibrium $x_e = 0$ of the DDS S is *uniformly asymptotically stable*.

Proof. This result is a direct consequence of Theorem 3.2.2. □

Theorem 6.4.3 Let $\{\mathbb{R}^+, \mathbb{R}^n, A, S\}$ be a finite-dimensional DDS and assume that $x_e = 0$ is an equilibrium. Assume that there exist a function $v\colon \mathbb{R}^n \times \mathbb{R}^+ \to \mathbb{R}^+$ and four positive constants c_1, c_2, c_3, and b such that

$$c_1|x|^b \le v(x,t) \le c_2|x|^b \tag{6.4.5}$$

for all $x \in \mathbb{R}^n$ and $t \in \mathbb{R}^+$.

Assume that there exists a neighborhood U of the origin $x_e = 0$ such that for all $x_0 \in U$ and $\varphi(\cdot, t_0, x_0) \in S$ $(t_0 = \tau_0)$, $v(\varphi(t,t_0,x_0),t)$ is continuous everywhere on $\mathbb{R}^+_{t_0}$ except on an unbounded subset $E = \{\tau_1, \tau_2, \dots : \tau_1 < \tau_2 < \cdots\}$ of $\mathbb{R}^+_{t_0}$. Furthermore, assume that there exists a function $f \in C[\mathbb{R}^+, \mathbb{R}^+]$ such that $f(0) = 0$ and

$$v(\varphi(t,t_0,x_0),t) \le f(v(\varphi(\tau_k,t_0,x_0),\tau_k)) \tag{6.4.6}$$

for all $t \in (\tau_k, \tau_{k+1})$, $k \in \mathbb{N}$, and that for some positive constant q, f satisfies

$$f(r) = o(r^q) \qquad \text{as } r \to 0^+ \tag{6.4.7}$$

(i.e., $\lim_{r\to 0^+} (f(r)/r^q) = 0$). Assume that for all $k \in \mathbb{N}$,

$$Dv(\varphi(\tau_k, t_0, x_0), \tau_k) \leq -c_3 \big|\varphi(\tau_k, t_0, x_0)\big|^b \tag{6.4.8}$$

for all $x_0 \in U$ and $\varphi(\cdot, t_0, x_0) \in S$, where Dv is given in (6.4.4).

Then the equilibrium $x_e = 0$ of the DDS S is *exponentially stable*.

Proof. This result is a direct consequence of Theorem 3.2.3. □

B. Global stability and boundedness results

Next, we address global results.

Theorem 6.4.4 Let $\{\mathbb{R}^+, \mathbb{R}^n, A, S\}$ be a finite-dimensional DDS and assume that $x_e = 0$ is an equilibrium. Assume that there exist a function $v\colon \mathbb{R}^n \times \mathbb{R}^+ \to \mathbb{R}^+$ and two strictly increasing functions $\psi_1, \psi_2 \in C[\mathbb{R}^+, \mathbb{R}^+]$ with $\lim_{r\to\infty} \psi_i(r) = \infty$, $i = 1, 2$, such that

$$\psi_1(|x|) \leq v(x,t) \leq \psi_2(|x|) \tag{6.4.9}$$

for all $x \in \mathbb{R}^n$ and $t \in \mathbb{R}^+$ whenever $|x| \geq \Omega$, where Ω is a positive constant.

Assume that for every $\varphi(\cdot, t_0, x_0) \in S$ $(t_0 = \tau_0)$, $v(\varphi(t, t_0, x_0), t)$ is continuous everywhere on $\mathbb{R}^+_{t_0}$ except on an unbounded subset $E = \{\tau_1, \tau_2, \ldots : \tau_1 < \tau_2 < \cdots\}$ of $\mathbb{R}^+_{t_0}$. Also, assume that

$$v(\varphi(\tau_{k+1}, t_0, x_0), \tau_{k+1}) \leq v(\varphi(\tau_k, t_0, x_0), \tau_k) \tag{6.4.10}$$

for all τ_k whenever $|\varphi(\tau_k, t_0, x_0)| \geq \Omega$.

Furthermore, assume that there exists a function $f \in C[\mathbb{R}^+, \mathbb{R}^+]$, independent of $\varphi \in S$, such that for all $k \in \mathbb{N}$ and $\varphi \in S$,

$$v(\varphi(t, t_0, x_0), t) \leq f(v(\varphi(\tau_k, t_0, x_0), \tau_k)) \tag{6.4.11}$$

for all $t \in (\tau_k, \tau_{k+1})$ whenever $|\varphi(t, t_0, x_0)| \geq \Omega$.

Assume that there exists a constant $\Gamma > 0$ such that $|\varphi(\tau_{k+1}, t_0, x_0)| \leq \Gamma$ whenever $|\varphi(\tau_k, t_0, x_0)| \leq \Omega$ for all $\varphi \in S$.

Then S is *uniformly bounded*.

Proof. This result is a direct consequence of Theorem 3.2.4. □

Theorem 6.4.5 If in addition to the assumptions in Theorem 6.4.4 there exists a function $\psi_3 \in \mathcal{K}$ defined on $\mathbb{R}^+$ such that for all $\varphi \in S$

$$Dv(\varphi(\tau_k, t_0, x_0), \tau_k) \leq -\psi_3(|\varphi(\tau_k, t_0, x_0)|) \tag{6.4.12}$$

for all τ_k whenever $|\varphi(\tau_k, t_0, x_0)| \geq \Omega$, where Dv in (6.4.12) is defined in (6.4.4), then S is *uniformly ultimately bounded*.

Proof. This result is a direct consequence of Theorem 3.2.5. □

Theorem 6.4.6 Let $\{\mathbb{R}^+, \mathbb{R}^n, A, S\}$ be a finite-dimensional DDS and assume that $x_e = 0$ is an equilibrium. Assume that there exist a function $v\colon \mathbb{R}^n \times \mathbb{R}^+ \to \mathbb{R}^+$ and functions $\psi_1, \psi_2 \in \mathcal{K}_\infty$ defined on $\mathbb{R}^+$ such that

$$\psi_1(|x|) \le v(x,t) \le \psi_2(|x|) \tag{6.4.13}$$

for all $x \in \mathbb{R}^n$ and $t \in \mathbb{R}^+$.

Assume that for any $\varphi(\cdot, t_0, x_0) \in S$ $(t_0 = \tau_0)$, $v(\varphi(t, t_0, x_0), t)$ is continuous everywhere on $\mathbb{R}^+_{t_0}$ except on an unbounded subset $E = \{\tau_1, \tau_2, \dots : \tau_1 < \tau_2 < \cdots\}$ of $\mathbb{R}^+_{t_0}$. Furthermore, assume that there exists a function $f \in C[\mathbb{R}^+, \mathbb{R}^+]$ with $f(0) = 0$ such that for any $\varphi \in S$,

$$v(\varphi(t, t_0, x_0), t) \le f(v(\varphi(\tau_k, t_0, x_0), \tau_k)) \tag{6.4.14}$$

for all $t \in (\tau_k, \tau_{k+1})$, $k \in \mathbb{N}$.

Assume that there exists a function $\psi_3 \in \mathcal{K}$ defined on $\mathbb{R}^+$ such that for any $\varphi \in S$,

$$Dv(\varphi(\tau_k, t_0, x_0), \tau_k) \le -\psi_3(|\varphi(\tau_k, t_0, x_0)|), \tag{6.4.15}$$

$k \in \mathbb{N}$, where Dv in (6.4.15) is defined in (6.4.4).

Then the equilibrium $x_e = 0$ of the DDS S is *uniformly asymptotically stable in the large*.

Proof. This result is a direct consequence of Theorem 3.2.6. □

Theorem 6.4.7 Let $\{\mathbb{R}^+, \mathbb{R}^n, A, S\}$ be a finite-dimensional DDS and assume that $x_e = 0$ is an equilibrium. Assume that there exist a function $v\colon \mathbb{R}^n \times \mathbb{R}^+ \to \mathbb{R}^+$ and four positive constants c_1, c_2, c_3, and b such that

$$c_1|x|^b \le v(x,t) \le c_2|x|^b \tag{6.4.16}$$

for all $x \in \mathbb{R}^n$ and $t \in \mathbb{R}^+$.

Assume that for every $\varphi(\cdot, t_0, x_0) \in S$ $(t_0 = \tau_0)$, $v(\varphi(t, t_0, x_0), t)$ is continuous everywhere on $\mathbb{R}^+_{t_0}$ except on an unbounded subset $E = \{\tau_1, \tau_2, \dots : \tau_1 < \tau_2 < \cdots\}$ of $\mathbb{R}^+_{t_0}$. Furthermore, assume that there exists a function $f \in C[\mathbb{R}^+, \mathbb{R}^+]$ with $f(0) = 0$ such that

$$v(\varphi(t, t_0, x_0), t) \le f(v(\varphi(\tau_k, t_0, x_0), \tau_k)) \tag{6.4.17}$$

for all $t \in (\tau_k, \tau_{k+1})$, $k \in \mathbb{N}$, and such that for some positive constant q, f satisfies

$$f(r) = \mathcal{O}(r^q) \qquad \text{as } r \to 0^+. \tag{6.4.18}$$

Assume that for all $k \in \mathbb{N}$

$$Dv(\varphi(\tau_k, t_0, x_0), \tau_k) \le -c_3\big|\varphi(\tau_k, t_0, x_0)\big|^b \tag{6.4.19}$$

for all $x_0 \in A$ and $\varphi \in S$, where Dv in (6.4.19) is defined in (6.4.4).

Then the equilibrium $x_e = 0$ of the DDS S is *exponentially stable in the large*.

Proof. This result is a direct consequence of Theorem 3.2.7. □

C. Instability results

Next, we address instability results.

Theorem 6.4.8 Let $\{\mathbb{R}^+, \mathbb{R}^n, A, S\}$ be a finite-dimensional DDS and assume that $x_e = 0$ is an equilibrium. Assume that there exist a function $v\colon \mathbb{R}^n \times \mathbb{R}^+ \to \mathbb{R}^+$ and a $t_0 = \tau_0 \in \mathbb{R}^+$ that satisfy the following conditions.

(i) There exists a function $\psi_2 \in \mathcal{K}$ defined on $\mathbb{R}^+$ such that

$$v(x,t) \le \psi_2(|x|) \tag{6.4.20}$$

for all $x \in \mathbb{R}^n$ and $t \in \mathbb{R}^+$.

(ii) In every neighborhood of $x_e = 0$ there are points x such that $v(x, t_0) > 0$.

(iii) For any $x_0 \in A$ such that $v(x_0, t_0) > 0$ and any $\varphi(\cdot, t_0, x_0) \in S$ ($t_0 = \tau_0$), $v(\varphi(t, t_0, x_0), t)$ is continuous everywhere on $\mathbb{R}^+_{t_0}$ except on an unbounded subset $E = \{\tau_1, \tau_2, \ldots : \tau_1 < \tau_2 < \cdots\}$ of $\mathbb{R}^+_{t_0}$. Assume that there exists a function $\psi \in \mathcal{K}$ defined on $\mathbb{R}^+$ such that

$$Dv(\varphi(\tau_k, t_0, x_0), \tau_k) \ge \psi\big(|v(\varphi(\tau_k, t_0, x_0), \tau_k)|\big), \tag{6.4.21}$$

$k \in \mathbb{N}$, where Dv is defined in (6.4.4).

Then the equilibrium $x_e = 0$ of the DDS S is *unstable*.

Proof. This result is a direct consequence of Theorem 3.2.8. □

Theorem 6.4.9 If in addition to the assumptions given in Theorem 6.4.8, $v(x, t_0) > 0$ for all $x \neq 0$, then $x_e = 0$ of the DDS S is *completely unstable*.

Proof. This result is a direct consequence of Theorem 3.2.9. □

D. Examples

We now consider several specific examples to demonstrate the applicability of the results of the present section. In all cases, we assume that the sets of discontinuities, $\{\tau_1, \tau_2, \ldots : \tau_1 < \tau_2 < \cdots\}$, are unbounded.

Example 6.4.1 We consider dynamical systems determined by equations of the form

$$\begin{cases} \dot{x}(t) = Ax(t), & \tau_k \le t < \tau_{k+1} \\ x(t) = Bx(t^-), & t = \tau_{k+1}, \quad k \in \mathbb{N} \end{cases} \tag{6.4.22}$$

where $x(t) \in \mathbb{R}^n$ for all $t \in \mathbb{R}^+$, $A, B \in \mathbb{R}^{n\times n}$, and $x(t^-) = \lim_{t'\to t, t'<t} x(t')$ denotes the left limit of $x(t')$ at $t' = t$. We assume that for (6.4.22) the following conditions hold.

(i) $\sup_{k\in\mathbb{N}}\{\tau_{k+1}-\tau_k\} \le \lambda < \infty$ where $\lambda > 0$ is a constant.

(ii) $\|B\|e^{\|A\|\lambda} \le \alpha < 1$, where $\alpha > 0$ is constant and $\|\cdot\|$ denotes the matrix norm induced by the vector norm $|\cdot|$.

We choose $v(x) = |x|$. Then clearly (6.4.13) (resp., (6.4.16)) is satisfied. Along the solutions $\varphi(t, t_0, x_0) \triangleq x(t)$ of (6.4.22) we have, for all $k \in \mathbb{N}$,

$$\begin{aligned} Dv(x(\tau_k)) &= \frac{|x(\tau_{k+1})| - |x(\tau_k)|}{\tau_{k+1}-\tau_k} \\ &\le \frac{\|Be^{A(\tau_{k+1}-\tau_k)}\|\,|x(\tau_k)| - |x(\tau_k)|}{\tau_{k+1}-\tau_k} \\ &\le \frac{\left(\|B\|e^{\|A\|\lambda}-1\right)|x(\tau_k)|}{\tau_{k+1}-\tau_k} \\ &\le \frac{\alpha-1}{\lambda}|x(\tau_k)|. \end{aligned}$$

Therefore, inequality (6.4.15) (resp., (6.4.19)) is satisfied. Also,

$$v(x(t)) \le f(v(x(\tau_k)))$$

is true for all $t \in (\tau_k, \tau_{k+1}), k \in \mathbb{N}$, where $f(r) = e^{\|A\|\lambda}r$, and (6.4.14) (resp., (6.4.17)) is satisfied. Also, $f(r) = o(r^q)$ as $r \to 0^+$ for $0 < q < 1$. It follows from Theorem 6.4.6 (resp., Theorem 6.4.7) that the equilibrium $x_e = 0$ of (6.4.22) is *uniformly asymptotically stable in the large*, in fact, *exponentially stable in the large.* □

We emphasize that in the above example, the matrix A may be unstable. In the case when A has eigenvalues in the right half of the complex plane, the function $v(x(t))$ may increase over the intervals (τ_k, τ_{k+1}), $k \in \mathbb{N}$. However, the jumps of $v(x(t)) = |x(t)|$ at $\tau_{k+1}, k \in \mathbb{N}$, offset these increases, with the consequence that $v(x(\tau_{k+1}))$, and hence the norms of the solutions of (6.4.22) tend to zero as $t \to \infty$.

Example 6.4.2 We consider dynamical systems determined by equations of the form

$$\begin{cases} \dot{x}(t) = Ax(t), & \tau_k \le t < \tau_{k+1} \\ x(t) = Bx(t^-) + u(t), & t = \tau_{k+1},\ \ k \in \mathbb{N} \end{cases} \tag{6.4.23}$$

where $x(t) \in \mathbb{R}^n$ for all $t \in \mathbb{R}^+$, $u(t) \in \mathbb{R}^n$, and $|u(t)| < K$ for $t = \tau_{k+1}$, $k \in \mathbb{N}$, where $K > 0$ is a constant, and $A, B \in \mathbb{R}^{n\times n}$. We assume that for (6.4.23) the following conditions hold.

(i) $\sup_{k\in\mathbb{N}}\{\tau_{k+1}-\tau_k\} \le \lambda < \infty$ where $\lambda > 0$ is a constant.

(ii) $\|B\|e^{\|A\|\lambda} \le \alpha < 1$, where $\alpha > 0$ is a constant.

We choose $v(x) = |x|$. Along the solutions $\varphi(t, t_0, x_0) \triangleq x(t)$ of (6.4.23) we have

$$\begin{aligned} Dv(x(\tau_k)) &= \frac{|x(\tau_{k+1})| - |x(\tau_k)|}{\tau_{k+1} - \tau_k} \\ &\leq \frac{\|Be^{A(\tau_{k+1}-\tau_k)}\| \, |x(\tau_k)| - |x(\tau_k)| + |u(\tau_{k+1})|}{\tau_{k+1} - \tau_k} \\ &\leq \frac{(\|B\|e^{\|A\|\lambda} - 1)|x(\tau_n)| + |u(\tau_{k+1})|}{\tau_{k+1} - \tau_k} \\ &\leq \frac{(\alpha - 1)|x(\tau_k)| + K}{\lambda}. \end{aligned}$$

In the last step we require that $|x(\tau_k)| > \Omega = K/(1-\alpha)$. Therefore $Dv(x(\tau_k))$ is negative when $|x(\tau_k)| > \Omega$ and condition (6.4.12) in Theorem 6.4.5 is satisfied. Also, $v(x(t)) \leq f(v(x(\tau_k)))$ is true for all $t \in (\tau_k, \tau_{k+1})$, $k \in \mathbb{N}$, where $f(r) = e^{\|A\|\lambda}r$, and relation (6.4.11) is satisfied. Finally, it is easily verified that when $|x(\tau_k)| \leq \Omega$, $|x(\tau_{k+1})| \leq K + \alpha K$. All conditions of Theorems 6.4.4 and 6.4.5 are satisfied. Therefore, the solutions of system (6.4.23) are *uniformly bounded*, in fact, *uniformly ultimately bounded.* □

The results given in Examples 6.4.1 and 6.4.2 can be improved by making use of the *matrix measure*, $\mu(A)$, of a real matrix $A \in \mathbb{R}^{n\times n}$, defined by

$$\mu(A) = \lim_{\varepsilon\to 0^+} \frac{|I + \varepsilon A| - 1}{\varepsilon}, \tag{6.4.24}$$

where $I \in \mathbb{R}^{n\times n}$ denotes the identity matrix. For $x \in \mathbb{R}^n$ and

$$|x|_p = \left(\sum_{i=1}^{n} |x_i|^p\right)^{1/p}, \qquad 1 \leq p < \infty$$

and

$$|x|_\infty = \max_{1\leq j\leq n} \{|x_j|\},$$

the matrix measure of $A = [a_{ij}]$ is given, for $p = 1, 2, \infty$, by

$$\mu_1(A) = \max_{1\leq j\leq n} \Big\{a_{jj} + \sum_{i\neq j} |a_{ij}|\Big\}, \tag{6.4.25}$$

$$\mu_2(A) = \frac{1}{2}\lambda_M(A^T + A),$$

where $\lambda_M(A^T + A)$ denotes the largest eigenvalue of $A + A^T$, and

$$\mu_\infty(A) = \max_{1\leq i\leq n} \Big\{a_{ii} + \sum_{i\neq j} |a_{ij}|\Big\}. \tag{6.4.26}$$

Of particular interest to us is the relation

$$\left|e^{At}x\right| \leq e^{\mu(A)t}\left|x\right|, \quad t \geq 0. \tag{6.4.27}$$

A moment's reflection makes it now clear that the results of Examples 6.4.1 and 6.4.2 can be improved substantially by replacing condition (ii) in these examples by the condition

$$\text{(ii}'\text{)} \quad \|B\|e^{\mu(A)\lambda} \leq \alpha < 1. \tag{6.4.28}$$

Example 6.4.3 For system (6.4.22) assume that the following conditions hold.

(i) $\sup_{k\in\mathbb{N}}\{\tau_{k+1} - \tau_k\} \leq \lambda < \infty$, where $\lambda > 0$ is a constant; and

(ii) B is nonsingular and $\|B^{-1}\|e^{\|A\|\lambda} \leq \beta < 1$, where $\beta > 0$ is a constant.

We now apply Theorem 6.4.9 to show that under the above assumptions, the equilibrium $x_e = 0$ of (6.4.22) is *unstable*, in fact, *completely unstable*.

Choose $v(x) = |x|$. Along any solution $\varphi(t, t_0, x_0) \triangleq x(t)$ of system (6.4.22) we have

$$\begin{aligned}
Dv(x(\tau_k)) &= \frac{\left|x(\tau_{k+1})\right| - \left|x(\tau_k)\right|}{\tau_{k+1} - \tau_k} \\
&= \frac{\left|Be^{A(\tau_{k+1}-\tau_k)}x(\tau_k)\right| - \left|x(\tau_k)\right|}{\tau_{k+1} - \tau_k} \\
&\geq \frac{\left(\left\|e^{-A(\tau_{k+1}-\tau_k)}B^{-1}\right\|\right)^{-1}\left|x(\tau_k)\right| - \left|x(\tau_k)\right|}{\tau_{k+1} - \tau_k} \\
&\geq \frac{\left(\|B^{-1}\|e^{\|A\|(\tau_{k+1}-\tau_k)}\right)^{-1} - 1}{\tau_{k+1} - \tau_k}\left|x(\tau_k)\right| \\
&\geq \frac{\beta^{-1} - 1}{\lambda}\left|x(\tau_k)\right|.
\end{aligned}$$

Therefore, (6.4.21) is satisfied. In fact, all the hypotheses of Theorems 6.4.8 and 6.4.9 are satisfied. Therefore, the equilibrium $x_e = 0$ of system (6.4.22) is *unstable*, in fact, *completely unstable*. □

Example 6.4.4 We consider dynamical systems determined by equations of the form

$$\begin{cases} \dot{x}(t) = f(x(t)), & \tau_k \leq t < \tau_{k+1} \\ x(t) = g(x(t^-)), & t = \tau_{k+1}, \quad k \in \mathbb{N} \end{cases} \tag{6.4.29}$$

and associated with (6.4.29), the initial value problems given by

$$\begin{cases} \dot{x}(t) = f(x(t)), \\ x(\tau_k) = x_k, \end{cases} \tag{6.4.30}$$

$k \in \mathbb{N}$. We assume that $f\colon \mathbb{R}^n \to \mathbb{R}^n$, $f(0) = 0$, $g\colon \mathbb{R}^n \to \mathbb{R}^n$, $g(0) = 0$, and that $|g(\eta)| \leq \gamma|\eta|$ for all $\eta \in \mathbb{R}^n$ for some constant $\gamma > 0$. We assume that for every $(\tau_k, x_k) \in \mathbb{R}^+ \times \mathbb{R}^n$, (6.4.30) possesses a unique solution $x^{(k)}(t, \tau_k, x_k)$ that exists for all $t \geq \tau_k$. Then (6.4.29) possesses for every $(t_0, x_0) \triangleq (\tau_0, x_0) \in \mathbb{R}^+ \times \mathbb{R}^n$ a unique solution $\varphi(t, t_0, x_0)$ that exists for all $t \geq t_0$, where

$$\varphi(t, t_0, x_0) = x^{(k)}(t, \tau_k, x_k), \qquad \tau_k \leq t < \tau_{k+1}, \quad k \in \mathbb{N}. \tag{6.4.31}$$

In particular, (6.4.30) possesses the trivial solution $x^{(k)}(t, \tau_k, 0) = 0$ for all $t \geq \tau_k$, $k \in \mathbb{N}$, (6.4.29) possesses the trivial solution $\varphi(t, t_0, 0) = 0$, $t \geq t_0 = \tau_0 \geq 0$, and (6.4.30) and (6.4.29) have an equilibrium at the origin $x_e = 0$.

We now assume that for every initial value problem (6.4.30) there exist a function $v \in C^1[\mathbb{R}^n, \mathbb{R}^+]$ and four positive constants c_1, c_2, c_3, and $b > 0$ such that

$$\begin{cases} c_1|x|^2 \leq v(x) \leq c_2|x|^2 \\ v'_{(6.4.30)}(x) \leq -c_3|x|^2. \end{cases} \tag{6.4.32}$$

Letting $c = -c_3/c_2$, we have

$$v'_{(6.4.30)}(x) \leq cv(x) \tag{6.4.33}$$

which yields for (6.4.30) the estimate

$$v(x^{(k)}(t, \tau_k, x_k)) \leq e^{c(t-\tau_k)}v(x_k), \qquad t \geq \tau_k, \quad k \in \mathbb{N}. \tag{6.4.34}$$

For (6.4.29) we choose the v-function

$$v(\varphi(t, t_0, x_0)) = v(x^{(k)}(t, \tau_k, x_k)), \qquad \tau_k \leq t < \tau_{k+1}, \quad k \in \mathbb{N}. \tag{6.4.35}$$

Then for (6.4.29) we still have

$$c_1|x|^2 \leq v(x) \leq c_2|x|^2 \tag{6.4.36}$$

for all $x \in \mathbb{R}^n$. Thus, (6.4.13) in Theorem 6.4.6 and (6.4.16) in Theorem 6.4.7 are satisfied with $\psi_i(s) = c_i s^2$, $i = 1, 2$, and $s \geq 0$.

Next, using (6.4.34) and (6.4.35), we have for the DDS given in (6.4.29) that

$$v(x^{(k)}(t, \tau_k, x_k)) \leq e^{c(t-\tau_k)}v(x_k), \qquad \tau_k \leq t < \tau_{k+1}, \quad k \in \mathbb{N}. \tag{6.4.37}$$

At $t = \tau_{k+1}$ we have in view of (6.4.29) and (6.4.32) that

$$\begin{aligned} v(x^{(k+1)}(\tau_{k+1}, \tau_{k+1}, x_{k+1})) &= v(x_{k+1}) \\ &\leq c_2\big|x_{k+1}\big|^2 \\ &= c_2\big|x^{(k+1)}(\tau_{k+1}, \tau_{k+1}, x_{k+1})\big|^2 \\ &\leq c_2\gamma^2\big|x^{(k)}(\tau_{k+1}^-, \tau_k, x_k)\big|^2 \\ &\leq (c_2/c_1)\gamma^2 v(x^{(k)}(\tau_{k+1}^-, \tau_k, x_k)). \end{aligned} \tag{6.4.38}$$

Letting $\delta_k = \tau_{k+1} - \tau_k$, $k \in \mathbb{N}$, and using (6.4.37) and (6.4.38), we have that

$$\begin{aligned} v(x_{k+1}) &\leq (c_2/c_1)\gamma^2 e^{-(c_3/c_2)\delta_k} v(x^{(k)}(\tau_k, \tau_k, x_k)) \\ &= (c_2/c_1)\gamma^2 e^{-(c_3/c_2)\delta_k} v(x_k), \qquad k \in \mathbb{N}. \end{aligned} \tag{6.4.39}$$

If we require that

$$(c_2/c_1)\gamma^2 e^{-(c_3/c_2)\delta_k} \leq 1 \tag{6.4.40}$$

then $v(x_k)$ is nonincreasing with increasing k, and if

$$(c_2/c_1)\gamma^2 e^{-(c_3/c_2)\delta_k} \leq \alpha_k < 1 \tag{6.4.41}$$

then $v(x_k)$ is strictly decreasing with k.

Next, from (6.4.39) and the definition of Dv, we have

$$Dv(\varphi(\tau_k, t_0, x_0)) \leq \frac{\alpha_k - 1}{\tau_{k+1} - \tau_k} c_1 \left|\varphi(\tau_k, t_0, x_0)\right|^2, \qquad k \in \mathbb{N}.$$

We assume that $\inf_{k\in\mathbb{N}} [(1 - \alpha_k)/(\tau_{k+1} - \tau_k)]\, c_1 = c_3 > 0$. Then (6.4.15) in Theorem 6.4.6 and (6.4.19) in Theorem 6.4.7 are satisfied with $\psi_3(s) = c_3|s|^2$, $s \geq 0$.

Next, assume that $\inf_{k\in\mathbb{N}}(c_3/c_2)\delta_k = \beta > 0$. Then in view of (6.4.37) we have that

$$v(\varphi(t, t_0, x_0)) \leq e^{-\beta} v(x_k) = f(v(x_k)), \qquad \tau_k \leq t < \tau_{k+1}, \quad k \in \mathbb{N},$$

where $f(s) = e^{-\beta}s$, $s \geq 0$. Thus, (6.4.17) in Theorem 6.4.7 holds. Furthermore, it is clear that $f(s) = o(s^q)$ as $s \to 0^+$ for any $q \in (0, 1)$. Therefore, condition (6.4.18) of Theorem 6.4.7 holds as well.

All the conditions of Theorems 6.4.1, 6.4.6, and 6.4.7 are satisfied and we have the following result.

Proposition 6.4.1 Let c_1, c_2, c_3, γ, and δ_k be the parameters for system (6.4.29), as defined above.

(a) If for all $k \in \mathbb{N}$, $(c_2/c_1)\gamma^2 e^{-(c_3/c_2)\delta_k} \leq 1$, then the equilibrium $x_e = 0$ of system (6.4.29) is *uniformly stable*.

(b) If for all $k \in \mathbb{N}$, $(c_2/c_1)\gamma^2 e^{-(c_3/c_2)\delta_k} \leq \alpha < 1$ $(\alpha > 0)$, then the equilibrium $x_e = 0$ of system (6.4.29) is *uniformly asymptotically stable in the large* and *exponentially stable in the large*. □

Example 6.4.5 We now consider a class of DDS determined by equations of the form

$$\begin{cases} \dot{x}(t) = f_k(t, x(t)), & \tau_k \leq t < \tau_{k+1} \\ x(t) = g_k(x(t^-)), & t = \tau_{k+1}, \quad k \in \mathbb{N} \end{cases} \tag{6.4.42}$$

and the associated family of initial value problems given by

$$\begin{cases} \dot{x}(t) = f_k(t, x(t)) \\ x(\tau_k) = x_k, \end{cases} \tag{6.4.43}$$

$k \in \mathbb{N}$, where $x(t) \in \mathbb{R}^n$, $f_k \in C[\mathbb{R}^+ \times \mathbb{R}^n, \mathbb{R}^n]$, $f_k(t, 0) = 0$ for all $t \geq 0$, $g_k \colon \mathbb{R}^n \to \mathbb{R}^n$, and $g(0) = 0$. We assume that there exists a constant $\gamma_k > 0$ such that $|g_k(\eta)| \leq \gamma_k|\eta|$ for all $\eta \in \mathbb{R}^n$.

We assume that for every $(\tau_k, x_k) \in \mathbb{R}^+ \times \mathbb{R}^n$ there exists a unique solution $x^{(k)}(t, \tau_k, x_k)$ for (6.4.43) that exists for all $t \geq \tau_k$. We note that for (6.4.43) we have $x^{(k)}(t, \tau_k, 0) = 0$ for all $t \geq \tau_k$ and that $x_e = 0$ is an equilibrium.

As a consequence of the above assumptions, we have that (6.4.42) possesses for every (t_0, x_0) a unique solution

$$\varphi(t, t_0, x_0) = x^{(k)}(t, \tau_k, x_k), \qquad \tau_k \leq t < \tau_{k+1}, \quad k \in \mathbb{N},$$

with $t_0 = \tau_0$ and that $x_e = 0$ is an equilibrium for (6.4.42).

Consistent with the above assumptions, we also assume that f_k in (6.4.43) satisfies the Lipschitz condition

$$\big|f_k(t, x) - f_k(t, y)\big| \leq K_k\big|x - y\big| \tag{6.4.44}$$

for all $x, y \in \mathbb{R}^n$ and $t \in [\tau_k, \tau_{k+1}]$, $k \in \mathbb{N}$, where $K_k > 0$ is a constant.

From (6.4.43) we have

$$\begin{aligned}
&\big|x^{(k)}(t, \tau_k, x_k) - y^{(k)}(t, \tau_k, y_k)\big| \\
&= \bigg|x_k - y_k + \int_{\tau_k}^{t} \Big(f_k(\eta, x^{(k)}(\eta, \tau_k, x_k)) - f_k(\eta, y^{(k)}(\eta, \tau_k, y_k))\Big) d\eta\bigg|.
\end{aligned}$$

Choosing $y_k = 0$ and recalling that $f_k(t, 0) = 0$, $t \geq \tau_k$, we have in view of (6.4.44) that

$$\big|x^{(k)}(t, \tau_k, x_k)\big| \leq \big|x_k\big| + \int_{\tau_k}^{t} K_k\big|x^{(k)}(\eta, \tau_k, x_k)\big| d\eta \tag{6.4.45}$$

for all $t \in [\tau_k, \tau_{k+1})$ and $k \in \mathbb{N}$. Applying the Gronwall inequality (see Problem 2.14.9) to (6.4.45), we obtain the estimate

$$\big|x^{(k)}(t, \tau_k, x_k)\big| \leq \big|x_k\big| e^{K_k(t - \tau_k)}, \qquad \tau_k \leq t < \tau_{k+1} \tag{6.4.46}$$

for all $x_k \in \mathbb{R}^n$ and $k \in \mathbb{N}$.

In what follows, we further assume that $\sup_{k \in \mathbb{N}} K_k \triangleq K < \infty$, and letting $\lambda_k = \tau_{k+1} - \tau_k$, that $\sup_{k \in \mathbb{N}} \lambda_k \triangleq \Lambda < \infty$, and that $\sup_{k \in \mathbb{N}} \gamma_k \triangleq \Gamma < \infty$.

Proposition 6.4.2 Let $K_k, \gamma_k, \lambda_k, K, \Gamma$, and Λ be the parameters for system (6.4.42), as defined above.

(a) If for every $k \in \mathbb{N}$, $\gamma_k e^{K_k \lambda_k} \leq 1$, then the equilibrium $x_e = 0$ of system (6.4.42) is *uniformly stable*.

(b) If for every $k \in \mathbb{N}$, $\gamma_k e^{K_k \lambda_k} \leq \alpha < 1$, where $\alpha > 0$ is a constant, then the equilibrium $x_e = 0$ of system (6.4.42) is *uniformly asymptotically stable in the large* and *exponentially stable in the large*.

Proof. We choose the Lyapunov function $v(x) = |x|$, $x \in \mathbb{R}^n$. When evaluated along the solutions of (6.4.42) we have

$$\begin{aligned}
v\big(\varphi(t, t_0, x_0)\big) &\triangleq v\big(x^{(k)}(t, t_0, x_0)\big) \\
&= \big|x^{(k)}(t, \tau_k, x_k)\big|, \qquad \tau_k \leq t < \tau_{k+1}, \quad k \in \mathbb{N}.
\end{aligned}$$

Clearly,

$$\psi_1(|x|) \le v(x) \le \psi_2(|x|) \tag{6.4.47}$$

for all $x \in \mathbb{R}^n$, where $\psi_1(s) = \psi_2(s) = s \ge 0$ and $\psi_1, \psi_2 \in \mathcal{K}_\infty$. Along the solutions of (6.4.42) we have, in view of (6.4.46),

$$\big|x^{(k)}(t, \tau_k, x_k)\big| \le e^{K_k(t-\tau_k)}\big|x_k\big| = e^{K_k(t-\tau_k)}\big|x^{(k)}(\tau_k, \tau_k, x_k)\big|, \tag{6.4.48}$$

for all $t \in [\tau_k, \tau_{k+1})$. At $t = \tau_{k+1}$ we have

$$\big|x^{(k+1)}(\tau_{k+1}, \tau_{k+1}, x_{k+1})\big| = \big|g(x^{(k)}(\tau_{k+1}^-, \tau_k, x_k))\big| \le \gamma_k\big|x^{(k)}(\tau_{k+1}^-, \tau_k, x_k)\big|. \tag{6.4.49}$$

Combining (6.4.48) and (6.4.49), we have

$$\big|x^{(k+1)}(\tau_{k+1}, \tau_{k+1}, x_{k+1})\big| \le \gamma_k e^{K_k\lambda_k}\big|x^{(k)}(\tau_k, \tau_k, x_k)\big| \tag{6.4.50}$$

and because by assumption, $\gamma_k e^{K_k\lambda_k} \le 1$, we have that

$$\begin{aligned} v\big(x^{(k+1)}(\tau_{k+1}, \tau_{k+1}, x_{k+1})\big) &= \big|x^{(k+1)}(\tau_{k+1}, \tau_{k+1}, x_{k+1})\big| \\ &\le \big|x^{(k)}(\tau_k, \tau_k, x_k)\big| = v\big(x^{(k)}(\tau_k, \tau_k, x_k)\big). \end{aligned}$$

The above relation holds for all $k \in \mathbb{N}$; thus it follows that $v\big(\varphi(\tau_k, t_0, x_0)\big)$, $k \in \mathbb{N}$, is nonincreasing.

Next, from (6.4.48) we have, recalling that $\sup_{k\in\mathbb{N}} K_k = K$ and $\sup_{k\in\mathbb{N}} \lambda_k = \Lambda$, that

$$\begin{aligned} v\big(x^{(k)}(t, \tau_k, x_k)\big) &= \big|x^{(k)}(t, \tau_k, x_k)\big| \\ &\le e^{K\Lambda}\big|x^{(k)}(\tau_k, \tau_k, x_k)\big| \\ &\triangleq f\big(v\big(x^{(k)}(\tau_k, \tau_k, x_k)\big)\big), \end{aligned} \tag{6.4.51}$$

$t \in [\tau_k, \tau_{k+1})$, $k \in \mathbb{N}$, where $f(s) = e^{K\Lambda}s$. Therefore, all the hypotheses of Theorem 6.4.1 are satisfied and we conclude that the equilibrium $x_e = 0$ of system (6.4.42) is *uniformly stable*.

If in (6.4.50) we assume that $\gamma_k e^{K_k\lambda_k} \le \alpha < 1$, $\alpha > 0$, we have that

$$v\big(x^{(k+1)}(\tau_{k+1}, \tau_{k+1}, x_{k+1})\big) < \alpha v\big(x^{(k)}(\tau_k, \tau_k, x_k)\big)$$

and

$$\begin{aligned} &\Big[v\big(x^{(k+1)}(\tau_{k+1}, \tau_{k+1}, x_{k+1})\big) - v\big(x^{(k)}(\tau_k, \tau_k, x_k)\big)\Big]\Big/(\tau_{k+1} - \tau_k) \\ &\quad\le \big[(\alpha - 1)/\Lambda\big]v\big(x^{(k)}(\tau_k, \tau_k, x_k)\big) \\ &\quad= -\psi_3\big(\big|x^{(k)}(\tau_k, \tau_k, x_k)\big|\big) \\ &\quad= -\psi_3\big(\big|\varphi(\tau_k, t_0, x_0)\big|\big) \end{aligned} \tag{6.4.52}$$

for all $x_0 \in \mathbb{R}^n$, $k \in \mathbb{N}$. In (6.4.52) we have $\psi_3(s) = [(1-\alpha)/\Lambda]s, s \geq 0$ (i.e., $\psi_3 \in \mathcal{K}_\infty$). Therefore, all the hypotheses of Theorem 6.4.2 are satisfied and we conclude that the equilibrium $x_e = 0$ of system (6.4.42) is *uniformly asymptotically stable*.

Because (6.4.47) holds for all $x \in \mathbb{R}^n$ and because actually $\psi_1, \psi_2 \in \mathcal{K}_\infty$ and because (6.4.52) is true for all $\varphi(\tau_k, t_0, x_0) \in \mathbb{R}^n$, it follows from Theorem 6.4.6 that the equilibrium $x_e = 0$ of system (6.4.42) is *uniformly asymptotically stable in the large*.

From (6.4.47) it is clear that in relation (6.4.16) of Theorem 6.4.7 we have that $c_1 = c_2 = b = 1$ and from (6.4.52) it is clear that in relation (6.4.19) in Theorem 6.4.7, $c_3 = (1-\alpha)/\Lambda$. We have already shown that relation (6.4.17) of Theorem 6.4.7 is true, and clearly, for $f(s) = e^{K\Lambda}s$, we have that $f(s) = \mathcal{O}(s^q)$ as $s \to 0^+$ for any $q \in (0,1)$. Therefore, all the hypotheses of Theorem 6.4.7 are satisfied and we conclude that the equilibrium $x_e = 0$ of system (6.4.42) is *exponentially stable in the large*. □

Example 6.4.6 We now consider the DDS determined by equations of the form

$$\begin{cases} \dot{x}(t) = A_k x(t), & \tau_k \leq t < \tau_{k+1} \\ x(t) = B_k x(t^-), & t = \tau_{k+1}, \quad k \in \mathbb{N} \end{cases} \tag{6.4.53}$$

and the associated family of initial value problems given by

$$\begin{cases} \dot{x}(t) = A_k x(t), \\ x(\tau_k) = x_k, \end{cases} \tag{6.4.54}$$

$k \in \mathbb{N}$, where $t \in \mathbb{R}^+$, $x(t) \in \mathbb{R}^n$, and $A_k, B_k \in \mathbb{R}^{n\times n}$. We denote the solutions of (6.4.54) by $x^{(k)}(t, \tau_k, x_k)$, $t \geq \tau_k$, $k \in \mathbb{N}$, and the solutions of (6.4.53) by

$$\varphi(t, t_0, x_0) = x^{(k)}(t, \tau_k, x_k), \qquad \tau_k \leq t < \tau_{k+1},$$

$k \in \mathbb{N}$, where $\tau_0 = t_0$. Note that $x_e = 0$ is an equilibrium for (6.4.53) and (6.4.54).

If all the eigenvalues λ of A_k satisfy the relation $\mathrm{Re}\lambda \leq -\alpha_0$, then for any positive $\alpha_k < \alpha_0$, there is a constant $M_k(\alpha_k) > 0$ such that the solutions of (6.4.54) satisfy the estimate

$$\left|x^{(k)}(t, \tau_k, x_k)\right| \leq M_k(\alpha_k)e^{-\alpha_k(t-\tau_k)}\left|x_k\right| \tag{6.4.55}$$

for all $t \geq \tau_k \geq 0$ and $x_k \in \mathbb{R}^n$ (refer to Theorem 2.9.5). When the aforementioned assumption is not true, then the solutions of (6.4.54) still allow the estimate

$$\left|x^{(k)}(t, \tau_k, x_k)\right| \leq e^{\|A_k\|(t-\tau_k)}\left|x_k\right| \tag{6.4.56}$$

for all $t \geq \tau_k \geq 0$ and $x_k \in \mathbb{R}^n$. Thus, in either case we have

$$\left|x^{(k)}(t, \tau_k, x_k)\right| \leq Q_k e^{w_k(t-\tau_k)}\left|x_k\right| \tag{6.4.57}$$

for all $t \geq \tau_k$ and $x_k \in \mathbb{R}^n$, where $Q_k = 1$ and $w_k = \|A_k\|$ when (6.4.56) applies and $Q_k = M_k(\alpha_k)$ and $w_k = -\alpha_k$, $\alpha_k > 0$, when (6.4.55) applies.

We assume that $\sup_{k\in\mathbb{N}} M_k(\alpha_k) = M < \infty$ and $\sup_{k\in\mathbb{N}} \lambda_k = \Lambda < \infty$.

Proposition 6.4.3 Let $w_k, M_k(\alpha_k), Q_k, \lambda_k, \Lambda$, and M be the parameters for system (6.4.53), as defined above.

(a) If for all $k \in \mathbb{N}$, $\|B_k\| Q_k e^{w_k \lambda_k} \leq 1$, then the equilibrium $x_e = 0$ of system (6.4.53) is *uniformly stable*.

(b) If for all $k \in \mathbb{N}$, $\|B_k\| Q_k e^{w_k \lambda_k} \leq \alpha < 1$, where $\alpha > 0$ is a constant, then the equilibrium $x_e = 0$ of system (6.4.53) is *uniformly asymptotically stable in the large* and *exponentially stable in the large*. □

The proof of Proposition 6.4.3 is similar to the proof of Proposition 6.4.2 and is left as an exercise for the reader.

6.5 Converse Theorems for Ordinary Differential Equations

In the present section we establish sample converse results for the principal Lyapunov stability and boundedness results for ordinary differential equations presented in Section 6.2. We recall the system of ordinary differential equations given by

$$\dot{x} = f(t, x) \tag{E}$$

where $t \in \mathbb{R}^+$, $x \in \Omega$, $0 \in \Omega$, Ω is an open connected subset of $\mathbb{R}^n$, and where it is now assumed that $f \in C[\mathbb{R}^+ \times \Omega, \mathbb{R}^n]$. In the appendix section (Section 6.8) it is shown that the continuity of $f(t, x)$ ensures the continuity of the solutions $\varphi(t, t_0, x_0)$ of (E) with respect to the initial conditions (t_0, x_0). We assume that $x_e = 0$ is an equilibrium for (E).

A. Local results

In our first result we address uniform stability.

Theorem 6.5.1 Assume that the equilibrium $x_e = 0$ for (E) is *uniformly stable*.

(a) Then there exist functions $\psi_1, \psi_2 \in \mathcal{K}$ and a function $v\colon B(r) \times \mathbb{R}^+ \to \mathbb{R}^+$ for some $r > 0$, where $B(r) \subset \Omega$, such that

$$\psi_1(|x|) \leq v(x, t) \leq \psi_2(|x|)$$

for all $(x, t) \in B(r) \times \mathbb{R}^+$, and $v'_{(E)}$ is nonpositive.

(b) If f is Lipschitz continuous, then there exists a positive definite and decrescent function $v \in C[B(r) \times \mathbb{R}^+, \mathbb{R}]$ for some $r > 0$, where $B(r) \subset \Omega$, such that $v'_{(E)}$ is negative semidefinite.

Proof. (a) This result is a direct consequence of Theorem 3.6.1.

(b) Let $r_0 > 0$ such that $B(r_0) \subset \Omega$. Because $x_e = 0$ is uniformly stable, there exists a $\delta(\varepsilon) > 0$ for any $0 < \varepsilon \le r_0$ such that $|x(t, t_0, x_0)| < \varepsilon$ for all $t \ge t_0$ whenever $|x_0| < \delta$. Let $\delta(0) = 0$. Without loss of generality, we may assume that $\delta \in \mathcal{K}$ and $\delta(\varepsilon) < \varepsilon$ for all $\varepsilon \le r_0$.

Let $r_1 < r_0$ and $r \triangleq \delta(\delta(r_1))$. Define the Lyapunov function $v(x, t)$ as

$$v(x,t) = \min\left\{|x(\tau, t, x)| \colon \tau \in [t^*, t]\right\}$$

for all $(x, t) \in B(r) \times \mathbb{R}^+$, where $t^* \in [0, t]$ is the smallest value to the left of t that $x(\tau, t, x)$ can be continued to such that $|x(\tau, t, x)| < r_0$ for all $\tau \in (t^*, t]$.

Fix $(x_0, t_0) \in B(r) \times \mathbb{R}^+$. If $|x(t_0^*, t_0, x_0)| < r_0$, then $t_0^* = 0$ and $|x(\tau, t_0, x_0)| < r_0$ for all $\tau \in [0, t_0]$. By the continuity of solutions with respect to initial conditions, there exists a neighborhood U of (x_0, t_0) such that for all (x, t) in this neighborhood, $t^* = 0$. If $|x(t_0^*, t_0, x_0)| = r_0$, let $\widehat{t_0} < t_0$ be the value of t for which $|x(t, t_0, x_0)| = r_1$ for the first time to the left of t_0. Because $r < \delta(r_1) < r_1$, there exists a neighborhood U of (x_0, t_0) such that all the solutions of (E) starting within this neighborhood stay within the $(r_1 - \delta(r_1))$-neighborhood of $x(\widehat{t_0}, t_0, x_0)$. Therefore, they are all continuable to $\widehat{t_0}$. Furthermore, if $(x_1, t_1) \in U$, then $|x(\widehat{t_0}, t_1, x_1)| > \delta(r_1)$. By the choice of $\delta(r_1)$, it must be true that $|x(\tau, t_1, x_1)| \ge \delta(\delta(r_1)) = r$ for all $\tau < \widehat{t_0}$. On the other hand $|x(t_1, t_1, x_1)| = |x_1| < r$. Hence, $v(x, t)$ takes place at some τ such that $\widehat{t_0} \le \tau \le t$ for all $(x, t) \in U$.

Because $(x_0, t_0) \in B(r)$, $x(t, t_0, x_0)$ is continuable to the right. Without loss of generality, we assume that $x(t, t_0, x_0)$ can be continued to $[t_0, t_0 + c], c > 0$ and $U \subset \{x \colon |x - x_0| < \varepsilon\} \times [t_0^*, t_0 + c]$ for some $\varepsilon > 0$. By the Lipschitz continuity of f, there exist an $L > 0$ such that $|f(t, x) - f(t, y)| \le L|x - y|$ for $(x, t), (y, t) \in B(r_1) \times [0, t_0 + c]$ and a $K = \max\{|f(t, x)| \colon (t, x) \in [0, t_0 + c] \times B(r_0)]$. For any $(x, t), (y, t) \in U$, subtract the integral equations satisfied by $x(s, t, x)$ and $x(s, t, y)$ and estimate as follows,

$$\begin{aligned} |x(s,t,x) - x(s,t,y)| &\le |x - y| + \left| \int_t^s f(\tau, x(\tau, t, x)) d\tau - \int_t^s f(\tau, x(\tau, t, y) d\tau \right| \\ &\le |x - y| + L \int_t^s \big|x(\tau, t, x) - x(\tau, t, y)\big| d\tau \end{aligned}$$

for all $s \ge t$ for which the solutions exist. Apply the Gronwall inequality to obtain $|x(s, t, x) - x(s, t, y)| < e^{L|t-s|}|x - y|$.

For any $(x, t), (y, t) \in U$, suppose $v(x, t) = |x(t_x, t, x)|$ and $v(y, t) = |x(t_y, t, y)|$. Then

$$v(x,t) - v(y,t) \le |x(t_y, t, x)| - |x(t_y, t, y)| \le e^{L(t_0+c)}|x - y|,$$

and

$$v(x,t) - v(y,t) \ge |x(t_x, t, x)| - |x(t_x, t, y)| \ge -e^{L(t_0+c)}|x - y|.$$

Thus, we have

$$|v(x,t) - v(y,t)| \le e^{L(t_0+c)}|x - y|. \tag{6.5.1}$$

We now are ready to establish the continuity of $v(x, t)$. Let $(x_1, t_1) \in U$ be arbitrarily close to (x_0, t_0). We have

$$\begin{aligned}|v(x_0, t_0) - v(x_1, t_1)| \leq & |v(x_0, t_0) - v(x(t_1, t_0, x_0), t_1)| \\ & + |v(x(t_1, t_0, x_0), t_1) - v(x_0, t_1)| \\ & + |v(x_0, t_1) - v(x_1, t_1)|. \end{aligned} \tag{6.5.2}$$

It follows from

$$|x(t_1, t_0, x_0) - x_0| \leq K|t_1 - t_0|$$

that when $|t_1 - t_0|$ is sufficiently small, we have $x(t_1, t_0, x_0) \in U$. Hence, the second and third terms in (6.5.2) are small in view of (6.5.1). To obtain an estimate for the first term, we first let $t_0 > t_1$ and set $v(x_0, t_0) = |x(t_{x_0}, t_0, x_0)|$. If $t_{x_0} \leq t_1$ then it is true that

$$\begin{aligned} v(x_0, t_0) &= \min\left\{|x(\tau, t_0, x_0)| : \tau \leq t_0\right\} \\ &= \min\left\{|x(\tau, t_0, x_0)| : \tau \leq t_1\right\} \\ &= v(x(t_1, t_0, x_0), t_1). \end{aligned}$$

If t_{x_0} is between t_1 and t_0, we have that

$$|x(t_{x_0}, t_0, x_0)| = v(x_0, t_0) \leq v(x(t_1, t_0, x_0), t_1) \leq |x(t_1, t_0, x_0)|.$$

Thus,

$$|v(x_0, t_0) - v(x(t_1, t_0, x_0), t_1)| \leq |x(t_{x_0}, t_0, x_0) - x(t_1, t_0, x_0)| \leq K|t_1 - t_0|. \tag{6.5.3}$$

When $t_0 < t_1$, it can be shown similarly as above that (6.5.3) holds. Thus, we have shown that $v(x, t)$ is continuous.

Clearly, $\delta(|x|) \leq v(x, t) \leq |x|$ and $v'_{(E)}$ is negative semidefinite due to the fact that $v(x(t, t_0, x_0), t)$ is nonincreasing.

The proof is completed. □

In the next result we address the uniform asymptotic stability of the equilibrium $x_e = 0$ for (E).

Theorem 6.5.2 Assume that for every initial condition resulting in a solution of (E), the solution is unique. Assume that the equilibrium $x_e = 0$ is *uniformly asymptotically stable*. Then there exists a positive definite and decrescent function $v \in C[B(r) \times \mathbb{R}^+, \mathbb{R}]$ for some $r > 0$, where $B(r) \subset \Omega$, such that $v'_{(E)}$ is negative definite.

Proof. This result is a consequence of Theorem 3.6.4 and the continuity of the solutions of (E) with respect to initial conditions. □

The next result, which addresses the exponential stability of the equilibrium $x_e = 0$ for (E), is not symmetric to the exponential stability theorem given in Theorem 6.2.4. Nevertheless, this result does provide a set of necessary conditions for exponential stability.

Theorem 6.5.3 Assume that for every initial condition resulting in a solution of (E), the solution is unique. Assume that the equilibrium $x_e = 0$ is *exponentially stable*. Then there exists a positive definite and decrescent function $v \in C[B(r) \times \mathbb{R}^+, \mathbb{R}]$ for some $r > 0$, where $B(r) \subset \Omega$, such that $v'_{(E)}(x,t) \le -cv(x,t)$ for all $(x,t) \in B(r) \times \mathbb{R}^+$, where $c > 0$ is a constant.

Proof. This result is a consequence of Theorem 3.6.5 and the continuity of the solutions of (E) with respect to initial conditions. □

We emphasize that converse theorems for ordinary differential equations for *uniform boundedness*, *uniform ultimate boundedness*, *uniform asymptotic stability in the large*, *exponential stability in the large*, and *instability* can also be established.

B. Some refinements

By imposing appropriate restrictions on the function f in (E), it is possible to refine the converse theorems. For example, when $f(t,x) \equiv f(x)$ then it turns out (similarly as in the case of Theorems 4.1.3 and 4.1.4) that the Lyapunov functions for the preceding results are time invariant (i.e., $v(x,t) \equiv v(x)$). (We address this in Chapter 7.) Similarly, for the case of periodic systems (where in (E), $f(t,x) = f(t+T,x)$ for all $t \in \mathbb{R}$, $x \in \mathbb{R}^n$ for some $T > 0$), the Lyapunov functions in the preceding converse theorems are also periodic with the same period T (i.e., $v(x,t) = v(x,t+T)$ for the same $T > 0$, $x \in \mathbb{R}^n$). (We address this also in Chapter 7.)

In the present subsection we first identify conditions on f in (E) that yield continuously differentiable v-functions in the converse theorems. We present only a sample result. In the proof of this result we require the following two preliminary results. (In these results, we use the notation $f_x(t,x) = (\partial f/\partial x)(t,x)$.)

Lemma 6.5.1 Let $f, \partial f/\partial x_i \in C[\mathbb{R}^+ \times \overline{B(r)}, \mathbb{R}^n]$, $i = 1, \dots, n$. Then there exists a function $\psi \in C^1[\mathbb{R}^+, \mathbb{R}^+]$ such that $\psi(0) = 0$, $d\psi(t)/dt > 0$, and such that $s = \psi(t)$ transforms the equation

$$\frac{dx}{dt} = f(t,x) \tag{E}$$

into

$$\frac{dx}{ds} = f^*(s,x) \tag{E^*}$$

where $|\nabla f^*(s,x)| \le 1$ for all $(s,x) \in \mathbb{R}^+ \times B(r)$ with

$$\nabla f^*(s,x) \triangleq \left[\frac{\partial f^*}{\partial x_1}(s,x), \dots, \frac{\partial f^*}{\partial x_n}(s,x)\right].$$

Moreover, if $v(x,s)$ is a C^1-smooth function such that $v'_{(E^*)}(x,s)$ is negative definite, then for $\widetilde{v}(x,t) = v(x,\psi(t))$, $\widetilde{v}'_{(E)}(x,t)$ is negative definite.

Proof. Pick a positive and continuous function F such that $|(\partial f/\partial x)(t,x)| \leq F(t)$ for all $(t,x) \in \mathbb{R}^+ \times B(r)$. We can assume that $F(t) \geq 1$ for all $t \geq 0$. Define

$$\psi(t) = \int_0^t F(v)dv$$

and define Ψ as the inverse function $\Psi = \psi^{-1}$. Define $s = \psi(t)$ so that (E) becomes (E^*) with

$$f^*(s,x) = \frac{f(\Psi(s),x)}{F(\Psi(s))}.$$

Clearly, for all $(t,x) \in \mathbb{R}^+ \times B(r)$ we have

$$\left|\frac{\partial f^*}{\partial x}(s,x)\right| = \left|\frac{\partial f}{\partial x}(\Psi(s),x)\right| \Big/ F(\Psi(s)) \leq \frac{F(\Psi(s))}{F(\Psi(s))} = 1.$$

If $v(x,s)$ has a negative definite derivative with respect to system (E^*), then define $\widetilde{v}(x,t) = v(x,\psi(t))$. There is a function $\psi_1 \in \mathcal{K}$ such that $v'_{(E^*)}(x,t) \leq -\psi_1(|x|)$. Thus

$$\begin{aligned}
\widetilde{v}'_{(E)}(x,t) &= v_s(x,\psi(t))\dot{\psi}(t) + \nabla v(x,\psi(t))f(t,x) \\
&= v_s(x,\psi(t))F(t) + \nabla v(x,\psi(t))\frac{f(t,x)}{F(t)}F(t) \\
&= F(t)v'_{(E^*)}(x,\psi(t)) \\
&\leq v'_{(E^*)}(x,\psi(t)) \\
&\leq -\psi_1(|x|).
\end{aligned}$$

Thus $\widetilde{v}'_{(E)}(t,x)$ is also negative definite. □

Lemma 6.5.2 Let $g(t)$ be a positive, continuous function defined for all $t \geq 0$ and satisfying $g(t) \to 0$ as $t \to \infty$. Let $h(t)$ be a positive, continuous, monotone nondecreasing function defined for all $t \geq 0$. Then there exists a function $G(u)$ defined for $u \geq 0$, positive for $u > 0$, continuous, increasing, having an increasing, continuous derivative $\dot{G}$, and such that $G(0) = \dot{G}(0) = 0$, and such that for any $a > 0$ and any continuous function $g^*(t)$ which satisfies $0 < g^*(t) \leq ag(t)$ the integrals

$$\int_0^\infty G(g^*(t))ds \quad \text{and} \quad \int_0^\infty \dot{G}(g^*(t))h(t)dt \tag{6.5.4}$$

converge uniformly in g^*.

Proof. We first construct a function $u(t)$ defined for $t > 0$ that is continuous and decreasing and satisfies $u(t) \to 0$ as $t \to \infty$, and $u(t) \to \infty$ as $t \to 0^+$ such that for any $a > 0$ there exists a $T(a)$ with the property that if $t \geq T(a)$ then $ag(t) \leq u(t)$.

Pick a sequence $\{t_m\}$ such that $t_1 \geq 1$, $t_{m+1} \geq t_m + 1$, and such that if $t \geq t_m$ then $g(t) \leq (m+1)^{-2}$. Define $u(t_m) = m^{-1}$, $u(t)$ linear between the t_ms and such that $u(t) = (t_1/t)^p$ on $0 < t < t_1$, where p is chosen so large that $\dot{u}(t_1^-) < \dot{u}(t_1^+)$. For $t_m \leq t \leq t_{m+1}$ we have

$$ag(t) \leq a(m+1)^{-2} \quad \text{and} \quad u(t) \geq (m+1)^{-1}$$

so that

$$ag(t) \leq u(t)a(m+1)^{-1} \leq u(t)$$

as soon as m is larger than $[a]$, the integer part of a. Thus we can take $T(a) = [a]$.

Define $F(u)$ to be the inverse function of $u(t)$ and define

$$G(u) = \int_0^u \frac{e^{-F(s)}}{h(F(s))} ds. \tag{6.5.5}$$

Because F is continuous and h is positive, the integrand in (6.5.5) is continuous on $0 < u < \infty$ and $F(u) \to \infty$ as $u \to 0^+$. Hence the integral exists and defines a function $G \in C^1[\mathbb{R}^+, \mathbb{R}^+]$.

Fix $a > 0$ and choose a continuous function g^* such that $0 < g^*(t) < ag(t)$. For $t \geq T(a)$ we have $0 < g^*(t) \leq u(t)$ or $F(g^*(u)) \geq t$. Thus

$$\dot{G}(g^*)) = \frac{e^{-F(g^*(t))}}{h(F(g^*(t)))} \leq \frac{e^{-t}}{h(t)}, \quad t \geq T(a).$$

Hence the uniform convergence of the second integral in (6.5.4) is clear.

The tail of the first integral in (6.5.4) can be estimated by

$$\int_{T(a)}^{\infty} \left(\int_0^{u(t)} \frac{e^{-F(s)}}{h(0)} ds \right) dt.$$

Because $u(t)$ is piecewise C^1 on $0 < t < \infty$, we can change variables from u to s in the inner integral to compute

$$\begin{aligned}
\int_{T(a)}^{\infty} \left(\int_{\infty}^{s} \frac{\dot{u}(s)e^{-s}}{h(s)} ds \right) dt &\leq \int_{T(a)}^{\infty} \left(\int_{\infty}^{t} \frac{\dot{u}(s)e^{-s}}{h(0)} ds \right) dt \\
&\leq h(0)^{-1} \int_{T(a)}^{\infty} \left(\int_{t}^{\infty} e^{-s} ds \right) dt \\
&< \infty
\end{aligned}$$

because $0 > \dot{u}(t) > -1$. Hence the uniform convergence of the first integral in (6.5.4) is also clear. □

In our next result we make use of the fact that if f is continuously differentiable, then it is Lipschitz continuous and if $|\partial f/\partial x| \leq L$ for some $L > 0$, then L is a Lipschitz constant for f.

Theorem 6.5.4 Assume that f, $\partial f/\partial x_i \in C[\mathbb{R}^+ \times B(r), \mathbb{R}^n]$, $i = 1, \dots, n$ for some $r > 0$. Assume that $x_e = 0$ is an equilibrium of (E) which is *uniformly asymptotically stable*. Then there exists a function $v \in C^1[B(r_1) \times \mathbb{R}^+, \mathbb{R}^+]$ for some $r_1 > 0$ such that v is positive definite and decrescent and such that $v'_{(E)}$ is negative definite.

Proof. By Lemma 6.5.1 we can assume without loss of generality that $|\partial f/\partial x| \le 1$ on $\mathbb{R}^+ \times B(r)$. For all $x_0, y_0 \in B(r)$, $t_0 \ge 0$, subtract the integral equations satisfied by $\varphi(t, t_0, x_0)$ and $\varphi(t, t_0, y_0)$ and estimate as follows,

$$\begin{aligned}\big|\varphi(t,t_0,x_0) - \varphi(t,t_0,y_0)\big| &\le \big|x_0 - y_0\big| \\ &\quad + \int_{t_0}^{t} \big|f(s,\varphi(t,t_0,x_0)) - f(s,\varphi(t,t_0,y_0))\big| ds \\ &\le \big|x_0 - y_0\big| + \int_{t_0}^{t} L\big|\varphi(t,t_0,x_0) - \varphi(t,t_0,y_0)\big| ds\end{aligned}$$

for all $t \ge t_0$ for which the solutions exist. Apply the Gronwall inequality to obtain

$$\big|\varphi(t,t_0,x_0) - \varphi(t,t_0,y_0)\big| \le \big|x_0 - y_0\big| e^{t-t_0}.$$

Define $h(t) = e^t$.

Pick r_1 such that $0 < r_1 \le r$ and such that if $(t_0, x_0) \in \mathbb{R}^+ \times B(r_1)$, then $\varphi(t, t_0, x_0) \in B(r)$ for all $t \ge t_0$ and such that

$$\lim_{t\to\infty} \varphi(t + t_0, t_0, x_0) = 0$$

uniformly for $(t_0, x_0) \in \mathbb{R}^+ \times B(r_1)$. This is possible because $x_e = 0$ is uniformly asymptotically stable. Let $g(s)$ be a positive continuous function such that $g(s) \to 0$ as $s \to \infty$, and such that $|\varphi(s + t, t, x)|^2 \le g(s)$ on $s \ge 0$, $t \ge 0$, $x \in B(r_1)$.

Let G be the function given by Lemma 6.5.2 and define

$$v(x,t) = \int_0^\infty G\big(|\varphi(s+t,t,x)|^2\big) ds.$$

Clearly v is defined on $B(r_1) \times \mathbb{R}^+$. The integral converges uniformly in $(x, t) \in B(r_1) \times \mathbb{R}^+$, therefore v is also continuous. If $D = \partial/\partial x_1$, $D\varphi(s + t, t, x)$ must satisfy the linear equation

$$\frac{dy}{ds} = f_x\big(s, \varphi(s+t,t,x)\big)y; \qquad y(t) = (1, 0, \dots, 0)^T$$

(refer to Miller and Michel [19, p. 69, Theorem 2.7.1]). Thus $|D\varphi(s+t, t, x)| \le ke^s$ for some constant $k \ge 1$. Thus

$$\frac{\partial v}{\partial x_1}(x,t) = \int_0^\infty \dot{G}\big(|\varphi(s+t,t,x)|^2\big) \left(2\varphi(s+t,t,x)\frac{\partial \varphi}{\partial x_1}(s+t,t,x)\right) ds$$

exists and is continuous and

$$\left|\frac{\partial v}{\partial x_1}(x,t)\right| \leq \int_0^\infty \dot{G}(g(s))k_1 e^s ds < \infty$$

for some constant $k_1 > 0$. A similar argument can be used on the other partial derivatives. Hence $v \in C^1[B(r_1) \times \mathbb{R}^+, \mathbb{R}^+]$.

Because v_x exists and is bounded by some number B whereas $v(0,t)$ is zero, then clearly

$$0 \leq v(x,t) = v(x,t) - v(0,t) \leq B|x|.$$

Thus, v is decrescent. To see that v is positive definite, first find $M_1 > 0$ such that $|f(t,x)| \leq M_1|x|$ for all $(t,x) \in \mathbb{R}^+ \times B(r_1)$. For $M = M_1 r_1$ we have

$$\big|\varphi(t+s,t,x) - x\big| \leq \int_t^{t+s} \big|f(u,\varphi(u,t,x))\big| du \leq Ms.$$

Thus, for $0 \leq s \leq |x|/(2M)$ we have $|\varphi(t+s,t,x)| \geq |x|/2$ and

$$v(x,t) \geq \int_0^{|x|/(2M)} G\big(|\varphi(t+s,t,x)|^2\big)ds \geq \big(|x|/(2M)\big)G(|x|^2/4).$$

This proves that v is positive definite.

To compute $v'_{(E)}$ we replace x by a solution $\varphi(t,t_0,x_0)$. Because by uniqueness $\varphi(t+s,t,\varphi(t,t_0,x_0)) = \varphi(t+s,t_0,x_0)$, then

$$v(\varphi(t,t_0,x_0),t) = \int_0^\infty G\big(|\varphi(t+s,t_0,x_0)|^2\big)ds = \int_t^\infty G\big(|\varphi(s,t_0,x_0)|^2\big)ds,$$

and

$$v'_{(E)}(\varphi(t,t_0,x_0),t) = -G\big(|\varphi(t,t_0,x_0)|^2\big).$$

Thus, $v'_{(E)}(x_0,t_0) = -G(|x_0|^2)$. □

Next, as we noted earlier in Subsection 6.5A, the converse theorem for exponential stability presented in Theorem 6.5.3 is not symmetric to the exponential stability theorem given in Theorem 6.2.4. However, by imposing additional restrictions, we are able to establish a converse result for exponential stability that is nearly symmetric to the stability result given in Theorem 6.2.4, as demonstrated in the last result of this subsection.

Theorem 6.5.5 Assume that for the system

$$\dot{x} = f(t,x) \tag{E}$$

$f \in C[\mathbb{R}^+ \times \Omega, \mathbb{R}^n]$, where Ω is a neighborhood of the origin in $\mathbb{R}^n$, and assume that f satisfies the Lipschitz condition

$$\big|f(t,x) - f(t,y)\big| \leq L\big|x - y\big|$$

for all $x, y \in B(r)$, $r > 0$, $B(r) \subset \Omega$, and for all $t \in \mathbb{R}^+$. Assume that the equilibrium $x_e = 0$ of (E) is *exponentially stable* in the sense that

$$|\varphi(t, t_0, x_0)| \leq B|x_0|e^{-\alpha(t-t_0)} \tag{6.5.6}$$

for all $t \geq t_0$, whenever $|x_0| < r$, where B and α are positive constants. Then there exist a function $v \in C[B(r) \times \mathbb{R}^+, \mathbb{R}]$, and three positive constants c_1, c_2, and c_3 such that

$$c_1|x|^2 \leq v(x,t) \leq c_2|x|^2$$
$$v'_{(E)}(x, t) \leq -c_3|x|^2.$$

Proof. Let the function $v(x, t)$ be given by

$$v(x_0, t_0) = \int_{t_0}^{t_0+T} |\varphi(t, t_0, x_0)|^2 dt, \tag{6.5.7}$$

for all $(x_0, t_0) \in B(r) \times \mathbb{R}^+$, where $T = \ln(B\sqrt{2}/\alpha)$ is a constant.

First we need to obtain a lower bound for $|\varphi(t, t_0, x_0)|$. To this end, we let $y(s) = x(-s)$, $s \in (-\infty, -t_0]$. Then

$$\dot{y}(s) = \dot{x}(-s) = -f(s, y(s)),$$

and for an arbitrary $t \geq t_0$,

$$y(s) = y(-t) + \int_{-t}^{s} -f(\tau, y(\tau))d\tau$$

for all $s \in [-t, -t_0]$. From the Lipschitz condition it is easily obtained that $|f(t, x)| = |f(t, x) - f(t, 0)| \leq L|x|$. Thus,

$$|y(s)| \leq |y(-t)| + \int_{-t}^{s} L|y(\tau)|d\tau.$$

By the Gronwall inequality, we have

$$|y(s)| \leq |y(-t)|e^{L(s+t)}.$$

In particular, at $s = -t_0$, we have

$$|x(t_0)| = |y(-t_0)| \leq |x(t)|e^{L(t-t_0)},$$

which in turn yields $|x(t)| \geq |x(t_0)|e^{-L(t-t_0)}$ for all $t \geq t_0$.

We now have the following estimates for the v-function,

$$v(x_0, t_0) \leq \int_{t_0}^{t_0+T} |x_0|^2 B^2 e^{-2\alpha(t-t_0)} dt = |x_0|^2 B^2 \int_0^T e^{-2\alpha t} dt = c_2|x_0|^2,$$

and

$$v(x_0, t_0) \geq \int_{t_0}^{t_0+T} |x_0|^2 e^{-2L(t-t_0)} dt = |x_0|^2 \int_0^T e^{-2Lt} dt = c_1 |x_0|^2.$$

Along the solution $\varphi(t, t_0, x_0)$ of (E), we have

$$\begin{aligned} v'_{(E)}(\varphi(t, t_0, x_0), t) &= -|\varphi(t, t_0, x_0)|^2 + |\varphi(t + T, t_0, x_0)|^2 \\ &\quad + \int_t^{t+T} \frac{d}{dt} |\varphi(\tau, t, \varphi(t, t_0, x_0))|^2 dt. \end{aligned}$$

Because

$$\varphi(\tau, t + \Delta t, \varphi(t + \Delta t, t_0, x_0)) = \varphi(\tau, t_0, x_0) = \varphi(\tau, t, \varphi(t, t_0, x_0)),$$

the last term in the above equation is zero. Additionally, it follows from (6.5.6) that

$$|\varphi(t + T, t_0, x_0)| = |\varphi(t + T, t, \varphi(t, t_0, x_0))| \leq B |\varphi(t, t_0, x_0)|.$$

Hence,

$$\begin{aligned} v'_{(E)}(\varphi(t, t_0, x_0), t) &\leq -|\varphi(t, t_0, x_0)|^2 + Be^{-2T\alpha} |\varphi(t + T, t_0, x_0)|^2 \\ &= -\frac{1}{2} |\varphi(t, t_0, x_0)|^2. \end{aligned}$$

This completes the proof of the theorem. □

6.6 Converse Theorems for Ordinary Difference Equations

In the present section we establish sample converse results for the principal Lyapunov stability and boundedness results for ordinary difference equations presented in Section 6.3. We recall the system of ordinary difference equations given by

$$x(k + 1) = f(k, x(k)) \tag{D}$$

where $k \in \mathbb{N}$, $x(k) \in \Omega \subset \mathbb{R}^n$, and $f\colon \mathbb{N} \times \Omega \to \Omega$. We assume that Ω is an open connected set and that $0 \in \Omega$. We assume that $x_e = 0$ is an equilibrium for (D).

In our first result we address uniform stability.

Theorem 6.6.1 Assume that the equilibrium $x_e = 0$ for (D) is *uniformly stable*. Then there exists a function $v\colon B(r) \times \mathbb{N} \to \mathbb{R}^+$ for some $r > 0$, $B(r) \subset \Omega$, which satisfies the following conditions.

(i) There exist functions $\psi_1, \psi_2 \in \mathcal{K}$ such that

$$\psi_1(|x|) \leq v(x, k) \leq \psi_2(|x|)$$

for all $(x, k) \in B(r) \times \mathbb{N}$.

(ii) $\Delta_{(D)}v(x,k) \le 0$ for all $(x,k) \in B(r) \times \mathbb{N}$.

Proof. This result is a direct consequence of Theorem 3.7.1. □

In the next result we address the uniform asymptotic stability of the equilibrium $x_e = 0$ for (D). We recall that the motions determined by (D) are unique.

Theorem 6.6.2 Assume that the equilibrium $x_e = 0$ of (D) is *uniformly asymptotically stable*. Then there exists a function $v\colon B(r) \times \mathbb{N} \to \mathbb{R}^+$ for some $r > 0$, $B(r) \subset \Omega$, that satisfies the following conditions.

(i) There exist functions $\psi_1, \psi_2 \in \mathcal{K}$ such that

$$\psi_1(|x|) \le v(x,k) \le \psi_2(|x|)$$

for all $(x,k) \in B(r) \times \mathbb{N}$.

(ii) There exists a function $\psi_3 \in \mathcal{K}$ such that

$$\Delta_{(D)}v(x,k) \le -\psi_3(|x|)$$

for all $(x,k) \in B(r) \times \mathbb{N}$.

Proof. This result is a direct consequence of Theorem 3.7.2. □

The next result, which address the exponential stability of the equilibrium $x_e = 0$ for (E), is not symmetric to the exponential stability theorem given in Theorem 6.3.1(d). Nevertheless, this result does provide a set of necessary conditions for exponential stability.

Theorem 6.6.3 Assume that the equilibrium $x_e = 0$ of (D) is *exponentially stable*. Then there exists a function $v\colon B(r) \times \mathbb{N} \to \mathbb{R}^+$ for some $r > 0$, $B(r) \subset \Omega$, that satisfies the following conditions.

(i) There exist functions $\psi_1, \psi_2 \in \mathcal{K}$ such that

$$\psi_1(|x|) \le v(x,k) \le \psi_2(|x|)$$

for all $(x,k) \in B(r) \times \mathbb{N}$.

(ii) There exists a positive constant c such that

$$\Delta_{(D)}v(x,k) \le -cv(x,k)$$

for all $(x,k) \in B(r) \times \mathbb{N}$.

Proof. This result is a direct consequence of Theorem 3.7.3. □

We emphasize that converse theorems for ordinary difference equations can also be established for *uniform boundedness*, *uniform ultimate boundedness*, *uniform asymptotic stability in the large*, *exponential stability in the large*, and *instability*.

6.7 Converse Theorems for Finite-Dimensional DDS

In this section we present sample converse theorems for the stability and boundedness results of Section 6.4 for finite-dimensional discontinuous dynamical systems. In the first subsection we present results involving Lyapunov functions that in general need not be continuous. In the second subsection we show that under reasonable additional assumptions, the Lyapunov functions for the converse theorems are continuous.

A. Local results

We first address the uniform stability of the equilibrium $x_e = 0$ for finite-dimensional DDS.

Theorem 6.7.1 Let $\{\mathbb{R}^+, \mathbb{R}^n, A, S\}$ be a finite-dimensional discontinuous dynamical system (for short, a finite-dimensional DDS S) for which Assumption 3.5.1 holds. Assume that the equilibrium $x_e = 0$ is *uniformly stable*. Then there exists a function $v\colon B(r) \times \mathbb{R}^+ \to \mathbb{R}^+$, $B(r) \subset \Omega$, for some $r > 0$, that satisfies the following conditions.

(i) There exist two functions $\psi_1, \psi_2 \in \mathcal{K}$ such that

$$\psi_1(|x|) \leq v(x,t) \leq \psi_2(|x|)$$

for all $(x,t) \in B(r) \times \mathbb{R}^+$.

(ii) For every $\varphi(\cdot, t_0, x_0) \in S$ with $x_0 \in B(r)$, $v(\varphi(t, t_0, x_0), t)$ is nonincreasing for all $t \geq t_0$.

Proof. This result is a direct consequence of Theorem 3.5.1. □

In the next result we address the uniform asymptotic stability of the equilibrium $x_e = 0$ of finite-dimensional DDS.

Theorem 6.7.2 Let $\{\mathbb{R}^+, \mathbb{R}^n, A, S\}$ be a finite-dimensional DDS for which Assumptions 3.5.1 and 3.5.2 hold. Assume that for every $(t_0, x_0) \in \mathbb{R}^+ \times A$ there exists a *unique* $\varphi(\cdot, t_0, x_0) \in S$. Assume that the equilibrium $x_e = 0$ is *uniformly asymptotically stable*. Then there exists a function $v\colon B(r) \times \mathbb{R}^+ \to \mathbb{R}^+$, $B(r) \subset \Omega$, for some $r > 0$, that satisfies the following conditions.

(i) There exist two functions $\psi_1, \psi_2 \in \mathcal{K}$ such that

$$\psi_1(|x|) \leq v(x,t) \leq \psi_2(|x|)$$

for all $(x,t) \in B(r) \times \mathbb{R}^+$.

(ii) There exists a function $\psi_3 \in \mathcal{K}$ such that for all $\varphi(\cdot, t_0, x_0) \in S$ ($t_0 = \tau_0$), we have

$$Dv(\varphi(\tau_k, t_0, x_0), \tau_k) \leq -\psi_3(|\varphi(\tau_k, t_0, x_0)|),$$

$k \in \mathbb{N}$, where $x_0 \in B(r)$ and Dv is defined in (6.4.4).

(iii) There exists a function $f \in C[\mathbb{R}^+, \mathbb{R}^+]$ such that $f(0) = 0$ and such that

$$v(\varphi(t, t_0, x_0), t) \leq f(v(\varphi(\tau_k, t_0, x_0), \tau_k))$$

for every $\varphi(\cdot, t_0, x_0) \in S$, $t \in [\tau_k, \tau_{k+1})$, $k \in \mathbb{N}$ with $x_0 \in B(r)$ and $t_0 \in \mathbb{R}^+$.

Proof. This result is a direct consequence of Theorem 3.5.2. □

Next, we consider the exponential stability of the equilibrium $x_e = 0$ of the finite-dimensional DDS.

Theorem 6.7.3 Let $\{\mathbb{R}^+, \mathbb{R}^n, A, S\}$ be a finite-dimensional DDS for which Assumptions 3.5.1 and 3.5.2 hold. Assume that for every $(t_0, x_0) \in \mathbb{R}^+ \times A$ there exists a *unique* $\varphi(\cdot, t_0, x_0) \in S$. Assume that the equilibrium $x_e = 0$ for system S is *exponentially stable*. Then there exists a function $v\colon B(r) \times \mathbb{R}^+ \to \mathbb{R}^+$, $B(r) \subset \Omega$, for some $r > 0$, that satisfies the following conditions.

(i) There exist two functions $\psi_1, \psi_2 \in \mathcal{K}$ such that

$$\psi_1(|x|) \leq v(x, t) \leq \psi_2(|x|)$$

for all $(x, t) \in B(r) \times \mathbb{R}^+$.

(ii) There exists a constant $c > 0$ such that for all $\varphi(\cdot, t_0, x_0) \in S$ $(t_0 = \tau_0)$,

$$Dv(\varphi(\tau_k, t_0, x_0), \tau_k) \leq -cv(\varphi(\tau_k, t_0, x_0), \tau_k),$$

for all $k \in \mathbb{N}$, $t_0 \in \mathbb{R}^+$, $x_0 \in B(r)$, and Dv is defined in (6.4.4).

(iii) There exists a function $f \in C[\mathbb{R}^+, \mathbb{R}^+]$ with $f(0) = 0$ and

$$f(r) = \mathcal{O}(r^q) \quad \text{as } r \to 0^+$$

for some constant $q > 0$ such that

$$v(\varphi(t, t_0, x_0), t) \leq f(v(\varphi(\tau_k, t_0, x_0), \tau_k))$$

for every $\varphi(\cdot, t_0, x_0) \in S$, $t \in [\tau_k, \tau_{k+1})$, $k \in \mathbb{N}$ with $x_0 \in B(r)$ and $t_0 \in \mathbb{R}^+$.

Proof. This result is a direct consequence of Theorem 3.5.3. □

We emphasize that converse theorems for finite-dimensional DDS for *uniform boundedness*, *uniform ultimate boundedness*, *uniform asymptotic stability in the large*, *exponential stability in the large*, and *instability* can also be established.

B. Some refinements

The converse theorems presented in the preceding subsection involve Lyapunov functions that need not necessarily be continuous. In the present subsection, we show that under some additional mild assumptions, the Lyapunov functions for converse theorems are continuous.

The following concept of continuous dependence of solution on initial conditions for finite-dimensional DDS is used as a sufficient condition for the continuity of the Lyapunov functions.

Definition 6.7.1 Suppose $\{x_{0m}\} \subset A \subset \mathbb{R}^n$, $\{\tau_{0m}\} \subset \mathbb{R}^+$, and $x_{0m} \to x_0 \in A$ and $\tau_{0m} \to \tau_0$ as $m \to \infty$. Assume that the motions of the dynamical system $\{\mathbb{R}^+, \mathbb{R}^n, A, S\}$ are given by

$$p(t, \tau_0, x_0) = p^{(k)}(t, \tau_k, x_k), \qquad t \in [\tau_k, \tau_{k+1}),$$

and

$$p_m(t, \tau_{0m}, x_{0m}) = p_m^{(k)}(t, \tau_{km}, x_{km}), \qquad t \in [\tau_{km}, \tau_{(k+1)m}),$$

$k \in \mathbb{N}$, where $p^{(k)}(t, \tau_k, x_k)$ and $p_m^{(k)}(t, \tau_{km}, x_{km})$ are continuous for all $t \in \mathbb{R}^+$ with $p^{(k)}(\tau_k, \tau_k, x_k) = p(\tau_k, \tau_0, x_0) = x_k$ and $p_m^{(k)}(\tau_{km}, \tau_{km}, x_{km}) = p_m(\tau_{km}, \tau_{0m}, \tau_{0m}) = x_{km}$.

The motions in S are said to be *continuous with respect to initial conditions* if

(1) $\tau_{km} \to \tau_k$ as $m \to \infty$, for all $k \in \mathbb{N}$; and
(2) for every compact set $K \subset \mathbb{R}^+$ and every $\varepsilon > 0$ there exists an $L = L(K, \varepsilon) > 0$ such that for all $t \in K$ and $k \in \mathbb{N}$ such that $K \cap [\tau_k, \tau_{k+1}) \neq \emptyset$,

$$\left| p_m^{(k)}(t, \tau_{km}, x_{km}) - p^{(k)}(t, \tau_k, x_k) \right| < \varepsilon$$

whenever $m > L$. □

Theorem 6.7.4 If in addition to the assumptions given in Theorem 6.7.2, the motions in S are continuous with respect to initial conditions (in the sense of Definition 6.7.1), then there exists a continuous Lyapunov function that satisfies the conditions of Theorem 6.7.2.

Proof. The proof of this result is a direct consequence of Theorem 3.5.5. □

Converse theorems for DDS with continuous Lyapunov functions for other Lyapunov stability and boundedness types, which are in the spirit of Theorem 6.7.4, can also be established.

6.8 Appendix: Some Background Material on Differential Equations

In this section we present results concerning the continuity of solutions with respect to initial conditions for ordinary differential equations. We require these results in establishing the continuity of v-functions in the converse theorems for continuous finite-dimensional dynamical systems and finite-dimensional DDS.

We consider systems of differential equations given by

$$\dot{x} = f(t, x) \tag{E}$$

where $(t, x) \in D$, D is a domain in the (t, x)-space ($t \in \mathbb{R}^+, x \in \mathbb{R}^n$), and $f \in C[D, \mathbb{R}^n]$. Associated with (E) we have the initial value problem

$$\dot{x} = f(t, x), \qquad x(\tau) = \xi, \tag{I_E}$$

which can equivalently be expressed as

$$x(t) = \xi + \int_{\tau}^{t} f(s, x(s))ds, \qquad (I)$$

with noncontinuable solution $\varphi(t)$ defined on interval J.

In our subsequent discussion we require "perturbed systems" characterized by a sequence of initial value problems

$$x(t) = \xi_m + \int_{\tau}^{t} f_m(s, x(s))ds, \qquad (I_m)$$

with noncontinuable solutions $\varphi_m(t)$ defined on intervals J_m. We assume that $f_m \in C[D, \mathbb{R}^n]$, that $\xi_m \to \xi$ as $m \to \infty$ and that $f_m \to f$ uniformly on compact subsets of D.

In the proof of the main result of the present section, we require the following preliminary result.

Lemma 6.8.1 Let D be bounded. Suppose a solution φ of (I) exists on an interval $J = [\tau, b)$, or on $[\tau, b]$, or on the "degenerate interval" $[\tau, \tau]$, and suppose that $(t, \varphi(t))$ does not approach ∂D as $t \to b^-$; that is,

$$\text{dist}((t, \varphi(t)), \partial D) \triangleq \inf \big\{ |t - s| + |\varphi(t) - x| \colon (s, x) \notin D \big\} \geq \eta > 0 \qquad (6.8.1)$$

for all $t \in J$. Suppose that $\{b_m\} \subset J$ is a sequence that tends to b and the solutions $\varphi_m(t)$ of (I_m) are defined on $[\tau, b_m] \subset J$ and satisfy

$$\Phi_m = \sup \big\{ |\varphi_m(t) - \varphi(t)| \colon \tau \leq t \leq b_m \big\} \to 0$$

as $m \to \infty$. Then there is a number $b' > b$, where b' depends only on η (in (6.8.1)) and there is a subsequence $\{\varphi_{m_j}\}$ such that φ_{m_j} and φ are defined on $[\tau, b']$ and $\varphi_{m_j} \to \varphi$ as $j \to \infty$ uniformly on $[\tau, b']$.

Proof. Define $G = \{(t, \varphi(t)) \colon t \in J\}$, the graph of φ over J. By hypothesis, the distance from G to ∂D is at least $\eta = 3A > 0$. Define

$$D(b) = \big\{ (t, x) \in D \colon \text{dist}((t, x), G) \leq b \big\}.$$

Then $D(2A)$ is a compact subset of D and f is bounded there, say $|f(t, x)| \leq M$ $(M > 1)$ on $D(2A)$. Because $f_m \to f$ uniformly on $D(2A)$, it may be assumed (by increasing the size of M) that $|f_m(t, x)| \leq M$ on $D(2A)$ for all $m \geq 1$. Choose m_0 such that for $m \geq m_0$, $\Phi_m < A$. This means that $(t, \varphi_m(t)) \in D(A)$ for all $m \geq m_0$ and $t \in [\tau, b_m]$. Choose $m_1 \geq m_0$ so that if $m \geq m_1$, then $b - b_m < A/(4M)$. Define $b' = b + A/(4M)$.

Fix $m \geq m_1$. Because $(t, \varphi_m(t)) \in D(A)$ on $[\tau, b_m]$, then $|\dot{\varphi}_m(t)| \leq M$ on $[\tau, b_m]$ and until such time as $(t, \varphi_m(t))$ leaves $D(2A)$. Hence

$$|\varphi_m(t) - \varphi_m(b_m)| \leq M|t - b_m| \leq MA/(2M) = A/2$$

for as long as both $(t, \varphi_m(t)) \in D(2A)$ and $|t - b_m| \leq A/(2M)$. Thus $(t, \varphi_m(t)) \in D(2A)$ on $\tau \leq t \leq b_m + A/(2M)$. Moreover, $b_m + A/(2M) > b'$ when m is large.

Thus, it has been shown that $\{\varphi_m \colon m \geq m_1\}$ is a uniformly bounded family of functions and each is Lipschitz continuous with Lipschitz constant M on $[\tau, b']$. By Ascoli's Lemma (see Problem 2.14.7), a subsequence $\{\varphi_{m_j}\}$ will converge uniformly to a limit φ. The arguments used at the end of the proof of Theorem 2.3.1 (refer to the hint in Problem 2.14.8) show that

$$\lim_{j \to \infty} \int_\tau^t f(s, \varphi_{m_j}(s))ds = \int_\tau^t f(s, \varphi(s))ds.$$

Thus, the limit of

$$\phi_{m_j}(t) = \xi_{m_j} + \int_\tau^t f(s, \varphi_{m_j}(s))ds + \int_\tau^t \left[f_{m_j}(s, \varphi_{m_j}(s)) - f(s, \varphi_{m_j}(s))\right]ds$$

as $j \to \infty$, is

$$\varphi(t) = \xi + \int_\tau^t f(s, \varphi(s))ds.$$

□

We are now in a position to prove the following result.

Theorem 6.8.1 Let $f, f_m \in C[D, \mathbb{R}^n]$, let $\xi_m \to \xi$, and let $f_m \to f$ uniformly on compact subsets of D. If $\{\varphi_m\}$ is a sequence of noncontinuable solutions of (I_m) defined on intervals J_m, then there is a subsequence $\{m_j\}$ and a noncontinuable solution φ of (I) defined on an interval J_0 containing τ such that

(i) $\lim_{j \to \infty} \inf J_{m_j} \supset J_0$; and

(ii) $\varphi_{m_j} \to \varphi$ uniformly on compact subsets of J_0 as $j \to \infty$.

If in addition, the solution of (I) is unique, then the entire sequence $\{\varphi_m\}$ tends to φ uniformly for t on compact subsets of J_0.

Proof. With $J = [\tau, \tau]$ (a single point) and $b_m = \tau$ for all $m \geq 1$ apply Lemma 6.8.1. (If D is not bounded, use a subdomain.) Thus, there is a subsequence of $\{\varphi_m\}$ that converges uniformly to a limit function φ on some interval $[\tau, b']$, $b' > \tau$. Let B_1 be the supremum of these numbers b'. If $B_1 = +\infty$, choose b_1 to be any fixed b'. If $B_1 < \infty$, let b_1 be a number $b_1 \geq \tau$ such that $B_1 - b' < 1$. Let $\{\varphi_{1m}\}$ be a subsequence of $\{\varphi_m\}$ that converges uniformly on $[\tau, b_1]$.

Suppose for induction that we are given $\{\varphi_{km}\}$, b_k, $B_k > b_k$ with $\varphi_{km} \to \varphi$ uniformly on $[\tau, b_k]$ as $m \to \infty$. Define B_{k+1} as the supremum of all numbers $b' > b_k$ such that a subsequence of $\{\varphi_{km}\}$ will converge uniformly on $[\tau, b']$. Clearly $b_k < B_{k+1} \leq B_k$. If $B_{k+1} = +\infty$, pick $b_{k+1} > b_k + 1$ and if $B_{k+1} < \infty$, pick b_{k+1} so that $b_k < b_{k+1} < B_{k+1}$ and $b_{k+1} > B_{k+1} - 1/(k+1)$. Let $\{\varphi_{k+1,m}\}$ be a subsequence of $\{\varphi_{km}\}$ that converges uniformly on $[\tau, b_{k+1}]$ to a limit φ. Clearly, by possibly deleting finitely many terms of the new subsequence, we can assume without loss of generality that $|\varphi_{k+1,m}(t) - \varphi(t)| < 1/(k+1)$ for $t \in [\tau, b_{k+1}]$ and $m \geq k+1$.

Because $\{b_k\}$ is monotonically increasing, it has a limit $b \leq +\infty$. Define $J_0 = [\tau, b)$. The diagonal sequence $\{\varphi_{mm}\}$ will eventually become a subsequence of each sequence $\{\varphi_{km}\}$. Hence $\varphi_{mm} \to \varphi$ as $m \to \infty$ with convergence uniform on compact subsets of J_0. By the argument used at the end of the proof of Lemma 6.8.1, the limit φ must be a solution of (I_E).

If $b = \infty$, then φ is clearly noncontinuable. If $b < \infty$, then this means that B_k tends to b from above. If φ could be continued to the right past b (i.e., if $(t, \varphi(t))$ stays in a compact subset of D as $t \to b^-$), then by Lemma 6.8.1 there would be a number $b' > b$, a continuation of φ, and a subsequence of $\{\varphi_{mm}\}$ that would converge uniformly on $[\tau, b']$ to φ. Because $b' > b$ and $B_k \to b^+$, then for sufficiently large k (i.e., when $b' > B_k$), this would contradict the definition of B_k. Hence, φ must be noncontinuable. A similar argument works for $t < \tau$, therefore parts (i) and (ii) are proved.

Now assume that the solution of (I_E) is unique. If the entire sequence $\{\varphi_m\}$ does not converge to φ uniformly on compact subsets of J_0, then there is a compact set $K \subset J_0$, an $\varepsilon > 0$, a sequence $\{t_k\} \subset K$, and a subsequence $\{\varphi_{mk}\}$ such that

$$|\varphi_{mk}(t_k) - \varphi(t_k)| \geq \varepsilon. \tag{6.8.2}$$

By the part of the present theorem that has already been proved, there is a subsequence, we still call it $\{\varphi_{mk}\}$ in order to avoid a proliferation of subscripts, that converges uniformly on compact subsets of an interval J' to a solution ψ of (I_E). By uniqueness $J' = J_0$ and $\psi = \varphi$. Thus $\varphi_{mk} \to \varphi$ as $k \to \infty$ uniformly on $K \subset J_0$ which contradicts (6.8.2). □

Using Theorem 6.8.1, we now can prove the following result.

Corollary 6.8.1 Consider the system

$$\dot{x} = f(t, x) \tag{E}$$

where $t \in \mathbb{R}^+, x \in \Omega, \Omega$ is an open connected subset of $\mathbb{R}^n$, and $f \in C[\mathbb{R}^+ \times \Omega, \mathbb{R}^n]$. Assume that for each $(t_0, x_0) \in \mathbb{R}^+ \times \Omega$, there exists a unique noncontinuable solution $\varphi(t, t_0, x_0)$ with initial condition $\varphi(t_0) = x_0$. Then φ is continuous for $(t, t_0, x_0) \in S$ where

$$S \triangleq \{(t, t_0, x_0) \in \mathbb{R}^+ \times \mathbb{R}^+ \times \Omega \colon \alpha(t_0, x_0) < t < \beta(t_0, x_0)\},$$

where $\varphi(\cdot, t_0, x_0)$ is defined on (α, β), $\alpha = \alpha(t_0, x_0)$ is upper semicontinuous in $(t_0, x_0) \in \mathbb{R}^+ \times \Omega$ and $\beta = \beta(t_0, x_0)$ is lower semicontinuous in $(t_0, x_0) \in \mathbb{R}^+ \times \Omega$.

Proof. Define $\psi(t, t_0, x_0) = \varphi(t + t_0, t_0, x_0)$ so that ψ solves

$$\dot{y} = f(t + t_0, y), \qquad y(0) = x_0.$$

Let (t_{1m}, t_{0m}, x_{0m}) be a sequence in S that tends to a limit $(t_1, t_0, x_0) \in S$. By Theorem 6.8.1 it follows that

$$\psi(t, t_{0m}, x_{0m}) \to \psi(t, t_0, x_0) \qquad \text{as } m \to \infty$$

uniformly for t in compact subsets of $\alpha(t_0, x_0) - t_0 < t < \beta(t_0, x_0) - t_0$ and in particular uniformly in m for $t = t_1$. Therefore, we see that

$$\begin{aligned}\left|\varphi(t_{1m}, t_{0m}, x_{0m}) - \varphi(t_1, t_0, x_0)\right| &\le \left|\varphi(t_{1m}, t_{0m}, x_{0m}) - \varphi(t_{1m}, t_0, x_0)\right| \\ &+ \left|\varphi(t_{1m}, t_0, x_0) - \varphi(t_1, t_0, x_0)\right| \to 0 \text{ as } m \to \infty.\end{aligned}$$

This proves that φ is continuous on S.

To prove the remainder of the conclusions, we note that by Theorem 6.8.1(i), if J_m is the interval $(\alpha(t_{0m}, x_{0m}), \beta(t_{0m}, x_{0m}))$, then

$$\lim_{m \to \infty} \inf J_m \supset J_0.$$

The remaining assertions follow immediately. □

6.9 Notes and References

The various concepts of stability of an equilibrium for systems determined by ordinary differential equations, without reference to uniformity, were originally formulated by A. M. Lyapunov in 1892 [12]. The distinction between stability and uniform stability (resp., asymptotic stability and uniform asymptotic stability) was introduced in the process of establishing converse theorems (e.g., Malkin [13] and Massera [14]).

There are many interesting and excellent texts and monographs dealing with the stability theory of dynamical systems determined by ordinary differential equations (e.g., Hahn [4], Hale [5], Krasovskii [8], Lakshmikantham and Leela [9], Yoshizawa [22], and Zubov [23]). Excellent references that emphasize engineering applications include Khalil [7] and Vidyasagar [20]. Our presentation in Sections 6.2 and 6.5 concerning the stability of an equilibrium and the boundedness of solutions was greatly influenced by the presentations in Hahn [4], Miller and Michel [19], and Michel *et al.* [18]. For more complete treatments of converse theorems for ordinary differential equations, refer to Hahn [4, Chapter 6] and Yoshizawa [22, Chapter 5].

Our treatment in Sections 6.3 and 6.6 of the stability of an equilibrium and the boundedness of solutions of discrete-time dynamical systems determined by ordinary difference equations is more complete than what is usually found in texts. We note here that in the converse theorems presented in Section 6.6 we do not have any restrictions on the function f in (D), whereas the results in the literature usually require f to be continuous (see, e.g., [6]), globally Lipschitz continuous (see, e.g., [1] and [10]), or bijective (see, e.g., [3]). A good source on the stability of discrete-time systems determined by difference equations is the monograph by LaSalle [11]. Refer also to Antsaklis and Michel [2] and Michel *et al.* [18].

The material given in Sections 6.4 and 6.7 is perhaps the first systematic presentation of the stability and boundedness results of finite-dimensional discontinuous dynamical systems in book form. The first results of the type presented in Sections 6.4 and 6.7 were first addressed in Ye *et al.* [21]. For subsequent results on this subject, refer to Michel [15], Michel and Hu [17], and Michel *et al.* [18].

The results presented in Section 6.8 concerning the continuity of solutions of ordinary differential equations with respect to initial conditions are based on similar results given in Miller and Michel [19]. For additional results concerning this topic, refer to [19].

6.10 Problems

Problem 6.10.1 Show that if the equilibrium $x_e = 0$ of (E) satisfies (6.1.6) for a single initial time $t_0 \geq 0$ when (6.1.7) is true, then it also satisfies this condition at every other initial time $t_0' > t_0$. □

Problem 6.10.2 Prove that if $f(t, x_e) = 0$ for all $t \in \mathbb{R}^+$, then x_e is an equilibrium for (E).

Prove that if (E) possesses a unique solution for every $(t_0, x_0) \in \mathbb{R}^+ \times \Omega$, where Ω is an open connected set and $0 \in \Omega$, then $x_e = 0$ is an equilibrium for (E) if and only if $f(t, 0) = 0$ for all $t \in \mathbb{R}^+$. □

Problem 6.10.3 Prove relation (6.1.3). Prove relation (6.1.4). □

Problem 6.10.4 Prove that $x_e \in \Omega$ is an equilibrium of (D) if and only if for all $k \in \mathbb{N}$, $x_e = f(k, x_e)$.

Similarly as in the case of ordinary differential equations, prove that if (D) has an equilibrium at x_e, we may assume without loss of generality that the equilibrium is at the origin. □

Problem 6.10.5 Prove Theorem 6.1.2. □

Problem 6.10.6 Prove Theorem 6.1.3. □

Problem 6.10.7 Determine all the equilibrium points of the following differential equations (or systems of differential equations).

(a) $\dot{y} = \sin y$.

(b) $\dot{y} = y^2(y^2 - 3y + 2)$.

(c) $\ddot{x} + (x^2 - 1)\dot{x} + x = 0$.

(d) $\begin{cases} \dot{x}_1 = x_2 + x_1 x_2 \\ \dot{x}_2 = -x_1 + 2x_2. \end{cases}$

(e) $\ddot{x} + \dot{x} + \sin x = 0$.

(f) $\ddot{x} + \dot{x} + x(x^2 - 4) = 0$.

(g) $\dot{x} = a(1 + t^2)^{-1}x, a > 0$ is a constant or $a < 0$ is a constant. □

Problem 6.10.8 Determine the stability properties of the systems given in Problem 6.10.7. □

Problem 6.10.9 Consider the scalar equation

$$\dot{x} = -x^{2n+1} \tag{6.10.1}$$

where $k \in \mathbb{N}$. Prove that for arbitrary n, the equilibrium $x_e = 0$ of (6.10.1) is uniformly asymptotically stable in the large. Prove that when $n = 0$, the equilibrium $x_e = 0$ is exponentially stable in the large. Prove that when $n \geq 1$, the equilibrium $x_e = 0$ of system (6.10.1) is not exponentially stable. □

Problem 6.10.10 Consider the system

$$\begin{cases} \dot{x}_1 = x_2 - x_1(x_1^2 + x_2^2) \\ \dot{x}_2 = -x_1 - x_2(x_1^2 + x_2^2). \end{cases} \tag{6.10.2}$$

Prove that the equilibrium $x_e = 0$ of (6.10.2) is not exponentially stable. □

Problem 6.10.11 Let $f \in C^1[\mathbb{R}^+ \times \mathbb{R}^n, \mathbb{R}^n]$ with $f(t, 0) = 0$ for all $t \geq 0$, and assume that the eigenvalues $\lambda_i(t, x)$, $i = 1, \dots, n$, of the symmetric matrix

$$J(t, x) = \frac{1}{2}[f_x(t, x) + f_x(t, x)^T]$$

satisfy $\lambda_i(t, x) \leq -c$, $i = 1, \dots, n$ for all $(t, x) \in \mathbb{R}^+ \times \mathbb{R}^n$.

(i) If $c = 0$, show that the trivial solution of (E) is *stable* and that the solutions of (E) are *uniformly bounded*.

(ii) If $c > 0$, show that the equilibrium $x_e = 0$ of (E) is *exponentially stable in the large*. □

Problem 6.10.12 Investigate the *boundedness*, *uniform boundedness*, and *uniform ultimate boundedness* of the solutions for each of the following systems.

(a) $\ddot{x} + \dot{x} + x(x^2 - 4) = 0$.

(b) $\ddot{x} + \dot{x} + x^3 = \sin t$.

(c) $\begin{cases} \dot{x}_1 = x_2 + (x_1 x_2)/(1 + x_1^2 + x_2^2) \\ \dot{x}_2^2 = -2x_1 + 2x_2 + \arctan x_1. \end{cases}$

(d) $\begin{cases} \dot{x}_1 = x_2^3 + x_1(x_3^2 + 1) \\ \dot{x}_2 = -x_1^3 + x_2(x_3^2 + 2) \\ \dot{x}_3 = -(x_3)^{2/3}. \end{cases}$ □

Problem 6.10.13 Analyze the stability of the equilibrium $(x, \dot{x}) = 0$ of the system

$$x^{(n)} + g(x) = 0$$

where $n > 2$ is odd and $xg(x) > 0$ when $x \neq 0$. □

Hint: For $n = 2m+1$, use the Lyapunov function

$$v = \sum_{k=1}^{m} (-1)^k x_k x_{2m+2-k} + (-1)^{m+1} x_{m+1}^2/2.$$

Problem 6.10.14 Prove Corollary 6.2.2

Problem 6.10.15 Determine all the equilibrium points of the following discrete-time systems given by

(a) $$\begin{cases} x_1(k+1) = x_2(k) + |x_1(k)| \\ x_2(k+1) = -x_1(k) + |x_2(k)|. \end{cases}$$

(b) $$\begin{cases} x_1(k+1) = x_1(k)x_2(k) - 1 \\ x_2(k+1) = 2x_1(k)x_2(k) + 1. \end{cases}$$

(c) $$\begin{cases} x_1(k+1) = \text{sat}(x_1(k) + 2x_2(k)) \\ x_2(k+1) = \text{sat}(-x_1(k) + 2x_2(k)). \end{cases}$$ □

Problem 6.10.16 Consider the system given by

$$\begin{cases} x_1(k+1) = \dfrac{a x_2(k)}{1 + x_1(k)^2} \\ x_2(k+1) = \dfrac{b x_1(k)}{1 + x_2(k)^2}, \end{cases}$$

where a and b are constants with $a^2 < 1$ and $b^2 < 1$. Show that the equilibrium $x_e = (x_1, x_2)^T = 0$ is uniformly asymptotically stable. □

Problem 6.10.17 Prove that the equilibrium $x_e = 0$ of (6.3.8) is not exponentially stable. □

Problem 6.10.18 Analyze the stability of the equilibrium $x_e = 0$ of the system

$$x(k+1) = \begin{bmatrix} \cos\theta & \sin\theta \\ -\sin\theta & \cos\theta \end{bmatrix} x(k)$$

where θ is fixed. □

Problem 6.10.19 Investigate the boundedness, uniform boundedness, and uniform ultimate boundedness of the solutions for the following system

$$\begin{cases} x_1(k+1) = -0.5x_1(k) + 0.5x_2(k) + \cos(k\omega_0) \\ x_2(k+1) = -0.5x_1(k) - 0.5x_2(k) + \sin(k\omega_0), \end{cases}$$

where ω_0 is fixed. □

Problem 6.10.20 Prove Proposition 6.4.3. □

Problem 6.10.21 Consider the discontinuous dynamical system given by

$$\begin{cases} \dot{x}(t) = A_k(t)x(t), & \tau_k \le t < \tau_{k+1} \\ x(t) = B_k(t^-)x(t^-), & t = \tau_{k+1} \end{cases} \tag{6.10.3}$$

where $t \in \mathbb{R}^+$, $x(t) \in \mathbb{R}^n$, $A_k \in C[\mathbb{R}^+, \mathbb{R}^{n\times m}]$, and $B_k \in C[\mathbb{R}^+, \mathbb{R}^{n\times n}]$. Assume that $\|A_k(t)\| \le M_k$ for all $t \ge 0$, where $M_k > 0$ is a constant, $k \in \mathbb{N}$, and $\|B_k(t)\| < L_k$ for all $t \ge 0$, where $L_k > 0$ is a constant.

Prove that $x_e = 0$ is an equilibrium of (6.10.3). Establish conditions for the uniform stability, uniform asymptotic stability in the large, and exponential stability in the large of the equilibrium $x_e = 0$ of (6.10.3). □

Problem 6.10.22 Without making reference to the results given in Chapter 3, prove Theorems 6.2.1–6.2.10 by invoking fundamental concepts. □

Problem 6.10.23 Without making reference to the results given in Chapter 3, prove Theorems 6.3.1–6.3.5 by invoking fundamental concepts. □

Bibliography

[1] R. P. Agarwal, *Difference Equations and Inequalities: Theory, Methods, and Applications*, New York: Marcel Dekker, 1992.

[2] P. J. Antsaklis and A. N. Michel, *Linear Systems*, Boston: Birkhäuser, 2005.

[3] S. P. Gordon, "On converse to the stability theorems for difference equations," *SIAM J. Control Optim.*, vol. 10, pp. 76–81, 1972.

[4] W. Hahn, *Stability of Motion*, Berlin: Springer-Verlag, 1967.

[5] J. K. Hale, *Ordinary Differential Equations*, New York: Wiley-Interscience, 1969.

[6] Z. P. Jiang and Y. Wang, "A converse Lyapunov theorem for discrete-time systems with disturbances," *Syst. Control Lett.*, vol. 45, pp. 49–58, 2002.

[7] H. K. Khalil, *Nonlinear Systems*, New York: Macmillan, 1992.

[8] N. N. Krasovskii, *Stability of Motion*, Stanford, CA: Stanford University Press, 1963.

[9] V. Lakshmikantham and S. Leela, *Differential and Integral Inequalities*, vol. I and II, New York: Academic Press, 1969.

[10] V. Lakshmikantham and D. Trigiante, *Theory of Difference Equations: Numerical Methods and Applications*, New York: Marcel Dekker, 1988.

[11] J. P. LaSalle, *The Stability and Control of Discrete Processes*, New York: Springer-Verlag, 1986.

[12] A. M. Liapounoff, "Problème générale de la stabilité de mouvement," *Annales de la Faculté des Sciences de l'Université de Toulouse*, vol. 9, pp. 203–474, 1907. (Translation of a paper published in *Comm. Soc. Math.*, Kharkow, 1892, reprinted in *Ann. Math. Studies*, vol. 17, Princeton, NJ: Princeton, 1949.)

[13] I. G. Malkin, "On the question of the reciprocal of Lyapunov's theorem on asymptotic stability," *Prikl. Mat. Mekh.*, vol. 18, pp. 129–138, 1954.

[14] J. L. Massera, "Contributions to stability theory," *Ann. Math.*, vol. 64, pp. 182–206, 1956.

[15] A. N. Michel, "Recent trends in the stability analysis of hybrid dynamical systems," *IEEE Trans. Circ. Syst. – Part I: Fund. Theor. Appl.*, vol. 46, pp. 120–134, 1999.

[16] A. N. Michel and C. J. Herget, *Algebra and Analysis for Engineers and Scientists*, Boston: Birkhäuser, 2007.

[17] A. N. Michel and B. Hu, "Towards a stability theory of general hybrid dynamical systems," *Automatica*, vol. 35, pp. 371–384, 1999.

[18] A. N. Michel, K. Wang, and B. Hu, *Qualitative Theory of Dynamical Systems – The Role of Stability Preserving Mappings*, Second Edition, New York: Marcel Dekker, 2001.

[19] R. K. Miller and A. N. Michel, *Ordinary Differential Equations*, New York: Academic Press, 1982.

[20] M. Vidyasagar, *Nonlinear Systems Analysis*, Second Edition, Englewood Cliffs, NJ: Prentice-Hall, 1993.

[21] H. Ye, A. N. Michel, and L. Hou, "Stability theory for hybrid dynamical systems," *IEEE Trans. Autom. Control*, vol. 43, pp. 461–474, 1998.

[22] T. Yoshizawa, *Stability Theory by Liapinov's Second Method*, Tokyo: Math. Soc. of Japan, 1966.

[23] V. I. Zubov, *Methods of A. M. Lyapunov and Their Applications*, Amsterdam: Noordhoff, 1964.

Chapter 7

Finite-Dimensional Dynamical Systems: Specialized Results

In Chapter 6 we presented the principal stability and boundedness results for continuous, discrete-time, and discontinuous finite-dimensional dynamical systems, including converse theorems. In the present chapter we continue our study of finite-dimensional dynamical systems with the presentation of some important specialized results for continuous and discrete-time dynamical systems. This chapter consists of eight sections.

In the first section we present some general stability results concerning autonomous and periodic systems for continuous systems and in the second section we present some of the results from the invariance theory for differential equations and difference equations. In the third section we consider some results that make it possible to estimate the domain of attraction of an asymptotically stable equilibrium for systems described by differential equations. In the fourth and fifth sections we concern ourselves with the stability of systems described by linear homogeneous differential equations and difference equations, respectively. Some of these results require knowledge of state transition matrices, whereas other results involve Lyapunov matrix equations. Also, in the fourth section we present stability results for linear periodic systems and we study in detail second-order systems described by differential equations. In the sixth section we investigate various aspects of the qualitative properties of perturbed linear systems, including Lyapunov's First Method (also called Lyapunov's Indirect Method) for continuous and discrete-time systems; existence of stable and unstable manifolds in continuous linear perturbed systems; and stability properties of periodic solutions in continuous perturbed linear systems. In the seventh section we present stability results for the comparison theory for continuous and discrete-time finite-dimensional systems. Finally, in the eighth section, we provide some background material on differential and difference equations.

7.1 Autonomous and Periodic Systems

In the present section we first show that in the case of autonomous systems,

$$\dot{x} = f(x) \tag{A}$$

and in the case of periodic systems (with period $T > 0$),

$$\dot{x} = f(t, x), \qquad f(t, x) = f(t + T, x) \tag{P}$$

the stability of the equilibrium $x_e = 0$ is equivalent to the uniform stability, and the asymptotic stability of the equilibrium $x_e = 0$ is equivalent to the uniform asymptotic stability. In (A), we assume that $f \in C[\Omega, \mathbb{R}^n]$ where $\Omega \subset \mathbb{R}^n$ is an open connected set, and we assume that $0 \in \Omega$ and $f(0) = 0$. In (P), Ω is defined as above and we assume that $f(t, 0) = 0$ for all $t \geq 0$ and that $f \in C[\mathbb{R}^+ \times \Omega, \mathbb{R}^n]$.

Because an autonomous system may be viewed as a periodic system with arbitrary period, it suffices to prove our first two results for the case of periodic systems.

Theorem 7.1.1 Assume that for every initial condition resulting in a solution of (P) (or of (A)), the solution is unique. If the equilibrium $x_e = 0$ of (P) (or of (A)) is stable, then it is *uniformly stable*.

Proof. Denote the solutions of (P) by $\varphi(t, t_0, \xi_0)$ with $\varphi(t_0, t_0, \xi_0) = \xi_0$. For purposes of contradiction, assume that the equilibrium $x_e = 0$ of (P) is not uniformly stable. Then there is an $\varepsilon > 0$ and sequences $\{t_{0m}\}$ with $t_{0m} \geq 0$, $\{\xi_m\}$, and $\{t_m\}$ such that $\xi_m \to 0$, $t_m \geq t_{0m}$, and $|\varphi(t_m, t_{0m}, \xi_m)| \geq \varepsilon$. Let $t_{0m} = k_m T + \tau_m$, where k_m is a nonnegative integer and $0 \leq \tau_m < T$, and define $t_m^* = t_m - k_m T \geq \tau_m$. Then by the uniqueness of solutions and periodicity of (P), we have $\varphi(t + k_m T, t_{0m}, \xi_m) \equiv \varphi(t, \tau_m, \xi_m)$ because both of these solve (P) and satisfy the initial condition $x(\tau_m) = \xi_m$. Thus,

$$|\varphi(t_m^*, \tau_m, \xi_m)| \geq \varepsilon. \tag{7.1.1}$$

We claim that the sequence $t_m^* \to \infty$. For if it did not, then by going to a convergent subsequence and relabeling, we could assume that $\tau_m \to \tau^*$ and $t_m^* \to t^*$. Then by continuity with respect to initial conditions, $\varphi(t_m^*, \tau_m, \xi_m) \to \varphi(t^*, \tau^*, 0) = 0$. This contradicts (7.1.1).

Because $x_e = 0$ is stable by assumption, then at $t_0 = T$ there is a $\delta > 0$ such that if $|\xi| < \delta$ then $|\varphi(t, T, \xi)| < \varepsilon$ for $t \geq T$. Because $\xi_m \to 0$, then by continuity with respect to initial conditions, $|\varphi(T, \tau_m, \xi_m)| < \delta$ for all $m \geq m(\delta)$. But then by the choice of δ and by (7.1.1), we have

$$\varepsilon > |\varphi(t_m^*, T, \varphi(T, \tau_m, \xi_m))| = |\varphi(t_m^*, \tau_m, \xi_m)| \geq \varepsilon.$$

This contradiction completes the proof. □

Theorem 7.1.2 If the equilibrium $x_e = 0$ of (P) (or of (A)) is asymptotically stable, then it is *uniformly asymptotically stable*.

Proof. The uniform stability is already proved in Theorem 7.1.1. We only need to prove uniform attractivity. Fix $\varepsilon > 0$. By hypothesis, there is an $\eta(T) > 0$ and a $t(\varepsilon, T) > 0$ such that if $|\xi| \leq \eta(T)$, then $|\varphi(t, T, \xi)| < \varepsilon$ for all $t \geq T + t(\varepsilon, T)$. Uniform stability and attractivity imply $t(\varepsilon, T)$ is independent of $|\xi| \leq \eta$. By continuity with respect to initial conditions, there is a $\delta' > 0$ such that $|\varphi(T, \tau, \xi)| < \eta(T)$ if $|\xi| < \delta'$ and $0 \leq \tau \leq T$. So $|\varphi(t + T, \tau, \xi)| < \varepsilon$ if $|\xi| < \delta'$, $0 \leq \tau \leq T$, and $t \geq t(\varepsilon, T)$. Thus for $0 \leq \tau \leq T$, $|\xi| < \delta'$, and $t \geq (T - \tau) + t(\varepsilon, T)$, we have $|\varphi(t + \tau, \tau, \xi)| < \varepsilon$. Put $\delta(\varepsilon) = \delta'$ and $t(\varepsilon) = t(\varepsilon, T) + T$. If $kT \leq \tau < (k+1)T$, then $\varphi(t, \tau, \xi) = \varphi(t - kT, \tau - kT, \xi)$. Thus, if $|\xi| < \delta(\varepsilon)$ and $t \geq \tau + t(\varepsilon)$, then $t - kT \geq \tau - kT + t(\varepsilon)$ and $|\varphi(t, \tau, \xi)| = |\varphi(t - kT, \tau - kT, \xi)| < \varepsilon$. □

Next we address sample converse theorems for systems (A) and (P).

Theorem 7.1.3 Assume that for every initial condition resulting in a solution of (A), the solution is unique. Assume that the equilibrium $x_e = 0$ of (A) is *asymptotically stable*. Then there exists a positive definite function $v \in C[B(r), \mathbb{R}]$ for some $r > 0$ where $B(r) \subset \Omega$ such that $v'_{(A)}$ is negative definite.

Proof. It follows from Theorem 7.1.2 that the asymptotic stability of the equilibrium $x_e = 0$ implies that it is also *uniformly asymptotically stable*. Furthermore, by Lemma 3.10.5 (refer to Problem 3.10.17), there exist a function $\psi \in \mathcal{K}$, defined on $[0, r]$ for some $r > 0$, and a function $\sigma \in \mathcal{L}$, defined on $\mathbb{R}^+$, such that

$$\left|\varphi(t, t_0, x_0)\right| < \psi(|x_0|)\sigma(t - t_0) \tag{7.1.2}$$

for all $\varphi(\cdot, t_0, x_0)$ and all $t \geq t_0$ whenever $|x_0| < r$.

Let

$$Z(x, t) = \int_t^\infty u\big(|\varphi(\tau, t, x)|\big)\, d\tau, \tag{7.1.3}$$

where $u(s) = \alpha(s)^2$ and $\alpha(\cdot)$ is chosen by applying Lemma 3.6.1 to $\beta(\tau) = \psi(r)\sigma(\tau)$ so that $\int_0^\infty \alpha(\beta(\tau))d\tau \leq 1$. Therefore,

$$Z(x, t) \leq [u(\psi(|x|))]^{1/2} \int_t^\infty [u(\psi(r)\sigma(\tau - t))]^{1/2} d\tau \leq \left[u(\psi(|x|))\right]^{1/2}, \tag{7.1.4}$$

which implies that the integral in (7.1.3) converges uniformly with respect to $|x|$. By Corollary 6.8.1, $Z(x, t)$ is continuous with respect to x. Furthermore, because the system is assumed to be autonomous, it is easily seen that $Z(x, t)$ is independent of t. We let the v-function be $v(x) = Z(x, t)$. Then $v(x) \in C[B(r), \mathbb{R}]$ is positive definite. Inequality (7.1.4) shows that $v(x)$ is decrescent. Also, $v'_{(A)}$ is clearly negative definite. The proof is completed. □

Theorem 7.1.4 Assume that for every initial condition resulting in a solution of (P), the solution is unique and that the equilibrium $x_e = 0$ of (P) is *asymptotically stable*. Then there exists a positive definite and decrescent function $v \in C[B(r) \times \mathbb{R}^+, \mathbb{R}]$ for some $r > 0$, where $B(r) \subset \Omega$, which is periodic in t with period T (i.e., $v(x, t) = v(x, t + T)$ for all (x, t), $(x, t + T) \in B(r) \times \mathbb{R}^+$) such that $v'_{(P)}$ is negative definite.

Proof. The proof proceeds similarly as in the proof of Theorem 7.1.3. Let the v-function be $v(x,t) = Z(x,t)$ which is given by (7.1.3). It can readily be verified that $v(x,t+T) = v(x,t)$ for all $(x,t), (x,t+T) \in B(r) \times \mathbb{R}^+$. It has been proved that $v(x,t)$ is decrescent and $v'_{(A)}$ is negative definite. We only need to show that v is positive definite.

Let $y(t) = |\varphi(t,t_0,x_0)|$. Then $\lim_{t\to\infty} y(t) = 0$ because $x_e = 0$ is asymptotically stable. Because

$$x^T \dot{x} = x^T f(t,x),$$

we have the following estimate for $|\dot{y}(t)|$,

$$|\dot{y}(t)| \leq \big|f(t,\varphi(t,t_0,x_0))\big|.$$

By the assumption that f is continuous and $f(t,x) = f(t+T,x)$ for all $x \in \Omega$ and $t \in \mathbb{R}^+$, there exists a $K > 0$ such that $\big|f(t,x)\big| < K$ for all $(t,x) \in \mathbb{R}^+ \times B(r)$. To obtain an estimate for $v(x,t)$ from below, we first assume that $y(t)$ is monotone decreasing. By change of variables, we have

$$v(x,t) = \int_{|x_0|}^{0} u(y) \left(\frac{dy}{dt}\right)^{-1} dy \geq \frac{1}{K}\int_0^{|x_0|} u(y)dy,$$

from which we conclude that v is positive definite. If $y(t)$ is increasing in certain intervals $a_j < t < b_j$, $j = 1,2,\ldots$, we omit them and restrict the integration to the remaining t-axis. Then the above estimate is still valid.

The proof is completed. □

Results for autonomous and periodic discrete-time dynamical systems described by ordinary difference equations that are in the spirit of Theorems 7.1.1 to 7.1.4 can also be established. Also, converse theorems of the type given in Theorems 7.1.3 and 7.1.4 for other types of stability and boundedness can be established as well.

7.2 Invariance Theory

In this section we first present some of the results that comprise the invariance theory for continuous dynamical systems described by autonomous ordinary differential equations (Subsection A). Next, we present some of the results that make up the invariance theory for discrete-time dynamical systems described by autonomous ordinary difference equations (Subsection B). At the end of this section we consider a couple of examples to demonstrate the applicability of these results.

A. Continuous-time systems

We consider again autonomous systems given by

$$\dot{x} = f(x) \tag{A}$$

where $f \in C[\Omega, \mathbb{R}^n]$, $\Omega \subset \mathbb{R}^n$ is an open connected set, $0 \in \Omega$, and $f(0) = 0$.

In the results that follow, we relax some of the conditions required in the stability and boundedness results given in Chapter 6 that broaden the applicability of the Direct Method of Lyapunov (the Second Method of Lyapunov) appreciably.

Theorem 7.2.1 Assume that there exists a function $v \in C[\Omega, \mathbb{R}]$ such that $v'_{(A)}(x) \leq 0$ for all $x \in \Omega$ and such that for some constant $c \in \mathbb{R}$, the set H_c is a closed and bounded component of the set $\{x \in \Omega\colon v(x) \leq c\}$. Let M be the largest invariant set in the set

$$Z = \{x \in \Omega\colon v'_{(A)}(x) = 0\}$$

with respect to (A). Then every solution $\varphi(t)$ of (A) with $\varphi(t_0) \in H_c$ approaches the set M as $t \to \infty$.

Proof. The proof of this result is a direct consequence of Theorem 4.2.1 for the case $T = \mathbb{R}^+$. □

Theorem 7.2.2 With $\Omega = \mathbb{R}^n$, assume that there exists a function $v \in C[\mathbb{R}^n, \mathbb{R}]$ such that $v'_{(A)}(x) \leq 0$ for all $x \in \mathbb{R}^n$. Let M be the largest invariant set with respect to (A) in the set

$$Z = \{x \in \Omega\colon v'_{(A)}(x) = 0\}.$$

Then every bounded solution $\varphi(t)$ of (A) approaches the set M as $t \to \infty$.

Proof. The proof of this theorem is a direct consequence of Theorem 4.2.1 for the case $T = \mathbb{R}^+$, where for every bounded solution $\varphi(t)$ of (A) we choose X_1 as a compact set that contains the trajectory of φ. □

Corollary 7.2.1 With $\Omega = \mathbb{R}^n$, assume that there exists a positive definite and radially unbounded function $v \in C[\mathbb{R}^n, \mathbb{R}]$ such that $v'_{(A)}(x) \leq 0$ for all $x \in \mathbb{R}^n$. Suppose that the origin $x_e = 0$ of $\mathbb{R}^n$ is the only invariant subset of the set

$$Z = \{x \in \Omega\colon v'_{(A)}(x) = 0\}.$$

Then the equilibrium $x_e = 0$ of (A) is *uniformly asymptotically stable in the large*.

Proof. The proof of this result is an immediate consequence of Theorems 7.2.2, 6.2.1(b), and 6.2.6. □

Note that in the above result for the *uniform asymptotic stability in the large* of the equilibrium $x_e = 0$ of (A) we require only that $v'_{(A)}$ be negative semidefinite whereas in the corresponding results given in Chapter 6, we require that $v'_{(A)}$ be negative definite.

B. Discrete-time systems

Next, we consider dynamical systems that are determined by systems of autonomous ordinary difference equations of the form

$$x(k+1) = f(x(k)) \tag{DA}$$

where $k \in \mathbb{N}$, $x(k) \in \Omega$, $f\colon \Omega \to \Omega$, and Ω is an open connected subset of $\mathbb{R}^n$ that contains the origin $x = 0$.

Theorem 7.2.3 Assume for (DA) that there exists a function $v \in C[\Omega, \mathbb{R}]$ such that $v(f(x)) \leq v(x)$ for all $x \in \Omega$. Assume that the set $S_c = \{x \in \Omega\colon v(x) \leq c\}$, for some $c \in \mathbb{R}$, is bounded. Let M be the largest invariant set with respect to (DA) contained in the set

$$Z = \{x \in \Omega\colon v(f(x)) = v(x)\}.$$

Then every solution $\varphi(k)$ of (DA) with $\varphi(k_0) \in S_c$ approaches the set M as $k \to \infty$.

Proof. The proof of this result is a direct consequence of Theorem 4.2.1 for the case $T = \mathbb{N}$. □

Theorem 7.2.4 With $\Omega = \mathbb{R}^n$, assume that there exists a radially unbounded function $v \in C[\mathbb{R}^n, \mathbb{R}]$ such that $v(f(x)) \leq v(x)$ for all $x \in \mathbb{R}^n$. Let M be the largest invariant set with respect to (DA) in the set

$$Z = \{x \in \Omega\colon v(f(x)) = v(x)\}.$$

Then every bounded solution $\varphi(k)$ of (DA) approaches the set M as $k \to \infty$.

Proof. The proof of this theorem is a direct consequence of Theorem 4.2.1 for the case $T = \mathbb{N}$. □

Corollary 7.2.2 With $\Omega = \mathbb{R}^n$, assume that there exists a positive definite and radially unbounded function $v \in C[\mathbb{R}^n, \mathbb{R}]$ such that $v(f(x)) \leq v(x)$ for all $x \in \mathbb{R}^n$. Suppose that the origin $x_e = 0$ of $\mathbb{R}^n$ is the only invariant subset of the set

$$Z = \{x \in \Omega\colon v(f(x)) = v(x)\}.$$

Then the equilibrium $x_e = 0$ of (DA) is *uniformly asymptotically stable in the large*.

Proof. The proof of this result is an immediate consequence of Theorems 7.2.3, 6.3.1(b), and 6.3.3. □

C. Examples

To demonstrate the applicability of the above results, we now consider two specific examples.

Example 7.2.1 (*Lienard equation*) Consider systems described by the equation

$$\ddot{x} + f(x)\dot{x} + g(x) = 0 \tag{7.2.1}$$

where $f \in C^1[\mathbb{R}, \mathbb{R}^+]$, $g \in C^1[\mathbb{R}, \mathbb{R}]$, $g(x) = 0$ if and only if $x = 0$, $xg(x) > 0$ for $x \in \mathbb{R} - \{0\}$, and $\lim_{|x|\to\infty} \int_0^x g(s)ds = \infty$. Equation (7.2.1), called the *Lienard equation*, has been used in the modeling of a variety of physical systems.

Letting $x_1 = x$ and $x_2 = \dot{x}$, we obtain from (7.2.1) the equivalent system

$$\begin{cases} \dot{x}_1 = x_2 \\ \dot{x}_2 = -f(x_1)x_2 - g(x_1). \end{cases} \tag{7.2.2}$$

The origin $(x_1, x_2)^T = (0,0)^T \in \mathbb{R}^2$ is clearly an equilibrium for (7.2.2). We show that this equilibrium is uniformly asymptotically stable in the large.

We choose as a v-function,

$$v(x_1, x_2) = \frac{1}{2}x_2^2 + \int_0^{x_1} g(s)ds \tag{7.2.3}$$

which is positive definite and radially unbounded. Along the solutions of (7.2.2) we have

$$v'_{(7.2.2)}(x_1, x_2) = -x_2^2 f(x_1) \leq 0$$

for all $(x_1, x_2)^T \in \mathbb{R}^2$.

In the notation of Corollary 7.2.1, the set

$$Z = \left\{(x_1, x_2)^T \in \mathbb{R}^2 \colon v'_{(7.2.2)}(x_1, x_2) = 0\right\} \tag{7.2.4}$$

is the x_1-axis. Let M be the largest invariant set in Z. At any point $(x_1, 0)^T \in M$ with $x_1 \neq 0$, equation (7.2.2) implies that $\dot{x}_2 = -g(x_1) \neq 0$. Therefore, the solution emanating from $(x_1, 0)^T$ must leave the x_1-axis. This means that $(x_1, 0) \notin M$ if $x_1 \neq 0$. However, the origin $(0,0)^T$ is clearly in M. Hence, $M = \{(0,0)^T\}$.

It follows from Corollary 7.2.1 that the origin in $\mathbb{R}^2$, which is an equilibrium for system (7.2.2), is uniformly asymptotically stable in the large. □

Example 7.2.2 Let us consider the Lienard equation (7.2.2) given in Example 7.2.1. We assume again that $f \in C^1[\mathbb{R}, \mathbb{R}^+]$, $g \in C^1[\mathbb{R}, \mathbb{R}]$, $g(x) = 0$ if and only if $x = 0$, and $xg(x) > 0$ for $x \in \mathbb{R} - \{0\}$. We also assume that $\lim_{|x_1| \to \infty} |\int_0^{x_1} f(s)ds| = \infty$. This is the case if, for example, $f(s) = k > 0$ for all $s \in \mathbb{R}$. However, *we no longer assume that* $\lim_{|x_1| \to \infty} \int_0^{x_1} g(s)ds = \infty$.

We choose again the v-function

$$v(x_1, x_2) = \frac{1}{2}x_2^2 + \int_0^{x_1} g(s)ds,$$

resulting again in

$$v'_{(7.2.2)}(x_1, x_2) = -x_2^2 f(x_1) \leq 0$$

for all $(x_1, x_2)^T \in \mathbb{R}^2$.

As before, v is positive definite. However, it is not necessarily radially unbounded and therefore, we cannot apply Corollary 7.2.1 to conclude that the equilibrium $(x_1, x_2)^T = (0,0)^T$ of system (7.2.2) is asymptotically stable in the large.

Because $v(x_1, x_2)$ is positive definite and because $v'_{(7.2.2)}$ is negative semidefinite, we can conclude that the equilibrium $(x_1, x_2)^T = (0,0)^T$ of system (7.2.2) is *stable*. We use Theorem 7.2.2 to prove that the equilibrium $(x_1, x_2)^T = (0,0)^T$ is globally attractive, and therefore, that the equilibrium $(x_1, x_2)^T = (0,0)^T$ of system (7.2.2) is *asymptotically stable in the large*.

From Example 7.2.1 we know that $M = \{0\}$ is the largest invariant set in Z given in (7.2.4). To apply Theorem 7.2.2, what remains to be shown is that all the solutions $\varphi(t)$ of system (7.2.2) are bounded.

To this end, let l and a be arbitrary given positive numbers and consider the region U defined by the inequalities

$$v(x) < l \quad \text{and} \quad \left(x_2 + \int_0^{x_1} f(s)ds\right)^2 < a^2. \tag{7.2.5}$$

For each pair of numbers (l, a), U is a bounded region as shown, for example, in Figure 7.2.1.

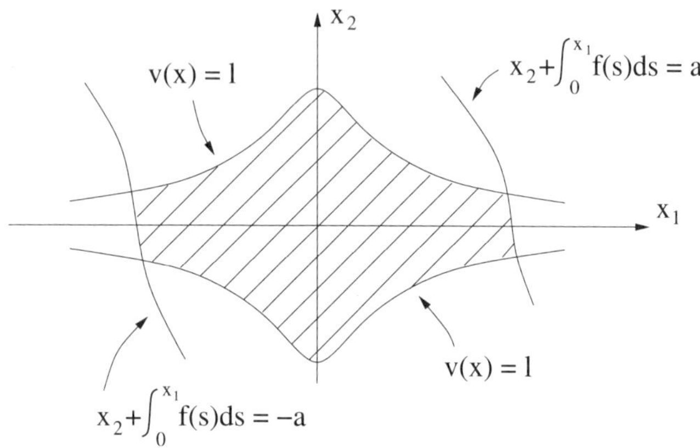

Figure 7.2.1: Region U determined by (7.2.5).

Now let $x_0^T = (x_{10}, x_{20}) = (x_1(0), x_2(0))$ be *any* point in $\mathbb{R}^2$. If we choose (l, a) properly, x_0 will be in the interior of U. Now let $\varphi(t, x_0)$ be a solution of (7.2.2) such that $\varphi(0, x_0) = x_0$. We show that $\varphi(t, x_0)$ cannot leave the bounded region U. This in turn shows that all solutions of (7.2.2) are bounded, inasmuch as $\varphi(t, x_0)$ was chosen arbitrarily.

In order to leave U, the solution $\varphi(t, x_0)$ must either cross the locus of points determined by $v(x) = l$ or one of the loci determined by $x_2 + \int_0^{x_1} f(s)ds = \pm a$. Here we choose, without loss of generality, $a > 0$ so large that the part of the curve determined by $x_2 + \int_0^{x_1} f(s)ds = a$ that is also the boundary of U corresponds to $x_1 > 0$ and the part of the curve determined by $x_2 + \int_0^{x_1} f(s)ds = -a$ corresponds to $x_1 < 0$. Now because $v'_{(7.2.2)}(\varphi(t, x_0)) \le 0$, the solution $\varphi(t, x_0)$ cannot cross the curve determined by $v(x) = l$. To show that it does not cross either of the curves determined by $x_2 + \int_0^{x_1} f(s)ds = \pm a$, we consider the function

$$w(t) = \left[\varphi_2(t, x_0) + \int_0^{\varphi_1(t,x_0)} f(s)ds\right]^2, \tag{7.2.6}$$

where $\varphi(t, x_0)^T = [\varphi_1(t, x_0), \varphi_2(t, x_0)]$. Then

$$w'(t) = -2\left[\varphi_2(t, x_0) + \int_0^{\varphi_1(t,x_0)} f(s)ds\right] g(\varphi_1(t, x_0)). \tag{7.2.7}$$

Now suppose that $\varphi(t, x_0)$ reaches that part of the boundary of U determined by the equation $x_2 + \int_0^{x_1} f(s)ds = a$, $x_1 > 0$. Then along this part of the boundary $w'(t) = -2ag(\varphi(t, x_0)) < 0$ because $x_1 > 0$ and $a > 0$. Therefore, the solution $\varphi(t, x_0)$ cannot cross outside of the set U through that part of the boundary determined by $x_2 + \int_0^{x_1} f(s)ds = a$. We apply the same argument to the part of the boundary determined by $x_2 + \int_0^{x_1} f(s)ds = -a$.

Therefore, every solution of (7.2.2) is bounded and we can apply Theorem 7.2.2 to conclude that the equilibrium $(x_1, x_2)^T = (0, 0)^T$ is globally attractive. □

We apply some of the results of this section in Chapter 8 in the qualitative analysis of a large class of artificial neural networks.

7.3 Domain of Attraction

Many practical systems possess more than one equilibrium point. In such cases, the concept of asymptotic stability in the large is no longer applicable and one is usually interested in knowing the extent of the domain of attraction of an asymptotically stable equilibrium. In the present section, we briefly address the problem of obtaining estimates of the domain of attraction of the equilibrium $x_e = 0$ of the autonomous system

$$\dot{x} = f(x) \tag{A}$$

where $f \in C[\Omega, \mathbb{R}^n]$, $\Omega \subset \mathbb{R}^n$ is an open connected set, $0 \in \Omega$, and $f(0) = 0$.

We assume that there exists a positive definite and time-independent function $v \in C[\Omega, \mathbb{R}^+]$ such that $v'_{(A)}$ is negative definite on some subset $\Omega_1 \subset \Omega$, $0 \in \Omega_1$. Let $D_c = \{x \in \Omega\colon v(x) \leq c\}$ for $c > 0$. If $D_c \subset \Omega_1$, then D_c is *contained in the domain of attraction of the equilibrium* $x_e = 0$ of (A). In fact, D_c is an invariant set for system (A): any trajectory of a solution of (A) starting within D_c will remain in D_c. As such it will remain in Ω_1 where $v'_{(A)}$ is negative definite. Therefore, it follows from the proof of Theorem 6.2.2 that all trajectories for (A) that start in D_c will approach the equilibrium $x_e = 0$. If c_M denotes the largest number for which the above discussion is true, then D_{c_M} is the best estimate of the domain of attraction of $x_e = 0$ for (A), *using the particular* v*-function employed*. Another choice of Lyapunov function will generally result in a different estimate for the domain of attraction.

Example 7.3.1 Consider the system

$$\begin{cases} \dot{x}_1 = -x_1 + x_1\left(x_1^2 + x_2^2\right) \\ \dot{x}_2 = -x_2 + x_2\left(x_1^2 + x_2^2\right). \end{cases} \tag{7.3.1}$$

This system clearly has an equilibrium at the origin $x_e = (x_1, x_2)^T = (0, 0)^T$.

We choose the function

$$v(x_1, x_2) = \frac{1}{2}\left(x_1^2 + x_2^2\right) \tag{7.3.2}$$

and we compute

$$v'_{(7.3.1)}(x_1, x_2) = -\left(x_1^2 + x_2^2\right) + \left(x_1^2 + x_2^2\right)^2. \tag{7.3.3}$$

The function v is positive definite and radially unbounded whereas $v'_{(7.3.1)}$ is negative definite on the set $\{x \in \mathbb{R}^2 \colon (x_1^2 + x_2^2) < 2c\}$, $c < 1/2$; that is, on the set

$$D_c = \left\{x \in \mathbb{R}^2 \colon v(x) < c\right\}, \qquad c < \frac{1}{2}. \tag{7.3.4}$$

We conclude that the equilibrium $x_e = 0$ of (7.3.1) is asymptotically stable and that the set $\{x \in \mathbb{R}^2 \colon (x_1^2 + x_2^2) < 1\}$ is contained in the domain of attraction of $x_e = 0$. Indeed, it is not hard to show that this set is the entire domain of attraction of the equilibrium $x_e = 0$. □

The procedure described above for determining estimates for the domain of attraction of an asymptotically stable equilibrium $x_e = 0$ applies, without substantial changes, to autonomous discrete-time systems described by difference equations

$$x(k+1) = f(x(k)) \tag{DA}$$

where $f \colon \Omega \to \Omega$, $\Omega \subset \mathbb{R}^n$ is an open connected set, and $0 \in \Omega$. We omit the details in the interests of brevity.

There are results that determine the domain of attraction of an asymptotically stable equilibrium $x_e = 0$ of (A) precisely. In the following, we let $G \subset \Omega$ and we assume that G is a simply connected domain containing a neighborhood of the origin.

Theorem 7.3.1 [21] (*Zubov*) Suppose there exist two functions $v \in C^1[G, \mathbb{R}]$ and $h \in C[\mathbb{R}^n, \mathbb{R}]$ satisfying the following hypotheses.

(i) v is positive definite in G and satisfies on G the inequality $0 < v(x) < 1$ when $x \neq 0$. For any $b \in (0, 1)$, the set $\{x \in G \colon v(x) \leq b\}$ is bounded.

(ii) $h(0) = 0$, and $h(x) > 0$ for $x \neq 0$.

(iii) For $x \in G$, we have

$$v'_{(A)}(x) = -h(x)\left[1 - v(x)\right]\left[1 + |f(x)|^2\right]^{1/2}. \tag{7.3.5}$$

(iv) As $x \in G$ approaches a point on the boundary of G, or in case of an unbounded region G, as $|x| \to \infty$, $\lim v(x) = 1$.

Then G is exactly the domain of attraction of the equilibrium $x_e = 0$.

Proof. Under the given hypotheses, it follows from Theorem 6.2.2 that $x_e = 0$ is uniformly asymptotically stable. Note also that if we introduce the change of variables

$$ds = \left[1 + |f(\varphi(t))|^2\right]^{1/2} dt,$$

then (7.3.5) reduces to

$$\frac{dv}{ds} = -h(x)\left[1 - v(x)\right],$$

but the stability properties of (A) remain unchanged. Let $V(s) = v(\varphi(s))$ for a given function $\varphi(s)$ such that $\varphi(0) = x_0$. Then

$$\frac{d}{ds}\log\big[1 - V(s)\big] = h(\varphi(s)),$$

or

$$1 - V(s) = \big[1 - V(0)\big]\exp\left(\int_0^s h(\varphi(u))du\right). \tag{7.3.6}$$

Let $x_0 \in G$ and assume that x_0 is not in the domain of attraction of the trivial solution. Then $h(\varphi(s)) \geq \delta > 0$ for some fixed δ and for all $s \geq 0$. Hence, in (7.3.6) as $s \to \infty$ the term on the left is at most one, whereas the term on the right tends to infinity, which is impossible. Thus, x_0 is in the domain of attraction of $x_e = 0$.

Suppose x_1 is in the domain of attraction but $x_1 \notin G$. Then $\varphi(s, x_1) \to 0$ as $s \to \infty$, so there must exist s_1 and s_2 such that $\varphi(s_1, x_1) \in \partial G$ and $\varphi(s_2, x_1) \in G$. Let $x_0 = \varphi(s_2, x_1)$ in (7.3.6). Take the limit in (7.3.6) as $s \to s_1^+$. We see that

$$\lim_{s\to s_1^+}\big[1 - V(s)\big] = 1 - 1 = 0,$$

and the limit on the right-hand side is

$$\big[1 - v(x_0)\big]\exp\left(\int_{s_2}^{s_1} h(\varphi(s, x_1))ds\right) > 0.$$

This is impossible. Hence x_1 must be in G. □

An immediate result of Theorem 7.3.1 is the following result.

Corollary 7.3.1 Assume that there exists a function h that satisfies the hypotheses of Theorem 7.3.1 and assume that there exists a continuously differentiable, positive definite function $v\colon G \to \mathbb{R}$ that satisfies the inequality $0 \leq v(x) \leq 1$ for all $x \in G$ as well as the differential equation

$$\nabla v(x)^T f(x) = -h(x)\big[[1 - v(x)\big]\big[1 + |f(x)|^2\big]^{1/2}. \tag{7.3.7}$$

Then the boundary of the domain of attraction is defined by the equation

$$v(x) = 1. \tag{7.3.8}$$

If the domain of attraction G is all of $\mathbb{R}^n$, then the equilibrium $x_e = 0$ of (A) is asymptotically stable in the large. In this case, we have

$$v(x) \to 1 \quad \text{as} \quad |x| \to \infty. \tag{7.3.9}$$

□

In the foregoing results, we can work with different v functions. For example, if we let

$$w(x) = -\log\left[1 - v(x)\right],$$

then (7.3.5) assumes the form

$$w'_{(A)}(x) = -h(x)\left[1 + |f(x)|^2\right]^{1/2}$$

and condition (7.3.8) defining the boundary becomes $w(x) \to \infty$.

Note that in Theorem 7.3.1, the function $h(x)$ is arbitrary. In specific applications it is chosen in a fashion that makes the solution of the partial differential equation (7.3.7) easy.

Example 7.3.2 Consider the system

$$\begin{cases} \dot{x}_1 = 2x_1 \dfrac{1 - x_1^2 + x_2^2}{(x_1+1)^2 + x_2^2} + x_1 x_2 \triangleq f_1(x_1, x_2) \\ \dot{x}_2 = \dfrac{1 - x_1^2 + x_2^2}{2} - \dfrac{4x_1^2 x_2}{(x_1+1)^2 + x_2^2} \triangleq f_2(x_1, x_2). \end{cases} \tag{7.3.10}$$

This system has an equilibrium at $x_1 = 1$, $x_2 = 0$. The partial differential equation (7.3.7) assumes the form

$$\frac{\partial v}{\partial x_1}(x_1, x_2) f_1(x_1, x_2) + \frac{\partial v}{\partial x_2}(x_1, x_2) f_2(x_1, x_2) = -2\frac{(x_1 - 1)^2 + x_2^2}{(x_1+1)^2 + x_2^2}\left[1 - v(x_1, x_2)\right]$$

where

$$h(x_1, x_2) = 2\frac{(x_1 - 1)^2 + x_2^2}{(x_1+1)^2 + x_2^2}\left[1 + f_1(x_1, x_2)^2 + f_2(x_1, x_2)^2\right]^{-(1/2)}.$$

It is readily verified that a solution of the above partial differential equation is

$$v(x_1, x_2) = \frac{(x_1 - 1)^2 + x_2^2}{(x_1+1)^2 + x_2^2}.$$

Because $v(x_1, x_2) = 1$ if and only if $x_1 = 0$, the domain of attraction is the set $\{(x_1, x_2)\colon 0 < x_1 < \infty, -\infty < x_2 < \infty\}$. □

7.4 Linear Continuous-Time Systems

In this section we study the stability properties of the equilibrium of linear homogeneous systems

$$\dot{x} = A(t)x, \qquad t \geq t_0,\ t_0 \geq 0 \tag{LH}$$

and linear autonomous homogeneous systems

$$\dot{x} = Ax, \qquad t \geq 0. \tag{L}$$

In (LH), $t \in \mathbb{R}^+$, $x \in \mathbb{R}^n$, and $A \in C[\mathbb{R}^+, \mathbb{R}^{n\times n}]$. In (L), $t \in \mathbb{R}^+$, $x \in \mathbb{R}^n$, and $A \in \mathbb{R}^{n\times n}$. Recall that $x_e = 0$ is always an equilibrium of (L) and (LH) and that $x_e = 0$ is the only equilibrium of (LH) if $A(t)$ is nonsingular for $t \geq 0$. Recall also that the solution of (LH) for $x(t_0) = x_0$ is of the form

$$\varphi(t, t_0, x_0) = \Phi(t, t_0)x_0, \qquad t \geq t_0$$

where Φ denotes the state transition matrix of $A(t)$. Recall further that the solution of (L) for $x(t_0) = x_0$ is given by

$$\begin{aligned}\varphi(t, t_0, x_0) &= \Phi(t, t_0)x_0\\ &= \Phi(t - t_0, 0)x_0\\ &\triangleq \Phi(t - t_0)x_0\\ &= e^{A(t-t_0)}x_0\end{aligned}$$

where in the preceding equation, a slight abuse in notation has been used.

For some of the properties of the transition matrix $\Phi(t, t_0)$ that are used in the proofs of the result that follow, the reader should refer to the appendix (Subsection 7.8A).

A. Linear homogeneous systems

In the next four results, we explore some of the basic qualitative properties of (LH).

Theorem 7.4.1 The equilibrium $x_e = 0$ of (LH) is *stable* if and only if the solutions of (LH) are bounded, or equivalently, if and only if

$$\sup_{t\geq t_0} \|\Phi(t, t_0)\| \triangleq c(t_0) < \infty$$

where $\|\Phi(t, t_0)\|$ denotes the matrix norm induced by the vector norm used on $\mathbb{R}^n$ and $c(t_0)$ denotes a constant that may depend on the choice of t_0.

Proof. Assume that the equilibrium $x_e = 0$ of (LH) is stable. Then for any $t_0 \geq 0$ and for $\varepsilon = 1$ there is a $\delta = \delta(1, t_0) > 0$ such that $|\varphi(t, t_0, x_0)| < 1$ for all $t \geq t_0$ and all x_0 with $|x_0| \leq \delta$. In this case

$$|\varphi(t, t_0, x_0)| = |\Phi(t, t_0)x_0| = \left|\frac{\Phi(t, t_0)(x_0\delta)}{|x_0|}\right| \left(\frac{|x_0|}{\delta}\right) < \frac{|x_0|}{\delta}$$

for all $x_0 \neq 0$ and all $t \geq t_0$. In the above inequality we have used the fact that

$$\left|\varphi\left(t, t_0, \frac{x_0\delta}{|x_0|}\right)\right| = \left|\frac{\Phi(t, t_0)(x_0\delta)}{|x_0|}\right| < 1.$$

Using the definition of matrix norm it follows that

$$\|\Phi(t, t_0)\| \leq \delta^{-1}, \qquad t \geq t_0.$$

We have proved that if the equilibrium $x_e = 0$ of system (LH) is stable, then the solutions of (LH) are bounded.

Conversely, suppose that all solutions $\varphi(t, t_0, x_0) = \Phi(t, t_0)x_0$ are bounded. Let $\{e_1, \dots, e_n\}$ denote the natural basis for n-space and let $|\varphi(t, t_0, e_j)| < \beta_j$ for all $t \geq t_0, j = 1, \dots, n$. Then for any vector $x_0 = \sum_{j=1}^{n} \alpha_j e_j$ we have that

$$\begin{aligned}
|\varphi(t, t_0, x_0)| &= \left| \sum_{j=1}^{n} \alpha_j \varphi(t, t_0, e_j) \right| \\
&\leq \sum_{j=1}^{n} |\alpha_j| \beta_j \\
&\leq \left(\max_{1 \leq j \leq n} \beta_j \right) \sum_{j=1}^{n} |\alpha_j| \\
&\leq K|x_0|
\end{aligned}$$

for some constant $K > 0$ and for $t \geq t_0$. For given $\varepsilon > 0$, we choose $\delta = \varepsilon/K$. Thus, if $|x_0| < \delta$, then $|\varphi(t, t_0, x_0)| < K|x_0| < \varepsilon$ for all $t \geq t_0$. We have proved that if the solutions of (LH) are bounded, then the equilibrium $x_e = 0$ of (LH) is stable. □

Theorem 7.4.2 The equilibrium $x_e = 0$ of (LH) is *uniformly stable* if and only if

$$\sup_{t_0 \geq 0} c(t_0) \triangleq \sup_{t_0 \geq 0} \left(\sup_{t \geq t_0} \|\Phi(t, t_0)\| \right) \triangleq c_0 < \infty.$$

The proof of the above result is similar to the proof of Theorem 7.4.1 and is left as an exercise.

Example 7.4.1 Consider the system

$$\begin{bmatrix} \dot{x}_1 \\ \dot{x}_2 \end{bmatrix} = \begin{bmatrix} e^{-2t} & (e^{-t} - e^{-2t}) \\ 0 & e^{-t} \end{bmatrix} \begin{bmatrix} x_1 \\ x_2 \end{bmatrix} \tag{7.4.1}$$

where $x(0) = x_0$. We transform (7.4.1) using the relation $x = Py$, where

$$P = \begin{bmatrix} 1 & 1 \\ 0 & 1 \end{bmatrix}, \qquad P^{-1} = \begin{bmatrix} 1 & -1 \\ 0 & 1 \end{bmatrix}$$

to obtain the equivalent system

$$\begin{bmatrix} \dot{y}_1 \\ \dot{y}_2 \end{bmatrix} = \begin{bmatrix} e^{-2t} & 0 \\ 0 & e^{-t} \end{bmatrix} \begin{bmatrix} y_1 \\ y_2 \end{bmatrix} \tag{7.4.2}$$

with $y(0) = y_0 = P^{-1}x_0$. System (7.4.2) has the solution $\psi(t, 0, y_0) = \Psi(t, 0)y_0$, where

$$\Psi(t, 0) = \begin{bmatrix} e^{(1/2)(1-e^{-2t})} & 0 \\ 0 & e^{(1-e^{-t})} \end{bmatrix}.$$

The solution of (7.4.1) is obtained as $\varphi(t,0,x_0) = P\Psi(t,0)P^{-1}x_0$. From this we obtain for $t_0 \neq 0$, $\varphi(t,t_0,x_0) = \Phi(t,t_0)x_0$, where

$$\Phi(t,t_0) = \begin{bmatrix} e^{(1/2)(e^{-2t_0}-e^{-2t})} & e^{(e^{-t_0}-e^{-t})} - e^{(1/2)(e^{-2t_0}-e^{-2t})} \\ 0 & e^{(e^{-t_0}-e^{-t})} \end{bmatrix}.$$

Letting $t \to \infty$, we obtain

$$\lim_{t\to\infty} \Phi(t,t_0) = \begin{bmatrix} e^{(1/2)e^{-2t_0}} & e^{e^{-t_0}} - e^{(1/2)e^{-2t_0}} \\ 0 & e^{e^{-t_0}} \end{bmatrix}. \tag{7.4.3}$$

We conclude that

$$\lim_{t_0\to\infty} \lim_{t\to\infty} \|\Phi(t,t_0)\| < \infty,$$

and therefore that

$$\sup_{t_0\geq 0} \Big(\sup_{t\geq t_0} \|\Phi(t,t_0)\| \Big) < \infty$$

because

$$\|\Phi(t,t_0)\| = \big\|[\phi_{ij}(t,t_0)]\big\| \leq \sqrt{\sum_{i,j=1}^{2} |\phi_{ij}(t,t_0)|^2} \leq \sum_{i,j=1}^{2} |\phi_{ij}(t,t_0)|.$$

Therefore, the equilibrium $x_e = 0$ of system (7.4.1) is *stable* by Theorem 7.4.1 and *uniformly stable* by Theorem 7.4.2. □

Theorem 7.4.3 The following statements are equivalent.

(i) The equilibrium $x_e = 0$ of (LH) is asymptotically stable.

(ii) The equilibrium $x_e = 0$ of (LH) is asymptotically stable in the large.

(iii) $\lim_{t\to\infty} \|\Phi(t,t_0)\| = 0$.

Proof. Assume that statement (i) is true. Then there is an $\eta(t_0) > 0$ such that when $|x_0| \leq \eta(t_0)$, then $\varphi(t,t_0,x_0) \to 0$ as $t \to \infty$. But then we have for any $x_0 \neq 0$ that

$$\varphi(t,t_0,x_0) = \varphi\left(t, t_0, \frac{\eta(t_0)x_0}{|x_0|}\right)\left(\frac{|x_0|}{\eta(t_0)}\right) \to 0$$

as $t \to \infty$. It follows that statement (ii) is true.

Next, assume that statement (ii) is true. Fix $t_0 \geq 0$. For any $\varepsilon > 0$ there must exist a $T(\varepsilon) > 0$ such that for all $t \geq t_0 + T(\varepsilon)$ we have that $|\varphi(t,t_0,x_0)| = |\Phi(t,t_0)x_0| < \varepsilon$. To see this, let $\{e_1,\dots,e_n\}$ be the natural basis for $\mathbb{R}^n$. Thus, for some fixed constant $K > 0$, if $x_0 = (\alpha_1,\dots,\alpha_n)^T$ and if $|x_0| \leq 1$, then $x_0 = \sum_{j=1}^n \alpha_j e_j$ and $\sum_{j=1}^n |\alpha_j| \leq K$. For each j there is a $T_j(\varepsilon)$ such that $|\Phi(t,t_0)e_j| < \varepsilon/K$ for all $t \geq t_0 + T_j(\varepsilon)$. Define $T(\varepsilon) = \max\{T_j(\varepsilon)\colon j = 1,\dots,n\}$. For $|x_0| \leq 1$ and $t \geq t_0 + T(\varepsilon)$, we have that

$$|\Phi(t,t_0)x_0| = \left|\sum_{j=1}^n \alpha_j \Phi(t,t_0)e_j\right| \leq \sum_{j=1}^n |\alpha_j| \left(\frac{\varepsilon}{K}\right) \leq \varepsilon.$$

By the definition of matrix norm, this means that $\|\Phi(t,t_0)\| \le \varepsilon$ for all $t \ge t_0 + T(\varepsilon)$. Therefore, statement (iii) is true.

Finally, assume that statement (iii) is true. Then $\|\Phi(t,t_0)\|$ is bounded in t for all $t \ge t_0$. By Theorem 7.4.1, the equilibrium $x_e = 0$ is stable. To prove asymptotic stability, fix $t_0 \ge 0$ and $\varepsilon > 0$. If $|x_0| < \eta(t_0) = 1$, then $|\varphi(t,t_0,x_0)| \le \|\Phi(t,t_0)\||x_0| \to 0$ as $t \to \infty$. Therefore, statement (i) is true. This completes the proof. □

Example 7.4.2 The equilibrium $x_e = 0$ of system (7.4.1) given in Example 7.4.1 is stable but not asymptotically stable because $\lim_{t\to\infty} \|\Phi(t,t_0)\| \ne 0$. □

Example 7.4.3 The solution of the system

$$\dot{x} = -e^{2t}x, \qquad x(t_0) = x_0 \tag{7.4.4}$$

is $\varphi(t,t_0,x_0) = \Phi(t,t_0)x_0$, where

$$\Phi(t,t_0) = e^{(1/2)(e^{2t_0} - e^{2t})}.$$

Because $\lim_{t\to\infty} \Phi(t,t_0) = 0$, it follows that the equilibrium $x_e = 0$ of system (7.4.4) is asymptotically stable in the large. □

Theorem 7.4.4 The equilibrium $x_e = 0$ of system (LH) is uniformly asymptotically stable if and only if it is *exponentially stable*.

Proof. The exponential stability of the equilibrium $x_e = 0$ implies the uniform asymptotic stability of the equilibrium $x_e = 0$ of system (E) in general, and hence, for system (LH) in particular.

Conversely, assume that the equilibrium $x_e = 0$ of system (LH) is uniformly asymptotically stable. Then there are a $\delta > 0$ and a $T > 0$ such that if $|x_0| \le \delta$, then $|\Phi(t + t_0 + T, t_0)x_0| \le (\delta/2)$ for all $t, t_0 \ge 0$. This implies that

$$\|\Phi(t + t_0 + T, t_0)\| \le \frac{1}{2} \quad \text{if } t, t_0 \ge 0. \tag{7.4.5}$$

From Theorem 7.8.6(iii) (Subsection 7.8A) we have that $\Phi(t,\tau) = \Phi(t,\sigma)\Phi(\sigma,\tau)$ for any t, σ, and τ. Therefore,

$$\|\Phi(t + t_0 + 2T, t_0)\| = \|\Phi(t + t_0 + 2T, t + t_0 + T)\Phi(t + t_0 + T, t_0)\| \le \frac{1}{4},$$

in view of (7.4.5). By induction, we obtain for $t, t_0 \ge 0$ that

$$\|\Phi(t + t_0 + nT, t_0)\| \le 2^{-n}. \tag{7.4.6}$$

Now let $\alpha = (\ln 2)/T$. Then (7.4.6) implies that for $0 \le t < T$ we have that

$$\begin{aligned} |\varphi(t + t_0 + nT, t_0, x_0)| &\le 2|x_0|2^{-(n+1)} \\ &= 2|x_0|e^{-\alpha(n+1)T} \\ &\le 2|x_0|e^{-\alpha(t+nT)}, \end{aligned}$$

which proves the result. □

Example 7.4.4 For system (7.4.4) given in Example 7.4.3 we have

$$|\varphi(t, t_0, x_0)| = \left| x_0 e^{(1/2)e^{2t_0}} e^{-(1/2)e^{2t}} \right| \leq |x_0| e^{(1/2)e^{2t_0}} e^{-t}, \qquad t \geq t_0 \geq 0,$$

because $e^{2t} > 2t$. Therefore, the equilibrium $x_e = 0$ of system (7.4.4) is uniformly asymptotically stable in the large, and exponentially stable in the large. □

Even though the preceding results require knowledge of the state transition matrix $\Phi(t, t_0)$ of (LH), they are quite useful in the qualitative analysis of linear systems.

B. Linear autonomous homogeneous systems

Revisiting Example 3.1.8, we now address the stability properties of system (L),

$$\dot{x} = Ax, \qquad t \geq 0. \tag{L}$$

To this end we transform matrix A into the Jordan canonical form, $J = P^{-1}AP$, using the transformation $x = Py$ to obtain from (L) the equivalent system

$$\dot{y} = P^{-1}APy = Jy. \tag{7.4.7}$$

It is easily verified (the reader is asked to do this in the exercise section) that the equilibrium $x_e = 0$ of (L) is stable (resp., asymptotically stable or unstable) if and only if $y_e = 0$ of system (7.4.7) is stable (resp., asymptotically stable or unstable). In view of this, we can assume without loss of generality that the matrix A in (L) is in Jordan canonical form given by

$$A = \text{diag}\big[J_0, J_1, \ldots, J_s\big]$$

where

$$J_0 = \text{diag}\big[\lambda_1, \ldots, \lambda_k\big] \quad \text{and} \quad J_i = \lambda_{k+i} I_i + N_i$$

for the Jordan blocks $J_1, \ldots, J_s$, where I_i denotes the $n_i \times n_i$ identity matrix and N_i denotes the $n_i \times n_i$ nilpotent matrix given by

$$N_i = \begin{bmatrix} 0 & 1 & 0 & \cdots & 0 \\ 0 & 0 & 1 & \cdots & 0 \\ \vdots & \vdots & \ddots & \ddots & \vdots \\ 0 & 0 & 0 & \ddots & 1 \\ 0 & 0 & 0 & \cdots & 0 \end{bmatrix},$$

and λ_j, $j = 1, \ldots n$, denote the eigenvalues of A. We have

$$e^{At} = \begin{bmatrix} e^{J_0 t} & & & 0 \\ & e^{J_1 t} & & \\ & & \ddots & \\ 0 & & & e^{J_s t} \end{bmatrix}$$

where

$$e^{J_0 t} = \text{diag}\left[e^{\lambda_1 t}, \dots, e^{\lambda_k t}\right] \tag{7.4.8}$$

and

$$e^{J_i t} = e^{\lambda_{k+i} t} \begin{bmatrix} 1 & t & t^2/2 & \cdots & t^{n_i - 1}/(n_i - 1)! \\ 0 & 1 & t & \cdots & t^{n_i - 2}/(n_i - 2)! \\ \vdots & \vdots & \vdots & \vdots & \vdots \\ 0 & 0 & 0 & \cdots & 1 \end{bmatrix} \tag{7.4.9}$$

for $i = 1, \dots, s$.

Now suppose that $\text{Re}\lambda_i \leq \beta$ for all $i = 1, \dots, k$. Then it is clear that

$$\lim_{t \to \infty} \frac{\|e^{J_0 t}\|}{e^{\beta t}} < \infty$$

where $\|e^{J_0 t}\|$ is the matrix norm induced by one of the equivalent vector norms defined on $\mathbb{R}^n$. We write this as $\|e^{J_0 t}\| = \mathcal{O}(e^{\beta t})$. Similarly, if $\beta = \text{Re}\lambda_{k+i}$, then for any $\varepsilon > 0$ we have that $\|e^{J_i t}\| = \mathcal{O}(t^{n_i - 1} e^{\beta t}) = \mathcal{O}(e^{(\beta + \varepsilon)t})$.

From the foregoing it is now clear that $\|e^{At}\| \leq K$ for some $K > 0$ if and only if all eigenvalues of A have nonpositive real parts, and the eigenvalues with zero real part occur in the Jordan form only in J_0 and not in any of the Jordan blocks J_i, $1 \leq i \leq s$. Hence, by Theorems 7.4.1 and 7.4.2, the equilibrium $x_e = 0$ of (L) is under these conditions *stable*, in fact *uniformly stable*.

Now suppose that all eigenvalues of A have negative real parts. From the preceding discussion it is clear that there is a constant $K > 0$ and an $\alpha > 0$ such that $\|e^{At}\| \leq Ke^{-\alpha t}$, and therefore, $|\varphi(t, t_0, x_0)| \leq Ke^{-\alpha(t - t_0)}|x_0|$ for all $t \geq t_0 \geq 0$ and for all $x_0 \in \mathbb{R}^n$. It follows that the equilibrium $x_e = 0$ is uniformly asymptotically stable in the large, in fact exponentially stable in the large. Conversely, assume that there is an eigenvalue λ_i with nonnegative real part. Then either one term in (7.4.8) does not tend to zero, or else a term in (7.4.9) is unbounded as $t \to \infty$. In either case, $e^{At}x(0)$ will not tend to zero when the initial condition $x(0) = x_0$ is properly chosen. Hence, the equilibrium $x_e = 0$ of (L) cannot be asymptotically stable (and hence, it cannot be exponentially stable).

Summarizing the above, we have proved the following result.

Theorem 7.4.5 The equilibrium $x_e = 0$ of (L) is *stable*, in fact, *uniformly stable*, if and only if all eigenvalues of A have nonpositive real parts, and every eigenvalue with zero real part has an associated Jordan block of order one. The equilibrium $x_e = 0$ of (L) is *uniformly asymptotically stable in the large*, in fact, *exponentially stable in the large*, if and only if all eigenvalues of A have negative real parts. □

A consequence of the above result is the following result.

Theorem 7.4.6 The equilibrium $x_e = 0$ of (L) is *unstable* if and only if at least one of the eigenvalues of A has either positive real part or has zero real part that is associated with a Jordan block of order greater than one. □

Before proceeding any further, it may be appropriate to take note of certain conventions concerning matrices that are used in the literature. Some of these are not consistent with the terminology used in Theorem 7.4.5. Thus, a real $n \times n$ matrix A is called *stable* or a *Hurwitz matrix* if all its eigenvalues have negative real parts. If at least one of the eigenvalues has a positive real part, then A is called *unstable*. A matrix A, which is neither stable nor unstable, is called *critical*, and the eigenvalues with zero real parts are called *critical eigenvalues*.

We conclude the present subsection with some examples.

Example 7.4.5 Consider system (L) with

$$A = \begin{bmatrix} 0 & 1 \\ -1 & 0 \end{bmatrix}.$$

The eigenvalues of A are $\lambda_1, \lambda_2 = \pm i$ $(i = \sqrt{-1})$. According to Theorem 7.4.5, the equilibrium $x_e = 0$ of this system is *stable*. This can also be verified by computing the solutions of this system for the given set of initial data $x(0) = (x_1(0), x_2(0))^T$,

$$\begin{cases} \varphi_1(t, 0, x_0) & = x_1(0)\cos t + x_2(0)\sin t \\ \varphi_2(t, 0, x_0) & = -x_1(0)\sin t + x_2(0)\cos t, \end{cases}$$

$t \geq 0$, and then applying Definition 6.1.1(a). □

Example 7.4.6 Consider system (L) with

$$A = \begin{bmatrix} 2.8 & 9.6 \\ 9.6 & -2.8 \end{bmatrix}.$$

The eigenvalues of A are $\lambda_1, \lambda_2 = \pm 10$. According to Theorem 7.4.6, the equilibrium $x_e = 0$ of this system is *unstable*. □

Example 7.4.7 Consider system (L) with

$$A = \begin{bmatrix} 0 & 1 \\ 0 & 0 \end{bmatrix}.$$

The eigenvalues of A are $\lambda_1 = 0$ and $\lambda_2 = 0$. According to Theorem 7.4.6, the equilibrium $x_e = 0$ of this system is *unstable*. This can also be verified by computing the solutions of this system for the given set of initial data $x(0) = (x_1(0), x_2(0))^T$,

$$\begin{cases} \varphi_1(t, 0, x_0) & = x_1(0) + x_2(0)t, \\ \varphi_2(t, 0, x_0) & = x_2(0), \end{cases}$$

$t \geq 0$, and then applying Definition 6.1.1(k). □

Example 7.4.8 Consider system (L) with

$$A = \begin{bmatrix} -1 & 0 \\ -1 & -2 \end{bmatrix}.$$

The eigenvalues of A are $\lambda_1, \lambda_2 = -1, -2$. According to Theorem 7.4.5, the equilibrium $x_e = 0$ of this system is *exponentially stable*. □

C. The Lyapunov matrix equation

The stability results that we established thus far in this section require *explicit* knowledge of the solutions of (L) and (LH). In the present subsection we develop stability criteria for system (L) with *arbitrary* matrix A. To accomplish this, we employ the Lyapunov stability results established in Chapter 6. We recall that these involve the existence of Lyapunov functions.

Lyapunov functions v for a system are sometimes viewed as "*generalized distance functions*" of the state x from the equilibrium x_e ($x_e = 0$) and the stability properties are then deduced directly from the properties of v and its time derivative v', along the solutions of the system on hand.

A logical choice of Lyapunov function for system (L) is $v(x) = x^T x = |x|^2$ which represents the square of the Euclidean distance of the state from the equilibrium $x_e = 0$. The stability properties of this equilibrium are then determined by examining the properties of $v'_{(L)}(x)$, the time derivative of $v(x)$ along the solutions of (L),

$$\dot{x} = Ax. \tag{L}$$

This derivative can be determined without explicitly solving for the solutions of system (L) as

$$\begin{aligned} v'_{(L)}(x) &= \dot{x}^T x + x^T \dot{x} \\ &= (Ax)^T x + x^T (Ax) \\ &= x^T (A^T + A)x. \end{aligned}$$

If the matrix A is such that $v'_{(L)}(x)$ is negative for all $x \neq 0$, then it is reasonable to expect that the distance of the state of (L) from the equilibrium $x_e = 0$ will decrease with increasing time, and that the state will therefore tend to the equilibrium of (L) with increasing time.

The above discussion is consistent with our earlier discussion of Subsection 6.1G. It turns out that the Lyapunov function used above is not sufficiently flexible. In the following we employ as a "generalized distance function" the *quadratic form*

$$v(x) = x^T P x, \qquad P = P^T \tag{7.4.10}$$

where $P \in \mathbb{R}^{n \times n}$. (Refer to Subsection 6.1F for a discussion of quadratic forms.) The derivative of $v(x)$ along the solutions of (L) is determined as

$$\begin{aligned} v'_{(L)}(x) &= \dot{x}^T P x + x^T P \dot{x} \\ &= x^T A^T P x + x^T P A x \\ &= x^T (A^T P + PA)x; \end{aligned}$$

that is,

$$v'_{(L)}(x) = x^T C x, \tag{7.4.11}$$

where

$$C = A^T P + PA. \tag{7.4.12}$$

Note that C is real and $C^T = C$. The system of equations given in (7.4.12) is called the *Lyapunov Matrix Equation*.

Before proceeding further, we recall that because P is real and symmetric, all of its eigenvalues are real. Also, we recall that P is said to be *positive definite* (resp., *positive semidefinite*) if all its eigenvalues are positive (resp., nonnegative) and is called *indefinite* if P has eigenvalues of opposite sign. The notions of *negative definite* and *negative semidefinite* for matrix P are defined similarly (refer, e.g., to Michel and Herget [14]). Thus (see Subsection 6.1F), the function $v(x)$ given in (7.4.10) is *positive definite*, *positive semidefinite*, and so forth, if the matrix P has the corresponding definiteness properties. Finally, we recall from Subsection 6.1F that instead of solving for the eigenvalues of a real symmetric matrix to determine its definiteness properties, there are more efficient and direct methods to accomplish this (refer to Theorem 6.1.4).

In view of the above discussion, the results below now follow readily by invoking the Lyapunov results established in Section 6.2.

Proposition 7.4.1 (a) The equilibrium $x_e = 0$ of (L) is *uniformly stable* if there exists a real, symmetric, and positive definite $n \times n$ matrix P such that the matrix C given in (7.4.12) is negative semidefinite.

(b) The equilibrium $x_e = 0$ of (L) is *exponentially stable in the large* if there exists a real, symmetric, and positive definite $n \times n$ matrix P such that the matrix C given in (7.4.12) is negative definite.

(c) The equilibrium $x_e = 0$ of (L) is *unstable* if there exists a real, symmetric $n \times n$ matrix P that is either negative definite or indefinite such that the matrix C given in (7.4.12) is negative definite. □

We leave the proofs of the above results as an exercise to the reader.

Example 7.4.9 For the system given in Example 7.4.5 we choose $P = I$, and we compute

$$C = A^T P + PA = A^T + A = 0.$$

According to Proposition 7.4.1(a), the equilibrium $x_e = 0$ of this system is *stable* (as expected in Example 7.4.5). □

Example 7.4.10 For the system given in Example 7.4.8 we choose

$$P = \begin{bmatrix} 1 & 0 \\ 0 & 0.5 \end{bmatrix}$$

and we compute the matrix

$$C = A^T P + PA = \begin{bmatrix} -2 & -0.5 \\ -0.5 & -2 \end{bmatrix}.$$

According to Proposition 7.4.1(b), the equilibrium $x_e = 0$ of this system is *exponentially stable in the large* (as expected in Example 7.4.8). □

Example 7.4.11 For the system given in Example 7.4.7 we choose

$$P = \begin{bmatrix} -0.28 & -0.96 \\ -0.96 & 0.28 \end{bmatrix}$$

and we compute

$$C = A^T P + PA = \begin{bmatrix} -20 & 0 \\ 0 & -20 \end{bmatrix}.$$

The eigenvalues of P are ± 1. According to Proposition 7.4.1(c), the equilibrium $x_e = 0$ of this system is *unstable* (as expected from Example 7.4.7). □

In applying the results given in Proposition 7.4.1, we start by guessing a matrix P that has certain desired properties. Next, we solve the Lyapunov matrix equation for C, using (7.4.12). If C possesses certain desired properties (it is negative definite), we draw appropriate conclusions, using Proposition 7.4.1; if this is not possible (i.e., Proposition 7.4.1 does not apply), we need to choose another matrix P. This points to the principal shortcomings of Lyapunov's Direct Method, when applied to general systems. However, in the case of the *special case* of linear system (L), it is possible to *construct Lyapunov functions* of the form $v(x) = x^T Px$ in a *systematic manner*. In the process of doing so, one first chooses the matrix C in (7.4.12), having desired properties, and then one solves (7.4.12) for P. Conclusions are then drawn by applying the appropriate results given in Proposition 7.4.1. In applying this construction procedure, *we need to know the conditions under which (7.4.12) possesses a unique solution P for a given C*. We address this topic next.

Once more, we consider the quadratic form

$$v(x) = x^T Px, \qquad P = P^T \tag{7.4.13}$$

and the time derivative of v along the solutions of (L), given by

$$v'_{(L)}(x) = x^T Cx, \qquad C = C^T \tag{7.4.14}$$

where

$$C = A^T P + PA, \tag{7.4.15}$$

where all symbols are defined as before. Our objective is to determine the as yet unknown matrix P in such a way that $v'_{(L)}$ becomes a preassigned negative definite quadratic form, that is, in such a way that C is a preassigned negative definite matrix.

We first note that (7.4.15) constitutes a system of $n(n+1)/2$ linear equations, because P is symmetric. We need to determine under what conditions we can solve for the $n(n+1)/2$ elements, p_{ik}, given the matrices A and C. To make things tractable, we choose a similarity transformation Q such that

$$QAQ^{-1} = \overline{A}, \tag{7.4.16}$$

or equivalently,

$$A = Q^{-1}\overline{A}Q, \tag{7.4.17}$$

where $\overline{A}$ is similar to A and Q is a real $n \times n$ nonsingular matrix. From (7.4.17) and (7.4.15) we obtain

$$(\overline{A})^T(Q^{-1})^T PQ^{-1} + (Q^{-1})^T PQ^{-1}\overline{A} = (Q^{-1})^T CQ^{-1} \tag{7.4.18}$$

or

$$(\overline{A})^T\overline{P} + \overline{P}\,\overline{A} = \overline{C} \tag{7.4.19}$$

where

$$\overline{P} = (Q^{-1})^T PQ^{-1}, \qquad \overline{C} = (Q^{-1})^T CQ^{-1}. \tag{7.4.20}$$

In (7.4.19) and (7.4.20), P and C are subjected to a congruence transformation and $\overline{P}$ and $\overline{C}$ have the same definiteness properties as P and C, respectively. Because every real $n \times n$ matrix can be *triangularized*, we can choose Q in such a manner that $\overline{A} = [\overline{a}_{ij}]$ is *triangular*; that is, $\overline{a}_{ij} = 0$ for $i > j$. Note that in this case the eigenvalues of A, $\lambda_1, \dots, \lambda_n$, appear in the main diagonal of $\overline{A}$. To simplify our subsequent notation, we rewrite (7.4.19), (7.4.20) in the form of (7.4.15) by dropping the bars, that is,

$$A^T P + PA = C, \qquad C = C^T \tag{7.4.21}$$

and *we assume without loss of generality that* $A = [a_{ij}]$ *has been triangularized*; that is, $a_{ij} = 0$ for $i > j$. Because the eigenvalues $\lambda_1, \dots, \lambda_n$ appear in the diagonal of A, we can rewrite (7.4.21) as

$$\begin{aligned} 2\lambda_1 p_{11} &= c_{11} \\ a_{12}p_{11} + (\lambda_1 + \lambda_2)p_{12} &= c_{12} \\ &\vdots \end{aligned} \tag{7.4.22}$$

Note that λ_1 may be a complex number, in which case c_{11} will also be complex. Because this system of equations is triangular, and because its determinant is equal to

$$2^n \lambda_1 \cdots \lambda_n \prod_{i<j}(\lambda_i + \lambda_j), \tag{7.4.23}$$

the matrix P can be determined uniquely if and only this determinant is not zero. This is true when all eigenvalues of A are nonzero and no two of them are such that $\lambda_i + \lambda_j = 0$. This condition is not affected by a similarity transformation and is therefore also valid for the original system of equations (7.4.15).

We summarize the above discussion as follows.

Lemma 7.4.1 Let $A \in \mathbb{R}^{n\times n}$ and let $\lambda_1, \dots, \lambda_n$ denote the (not necessarily distinct) eigenvalues of A. Then (7.4.21) has a unique solution for P corresponding to each $C \in \mathbb{R}^{n\times n}$ if and only if

$$\lambda_i \neq 0 \quad \text{for all } i = 1, \dots, n \quad \text{and} \quad \lambda_i + \lambda_j \neq 0 \quad \text{for all } i, j = 1, \dots, n. \tag{7.4.24}$$

□

In order to construct $v(x)$, we must still check the definiteness of P. This can be done in a purely algebraic way; however, it is much easier to invoke the stability results of the present section and argue as follows.

(a) If all the eigenvalues λ_i of A have negative real parts, then the equilibrium $x_e = 0$ of (L) is exponentially stable in the large and if C in (7.4.15) is negative definite, then P must be positive definite. To prove this, we note that if P is not positive definite, then for $\delta > 0$ and sufficiently small, $(P - \delta I)$ has at least one negative eigenvalue and the function $v(x) = x^T(P - \delta I)x$ has a negative definite derivative; that is,

$$v'_{(L)}(x) = x^T[C - \delta(A + A^T)]x < 0$$

for all $x \neq 0$. By Theorem 6.2.8 (resp., Proposition 7.4.1(c)), the equilibrium $x_e = 0$ of (L) is unstable. We have arrived at a contradiction. Therefore, P must be positive definite.

(b) If A has eigenvalues with positive real parts and no eigenvalues with zero real parts we can use a similarity transformation $x = Qy$ in such a way that $Q^{-1}AQ$ is a block diagonal matrix of the form diag$[A_1, A_2]$, where all the eigenvalues of A_1 have positive real parts and all eigenvalues of A_2 have negative real parts. (If A does not have any eigenvalues with negative real parts, then we take $A = A_1$). By the result established in (a), noting that all eigenvalues of $-A_1$ have negative real parts, given any negative definite matrices C_1 and C_2, there exist positive definite matrices P_1 and P_2 such that

$$(-A_1^T)P_1 + P_1(-A_1) = C_1, \qquad A_2^T P_2 + P_2 A_2 = C_2.$$

Then $w(y) = y^T Py$, with $P = \text{diag}[-P_1, P_2]$ is a Lyapunov function for the system $\dot{y} = Q^{-1}AQy$ (and hence, for the system $\dot{x} = Ax$) that satisfies the hypotheses of Theorem 6.2.8 (resp., Proposition 7.4.1(c)). Therefore, the equilibrium $x_e = 0$ of system (L) is unstable. If A does not have any eigenvalues with negative real parts, then the equilibrium $x_e = 0$ of (L) is completely unstable.

In the above proof, we did not invoke Lemma 7.4.1. We note, however, that if additionally, (7.4.24) is true, then we can construct the Lyapunov function for (L) in a systematic manner.

Summarizing the above discussion, we now can state the main result of this subsection.

Theorem 7.4.7 Assume that the matrix A (for system (L)) has no eigenvalues with real part equal to zero. If all the eigenvalues of A have negative real parts, or if at least one of the eigenvalues of A has a positive real part, then there exists a quadratic Lyapunov function

$$v(x) = x^T Px, \qquad P = P^T$$

whose derivative along the solutions of (L) is definite (i.e., either negative definite or positive definite). □

This result shows that *when A is a stable matrix (i.e., all the eigenvalues of A have negative real parts), then for system (L), the conditions of Theorem 6.2.3 are* *also necessary conditions for asymptotic stability.* Moreover, in the case *when the matrix A has at least one eigenvalue with positive real part and no eigenvalues on the imaginary axis, then the conditions of Theorem 6.2.8 are also necessary conditions for instability.*

Example 7.4.12 We consider the system (L) with

$$A = \begin{bmatrix} 0 & 1 \\ -1 & 0 \end{bmatrix}.$$

The eigenvalues of A are $\lambda_1, \lambda_2 = \pm i$ $(i = \sqrt{-1})$ and therefore, condition (7.4.24) is violated. According to Lemma 7.4.1, the Lyapunov matrix equation

$$A^T P + PA = C$$

does not possess a unique solution for a given C. We demonstrate this for two specific cases:

(i) If $C = 0$, we obtain

$$\begin{bmatrix} 0 & -1 \\ 1 & 0 \end{bmatrix}\begin{bmatrix} p_{11} & p_{12} \\ p_{12} & p_{22} \end{bmatrix} + \begin{bmatrix} p_{11} & p_{12} \\ p_{12} & p_{22} \end{bmatrix}\begin{bmatrix} 0 & 1 \\ -1 & 0 \end{bmatrix} = \begin{bmatrix} -2p_{12} & p_{11} - p_{22} \\ p_{11} - p_{22} & 2p_{12} \end{bmatrix} = \begin{bmatrix} 0 & 0 \\ 0 & 0 \end{bmatrix},$$

or $p_{12} = 0$ and $p_{11} = p_{22}$. Therefore, for any $c \in \mathbb{R}$, the matrix $P = cI$ is a solution of the Lyapunov matrix equation. Thus, for $C = 0$, the Lyapunov matrix equation has in this case denumerably many solutions.

(ii) If $C = -2I$, we have

$$\begin{bmatrix} -2p_{12} & p_{11} - p_{22} \\ p_{11} - p_{22} & 2p_{12} \end{bmatrix} = \begin{bmatrix} -2 & 0 \\ 0 & -2 \end{bmatrix},$$

or $p_{11} = p_{22}$ and $p_{12} = 1$ and $p_{12} = -1$, which is impossible. Therefore, for $C = -2I$, the Lyapunov matrix equation has in this example no solution at all. □

We conclude the present section with a result which shows that when all the eigenvalues of matrix A for system (L) have negative real parts, then the matrix P in (7.4.15) can be computed explicitly.

Theorem 7.4.8 If all the eigenvalues of a real $n \times n$ matrix A have negative real parts, then for each matrix $C \in \mathbb{R}^{n\times n}$, the unique solution of (7.4.15) is given by

$$P = \int_0^\infty e^{A^T s}(-C)e^{As}\,ds. \tag{7.4.25}$$

Proof. If all eigenvalues of A have negative real parts, then (7.4.24) is satisfied and therefore (7.4.15) has a unique solution for every $C \in \mathbb{R}^{n\times n}$. To verify that (7.4.25) is indeed this solution, we first note that the right-hand side of (7.4.25) is well defined, because all eigenvalues of A have negative real parts. Substituting the right-hand side of (7.4.25) for P into (7.4.15), we obtain

$$\begin{aligned} A^T P + PA &= \int_0^\infty A^T e^{A^T t}(-C)e^{At}dt + \int_0^\infty e^{A^T t}(-C)e^{At}Adt \\ &= \int_0^\infty \frac{d}{dt}\left(e^{A^T t}(-C)e^{At}\right)dt \\ &= e^{A^T t}(-C)e^{At}\Big|_0^\infty \\ &= C, \end{aligned}$$

which proves the theorem. □

D. Periodic systems

We now briefly consider linear periodic systems given by

$$\dot{x} = A(t)x \qquad (LP)$$

where $A \in C[\mathbb{R}, \mathbb{R}^{n\times n}]$ and $A(t) = A(t+T)$ for all $t \in \mathbb{R}$, where $T > 0$ denotes the *period* for (LP). Making reference to the appendix section (Subsection 7.8B), we recall that if $\Phi(t, t_0)$ is the state transition matrix for (LP), then there exists a constant matrix $R \in \mathbb{R}^{n\times n}$ and a nonsingular $n \times n$ matrix $\Psi(t, t_0)$ such that

$$\Phi(t, t_0) = \Psi(t, t_0)e^{R(t-t_0)}, \qquad (7.4.26)$$

where

$$\Psi(t, t_0) = \Psi(t+T, t_0)$$

for all $t \in \mathbb{R}$. In Section 7.8 it is shown that the change of variables given by

$$x = \Psi(t, t_0)y$$

transforms system (LP) into the system

$$\dot{y} = Ry, \qquad (7.4.27)$$

where R is given in (7.4.26). Moreover, because $\Psi(t, t_0)^{-1}$ exists over $t_0 \le t \le t+T$, the equilibrium $x_e = 0$ is uniformly stable (resp., uniformly asymptotically stable) if and only if $y_e = 0$ is also uniformly stable (resp., uniformly asymptotically stable). Applying Theorem 7.4.5 to system (7.4.27), we obtain the following results.

Theorem 7.4.9 The equilibrium $x_e = 0$ of (LP) is *uniformly stable* if and only if all eigenvalues of the matrix R (given in (7.4.26)) have nonpositive real parts, and every eigenvalue with a zero real part has an associated Jordan block of order one. The equilibrium $x_e = 0$ of (LP) is *uniformly asymptotically stable in the large* if and only if all the eigenvalues of R have negative real parts. □

E. Second-order systems

At this point it might be appropriate to investigate the qualitative behavior of the solutions of second-order linear autonomous homogeneous systems in the vicinity of the equilibrium $x_e = 0$. In the process of doing this, we establish a classification of equilibrium points for second-order systems. Knowledge of the qualitative behavior of the solutions of second-order linear systems frequently provides motivation and guidelines for the study of higher-dimensional and more complex systems.

We consider systems given by

$$\begin{cases} \dot{x}_1 = a_{11}x_1 + a_{12}x_2 \\ \dot{x}_2 = a_{21}x_1 + a_{22}x_2 \end{cases} \tag{7.4.28}$$

that can be expressed by

$$\dot{x} = Ax, \tag{7.4.29}$$

where

$$A = \begin{bmatrix} a_{11} & a_{12} \\ a_{21} & a_{22} \end{bmatrix}. \tag{7.4.30}$$

When $\det A \neq 0$, system (7.4.28) has only one equilibrium point, $x_e = 0$. We classify this equilibrium point (resp., system (7.4.28)) according to the following properties of the eigenvalues λ_1, λ_2 of A.

(a) If λ_1, λ_2 are real and negative, then $x_e = 0$ is called a *stable node*.

(b) If λ_1, λ_2 are real and positive, then $x_e = 0$ is called an *unstable node*.

(c) If λ_1, λ_2 are real and if $\lambda_1\lambda_2 < 0$, then $x_e = 0$ is called a *saddle*.

(d) If λ_1, λ_2 are complex conjugates and $\text{Re}\lambda_1 = \text{Re}\lambda_2 < 0$, then $x_e = 0$ is called a *stable focus*.

(e) If λ_1, λ_2 are complex conjugates and $\text{Re}\lambda_1 = \text{Re}\lambda_2 > 0$, then $x_e = 0$ is called an *unstable focus*.

(f) If λ_1, λ_2 are complex conjugates and $\text{Re}\lambda_1 = \text{Re}\lambda_2 = 0$, then $x_e = 0$ is called a *center*.

In accordance with the results of the present section, stable nodes and stable foci are exponentially stable equilibrium points; centers are stable equilibrium points; and saddles, unstable nodes, and unstable foci are unstable equilibrium points.

To simplify our subsequent discussion, we transform system (7.4.29) into special forms, depending on the situation on hand. To this end, we let

$$y = P^{-1}x \tag{7.4.31}$$

where $P \in \mathbb{R}^{2\times 2}$ is nonsingular. Under this similarity transformation, (7.4.29) assumes the form

$$\dot{y} = \Lambda y \tag{7.4.32}$$

where

$$\Lambda = P^{-1}AP. \tag{7.4.33}$$

Corresponding to an initial condition $x(0) = x_0$ for (7.4.29) we have the initial condition

$$y(0) = y_0 = P^{-1}x_0 \tag{7.4.34}$$

for system (7.4.32).

In the following, we assume without loss of generality that when λ_1, λ_2 are real and not equal, then $\lambda_1 > \lambda_2$.

We first assume that λ_1 *and* λ_2 *are real and that* A *can be diagonalized*, so that

$$\Lambda = \begin{bmatrix} \lambda_1 & 0 \\ 0 & \lambda_2 \end{bmatrix}, \tag{7.4.35}$$

where λ_1, λ_2 are not necessarily distinct. Then (7.4.32) assumes the form

$$\begin{cases} \dot{y}_1 = \lambda_1 y_1 \\ \dot{y}_2 = \lambda_2 y_2. \end{cases} \tag{7.4.36}$$

For a given set of initial conditions $(y_{10}, y_{20})^T = (y_1(0), y_2(0))^T$, the solution of (7.4.36) is given by

$$\begin{cases} \varphi_1(t, 0, y_{10}) \triangleq y_1(t) = e^{\lambda_1 t} y_{10} \\ \varphi_2(t, 0, y_{20}) \triangleq y_2(t) = e^{\lambda_2 t} y_{20}. \end{cases} \tag{7.4.37}$$

Eliminating t, we can express (7.4.37) equivalently as

$$y_2(t) = y_{20}\left[y_1(t)/y_{10}\right]^{\lambda_2/\lambda_1}. \tag{7.4.38}$$

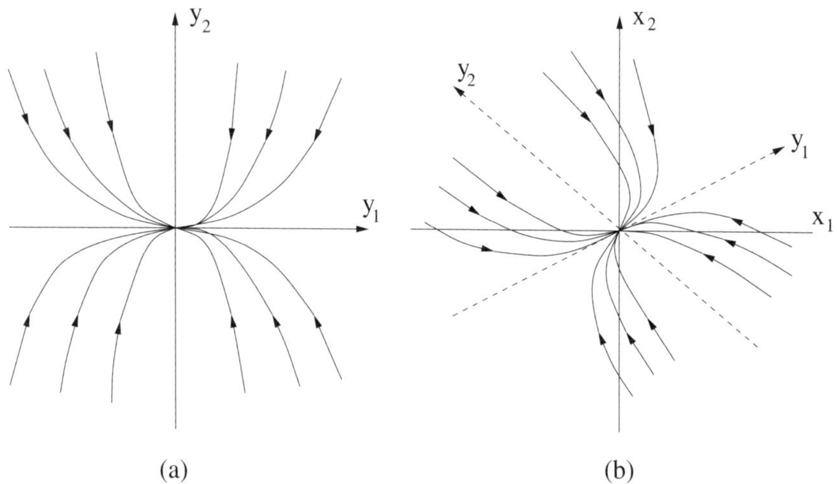

Figure 7.4.1: Trajectories near a stable node.

Using either (7.4.37) or (7.4.38), we now can sketch families of trajectories in the y_1–y_2 plane for a *stable node* (Figure 7.4.1(a)), for an *unstable node* (Figure 7.4.2(a)), and for a *saddle* (Figure 7.4.3(a)). Using (7.4.31) and (7.4.37) or (7.4.38), we can sketch corresponding families of trajectories in the $x_1 - x_2$ plane. In all figures, the arrows signify increasing time t. Note that in all cases, the qualitative properties of the trajectories have been preserved under the similarity transformation (7.4.31) (refer to Figures 7.4.1(b), 7.4.2(b), and 7.4.3(b)).

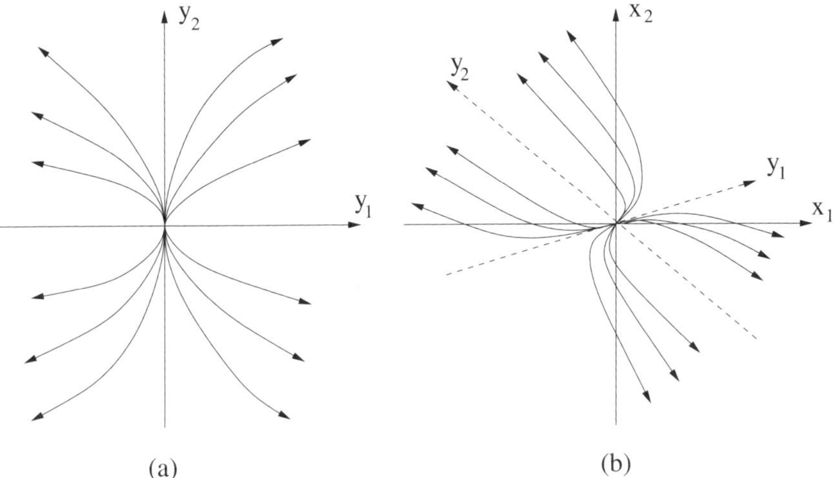

Figure 7.4.2: Trajectories near an unstable node.

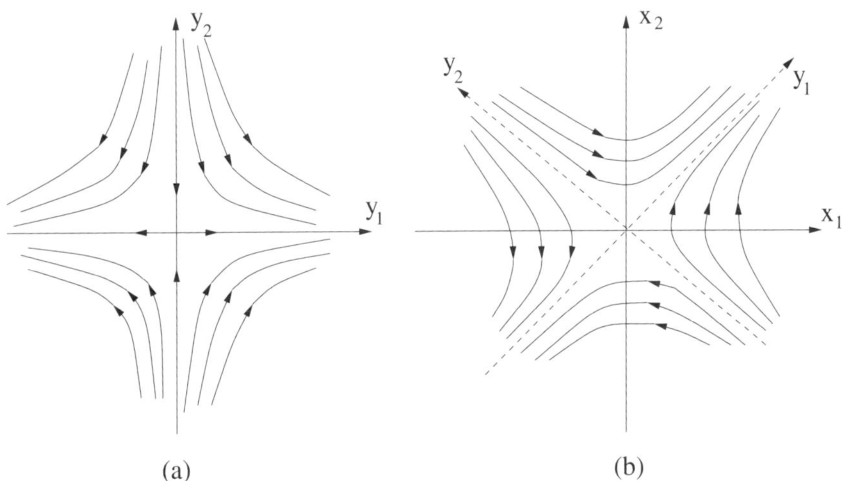

Figure 7.4.3: Trajectories near a saddle.

Next, we assume that *matrix A has two real repeated eigenvalues*, $\lambda_1 = \lambda_2 = \lambda$, and that Λ is in Jordan canonical form,

$$\Lambda = \begin{bmatrix} \lambda & 1 \\ 0 & \lambda \end{bmatrix}.$$

In this case (7.4.32) assumes the form

$$\begin{cases} \dot{y}_1 = \lambda y_1 + y_2 \\ \dot{y}_2 = \lambda y_2. \end{cases} \tag{7.4.39}$$

For an initial point, we obtain for (7.4.39) the solution

$$\begin{cases} \varphi_1(t,0,y_{10},y_{20}) = y_1(t) = e^{\lambda t} y_{10} + t e^{\lambda t} y_{20} \\ \varphi_2(t,0,y_{20}) = e^{\lambda t} y_{20}. \end{cases} \tag{7.4.40}$$

Eliminating the parameter t, we can plot trajectories in the $y_1 - y_2$ plane (resp., in the $x_1 - x_2$ plane) for different sets of initial data near the origin. In Figure 7.4.4, we depict typical trajectories near a stable node ($\lambda < 0$) for repeated eigenvalues.

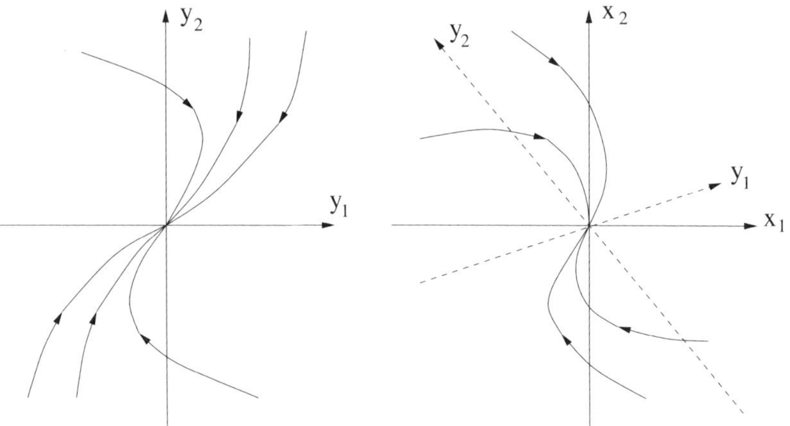

Figure 7.4.4: Trajectories near a stable node (repeated eigenvalues).

Next, we consider the case when *matrix A has two complex conjugate eigenvalues*,

$$\lambda_1 = \delta + i\omega, \qquad \lambda_2 = \delta - i\omega.$$

In this case there exists a similarity transformation P such that the matrix $\Lambda = P^{-1}AP$ assumes the form

$$\Lambda = \begin{bmatrix} \delta & \omega \\ -\omega & \delta \end{bmatrix} \tag{7.4.41}$$

and therefore

$$\begin{cases} \dot{y}_1 = \delta y_1 + \omega y_2 \\ \dot{y}_2 = -\omega y_1 + \delta y_2. \end{cases} \tag{7.4.42}$$

The solution for the case $\delta > 0$, for a set of initial data $(y_{10}, y_{20})^T$, is given by

$$\begin{cases} \varphi_1(t,0,y_{10},y_{20}) = y_1(t) = e^{\delta t}(y_{10}\cos\omega t + y_{20}\sin\omega t) \\ \varphi_2(t,0,y_{10},y_{20}) = y_2(t) = e^{\delta t}(-y_{10}\sin\omega t + y_{20}\cos\omega t). \end{cases} \tag{7.4.43}$$

Letting $\rho = (y_{10}^2 + y_{20}^2)^{1/2}$, $\cos\alpha = y_{10}/\rho$, and $\sin\alpha = y_{20}/\rho$, we can rewrite (7.4.43) as

$$\begin{cases} \varphi_1(t, 0, y_{10}, y_{20}) = y_1(t) = e^{\delta t}\rho\cos(\omega t - \alpha) \\ \varphi_2(t, 0, y_{10}, y_{20}) = y_2(t) = -e^{\delta t}\rho\sin(\omega t - \alpha). \end{cases} \tag{7.4.44}$$

Letting r and θ denote the polar coordinates, $y_1 = r\cos\theta$ and $y_2 = r\sin\theta$, we may rewrite the solution (7.4.44) as

$$r(t) = \rho e^{\delta t}, \qquad \theta(t) = -(\omega t - \alpha). \tag{7.4.45}$$

Eliminating the parameter t, we obtain

$$r = c e^{-(\delta/\omega)\theta}, \qquad c = \rho e^{(\delta/\omega)\alpha}. \tag{7.4.46}$$

In the present case, the origin is an *unstable focus*. For different initial conditions, (7.4.45) and (7.4.46) yield a family of trajectories in the form of spirals tending away from the origin with increasing t, as shown in Figure 7.4.5 (for $\omega > 0$).

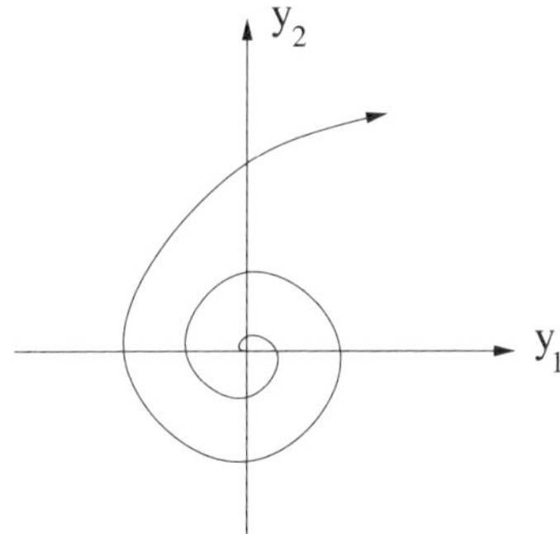

Figure 7.4.5: Trajectories near an unstable focus.

When $\delta < 0$, we obtain in a similar manner, for different initial conditions near the origin, a family of trajectories as shown in Figure 7.4.6 (for $\omega > 0$). In the present case, the origin is a *stable focus* and the trajectories are in the form of spirals that tend towards the origin with increasing t.

Finally, if $\delta = 0$, the origin is a *center* and the preceding expressions ((7.4.45) and (7.4.46)) yield for different initial conditions near the origin, a family of concentric circles of radius ρ, as shown in Figure 7.4.7 (for $\omega > 0$).

7.5 Linear Discrete-Time Systems

In the present section we study the stability properties of the equilibrium of linear homogeneous systems

$$x(k+1) = A(k)x(k), \qquad k \geq k_0 \geq 0 \tag{LH_D}$$

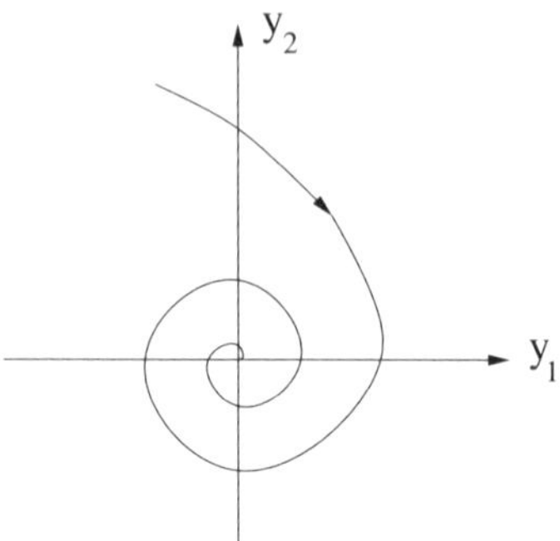

Figure 7.4.6: Trajectories near a stable focus.

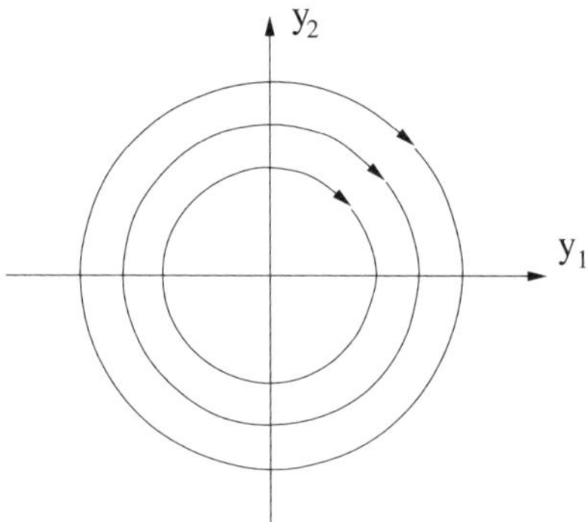

Figure 7.4.7: Trajectories near a center.

$k, k_0 \in \mathbb{N}$, and linear autonomous homogeneous systems

$$x(k+1) = Ax(k), \qquad k \geq 0. \tag{L_D}$$

In (LH_D), $k \in \mathbb{N}$, $x \in \mathbb{R}^n$, and $A\colon \mathbb{N} \to \mathbb{R}^{n\times n}$. In (L_D), $k \in \mathbb{N}$, $x \in \mathbb{R}^n$, and $A \in \mathbb{R}^{n\times n}$. Recall that $x_e = 0$ is always an equilibrium of (L_D) and (LH_D) and that $x_e = 0$ is the only equilibrium of (LH_D) if $A(k)$ is nonsingular for $k \in \mathbb{N}$. The solutions of (LH_D) are of the form

$$\varphi(k, k_0, x_0) = \Phi(k, k_0)x_0, \qquad k \geq k_0,$$

where Φ denotes the state transition matrix of $A(k)$ and $\varphi(k_0, k_0, x_0) = x_0$. Also, the solutions of (L_D) are of the form

$$\begin{aligned}\varphi(k, k_0, x_0) &= \Phi(k, k_0)x_0\\ &= \Phi(k - k_0, 0)x_0\\ &\triangleq \Phi(k - k_0)x_0\end{aligned}$$

where in the preceding equation, a slight abuse in notation has been used.

For some of the properties of the transition matrix $\Phi(k, k_0)$ that are required in the proofs of the results that follow, the reader should refer to the appendix (Subsection 7.8D).

A. Linear homogeneous systems

In the next four results, we provide some of the basic qualitative properties of (LH_D).

Theorem 7.5.1 The equilibrium $x_e = 0$ of (LH_D) is *stable* if and only if the solutions of (LH_D) are bounded, or equivalently, if and only if

$$\sup_{k \geq k_0} \|\Phi(k, k_0)\| \triangleq c(k_0) < \infty,$$

where $\|\Phi(k, k_0)\|$ denotes the matrix norm induced by the vector norm used on $\mathbb{R}^n$ and $c(k_0)$ denotes a constant that may depend on the choice of k_0. □

The proof of the above result is similar to the proof of Theorem 7.4.1 and is left as an exercise for the reader.

Theorem 7.5.2 The equilibrium $x_e = 0$ of (LH_D) is *uniformly stable* if and only if the solutions of (LH_D) are uniformly bounded, or equivalently, if and only if

$$\sup_{k_0 \geq 0} c(k_0) \triangleq \sup_{k_0 \geq 0} \left(\sup_{k \geq k_0} \|\Phi(k, k_0)\| \right) \triangleq c_0 < \infty. \qquad \square$$

The proof of the above result is similar to the proof of Theorem 7.4.2 and is left as an exercise.

Theorem 7.5.3 The following statements are equivalent.

(i) The equilibrium $x_e = 0$ of (LH_D) is asymptotically stable.

(ii) The equilibrium $x_e = 0$ of (LH_D) is asymptotically stable in the large.

(iii) $\lim_{k \to \infty} \|\Phi(k, k_0)\| = 0$. □

The proof of the above result is similar to the proof of Theorem 7.4.3 and is left as an exercise.

Theorem 7.5.4 The equilibrium $x_e = 0$ of (LH_D) is uniformly asymptotically stable if and only if it is *exponentially stable*. □

The proof of the above result is similar to the proof of Theorem 7.4.4 and is left as an exercise.

B. Linear autonomous homogeneous systems

Revisiting Example 3.1.9, we now address the stability properties of system (L_D),

$$x(k+1) = Ax(k), \qquad k \geq 0. \tag{L_D}$$

To this end we transform the matrix A into the Jordan canonical form, $J = P^{-1}AP$, using the transformation $x = Py$ to obtain the equivalent system

$$y(k+1) = P^{-1}APy(k) = Jy(k). \tag{7.5.1}$$

Because the equilibrium $x_e = 0$ of (L_D) possesses the same stability properties as the equilibrium $y_e = 0$ of (7.5.1), we may assume without loss of generality that the matrix A in (L_D) is in Jordan canonical form. We may now use the same reasoning as was done in Subsection 7.4B (for system (L)), to arrive at the following result.

Theorem 7.5.5 The equilibrium $x_e = 0$ of system (L_D) is *stable*, in fact, *uniformly stable*, if and only if all eigenvalues of A are within or on the unit circle of the complex plane, and every eigenvalue that is on the unit circle has an associated Jordan block of order one. The equilibrium $x_e = 0$ of (L_D) is *uniformly asymptotically stable in the large*, in fact, *exponentially stable in the large*, if and only if all eigenvalues of A are within the unit circle of the complex plane. □

The proof of this result proceeds along similar lines as the proof of Theorem 7.4.5 and is left to the reader as an exercise. As a consequence of Theorem 7.5.5, we have the following result.

Theorem 7.5.6 The equilibrium $x_e = 0$ of (L_D) is *unstable* if and only if at least one of the eigenvalues of A is outside of the unit circle in the complex plane or at least one of the eigenvalues of A is on the unit circle in the complex plane and is associated with a Jordan block of order greater than one. □

The proof of the above result is similar to the proof of Theorem 7.4.6 and is left as an exercise.

As in the case of linear system (L), it may be appropriate to take note of certain conventions concerning matrices for system (L_D) that are used in the literature. Again, these are not consistent with the terminology used in the preceding results. Thus, a real $n \times n$ matrix A is called *Schur stable* or just *stable*, if all of its eigenvalues are within the unit circle of the complex plane. If at least one of the eigenvalues of A is outside the unit circle, then A is called *unstable*. A matrix A that is neither stable nor unstable is called *critical*, and the eigenvalues on the unit circle in the complex plane are called *critical eigenvalues*.

Example 7.5.1 For system (L_D), we let

$$A = \begin{bmatrix} 0 & 1 \\ -1 & 0 \end{bmatrix}.$$

The eigenvalues of A are $\lambda_1, \lambda_2 = \pm\sqrt{-1}$. According to Theorem 7.5.5, the equilibrium $x_e = 0$ of the system is *stable*, and according to Theorems 7.5.1 and 7.5.2, the matrix A^k is bounded (resp., uniformly bounded) for all $k \geq 0$. □

Example 7.5.2 For system (L_D), we let

$$A = \begin{bmatrix} 0 & -1/2 \\ -1 & 0 \end{bmatrix}.$$

The eigenvalues of A are $\lambda_1, \lambda_2 = \pm 1/\sqrt{2}$. According to Theorem 7.5.5, the equilibrium $x_e = 0$ of the system is *asymptotically stable* and according to Theorem 7.5.3, $\lim_{k\to\infty} A^k = 0$. □

Example 7.5.3 For system (L_D), we let

$$A = \begin{bmatrix} 0 & -1/2 \\ -3 & 0 \end{bmatrix}.$$

The eigenvalues of A are $\lambda_1, \lambda_2 = \pm\sqrt{3/2}$. According to Theorem 7.5.6, the equilibrium $x_e = 0$ of system (L_D) is *unstable*, and according to Theorems 7.5.1 and 7.5.2, the matrix A^k is not bounded with increasing k. □

Example 7.5.4 For system (L_D), we let

$$A = \begin{bmatrix} 1 & 1 \\ 0 & 1 \end{bmatrix}.$$

The matrix A is a Jordan block of order 2 for the eigenvalue $\lambda = 1$. According to Theorem 7.5.6, the equilibrium $x_e = 0$ of the system is *unstable*. □

C. The Lyapunov matrix equation

In the present section we employ quadratic forms

$$v(x) = x^T Bx, \qquad B = B^T \tag{7.5.2}$$

to establish stability criteria for linear systems

$$x(k+1) = Ax(k). \tag{L_D}$$

Evaluating v along the solutions of system (L_D), we obtain the first forward difference of v as

$$\begin{aligned}\Delta_{L_D} v(x(k)) &= v(x(k+1)) - v(x(k)) \\ &= x(k+1)^T Bx(k+1) - x(k)^T Bx(k) \\ &= x(k)^T A^T BAx(k) - x(k)^T Bx(k) \\ &= x(k)^T (A^T BA - B)x(k),\end{aligned}$$

and therefore

$$\Delta_{L_D} v(x) = x^T (A^T BA - B)x \triangleq x^T Cx \tag{7.5.3}$$

where

$$A^T BA - B = C, \qquad C^T = C. \tag{7.5.4}$$

Equation (7.5.4) is called the *Lyapunov Matrix Equation* for system (L_D).

Invoking the Lyapunov stability results of Section 6.3, the following results follow readily.

Proposition 7.5.1 (a) The equilibrium $x_e = 0$ of system (L_D) is *stable* if there exists a real, symmetric, and positive definite matrix B such that the matrix C given in (7.5.4) is negative semidefinite.

(b) The equilibrium $x_e = 0$ of system (L_D) is *asymptotically stable in the large*, in fact, *exponentially stable in the large*, if there exists a real, symmetric, and positive definite matrix B such that the matrix C given in (7.5.4) is negative definite.

(c) The equilibrium $x_e = 0$ of system (L_D) is *unstable* if there exists a real, symmetric matrix B that is either negative definite or indefinite such that the matrix C given in (7.5.4) is negative definite. □

We leave the proofs of the above results as an exercise for the reader.

In applying Proposition 7.5.1, we start by guessing a matrix B having certain properties and we then solve for the matrix C in (7.5.4). If C possesses desired properties, we can apply Proposition 7.5.1 to draw appropriate conclusions; if not, we need to choose another matrix B. This is not a very satisfactory approach, and in the following, we derive results that, similarly as in the case of linear continuous-time systems (L), enable us to *construct* Lyapunov functions of the form $v(x) = x^T Bx$ in a systematic manner. In this approach we first choose a matrix C in (7.5.4) which is either negative definite or positive definite, then we solve (7.5.4) for B, and then we draw appropriate conclusions by invoking existing Lyapunov results (e.g., Proposition 7.5.1). In applying this approach of constructing Lyapunov functions, we need to know under what conditions equation (7.5.4) possesses a unique solution B for *any* definite (i.e., positive definite or negative definite) matrix C. We address this issue next.

We first show that if all the eigenvalues of A are within the unit circle of the complex plane (i.e., A is stable), then we can compute B in (7.5.4) *explicitly*. To see this, assume that in (7.5.4) C is given and that A is stable. Then

$$(A^T)^{k+1} BA^{k+1} - (A^T)^k BA^k = (A^T)^k CA^k,$$

and summing from $k = 0$ to l yields

$$A^T BA - B + (A^T)^2 BA^2 - A^T BA + \cdots + (A^T)^{l+1} BA^{l+1} - (A^T)^l BA^l = \sum_{k=0}^{l} (A^T)^k CA^k$$

or

$$(A^T)^{l+1}BA^{l+1} - B = \sum_{k=0}^{l}(A^T)^k CA^k.$$

Letting $l \to \infty$, we have

$$B = -\sum_{k=0}^{\infty}(A^T)^k CA^k. \tag{7.5.5}$$

It is easily verified that (7.5.5) is a solution of (7.5.4). Indeed, we have

$$-A^T\left[\sum_{k=0}^{\infty}(A^T)^k CA^k\right]A + \sum_{k=0}^{\infty}(A^T)^k CA^k = C$$

or

$$-A^T CA + C - (A^T)^2CA^2 + A^TCA - (A^T)^3CA^3 + (A^T)^2CA^2 - \cdots = C.$$

Furthermore, if C is negative definite, then B is positive definite.

Combining the above discussion with Proposition 7.5.1(b), we have proved the following result.

Theorem 7.5.7 If there is a positive definite and symmetric matrix B and a negative definite and symmetric matrix C satisfying the Lyapunov matrix equation (7.5.4), then the matrix A is stable. Conversely, if A is stable, then given *any* symmetric matrix C, equation (7.5.4) has a unique solution B, and if C is negative definite, then B is positive definite. □

Next, we determine conditions under which the Lyapunov matrix equation (7.5.4) has a unique solution $B = B^T \in \mathbb{R}^{n\times n}$ for a given matrix $C = C^T \in \mathbb{R}^{n\times n}$. In doing so, we consider the more general system of equations

$$A_1XA_2 - X = C \tag{7.5.6}$$

where $A_1 \in \mathbb{R}^{m\times m}$, $A_2 \in \mathbb{R}^{n\times n}$, $X \in \mathbb{R}^{m\times n}$, and $C \in \mathbb{R}^{m\times n}$.

Lemma 7.5.1 Let $A_1 \in \mathbb{R}^{m\times m}$ and $A_2 \in \mathbb{R}^{n\times n}$. Then equation (7.5.6) has a unique solution $X \in \mathbb{R}^{m\times n}$ for a given $C \in \mathbb{R}^{m\times n}$ if and only if no eigenvalue of A_1 is a reciprocal of an eigenvalue of A_2.

Proof. We need to show that the condition on A_1 and A_2 is equivalent to the condition that $A_1XA_2 = X$ implies $X = 0$. Once we have proved that $A_1XA_2 = X$ has the unique solution $X = 0$, then it can be shown that (7.5.6) has a unique solution for every C, because (7.5.6) is a linear equation.

Assume first that the condition on A_1 and A_2 is satisfied. Now $A_1XA_2 = X$ implies that $A_1^{k-j}XA_2^{k-j} = X$ and

$$A_1^j = A_1^kXA_2^{k-j} \qquad \text{for } k \geq j \geq 0.$$

Now for the polynomial of degree k,

$$p(\lambda) = \sum_{j=0}^{k} a_j \lambda^j,$$

we define the polynomial of degree k,

$$p^*(\lambda) = \sum_{j=0}^{k} a_j \lambda^{k-j} = \lambda^k p\left(\frac{1}{\lambda}\right),$$

from which it follows that

$$p(A_1)X = A_1^k X p^*(A_2). \tag{7.5.7}$$

Now let $\varphi_i(\lambda)$ be the characteristic polynomial of A_i, $i = 1, 2$. It follows from the assumption that $\varphi_1(\lambda)$ and $\varphi_2^*(\lambda)$ do not have common roots. Thus, $\varphi_1(\lambda)$ and $\varphi_2^*(\lambda)$ are relatively prime, which in turn yields that there are polynomials $p(\lambda)$ and $q(\lambda)$ such that

$$p(\lambda)\varphi_1(\lambda) + q(\lambda)\varphi_2^*(\lambda) = 1.$$

Now define $\varphi(\lambda) = q(\lambda)\varphi_2^*(\lambda)$ and note that $\varphi^*(\lambda) = q^*(\lambda)\varphi_2(\lambda)$. It follows that $\varphi^*(A_2) = 0$ and $\varphi(A_1) = I$. Replacing $p(\lambda)$ in (7.5.7) by $\varphi(\lambda)$, we obtain

$$X = \varphi(A_1)X = A_1^k X \varphi^*(A_2) = 0.$$

From this it follows that $A_1 X A_2 = X$ implies $X = 0$.

To prove the converse, we assume that λ is an eigenvalue of A_1 and λ^{-1} is an eigenvalue of A_2 (and hence, is also an eigenvalue of A_2^T). Let $A_1 w = \lambda w$ and $A_2^T z = \lambda^{-1} z$, $0 \neq w \in \mathbb{R}^{n \times 1}$, and $0 \neq z \in \mathbb{R}^{m \times 1}$. Define $X = wz^T$. Then $X \neq 0$ and $A_1 X A_2 = X$. □

To construct the Lyapunov function $v(x)$, using Lemma 7.5.1, we must still check the definiteness of B, using the results of the present section (e.g., Proposition 7.5.1).

(a) If all the eigenvalues of A for system (L_D) are within the unit circle of the complex plane, then no reciprocal of an eigenvalue of A is an eigenvalue, and Lemma 7.5.1 gives another way of showing that equation (7.5.4) has a unique solution B for every C if A is stable. If C is negative definite, then B is positive definite. This can be shown similarly as was done for the corresponding case of linear ordinary differential equations (Subsection 7.4C).

(b) Suppose that at least one of the eigenvalues of A is outside the unit circle in the complex plane and that A has no eigenvalues on the unit circle. As in the case of linear differential equations (L) (Subsection 7.4C), we use a similarity transformation $x = Qy$ in such a way that $Q^{-1}AQ = \text{diag}[A_1, A_2]$, where all eigenvalues of A_1 are outside the unit circle and all eigenvalues of A_2 are within the unit circle. We then proceed identically as in the case of linear differential equations to show that under the present assumptions there exists for system (L_D) a Lyapunov function that satisfies the hypotheses of Proposition 7.5.1(c). Therefore, the equilibrium $x_e = 0$ of system (L_D) is unstable. If A does not have any eigenvalues within the unit circle, then the equilibrium $x_e = 0$ of (L) is completely unstable. In this proof, Lemma 7.5.1 has not been invoked. If additionally, the hypotheses of Lemma 7.5.1 are true (i.e., no reciprocal of an eigenvalue of A is an eigenvalue of A), then we can construct the Lyapunov function for (L_D) in a systematic manner.

Summarizing the above discussion, we have arrived at the following result.

Theorem 7.5.8 Assume that the matrix A (for system (L_D)) has no eigenvalues on the unit circle. If all the eigenvalues of A are within the unit circle of the complex plane, or if at least one eigenvalue is outside the unit circle of the complex plane, then there exists a Lyapunov function of the form $v(x) = x^T Bx$, $B = B^T$, whose first forward difference along the solutions of system (L_D) is definite (i.e., $\Delta_{L_D} v(x)$ is either negative definite or positive definite). □

Theorem 7.5.8 shows that *when all the eigenvalues of A are within the unit circle, then for system (L_D), the conditions of Theorem 6.3.2 are also necessary conditions for asymptotic stability*. Furthermore, *when at least one eigenvalue is outside the unit circle and no eigenvalues are on the unit circle, then the conditions of Theorem 6.3.4 are also necessary conditions for instability*.

We conclude the present section by considering several specific examples.

Example 7.5.5 For system (L_D), let

$$A = \begin{bmatrix} 0 & 1 \\ -1 & 0 \end{bmatrix}.$$

Choose $B = I$, which is positive definite. From (7.5.4) we have

$$C = A^T A - I = \begin{bmatrix} 0 & -1 \\ 1 & 0 \end{bmatrix} \begin{bmatrix} 0 & 1 \\ -1 & 0 \end{bmatrix} - \begin{bmatrix} 1 & 0 \\ 0 & 1 \end{bmatrix} = \begin{bmatrix} 0 & 0 \\ 0 & 0 \end{bmatrix}.$$

It follows from Proposition 7.5.1(a) that the equilibrium $x_e = 0$ of (L_D) is *stable*. This is the same conclusion as the one we arrived at in Example 7.5.1. □

Example 7.5.6 For system (L_D), let

$$A = \begin{bmatrix} 0 & -1/2 \\ -1 & 0 \end{bmatrix}.$$

We choose

$$B = \begin{bmatrix} 8/3 & 0 \\ 0 & 5/3 \end{bmatrix},$$

which is positive definite. From (7.5.4) we obtain

$$\begin{aligned} C &= A^T BA - B \\ &= \begin{bmatrix} 0 & -1 \\ -1/2 & 0 \end{bmatrix} \begin{bmatrix} 8/3 & 0 \\ 0 & 5/3 \end{bmatrix} \begin{bmatrix} 0 & -1/2 \\ -1 & 0 \end{bmatrix} - \begin{bmatrix} 8/3 & 0 \\ 0 & 5/3 \end{bmatrix} \\ &= \begin{bmatrix} -1 & 0 \\ 0 & -1 \end{bmatrix}, \end{aligned}$$

which is negative definite. It follows from Proposition 7.5.1(b) that the equilibrium $x_e = 0$ of system (L_D) is *asymptotically stable in the large*, in fact, *exponentially stable in the large*. This is the same conclusion that was drawn in Example 7.5.2. □

Example 7.5.7 For system (L_D), let

$$A = \begin{bmatrix} 0 & -1/2 \\ -3 & 0 \end{bmatrix}.$$

We choose

$$C = \begin{bmatrix} -1 & 0 \\ 0 & -1 \end{bmatrix}$$

which is negative definite. From (7.5.4) we obtain

$$C = A^T BA - B = \begin{bmatrix} 0 & -3 \\ -1/2 & 0 \end{bmatrix} \begin{bmatrix} b_{11} & b_{12} \\ b_{12} & b_{22} \end{bmatrix} \begin{bmatrix} 0 & -1/2 \\ -3 & 0 \end{bmatrix} - \begin{bmatrix} b_{11} & b_{12} \\ b_{12} & b_{22} \end{bmatrix}$$

or

$$\begin{bmatrix} (9b_{22} - b_{11}) & b_{12}/2 \\ b_{12}/2 & (b_{11}/4 - b_{22}) \end{bmatrix} = \begin{bmatrix} -1 & 0 \\ 0 & -1 \end{bmatrix},$$

which yields

$$B = \begin{bmatrix} -8 & 0 \\ 0 & -1 \end{bmatrix},$$

which is also negative definite. It follows from Proposition 7.5.1(c) that the equilibrium $x_e = 0$ of this system is *unstable*. This conclusion is consistent with the conclusion made in Example 7.5.3. □

Example 7.5.8 For system (L_D), let

$$A = \begin{bmatrix} 1/3 & 1 \\ 0 & 3 \end{bmatrix}.$$

The eigenvalues of A are $\lambda_1 = 1/3$ and $\lambda_2 = 3$. According to Lemma 7.5.1, for a given matrix $C = C^T$, equation (7.5.4) *does not have a unique solution*, because in this case $\lambda_1 = 1/\lambda_2$. For purposes of illustration, we choose $C = -I$. Then

$$-I = A^T BA - B = \begin{bmatrix} 1/3 & 0 \\ 1 & 3 \end{bmatrix} \begin{bmatrix} b_{11} & b_{12} \\ b_{12} & b_{22} \end{bmatrix} \begin{bmatrix} 1/3 & 1 \\ 0 & 3 \end{bmatrix} - \begin{bmatrix} b_{11} & b_{12} \\ b_{12} & b_{22} \end{bmatrix}$$

or

$$\begin{bmatrix} -8/9 & b_{11}/3 \\ b_{11}/3 & (b_{11} + 6b_{12} + 8b_{22}) \end{bmatrix} = \begin{bmatrix} -1 & 0 \\ 0 & -1 \end{bmatrix},$$

which shows that for $C = -I$, equation (7.5.4) does not have any solution for B at all. □

7.6 Perturbed Linear Systems

Perturbed linear systems come about because of uncertainties incurred during the modeling process; because of errors made in measurements; because of errors introduced when linearizing nonlinear systems; and the like. A natural question of fundamental importance is to identify conditions under which linear systems and their perturbations have similar qualitative properties. We answer this question in pieces, by addressing different aspects of system behavior.

We first determine conditions under which the stability properties of the equilibrium $x_e = 0$ of a nonlinear system can be deduced from the stability properties of the equilibrium $w = 0$ of its linearization, for noncritical cases. This is known as *Lyapunov's First Method* or *Lyapunov's Indirect Method.* Next, for noncritical cases, we determine the qualitative properties of the solutions of perturbed linear systems by proving the existence of *stable* and *unstable manifolds* near the equilibrium $x_e = 0$ of such systems. Finally, by introducing the notion of *orbital stability*, we study the stability and instability properties of periodic solutions of perturbed linear periodic systems.

A. Preliminaries

The present subsection consists of several parts.

Some Notation

We recall that for a function $g\colon \mathbb{R}^l \to \mathbb{R}^k$, the notation $g(x) = \mathcal{O}(|x|^\beta)$ as $|x| \to \alpha$ means that

$$\lim_{|x|\to\alpha} \sup \frac{|g(x)|}{|x|^\beta} < \infty,$$

$\beta \geq 0$, with the interesting cases including $\alpha = 0$ and $\alpha = \infty$. Above, $|\cdot|$ denotes any one of the equivalent norms on $\mathbb{R}^l$. Also, when $g\colon \mathbb{R} \times \mathbb{R}^l \to \mathbb{R}^k$, $g(t,x) = \mathcal{O}(|x|^\beta)$ as $|x| \to \alpha$ uniformly for t in an interval I means that

$$\lim_{|x|\to\alpha} \sup \left(\sup_{t\in I} \frac{|g(t,x)|}{|x|^\beta} \right) < \infty.$$

We also recall that $g(x) = o(|x|^\beta)$ as $|x| \to \alpha$ means that

$$\lim_{|x|\to\alpha} \frac{|g(x)|}{|x|^\beta} = 0$$

and $g(t,x) = o(|x|^\beta)$ as $|x| \to \alpha$ uniformly for t in an interval is defined in the obvious way.

The Implicit Function Theorem

In Subsection 7.6D we make use of the *Implicit Function Theorem* which we present next, without proof. To this end, we consider a system of functions

$$g_i(x,y) = g_i(x_1,\ldots,x_n,y_1,\ldots,y_r), \qquad i = 1,\ldots,r.$$

We assume that these functions have continuous first derivatives in an open set containing a point (x_0, y_0). We define the *Jacobian matrix* of $g(\cdot) = (g_1(\cdot), \ldots, g_r(\cdot))$ with respect to $(y_1, \ldots, y_r)$ by

$$g_y(\cdot) = \frac{\partial g}{\partial y}(\cdot) = \begin{bmatrix} \partial g_1/\partial y_1 & \cdots & \partial g_1/\partial y_r \\ \partial g_2/\partial y_1 & \cdots & \partial g_2/\partial y_r \\ \vdots & \ddots & \vdots \\ \partial g_r/\partial y_1 & \cdots & \partial g_r/\partial y_r \end{bmatrix}.$$

The determinant of this matrix is called the *Jacobian* of $g(\cdot)$ with respect to $(y_1, \ldots, y_r)$ and is denoted

$$J = \det(\partial g/\partial y).$$

Theorem 7.6.1 (*Implicit Function Theorem*) Let $g_1(\cdot), \ldots, g_r(\cdot)$ have continuous first derivatives in a neighborhood of a point (x_0, y_0). Assume that $g_i(x_0, y_0) = 0, i = 1, \ldots, r$ and that $J \neq 0$ at (x_0, y_0). Then there is a δ-neighborhood U of x_0 and a γ-neighborhood S of y_0 such that for any $x \in U$ there is a unique solution y of $g_i(x, y) = 0, i = 1, \ldots, r$ in S. The vector-valued function $y(x) = (y_1(x), \ldots, y_r(x))^T$ defined in this way has continuous first derivatives. If the functions $g_i(\cdot)$, $i = 1, \ldots, r$, have a continuous kth derivative, or if they are analytic, then so are the functions $y_i(\cdot)$, $i = 1, \ldots, r$. □

Hypersurfaces

We characterize *stable* and *unstable manifolds* by means of hypersurfaces in $\mathbb{R}^n$.

Definition 7.6.1 A *local hypersurface* S *of dimension* $k + 1$ *located along a curve* $v(t)$ is determined as follows. There is a neighborhood V of the origin $x = 0$ in $\mathbb{R}^n$ and there are $(n - k)$ functions $H_i \in C^1[\mathbb{R} \times V, \mathbb{R}]$ such that

$$S = \big\{(t, x) \colon t \in \mathbb{R}, x - v(t) \in V \text{ and } H_i(t, x + v(t)) = 0, \ i = k + 1, \ldots, n\big\}. \tag{7.6.1}$$

Here $H_i(t, v(t)) = 0$, $i = k + 1, \ldots, n$, for all $t \in \mathbb{R}$. Moreover, if ∇ denotes the gradient with respect to x, then for each $t \in \mathbb{R}$, $\{\nabla H_i(t, v(t)) \colon i = k + 1, \ldots, n\}$ is a set of $(n - k)$ linearly independent vectors. A *tangent hypersurface* to S at a point (t, x) is determined by $\{y \in \mathbb{R}^n \colon \langle y, \nabla H_i(t, v(t))\rangle = 0, \ i = k + 1, \ldots, n\}$. We say that S is C^m*-smooth* if $v \in C^m[\mathbb{R}, \mathbb{R}^n]$ and $H_i \in C^m[\mathbb{R} \times V, \mathbb{R}]$ and we say that S is *analytic* if v and H_i are holomorphic in t and (t, x), respectively. □

In the present section, $v(t)$ is usually a constant (usually, $v(t) \equiv 0$) or it is a periodic function. Moreover, there is typically a constant $n \times n$ matrix Q, a neighborhood U of the origin in the $\hat{y} = (y_1, \ldots, y_k)^T$-space, and a function $G \in C^1[\mathbb{R} \times U, \mathbb{R}^{n-k}]$ such that $G(t, 0) \equiv 0$ and such that

$$S = \big\{(t, x) \colon y = Q(x - v) \in U \ \text{ and } \ (y_{k+1}, \ldots, y_n)^T = G(t, y_1, \ldots, y_k)\big\}. \tag{7.6.2}$$

The functions $H_i(t, x)$ can be determined immediately from $G(t, y)$ and Q.

Positively and Negatively Invariant Sets

In Subsection 7.6D where we study the qualitative properties of the solutions of perturbed linear equations using stable and unstable manifolds, we need to allow the solutions of the equations to evolve forward and backward in time. As a consequence of this, as pointed out in Subsection 3.1A, we require not only *positively invariant sets* but *negatively invariant sets* as well. Thus, a set $M \subset \mathbb{R}^n$ is *positively invariant* with respect to (E) if for every solution $\varphi(\cdot, t_0, x_0)$ of (E), $x_0 \in M$ implies that $\varphi(t, t_0, x_0) \in M$ for all $t \geq t_0$ and *negatively invariant* with respect to (E) if for every solution $\varphi(\cdot, t_0, x_0)$ of (E), $x_0 \in M$ implies that $\varphi(t, t_0, x_0) \in M$ for all $t \leq t_0$.

B. Stability of an equilibrium (continuous-time systems)

To fix some of the ideas involved, we consider systems of equations

$$\dot{x} = g(t, x) \tag{G}$$

where $g \in C^1[\mathbb{R}^+ \times \Omega, \mathbb{R}^n]$ and Ω is an open connected set. Let φ denote a given solution of (G) that is defined for all $t \geq t_0 \geq 0$. We can *linearize* (G) *about* φ in the following manner. Define $y = x - \varphi(t)$ so that

$$\begin{aligned}\dot{y} &= g(t, x) - g(t, \varphi(t)) \\ &= g(t, y + \varphi(t)) - g(t, \varphi(t)) \\ &= \frac{\partial g}{\partial x}(t, \varphi(t))y + G(t, y)\end{aligned}$$

where

$$G(t, y) \triangleq [g(t, y + \varphi(t)) - g(t, \varphi(t))] - \frac{\partial g}{\partial x}(t, \varphi(t))y$$

which is $o(|y|)$ as $|y| \to 0$, uniformly in t on compact subsets of $[t_0, \infty)$.

Of special interest is the case when $g(t, x) \equiv g(x)$ and $\varphi(t) = x_0$ is a constant (i.e., an equilibrium point). Under these conditions, we have

$$\dot{y} = Ay + G(y)$$

where $A = (\partial g / \partial x)(x_0)$.

Also, of special interest is the case in which $g(t, x)$ is T periodic in t (or is independent of t) and $\varphi(t)$ is T periodic. We consider this case in Subsection E.

We now consider systems of equations given by

$$\dot{x} = Ax + F(t, x) \tag{PE}$$

where $F \in C[\mathbb{R}^+ \times B(r), \mathbb{R}^n]$, $B(r) \subset \Omega \subset \mathbb{R}^n$ for some $r > 0$, where Ω is a connected set containing the origin $x = 0$ and $A \in \mathbb{R}^{n \times n}$. We call Ax the *linear part* of the right-hand side of (PE) and $F(t, x)$ represents the remaining terms of order higher than one in the various components of x. System (PE) constitutes a *perturbed linear system* corresponding to the *unperturbed linear system*

$$\dot{w} = Aw. \tag{L}$$

Theorem 7.6.2 Let $A \in \mathbb{R}^{n\times n}$ be stable, let $F \in C[\mathbb{R}^+ \times B(r), \mathbb{R}^n]$, and assume that

$$F(t,x) = o(|x|) \quad \text{as } |x| \to 0, \tag{7.6.3}$$

uniformly in $t \in \mathbb{R}^+$. Then the equilibrium $x_e = 0$ of (PE) is *uniformly asymptotically stable*, in fact, *exponentially stable*.

Proof. Because (L) is an autonomous linear system, Theorem 7.4.7 applies. In view of that theorem, there exists a symmetric, real, positive definite $n \times n$ matrix P such that $A^T P + PA = -C$, where C is positive definite. Consider the Lyapunov function $v(x) = x^T Px$. The derivative of v with respect to t along the solutions of (PE) is given by

$$v'_{(PE)}(t,x) = -x^T Cx + 2x^T PF(t,x). \tag{7.6.4}$$

Now pick $\gamma > 0$ such that $x^T Cx \geq 3\gamma|x|^2$ for all $x \in \mathbb{R}^n$. By (7.6.3) there is a δ with $0 < \delta < r$ such that if $|x| \leq \delta$, then $|PF(t,x)| \leq \gamma|x|$ for all $(t,x) \in \mathbb{R}^+ \times \overline{B(\delta)}$. For all $(t,x) \in \mathbb{R}^+ \times \overline{B(\delta)}$ we obtain, in view of (7.6.4), the estimate

$$v'_{(PE)}(t,x) \leq -3\gamma|x|^2 + 2\gamma|x|^2 = -\gamma|x|^2.$$

It follows that $v'_{(PE)}(t,x)$ is negative definite in a neighborhood of the origin. By Theorem 6.2.2 the trivial solution of (PE) is uniformly asymptotically stable and by Theorem 6.2.4, it is exponentially stable, because $c_1|x|^2 \leq v(x) \leq c_2|x|^2$ for some $c_2 > c_1 > 0$ and for all $x \in \mathbb{R}^n$. □

Example 7.6.1 We consider the *Lienard Equation*

$$\ddot{x} + f(x)\dot{x} + x = 0 \tag{7.6.5}$$

where $f \in C[\mathbb{R}, \mathbb{R}]$. Assume that $f(0) > 0$. We can rewrite (7.6.5) (letting $x = x_1$ and $\dot{x} = x_2$) as

$$\begin{cases} \dot{x}_1 = x_2 \\ \dot{x}_2 = -x_1 - f(0)x_2 + \big(f(0) - f(x_1)\big)x_2 \end{cases} \tag{7.6.6}$$

and we can apply Theorem 7.6.2 with $x^T = (x_1, x_2)$,

$$A = \begin{bmatrix} 0 & 1 \\ -1 & -f(0) \end{bmatrix} \quad \text{and} \quad F(t,x) = \begin{bmatrix} 0 \\ (f(0) - f(x_1))x_2 \end{bmatrix}.$$

Because A is a stable matrix and $F(t,x)$ satisfies (7.6.3), we conclude that the equilibrium $x_e = 0$ of (7.6.5) is *uniformly asymptotically stable*. □

We emphasize that the results one obtains by applying Theorem 7.6.2 are local, and no information concerning the extent of the uniform asymptotic stability of the equilibrium $x_e = 0$ (domain of attraction) is provided.

Theorem 7.6.3 Assume that $A \in \mathbb{R}^{n\times n}$ has at least one eigenvalue with positive real part and no eigenvalues with real part equal to zero. If $F \in C[\mathbb{R}^+ \times B(r), \mathbb{R}^n]$ and if F satisfies (7.6.3), then the equilibrium $x_e = 0$ of (PE) is *unstable*.

Proof. We use Theorem 7.4.7 to choose a real, symmetric $n \times n$ matrix P such that $A^T P + PA = -C$ is negative definite. The matrix P is not positive definite or even positive semidefinite. Hence, the function $v(x) \triangleq x^T Px$ is negative at points arbitrarily close to the origin. Evaluating the derivative of v with respect to t along the solutions of (PE), we obtain

$$v'_{(PE)}(t, x) = -x^T Cx + 2x^T PF(t, x).$$

Pick $\gamma > 0$ such that $x^T Cx \geq 3\gamma|x|^2$ for all $x \in \mathbb{R}^n$. In view of (7.6.3) we can pick δ such that $0 < \delta < r$ and $|PF(t, x)| \leq \gamma|x|$ for all $(t, x) \in \mathbb{R}^+ \times B(\delta)$. Thus, for all $(t, x) \in \mathbb{R}^+ \times B(\delta)$, we obtain

$$v'_{(PE)}(t, x) \leq -3\gamma|x|^2 + 2\gamma|x|^2 = -\gamma|x|^2,$$

so that $v'_{(PE)}(t, x)$ is negative definite. By Theorem 6.2.8 the trivial solution of (PE) is unstable. □

Example 7.6.2 Consider the *simple pendulum* described by the equation

$$\ddot{x} + a \sin x = 0 \tag{7.6.7}$$

where $a > 0$ is a constant. Note that $(x_e, \dot{x}_e)^T = (\pi, 0)^T$ is an equilibrium for (7.6.7). Let $y = x - x_e$. Then

$$\ddot{y} + a \sin(y + \pi) = \ddot{y} - ay + a(\sin(y + \pi) + y) = 0.$$

This equation can be put into the form of (PE) with

$$A = \begin{bmatrix} 0 & 1 \\ a & 0 \end{bmatrix} \quad \text{and} \quad F(t, x) = \begin{bmatrix} 0 \\ a(\sin(y + \pi) + y) \end{bmatrix}.$$

The eigenvalues of A are $\lambda_1, \lambda_2 = \pm\sqrt{a}$ and F satisfies condition (7.6.3). Thus, Theorem 7.6.3 is applicable and we can conclude that the equilibrium $(x_e, \dot{x}_e) = (\pi, 0)$ is *unstable*. □

Next, we consider *periodic systems* given by

$$\dot{x} = P(t)x + F(t, x) \tag{7.6.8}$$

where $P \in C[\mathbb{R}, \mathbb{R}^{n\times n}]$ is periodic with period $T > 0$ and where F has the properties enumerated in Theorem 7.6.2. As in the case of system (PE), system (7.6.8) may arise in the process of linearizing equations of the form (E) or they may arise in the process of modeling a physical system. Thus, system (7.6.8) constitutes a *perturbed linear periodic system* corresponding to the *unperturbed linear periodic system*

$$\dot{w} = P(t)w. \tag{LP}$$

Corollary 7.6.1 Let $P(t)$ be defined as above and let F satisfy the hypotheses of Theorem 7.6.2.

(i) If all the characteristic exponents of the linear system (LP) have negative real parts, then the equilibrium $x_e = 0$ of system (7.6.8) is *uniformly asymptotically stable*.

(ii) If at least one characteristic exponent of (LP) has positive real part and no characteristic exponent has zero real part, then the equilibrium $x_e = 0$ of system (7.6.8) is *unstable*.

Proof. By Theorem 7.8.8 (in the appendix section, Section 7.8), the fundamental matrix Φ for (LP) satisfying $\Phi(0) = I$ has the form $\Phi(t) = U(t)e^{Rt}$, where $U(t)$ is a continuous, periodic, and nonsingular matrix. Now define $x = U(t)y$, where x solves (7.6.8), so that

$$\dot{U}(t)y + U(t)\dot{y} = P(t)U(t)y + F(t, U(t)y),$$

and $\dot{U} = PU - UR$. Thus y solves the equation

$$\dot{y} = Ry + U^{-1}(t)F(t, U(t)y),$$

and $U^{-1}(t)F(t, U(t)y)$ satisfies (7.6.3). Now apply Theorem 7.6.2 or 7.6.3 to determine the stability of the equilibrium $y_e = 0$. Because $U(t)$ and $U^{-1}(t)$ are both bounded on $\mathbb{R}$, the trivial solution $y_e = 0$ and $x_e = 0$ have the same stability properties. □

It is clear from the preceding results that the stability properties of the trivial solution of many nonlinear systems can be deduced from their linearizations. As mentioned earlier, these results comprise what is called *Lyapunov's First Method* or *Lyapunov's Indirect Method* for systems described by ordinary differential equations.

C. Stability of an equilibrium (discrete-time systems)

We now establish conditions under which the stability properties of the equilibrium $x_e = 0$ of the *perturbed linear system*

$$x(k+1) = Ax(k) + F(k, x(k)) \tag{7.6.9}$$

can be deduced from the stability properties of the equilibrium $w_e = 0$ of the *linear system*

$$w(k+1) = Aw(k) \tag{7.6.10}$$

under the assumption that $F(k, x) = o(|x|)$ as $|x| \to 0$, uniformly in $k \in N_1 \subset \mathbb{N}$. In (7.6.9), $A \in \mathbb{R}^{n \times n}$ and $F \in C[\mathbb{N} \times \Omega, \Omega]$ where $\Omega \subset \mathbb{R}^n$ is a connected set containing the origin $x = 0$.

Theorem 7.6.4 Assume that $F \in C[\mathbb{N} \times \Omega, \Omega]$ where $\Omega \subset \mathbb{R}^n$ is an open connected set containing the origin $x_e = 0$ and assume that $F(k, x) = o(|x|)$ as $|x| \to 0$, uniformly in $k \in N_1$ where N_1 is a subset of $\mathbb{N}$.

(i) If A is Schur stable (i.e., all eigenvalues of A are within the unit circle of the complex plane), then the equilibrium $x_e = 0$ of system (7.6.9) is *uniformly asymptotically stable* (in fact, *exponentially stable*).

(ii) If at least one eigenvalue of A is outside the unit circle of the complex plane and if A has no eigenvalues on the unit circle in the complex plane, then the equilibrium $x_e = 0$ of system (7.6.9) is *unstable*. □

The proofs of the results in Theorem 7.6.4 are similar to the proofs of Theorems 7.6.2 and 7.6.3 and are left as an exercise to the reader.

Example 7.6.3 Consider the system

$$\begin{cases} x_1(k+1) = -0.5x_2(k) + x_1(k)^2 + x_2(k)^2 \\ x_2(k+1) = -x_1(k) + x_1(k)^2 + x_2(k)^2. \end{cases} \tag{7.6.11}$$

System (7.6.11) has an equilibrium at the origin, $x_e = (x_1, x_2)^T = (0, 0)^T$. Using the notation of (7.6.9) we have

$$A = \begin{bmatrix} 0 & -1/2 \\ -1 & 0 \end{bmatrix} \quad \text{and} \quad F(k, x) \equiv F(x) = \begin{bmatrix} x_1^2 + x_2^2 \\ x_1^2 + x_2^2 \end{bmatrix}.$$

The eigenvalues of A are $\lambda_1, \lambda_2 = \pm\sqrt{1/2}$. Also, it is clear that $F(x) = o(|x|)$ as $|x| \to 0$. All the hypotheses of Theorem 7.6.4(i) are satisfied. Therefore, the equilibrium $x_e = 0$ of system (7.6.11) is *asymptotically stable*. □

Example 7.6.4 Consider the system

$$\begin{cases} x_1(k+1) = -0.5x_2(k) + x_1(k)^3 + x_2(k)^2 \\ x_2(k+1) = -3x_1(k) + x_1(k)^4 - x_2(k)^5. \end{cases} \tag{7.6.12}$$

Using the notation of (7.6.9), we have

$$A = \begin{bmatrix} 0 & -1/2 \\ -3 & 0 \end{bmatrix} \quad \text{and} \quad F(k, x) \equiv F(x) = \begin{bmatrix} x_1^3 + x_2^2 \\ x_1^4 - x_2^5 \end{bmatrix}.$$

The eigenvalues of A are $\lambda_1, \lambda_2 = \pm\sqrt{3/2}$. Also, it is clear that $F(x) = o(|x|)$ as $|x| \to 0$. All the hypotheses of Theorem 7.6.4(ii) are satisfied. Therefore, the equilibrium $x_e = 0$ of system (7.6.12) is *unstable*. □

D. Stable and unstable manifolds

In the present subsection we consider systems described by equations of the form

$$\dot{x} = Ax + F(t, x) \tag{PE}$$

under the assumption that the matrix A does not have any critical eigenvalues. We wish to study in some detail the properties of the solutions of (PE) in a neighborhood of the origin $x_e = 0$. To accomplish this, we establish the existence of stable and unstable manifolds (defined shortly). In doing so, we need to strengthen hypothesis (7.6.3) by making the following assumption.

Assumption 7.6.1 Let $F \in C[\mathbb{R} \times \Omega, \mathbb{R}^n]$ where $\Omega \subset \mathbb{R}^n$ is an open connected set containing the origin $x_e = 0$. Assume that $F(t, 0) = 0$ for all $t \in \mathbb{R}$ and that for any $\varepsilon > 0$ there is a $\delta > 0$ such that $B(\delta) \subset \Omega$ and such that if $(t, x), (t, y) \in \mathbb{R} \times B(\delta)$, then $|F(t, x) - F(t, y)| \leq \varepsilon |x - y|$. □

Assumption 7.6.1 is satisfied if, e.g., $F(t, x)$ is periodic in t, or if it is independent of t (i.e., $F(t, x) \equiv F(x)$), or if $F \in C^1[\mathbb{R} \times \Omega, \mathbb{R}^n]$ and both $F(t, 0) = 0$ and $F_x(t, 0) = 0$ for all $t \in \mathbb{R}$.

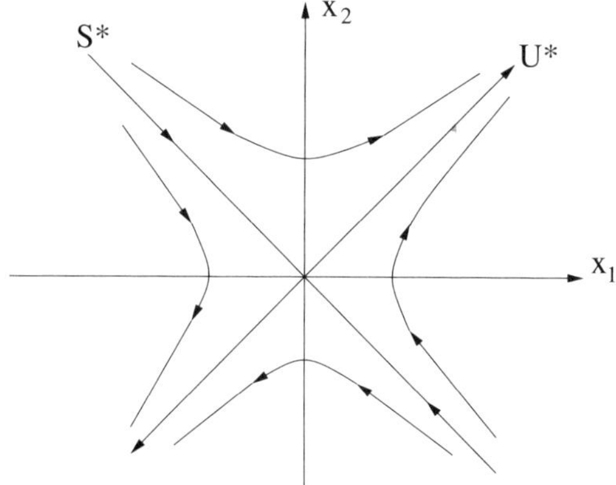

Figure 7.6.1: Stable and unstable manifolds for a linear system.

Before proceeding further, it may be useful to first present some motivation and insight for the principal results of the present subsection. To this end, we make reference to the trajectory portraits of the two-dimensional systems considered in Subsection 7.4E. We single out noncritical cases, and we are specifically interested in Figure 7.4.3 which depicts the qualitative behavior of the trajectories in the vicinity of a saddle. Making reference to Figure 7.6.1, we see that there is a one-dimensional linear subspace S^* such that the solutions starting in S^* tend to the origin as $t \to \infty$. This set is called a *stable manifold*. There is also a linear subspace U^*, called an *unstable manifold*, consisting of those trajectories that tend to the origin as $t \to -\infty$. If time is reversed, then S^* and U^* change roles. In the principal results of this subsection we prove that if the linear system is perturbed by terms that satisfy Assumption 7.6.1, then the resulting trajectory portrait for the perturbed linear system (PE) remains essentially unchanged, as shown in Figure 7.6.2. In this case, the stable manifold S and the unstable manifold U may become slightly distorted, but their essential qualitative properties persist.

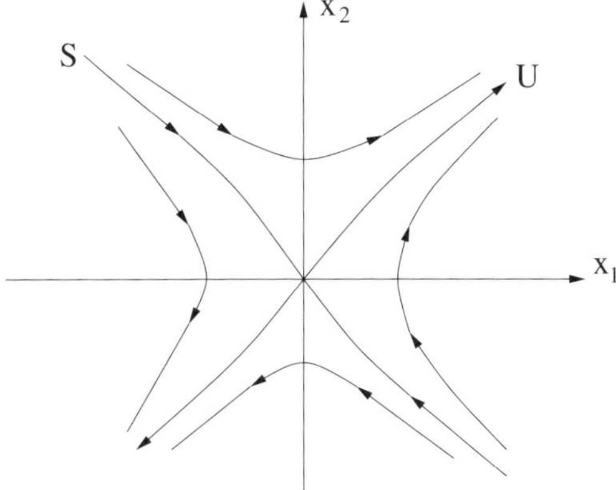

Figure 7.6.2: Stable and unstable manifolds for a perturbed linear system.

Our analysis is local, valid in a small neighborhood of the origin of $\mathbb{R}^n$. For n-dimensional systems (PE), we allow k eigenvalues of matrix A with negative real parts and $(n-k)$ eigenvalues with positive real parts. We allow $k = 0$ or $k = n$ as special cases and we allow F to depend on time t. In the (t, x)-space, we show that there is a $(k+1)$-dimensional stable manifold and an $(n-k+1)$-dimensional unstable manifold in a sufficiently small neighborhood of the line determined by $(t, 0)$, $t \in \mathbb{R}$.

Theorem 7.6.5 For system (PE), let $F \in C^1[\mathbb{R} \times \Omega, \mathbb{R}^n]$ and satisfy Assumption 7.6.1 and assume that $A \in \mathbb{R}^{n \times n}$ has k eigenvalues with negative real parts and $(n-k)$ eigenvalues with positive real parts. Then there exists a $(k+1)$-dimensional local hypersurface S (refer to Definition 7.6.1), located at the origin, called the *stable manifold* of (PE), such that S is positively invariant with respect to (PE), and for any solution φ of (PE) and any τ such that $(\tau, \varphi(\tau)) \in S$, we have $\varphi(t) \to 0$ as $t \to \infty$. Moreover, there is a $\delta > 0$ such that if $(\tau, \varphi(\tau)) \in \mathbb{R} \times B(\delta)$ for some solution φ of (PE) but $(\tau, \varphi(\tau)) \notin S$, then $\varphi(t)$ must leave the ball $B(\delta)$ at some finite time $t_1 > \tau$.

If $F \in C^l[\mathbb{R} \times \Omega, \mathbb{R}^n]$ for $l = 1, 2, 3, \ldots$ or $l = \infty$, or if F is holomorphic in (t, x), then S has the same degree of smoothness as F. Moreover, S is tangent at the origin to the stable manifold S^* for the linear system (L).

Proof. Choose a linear transformation $x = Qy$ such that (PE) becomes

$$\dot{y} = By + g(t, y), \qquad (PE')$$

where $B = Q^{-1}AQ = \text{diag}[B_1, B_2]$ and $g(t, y) = Q^{-1}F(t, Qy)$. The matrix Q can be chosen so that B_1 is a $k \times k$ stable matrix and $-B_2$ is an $(n-k) \times (n-k)$ stable matrix. Clearly g will satisfy Assumption 7.6.1. Moreover, if we define

$$U_1(t) = \begin{bmatrix} e^{B_1 t} & 0 \\ 0 & 0 \end{bmatrix}, \qquad U_2(t) = \begin{bmatrix} 0 & 0 \\ 0 & e^{B_2 t} \end{bmatrix},$$

then $e^{Bt} = U_1(t) + U_2(t)$ and for some positive constants K and σ we have

$$|U_1(t)| \leq Ke^{-2\sigma t}, \quad t \geq 0, \quad \text{and} \quad |U_2(t)| \leq Ke^{\sigma t}, \quad t \leq 0.$$

Let φ be a bounded solution of (PE') with $\varphi(\tau) = \xi$. Then by the variation of constants formula (see (7.8.10)), we have

$$\begin{aligned}
\varphi(t) &= e^{B(t-\tau)}\xi + \int_\tau^t e^{B(t-s)} g(s, \varphi(s))ds \\
&= U_1(t-\tau)\xi + \int_\tau^t U_1(t-s)g(s,\varphi(s))ds + U_2(t-\tau)\xi \\
&\quad + \int_\tau^\infty U_2(t-s)g(s,\varphi(s))ds - \int_t^\infty U_2(t-s)g(s,\varphi(s))ds.
\end{aligned}$$

Because $U_2(t-s) = U_2(t)U_2(-s)$, the bounded solution φ of (PE') must satisfy

$$\begin{aligned}
\varphi(t) = U_1(t-\tau)\xi &+ \int_\tau^t U_1(t-s)g(s,\varphi(s))ds - \int_t^\infty U_2(t-s)g(s,\varphi(s))ds \\
&+ U_2(t)\left(U_2(-\tau)\xi + \int_\tau^\infty U_2(-s)g(s,\varphi(s))ds\right). \qquad (7.6.13)
\end{aligned}$$

Conversely, any solution φ of (7.6.13) that is bounded and continuous on $[\tau, \infty)$ must solve (PE').

In order to satisfy (7.6.13) it is sufficient to find bounded and continuous solutions of the integral equation

$$\begin{aligned}
\psi(t,\tau,\xi) = U_1(t-\tau)\xi &+ \int_\tau^t U_1(t-s)g(s,\psi(s,\tau,\xi))ds \\
&- \int_t^\infty U_2(t-s)g(s,\psi(s,\tau,\xi))ds \qquad (7.6.14)
\end{aligned}$$

that also satisfy the condition

$$U_2(-\tau)\xi + \int_\tau^\infty U_2(-s)g(s,\psi(s,\tau,\xi))ds = 0. \qquad (7.6.15)$$

Successive approximation is used (refer, e.g., to Theorem 7.8.10) to solve (7.6.14) starting with $\psi_0(t,\tau,\xi) \equiv 0$. Choose $\varepsilon > 0$ such that $4\varepsilon K < \sigma$, pick $\delta = \delta(\varepsilon)$ using Assumption 7.6.1, and choose ξ with $|\xi| < \delta/(2K)$. Define

$$\|\psi\| \triangleq \sup\{|\psi(t)|: t \geq \tau\}.$$

If $\|\psi_j\| \leq \delta$, then ψ_{j+1} must satisfy

$$\begin{aligned}
|\psi_{j+1}(t,\tau,\xi)| &\leq K|\xi| + \int_\tau^t Ke^{-\sigma(t-s)}\varepsilon\|\psi_j\|ds + \int_t^\infty Ke^{\sigma(t-s)}\varepsilon\|\psi_j\|ds \\
&\leq \frac{1}{2}\delta + (2\varepsilon K/\sigma)\|\psi_j\| \\
&\leq \delta.
\end{aligned}$$

Because $\psi_0 \equiv 0$, then the ψ_j are well defined and satisfy $\|\psi_j\| \le \delta$ for all j. Thus

$$\begin{aligned}|\psi_{j+1}(t,\tau,\xi) - \psi_j(t,\tau,\xi)| &\le \int_\tau^t Ke^{-\sigma(t-s)}\varepsilon\|\psi_j - \psi_{j-1}\|ds \\ &\quad + \int_t^\infty Ke^{\sigma(t-s)}\varepsilon\|\psi_j - \psi_{j-1}\|ds \\ &\le (2\varepsilon K/\sigma)\|\psi_j - \psi_{j-1}\| \\ &\le \frac{1}{2}\|\psi_j - \psi_{j-1}\|.\end{aligned}$$

By induction, we have $\|\psi_{k+l+1} - \psi_{k+l}\| \le 2^{-l}\|\psi_{k+1} - \psi_k\|$ and

$$\begin{aligned}\|\psi_{k+j} - \psi_k\| &\le \|\psi_{k+j} - \psi_{k+j-1}\| + \cdots + \|\psi_{k+1} - \psi_k\| \\ &\le (2^{-j+1} + \cdots + 2^{-1} + 1)\,\|\psi_{k+1} - \psi_k\| \\ &\le 2\|\psi_{k+1} - \psi_k\| \\ &\le 2^{-k+1}\|\psi_1\|.\end{aligned}$$

From this estimate, it follows that $\{\psi_j\}$ is a Cauchy sequence uniformly in (t,τ,ξ) over $\tau \in \mathbb{R}$, $t \in [\tau,\infty)$, and $\xi \in B(\delta/(2K))$. Thus $\psi_j(t,\tau,\xi)$ tends to a limit $\psi(t,\tau,\xi)$ uniformly on (t,τ,ξ) on compact subsets of $(\tau,\xi) \in \mathbb{R} \times B(\delta/(2K))$, $t \in [\tau,\infty)$. The limit function ψ must be continuous in (t,τ,ξ) and it must satisfy $\|\psi\| \le \delta$.

The limit function ψ must satisfy (7.6.14). This is argued as follows. Note first that

$$\begin{aligned}&\left|\int_t^\infty U_2(t-s)g(s,\psi(s,\tau,\xi))ds - \int_t^\infty U_2(t-s)g(s,\psi_j(s,\tau,\xi))ds\right| \\ &\qquad \le \int_t^\infty Ke^{\sigma(t-s)}\varepsilon|\psi(s,\tau,\xi) - \psi_j(s,\tau,\xi)|ds \to 0, \qquad j \to \infty.\end{aligned}$$

A similar procedure applies to the other integral term in (7.6.14). Thus we can take the limit as $j \to \infty$ in the equation

$$\begin{aligned}\psi_{j+1}(t,\tau,\xi) = U_1(t-\tau)\xi &+ \int_\tau^t U_1(t-s)g(s,\psi_j(s,\tau\xi))ds \\ &- \int_t^\infty U_2(t-s)g(s,\psi_j(s,\tau,\xi))ds\end{aligned}$$

to obtain (7.6.14). Note that the solution of (7.6.14) is unique for given τ and ξ inasmuch as a second solution $\widetilde{\psi}$ would have to satisfy $\|\psi - \widetilde{\psi}\| \le 0.5\|\psi - \widetilde{\psi}\|$.

The stable manifold S is the set of all points (τ,ξ) such that equation (7.6.15) is true. It will be clear that S is a local hypersurface of dimension $(k+1)$. If $\xi = 0$, then by uniqueness $\psi(t,\tau,0) \equiv 0$ for $t \ge \tau$ and so $g(t,\psi(t,\tau,0)) \equiv 0$. Hence, $(\tau,0) \in S$ for all $\tau \in \mathbb{R}$. To see that S is positively invariant, let $(\tau,\xi) \in S$. Then $\psi(t,\tau,\xi)$ will solve (7.6.13), and hence it will solve (PE'). For any $\tau_1 > \tau$ let $\xi_1 = \psi(\tau_1,\tau,\xi)$

and define $\varphi(t,\tau_1,\xi_1) \triangleq \psi(t,\tau,\xi)$. Then $\varphi(t,\tau_1,\xi_1)$ solves (PE') and hence it also solves (7.6.13) with (τ,ξ) replaced by (τ_1,ξ_1). Hence

$$\begin{aligned}
&\left|U_2(t)\left(U_2(-\tau_1)\xi_1 + \int_{\tau_1}^{\infty} U_2(-s)g(s,\varphi(s,\tau_1,\xi_1))ds\right)\right| \\
&= \left|\varphi(t,\tau_1,\xi_1) - U_1(t-\tau_1)\xi_1 - \int_{\tau_1}^{t} U_1(t-s)g(s,\varphi(s,\tau_1,\xi_1))ds\right. \\
&\qquad \left. + \int_{t}^{\infty} U_2(t-s)g(s,\varphi(s,\tau_1,\xi_1))ds\right| \\
&\le \delta + Ke^{-\sigma(t-\tau_1)}|\xi_1| + \int_{\tau_1}^{t} Ke^{-\sigma(t-s)}\varepsilon\delta ds + \int_{t}^{\infty} Ke^{\sigma(t-s)}\varepsilon\delta ds \\
&\le \delta + \delta + (2K\varepsilon\delta/\sigma) \\
&\le 3\delta \\
&< \infty.
\end{aligned} \tag{7.6.16}$$

Because $U_2(t) = \text{diag}[0, e^{B_2 t}]$ and $-B_2$ is a stable matrix, this is only possible when $(\tau_1,\xi_1) \in S$. Hence S is positively invariant.

To see that any solution starting on S tends to the origin as $t \to \infty$, let $(\tau,\xi) \in S$ and let ψ_j be the successive approximation defined above. Then clearly

$$|\psi_1(t,\tau,\xi)| \le K|\xi|e^{-2\sigma(t-\tau)} \le 2K|\xi|e^{-\sigma(t-\tau)}.$$

If $|\psi_j(t,\tau,\xi)| \le 2K|\xi|e^{-\sigma(t-\tau)}$, then

$$\begin{aligned}
|\psi_{j+1}(t,\tau,\xi)| &\le K|\xi|e^{-\sigma(t-\tau)} + \int_{\tau}^{t} Ke^{-2\sigma(t-s)}\varepsilon\left(2K|\xi|e^{-\sigma(s-\tau)}\right)ds \\
&\quad + \int_{t}^{\infty} Ke^{\sigma(t-s)}\varepsilon\left(2K|\xi|e^{-\sigma(s-\tau)}\right)ds \\
&\le K|\xi|e^{-\sigma(t-\tau)} + 2K|\xi|(\varepsilon K/\sigma)e^{-\sigma(t-\tau)} + 2K|\xi|(\varepsilon K/2\sigma)e^{-\sigma(t-\tau)} \\
&\le 2K|\xi|e^{-\sigma(t-\tau)}
\end{aligned}$$

because $(4\varepsilon K/\sigma) < 1$. Hence in the limit as $j \to \infty$ we have

$$|\psi(t,\tau,\xi)| \le 2K|\xi|e^{-\sigma(t-\tau)}$$

for all $t \ge \tau$ and for all $\xi \in B(\delta/(2K))$.

Suppose that $\varphi(t,\tau,\xi)$ solves (PE') but (τ,ξ) does not belong to S. If $\varphi(t)$ stays in the ball $B(\delta)$ (i.e., $|\varphi(t,\tau,\xi)| \le \delta$, for all $t \ge \tau$), then (7.6.16) is true. Hence $(\tau,\xi) \in S$, which is a contradiction.

Equation (7.6.15) can be rearranged as

$$(\xi_{k+1},\ldots,\xi_n)^T = P\left(-\int_{t}^{\infty} U(\tau-s)g(s,\psi(s,\tau,\xi))ds\right), \tag{7.6.17}$$

where P denotes the projection on the last $n-k$ components. Utilizing estimates of the type used above, we see that the function on the right side of (7.6.17) is Lipschitz continuous in ξ with Lipschitz constant $L \leq 1$. Hence, successive approximations can be used to solve (7.6.17), say

$$(\xi_{k+1}, \ldots, \xi_n)^T = h(\tau, \xi_1, \ldots, \xi_k) \tag{7.6.18}$$

with h continuous. If F is of class C^1 in (t, x), then the partial derivatives of the right-hand side of (7.6.17) with respect to $\xi_1, \ldots, \xi_n$ all exist and are zero at $\xi_1 = \cdots = \xi_n = 0$. The Jacobian with respect to $(\xi_{k+1}, \ldots, \xi_n)$ on the left side of (7.6.17) is one. By the implicit function theorem (see Theorem 7.6.1), the solution of (7.6.18) is C^1 smooth; indeed h is at least as smooth as F is. Inasmuch as

$$\frac{\partial h}{\partial \xi_j} = 0 \quad \text{for } k < j \leq n \text{ at } \xi_1 = \cdots = \xi_n = 0,$$

then S is tangent to the hyperplane $\xi_{k+1} = \cdots = \xi_n = 0$ at $\xi = 0$; that is, S is tangent to the stable manifold of the linear system (L) at $\xi = 0$. □

If in (PE) we reverse time, we obtain the system

$$\dot{y} = -Ay - F(-t, y). \tag{7.6.19}$$

Applying Theorem 7.6.5 to system (7.6.19), we obtain the following result.

Theorem 7.6.6 If the hypotheses of Theorem 7.6.5 are satisfied, then there is an $(n-k+1)$-dimensional local hypersurface U based at the origin, called the *unstable manifold* of (PE), such that U is negatively invariant with respect to (PE), and for any solution φ of (PE) and any $\tau \in \mathbb{R}$ such that $(\tau, \varphi(\tau)) \in U$, we have $\varphi(t) \to 0$ as $t \to -\infty$. Moreover, there is a $\delta > 0$ such that if $(\tau, \varphi(\tau)) \in \mathbb{R} \times B(\delta)$ but $(\tau, \varphi(\tau)) \notin U$, then $\varphi(t)$ must leave the ball $B(\delta)$ at some finite time $t_1 < \tau$.

The surface U has the same degree of smoothness as F and is tangent at the origin to the unstable manifold U^* of the linear system (L). □

If F in (PE) is independent of time t, that is, if $F(t, x) \equiv F(x)$, then it is not necessary to keep track of initial time in Theorems 7.6.5 and 7.6.6. Thus, in this case one dispenses with time and one defines S and U in the x-space, $\mathbb{R}^n$. This was done in our discussion concerning Figures 7.6.1 and 7.6.2.

Example 7.6.5 Consider equations of the form

$$\begin{cases} \dot{x}_1 = ax_1 - bx_1x_2 \\ \dot{x}_2 = cx_2 - dx_1x_2 \end{cases} \tag{7.6.20}$$

where $a, b, c, d > 0$ are constants, where $x_1 \geq 0$ and $x_2 \geq 0$, and where nonnegative initial data $x_1(0) = x_{10}$ and $x_2(0) = x_{20}$ must be specified.

Equation (7.6.20), which is an example of a *Volterra competition equation*, can be used to describe the growth of two competing species (e.g., of small fish) that prey on each other (e.g., the adult members of specie A prey on the young members of specie B, and vice versa).

System (7.6.20) has two equilibrium points, $x_{e1} = (0,0)$ and $x_{e2} = (c/d, a/b)$. The eigenvalues of the linear part of system (7.6.20) at the equilibrium x_{e1} are $\lambda_1 = a$ and $\lambda_2 = c$. Both are positive, therefore this equilibrium is completely unstable. The eigenvalues of the linear part of system (7.6.20) at the equilibrium x_{e2} are $\lambda_1 = \sqrt{ac} > 0$ and $\lambda_2 = -\sqrt{ac} < 0$. The right-hand side of equation (7.6.20) is time-invariant, so we may ignore time, and the stable manifold S and the unstable manifold U each have dimension one. These manifolds are tangent at x_{e2} to the lines

$$\sqrt{ac}x_1 + (bc/d)x_2 = 0 \quad \text{and} \quad -\sqrt{ac}x_1 + (bc/d)x_2 = 0.$$

If $x_2 = a/b$ and $0 < x_1 < c/d$, then $\dot{x}_1 = 0$ and $\dot{x}_2 > 0$; if $x_2 > a/b$ and $0 < x_1 < c/d$, then $\dot{x}_1 < 0$ and $\dot{x}_2 > 0$; and if $x_1(0) = 0$, then $x_1(t) = 0$ for all $t \geq 0$. Therefore, the set $G_1 = \{(x_1, x_2)\colon 0 < x_1 < c/d, x_2 > a/b\}$ is positively invariant and all solutions $(x_1(t), x_2(t))$ that enter this set must satisfy the condition that $x_2(t) \to \infty$ as $t \to \infty$. In a similar manner we can conclude that the set $G_2 = \{(x_1, x_2)\colon x_1 > c/d, 0 < x_2 < a/b\}$ is also positively invariant and all solutions that enter G_2 must satisfy the condition that $x_1(t) \to \infty$ as $t \to \infty$.

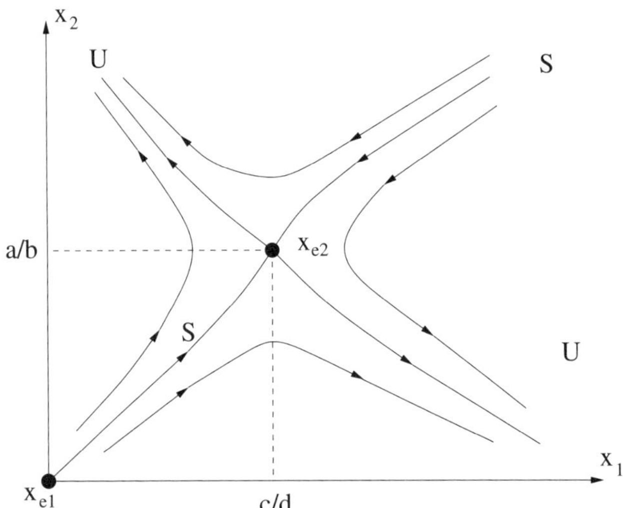

Figure 7.6.3: Trajectory portrait for system (7.6.20).

Because the unstable manifold U of the equilibrium x_{e2} is tangent to the line

$$\sqrt{ac}x_1 + (bc/d)x_2 = 0,$$

then, as shown in Figure 7.6.3, one branch of U enters G_1 and one enters G_2. The stable manifold S of x_{e2} cannot meet either G_1 or G_2. Hence, the trajectory portrait for system (7.6.20) is completely determined, as shown in Figure 7.6.3. From this portrait we can conclude that for almost all initial conditions one of the competing species will eventually die out and the second will grow. However, the outcome is unpredictable in the sense that near the manifold S, a slight change in initial conditions can radically alter the outcome. □

E. Stability of periodic solutions

We first consider T-periodic systems described by

$$\dot{x} = f(t, x) \tag{P}$$

where $f \in C^1[\mathbb{R} \times \Omega, \mathbb{R}^n]$, $\Omega \subset \mathbb{R}^n$ is a domain and $f(t,x) = f(t+T, x)$ for all $(t,x) \in \mathbb{R} \times \Omega$. Now let φ be a T-periodic solution of (P) with the property that $\varphi(t) \in \Omega$ for all $t \in \mathbb{R}$. Defining $y = x - \varphi(t)$, we obtain from (P) that

$$\dot{y} = f_x(t, \varphi(t))y + h(t, y) \tag{7.6.21}$$

where

$$h(t,y) = f(t, y + \varphi(t)) - f(t, \varphi(t)) - f_x(t, \varphi(t))y$$

satisfies Assumption 7.6.1. Corresponding to the perturbed linear system (7.6.21), we have the linear system

$$\dot{y} = f_x(t, \varphi(t))y. \tag{7.6.22}$$

By the Floquet theory (refer to Subsection 7.8B), there is a periodic nonsingular matrix $V(t)$ that transforms (7.6.21), using $y = V(t)z$, into a system of the form

$$\dot{z} = Az + \left[V(t)\right]^{-1}\left(h(t, V(t)z)\right).$$

If A is noncritical, then this system satisfies all the hypotheses of Theorems 7.6.5 and 7.6.6 to yield the following result.

Theorem 7.6.7 Assume that $f \in C^1[\mathbb{R} \times \Omega, \mathbb{R}^n]$ and let φ be a periodic solution of (P) with period T. Suppose that the linear variational system (7.6.22) for $\varphi(t)$ has k characteristic exponents with negative real parts and $(n-k)$ characteristic exponents with positive real parts. Then there exist two hypersurfaces S and U for (P), each containing $(t, \varphi(t))$ for all $t \in \mathbb{R}$, where S is positively invariant and U is negatively invariant with respect to (P), and where S has dimension $(k+1)$ and U has dimension $(n-k+1)$ such that for any solution ψ of (P) in a δ neighborhood of φ and any $\tau \in \mathbb{R}$ we have

(i) $\psi(t) - \varphi(t) \to 0$ as $t \to \infty$ if $(\tau, \psi(\tau)) \in S$;
(ii) $\psi(t) - \varphi(t) \to 0$ as $t \to -\infty$ if $(\tau, \psi(\tau)) \in U$; and
(iii) ψ must leave the δ neighborhood of φ in finite time as t increases from τ and as t decreases from τ if $(\tau, \psi(\tau))$ is not in S and not in U.

The sets S and U are the stable and the unstable manifolds associated with φ. When $k = n$, then S is $(n+1)$-dimensional, U consists only of the points $(t, \varphi(t))$ for $t \in \mathbb{R}$, and φ is asymptotically stable. If $k < n$, then clearly φ is unstable. □

Next, assume that φ is a T-periodic solution of an autonomous system described by

$$\dot{x} = f(x) \tag{A}$$

where $f \in C^1[\Omega, \mathbb{R}^n]$. Using the transformation $y = x - \varphi(t)$, we obtain in this case the variational equation

$$\dot{y} = f_x(\varphi(t))y + h(t, y) \tag{7.6.23}$$

where $h(t, y) \triangleq f(y + \varphi(t)) - f(\varphi(t)) - f_x(\varphi(t))y$ which satisfies Assumption 7.6.1. Corresponding to (7.6.23) we have the linear first approximation given by

$$\dot{y} = f_x(\varphi(t))y. \tag{7.6.24}$$

Because $\varphi(t)$ solves (A), $\dot{\varphi}(t)$ is a T-periodic solution of (7.6.24). Therefore equation (7.6.24) cannot possibly satisfy the hypothesis that no characteristic exponent has zero real part. Indeed, one Floquet multiplier is one. Thus, the hypotheses of Theorem 7.6.7 can never be satisfied. Even if the remaining $(n - 1)$ characteristic exponents are all negative, φ cannot possibly be asymptotically stable. This can be seen by noting that for small τ, $\varphi(t + \tau)$ is near $\varphi(t)$ at $t = 0$, but $|\varphi(t + \tau) - \varphi(t)|$ does not tend to zero as $t \to \infty$. However, φ will satisfy the following more general notion of stability.

Definition 7.6.2 A T-periodic solution φ of system (A) is called *orbitally stable* if there is a $\delta > 0$ such that any solution ψ of (A) with $|\psi(\tau) - \varphi(\tau)| < \delta$ for some τ tends to the orbit

$$C(\varphi(\tau)) = \{\varphi(t) \colon 0 \le t \le T\}$$

as $t \to \infty$. If in addition for each such ψ there is a constant $\alpha \in [0, T)$ such that $\psi(t) - \varphi(t + \alpha) \to 0$ as $t \to \infty$, then ψ is said to have *asymptotic phase* α. □

We are now in a position to prove the following result.

Theorem 7.6.8 Let φ be a periodic solution of (A) with least period $T > 0$ and let $f \in C^1[\Omega, \mathbb{R}^n]$ where $\Omega \subset \mathbb{R}^n$ is a domain. If the linear system (7.6.24) has $(n - 1)$ characteristic exponents with negative real parts, then φ is *orbitally stable* and nearby solutions of (A) possess an *asymptotic phase*.

Proof. By a change of variables of the form $x = Qw + \varphi(0)$, where Q is assumed to be nonsingular, so that

$$\dot{w} = Q^{-1}f(Qw + \varphi(0)),$$

Q can be arranged so that $w(0) = 0$ and $\dot{w}(0) = Q^{-1}f(\varphi(0)) = (1, 0, \dots, 0)^T$. Hence, without loss of generality, we may assume in the original problem (A) that $\varphi(0) = 0$ and $\dot{\varphi}(0) = e_1 \triangleq (1, 0, \dots, 0)^T$.

Let Φ_0 be a real fundamental matrix solution of (7.6.24). There is a real nonsingular matrix C such that $\Phi_0(t+T) = \Phi_0(t)C$ all $t \in \mathbb{R}$ (refer to Subsection 7.8B). Because

$\dot{\varphi}(t)$ is a solution of (7.6.24), one eigenvalue of C is equal to one (refer to (7.8.8)). By hypothesis, all other eigenvalues of C have magnitude less than one; that is, all other characteristic exponents of (7.6.24) have negative real parts. Thus, there is a real $n \times n$ matrix R such that

$$R^{-1}CR = \begin{bmatrix} 1 & 0 \\ 0 & D_0 \end{bmatrix},$$

where D_0 is an $(n-1) \times (n-1)$ matrix and all eigenvalues of D_0 have absolute value less than one.

Now define $\Phi_1(t) = \Phi_0(t)R$ so that Φ_1 is a fundamental matrix for (7.6.24) and

$$\Phi_1(t+T) = \Phi_0(t+T)R = \Phi_0(t)CR = \Phi_0(t)R(R^{-1}CR) = \Phi_1(t)\begin{bmatrix} 1 & 0 \\ 0 & D_0 \end{bmatrix}.$$

The first column $\varphi_1(t)$ of $\Phi_1(t)$ necessarily must satisfy the relation

$$\varphi_1(t+T) = \varphi_1(t) \quad \text{for all } t \in \mathbb{R};$$

that is, it must be T periodic. Because $(n-1)$ characteristic exponents of (7.6.24) have negative real parts, there cannot be two linearly independent T periodic solutions of (7.6.24). Thus, there is a constant $k \neq 0$ such that $\varphi_1 = k\dot{\varphi}$. If $\Phi_1(t)$ is replaced by

$$\Phi(t) \stackrel{\triangle}{=} \Phi_1(t)\text{diag}\big[k^{-1}, 1, \ldots, 1\big],$$

then Φ satisfies the same conditions as Φ_1 but now $k = 1$.

There is a T periodic matrix $P(t)$ and a constant matrix B such that

$$e^{TB} = \begin{bmatrix} 1 & 0 \\ 0 & D_0 \end{bmatrix}, \qquad \Phi(t) = P(t)e^{Bt}.$$

(Both $P(t)$ and B may be complex valued.) The matrix B can be taken in the block diagonal form

$$B = \begin{bmatrix} 0 & 0 \\ 0 & B_1 \end{bmatrix}$$

where $e^{B_1 T} = D_0$ and B_1 is a stable $(n-1) \times (n-1)$ matrix. Define

$$U_1(t,s) = P(t)\begin{bmatrix} 1 & 0 \\ 0 & 0 \end{bmatrix} P^{-1}(s)$$

and

$$U_2(t,s) = P(t)\begin{bmatrix} 0 & 0 \\ 0 & e^{B_1(t-s)} \end{bmatrix} P^{-1}(s)$$

so that

$$U_1(t,s) + U_2(t,s) = P(t)e^{B(t-s)}P^{-1}(s) = \Phi(t)\Phi^{-1}(s).$$

Clearly $U_1 + U_2$ is real-valued. Because

$$P(t)\begin{bmatrix}1 & 0\\ 0 & 0\end{bmatrix} = (\varphi_1, 0, \dots, 0),$$

this matrix is real. Similarly, the first row of

$$\begin{bmatrix}1 & 0\\ 0 & 0\end{bmatrix} P^{-1}(s)$$

is the first row of $\Phi^{-1}(s)$ and the remaining rows are zero. Thus,

$$U_1(t,s) = P(t)\begin{bmatrix}1 & 0\\ 0 & 0\end{bmatrix}\begin{bmatrix}1 & 0\\ 0 & 0\end{bmatrix} P^{-1}(s)$$

is a real matrix. Hence,

$$U_2(t,s) = \Phi(t)\Phi^{-1}(s) - U_1(t,s)$$

is also real.

Choose constants $K > 1$ and $\sigma > 0$ such that $|U_1(t,s)| \leq K$ and $|U_2(t,s)| \leq Ke^{-2\sigma(t-s)}$ for all $t \geq s \geq 0$. As in the proof of Theorem 7.6.5, we utilize an integral equation. In the present case, it assumes the form

$$\psi(t) = U_2(t,\tau)\xi + \int_\tau^t U_2(t,s)h(s,\psi(s))ds - \int_t^\infty U_1(t,s)h(s,\psi(s))ds, \quad (7.6.25)$$

where h is the function defined in (7.6.23). This integral equation is again solved by successive approximations to obtain a unique, continuous solution $\psi(t,\tau,\xi)$ for $t \geq \tau$, $\tau \in \mathbb{R}$, and $|\xi| \leq \delta$ and with

$$\big|\psi(t+\tau,\tau,\xi)\big| \leq 2K|\xi|e^{-\sigma t}.$$

Solutions of (7.6.25) will be solutions of (7.6.23) provided that the condition

$$U_1(t,\tau)\xi + \int_\tau^\infty U_1(t,s)h(s,\psi(s,\tau,\xi))ds = 0 \quad (7.6.26)$$

is satisfied. Because

$$U_1(t,s) = P(t)\begin{bmatrix}1 & 0\\ 0 & 0\end{bmatrix} P^{-1}(s),$$

one can write equivalently

$$\begin{bmatrix}1 & 0\\ 0 & 0\end{bmatrix}\left(P^{-1}(\tau)\xi + \int_\tau^\infty P^{-1}(s)h(s,\psi(s,\tau,\xi))ds\right) = 0.$$

Because h_x and ψ_ξ exist and are continuous with $h_x(t,0) = 0$, then by the implicit function theorem (Theorem 7.6.1) one can solve for some ξ_j in terms of τ and the other ξ_ms. Hence, the foregoing equation determines a local hypersurface. For any τ, let G_τ be the set of all points ξ such that (τ, ξ) is on this hypersurface.

The set of points (τ, ξ) that satisfy (7.6.26) is positively invariant with respect to (7.6.23). Hence G_τ is mapped to $G_{\tau'}$ under the transformation determined by (A) as t varies from τ to τ'. As τ varies over $0 \leq \tau \leq T$, the surface G traces out a neighborhood N of the orbit $C(\varphi(0))$. Any solution that starts within N will tend to $C(\varphi(0))$ as $t \to \infty$. Indeed, for $|\widetilde{\varphi}(\tau) - \varphi(\tau')|$ sufficiently small, we define $\widetilde{\varphi}_1(t) = \widetilde{\varphi}(t + \tau - \tau')$. Then $\widetilde{\varphi}_1$ solves (A), $|\widetilde{\varphi}_1(\tau') - \varphi(\tau')|$ is small, and so, by continuity with respect to initial conditions, $\widetilde{\varphi}_1(t)$ will remain near $\varphi(t)$ long enough to intersect G_τ at $\tau = 0$ at some t_1. Then as $t \to \infty$,

$$\widetilde{\varphi}_1(t + t_1) - \varphi(t) \to 0,$$

or

$$\widetilde{\varphi}(t - \tau' + \tau + t_1) - \varphi(t) \to 0.$$

This completes the proof. □

The above result can be extended to obtain stable and unstable manifolds about a periodic solution, as shown next. The reader may find it helpful to make reference to Figure 7.6.4.

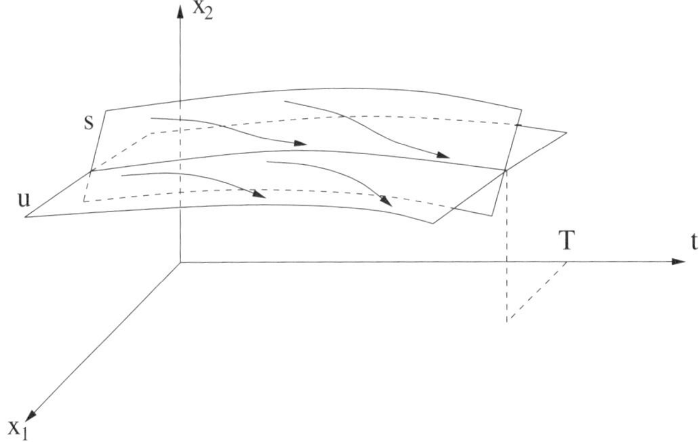

Figure 7.6.4: Stable and unstable manifolds about a periodic solution.

Theorem 7.6.9 Let $f \in C^1[\Omega, \mathbb{R}^n]$ where $\Omega \subset \mathbb{R}^n$ is a domain and let φ be a T-periodic solution of (A). Suppose k characteristic exponents of (7.6.24) have negative real parts and $(n-k-1)$ characteristic exponents of (7.6.24) have positive real parts. Then there exist T-periodic C^1-smooth manifolds S and U based at $\varphi(t)$ such that S has dimension $(k+1)$ and is positively invariant, U has dimension $(n-k)$ and is negatively invariant, and if ψ is a solution of (A) with $\psi(0)$ sufficiently close to $C(\varphi(0))$, then the following statements are true.

(i) $\psi(t)$ tends to $C(\varphi(0))$ as $t \to \infty$ if $(0, \psi(0)) \in S$.

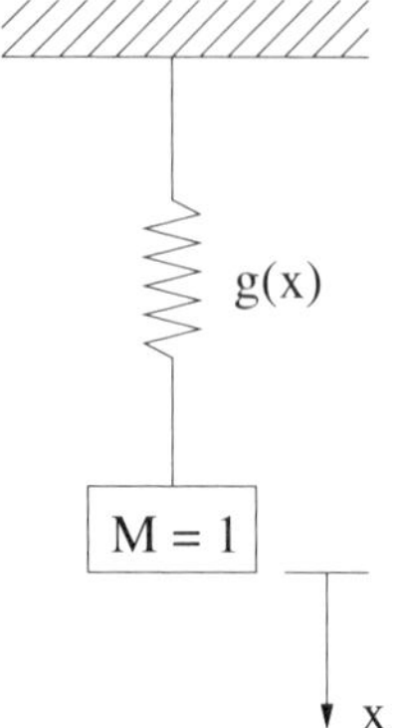

Figure 7.6.5: Nonlinear spring–unit mass system (7.6.27).

(ii) $\psi(t)$ tends to $C(\varphi(0))$ as $t \to -\infty$ if $(0, \psi(0)) \in U$.

(iii) $\psi(t)$ must have a neighborhood of $C(\varphi(0))$ as t increases and as t decreases if $(0, \psi(0)) \notin S \cup U$.

Proof. The proof of this theorem is very similar to the proof of Theorem 7.6.8. The matrix R can be chosen so that

$$R^{-1}CR = \begin{bmatrix} 1 & 0 & 0 \\ 0 & D_2 & 0 \\ 0 & 0 & D_3 \end{bmatrix}$$

where D_2 is a $k \times k$ matrix with eigenvalues that satisfy $|\lambda| < 1$ and D_3 is an $(n-k-1) \times (n-k-1)$ matrix whose eigenvalues satisfy $|\lambda| > 1$. Define B so that

$$B = \begin{bmatrix} 0 & 0 & 0 \\ 0 & B_2 & 0 \\ 0 & 0 & B_3 \end{bmatrix}, \qquad e^{BT} = R^{-1}CR.$$

Define U_1 as before and define U_2 and U_3 using $e^{B_2 t}$ and $e^{B_3 t}$. The rest of the proof involves similar modifications. □

Except in special cases, such as second-order systems and certain classes of Hamiltonian systems, the determination of Floquet multipliers of periodic linear systems is in general difficult. Nevertheless, results such as Theorems 7.6.8 and 7.6.9 are of great theoretical importance.

Example 7.6.6 An important class of conservative dynamical systems is described by equations of the form

$$\ddot{x} + g(x) = 0 \tag{7.6.27}$$

where $g \in C^1[\mathbb{R}, \mathbb{R}]$ and $xg(x) > 0$ for all $x \neq 0$. Equation (7.6.27) can be used to represent, for example, a mechanical system consisting of a unit mass and a nonlinear spring, as shown in Figure 7.6.5. Here, x denotes displacement and $g(x)$ denotes the restoring force due to the spring.

Letting $x_1 = x$ and $x_2 = \dot{x}$, we can express (7.6.27) equivalently as

$$\begin{cases} \dot{x}_1 = x_2 \\ \dot{x}_2 = -g(x_1). \end{cases} \tag{7.6.28}$$

The total energy for this system is given by

$$v(x) = \frac{1}{2}x_2^2 + \int_0^{x_1} g(\eta)d\eta = \frac{1}{2}x_2^2 + G(x_1) \tag{7.6.29}$$

where $G(x_1) = \displaystyle\int_0^{x_1} g(\eta)d\eta$. Note that v is positive definite and

$$v'_{(7.6.28)}(x) = 0. \tag{7.6.30}$$

Therefore, (7.6.28) is a conservative dynamical system and $(x_1, x_2)^T = (0, 0)^T$ is a stable equilibrium. Note that because $v'_{(7.6.28)} = 0$, it follows that

$$\frac{1}{2}x_2^2 + G(x_1) = c \tag{7.6.31}$$

where c is determined by the initial conditions (x_{10}, x_{20}). For different values of c we obtain different trajectories, as shown in Figure 7.6.6. The exact shapes of these trajectories depend on the function G. Note, however, that the curves determined by (7.6.31) will always be symmetric with respect to the x_1-axis. Furthermore, if $G(x) \to \infty$ as $|x| \to \infty$ then the entire x_1–x_2 plane can be covered by closed trajectories, each of which is an invariant set with respect to (7.6.28).

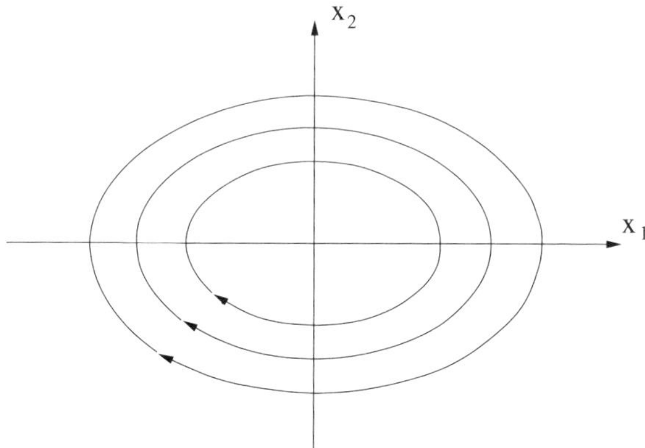

Figure 7.6.6: Trajectory portrait for system (7.6.27).

Now notice that no two periodic solutions of system (7.6.28) will approach each other or recede from each other. From this we see that the Floquet multipliers of a given periodic solution φ of system (7.6.28) must both be one. □

7.7 Comparison Theory

In this section we establish stability and boundedness results for ordinary differential equations (E) and ordinary difference equations (D) using comparison results developed in Section 4.3.

A. Continuous-time systems

Our *object of inquiry* is systems described by differential equations

$$\dot{x} = f(t, x) \tag{E}$$

where $x \in \mathbb{R}^n$, $f \in C[\mathbb{R}^+ \times \Omega, \mathbb{R}^n]$, $\Omega \subset \mathbb{R}^n$ is an open connected set, $0 \in \Omega$, and $f(t, 0) = 0$ for all $t \in \mathbb{R}^+$. For (E), we use *comparison systems* given by

$$\dot{y} = g(t, y) \tag{C}$$

where $y \in \mathbb{R}^l$, $g \in C[\mathbb{R}^+ \times B(r_1), \mathbb{R}^l]$ for some $r_1 > 0$, $B(r_1) \subset (\mathbb{R}^+)^l$, and $g(t, 0) = 0$ for all $t \in \mathbb{R}^+$.

For a vector-valued function $V\colon B(r) \times \mathbb{R}^+ \to \mathbb{R}^l$, where $B(r) \subset \Omega$, $r > 0$, we use the notation

$$V(x, t) = [v_1(x, t), \ldots, v_l(x, t)]^T$$

and

$$V'_{(E)}(x, t) = [v'_{1(E)}(x, t), \ldots, v'_{l(E)}(x, t)]^T.$$

In the results that follow, $|\cdot|$ denotes the Euclidean norm in $\mathbb{R}^l$.

Theorem 7.7.1 Assume that there exists a function $V \in C[B(r) \times \mathbb{R}^+, (\mathbb{R}^+)^l]$, where $B(r) \subset \Omega \subset \mathbb{R}^n$, $r > 0$, such that $|V(x, t)|$ is positive definite and decrescent, and that there exists a function $g \in C[\mathbb{R}^+ \times B(r_1), \mathbb{R}^l]$, where $B(r_1) \subset (\mathbb{R}^+)^l$, $r_1 > 0$, which is quasi-monotone nondecreasing (refer to Definition 3.8.1) and has the property that $g(t, 0) = 0$ for all $t \in \mathbb{R}^+$, and that

$$V'_{(E)}(x, t) \leq g(t, V(x, t))$$

holds componentwise for all $(x, t) \in B(r) \times \mathbb{R}^+$. Then the following statements are true.

(a) The *stability*, *asymptotic stability*, *uniform stability*, and *uniform asymptotic stability* of the equilibrium $y_e = 0$ of (C) imply the same corresponding types of stability of the equilibrium $x_e = 0$ of (E).

(b) If in addition to the above assumptions,

$$|V(x, t)| \geq a|x|^b \text{ for } (x, t) \in B(r) \times \mathbb{R}^+,$$

where $a > 0$ and $b > 0$ are constants, then the *exponential stability* of the equilibrium $y_e = 0$ of (C) implies the exponential stability of the equilibrium $x_e = 0$ of (E).

Proof. This theorem is a direct consequence of Theorem 4.3.2(b) and (c). □

Theorem 7.7.2 With $\Omega = \mathbb{R}^n$, assume that there exists a vector Lyapunov function $V \in C[\mathbb{R}^n \times \mathbb{R}^+, (\mathbb{R}^+)^l]$ such that $|V(x,t)|$ is positive definite, decrescent, and radially unbounded, and that there exists a quasi-monotone nondecreasing function $g \in C[\mathbb{R}^+ \times (\mathbb{R}^+)^l, \mathbb{R}^l]$ such that $g(t,0) = 0$ for all $t \in \mathbb{R}^+$ and such that the inequality

$$V'_{(E)}(x,t) \leq g(t, V(x,t))$$

holds componentwise for all $(x,t) \in \mathbb{R}^n \times \mathbb{R}^+$. Then the *uniform asymptotic stability in the large* of the equilibrium $y_e = 0$ of (C) implies the uniform asymptotic stability in the large of the equilibrium $x_e = 0$ of (E). Also, the *uniform boundedness* and the *uniform ultimate boundedness* of the set of solutions of (C), imply the same corresponding types of boundedness of the set of solutions of (E).

Proof. The proof is a direct consequence of Theorem 4.3.2(d) and (f). □

In the special case when $l = 1$, g is a scalar function that automatically satisfies the quasi-monotone nondecreasing property. Accordingly, Theorems 7.7.1 and 7.7.2 are applicable to any scalar comparison system (with $l = 1$).

Next, we consider comparison systems given by

$$\dot{y} = Py + m(y,t) \tag{LC}$$

where $P = [p_{ij}] \in \mathbb{R}^{l \times l}$ and $m\colon B(r_1) \times \mathbb{R}^+ \to \mathbb{R}^l$ is assumed to satisfy the condition

$$\lim_{|y| \to 0} \frac{|m(y,t)|}{|y|} = 0, \quad \text{uniformly for } t \in \mathbb{R}^+.$$

Applying *Lyapunov's First Method* (i.e., Theorem 7.6.2) to (LC), we obtain the following comparison results.

Corollary 7.7.1 Assume that there exists a function $V \in C[B(r) \times \mathbb{R}^+, (\mathbb{R}^+)^l]$, where $B(r) \subset \Omega \subset \mathbb{R}^n$, $r > 0$, such that $|V(x,t)|$ is positive definite and decrescent, and that there exist a real $l \times l$ matrix $P = [p_{ij}]$ and a quasi-monotone nondecreasing function $m \in C[B(r_1) \times \mathbb{R}^+, \mathbb{R}^l]$, where $B(r_1) \subset (\mathbb{R}^+)^l$, $r_1 > 0$, such that the inequality

$$V'_{(E)}(x,t) \leq PV(x,t) + m(V(x,t),t) \tag{7.7.1}$$

holds componentwise for $(x,t) \in B(r) \times \mathbb{R}^+$, and that

$$\lim_{|y| \to 0} \frac{|m(y,t)|}{|y|} = 0, \quad \text{uniformly for } t \in \mathbb{R}^+$$

where $p_{ij} \geq 0$ for $1 \leq i \neq j \leq l$. Then the following statements are true:

(a) If all eigenvalues of P have negative real parts, then the equilibrium $x_e = 0$ of (E) is *uniformly asymptotically stable*.

(b) If in addition to (a), $|V(x,t)| \geq a|x|^b$ for $(x,t) \in B(r) \times \mathbb{R}^+$, where $a > 0$ and $b > 0$ are constants, then the equilibrium $x_e = 0$ of (E) is *exponentially stable*.

Proof. The proofs of the above results are a direct consequence of Theorems 7.6.2 and 7.7.1. □

In connection with Corollary 7.7.1, we find the concept of the M-matrix very useful. Before proceeding any further, it might be useful to recall the definition of the M-matrix, along with some of the properties of such matrices (see, e.g., [3]).

Definition 7.7.1 A matrix $B = [b_{ij}] \in \mathbb{R}^{l \times l}$ is called an *M-matrix* if $b_{ij} \leq 0$ for all $1 \leq i \neq j \leq l$ and the real parts of all eigenvalues of B are positive. □

In the following we enumerate several useful *equivalent* characterizations of M-matrices.

(i) B is an M-matrix.

(ii) The principal minors of B are all positive.

(iii) The successive principal minors of B are all positive.

(iv) There is a vector $u \in (\mathbb{R}_+)^l$ such that $Bu \in (\mathbb{R}_+)^l$ (recall that $\mathbb{R}_+ = (0,\infty)$).

(v) There is a vector $v \in (\mathbb{R}_+)^l$ such that $B^T v \in (\mathbb{R}_+)^l$.

(vi) B is nonsingular and all elements of B^{-1} are nonnegative (in fact, all diagonal elements of B^{-1} are positive).

Clearly, the condition in part (a) of Corollary 7.7.1 is equivalent to saying that $-P$ is an M-matrix, and thus, the equivalent conditions for M-matrix enumerated above, apply to Corollary 7.7.1(a) as well.

Example 7.7.1 Consider the system

$$\dot{x} = B(x,t)x \tag{7.7.2}$$

where $x \in \mathbb{R}^n$, $t \in \mathbb{R}^+$, and $B(x,t) = [b_{ij}(x,t)] \in C[B(r) \times \mathbb{R}^+, \mathbb{R}^{n \times n}]$, where $B(r) \subset \mathbb{R}^n$, $r > 0$. Assume that

$$b_{ii}(x,t) \leq a_{ii}, \qquad i = 1,\dots,n$$

and

$$|b_{ij}(x,t)| \leq a_{ij}, \qquad 1 \leq i \neq j \leq n,$$

for all $(x,t) \in B(r) \times \mathbb{R}^+$. Assume that $-A = -[a_{ij}] \in \mathbb{R}^{n \times n}$ is an M-matrix. Clearly, $x_e = 0$ is an equilibrium of (7.7.2).

Choose the vector Lyapunov function $V(x) = (|x_1|,\dots,|x_n|)^T$ where $x = (x_1,\dots,x_n)^T$. The upper-right Dini derivative of $|x_i|$ with respect to time is given by

$$D^+|x_i| = \begin{cases} \dot{x}_i & \text{if } x_i > 0 \\ |\dot{x}_i| & \text{if } x_i = 0 \\ -\dot{x}_i & \text{if } x_i < 0. \end{cases}$$

In all three cases ($x_i > 0$, $x_i < 0$, $x_i = 0$) we have along the solutions of (7.7.2),

$$\begin{aligned} D^+|x_i| &\le b_{ii}(x,t)|x_i| + \sum_{i\neq j} |b_{ij}(x,t)||x_j| \\ &\le a_{ii}|x_i| + \sum_{i\neq j} a_{ij}|x_j| \\ &= \sum_{j=1}^{n} a_{ij}|x_j|. \end{aligned}$$

We have

$$V'_{(7.7.2)}(x) \le AV(x)$$

for all $x \in B(r)$. Because by assumption $-A$ is an M-matrix, it follows that all eigenvalues of A have negative real parts. Hence, all conditions of Corollary 7.7.1(b) are satisfied. Therefore, the equilibrium $x_e = 0$ of (7.7.2) is *exponentially stable*. □

B. Discrete-time systems

The *object of inquiry* is systems described by difference equations

$$x(k+1) = f(k, x(k)) \tag{D}$$

where $k \in \mathbb{N}$, $f\colon \mathbb{N} \times \Omega \to \Omega$, and Ω is an open connected subset of $\mathbb{R}^n$ that contains the origin. We assume that $f(k,0) = 0$ for all $k \in \mathbb{N}$. For (D) we use *comparison systems* given by

$$y(k+1) = h(k, y(k)) \tag{DC}$$

where $y \in \mathbb{R}^l$ and $h\colon \mathbb{N} \times \Omega_1 \to (\mathbb{R}^+)^l$, where Ω_1 is an open neighborhood of the origin in $(\mathbb{R}^+)^l$. We assume that $h(k,0) = 0$ for all $k \in \mathbb{N}$.

Similarly as in the case of continuous-time systems, we employ vector-valued Lyapunov functions $V\colon \Omega \times \mathbb{N} \to (\mathbb{R}^+)^l$.

Theorem 7.7.3 Assume that there exists a function $V \in C[\Omega \times \mathbb{N}, (\mathbb{R}^+)^l]$ such that $|V(x,k)|$ is positive definite and decrescent on $\Omega \times \mathbb{N}$, and that there exists a function $h\colon \mathbb{N} \times \Omega_1 \to (\mathbb{R}^+)^l$ that is monotone nondecreasing (refer to Definition 4.3.1), that

$$V(f(k,x), k+1) \le h(k, V(x,k))$$

holds componentwise for all $(x,k) \in \Omega \times \mathbb{N}$, and that $V(x,k) \in \Omega_1$ for all $(x,k) \in \Omega \times \mathbb{N}$. Then the following statements are true.

(a) The *stability*, *asymptotic stability*, *uniform stability*, and *uniform asymptotic stability* of the equilibrium $y_e = 0$ of system (DC) imply the same corresponding types of stability of the equilibrium $x_e = 0$ of system (D).

(b) If in addition to the above assumptions, $|V(x,k)| \ge a|x|^b$ for $(x,k) \in \Omega \times \mathbb{N}$, where $a > 0$ and $b > 0$ are constants, then the *exponential stability* of the equilibrium $y_e = 0$ of (DC) implies the exponential stability of the equilibrium $x_e = 0$ of (D).

(c) Let $\Omega = \mathbb{R}^n$. If $|V(x,k)|$ is radially unbounded and decresent, then the *uniform asymptotic stability in the large* of $y_e = 0$ of (DC) implies the uniform asymptotic stability in the large of $x_e = 0$ of (D). Furthermore, the *uniform boundedness* and the *uniform ultimate boundedness* of the set of solutions of (DC) imply the same corresponding types of boundedness of the set of solutions of (D).

(d) In the case $\Omega = \mathbb{R}^n$, if $a_1|x|^b \leq |V(x,k)| \leq a_2|x|^b$ for all $(x,k) \in \mathbb{R}^n \times \mathbb{N}$, where $a_2 > a_1 > 0$ and $b > 0$ are constants, then the *exponential stability in the large* of the equilibrium $y_e = 0$ of (DC) implies the exponential stability in the large of the equilibrium $x_e = 0$ of (D). □

The proofs of the above results can be accomplished by following similar steps to the corresponding proofs for continuous-time systems given in Theorems 7.7.1 and 7.7.2, and are left to the reader as an exercise.

By applying the *First Method of Lyapunov* (Theorem 7.6.4) to Theorem 7.7.3, we can readily establish the following results.

Corollary 7.7.2 Assume for (D) that there exists a function $V \in C[\Omega \times \mathbb{N}, (\mathbb{R}^+)^l]$ such that $|V(x,k)|$ is positive definite and decrescent on $\Omega \times \mathbb{N}$, and that there exist a real $l \times l$ matrix $P = [p_{ij}] \in (\mathbb{R}^+)^{l \times l}$, and a monotone nondecreasing function $m \in C[B(r_1) \times \mathbb{N}, (\mathbb{R}^+)^l]$, where $B(r_1) \subset (\mathbb{R}^+)^l$, $r_1 > 0$, such that $V(x,k) \in B(r_1)$ for all $(x,k) \in \Omega \times \mathbb{N}$ and such that the inequality

$$V(f(k,x),k) \leq PV(x,k) + m(V(x,k),k) \tag{7.7.3}$$

holds componentwise for all $(x,k) \in \Omega \times \mathbb{N}$, and that

$$\lim_{|y| \to 0} \frac{|m(y,k)|}{|y|} = 0, \quad \text{uniformly for } k \in \mathbb{N}.$$

Under the above assumptions, the following statements are true.

(a) If all eigenvalues of P are within the unit circle of the complex plane, then the equilibrium $x_e = 0$ of (D) is *uniformly asymptotically stable.*

(b) If in addition to (a), $|V(x,k)| \geq a|x|^b$ for all $(x,k) \in \Omega \times \mathbb{N}$, where $a > 0$ and $b > 0$ are constants, then the equilibrium $x_e = 0$ of (D) is *exponentially stable.* □

7.8 Appendix: Background Material on Differential Equations and Difference Equations

In this section we address some background material concerning linear homogeneous systems of ordinary differential equations and ordinary difference equations, linear nonhomogeneous systems of ordinary differential equations, and successive approximations of the solutions of ordinary differential equations. Because this material is standard fare in ordinary differential equations and linear systems, we do not include proofs for the presented results. However, we point to sources where these proofs can be found.

A. Linear homogeneous systems of differential equations

We consider linear homogeneous systems

$$\dot{x} = A(t)x \tag{LH}$$

where $t \in J = (a, b)$, $x \in \mathbb{R}^n$, and $A \in C[J, \mathbb{R}^{n \times n}]$ ($J = (-\infty, \infty)$ is allowed). We let

$$\Omega = \{(t, x)\colon t \in J \text{ and } x \in \mathbb{R}^n\}$$

and we recall (see Chapter 2) that for every $(t_0, x_0) \in \Omega$, (LH) possesses a unique solution $\varphi(t, t_0, x_0) \triangleq \varphi(t)$ that exists over the entire interval J.

Theorem 7.8.1 The set of all solutions of (LH) on the interval J forms an n-dimensional vector space. □

For a proof of Theorem 7.8.1, refer, for example, to [17, p. 89]. This result enables us to introduce the concept of *fundamental matrix*.

Definition 7.8.1 A set of n linearly independent solutions of system (LH) on J, $\{\varphi_1, \dots, \varphi_n\}$ is called a *fundamental set* of solutions of (LH) and the $n \times n$ matrix $\Phi = [\varphi_1 \ \cdots \ \varphi_n]$ is called *a fundamental matrix of* (LH). □

Note that there are infinitely many different fundamental matrices for (LH). In the following result we let $X = [x_{ij}]$ denote an $n \times n$ matrix and we let $\dot{X} = [\dot{x}_{ij}]$. If $A(t)$ is the matrix given in (LH), then we call the system of n^2 equations,

$$\dot{X} = A(t)X \tag{7.8.1}$$

a *matrix differential equation* for (LH).

Theorem 7.8.2 A fundamental matrix Φ of (LH) satisfies the matrix equation (7.8.1) on the interval J. □

The proof of Theorem 7.8.2 follows trivially from definitions. The next result, called *Abel's formula*, is used in the proofs of several of the subsequent results of this section.

Theorem 7.8.3 If Φ is a solution of the matrix equation (7.8.1) on an interval J and if τ is any point of J, then

$$\det \Phi(t) = \det \Phi(\tau) \exp\left[\int_\tau^t \operatorname{tr} A(s)ds\right]$$

for every t. □

For a proof of Theorem 7.8.3, refer to [17, pp. 91, 92]. It follows from Theorem 7.8.3, because $\tau \in J$ is arbitrary, that either $\det \Phi(t) \neq 0$ for each $t \in J$ or that $\det \Phi(t) = 0$ for every $t \in J$.

Theorem 7.8.4 A solution Φ of the matrix equation (7.8.1) is a fundamental matrix of (LH) if and only if its determinant is nonzero for all $t \in J$. □

For a proof of Theorem 7.8.4, refer to [17, p. 93]. The next result is also required in the development of some of the subsequent results.

Theorem 7.8.5 If Φ is a fundamental matrix of (LH) and if C is any nonsingular constant $n \times n$ matrix, then ΦC is also a fundamental matrix of (LH). Moreover, if Ψ is any other fundamental matrix of (LH), then there exists a constant $n \times n$ nonsingular matrix P such that $\Psi = \Phi P$. □

For a proof of Theorem 7.8.5, refer, for example, to [17, pp. 94, 95].

In what follows, we make use of the *natural basis for* $\mathbb{R}^n$, $\{e_1, \dots, e_n\}$ where $e_1 = (1, 0, \dots, 0)^T$, $e_2 = (0, 1, 0, \dots, 0)^T, \dots, e_n = (0, \dots, 0, 1)^T$.

Definition 7.8.2 A fundamental matrix Φ of (LH) whose columns are determined by the linearly independent solutions $\varphi_1, \dots, \varphi_n$ with

$$\varphi_1(\tau) = e_1, \dots, \varphi_n(\tau) = e_n, \qquad \tau \in J,$$

is called *the state transition matrix* Φ *for* (LH). Equivalently, if Ψ is *any* fundamental matrix of (LH), then the matrix Φ determined by

$$\Phi(t, \tau) \triangleq \Psi(t)\Psi(\tau)^{-1} \qquad \text{for all } t, \tau \in J$$

is said to be *the state transition matrix of* (LH). □

Note that the state transition matrix of (LH) is uniquely determined by the matrix $A(t)$ and is independent of the particular choice of the fundamental matrix. For example, let Ψ_1 and Ψ_2 be two different fundamental matrices for (LH). By Theorem 7.8.5, there exists a constant $n \times n$ nonsingular matrix P such that $\Psi_2 = \Psi_1 P$. By the definition of the state transition matrix, we have

$$\Phi(t, \tau) = \Psi_2(t)[\Psi_2(\tau)]^{-1} = \Psi_1(t)PP^{-1}[\Psi_1(\tau)]^{-1} = \Psi_1(t)[\Psi_1(\tau)]^{-1}.$$

This shows that $\Phi(t, \tau)$ is independent of the fundamental matrix chosen.

In the next result, we summarize the principal properties of the state transition matrix for (LH).

Theorem 7.8.6 Let $\tau \in J$, let $\varphi(\tau) = \xi$, and let $\Phi(t, \tau)$ denote the state transition matrix for (LH) for all $t \in J$. Then

(i) $\Phi(t, \tau)$ is the unique solution of the matrix equation

$$\frac{\partial}{\partial t}\Phi(t, \tau) \triangleq \dot{\Phi}(t, \tau) = A(t)\Phi(t, \tau)$$

with $\Phi(\tau, \tau) = I$, the $n \times n$ identity matrix;

(ii) $\Phi(t, \tau)$ is nonsingular for all $t \in J$;

(iii) for any $t, \sigma, \tau \in J$, we have $\Phi(t, \tau) = \Phi(t, \sigma)\Phi(\sigma, \tau)$;

(iv) $[\Phi(t, \tau)]^{-1} \triangleq \Phi^{-1}(t, \tau) = \Phi(\tau, t)$ for all $t, \tau \in J$; and

(v) the unique solution $\varphi(t, \tau, \xi)$ of (LH), with $\varphi(\tau, \tau, \xi) = \xi$ specified, is given by

$$\varphi(t, \tau, \xi) = \Phi(t, \tau)\xi \quad \text{for all } t \in J. \tag{7.8.2}$$

□

For a proof of Theorem 7.8.6, refer, for example, to [17, pp. 96 and 97].

B. Linear systems with periodic coefficients

In this subsection we consider linear homogeneous systems

$$\dot{x} = A(t)x, \qquad -\infty < t < \infty, \tag{LP}$$

where $A \in C[\mathbb{R}, \mathbb{R}^{n \times n}]$ and where

$$A(t) = A(t + T) \tag{7.8.3}$$

for some $T > 0$. System (LP) is called a *linear periodic system* and T is called a *period* of $A(t)$.

Theorem 7.8.7 Let B be a nonsingular $n \times n$ matrix. Then there exists an $n \times n$ matrix C, called the *logarithm of* B, such that

$$e^C = B. \tag{7.8.4}$$

□

For a proof of Theorem 7.8.7, refer to [17, pp. 112 and 113]. The matrix C in the above result is not unique, because, for example, $e^{C+2\pi kiI} = e^C e^{2\pi ki} = e^C$ for all integers k.

Theorem 7.8.8 Let (7.8.3) be true and let $A \in C[\mathbb{R}, \mathbb{R}^{n \times n}]$. If $\Phi(t)$ is a fundamental matrix for (LP), then so is $\Phi(t + T)$, $t \in \mathbb{R}$. Moreover, corresponding to every Φ, there exist a nonsingular matrix P that is also periodic with period T and a constant matrix R such that

$$\Phi(t) = P(t)e^{tR}. \tag{7.8.5}$$

□

For a proof of Theorem 7.8.8, refer to [17, pp. 113 and 114].

Now let us suppose that $\Phi(t)$ is known only over the interval $[t_0, t_0 + T]$. Because $\Phi(t + T) = \Phi(t)C$, we have by setting $t = t_0$, $C = \Phi(t_0)^{-1}\Phi(t_0 + T)$ and R is given by $T^{-1}\log C$. $P(t) = \Phi(t)e^{-tR}$ is now determined over $[t_0, t_0 + T]$. However, $P(t)$ is periodic over $(-\infty, \infty)$. Therefore, $\Phi(t)$ is given over $(-\infty, \infty)$

by $\Phi(t) = P(t)e^{tR}$. In other words, Theorem 7.8.8 allows us to conclude that the determination of a fundamental matrix Φ for (LP) over any interval of length T, leads at once to the determination of Φ over $(-\infty, \infty)$.

Next, let Φ_1 be any other fundamental matrix for (LP) with $A(t+T) = A(t)$. Then $\Phi = \Phi_1 S$ for some constant nonsingular matrix S. Because $\Phi(t+T) = \Phi(t)e^{TR}$, we have $\Phi_1(t+T)S = \Phi_1(t)Se^{tR}$, or

$$\Phi_1(t+T) = \Phi_1(t)\big(Se^{TR}S^{-1}\big) = \Phi_1(t)e^{T(SRS^{-1})}. \tag{7.8.6}$$

Therefore, every fundamental matrix Φ_1 determines a matrix $Se^{TR}S^{-1}$ which is similar to the matrix e^{TR}.

Conversely, let S be any constant nonsingular matrix. Then there exists a fundamental matrix of (LP) such that (7.8.6) holds. Thus, although Φ does not determine R uniquely, the set of all fundamental matrices of (LP), and hence of $A(t)$, determines uniquely all quantities associated with e^{TR} that are invariant under a similarity transformation. Specifically, the set of all fundamental matrices of $A(t)$ determines a unique set of eigenvalues of the matrix e^{TR}, $\lambda_1, \ldots, \lambda_n$, which are called the *Floquet multipliers* associated with $A(t)$. None of these vanishes because $\Pi\lambda_i = \det e^{TR} \neq 0$. Also, the eigenvalues of R are called the *characteristic exponents*.

Next, we let Q be a constant nonsingular matrix such that $J = Q^{-1}RQ$ where J is the Jordan canonical form of R; that is,

$$J = \begin{bmatrix} J_0 & 0 & \cdots & 0 \\ 0 & J_1 & \cdots & 0 \\ \vdots & \vdots & \ddots & \vdots \\ 0 & 0 & \cdots & J_s \end{bmatrix}.$$

Let $\Phi_1 = \Phi Q$ and $P_1 = PQ$. From Theorem 7.8.8 we have

$$\Phi_1(t) = P_1(t)e^{tJ} \quad \text{and} \quad P_1(t+T) = P_1(t). \tag{7.8.7}$$

Let the eigenvalues of R be $\rho_1, \ldots, \rho_n$. Then

$$e^{tJ} = \begin{bmatrix} e^{tJ_0} & 0 & \cdots & 0 \\ 0 & e^{tJ_1} & \cdots & 0 \\ \vdots & \vdots & \ddots & \vdots \\ 0 & 0 & \cdots & e^{tJ_s} \end{bmatrix}$$

where

$$e^{tJ_0} = \begin{bmatrix} e^{t\rho_1} & 0 & \cdots & 0 \\ 0 & e^{t\rho_2} & \cdots & 0 \\ \vdots & \vdots & \ddots & \vdots \\ 0 & 0 & \cdots & e^{t\rho_q} \end{bmatrix}$$

and

$$e^{tJ_i} = e^{t\rho_{q+i}} \begin{bmatrix} 1 & t & t^2/2 & \cdots & t^{r_i-1}/(r_i-1)! \\ 0 & 1 & t & \cdots & t^{r_i-2}/(r_i-2)! \\ 0 & 0 & 1 & \cdots & t^{r_i-3}/(r_i-3)! \\ \vdots & \vdots & \vdots & \ddots & \vdots \\ 0 & 0 & 0 & \cdots & 1 \end{bmatrix}, \; i = 1, \dots, s, \; q + \sum_{i=1}^{s} r_i = n.$$

Now $\lambda_i = e^{T\rho_i}$. Thus, even though the ρ_i are not uniquely determined, their real parts are. In view of (7.8.7), the columns $\varphi_1, \dots, \varphi_n$ of Φ_1 are linearly independent solutions of (LP). Let $p_1, \dots, p_n$ denote the periodic column vectors of P_1. Then

$$\begin{aligned}
\varphi_1(t) &= e^{t\rho_1} p_1(t), \\
\varphi_2(t) &= e^{t\rho_2} p_2(t), \\
&\vdots \\
\varphi_q(t) &= e^{t\rho_q} p_q(t), \\
\varphi_{q+1}(t) &= e^{t\rho_{q+1}} p_{q+1}(t), \\
\varphi_{q+2}(t) &= e^{t\rho_{q+1}} (t p_{q+1}(t) + p_{q+2}(t)), \\
&\vdots \\
\varphi_{q+r_1}(t) &= e^{t\rho_{q+1}} \left(\frac{t^{r_1-1}}{(r_1-1)!} p_{q+1}(t) + \cdots + t p_{q+r_1-1}(t) + p_{q+r_1}(t) \right), \\
&\vdots \\
\varphi_{n-r_s+1}(t) &= e^{t\rho_{q+s}} p_{n-r_s+1}(t), \\
&\vdots \\
\varphi_n(t) &= e^{t\rho_{q+s}} \left(\frac{t^{r_s-1}}{(r_s-1)!} p_{n-r_s+1}(t) + \cdots + t p_{n-1}(t) + p_n(t) \right).
\end{aligned} \tag{7.8.8}$$

From (7.8.8) it is now clear that when $\text{Re}\rho_i \stackrel{\triangle}{=} \alpha_i < 0$, or equivalently, when $|\lambda_i| < 1$, then there exists a $K > 0$ such that

$$|\varphi_i(t)| \leq K e^{\alpha_i t} \to 0 \quad \text{as } t \to \infty.$$

In other words, if the eigenvalues ρ_i, $i = 1, \dots, n$, of R have negative real parts, then the norm of any solution of (LP) tends to zero as $t \to \infty$ at an exponential rate.

From (7.8.5) we have $P(t) = \Phi(t)e^{-tR}$ and therefore it is easy to see that $AP - \dot{P} = PR$. Thus, for the transformation

$$x = P(t)y \tag{7.8.9}$$

we compute

$$\dot{x} = A(t)x = A(t)P(t)y = \dot{P}(t)y + P(t)\dot{y} = \frac{d}{dt}(P(t)y)$$

or

$$\dot{y} = P^{-1}(t)(A(t)P(t) - \dot{P}(t))y = P^{-1}(t)(P(t)R)y = Ry.$$

This shows that the transformation (7.8.9) reduces the linear, homogeneous, periodic system (LP) to

$$\dot{y} = Ry,$$

a linear homogeneous system with constant coefficients.

C. Linear nonhomogeneous systems of differential equations

We consider linear nonhomogeneous systems of differential equations given by

$$\dot{x} = A(t)x + g(t) \tag{LN}$$

where $g \in C[J, \mathbb{R}^n]$ and all other symbols are as defined in (LH).

Theorem 7.8.9 Let $\tau \in J$, let $(\tau, \xi) \in J \times \mathbb{R}^n$, and let $\Phi(t, \tau)$ denote the state transition matrix for (LH) for all $t \in J$. Then the unique solution $\varphi(t, \tau, \xi)$ of (LN) satisfying $\varphi(\tau, \tau, \xi) = \xi$ is given by the *variation of constants formula*

$$\varphi(t, \tau, \xi) = \Phi(t, \tau)\xi + \int_{\tau}^{t} \Phi(t, \eta)g(\eta)d\eta. \tag{7.8.10}$$

□

For a proof of Theorem 7.8.9, refer, for example, to [17, p. 99].

D. Linear homogeneous systems of difference equations

We consider systems of linear homogeneous difference equations

$$x(k+1) = A(k)x(k), \qquad x(k_0) = x_0, \qquad k \geq k_0 \geq 0 \tag{LH_D}$$

where $A \colon \mathbb{N} \to \mathbb{R}^{n \times n}$, $x(k) \in \mathbb{R}^n$, and $k, k_0 \in \mathbb{N}$. We denote the solutions of (LH_D) by $\varphi(k, k_0, x_0)$ with $\varphi(k_0, k_0, x_0) = x_0$.

For system (LH_D), several results that are analogous to corresponding results given in Subsection 7.8A for system (LH), are still true. Thus, the set of the solutions of system (LH_D) over some subset J of $\mathbb{N}$ (say, $J = \{k_0, k_0 + 1, \ldots, k_0 + n_J\}$) forms an n-dimensional vector space. To prove this, we note that the linear combination of solutions of system (LH_D) is also a solution of system (LH_D), and hence, this set of solutions forms a vector space. The dimension of this vector space is n. To show this, we choose a set of linearly independent vector $x_0^1, \ldots, x_0^n$ in the n-dimensional x-space and we show that the set of solutions $\varphi(k, k_0, x_0^i)$, $i = 1, \ldots, n$, is linearly independent and spans the set of the solutions of (LH_D) over the set J.

If in particular, we choose $\varphi(k, k_0, e^i)$, $i = 1, \ldots, n$, where e^i, $i = 1, \ldots, n$ denotes the natural basis for $\mathbb{R}^n$, and if we let

$$\Phi(k, k_0) = [\varphi(k, k_0, e^1), \ldots, \varphi(k, k_0, e^n)],$$

then it is easily verified that the $n \times n$ matrix $\Phi(k, k_0)$ satisfies the *matrix equation*

$$\Phi(k+1, k_0) = A(k)\Phi(k, k_0), \qquad \Phi(k_0, k_0) = I,$$

where I denotes the $n \times n$ identity matrix. Furthermore,

$$\Phi(k, k_0) = \prod_{j=k_0}^{k-1} A(j)$$

and

$$\varphi(k, k_0, x_0) = \Phi(k, k_0)x_0, \qquad k > k_0.$$

Other important properties that carry over from system (LH) include, for example, the semigroup property,

$$\Phi(k, l) = \Phi(k, m)\Phi(m, l), \qquad k \geq m \geq l.$$

However, whereas in the case of system (LH) it is possible to reverse time, this is in general not valid for system (LH_D). For example, in the case of system (LH), if $\varphi(t) = \Phi(t, \tau)\varphi(\tau)$, then we can compute $\varphi(\tau) = \Phi^{-1}(t, \tau)\varphi(t) = \Phi(\tau, t)\varphi(t)$. For (LH_D), this does not apply, unless $A^{-1}(k)$ exists for all $k \in \mathbb{N}$.

E. Successive approximations of solutions of initial value problems

We consider initial value problems given by

$$\dot{x} = f(t, x), \qquad x(\tau) = \xi \tag{I}$$

where $f \in C[D, \mathbb{R}^n]$, $D = J \times \Omega$ (where $J = (a, b) \subset \mathbb{R}$ is an interval and $\Omega \subset \mathbb{R}^n$ is a domain), $\tau \in J$, and $x(\tau) \in \Omega$. For (I) we define the *successive approximations*

$$\begin{cases} \varphi_0(t) = \xi \\ \varphi_{j+1} = \xi + \displaystyle\int_\tau^t f(s, \varphi_j(s))ds, \qquad j = 0, 1, 2, \ldots \end{cases} \tag{7.8.11}$$

for $|t - \tau| \leq c$ for some $c > 0$.

Theorem 7.8.10 If $f \in C[D, \mathbb{R}^n]$, if f is Lipschitz continuous on a compact set $S \subset D$ with Lipschitz constant L, and if S contains a neighborhood of (τ, ξ), then the successive approximations φ_j, $j = 0, 1, 2, \ldots$ given in (7.8.11) exist on $|t - \tau| \leq c$ for some $c > 0$, are continuous there, and converge uniformly to the unique solution $\varphi(t, \tau, \xi)$ of (I) as $j \to \infty$. □

For a proof of Theorem 7.8.10, refer, for example, to [17, pp. 56–58].

7.9 Notes and References

There are many excellent texts on the stability of finite dimensional dynamical systems determined by ordinary differential equations that treat the topics addressed in this chapter, including Hahn [5], Hale [6], Krasovskii [8], LaSalle and Lefschetz [12], Yoshizawa [20], and Zubov [21]. Texts on these topics that emphasize engineering applications include Khalil [7] and Vidyasagar [19]. Our presentation in this chapter was greatly influenced by Antsaklis and Michel [1], Hahn [5], Michel *et al.* [16], and Miller and Michel [17].

There are fewer sources dealing with the stability analysis of discrete-time systems described by difference equations. In our presentation in this chapter, we found the texts by LaSalle [11], Antsaklis and Michel [1], Hahn [5], and Michel *et al.* [16] especially useful.

The results in Subsection 7.2A, along with other results that comprise the invariance theory for systems described by ordinary differential equations are due to Barbashin and Krasovskii [2] and LaSalle [10]. Extensions of these results to other types of dynamical system (e.g., systems described by difference equations, as in Subsection 7.2B) have been reported, for example, in Michel *et al.* [16].

The necessary and sufficient conditions for the various Lyapunov stability types presented in Subsection 7.4C involving the Lyapunov matrix equation were originally established by Lyapunov [13] for ordinary differential equations. Our presentation in Subsection 7.5C of the analogous results for systems described by difference equations are in the spirit of similar results given in LaSalle [11].

The results in Subsections 7.6B and 7.6C comprise the First Method of Lyapunov (also called the Indirect Method of Lyapunov). For the case of ordinary differential equations (Subsection 7.6B) these results were originally established by Lyapunov [13]. The results that we present in Subsection 7.6C for systems described by difference equations are along similar lines as the results given in Antsaklis and Michel [1].

The stability results for autonomous systems (A), periodic systems (P), linear homogeneous systems (LH), linear autonomous systems (L), linear periodic systems (LP), and linear second-order differential equations with constant coefficients (Section 7.1, Subsections 7.4A, 7.4B, 7.4D, and 7.4E, resp.) are standard fare in texts on stability of systems described by ordinary differential equations (e.g., [5]–[8], [12], [16], [17], [19]–[21]). Sources for the analogous results for linear systems described by difference equations (L_D) and (LH_D) (Subsections 7.5A and 7.5B) include, for example, [1] and [11]. Results to estimate the domain of attraction of an equilibrium (Subsection 7.3) are also included in most texts on stability theory of differential equations ([5], [7], [8], [12], [17], [19]–[21]). The results concerning stable and unstable manifolds and stability properties of periodic solutions in perturbed linear systems (Subsections 7.6D and 7.6E) are addressed in the usual texts on ordinary differential equations (e.g., [6], [17]). A good source on the comparison theory for differential equations (Subsection 7.7A) includes Lakshmikantham [9] and on difference equations (Subsection 7.7B), Michel *et al.* [16]. For applications of the comparison theory to large-scale dynamical systems, refer to Grujic *et al.* [4], Michel and Miller [15], and Siljak [18].

7.10 Problems

Problem 7.10.1 Consider the systems

$$\dot{x} = Ax \tag{L}$$

and

$$\dot{y} = P^{-1}APy \tag{7.10.1}$$

where $A, P \in \mathbb{R}^{n\times n}$ and where P is assumed to be nonsingular. Show that the equilibrium $x_e = 0$ of (L) is stable, exponentially stable, unstable, and completely unstable if and only if the equilibrium $y_e = 0$ of (7.10.1) has the same corresponding stability properties. □

Problem 7.10.2 There are several variants to the results that make up the Invariance Theory. Corollary 7.2.1 provides conditions for *global* asymptotic stability of the equilibrium $x_e = 0$ of system (A). In the following we ask the reader to prove a *local* result for asymptotic stability. □

Corollary 7.10.1 Assume that for system (A) there exists a function $v \in C[\Omega, \mathbb{R}]$ where $\Omega \subset \mathbb{R}^n$ is an open connected set containing the origin. Assume that v is positive definite. Assume that $v'_{(A)}(x) \leq 0$ on Ω. Suppose that the origin is the only invariant subset with respect to (A) of the set $Z = \{x \in \Omega\colon v'_{(A)}(x) = 0\}$. Then the equilibrium $x_e = 0$ of (A) is *asymptotically stable.* □

Problem 7.10.3 Consider the system

$$\begin{cases} \dot{x}_1 = x_2 - \varepsilon(x_1 - x_1^3/3) \\ \dot{x}_2 = -x_1 \end{cases} \tag{7.10.2}$$

where $\varepsilon > 0$. This system has an equilibrium at the origin $x_e = 0 \in \mathbb{R}^2$.

First show that the equilibrium $x_e = 0$ of system (7.10.2) is asymptotically stable, choosing

$$v(x_1, x_2) = \frac{1}{2}(x_1^2 + x_2^2)$$

and applying Corollary 7.10.1. Next, show that the region $\{x \in \mathbb{R}^2\colon x_1^2 + x_2^2 < 3\}$ is contained in the domain of attraction of the equilibrium $x_e = 0$ of (7.10.2). □

Problem 7.10.4 Consider the linear system

$$\dot{x} = Ax \tag{L}$$

where $x \in \mathbb{R}^n$ and $A \in \mathbb{R}^{n\times n}$. Assume that there exists a positive definite matrix G such that the matrix

$$B = A^T G + GA$$

is negative semidefinite. Prove that the equilibrium $x_e = 0$ of (L) is exponentially stable if and only if

(a) the pair (B, A) is *observable*; that is, the $n \times n^2$ matrix

$$[B \ \ BA \ \cdots \ BA^{n-1}]$$

has full rank; or

(b) the pair $(C, A - D)$ is observable, where $C = P_1 B$, $D = P_2 B$, $P_1 \in \mathbb{R}^{n \times n}$ is nonsingular and $P_2 \in \mathbb{R}^{n \times n}$ is any matrix.

Hint: Apply Corollary 7.2.1, letting $v(x) = x^T G x$. Then $v'_{(L)}(x) = x^T B x$. Show that $Z = \{x \in \mathbb{R}^n : x^T B x = 0\} = \{x \in \mathbb{R}^n : Bx = 0\}$, using the fact that B is negative semidefinite. Next, show that $\{0\} \subset \mathbb{R}^n$ is the largest invariant set in Z.

For further details, refer to Miller and Michel [16a]. □

Problem 7.10.5 Consider a mechanical system consisting of n rigid bodies with masses m_i, $i = 1, \ldots, n$, that are interconnected by springs and are subjected to viscous damping, and are described by the equations

$$\begin{cases} \dot{q} = M^{-1} p \\ \dot{p} = -Hq + KM^{-1} p \end{cases} \tag{7.10.3}$$

where $q \in \mathbb{R}^n$ denotes the position vector, $p \in \mathbb{R}^n$ is the momentum vector, $M = \operatorname{diag}[m_1, \ldots, m_n]$, $K = K^T \in \mathbb{R}^{n \times n}$, and $H = H^T \in \mathbb{R}^{n \times n}$. We assume that M and H are positive definite and that K is negative semidefinite. Prove that the equilibrium $(q^T, p^T) = (0^T, 0^T)$ of system (7.10.3) is exponentially stable if and only if $(K, M^{-1}H)$ is observable.

Hint: Apply Problem 7.10.4(b) with

$$A = \begin{bmatrix} 0 & M^{-1} \\ -H & KM^{-1} \end{bmatrix}, \qquad B = \begin{bmatrix} 0 & 0 \\ 0 & M^{-1}KM^{-1} \end{bmatrix},$$

and

$$C = D = \begin{bmatrix} 0 & 0 \\ 0 & KM^{-1} \end{bmatrix}.$$

For further details, consult Miller and Michel [16a]. □

Problem 7.10.6 In the mechanical system depicted in Figure 7.10.1, x_i denotes displacement for mass m_i, $i = 1, 2, k_1, k_2, k$ denote linear spring constants, and B_1, B_2, B denote viscous damping coefficients. We assume that $m_i > 0$, $k_i > 0$, $i = 1, 2$, $k > 0$, $B_1 \geq 0$, $B_2 \geq 0$, $B \geq 0$, and $B_1 + B_2 + B > 0$. This system is governed by the equations

$$\begin{cases} m_1 \ddot{x}_1 + k_1 x_1 + k(x_1 - x_2) + B_1 \dot{x}_1 + B(\dot{x}_1 - \dot{x}_2) = 0 \\ m_2 \ddot{x}_2 + k_2 x_2 + k(x_2 - x_1) + B_2 \dot{x}_2 + B(\dot{x}_2 - \dot{x}_1) = 0. \end{cases} \tag{7.10.4}$$

System (7.10.4) is a special case of system (7.10.3) with

$$D = \begin{bmatrix} (-B_1 - B) & B \\ B & (-B_2 - B) \end{bmatrix}, \qquad H = \begin{bmatrix} (k_1 + k) & -k \\ -k & (k_2 + k) \end{bmatrix},$$

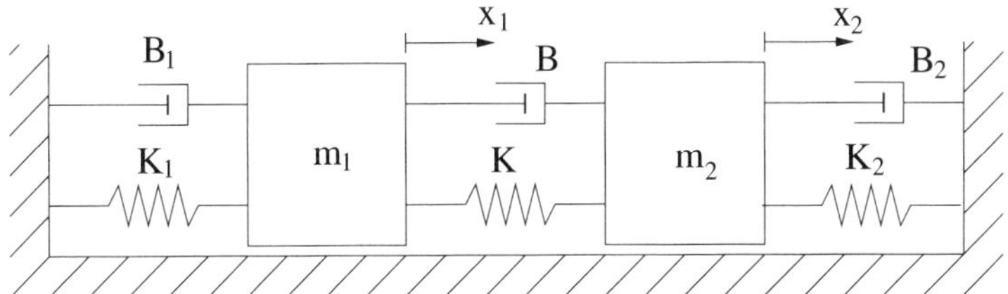

Figure 7.10.1: A mechanical system.

$M = \mathrm{diag}[m_1, m_2]$. Prove that a necessary and sufficient condition for the exponential stability of the equilibrium $x_e = 0 \in \mathbb{R}^4$ of system (7.10.4) is that

$$|\det D| + |B_1| + |B_2| + |(k_1/m_1) - (k_2/m_2)| \neq 0.$$

Hint: Use the result given in Problem 7.10.5, taking into account the following possibilities:

(i) $\det D \neq 0$.
(iia) $\det D = 0$ and $B_1 = B_2 = 0$.
(iib) $\det D = 0$ and $B_1 = B = 0$.
(iic) $\det D = 0$ and $B_2 = B = 0$.

For case (i), the pair $(D, M^{-1}H)$ is observable; for case (iia), the pair $(D, M^{-1}H)$ is observable if and only if $(k_1/m_1) \neq (k_2/m_2)$; for case (iib), the pair $(D, M^{-1}H)$ is observable if $B_1 k > 0$, which is satisfied by assumption; and for case (iic), the pair $(D, M^{-1}H)$ is observable if $B_2 k > 0$, which is true by assumption.

For further details, refer to Miller and Michel [16a]. □

Problem 7.10.7 Determine the state transition matrix $\Phi(t, t_0)$ of the system

$$\begin{bmatrix} \dot{x}_1 \\ \dot{x}_2 \end{bmatrix} = \begin{bmatrix} -t & 0 \\ (2t - t) & -2t \end{bmatrix} \begin{bmatrix} x_1 \\ x_2 \end{bmatrix}. \tag{7.10.5}$$

Use the results of Subsection 7.4A to determine the stability properties of the equilibrium $(x_1, x_2)^T = (0, 0)^T$ of system (7.10.5). □

Problem 7.10.8 Consider the equations

$$\dot{x} = (\cos t)x \tag{7.10.6}$$

and

$$\dot{x} = (4t \sin t - 2t)x. \tag{7.10.7}$$

Solve equations (7.10.6) and (7.10.7) and determine their stability properties. □

Problem 7.10.9 Prove Proposition 7.4.1. Prove Theorem 7.4.2. □

Problem 7.10.10 Show that the trivial solution of an nth-order, linear autonomous differential equation

$$a_n y^{(n)} + a_{n-1} y^{(n-1)} + \cdots + a_1 y^1 + a_0 y = 0, \qquad a_n \neq 0 \tag{7.10.8}$$

is stable if and only if all roots of

$$p(\lambda) = a_n \lambda^n + \cdots + a_1 \lambda + a_0 \tag{7.10.9}$$

have nonpositive real parts and all roots with zero real parts are simple roots. (In (7.10.8), $y^{(n)} = d^{(n)} y / dt^{(n)}$.) □

In the next few results, we use the following notation concerning quadratic forms. If $S = S^T \in \mathbb{R}^{n \times n}$ and $R = R^T \in \mathbb{R}^{n \times n}$ then we write $R > S$ if $x^T R x > x^T S x$ for all $x \in \mathbb{R}^n - \{0\}$; $R \geq S$ if $x^T R x \geq x^T S x$ for all $x \in \mathbb{R}^n$; $R > 0$ if $x^T R x > 0$ for all $x \in \mathbb{R}^n - \{0\}$; $R \geq 0$ if $x^T R x \geq 0$ for all $x \in \mathbb{R}^n$; and so forth.

Problem 7.10.11 Let $A \in C[\mathbb{R}^+, \mathbb{R}^{n \times n}]$ and $x \in \mathbb{R}^n$ and consider the system

$$\dot{x} = A(t)x. \tag{LH}$$

Show that the equilibrium $x_e = 0$ of (LH) is *uniformly stable* if there exists a $Q \in C^1[\mathbb{R}^+, \mathbb{R}^{n \times n}]$ such that $Q(t) = [Q(t)]^T$ for all t and if there exist constants $c_2 \geq c_1 > 0$ such that

$$c_1 I \leq Q(t) \leq c_2 I, \qquad t \in \mathbb{R} \tag{7.10.10}$$

and such that

$$[A(t)]^T Q(t) + Q(t) A(t) + \dot{Q} \leq 0, \qquad t \in \mathbb{R} \tag{7.10.11}$$

where I is the $n \times n$ identity matrix. □

Problem 7.10.12 Show that the equilibrium $x_e = 0$ of (LH) is *exponentially stable* if there exists a $Q \in C^1[\mathbb{R}^+, \mathbb{R}^{n \times n}]$ such that $Q(t) = [Q(t)]^T$ for all t and if there exist constants $c_2 \geq c_1 > 0$ and $c_3 > 0$ such that (7.10.10) holds and such that

$$[A(t)]^T Q(t) + Q(t) A(t) + \dot{Q}(t) \leq -c_3 I, \qquad t \in \mathbb{R}. \tag{7.10.12}$$

□

Problem 7.10.13 For (LH) let $\lambda_m(t)$ and $\lambda_M(t)$ denote the smallest and largest eigenvalues of $A(t) + [A(t)]^T$ at $t \in \mathbb{R}$, respectively. Let $\varphi(t, t_0, x_0)$ denote the unique solution of (LH) for the initial conditions $x(t_0) = x_0 = \varphi(t_0, t_0, x_0)$.

(a) Show that for any $x_0 \in \mathbb{R}^n$ and any $t_0 \in \mathbb{R}$, the unique solution of (LH) satisfies the estimate

$$|x_0| e^{(1/2) \int_{t_0}^t \lambda_m(s) ds} \leq |\varphi(t, t_0, x_0)| \leq |x_0| e^{(1/2) \int_{t_0}^t \lambda_M(s) ds}, \qquad t \geq t_0. \tag{7.10.13}$$

(b) Use the above result to show that the equilibrium $x_e = 0$ of (LH) is *uniformly stable* if there exists a constant c such that

$$\int_{\sigma}^{t} \lambda_M(s)ds \leq c \tag{7.10.14}$$

for all t, σ such that $t \geq \sigma$.

(c) Use the result in item (a) to show that the equilibrium $x_e = 0$ of (LH) is *exponentially stable* if there exist constants $\varepsilon > 0$, $\alpha > 0$ such that

$$\int_{\sigma}^{t} \lambda_M(s)ds \leq -\alpha(t-\sigma) + \varepsilon \tag{7.10.15}$$

for all t, σ such that $t \geq \sigma$. □

Problem 7.10.14 Show that if the equilibrium $x_e = 0$ of the system

$$x(k+1) = e^{A}x(k) \tag{7.10.16}$$

is asymptotically stable, then the equilibrium $x_e = 0$ of the system

$$\dot{x} = Ax \tag{L}$$

is also asymptotically stable. In systems (7.10.16) and (L), $x \in \mathbb{R}^n$, $A \in \mathbb{R}^{n\times n}$, and $k \in \mathbb{N}$. □

Problem 7.10.15 Prove Theorem 7.5.1. Prove Theorem 7.5.2. □

Problem 7.10.16 Prove Theorem 7.5.3. Prove Theorem 7.5.4. □

Problem 7.10.17 Prove Theorem 7.5.5. □

Problem 7.10.18 Prove Theorem 7.5.6. □

Problem 7.10.19 Prove Proposition 7.5.1. □

Problem 7.10.20 Prove Theorem 7.6.4. □

Problem 7.10.21 Consider the system

$$\dot{x} = f(x) \tag{A}$$

where $f \in C^1[\Omega, \mathbb{R}^n]$ and assume that x_e is an equilibrium of (A) (i.e., $f(x_e) = 0$). Define $A \in \mathbb{R}^{n\times n}$ by

$$A = \frac{\partial f}{\partial x}(x_e).$$

Prove the following.

(a) If A is a stable matrix, then the equilibrium x_e is exponentially stable.

(b) If A has an eigenvalue with a positive real part and no eigenvalue with zero real part, then the equilibrium x_e is unstable.

(c) If A is critical, then x_e can be either stable or unstable. (Show this by using specific examples.) □

Problem 7.10.22 Using Problem 7.10.21, analyze the stability properties of each equilibrium point of the following equations:

(a) $\ddot{x} + \varepsilon(x^2 - 1)\dot{x} + x = 0,\ \varepsilon \neq 0.$

(b) $\ddot{x} + \dot{x}\sin x = 0.$

(c) $\ddot{x} + \dot{x} + x(x^2 - 4) = 0.$

(d) $3x^{(3)} - 7\ddot{x} + 3\dot{x} + e^x - 1 = 0.$

(e) $\ddot{x} + c\dot{x} + \sin x = x^3,\ c \neq 0.$

(f) $\ddot{x} + 2\dot{x} + x = x^3.$ □

Problem 7.10.23 Ignoring the time dimension, for each equilibrium point in Problems 7.10.22(a)–(d), determine the dimensions of the stable manifold and the unstable manifold. □

Problem 7.10.24 Analyze the stability properties of the trivial solution (origin) of the following equations

(a) $\begin{bmatrix} \dot{x}_1 \\ \dot{x}_2 \end{bmatrix} = \begin{bmatrix} 2 & 1 \\ 7 & 3 \end{bmatrix} \begin{bmatrix} x_1 \\ x_2 \end{bmatrix} + \begin{bmatrix} (e^{x_1} - 1)\sin(x_2 t) \\ e^{-t}x_1 x_2 \end{bmatrix}.$

(b) $\begin{bmatrix} \dot{x}_1 \\ \dot{x}_2 \end{bmatrix} = \begin{bmatrix} \arctan x_1 + x_2 \\ \sin(x_1 - x_2) \end{bmatrix}.$

(c) $\begin{bmatrix} \dot{x}_1 \\ \dot{x}_2 \\ \dot{x}_3 \end{bmatrix} = \begin{bmatrix} -3 & -1 & 1 \\ -1 & -4 & 0 \\ 1 & 0 & -4 \end{bmatrix} \begin{bmatrix} x_1 \\ x_2 \\ x_3 \end{bmatrix} + \begin{bmatrix} x_1 x_2 \\ x_1 x_3 \\ \sin(x_1 x_2 x_3) \end{bmatrix}.$

(d) $\begin{cases} \dot{x}_1 = -a_0 x_2 - a_1 x_3 \\ \dot{x}_2 = b_0(e^{x_1} - 1) \\ \dot{x}_3 = -\lambda x_3 + b_1(e^{x_1} - 1) \end{cases}$

where $\lambda > 0$, $b_i \neq 0$, and $a_i/b_i > 0$, $i = 0, 1$. □

Problem 7.10.25 In Problem 7.10.24, when possible, determine a set of basis vectors for the stable manifold for each associated linearized equation. □

Problem 7.10.26 Let F satisfy Assumption 7.6.1, let $T = 2\pi$, and consider the system

$$\begin{aligned} \begin{bmatrix} \dot{x}_1 \\ \dot{x}_2 \end{bmatrix} &= \begin{bmatrix} -1 + 3(\cos^2 t)/2 & 1 - 3(\sin t \cos t)/2 \\ -1 - 3(\sin t \cos t)/2 & -1 + 3(\sin^2 t)/2 \end{bmatrix} \begin{bmatrix} x_1 \\ x_2 \end{bmatrix} + F(t, x_1, x_2) \\ &= P_0(t)x + F(t, x). \end{aligned} \tag{7.10.17}$$

(a) Show that $y(t) = (\cos t, -\sin t)^T e^{t/2}$ is a solution of

$$\dot{y} = P_0(t)y. \tag{7.10.18}$$

(b) Compute the Floquet multipliers of (7.10.18).

(c) Determine the stability properties of the trivial solution of (7.10.17).

(d) Compute the eigenvalues of $P_0(t)$. Discuss the possibility of using the eigenvalues of (7.10.18), rather than the Floquet multipliers, to determine the stability properties of the trivial solution of (7.10.17). □

Problem 7.10.27 The system described by the differential equations

$$\begin{cases} \dot{x}_1 = x_2 + x_1(x_1^2 + x_2^2) \\ \dot{x}_2 = -x_1 + x_2(x_1^2 + x_2^2) \end{cases} \tag{7.10.19}$$

has an equilibrium at the origin $(x_1, x_2)^T = (0, 0)^T$. Show that the trivial solution of the *linearization* of system (7.10.19) is *stable*. Prove that the equilibrium $x_e = 0$ of system (7.10.19) is *unstable*. (This example shows that the assumptions on the matrix A in Theorems 7.6.2 and 7.6.3 are essential.) □

Problem 7.10.28 Use the results of Section 7.7 (Comparison Theory) to show that the trivial solution of the system

$$\begin{cases} \dot{x}_1 = -x_1 - 2x_2^2 + 2kx_4 \\ \dot{x}_2 = -x_2 + 2x_1x_2 \\ \dot{x}_3 = -3x_3 + x_4 + kx_1 \\ \dot{x}_4 = -2x_4 - x_3 - kx_2 \end{cases} \tag{7.10.20}$$

is uniformly asymptotically stable when $|k|$ is small.

Hint: Choose $v_1(x_1, x_2) = x_1^2 + x_2^2$ and $v_2(x_3, x_4) = x_3^2 + x_4^2$. □

Problem 7.10.29 Prove Theorem 7.7.3. □

Bibliography

[1] P. J. Antsaklis and A. N. Michel, *Linear Systems*, Boston: Birkhäuser, 2006.

[2] E. A. Barbashin and N. N. Krasovskii, "On the stability of motion in the large," *Dokl. Akad. Nauk.*, vol. 86, pp. 453–456, 1952.

[3] M. Fiedler and V. Ptak, "On matrices with nonpositive off-diagonal elements and positive principal minors," *Czechoslovak Math. J.*, vol. 12, pp. 382–400, 1962.

[4] L. T. Grujić, A. A. Martynyuk, and M. Ribbens-Pavella, *Large Scale Systems Under Structural and Singular Perturbations*, Berlin: Springer-Verlag, 1987.

[5] W. Hahn, *Stability of Motion*, Berlin: Springer-Verlag, 1967.

[6] J. K. Hale, *Ordinary Differential Equations*, New York: Wiley, 1969.

[7] H. K. Khalil, *Nonlinear Systems*, New York: Macmillan, 1992.

[8] N. N. Krasovskii, *Stability of Motion*, Stanford, CA: Stanford University Press, 1963.

[9] V. Lakshmikantham and S. Leela, *Differential and Integral Inequalities*, vol. I and vol. II, New York: Academic Press, 1969.

[10] J. P. LaSalle, "The extent of asymptotic stability," *Proc. Nat. Acad. Sci.*, vol. 48, pp. 363–365, 1960.

[11] J. P. LaSalle, *The Stability and Control of Discrete Processes*, New York: Springer-Verlag, 1986.

[12] J. P. LaSalle and S. Lefschetz, *Stability by Liapunov's Direct Method*, New York: Academic Press, 1961.

[13] A. M. Liapounoff, "Problème générale de la stabilité de mouvement," *Annales de la Faculté des Sciences de l'Université de Toulouse*, vol. 9, pp. 203–474, 1907. (Translation of a paper published in *Comm. Soc. Math.*, Kharkow, 1892, reprinted in *Ann. Math. Studies*, vol. 17, Princeton, NJ: Princeton, 1949.)

[14] A. N. Michel and C. J. Herget, *Algebra and Analysis for Engineers and Scientists*, Boston, Birkhäuser, 2007.

[15] A. N. Michel and R. K. Miller, *Qualitative Analysis of Large Scale Dynamical Systems*, New York: Academic Press, 1977.

[16] A. N. Michel, K. Wang, and B. Hu, *Qualitative Theory of Dynamical Systems-The Role of Stability Preserving Mappings*, 2nd Edition, New York: Marcel Dekker, 2001.

[16a] R. K. Miller and A. N. Michel, "Asymptotic stability of systems: Results involving the system topology," *SIAM J. Optim. Control*, vol. 18, pp. 181–190, 1980.

[17] R. K. Miller and A. N. Michel, *Ordinary Differential Equations*, New York: Academic Press, 1982.

[18] D. D. Siljak, *Large-Scale Dynamical Systems: Stability and Structure*, New York: North Holland, 1978.

[19] M. Vidyasagar, *Nonlinear Systems Analysis*, Englewood Cliffs, NJ: Prentice Hall, 1993.

[20] T. Yoshizawa, *Stability Theory by Liapunov's Second Method*, Tokyo: Math. Soc. of Japan, 1966.

[21] V. I. Zubov, *Methods of A. M. Lyapunov and Their Applications*, Amsterdam: Noordhoff, 1964.

Chapter 8

Applications to Finite-Dimensional Dynamical Systems

In the present chapter we apply several of the results developed in Chapters 6 and 7 in the qualitative analysis of several important classes of dynamical systems, including specific classes of *continuous dynamical systems*, *discrete-time dynamical systems*, and *discontinuous dynamical systems* (DDS). The chapter is organized into five parts. First, we address the stability analysis of *nonlinear regulator systems*, using stability results for continuous dynamical systems. Next, we study the stability properties of two important classes of neural networks, *analog Hopfield neural networks* and *synchronous discrete-time Hopfield neural networks*, using stability results for continuous and discrete-time dynamical systems. In the third section we address the stability analysis of an important class of discontinuous dynamical systems, *digital control systems*, using stability results for DDS. In the fourth part we conduct a stability analysis of an important class of *pulse-width-modulated feedback control systems*. Systems of this type are continuous dynamical systems whose motions have discontinuous derivatives. We demonstrate in this section that the stability results for DDS are also well suited in the analysis of certain types of continuous dynamical systems (such as pulse-width-modulated feedback control systems). Finally, in the fifth section we address the stability analysis of an important class of dynamical *systems with saturation nonlinearities* with an application to a class of *digital filters*, using stability results for discrete-time dynamical systems.

8.1 Absolute Stability of Regulator Systems

An important class of systems that arise in control theory is regulator systems described by equations of the form

$$\begin{cases} \dot{x} = Ax + bu \\ \sigma = c^T x + du \\ u = -\varphi(\sigma) \end{cases} \tag{8.1.1}$$

where $A \in \mathbb{R}^{n\times n}$; $b, c, x \in \mathbb{R}^n$; and $d, \sigma, u \in \mathbb{R}$. We assume that $\varphi \in C[\mathbb{R}, \mathbb{R}]$ and $\varphi(0) = 0$, and is such that (8.1.1) possesses unique solutions for all $t \geq 0$ and for every $x(0) \in \mathbb{R}^n$ that depend continuously on $x(0)$.

System (8.1.1) can be represented in block diagram form as shown in Figure 8.1.1. As can be seen from this figure, system (8.1.1) may be viewed as an interconnection of a linear component with input u and output σ, and a nonlinear component with input σ and output $\varphi(\sigma)$.

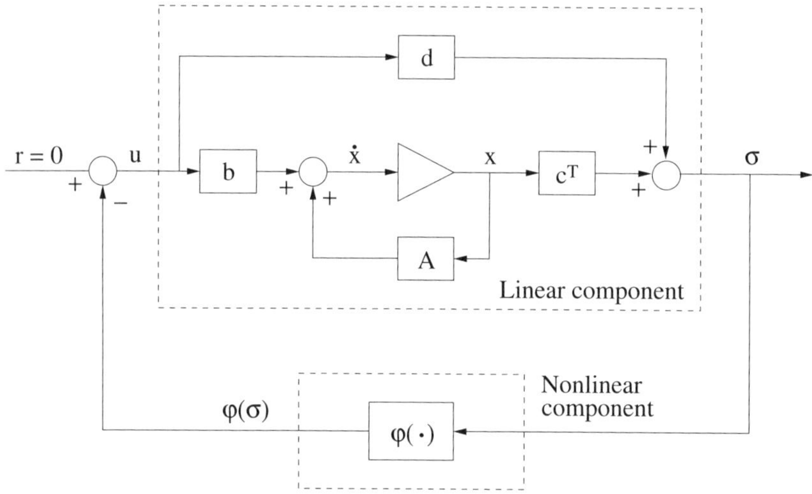

Figure 8.1.1: Block diagram of system (8.1.1).

In Figure 8.1.1 r denotes "reference input." Because we are interested in studying the stability properties of the equilibrium $x_e = 0$ of system (8.1.1), we choose $r \equiv 0$.

Assuming $x(0) = 0$ and using the Laplace transform, we can easily obtain the *transfer function* of the linear component of system (8.1.1) as

$$\hat{g}(s) = \frac{\hat{\sigma}(s)}{\hat{u}(s)} = c^T(sI - A)^{-1}b + d. \tag{8.1.2}$$

This in turn enables us to represent system (8.1.1) in block diagram form as shown in Figure 8.1.2.

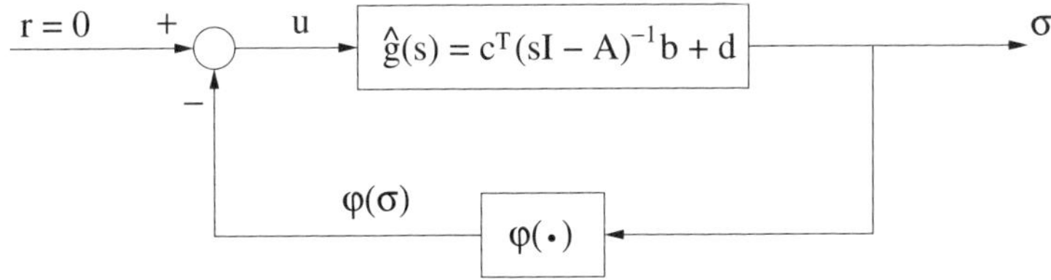

Figure 8.1.2: Block diagram of system (8.1.1).

In addition to the preceding assumptions concerning the nonlinearity $\varphi(\cdot)$, we assume, for example, that

$$k_1\sigma^2 \leq \sigma\varphi(\sigma) \leq k_2\sigma^2 \tag{8.1.3}$$

where k_1, k_2 are real constants. In this case we say that φ *belongs to the sector* $[k_1, k_2]$. Similarly, if we require that $k_1\sigma^2 < \sigma\varphi(\sigma) < k_2\sigma^2$, we say that φ belongs to the sector (k_1, k_2). Other sectors, such as $(k_1, k_2]$ and $[k_1, k_2)$ are defined similarly. Also, when $\sigma\varphi(\sigma) \geq 0$ for all $\sigma \in \mathbb{R}$, we say that φ belongs to the sector $[0, \infty)$.

Now let $d = 0$ and replace $\varphi(\sigma)$ by $k\sigma$, where $k_1 \leq k \leq k_2$. Then system (8.1.1) reduces to the *linear system*

$$\dot{x} = (A - kbc^T)x. \tag{8.1.4}$$

In 1949, Aizerman conjectured that if $d = 0$, if φ belongs to the sector $[k_1, k_2]$, and if for each $k \in [k_1, k_2]$ the matrix $(A - kbc^T)$ is a Hurwitz matrix, so that system (8.1.4) is exponentially stable in the large, then the equilibrium $x_e = 0$ of the nonlinear system (8.1.1) is uniformly asymptotically stable in the large. It turns out that this conjecture, called *Aizerman's conjecture*, is false. Nevertheless, it is still useful, because it serves as a benchmark in assessing how conservative some of the subsequent results are in particular applications.

In the present section we address the following problem: find conditions on A, b, c, d that ensure the equilibrium $x_e = 0$ of system (8.1.1) is uniformly asymptotically stable in the large for *any* nonlinearity φ belonging to some specified sector. A system (8.1.1) satisfying this property is said to be *absolutely stable.*

In the present section we address the absolute stability problem of regulator systems using two different methods: *Luré's criterion* and the *Popov criterion.*

A. Luré's result

In the following result, we assume that $d = 0$, that A is a Hurwitz matrix, and that φ belongs to the sector $[0, \infty)$. We use a Lyapunov function of the form

$$v(x) = x^T P x + \beta \int_0^\sigma \varphi(\xi)d\xi, \tag{8.1.5}$$

where $P = P^T \in \mathbb{R}^{n\times n}$ is positive definite and $\beta \geq 0$. We require that P is a solution of the Lyapunov matrix equation

$$A^T P + PA = -C \tag{8.1.6}$$

where $C = C^T$ is a positive definite matrix of our choice.

Theorem 8.1.1 (*Luré*) For system (8.1.1) assume that $d = 0$, that all eigenvalues of A have negative real parts (i.e., A is Hurwitz), and that there exist positive definite matrices P and C that satisfy (8.1.6). Also, assume that

$$\beta c^T b - w^T C^{-1} w > 0 \tag{8.1.7}$$

where

$$w = Pb - \frac{1}{2}\beta A^T c \tag{8.1.8}$$

and $\beta \geq 0$ is some constant (see (8.1.5)).

Then the equilibrium $x_e = 0$ of system (8.1.1) is asymptotically stable in the large for any φ satisfying $\sigma\phi(\sigma) \geq 0$ for all $\sigma \in \mathbb{R}$.

Proof. We choose as a Lyapunov function (8.1.5) which is continuous, positive definite, and radially unbounded (because P is positive definite and because $\sigma\varphi(\sigma) \geq 0$ for all $\sigma \in \mathbb{R}$ implies that $\int_0^\sigma \varphi(s)ds \geq 0$ for all $\sigma \in \mathbb{R}$). Along the solutions of (8.1.1) we have

$$\begin{aligned}
v'_{(8.1.1)}(x) &= x^T P(Ax - b\varphi(\sigma)) + (x^T A^T - b^T \varphi(\sigma))Px + \beta\varphi(\sigma)\dot{\sigma} \\
&= x^T(PA + A^T P)x - 2x^T Pb\varphi(\sigma) + \beta\varphi(\sigma)c^T(Ax - b\varphi(\sigma)) \\
&= -x^T Cx - 2x^T Pb\varphi(\sigma) + \beta x^T A^T c\varphi(\sigma) - \beta(c^T b)\varphi(\sigma)^2 \\
&= -x^T Cx - 2\varphi(\sigma)x^T w - \beta(c^T b)\varphi(\sigma)^2 \\
&= -(x + C^{-1}w\varphi(\sigma))^T C(x + C^{-1}w\varphi(\sigma)) - (\beta c^T b - w^T C^{-1} w)\varphi(\sigma)^2.
\end{aligned}$$

Invoking (8.1.7) and the positive definiteness of C, it follows that $v'_{(8.1.1)}(x)$ is negative definite for all $x \in \mathbb{R}^n$. Indeed, if $v'_{(8.1.1)}(x) = 0$, then $\varphi(\sigma) = 0$ and

$$x + C^{-1}w\varphi(\sigma) = x + C^{-1}w \cdot 0 = x = 0.$$

It now follows from Theorem 6.2.3 that the equilibrium $x_e = 0$ of system (8.1.1) is asymptotically stable in the large for any φ satisfying $\sigma\varphi(\sigma) \geq 0$ for all $\sigma \in \mathbb{R}$. □

B. The Popov criterion

System (8.1.1) which we considered in the preceding subsection is referred to in the literature as a *direct control system*. We now consider *indirect control systems* described by equations of the form

$$\begin{cases} \dot{x} = Ax - b\varphi(\sigma) \\ \sigma = c^T x + d\xi \\ \dot{\xi} = -\varphi(\sigma) \end{cases} \tag{8.1.9}$$

where $x, b, c \in \mathbb{R}^n$, $\sigma, \xi, d \in \mathbb{R}$, and $A \in \mathbb{R}^{n\times n}$ is assumed to be Hurwitz (i.e., all the eigenvalues of A have negative real parts). We assume that $d \neq 0$, for otherwise, (8.1.9) reduces to (8.1.1).

System (8.1.9) can be rewritten as

$$\begin{cases} \begin{bmatrix} \dot{x} \\ \dot{\xi} \end{bmatrix} = \begin{bmatrix} A & 0 \\ 0 & 0 \end{bmatrix} \begin{bmatrix} x \\ \xi \end{bmatrix} + \begin{bmatrix} b \\ 1 \end{bmatrix} \eta, \\ \sigma = \begin{bmatrix} c^T & d \end{bmatrix} \begin{bmatrix} x \\ \xi \end{bmatrix}, \\ \eta = -\varphi(\sigma). \end{cases} \tag{8.1.10}$$

Equation (8.1.10) is clearly of the same form as equation (8.1.1). However, in the present case, the matrix of the linear system component is given by

$$\widetilde{A} = \begin{bmatrix} A & 0 \\ 0 & 0 \end{bmatrix}$$

and satisfies the assumption that one eigenvalue is equal to zero because all the eigenvalues of A have negative real parts. We note that Theorem 8.1.1 (for the direct control problem) is not applicable to the indirect control problem (8.1.9).

In the following, we present the Popov stability criterion for system (8.1.9), assuming that φ belongs to the sector $(0, k]$, so that

$$0 < \sigma\varphi(\sigma) \leq k\sigma^2 \tag{8.1.11}$$

for all $\sigma \in \mathbb{R}$, $\sigma \neq 0$. In establishing his result, Popov relied heavily on results from functional analysis. Presently, we make use of the Yacubovich–Kalman Lemma to establish the absolute stability of system (8.1.9). In this lemma, which we state next, without proof, we assume that the pair (A, b) is *controllable*, that is, the matrix $[b \;\; Ab \;\; \cdots \;\; A^{n-1}b]$ has full rank.

Lemma 8.1.1 (*Yacubovich–Kalman*) Assume that $A \in \mathbb{R}^{n\times n}$ is a Hurwitz matrix and that $b \in \mathbb{R}^n$ is such that the pair (A, b) is controllable. Assume that $Q = Q^T$ is a positive definite matrix. Let $\gamma \geq 0$ and $\varepsilon > 0$. Then there exists an $n \times n$ positive definite matrix $P = P^T$ and a vector $q \in \mathbb{R}^n$ satisfying the equations

$$PA + A^T P = -qq^T - \varepsilon Q \tag{8.1.12}$$

and

$$Pb - w = \sqrt{\gamma}q \tag{8.1.13}$$

if and only if ε is sufficiently small and

$$\gamma + 2\mathrm{Re}\left[w^T(i\omega I - A)^{-1}b\right] > 0 \tag{8.1.14}$$

for all $\omega \in \mathbb{R}$, where $i = \sqrt{-1}$ and $I \in \mathbb{R}^{n\times n}$ denotes the identity matrix. □

For a proof of the Yacubovich–Kalman Lemma, please refer, for example, to Lefschetz [40, pp. 114–118].

We can rewrite system (8.1.9) as

$$\begin{cases} \dot{x} = Ax + bu \\ \sigma = c^T x + d\int u \\ u = -\varphi(\sigma) \end{cases} \tag{8.1.15}$$

where $\int u$ denotes a primary function of u. Similarly as in system (8.1.1), we may view (8.1.15) as an interconnection of a linear system component with input u and output σ, and a nonlinear component (refer to Figures 8.1.1 and 8.1.2). Assuming $x(0) = 0$ and making use of the Laplace transform, we obtain in the present case the transfer function

$$\frac{\hat{\sigma}(s)}{\hat{u}(s)} = \hat{g}(s) = \frac{d}{s} + c^T(sI - A)^{-1}b. \tag{8.1.16}$$

Theorem 8.1.2 (*Popov*) For system (8.1.9) assume that $d > 0$, that A is a Hurwitz matrix, and that there exists a nonnegative constant δ such that

$$\operatorname{Re}\big[(1 + i\omega\delta)\hat{g}(i\omega)\big] + \frac{1}{k} > 0 \tag{8.1.17}$$

for all $\omega \in \mathbb{R}$, $\omega \neq 0$, where $i = \sqrt{-1}$ and $\hat{g}(\cdot)$ is given in (8.1.16).

Then the equilibrium $(x, \xi) = (0, 0)$ of system (8.1.9) is *asymptotically stable in the large* for any φ belonging to the sector $(0, k]$.

Proof. In proving this result, we make use of Lemma 8.1.1. Choose $\alpha > 0$ and $\beta \geq 0$ such that $\delta = \beta(2\alpha d)^{-1}$. Also, choose $\gamma = \beta(c^T b + d) + (2\alpha d)/k$ and $w = \alpha dc + \beta A^T c/2$. We must show that $\gamma > 0$ and that (8.1.14) is satisfied.

Using (8.1.17) and the identity

$$s(sI - A)^{-1} = I + A(sI - A)^{-1}, \tag{8.1.18}$$

we obtain

$$\begin{aligned} 0 &< \operatorname{Re}[(1 + i\omega\delta)\hat{g}(i\omega)] + k^{-1} \\ &= k^{-1} + \delta d + \operatorname{Re}\{c^T[i\omega(i\omega I - A)^{-1}\delta + (i\omega I - A)^{-1}]b\} \\ &= k^{-1} + \delta d + \operatorname{Re}\{c^T[\delta I + \delta A(i\omega I - A)^{-1} + (i\omega I - A)^{-1}]b\} \\ &= k^{-1} + \delta(d + c^T b) + \operatorname{Re}\{c^T[(\delta A + I)(i\omega I - A)^{-1}]b\} \end{aligned}$$

for all $\omega > 0$. Let $\lambda = 1/\omega$. Then

$$\lim_{\omega\to\infty} \operatorname{Re}\{c^T[(\delta A + I)(i\omega I - A)^{-1}]b\} = \lim_{\lambda\to 0} \operatorname{Re}\{c^T[(\delta A + I)(iI - \lambda A)^{-1}]b\} = 0.$$

Therefore there exists an $\eta > 0$ such that

$$\eta \leq k^{-1} + \delta(d + c^T b) + \operatorname{Re}\{c^T[(\delta A + I)(i\omega I - A)^{-1}]b\}.$$

Letting $\omega \to \infty$, we have

$$0 < \eta \leq k^{-1} + \delta(d + c^T b) = k^{-1} + \frac{\beta}{2\alpha d(d + c^T b)} = \frac{\gamma}{2\alpha d}.$$

Therefore, $\gamma > 0$.

Next, using the identity (8.1.18) and $\delta = \beta/(2\alpha d)$, a straightforward computation shows that inequality (8.1.17) implies inequality (8.1.14) with the given choices of γ and w.

We now invoke Lemma 8.1.1 to choose P, q and $\varepsilon > 0$. Define

$$v(x,\xi) = x^T P x + \alpha d^2 \xi^2 + \beta \int_0^\sigma \varphi(s) ds$$

for the given choices of P, α, and β. Along the solutions of (8.1.9) we have

$$\begin{aligned}
&v'_{(8.1.9)}(x,\xi) \\
&\quad = x^T P(Ax - b\varphi(\sigma)) + (x^T A^T - b^T\varphi(\sigma))Px - 2d^2\alpha\xi\varphi(\sigma) + \beta\varphi(\sigma)\dot{\sigma} \\
&\quad = x^T(PA + A^T P)x - 2x^T Pb\varphi(\sigma) - 2\alpha d^2 \xi\varphi(\sigma) \\
&\qquad + \beta\varphi(\sigma)\big[c^T(Ax - b\varphi(\sigma)) - d\varphi(\sigma)\big] \\
&\quad = x^T(-qq^T - \varepsilon Q)x - 2x^T(Pb - w)\varphi(\sigma) - \beta(c^T b + d)\varphi(\sigma)^2 - 2\alpha d\sigma\varphi(\sigma) \\
&\quad = -\varepsilon x^T Q x - x^T qq^T x - 2x^T\sqrt{\gamma} q\varphi(\sigma) - \gamma\varphi(\sigma)^2 - 2\alpha d\Big[\sigma - \frac{\varphi(\sigma)}{k}\Big]\varphi(\sigma) \\
&\quad \le -\varepsilon x^T Q x - x^T qq^T x - 2x^T\sqrt{\gamma}\varphi(\sigma) - \gamma\varphi(\sigma)^2 \\
&\quad = -\varepsilon x^T Q x - \big[x^T q + \sqrt{\gamma}\varphi(\sigma)\big]^2 \\
&\quad \le 0
\end{aligned}$$

where in the preceding computations we have used the relations $w = \alpha dc + \frac{1}{2}\beta A^T c$, $Pb - w = \sqrt{\gamma} q$, $\gamma = \beta(c^T b + d) + (2\alpha d)/k$, and

$$2\alpha d\Big[\sigma - \frac{\varphi(\sigma)}{k}\Big]\varphi(\sigma) \ge 0.$$

The above inequality is true inasmuch as φ belongs to the sector $(0, k]$.

Next, we note that $v'_{(8.1.9)}(x,\xi) = 0$ implies that $x = 0$, because Q is positive definite, and that $\varphi(\sigma) = 0$. Because $\varphi(\sigma) = 0$ if and only if $\sigma = 0$ and because $\sigma = c^T x + d\xi$, where $d > 0$, it follows that $x = 0$ and $\varphi(\sigma) = 0$ implies that $(x, \xi) = 0$. Therefore, $v'_{(8.1.9)}(x,\xi)$ is negative definite.

Finally, it is clear that v is positive definite and radially unbounded. Therefore, it follows from Theorem 6.2.3 that the equilibrium $(x, \xi) = 0$ of system (8.1.9) is asymptotically stable in the large for any φ belonging to the sector $(0, k]$. □

Theorem 8.1.2 has a very useful geometric interpretation. If we plot in the complex plane, Re$[\hat{g}(i\omega)]$ versus ωIm$[\hat{g}(i\omega)]$, with ω as a parameter (such a plot is called a *Popov plot* or a *modified Nyquist plot*), then the condition (8.1.17) requires that there exists a number $\delta > 0$ such that the Popov plot of $\hat{g}(\cdot)$ lies to the right of a straight line with slope $1/\delta$ and passing through the point $-1/k + i \cdot 0$. In Figure 8.1.3 we depict a typical situation for which condition (8.1.17) is satisfied, using this interpretation.

Note that it suffices to consider only $\omega \ge 0$ in generating a Popov plot, because both Re$[\hat{g}(i\omega)]$ and ωIm$[\hat{g}(i\omega)]$ are even functions. In Figure 8.1.3, the arrow indicates the direction of increasing ω.

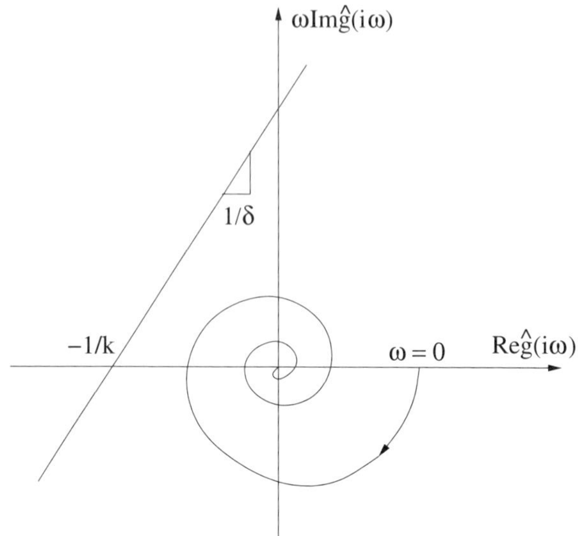

Figure 8.1.3: Geometric interpretation of inequality (8.1.17).

We conclude by noting that Popov-type results, such as Theorem 8.1.2, have also been established for direct control problems (such as system (8.1.1)) and Luré-type results, such as Theorem 8.1.1, have also been established for indirect control problems (such as system (8.1.9)). There is a large body of literature concerning the absolute stability of regulator systems, including, as pointed out in Section 8.6, several monographs.

8.2 Hopfield Neural Networks

An important class of artificial recurrent neural networks are *Hopfield neural networks* described by equations of the form

$$\dot{x} = -Bx + TS(x) + I \tag{H}$$

where $x = (x_1, \dots, x_n)^T \in \mathbb{R}^n$, $B = \text{diag}[b_1, \dots, b_n] \in \mathbb{R}^{n\times n}$ with $b_i > 0$, $1 \leq i \leq n$, $T = [T_{ij}] \in \mathbb{R}^{n\times n}$, $S(x) = [s_1(x_1), \dots, s_n(x_n)]^T : \mathbb{R}^n \to \mathbb{R}^n$, $s_i \in C^1[\mathbb{R}, (-1,1)]$ is strictly monotonically increasing with $s_i(0) = 0$ and $x_i s_i(x_i) > 0$ for all $x_i \neq 0$, and $I = [I_1, \dots, I_n]^T$, where $I_i \in C[\mathbb{R}^+, \mathbb{R}]$. Such networks, which have been popularized by Hopfield [20], have been applied in several areas, including image processing, pattern recognition, and optimization. In the application to associative memories, the external inputs I_i, $i = 1, \dots, n$, are frequently constant functions, used as bias terms. In the present section we assume that the I_i are constant functions.

Hopfield neural networks have been realized in a variety of ways, including by analog circuits, specialized digital hardware, and simulations on digital computers. In the case of the latter two, (H) is replaced by difference equations that comprise the synchronous discrete-time Hopfield neural network model. In Figure 8.2.1 we depict symbolically the realization of (H) by an analog circuit, using resistors, capacitors, operational amplifiers (capable of signal sign inversions, as required), and external inputs (bias terms). In Figure 8.2.1, dots indicate the presence of connections and the T_{ij}s denote conductances. It is easily shown that application of Kirchhoff's current law to the circuit in Figure 8.2.1 results in the system description (H) where $x_i = C_i u_i$, B and T are determined by the resistors R_i and the conductances T_{ij}, and the nonlinearities $s_i(x_i)$ are realized by the operational amplifiers.

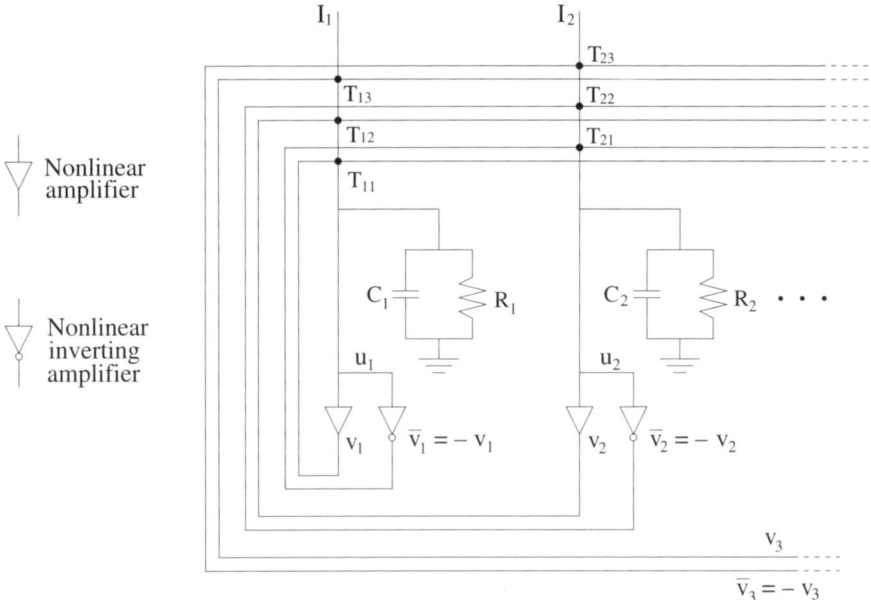

Figure 8.2.1: Hopfield neural network circuit.

In Chapter 9, we revisit the qualitative analysis of recurrent neural networks by establishing global and local stability results for Cohen–Grossberg neural networks endowed with time delays.

The present section consists of four parts. We first show that under reasonable assumptions, all states of system (H) eventually approach an equilibrium, using the Invariance Theory developed in Section 7.2 (Theorem 7.2.2). Next, we establish sufficient conditions under which an equilibrium of (H) is asymptotically stable, using the Comparison Theory developed in Section 7.7 (Corollary 7.7.1). In the third subsection we use the results of Section 7.3 to obtain an estimate for the domain of attraction of an asymptotically stable equilibrium of (H). In the final subsection we use the results of Section 6.3 (Theorem 6.3.1(c)) to establish a set of sufficient conditions for the asymptotic stability of an equilibrium of a class of synchronous discrete-time Hopfield neural networks.

A. A global result

In this subsection we present a result that serves as the basis for the application of Hopfield neural networks in the areas enumerated earlier. We require the following hypotheses.

(A–1) T is a symmetric matrix. □

(A–2) $s_i \in C^1[\mathbb{R}, (-1, 1)]$ and $\dfrac{ds_i}{d\eta}(\eta) > 0$ for all $\eta \in \mathbb{R}$. □

(A–3) System (H) has a finite number of equilibrium points. □

In [41], sufficient conditions are established which show that hypothesis (A–3) is satisfied under reasonable assumptions for (H).

Theorem 8.2.1 Assume that hypotheses (A–1), (A–2), and (A–3) are true. Then for every solution $\varphi(t)$ of (H), there exists an equilibrium x_e of (H) such that $\lim_{t\to\infty} \varphi(t) = x_e$.

Proof. In view of the results given in Chapter 2 (refer to Subsection 2.3B) it is easily established that for every $x(0) \in \mathbb{R}^n$, (H) possesses a unique solution $\varphi(t)$ with $\varphi(0) = x(0)$, which exists for all $t \geq 0$.

To apply Theorem 7.2.2 in the present case, we first need to show that every solution of (H) is bounded. To this end, let

$$c_i = \sum_{j=1}^{n} |T_{ij}| + |I_i|,$$

$i = 1, \dots, n$. Then in view of (A–2), every solution of (H), $\varphi(t) = [\varphi_1(t), \dots, \varphi_n(t)]^T$, satisfies the inequalities

$$\dot{\varphi}_i(t) \leq -b_i \varphi_i(t) + c_i,$$

$i = 1, \dots, n$. By the comparison theorem for ordinary differential equations, Theorem 3.8.1 (for the case $l = 1$), we have that $\varphi_i(t) \leq y_i(t)$ for all $t \geq 0$, where $y_i(0) = x_i(0)$ and where $y_i(t)$ is a solution of the differential equation

$$\dot{y}_i(t) = -b_i y_i(t) + c_i,$$

$i = 1, \dots, n$. Solving these equations, we have that

$$\varphi_i(t) \leq \frac{c_i}{b_i} - \Big(\frac{c_i}{b_i} - x_i(0)\Big) e^{-b_i t} \leq \frac{c_i}{b_i} + \Big|\frac{c_i}{b_i} - x_i(0)\Big|,$$

$i = 1, \dots, n$, for all $t \in \mathbb{R}^+$. Therefore, all the solutions of (H) are bounded from above. In a similar manner, we can show that all the solutions $\varphi(t)$ of (H) are bounded from below for all $t \in \mathbb{R}^+$.

Next, we choose as a Lyapunov function the "energy function" given by

$$v(x) = -\frac{1}{2}S(x)^T TS(x) - S(x)^T I + \sum_{i=1}^{n} b_i \int_0^{s_i(x_i)} s_i^{-1}(\eta)d\eta.$$

Then $v \in C^1[\mathbb{R}^n, \mathbb{R}]$ and the gradient of v is given by

$$\nabla v(x) = \nabla_x S(x)[-TS(x) - I + Bx]$$

where

$$\nabla_x S(x) = \text{diag}\left[\frac{ds_1}{dx_1}(x_1), \ldots, \frac{ds_n}{dx_n}(x_n)\right]$$

and hypothesis (A–1) has been used. Then

$$v'_{(H)}(x) = -(-Bx + TS(x) + I)^T \nabla_x S(x)(-Bx + TS(x) + I) \leq 0$$

for all $x \in \mathbb{R}^n$.

The set of all $x \in \mathbb{R}^n$ such that $v'_{(H)}(x) = 0$, given by

$$Z = \{x \in \mathbb{R}^n : v'_{(H)}(x) = 0\} = \{x \in \mathbb{R}^n : \; -Bx + TS(x) + I = 0\},$$

is an invariant set of (H) because it is precisely equal to the set of all equilibrium points of (H). By hypothesis (A–3), Z consists of a finite number of points. All the hypotheses of Theorem 7.2.2 are now satisfied and we conclude that every solution $\varphi(t)$ of (H) approaches an equilibrium of (H) as $t \to \infty$. □

B. A local result

In applications (e.g., to associative memories), asymptotically stable equilibria (called *stable memories*) are used to store information. It is important in such applications to establish conditions under which a given equilibrium x_e of (H) is asymptotically stable. We address this issue next.

Using the transformation $w = x - x_e$, system (H) assumes the form

$$\dot{w} = -Bw + T\widetilde{S}(w) \tag{8.2.1}$$

where $\widetilde{S}(w) = S(w + x_e) - S(x_e)$, $\widetilde{S}(0) = 0$, and $w_e = 0$ is an equilibrium of (8.2.1). We conclude that we may assume without loss of generality that $x_e = 0$ is an equilibrium of the Hopfield neural network and that the network has the form

$$\dot{x} = -Bx + TS(x), \tag{$\widetilde{H}$}$$

or equivalently,

$$\dot{x}_i = -b_i x_i + \sum_{j=1}^{n} T_{ij} s_j(x_j), \tag{$\widetilde{H}_i$}$$

$i = 1, \ldots, n$, where B and T are the same as in (H), $S(x) = [s_1(x_1), \ldots, s_n(x_n)]^T$ for $x = (x_1, \ldots, x_n)^T$, $s_i \in C^1[\mathbb{R}, (-2, 2)]$, s_i is strictly monotonically increasing, $s_i(0) = 0$, and $x_i s_i(x_i) > 0$ for all $x_i \neq 0$. In what follows, we make the realistic assumption that for all $i = 1, \ldots, n$,

$$0 < \frac{s_i(x_i)}{x_i} < \delta_i, \quad \text{for } 0 < |x_i| < \varepsilon_i. \tag{8.2.2}$$

System $(\widetilde{H}_i)$ (resp., $(\widetilde{H})$) can be rewritten as

$$\dot{x} = \begin{cases} F(x)x, & x \neq 0, \\ 0, & x = 0 \end{cases} \tag{8.2.3}$$

where $F(x) = [f_{ij}(x_j)]$ and

$$\begin{cases} f_{ii}(x_i) = -b_i + T_{ii}\dfrac{s_i(x_i)}{x_i}, & x_i \neq 0, \ i = 1, \ldots, n, \\ f_{ij}(x_j) = T_{ij}\dfrac{s_j(x_j)}{x_j}, & x_j \neq 0, \ 1 \leq i \neq j \leq n. \end{cases} \tag{8.2.4}$$

Now let $A = [a_{ij}]$, where

$$\begin{cases} a_{ii} = -b_i + T_{ii}\delta_i, & 1 \leq i \leq n, \\ a_{ij} = |T_{ij}|\delta_j, & 1 \leq i \neq j \leq n \end{cases} \tag{8.2.5}$$

where δ_i, $i = 1, \ldots, n$, is defined in (8.2.2).

Choosing the *vector Lyapunov function* $V(x) = \left[|x_1|, \ldots, |x_n|\right]^T$, where $x = (x_1, \ldots, x_n)^T$, and proceeding identically as in Example 7.7.1, we obtain along the solutions of $(\widetilde{H})$ the vector inequality

$$V'_{(\widetilde{H})}(x) \leq AV(x) \tag{8.2.6}$$

for all $x \in B(\varepsilon)$, where $\varepsilon = \min_{1 \leq i \leq n}\{\varepsilon_i\}$ and $\varepsilon_i > 0$, $i = 1, \ldots, n$, is given in (8.2.2). Identically as in Example 7.7.1, we now apply Corollary 7.7.1(b) to obtain the following result.

Proposition 8.2.1 The equilibrium $x_e = 0$ of $(\widetilde{H})$ is *exponentially stable* if $-A$ is an M-matrix (where A is defined in (8.2.5)). □

Remark 8.2.1 In view of Definition 7.7.1, because the matrix $D = -A$ given in Proposition 8.2.1 is an M-matrix, the following equivalent statements are true.

(a) The principal minors of $-A$ are all positive.

(b) The successive principal minors of $-A$ are all positive.

(c) There is a vector $u \in (\mathbb{R}_+)^n$ such that $-Au \in (\mathbb{R}_+)^n$ (recall that $\mathbb{R}_+ = (0, \infty)$).

(d) There is a vector $v \in (\mathbb{R}_+)^n$ such that $-A^T v \in (\mathbb{R}_+)^n$.

(e) $-A$ is nonsingular and all elements of $-A^{-1}$ are nonnegative (in fact, all diagonal elements of $-A^{-1}$ are positive). □

C. Domain of attraction

In applications (e.g., to associative memories), estimates for the domain of attraction of an asymptotically stable equilibrium are also of great interest. In the present subsection we apply the method developed in Section 7.3 to obtain estimates for the domain of attraction of the equilibrium $x_e = 0$ of system $(\widetilde{H})$.

We assume that all the hypotheses of Proposition 8.2.1 are still in place. In particular, we assume that $-A$ is an M-matrix where $A = [a_{ij}]$ is given in (8.2.5). In accordance with the property of M-matrices given in Remark 8.2.1(c), there exist $\lambda_j > 0$, $j = 1, \ldots, n$, such that

$$\sum_{j=1}^{n} \frac{a_{ij}}{\lambda_j} < 0, \qquad i = 1, \ldots, n. \tag{8.2.7}$$

Proposition 8.2.2 Assume that all conditions of Proposition 8.2.1 are true. Let

$$S = \left\{ x \in \mathbb{R}^n \colon \max_{1 \le i \le n} \lambda_i |x_i| < \min_{1 \le i \le n} \lambda_i \varepsilon_i \right\}$$

where λ_i and ε_i, $i = 1, \ldots, n$, are defined in (8.2.2) and (8.2.7), respectively. Then S is a subset of the domain of attraction of the equilibrium $x_e = 0$ of system $(\widetilde{H})$.

Proof. We choose as a Lyapunov function for $(\widetilde{H})$

$$v(x) = \max_{1 \le i \le n} \lambda_i |x_i|.$$

Along the solutions of $(\widetilde{H})$, we have, in view of (8.2.7),

$$\begin{aligned} v'_{(\widetilde{H})}(x) &\le \max_{1 \le i \le n} \lambda_i D^+ |x_i| \\ &\le \max_{1 \le i \le n} \left\{ \lambda_i a_{ii} |x_i| + \sum_{i \ne j} \lambda_i a_{ij} |x_j| \right\} \\ &\le a_{ii} v(x) + \sum_{i \ne j} \frac{\lambda_i}{\lambda_j} a_{ij} v(x) \\ &= \lambda_i \left(\sum_{j=1}^{n} \frac{a_{ij}}{\lambda_j} \right) v(x) \\ &\le 0 \end{aligned} \tag{8.2.8}$$

for all $|x_i| < \varepsilon_i$, $i = 1, \ldots, n$. In the above inequalities we have evaluated the Dini derivative $D^+|x_i|$ similarly as was done in Example 7.7.1.

For $c = \min_{1 \le i \le n} \lambda_i \varepsilon_i$, any $x \in \mathbb{R}^n$ satisfying $v(x) < c$ must satisfy $|x_i| < \varepsilon_i$ and therefore, (8.2.8) must be satisfied. From the method developed in Section 7.3 it is now clear that the set

$$S = \{x \in \mathbb{R}^n \colon v(x) < c\} = \left\{ x \in \mathbb{R}^n \colon \max_{1 \le i \le n} \lambda_i |x_i| < \min_{1 \le i \le n} \lambda_i \varepsilon_i \right\}$$

is contained in the domain of attraction of the equilibrium $x_e = 0$ of system $(\widetilde{H})$. □

D. Synchronous discrete-time Hopfield neural networks

In this subsection we establish local stability results for synchronous discrete-time Hopfield-type neural networks described by scalar-valued equations of the form

$$\begin{cases} u_i(k+1) = \sum_{j=1}^{n} T_{ij} v_j(u_j(k)) + (1 - \Delta t \cdot b_i) u_i(k) + I_i \\ \qquad\quad\;\; = \sum_{j=1}^{n} T_{ij} v_j(u_j(k)) - a_i u_i(k) + I_i \\ v_i(u_i(k)) = g_i(u_i(k)), \end{cases} \tag{W_i}$$

$i = 1, \ldots, n$, $k \in \mathbb{N}$, where $a_i = 1 - \Delta t \cdot b_i$, Δt denotes computation step size, $g_i \in C^1[\mathbb{R}, \mathbb{R}]$, $g_i(0) = 0$, $u_i g_i(u_i) > 0$ for all $u_i \neq 0$,

$$\frac{dg_i}{du_i}(u_i) > 0,$$

and $g_i(\cdot)$ satisfies the sector condition

$$d_{i1} \leq \frac{g_i(\sigma)}{\sigma} \leq d_{i2} \tag{8.2.9}$$

for all $\sigma \in B(r_i) - \{0\}$ for some $r_i > 0$, $i = 1, \ldots, n$, where $d_{i1}, d_{i2} > 0$, $i = 1, \ldots, n$, are constants.

Letting $u = (u_1, \ldots, u_n)^T \in \mathbb{R}^n$, $v = (v_1, \ldots, v_n)^T \in \mathbb{R}^n$, $T = [T_{ij}] \in \mathbb{R}^{n \times n}$, $A = \text{diag}[a_1, \ldots, a_n]$, $I = (I_1, \ldots, I_n)^T$, and $g(\cdot) = [g_1(\cdot), \ldots, g_n(\cdot)]^T \colon \mathbb{R}^n \to \mathbb{R}^n$, system (W_i) assumes the form

$$\begin{cases} u(k+1) = Tv(u(k)) + Au(k) + I, \\ v(u(k)) = g(u(k)), \end{cases}$$

$k \in \mathbb{N}$. Any point $u_e \in \mathbb{R}^n$ that satisfies the condition

$$u_e = Tg(u_e) + Au_e + I$$

or

$$0 = Tg(u_e) + Bu_e + I \tag{8.2.10}$$

where $B = A - I$, is an equilibrium for (W). Letting

$$\begin{cases} p(k) = u(k) - u_e \\ G(p(k)) = g(u(k)) - g(u_e) \end{cases} \tag{8.2.11}$$

where u_e satisfies (8.2.10) and $G(\cdot) = [G_1(\cdot), \ldots, G_n(\cdot)]^T$, system (W) reduces to the system

$$p(k+1) = TG(p(k)) + Ap(k) \tag{$\widetilde{W}$}$$

which has an equilibrium at $p_e = 0$.

System $(\widetilde{W})$ can be rewritten in component form as

$$p_i(k+1) = \sum_{j=1}^{n} T_{ij} G_j(p_j(k)) + a_i p_i(k), \qquad (\widetilde{W}_i)$$

$i = 1, \ldots, n$. From the relationship (8.2.11) it follows that the functions $G_i(\cdot)$ have the same qualitative properties as the functions $g_i(\cdot)$, $i = 1, \ldots, n$. In particular, $G_i(0) = 0$ and

$$c_{i1} \leq \frac{G_i(\sigma)}{\sigma} \leq c_{i2} \qquad (8.2.12)$$

for all $\sigma \in B(\delta_i) - \{0\}$ for some $\delta_i > 0$, $i = 1, \ldots, n$, where $c_{i1}, c_{i2} > 0$, $i = 1, \ldots, n$, are constants (in general not equal to d_{i1}, d_{i2} given in (8.2.9)).

As in [19], [51], and [64], we can view $(\widetilde{W}_i)$ as an *interconnection of n subsystems* described by equations of the form

$$x_i(k+1) = T_{ii} G_i(x_i(k)) + a_i x_i(k), \qquad (\Omega_i)$$

$i = 1, \ldots, n$, with the interconnecting structure specified by

$$h_i(x_1, \ldots, x_n) \triangleq \sum_{j=1, i \neq j}^{n} T_{ij} G_j(x_j(k)), \qquad (8.2.13)$$

$i = 1, \ldots, n$. Under this viewpoint, the stability properties of the equilibrium $p_e = 0$ of $(\widetilde{W})$ are established in terms of the qualitative properties of the subsystems (Ω_i), $i = 1, \ldots, n$, and the constraints imposed on the interconnecting structure of system $(\widetilde{W})$.

In the following, we assume that system $(\widetilde{W})$ satisfies the following hypotheses.

(B–1) For subsystem (Ω_i),

$$0 < \sigma_i \triangleq (|a_i| + |T_{ii}| c_{i2}) < 1$$

where c_{i2} is defined in (8.2.12). □

(B–2) Given σ_i in (B–1), the successive principal minors of the matrix $Q = [q_{ij}]$ are all positive, where

$$q_{ij} = \begin{cases} -(\sigma_i - 1), & 1 \leq i = j \leq n, \\ -\sigma_{ij}, & 1 \leq i \neq j \leq n \end{cases}$$

where $\sigma_{ij} = |T_{ij}| c_{j2}$ and c_{j2} is defined in (8.2.12). □

We are now in a position to prove the last result of this section.

Theorem 8.2.2 If Assumptions (B–1) and (B–2) are true, then the equilibrium $p_e = 0$ of system $(\widetilde{W})$ (resp., $(\widetilde{W}_i)$) is *asymptotically stable*.

Proof. We choose as a Lyapunov function for $(\widetilde{W})$,

$$v(p(k)) = \sum_{i=1}^{n} \lambda_i |p_i(k)|,$$

for some constants $\lambda_i > 0$, $i = 1, \ldots, n$. The first forward difference of v evaluated along the solutions of $(\widetilde{W}_i)$ is given by

$$\begin{aligned}
D_{(\widetilde{W}_i)} v(p(k)) &= v(p(k+1)) - v(p(k)) \\
&= \sum_{i=1}^{n} [|p_i(k+1)| - |p_i(k)|] \\
&= \sum_{i=1}^{n} \lambda_i \left[\left| a_i p_i(k) + \sum_{j=1}^{n} T_{ij} G_j(p_j(k)) \right| - |p_i(k)| \right] \\
&\leq \sum_{i=1}^{n} \lambda_i \left[|a_i| \, |p_i(k)| + \sum_{j=1}^{n} |T_{ij}| \, |G_j(p_j(k))| - |p_i(k)| \right] \\
&\leq \sum_{i=1}^{n} \lambda_i \left[(|a_i| - 1)|p_i(k)| + \sum_{j=1}^{n} |T_{ij}| \, |G_j(p_j(k))| \right] \\
&\leq \sum_{i=1}^{n} \lambda_i \left[(|a_i| - 1)|p_i(k)| + \sum_{j=1}^{n} |T_{ij}| c_{j2} |p_j(k)| \right] \\
&= \sum_{i=1}^{n} \lambda_i \left[(|a_i| - 1) + |T_{ii}| c_{i2}) |p_i(k)| + \sum_{j=1, i \neq j}^{n} |T_{ij}| c_{j2} |p_j(k)| \right] \\
&= \sum_{i=1}^{n} \lambda_i (\sigma_i - 1)|p_i(k)| + \sum_{i=1}^{n} \lambda_i \sum_{j=1, i \neq j}^{n} \sigma_{ij} |p_j(k)| \\
&= -\lambda^T Q w
\end{aligned}$$

where $\lambda = (\lambda_1, \ldots, \lambda_n)^T$ and $w = (|p_1|, \ldots, |p_n|)^T$. Because by (B–2), $q_{ij} \leq 0$ when $i \neq j$, and because the successive principal minors of matrix Q are all positive, it follows that Q is an M-matrix (refer to Definition 7.7.1(iii)). Moreover, Q^{-1} exists and each element of Q^{-1} is nonnegative (refer to Definition 7.7.1(vi)). Hence, there exists a vector $y = (y_1, \ldots, y_n)^T$ with $y_i > 0$, $i = 1, \ldots, n$, such that

$$-y^T w < 0 \quad \text{where } y^T = \lambda^T Q$$

and

$$\lambda = (Q^{-1})^T y > 0.$$

We have shown that $D_{(\widetilde{W}_i)} v(p(k))$ is negative for all $p(k) \in B(\delta) - \{0\}$ where $\delta = \min_{1 \leq i \leq n} \delta_i$ with δ_i defined in (8.2.12). Because v is positive definite, it follows from Theorem 6.3.1(c) that the equilibrium point $p_e = 0$ of $(\widetilde{W}_i)$ is *asymptotically stable*. □

We conclude by noting that similarly as in Subsection 8.2C, it is possible to determine estimates for the domain of attraction of the equilibrium $p_e = 0$ of $(\widetilde{W}_i)$, utilizing the method developed in Section 7.3.

8.3 Digital Control Systems

In the present section we apply the stability and boundedness results for DDS to investigate the qualitative behavior of digital feedback control systems with continuous-time plants and with digital controllers and interface elements with or without signal quantization.

A. Introduction and formulation of the problem

Digital feedback control systems, as shown in Figure 8.3.1, are hybrid dynamical systems that usually consist of an interconnection of a continuous-time plant (which can be described by a set of ordinary differential equations), a digital controller (which can be described by a set of ordinary difference equations), and interface elements (A/D and D/A converters).

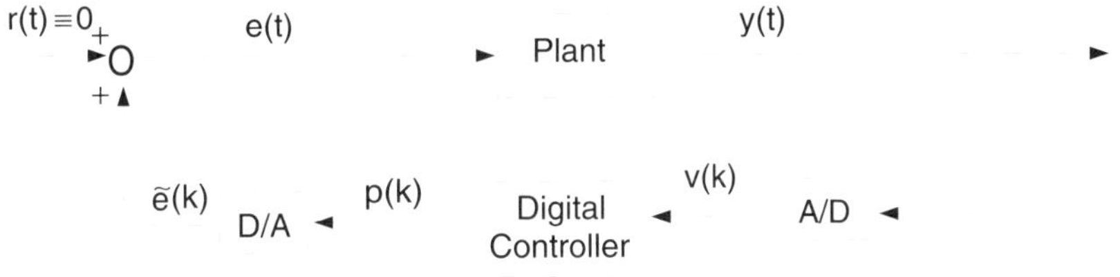

r(t) ≡ 0
e(t)
Plant
y(t)
ẽ(k)
D/A
p(k)
Digital Controller
v(k)
A/D

Figure 8.3.1: Nonlinear digital feedback control system.

The (nonlinear) plant is assumed to be given by equations of the form

$$\begin{cases} \dot{x}(t) = f(x(t)) + Fe(t), \qquad e(t) \equiv \widetilde{e}(k), \qquad t \in [k, k+1), \\ y(t) = Gx(t) \end{cases} \tag{8.3.1}$$

$k \in \mathbb{N}$, and the digital controller *without quantizers* is described by equations of the form

$$\begin{cases} u(k+1) = Cu(k) + \widetilde{F}v(k), \\ p(k) = \widetilde{G}u(k) \end{cases} \tag{8.3.2}$$

$k \in \mathbb{N}$, where $x \in \mathbb{R}^n$; $y, v \in \mathbb{R}^l$; $u \in \mathbb{R}^s$; $e, \widetilde{e}, p \in \mathbb{R}^m$; $F, G, \widetilde{F}, \widetilde{G}$, and C are real matrices of appropriate dimensions; and $f\colon \mathbb{R}^n \to \mathbb{R}^n$ is assumed to be continuously differentiable (i.e., $f \in C^1[\mathbb{R}^n, \mathbb{R}^n]$) with $f(0) = 0$. The interconnecting elements that make up the interfaces between the digital controller and the plant are A/D and D/A converters (with or without quantization). It is usually assumed that the A/D and D/A converters are synchronized in time. Also, as in Figure 8.3.1, we assume throughout that the sampling period is fixed at $T = 1$.

B. Stability analysis of systems without quantizers

In the present subsection we assume *ideal* A/D and D/A converters (i.e., there are no quantizers in the converters) and we assume infinite wordlength digital controllers (i.e., there is no quantization in the digital controller). Then $\widetilde{e}(k) = p(k) = \widetilde{G}u(k)$, $v(k) = y(k) = Gx(k)$, and the nonlinear digital feedback control system of Figure 8.3.1 is described by equations of the form

$$\begin{cases} \dot{x}(t) = f(x(t)) + Bu(k), & t \in [k, k+1) \\ u(k+1) = Cu(k) + Dx(k), \end{cases} \tag{8.3.3}$$

$k \in \mathbb{N}$, where $B = F\widetilde{G}$ and $D = \widetilde{F}G$. We note that because $f(0) = 0$, $(x^T, u^T)^T = 0$ is an equilibrium of system (8.3.3). We show that the stability (resp., instability) properties of (8.3.3) can under reasonable conditions be deduced from the associated linear system given by

$$\begin{cases} \dot{x}(t) = Ax(t) + Bu(k), & t \in [k, k+1) \\ u(k+1) = Cu(k) + Dx(k), \end{cases} \tag{8.3.4}$$

$k \in \mathbb{N}$, where $A \in \mathbb{R}^{n \times n}$ denotes the Jacobian of f evaluated at $x = 0$; that is,

$$A = \left[\frac{\partial f}{\partial x}(0)\right]_{n \times n}. \tag{8.3.5}$$

For the linear digital control system (8.3.4), the following results are well known (refer, e.g., to [10], [14], [30]).

Lemma 8.3.1 The equilibrium $(x^T, u^T)^T = (0^T, 0^T)^T$ of the linear digital control system (8.3.4) is *uniformly asymptotically stable in the large* if and only if the matrix

$$H \triangleq \begin{bmatrix} H_1 & H_2 \\ D & C \end{bmatrix} \tag{8.3.6}$$

is Schur stable, where $H_1 = e^A$ and $H_2 = \int_0^1 e^{A(1-\tau)} d\tau B$.

Lemma 8.3.2 Assume that the matrix H given in (8.3.6) has at least one eigenvalue outside the unit circle. Then the equilibrium $(x^T, u^T)^T = (0^T, 0^T)^T$ of the linear digital control system (8.3.4) is *unstable*.

We now prove the first stability result for system (8.3.3).

Theorem 8.3.1 The equilibrium $(x^T, u^T)^T = (0^T, 0^T)^T$ of the nonlinear digital control system (8.3.3) is *uniformly asymptotically stable* if the equilibrium $(x^T, u^T)^T = (0^T, 0^T)^T$ of the linear digital control system (8.3.4) is uniformly asymptotically stable, or equivalently, if the matrix H given in (8.3.6) is Schur stable.

Proof. To show that the trivial solution of system (8.3.3) is uniformly asymptotically stable, we verify that the conditions of Theorem 6.4.2 are satisfied.

Because $f \in C^1[\mathbb{R}^n, \mathbb{R}^n]$ and because $f(0) = 0$, we can represent f as

$$f(x) = Ax + g(x), \tag{8.3.7}$$

where $A \in \mathbb{R}^{n \times n}$ is given in (8.3.5) and $g \in C^1[\mathbb{R}^n, \mathbb{R}^n]$ satisfies the condition

$$\lim_{x \to 0} \frac{|g(x)|}{|x|} = 0. \tag{8.3.8}$$

The first equation in (8.3.3) now assumes the form

$$\dot{x}(t) = Ax(t) + g(x(t)) + Bu(k) \tag{8.3.9}$$

for $t \in [k, k+1)$. By the continuity of $x(t)$, the solution of equation (8.3.9) is given by

$$x(t) = e^{A(t-k)}x(k) + \int_k^t e^{A(t-\tau)}Bu(k)d\tau + \int_k^t e^{A(t-\tau)}g(x(\tau))d\tau \tag{8.3.10}$$

for all $t \in [k, k+1]$. Specifically, at $t = k+1$, we have

$$x(k+1) = e^A x(k) + \int_k^{k+1} e^{A(k+1-\tau)}d\tau Bu(k) + \int_k^{k+1} e^{A(k+1-\tau)}g(x(\tau))d\tau. \tag{8.3.11}$$

Combining (8.3.11) and the second equation in (8.3.3), we obtain

$$\begin{bmatrix} x(k+1) \\ u(k+1) \end{bmatrix} = H \begin{bmatrix} x(k) \\ u(k) \end{bmatrix} + \begin{bmatrix} \Delta(k) \\ 0 \end{bmatrix} \tag{8.3.12}$$

where

$$\Delta(k) \triangleq \int_k^{k+1} e^{A(k+1-\tau)}g(x(\tau))d\tau. \tag{8.3.13}$$

By assumption H is Schur stable. Thus there exists a positive definite symmetric matrix P such that $H^T PH - P = -I$, where $I \in \mathbb{R}^{(n+m)\times(n+m)}$ denotes the identity matrix (refer to Theorem 7.5.8). Define a Lyapunov function as

$$v(w) = w^T Pw \tag{8.3.14}$$

where $w \in \mathbb{R}^{n+m}$. Letting $w(t) = (x(t)^T, u(k)^T)^T$ when $t \in [k, k+1)$, and $m(k) \triangleq (\Delta(k)^T, 0^T)^T$, equation (8.3.12) can be written more concisely as

$$w(k+1) = Hw(k) + m(k). \tag{8.3.15}$$

The first forward difference of v evaluated along the solutions of the discrete-time system (8.3.15) yields

$$\begin{aligned}
Dv(w(k)) &\stackrel{\triangle}{=} v(w(k+1)) - v(w(k)) \\
&= w(k+1)^T P w(k+1) - w(k)^T P w(k) \\
&= [Hw(k) + m(k)]^T P [Hw(k) + m(k)] - w(k)^T P w(k) \\
&= w(k)^T [H^T P H - P] w(k) + 2m(k)^T P H w(k) + m(k)^T P m(k) \\
&= -|w(k)|^2 + 2m(k)^T P H w(k) + m(k)^T P m(k) \\
&\leq -|w(k)|^2 + 2|\Delta(k)| \|PH\| |w(k)| + |\Delta(k)|^2 \|P\|. \qquad (8.3.16)
\end{aligned}$$

Before proceeding further, we need the following result.

Proposition 8.3.1 For any given $\mu > 0$, there exists a $\delta = \delta(\mu) > 0$, such that

$$|\Delta(k)| \leq \int_k^{k+1} e^{\|A\|} |g(x(\tau))| d\tau < \mu |w(k)| \qquad (8.3.17)$$

whenever $|w(k)| < \delta$, for any $k \in \mathbb{N}$.

The proof of this result is presented at the end of this subsection. If we now choose a $\mu_0 > 0$ such that $c(\mu_0) = 1 - 2\mu_0 \|PH\| - \mu_0^2 \|P\| > 0$, then there exists a $\delta(\mu_0) > 0$ such that

$$\begin{aligned}
Dv(w(k)) &< -|w(k)|^2 + 2\mu_0 \|PH\| |w(k)|^2 + \mu_0^2 \|P\| |w(k)|^2 \\
&= -c(\mu_0) |w(k)|^2 \qquad (8.3.18)
\end{aligned}$$

whenever $|w(k)| < \delta(\mu_0)$. It follows from (8.3.14) and (8.3.18) that

$$\lambda_m(P) |w(k+1)|^2 \leq v(w(k+1)) < v(w(k)) \leq \lambda_M(P) |w(k)|^2 \qquad (8.3.19)$$

where $\lambda_m(P)$ and $\lambda_M(P)$ denote the smallest and largest eigenvalues of P, respectively. Let $d \stackrel{\triangle}{=} \sqrt{\lambda_m(P)/\lambda_M(P)}\delta(\mu_0)$. If $|w(k_0)| < d$ for some k_0, then (8.3.19) yields $|w(k_0+1)| < \delta(\mu_0)$. Thus, (8.3.18) is applicable for $k = k_0 + 1$, which yields $v(w(k_0+2)) < v(w(k_0+1)) < v(w(k_0))$. Replacing $(k+1)$ in (8.3.19) by $(k+2)$ yields $|w(k_0+2)| < \delta(\mu_0)$. By induction, it follows that $|w(k)| < \delta(\mu_0)$ for all $k \geq k_0$. Hence, (8.3.18) is satisfied for $k \geq k_0$ whenever $|w(k_0)| < d$. Therefore, (6.4.3) of Theorem 6.4.2 is satisfied.

Next, we note that for $t \in [k, k+1)$, it follows from (8.3.10) and (8.3.17) that

$$\begin{aligned}
|x(t)| &\leq e^{\|A\|} |x(k)| + e^{\|A\|} \|B\| |u(k)| + \int_k^{k+1} e^{\|A\|} \cdot |g(x(\tau))| d\tau \\
&\leq e^{\|A\|} \sqrt{1 + \|B\|^2} |w(k)| + \mu_0 |w(k)| \\
&= \left(e^{\|A\|} \sqrt{1 + \|B\|^2} + \mu_0 \right) |w(k)|. \qquad (8.3.20)
\end{aligned}$$

Hence for $k \le t < k+1$, we obtain that

$$
\begin{aligned}
v(w(t)) &\le \lambda_M(P)\big(|x(t)|^2 + |u(k)|^2\big) \\
&\le \lambda_M(P)\big(e^{\|A\|}\sqrt{1+\|B\|^2} + \mu_0 + 1\big)|w(k)|^2 \\
&\le \frac{\lambda_M(P)}{\lambda_m(P)}\big(e^{\|A\|}\sqrt{1+\|B\|^2} + \mu_0 + 1\big)v(w(k)).
\end{aligned}
$$

Let $c_1(\mu_0) = ((\lambda_M(P))/(\lambda_m(P)))\big(e^{\|A\|}\sqrt{1+\|B\|^2} + \mu_0 + 1\big)$. Then, (6.4.2) of Theorem 6.4.1 is satisfied with $f(r) = c_1(\mu_0)r$. Noting that d is independent of k_0, we conclude from Theorem 6.4.2 that the trivial solution of system (8.3.3) is uniformly asymptotically stable if H is Schur stable. □

Theorem 8.3.2 Assume that the matrix H given in (8.3.6) has no eigenvalues on the unit circle and has at least one eigenvalue outside the unit circle in the complex plane. Then the equilibrium $w_e = (x^T, u^T)^T = (0^T, 0^T)^T$ of the nonlinear digital feedback control system (8.3.3) is *unstable*.

Proof. The proof is similar to the proof of Theorem 8.3.1. By assumption, there exists a symmetric matrix P such that $H^T PH - P = I$, where $I \in \mathbb{R}^{(n+m)\times(n+m)}$ denotes the identity matrix (refer to Theorem 7.5.8). As before, we consider a Lyapunov function of the form $v(w) = w^T Pw$. Because in the present case P has at least one positive eigenvalue, there must exist points in every neighborhood of the origin where v is positive. Using a similar argument as in the proof of Theorem 8.3.1, we can show that there exists a $d > 0$ such that $v(w(k+1)) - v(w(k)) > c|w(k)|^2$ for a certain positive constant c whenever $|w(k)| < d$. Therefore, all the hypotheses of Theorem 6.4.8 are satisfied. Hence, the equilibrium $w_e = 0$ of (8.3.3) is unstable. □

Proof of Proposition 8.3.1. From (8.3.8) it follows that there exists a $\delta_1 > 0$ such that $|g(x)| \le |x|$ whenever $|x| \le \delta_1$. If we let

$$
\delta_2 = \frac{e^{-(\|A\|+1)}}{\sqrt{1+\|B\|^2}}\delta_1,
$$

then we can conclude that $|x(t)| \le \delta_1$ for all $t \in [k, k+1]$, whenever $|w(k)| \le \delta_2$. Otherwise, there must exist a $t_0 \in (k, k+1)$ such that $|x(t_0)| = \delta_1$ and $|x(t)| \le \delta_1$ for all $t \in [k, t_0]$. We show that this is impossible. For any $t \in [k, k+1]$, we have that

$$
x(t) = x(k) + \int_k^t \Big(Ax(\tau) + g(x(\tau)) + Bu(k)\Big)d\tau, \tag{8.3.21}
$$

and therefore, when $t \in [k, t_0]$, it is true that

$$
\begin{aligned}
|x(t)| &\le \Big(|x(k)| + (t-k)\|B\||u(k)|\Big) + \int_k^t \Big(\|A\||x(\tau)| + |g(x(\tau))|\Big)d\tau \\
&\le \sqrt{1+\|B\|^2}|w(k)| + \int_k^t (\|A\|+1)|x(\tau)|d\tau \qquad (8.3.22)
\end{aligned}
$$

where we have used in the last step of (8.3.22) the fact that $|g(x(\tau))| \leq |x(\tau)|$, because $|x(\tau)| \leq \delta_1$ for all $\tau \in [k, t_0]$, by assumption. By the Gronwall inequality (see, e.g., Problem 2.14.9), relation (8.3.22) implies that

$$|x(t)| \leq \sqrt{1 + \|B\|^2}|w(k)|e^{(\|A\|+1)(t-k)} \tag{8.3.23}$$

for all $t \in [k, t_0]$. Hence,

$$|x(t_0)| \leq \sqrt{1 + \|B\|^2}|w(k)|e^{(\|A\|+1)(t_0-k)} < \delta_1 \tag{8.3.24}$$

because $t_0 < k+1$. Inequality (8.3.24) contradicts the assumption that $|x(t_0)| = \delta_1$. We have shown that for any k, $|x(t)| \leq \delta_1$ for all $t \in [k, k+1]$ whenever $|w(k)| \leq \delta_2$.

For any given $\mu > 0$, we choose $\mu_1 > 0$ such that $\mu = \mu_1 \cdot e^{(2\|A\|+1)}\sqrt{1 + \|B\|^2}$. There exists a $\delta_3 > 0$ such that $|g(x)| < \mu_1|x|$, whenever $|x| < \delta_3$. Let

$$\delta \triangleq \min\left\{\delta_2, \frac{\delta_3}{\sqrt{1 + \|B\|^2}e^{(\|A\|+1)}}\right\}.$$

It now follows from (8.3.23) that whenever $|w(k)| \leq \delta$, then

$$|x(t)| \leq \delta\sqrt{1 + \|B\|^2}e^{(\|A\|+1)} \leq \delta_3$$

for all $t \in [k, k+1]$. Hence, for $\Delta(k)$ given by (8.3.13), we obtain

$$\begin{aligned}|\Delta(k)| &\leq \int_k^{k+1} e^{\|A\|}|g(x(\tau))|d\tau \\ &\leq e^{\|A\|}\mu_1\sqrt{1 + \|B\|^2}|w(k)|e^{(\|A\|+1)} \\ &= \mu \cdot |w(k)|\end{aligned}$$

whenever $|w(k)| \leq \delta$. □

C. Analysis of systems with quantization nonlinearities

In the implementation of digital controllers, quantization is unavoidable. This is due to the fact that computers store numbers with finite bits. In the present subsection, we investigate the nonlinear effects caused by quantization.

There are many types of quantization (see, e.g., [13], [55], and [56]). Presently, we concern ourselves primarily with the most commonly used fixed-point quantization which can be characterized by the relation

$$Q(\theta) = \theta + q(\theta) \tag{8.3.25}$$

where $|q(\theta)| < \varepsilon$, for all $\theta \in \mathbb{R}$ and ε depends on the desired precision.

If we add fixed-point quantization to both the sampler (A/D converter) and to the digital controller of the nonlinear digital feedback control system of Figure 8.3.1, assuming $r \equiv 0$, we obtain

$$v(k) = Q(y(k)) = y(k) + q_1(y(k))$$

$$u(k+1) = Q(Cu(k) + \widetilde{F}v(k)) = Cu(k) + \widetilde{F}Gx(k) + q_2(Cu(k) + \widetilde{F}v(k))$$
$$p(k) = Q(\widetilde{G}u(k)) = \widetilde{G}u(k) + q_3(\widetilde{G}u(k)),$$

$k \in \mathbb{N}$, where q_1, q_2, and q_3 should be interpreted as vectors whose components contain quantization terms. By a slight abuse of notation, we henceforth write $q_1(k)$ in place of $q_1(y(k))$, $q_2(k)$ in place of $q_2(Cu(k) + \widetilde{F}v(k))$, and so forth. It is easily verified that there exist positive constants J_i that are independent of ε such that $|q_i(k)| \leq J_i\varepsilon$, $i = 1, 2, 3$, $k \in \mathbb{N}$. (For further details concerning the inclusion of quantizers into digital controllers, refer to [13], [55], [56], and [69].)

In the presence of the quantizer nonlinearities, we can no longer expect that the system of Figure 8.3.1 will have a uniformly asymptotically stable equilibrium at the origin; in fact, there may not even be an equilibrium at the origin. In view of this, we investigate the (ultimate) boundedness of the solutions of the system of Figure 8.3.1, including the dependence of the bounds on the quantization size.

In the following, we represent the system of Figure 8.3.1 by the equations

$$\begin{cases} \dot{x}(t) = Ax(t) + g(x(t)) + F\widetilde{G}u(k) + Fq_3(k), & t \in [k, k+1) \\ u(k+1) = Cu(k) + \widetilde{F}Gx(k) + \widetilde{F}q_1(k) + q_2(k), \end{cases} \tag{8.3.26}$$

$k \in \mathbb{N}$. Letting $w(t) = (x(k)^T, u(k)^T)^T$ when $t \in [k, k+1)$, we obtain similarly as in Subsection B, the equivalent representation of (8.3.26), valid at sampling instants, as

$$w(k+1) = Hw(k) + m(k) \tag{8.3.27}$$

where H is defined as in (8.3.6) and where

$$m(k) = \begin{bmatrix} \int_k^{k+1} e^{A(k+1-\tau)} g(x(\tau))d\tau + \int_0^1 e^{A\tau}d\tau F q_3(k) \\ \widetilde{F}q_1(k) + q_2(k) \end{bmatrix}. \tag{8.3.28}$$

Now assume that H is Schur stable. Then there exists a symmetric positive definite matrix P such that $H^T PH - P = -I$. As in Subsection B, we choose as a Lyapunov function $v\colon \mathbb{R}^{n+m} \to \mathbb{R}^+$,

$$v(w) = w^T Pw. \tag{8.3.29}$$

Lemma 8.3.3 For any $d > 0$ that satisfies the relation

$$1 - 2d\|PH\| - d^2\|P\| > 0, \tag{8.3.30}$$

there exists a $\delta = \delta(d) > 0$ such that the estimate

$$|m(k)| \leq d|w(k)| + J\varepsilon, \qquad k \in \mathbb{N} \tag{8.3.31}$$

holds whenever $\sqrt{1 + \|B\|^2}|w(k)| + \left\|\int_0^1 e^{A\tau}d\tau\right\|\|F\|J_3\varepsilon < \delta$, where J is a positive constant independent of ε. (Recall that ε denotes a bound for the quantization size (see (8.3.25)) and J_3 is obtained from the estimate $|q_3(k)| \leq J_3\varepsilon$.)

Proof. The existence of d satisfying (8.3.30) is clear. By a similar argument as in the proof of Proposition 8.3.1, there exists a $\delta > 0$ such that $\int_k^{k+1} e^{\|A\|}|g(x(\tau))|d\tau < d|w(k)|$, whenever $\sqrt{1+\|B\|^2}|w(k)| + \|\int_0^1 e^{A\tau}d\tau\|\|F\|J_3\varepsilon < \delta$. Therefore,

$$\begin{aligned}|m(k)| &\le \left|\int_k^{k+1} e^{A(k+1-\tau)}g(x(\tau))d\tau + \int_0^1 e^{A\tau}d\tau F q_3(k)\right| + \left|\widetilde{F}q_1(k) + q_2(k)\right| \\ &\le d|w(k)| + e^{\|A\|}\|F\|J_3\varepsilon + \|\widetilde{F}\|J_1\varepsilon + J_2\varepsilon \\ &= d|w(k)| + J\varepsilon \qquad (8.3.32)\end{aligned}$$

where $J = e^{\|A\|}\|F\|J_3 + \|\widetilde{F}\|J_1 + J_2$ whenever

$$\sqrt{1+\|B\|^2}|w(k)| + \left\|\int_0^1 e^{A\tau}d\tau\right\| \|F\|J_3\varepsilon < \delta. \qquad \square$$

Now let us consider the Lyapunov function $v(w)$ given in (8.3.14). We compute the first forward difference of v along the solutions of the discrete-time system (8.3.26) to obtain

$$\begin{aligned}Dv(w(k)) &\triangleq v(w(k+1)) - v(w(k)) \\ &= w(k+1)^T P w(k+1) - w(k)^T P w(k) \\ &= [Hw(k) + m(k)]^T P[Hw(k) + m(k)] - w(k)^T P w(k) \\ &= w(k)^T[H^T P H - P]w(k) + 2m(k)^T P H w(k) + m(k)^T P m(k) \\ &= -|w(k)|^2 + 2m(k)^T P H w(k) + m(k)^T P m(k) \\ &\le -a_1|w(k)|^2 + a_2|w(k)|\varepsilon + a_3\varepsilon^2, \qquad (8.3.33)\end{aligned}$$

where $a_1 = 1 - 2d\|PH\| - d^2\|P\|$, $a_2 = 2(\|PH\| + d\|P\|)J$, and $a_3 = J^2\|P\|$. Let

$$R = a_2 + \frac{\sqrt{a_2^2 + 4a_1a_3}}{2a_1}.$$

We are now in a position to prove the following result.

Theorem 8.3.3 (i) If the matrix H defined in (8.3.6) is Schur stable, then the solutions of system (8.3.26) are *uniformly bounded*, provided that

$$\sqrt{1+\|B\|^2}|w(k_0)| + \left\|\int_0^1 e^{A\tau}d\tau\right\| \|F\|J_3\varepsilon < \delta,$$

for some $\delta > 0$. (ii) Let $L \triangleq \max\left\{\left(R\sqrt{1+\|B\|^2} + J_3\|F\|\right)e^{\|A\|+b}, \lambda_M(P)R\right\}$, where b is chosen such that $|g(x)| < b|x|$ for all $|x| \le R\varepsilon$. Then for sufficiently large k, the estimates

$$|w(k)| \le L\varepsilon \qquad (8.3.34)$$

and

$$|x(t)| \le \varepsilon\left(L\sqrt{1+\|B\|^2} + J_3\|F\|\right)e^{\|A\|+b_1} \qquad (8.3.35)$$

hold, where b_1 is such that $|g(x)| < b_1|x|$ for all $|x| \le L$.

Proof. We apply Theorem 6.4.4 in the present proof.

It is readily verified that under the present assumptions $Dv(w(k))$ is negative whenever $|w(k)| > R\varepsilon$. Hence, (6.4.10) of Theorem 6.4.4 is satisfied with $\Omega = R\varepsilon$. Furthermore, if $|w(k)| > R\varepsilon$, then $v(w(k+1)) < v(w(k))$ and thus $|w(k+1)| \leq \lambda_M(P)R\varepsilon \leq L\varepsilon$.

If $|w(k)| \leq R\varepsilon$, then by applying the Gronwall inequality to equation (8.3.26) when $t = k+1$, we obtain that $|w(k+1)| \leq L\varepsilon$. Thus, the last hypothesis of Theorem 6.4.4 is satisfied with $\Gamma = L\varepsilon$.

Solving the first equation in (8.3.26), we obtain

$$
\begin{aligned}
x(t) = e^{A(t-k)}x(k) &+ \int_k^t e^{A(t-\tau)}F\widetilde{G}u(k)d\tau + \int_k^t e^{A(t-\tau)}g(x(\tau))d\tau \\
&+ \int_k^t e^{A(t-\tau)}Fq_3(k)d\tau
\end{aligned}
$$

for $t \in (k,\ k+1)$ and therefore, when $|w(k)| \leq R\varepsilon$ it is true that

$$
\begin{aligned}
|x(t)| &\leq e^{\|A\|}|x(k)| + e^{\|A\|}\|B\||u(k)| + \int_k^{k+1} e^{\|A\|}|g(x(\tau))|d\tau + e^{\|A\|}J_3\|F\|\varepsilon \\
&\leq e^{\|A\|}\sqrt{1+\|B\|^2}|w(k)| + d|w(k)| + e^{\|A\|}J_3\|F\|\varepsilon.
\end{aligned}
$$

We have used the fact that $\int_k^{k+1} e^{\|A\|}|g(x(\tau))|d\tau < d|w(k)|$, whenever (refer to the proof of Lemma 8.3.3)

$$
\sqrt{1+\|B\|^2}|w(k)| + \left\|\int_0^1 e^{A\tau}d\tau\right\| \|F\|J_3\varepsilon < \delta.
$$

Therefore, (6.4.11) of Theorem 6.4.4 is satisfied with

$$
f(r) = \left(e^{\|A\|}\sqrt{1+\|B\|^2} + d\right)r + e^{\|A\|}J_3\|F\|\varepsilon.
$$

It now follows from Theorem 6.4.4 that the solutions of system (8.3.26) are *uniformly bounded.*

We have also shown above that for sufficiently large k, $|w(k)| \leq L\varepsilon$ holds. Finally, for $t \in (k, k+1)$, we apply the same argument as in the proof of Theorem 8.3.1 to obtain the bound (8.3.35) for $|x(t)|$. This concludes the proof of the theorem. □

In our final result we consider the difference in the response of the nonlinear digital control system with ideal samplers, given by equation (8.3.3), and the nonlinear digital control system with quantizers, given by (8.3.26). For our present purposes we rewrite (8.3.15) as

$$
\widetilde{w}(k+1) = H\widetilde{w}(k) + \widetilde{m}(k), \tag{8.3.36}
$$

where $\widetilde{w}(k) = (\widetilde{x}(k)^T, \widetilde{u}(k)^T)^T$ and

$$
\widetilde{m}(k) \triangleq \begin{bmatrix} \int_k^{k+1} e^{A(k+1-\tau)}g(\widetilde{x}(\tau))d\tau \\ 0 \end{bmatrix}. \tag{8.3.37}
$$

Letting $z(k) = w(k) - \widetilde{w}(k)$, we obtain the relation

$$z(k+1) = Hz(k) + \left[\begin{array}{c} \int_k^{k+1} e^{A(k+1-\tau)} \Big(g(x(\tau)) - g(\widetilde{x}(\tau)) \Big) d\tau + \int_0^1 e^{A\tau} d\tau F q_3(k) \\ \widetilde{F} q_1(k) + q_2(k) \end{array} \right]. \tag{8.3.38}$$

This equation is in the same form as equation (8.3.26), except that in (8.3.38) the nonlinearity includes the term $\int_k^{k+1} e^{A(k+1-\tau)} \big(g(x(\tau)) - g(\widetilde{x}(\tau))\big) d\tau$, rather than the term $\int_k^{k+1} e^{A(k+1-\tau)} g(x(\tau)) d\tau$. Now suppose that $g(\cdot)$ has the property

$$\lim_{x \to 0, \widetilde{x} \to 0} \frac{|g(x) - g(\widetilde{x})|}{|x - \widetilde{x}|} = 0, \tag{8.3.39}$$

which plays a similar role for system (8.3.38) as (8.3.8) does for system (8.3.26). Using similar arguments as in the proof of Theorem 8.3.3, we obtain the following result for the boundedness of $z(k)$, $k = 0, 1, \ldots$.

Theorem 8.3.4 Assume that H defined in (8.3.6) is Schur stable and $g(\cdot)$ satisfies (8.3.39). Then there exist a $d > 0$, a $K > 0$, and an $\varepsilon_0 > 0$ such that

$$|z(k)| \leq K\varepsilon \quad \text{when } k \text{ is sufficiently large} \tag{8.3.40}$$

whenever $\varepsilon < \varepsilon_0$, $|w(k_0)| < d$, and $|\widetilde{w}(k_0)| < d$, for some k_0, where ε is the quantization level. □

D. Examples

The purpose of the following specific example is to show that all conditions of Theorem 8.3.1 can be satisfied.

Example 8.3.1 In system (8.3.3) (resp., (8.3.4)) take

$$A = \begin{bmatrix} -0.6 & -1 \\ 0.8 & 0 \end{bmatrix}, \qquad B = \begin{bmatrix} 0 \\ 0.6 \end{bmatrix}, \qquad D = [-0.8 \ -0.3], \qquad C = [-1],$$

and in (8.3.7), take

$$g(x) = \begin{bmatrix} 0.013 \sin(x_1) \\ 0.008 x_2 \cos(x_2) \end{bmatrix}.$$

Then $|g(x)| \leq \alpha |x|$ for all x, where $\alpha = 0.0083$. We also compute that

$$H = \begin{bmatrix} 0.2962 & -0.6562 & 0.5174 \\ 0.5250 & 0.6899 & 0.0833 \\ -0.8000 & 0.3000 & -1.0000 \end{bmatrix}, \qquad P = \begin{bmatrix} 3.2515 & 0.6514 & 1.8906 \\ 0.6514 & 2.4873 & -0.0755 \\ 1.8906 & -0.0755 & 2.9856 \end{bmatrix}.$$

μ_0 is computed to be 0.1225, and $\mu_0^2\|P\| + 2\mu_0\|PH\| = 0.9035 < 1$. It follows from Theorem 8.3.1 that the equilibrium $(x^T, u^T)^T = (0^T, 0^T)^T$ of this system is uniformly asymptotically stable; in fact, it is uniformly asymptotically stable in the large because the conditions of Theorem 8.3.1 are satisfied for all x (i.e., $\delta(\mu_0) = +\infty$, where $\delta(\mu_0)$ is given in Proposition 8.3.1). □

Example 8.3.2 The present case is an example of the digital control of a nonlinear plant (whose linearization is a double-integrator) adopted from [13]. The system is given by

$$\begin{bmatrix} \dot{x}_1 \\ \dot{x}_2 \end{bmatrix} = \begin{bmatrix} 0 & 1 \\ 0 & 0 \end{bmatrix} \begin{bmatrix} x_1 \\ x_2 \end{bmatrix} + \begin{bmatrix} 0 \\ x_1^2 \end{bmatrix} - \begin{bmatrix} 0 \\ 1 \end{bmatrix} e, \qquad y = x_1.$$

The controller is given by

$$u(k+1) = \begin{bmatrix} 0 & 1 \\ b & a \end{bmatrix} u(k) + \begin{bmatrix} 1 \\ 0 \end{bmatrix} v(k), \qquad w(k) = [d \;\; c]u(k).$$

We choose $a = 0$, $b = -0.3$, $c = 4.4$, $d = -4.0$, and $T = .25$. Also, we assume fixed-point magnitude truncation quantization with $\varepsilon = 0.01$.

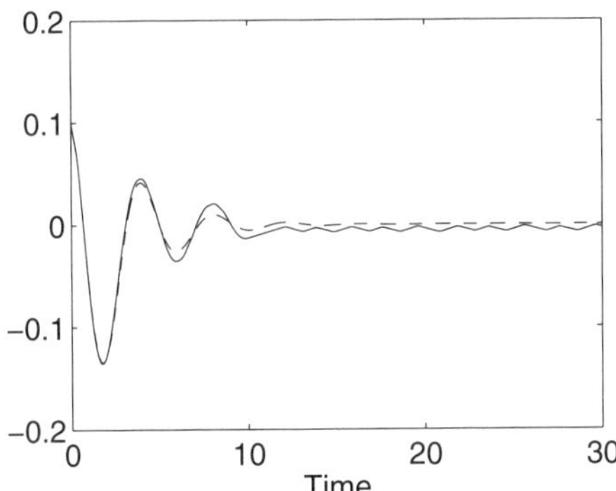

Figure 8.3.2: Output $y(t)$: dashed line, without quantization; solid line, with quantization (Example 8.3.2).

In Figure 8.3.2 we depict the output $y(t)$ of the above system with and without quantization effects. As shown in the figure, the output in the presence of quantization follows the ideal output (i.e., without quantization). However, as depicted in Figure 8.3.3, the difference between the ideal output and the output in the presence of quantization does not diminish as t increases. The difference stays within a certain bound. □

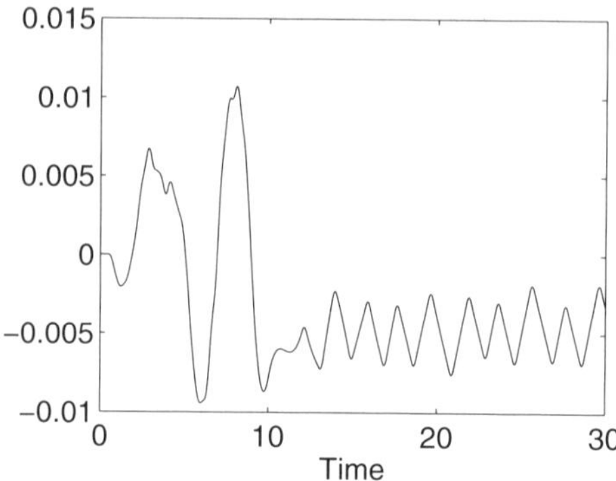

Figure 8.3.3: The difference between the ideal output and the output in the presence of quantization (Example 8.3.2).

8.4 Pulse-Width-Modulated Feedback Control Systems

In the present section we apply the stability and boundedness results for DDS to establish stability results for pulse-width-modulated (PWM) feedback systems with type II modulation.

A. Introduction and formulation of the problem

Pulse-width modulation has extensively been used in electronic, electrical, and electromechanical systems including attitude control systems, adaptive control systems, signal processing, power control systems, modeling of neuron behavior, and the like. The classical example of PWM control is the constant temperature oven suggested by Gouy in 1897 [17] and the most well-known modern application is the attitude control of satellites and space vehicles (see, e.g., [59]). In the latter it is usually required that power (i.e., engine thrust) be modulated in an on–off fashion and that the control computer be time-shared, thus almost always necessitating the use of pulse-width modulation if anything more than simple relay control is desired. Other applications where PWM feedback systems are used include the control of large electric furnaces, the control of electrolytic metal refining plants [18], and radar rendezvous systems [4]. Another interesting application is in the modeling of how information is transmitted in human beings. Specifically, the cardiac pulsatory system and the nervous system communication networks (see, e.g., [8], [33], and [46]) are believed to operate under a combination of pulse duration control and pulse repetition control. Indeed, such systems include one of the most important specific classes of practical nonlinear control systems (see, among others, [34], [35], [67], [68], and [70]) using pulse-width modulation.

One advantage of PWM control is the simplicity of its realization: the control variable typically assumes only two or three constant values, say $+M, -M$, and 0, and hence, the control action is realized through the operation of a switch. In many cases it provides a finer and more precise response than does simple relay control. Another reason for their wide applicability is that pulse-width modulators make it possible to process large signals with high efficiency and low sensitivity to noise. The advantages of PWM control also include the ability to regulate steady-state ripple oscillation frequency, the elimination of dead zone, and the possibility for time sharing of the control computer.

The PWM feedback control system considered is shown in Figure 8.4.1.

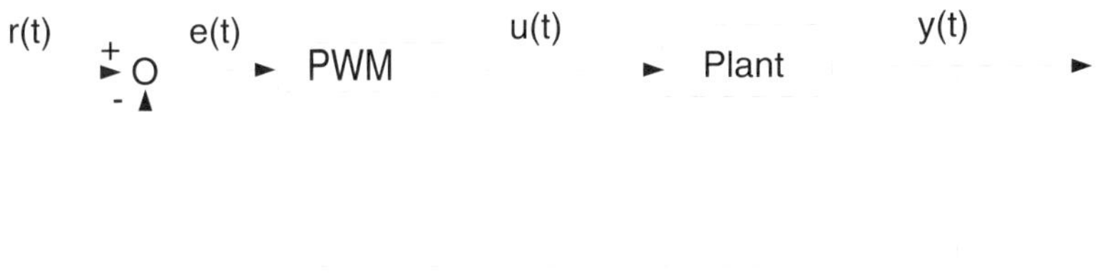

r(t) + - e(t) PWM u(t) Plant y(t)

Figure 8.4.1: PWM feedback system.

We assume that the plant is linear and has a state–space representation of the form

$$\begin{cases} \dot{x} = Ax + Bu, \\ y = Cx \end{cases} \tag{8.4.1}$$

where $x \in \mathbb{R}^n, y \in \mathbb{R}, u \in \mathbb{R}$, and A, B, and C are real matrices of appropriate dimensions.

The output of the pulse-width modulator is given by

$$u(t) = m(e(t)) = \begin{cases} M\sigma(e(k\mathsf{T})), & t \in [k\mathsf{T}, k\mathsf{T} + \mathsf{T}_k], \\ 0, & \text{otherwise} \end{cases} \tag{8.4.2}$$

where T is the sampling period, $k = 0, 1, 2, \ldots$, M is the amplitude of the pulse, T_k is the pulse width, and the signum function $\sigma(\cdot)$ is defined as

$$\sigma(r) = \begin{cases} 1, & r > 0, \\ 0, & r = 0, \\ -1, & r < 0. \end{cases}$$

The sampling period T, the amplitude of the pulse M, and the positive value β (defined below) are all assumed to be constant.

The pulse-width modulator yields piecewise continuous outputs, as illustrated in Figure 8.4.2. The amplitude of the pulses is fixed whereas their duration varies, depending on the error signal $e(t)$ and the type of modulation method being used. There are two types of pulse-width modulators. In a type II pulse-width modulator

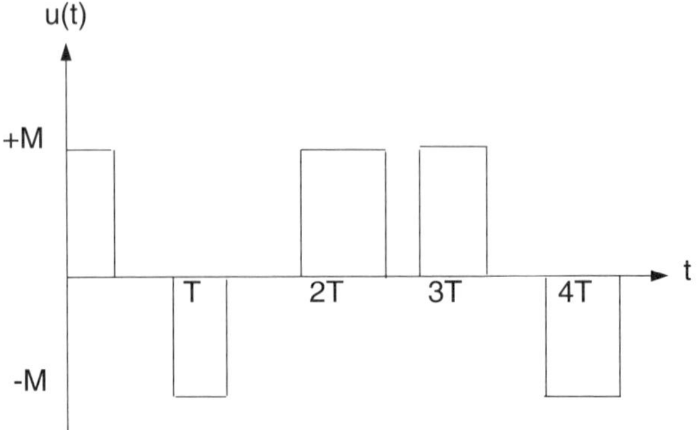

Figure 8.4.2: Example of the outputs of the pulse-width modulator.

(also called *pulse-width modulator with type II modulation* or *with natural sampling*), the pulse width T_k is the smallest value in $[0, \mathsf{T}]$ to satisfy all of the three conditions:

$$\begin{cases} \mathsf{T}_k = \beta|e(k\mathsf{T} + \mathsf{T}_k)| \\ |e(k\mathsf{T} + \mathsf{T}_k)| \leq \dfrac{\mathsf{T}}{\beta} \\ \sigma(e(k\mathsf{T} + \mathsf{T}_k)) = \sigma(e(k\mathsf{T})) \end{cases}$$

and $\mathsf{T}_k = \mathsf{T}$ if no such T_k exists. Graphically, T_k can be interpreted as the first intersection of the plot $\beta|e(t)|$ versus t and the sawtooth signal in each interval $[k\mathsf{T}, (k+1)\mathsf{T})$, as shown in Figure 8.4.3. If there are no intersections, then $\mathsf{T}_k = \mathsf{T}$. In a type I pulse-

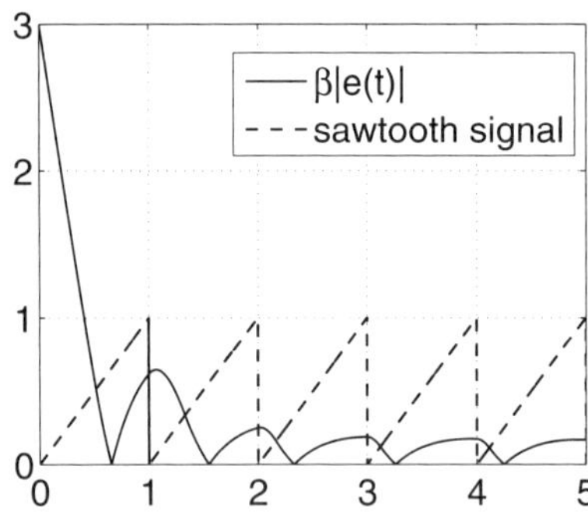

Figure 8.4.3: Determining pulse widths using the sawtooth signal.

width modulator (also called *pulse-width modulator with type I modulation* or *with uniform sampling*), the pulse duration T_k is solely determined by the error signal at the sampling instant $k\mathsf{T}$,

$$\mathsf{T}_k = \begin{cases} \beta|e(k\mathsf{T})|, & |e(k\mathsf{T})| \leq \mathsf{T}/\beta, \\ \mathsf{T}, & |e(k\mathsf{T})| > \mathsf{T}/\beta. \end{cases}$$

In this section, we investigate PWM feedback systems with *type II modulation*. We present sufficient conditions for *uniform asymptotic stability in the large* of the trivial solution and *necessary and sufficient* conditions for *uniform ultimate boundedness* of the solutions, respectively, for PWM feedback systems with *Hurwitz stable linear plants*. We also incorporate a procedure to compute and optimize the sufficient conditions for uniform asymptotic stability of the trivial solution presented herein. We demonstrate the applicability of our results by means of two specific examples.

B. Type II PWM feedback systems with Hurwitz stable plants

In the present subsection, we assume that A in (8.4.1) is *Hurwitz stable*.

Throughout this section, we let $\tau_0 = 0$ and $\tau_{k+1} \triangleq k\mathsf{T} + \mathsf{T}_k$, $k \in \mathbb{N}$. Combining (8.4.1) and (8.4.2), the PWM feedback system of Figure 8.4.1 assumes the form (with $r(t) \equiv 0$ and $e(t) \equiv -y(t)$)

$$\dot{x}(t) = \begin{cases} Ax(t) - BM\sigma(Cx(k\mathsf{T})), & t \in [k\mathsf{T}, \tau_{k+1}), \\ Ax(t), & t \in [\tau_{k+1},\ k\mathsf{T} + \mathsf{T}). \end{cases} \tag{8.4.3}$$

Over the time intervals $[k\mathsf{T}, \tau_{k+1}]$ and $[\tau_{k+1}, k\mathsf{T} + \mathsf{T}]$, $k \in \mathbb{N}$, equation (8.4.3) can be solved to yield the exact solution

$$x(t) = \begin{cases} e^{A(t-k\mathsf{T})}x(k\mathsf{T}) - \displaystyle\int_{k\mathsf{T}}^{t} e^{A(t-\tau)}d\tau BM\sigma(Cx(k\mathsf{T})), & t \in [k\mathsf{T}, \tau_{k+1}], \\ e^{A(t-\tau_{k+1})}x(\tau_{k+1}), & t \in [\tau_{k+1}, k\mathsf{T} + \mathsf{T}]. \end{cases} \tag{8.4.4}$$

We note that the trivial solution $x_e = 0$ is an equilibrium of system (8.4.3).

The first result is concerned with the ultimate boundedness of the solutions of the PWM feedback system (8.4.3).

Theorem 8.4.1 The solutions of system (8.4.3) are *uniformly bounded and uniformly ultimately bounded* for *any choice* of M and β.

Proof. To show that the solutions of system (8.4.3) are uniformly bounded and uniformly ultimately bounded, we verify that the hypotheses of Theorems 6.4.4 and 6.4.5 are satisfied, respectively. In doing so, the set E is chosen to be $E = \{k_0\mathsf{T}, k_0\mathsf{T} + \mathsf{T}, \dots\}$.

We recall that if A is Hurwitz stable, then $e^{A\mathsf{T}}$ is Schur stable and there exists a positive definite matrix Q such that

$$\left(e^{A\mathsf{T}}\right)^T Q\left(e^{A\mathsf{T}}\right) - Q = -I$$

where I is the identity matrix of appropriate dimensions. Choose $v\colon \mathbb{R}^n \to \mathbb{R}^+$ as

$$v(x) = x^T Q x.$$

The solutions of system (8.4.3) at $t = k\mathsf{T} + \mathsf{T}$ are given by

$$x(k\mathsf{T} + \mathsf{T}) = e^{A\mathsf{T}}\big(x(k\mathsf{T}) + \Delta(k\mathsf{T})\big) \tag{8.4.5}$$

where

$$\Delta(k\mathsf{T}) \triangleq -\int_0^{\mathsf{T}_k} e^{-A\tau} d\tau BM\sigma(Cx(k\mathsf{T}))$$

is bounded, because

$$|\Delta(k\mathsf{T})| = \left|\int_0^{\mathsf{T}_k} e^{-A\tau} d\tau BM\right| \le \mathsf{T}Me^{\|A\|\mathsf{T}}\|B\|. \tag{8.4.6}$$

Along the solutions of system (8.4.3) we have

$$\begin{aligned}
&\Delta v(x(k\mathsf{T})) \\
&\quad \triangleq v(x(k\mathsf{T}+\mathsf{T})) - v(x(k\mathsf{T})) \\
&\quad = \left(x(k\mathsf{T})^T + \Delta(k\mathsf{T})^T\right)(Q-I)\left(x(k\mathsf{T}) + \Delta(k\mathsf{T})\right) - x(k\mathsf{T})^T Q x(k\mathsf{T}) \\
&\quad = -|x(k\mathsf{T})|^2 + 2\Delta(k\mathsf{T})^T (Q-I)x(k\mathsf{T}) + \Delta(k\mathsf{T})^T (Q-I)\Delta(k\mathsf{T}) \\
&\quad \le -|x(k\mathsf{T})|^2 + 2\mathsf{T}Me^{\|A\|\mathsf{T}}\|B\|\|Q-I\||x(k\mathsf{T})| \\
&\qquad + \left(\mathsf{T}Me^{\|A\|\mathsf{T}}\|B\|\right)^2 \|Q-I\|.
\end{aligned}$$

It is readily verified that $Dv(x(k\mathsf{T})) = \Delta v(x(k\mathsf{T}))/\mathsf{T} < 0$ whenever $|x(k\mathsf{T})| > \Omega \triangleq \mathsf{T}Me^{\|A\|\mathsf{T}}\|B\|\|Q-I\|(1+\sqrt{1+1/\|Q-I\|})$.

If $|x(k\mathsf{T})| < \Omega$, we have that

$$\begin{aligned}
\left|x(k\mathsf{T}+\mathsf{T})\right|^2 &\le \frac{v(x(k\mathsf{T}+\mathsf{T}))}{\lambda_m(Q)} \\
&\le \frac{v(x(k\mathsf{T})) + \Delta v(x(k\mathsf{T}))}{\lambda_m(Q)} \\
&\le \frac{1}{\lambda_m(Q)}\Big(\|Q-I\|\Omega^2 + 2\mathsf{T}Me^{\|A\|\mathsf{T}}\|B\|\|Q-I\|\Omega \\
&\qquad + \left(\mathsf{T}Me^{\|A\|\mathsf{T}}\|B\|\right)^2\|Q-I\|\Big)
\end{aligned}$$

where $\lambda_m(\cdot)$ is the smallest eigenvalue of a matrix.

Next, we obtain an estimate for $x(t)$, $t \in [k\mathsf{T}, k\mathsf{T}+\mathsf{T})$. It follows from (8.4.4) that

$$|x(t)| \le e^{\|A\|\mathsf{T}}|x(k\mathsf{T})| + \mathsf{T}e^{\|A\|\mathsf{T}}\|B\|M,$$

and thus

$$\begin{aligned}
v(x(t)) &\le \lambda_M(Q)|x(t)|^2 \\
&\le \lambda_M(Q)\left(e^{\|A\|\mathsf{T}}|x(k\mathsf{T})| + \mathsf{T}e^{\|A\|\mathsf{T}}\|B\|M\right)^2 \\
&\le \lambda_M(Q)\left(e^{\|A\|\mathsf{T}}\sqrt{v(x(k\mathsf{T}))/\lambda_m(Q)} + \mathsf{T}e^{\|A\|\mathsf{T}}\|B\|M\right)^2
\end{aligned}$$

where $\lambda_M(\cdot)$ is the largest eigenvalue of a matrix.

Therefore, it follows from Theorems 6.4.4 and 6.4.5 that the solutions of system (8.4.3) are uniformly bounded and uniformly ultimately bounded, respectively. □

Remark 8.4.1 Note that the above proof does not rely on the modulation type. This result is also true for PWM feedback systems with type I modulation (see [25]). □

In the above proof, we utilized equation (8.4.5), which describes system (8.4.3) at discrete instants $k\mathsf{T}$, $k = 1, 2, 3, \ldots$. This representation, however, does not aid the analysis of the Lyapunov stability properties of system (8.4.3) due to the fact that we cannot obtain an explicit estimate for $\Delta(k\mathsf{T})$ in terms of $x(k\mathsf{T})$. We observe that the magnitude of $\Delta(k\mathsf{T})$ is determined by T_k, which is directly related to $x(\tau_{k+1})$ rather than $x(k\mathsf{T})$, and we easily can obtain an estimate of T_k in terms of $x(\tau_{k+1})$. These observations prompt us to consider system (8.4.3) at τ_k, $k \in \mathbb{N}$.

Noting that $\sigma(Cx(k\mathsf{T})) = \sigma(Cx(\tau_{k+1}))$, we have at $t = \tau_{k+1}$,

$$x(\tau_{k+1}) = e^{A\mathsf{T}_k}x(k\mathsf{T}) - \int_0^{\mathsf{T}_k} e^{A(\mathsf{T}_k-\tau)}d\tau BM\sigma(Cx(\tau_{k+1})). \tag{8.4.7}$$

To simplify equation (8.4.7), we let

$$\begin{aligned}\widetilde{x}(\tau_{k+1}) &\triangleq -\int_0^{\mathsf{T}_k} e^{A(\mathsf{T}_k-\tau)}d\tau BM\sigma(Cx(\tau_{k+1}))\\ &= -e^{A\mathsf{T}_k}(I - e^{-A\mathsf{T}_k})A^{-1}BM\sigma(Cx(\tau_{k+1}))\\ &= -M\beta e^{A\mathsf{T}_k}W(\delta_k)e^{-A\mathsf{T}_k}x(\tau_{k+1})\end{aligned}$$

where

$$\delta_k \triangleq \beta\left|Cx(\tau_{k+1})\right| \begin{cases} = \mathsf{T}_k, & \mathsf{T}_k < \mathsf{T},\\ \geq \mathsf{T}, & \mathsf{T}_k = \mathsf{T},\end{cases}$$

and $W(\cdot)$ is defined as

$$W(\delta) \triangleq \begin{cases} 0, & \delta = 0,\\ \dfrac{I - e^{-A\delta}}{\delta}A^{-1}BCe^{A\delta}, & 0 < \delta < \mathsf{T},\\ \dfrac{\mathsf{T}}{\delta}W(\mathsf{T}), & \delta \geq \mathsf{T}.\end{cases} \tag{8.4.8}$$

Equation (8.4.7) is then reduced to

$$\begin{aligned}x(\tau_{k+1}) &= e^{A\mathsf{T}_k}x(k\mathsf{T}) - M\beta e^{A\mathsf{T}_k}W(\delta_k)e^{-A\mathsf{T}_k}x(\tau_{k+1})\\ &= e^{A\mathsf{T}_k}\Big(x(k\mathsf{T}) - M\beta W(\delta_k)e^{-A\mathsf{T}_k}x(\tau_{k+1})\Big).\end{aligned}$$

Substituting $x(k\mathsf{T}) = e^{A(\mathsf{T}-\mathsf{T}_{k-1})}x(\tau_k)$ for $k = 1, 2, \ldots$, we further can obtain

$$\big(I + M\beta W(\delta_k)\big)e^{-A\mathsf{T}_k}x(\tau_{k+1}) = e^{A\mathsf{T}}e^{-A\mathsf{T}_{k-1}}x(\tau_k), \qquad k = 1, 2, \ldots. \tag{8.4.9}$$

To simplify the analysis below, let $z(\tau_{k+1}) \triangleq e^{-A\mathsf{T}_k}x(\tau_{k+1}), k \in \mathbb{N}$ and let $z(\tau_0) = e^{-A\mathsf{T}}x(0)$. At the discrete time instants in the set $E = \{\tau_0, \tau_1, \ldots\}$, system (8.4.3) is governed by the following equation

$$\big(I + M\beta W(\delta_k)\big)z(\tau_{k+1}) = e^{A\mathsf{T}}z(\tau_k), \qquad k = 1, 2, \ldots. \tag{8.4.10}$$

Note that the above equation is also true when $k = 0$.

We use a quadratic Lyapunov function which is constructed using a positive definite matrix P such that

$$(e^{A\mathsf{T}})^T P(e^{A\mathsf{T}}) - P = -(e^{A\mathsf{T}})^T e^{A\mathsf{T}}. \tag{8.4.11}$$

P is chosen such that $P + I = (e^{-A\mathsf{T}})^T P(e^{-A\mathsf{T}})$.

We now are in a position to prove the following result.

Theorem 8.4.2 The trivial solution $x_e = 0$ of the PWM feedback system (8.4.3) is *uniformly asymptotically stable in the large* whenever $M\beta$ satisfies

$$\Theta_{M\beta} \triangleq \inf_{\delta \in (0,\infty)} \lambda_m\big(\Phi(\delta, M\beta)\big) > 0 \tag{8.4.12}$$

where

$$\Phi(\delta, M\beta) = I + M\beta G_1(\delta) + M^2\beta^2 G_2(\delta),$$
$$G_1(\delta) = W(\delta)^T(P + I) + (P + I)W(\delta),$$
$$G_2(\delta) = W(\delta)^T(P + I)W(\delta),$$

$W(\cdot)$ is given by (8.4.8), and P is given in (8.4.11).

Proof. Choosing the Lyapunov function $v\colon \mathbb{R}^n \to \mathbb{R}^+$, $v(z) = z^T P z$, we obtain for the first forward difference of v along the solutions of system (8.4.10), the expression

$$\begin{aligned}
&v(z(\tau_{k+1})) - v(z(\tau_k))\\
&\quad = z(\tau_{k+1})^T P z(\tau_{k+1}) - z(\tau_{k+1})^T \big(I + M\beta W(\delta_k)\big)^T (e^{-A\mathsf{T}})^T P e^{-A\mathsf{T}}\\
&\qquad \times \big(I + M\beta W(\delta_k)\big) z(\tau_{k+1})\\
&\quad = z(\tau_{k+1})^T \Big(P - \big(I + M\beta W(\delta_k)\big)^T (P + I)\big(I + M\beta W(\delta_k)\big)\Big) z(\tau_{k+1})\\
&\quad = -z(\tau_{k+1})^T \Phi(\delta_k, M\beta) z(\tau_{k+1})
\end{aligned} \tag{8.4.13}$$

for all $k \in \mathbb{N}$. It follows from (8.4.13) that when $M\beta$ satisfies (8.4.12) , it is true that

$$Dv(z(\tau_k)) \triangleq \frac{v(z(\tau_{k+1})) - v(z(\tau_k))}{\tau_{k+1} - \tau_k} \le -c_1 v(z(\tau_{k+1})). \tag{8.4.14}$$

where

$$c_1 \triangleq \frac{\Theta_{M\beta}}{2\mathsf{T}\lambda_m(P)} > 0.$$

Next, we obtain an estimate for $x(t)$ when $t \in [k\mathsf{T}, \tau_{k+1})$. It is easily seen from the definition of T_k that

$$\mathsf{T}_k \le \beta|e(\tau_{k+1})| \le \beta\|C\||x(\tau_{k+1})|.$$

Because $\left\|e^{A\tau}\right\| \le e^{\|A\|\mathsf{T}}$ for all $\tau \in [0, \mathsf{T}]$, it follows from (8.4.4) that

$$|x(t)| \le e^{\|A\|\mathsf{T}}|x(\tau_{k+1})| \tag{8.4.15}$$

for $t \in [\tau_{k+1}, k\mathsf{T} + \mathsf{T})$. When $t \in [k\mathsf{T}, \tau_{k+1})$, we solve the first equation in (8.4.3) to yield

$$x(t) = e^{A(t-\tau_{k+1})}x(\tau_{k+1}) - \int_0^{t-\tau_{k+1}} e^{A(t-\tau_{k+1}-\tau)}d\tau BM\sigma(Cx(\tau_k)).$$

Thus, when $t \in [k\mathsf{T}, \tau_{k+1})$

$$|x(t)| \le e^{\|A\|\mathsf{T}}|x(\tau_{k+1})| + \mathsf{T}_k e^{\|A\|\mathsf{T}}\|B\|M \le c_2 e^{\|A\|\mathsf{T}}|x(\tau_{k+1})| \tag{8.4.16}$$

where $c_2 \triangleq 1 + M\beta\|C\|\|B\|$. In view of (8.4.15), (8.4.16) is true for all $t \in [k\mathsf{T}, k\mathsf{T} + \mathsf{T})$.

We now conclude from Theorem 6.4.1 that $x_e = 0$ is uniformly stable. However, we cannot apply Theorem 6.4.6 directly to conclude that $x_e = 0$ is uniformly asymptotically stable in the large because relation (8.4.14) is slightly different from (6.4.15) in Theorem 6.4.6. Nevertheless, in the following we can prove along similar lines the global uniform attractivity of the equilibrium $x_e = 0$ and hence the uniform asymptotic stability in the large of $x_e = 0$.

It follows from (8.4.14) that $v(z(\tau_k))$ is nonincreasing and that for all $k^* \le k$, $k^* > 0$,

$$\begin{aligned} v(z(\tau_{k^*})) - v(z(\tau_{k^*-1})) &\le -c_1 v(z(\tau_{k^*}))(\tau_{k^*} - \tau_{k^*-1}) \\ &\le -c_1 v(z(\tau_k))(\tau_{k^*} - \tau_{k^*-1}). \end{aligned}$$

The above inequality yields

$$v(z(\tau_k)) - v(z(\tau_0)) \le -c_1 v(z(\tau_k))(\tau_k - \tau_0) = -c_1 v(z(\tau_k))\tau_k.$$

Thus it is true for all $k > 0$ that

$$v(z(\tau_k)) \le \frac{v(z(\tau_0)) - v(z(\tau_k))}{c_1\tau_k} \le \frac{v(z(0))}{c_1\tau_k}. \tag{8.4.17}$$

For any $\varepsilon > 0$ and $\alpha > 0$, let

$$\Gamma = \frac{c_2^2\lambda_M(P)e^{2\|A\|\mathsf{T}}\alpha^2}{\varepsilon^2 c_1\lambda_m(P)}.$$

For any $x(0)$ such that $|x(0)| < \alpha$, and for all $k \ge \Gamma/\mathsf{T}$, we have $\tau_k \ge \Gamma$ and

$$v(z(\tau_k)) \le \frac{v(z(0))}{c_1\Gamma} < \frac{\lambda_M(P)e^{2\|A\|\mathsf{T}}\alpha^2}{c_1\Gamma}.$$

Hence,

$$|x(\tau_k)|^2 \le \frac{e^{2\|A\|\mathsf{T}}}{\lambda_m(P)}v(z(\tau_k)) < \frac{\varepsilon^2}{c_2^2 e^{2\|A\|\mathsf{T}}}.$$

Now applying the estimates established in (8.4.16), we have for $t \in [k\mathsf{T}, k\mathsf{T} + \mathsf{T})$ that

$$|x(t)| \le c_2 e^{\|A\|\mathsf{T}}|x(\tau_k)| < \varepsilon.$$

Therefore, we have shown that the trivial solution of (8.4.3) is uniformly asymptotically stable in the large. □

To obtain the least conservative stability results given by Theorem 8.4.2, we need to determine the largest upper bound of $M\beta$ such that $\Theta_{M\beta} < 0$ is satisfied for *all state representations* of (8.4.3). We denote this value by $(M\beta)_{\text{opt}}$. In Remark 8.4.2 given below, we outline a procedure for computing an *estimate* of the optimal value of $M\beta$ such that $\Theta_{M\beta} < 0$ for a *given state representation*. We call this $(M\beta)^*_{\text{opt}}$. The values of $(M\beta)^*_{\text{opt}}$ for different but equivalent state representations will in general vary. In Remark 8.4.3 given below, we outline a procedure for determining an *estimate* of $(M\beta)_{\text{opt}}$ using the different values of $(M\beta)^*_{\text{opt}}$ obtained by employing different state representations of (8.4.3). We denote the estimate of $(M\beta)_{\text{opt}}$ by $\overline{(M\beta)}_{\text{opt}}$.

Remark 8.4.2 To obtain $(M\beta)^*_{\text{opt}}$ for a given state representation, we proceed as follows. Denote $\alpha \triangleq \inf_{\delta\in[0,\mathsf{T}]} \lambda_m(G_1(\delta))$. Because $G_1(\delta) = G_1(\mathsf{T})/\delta$ for $\delta > \mathsf{T}$, it is easily seen that

$$\inf_{\delta\in[0,\infty)} \lambda_m(G_1(\delta)) = \alpha \qquad \text{if} \quad \alpha \le 0,$$

otherwise

$$\inf_{\delta\in[0,\mathsf{T}]} \lambda_m(G_1(\delta)) = 0.$$

The matrix $G_2(\delta)$ is positive semidefinite for all δ. Thus, $\Theta_{M\beta} > 0$ whenever $M\beta < -1/\alpha$ if $\alpha < 0$. When $\alpha \ge 0$, $\Theta_{M\beta} > 0$ for any choice of $M\beta$.

We now assume that $\alpha < 0$. Let $m_0 > 0$ be such that $\Theta_{M\beta} > 0$ is true for all $M\beta < m_0$ (m_0 can be initialized by choosing, for example, $-1/\alpha$). Notice that when $\delta > \mathsf{T}$, it is true that

$$\begin{aligned}\Phi(\delta, M\beta) &= I + M\beta G_1(\delta) + M^2\beta^2 G_2(\delta)\\ &= I + M\beta\frac{\mathsf{T}}{\delta}G_1(\mathsf{T}) + \left(M\beta\frac{\mathsf{T}}{\delta}\right)^2 G_2(\mathsf{T})\\ &= \Phi\left(\mathsf{T}, M\beta\frac{\mathsf{T}}{\delta}\right).\end{aligned} \tag{8.4.18}$$

Therefore, if we can show that the matrix $\Phi(\mathsf{T}, M\beta)$ is positive definite for all $M\beta$ less than a certain value, say $m_0 > 0$, then in view of (8.4.18) the matrix $\Phi(\delta, M\beta)$ is positive definite for all $\delta > \mathsf{T}$ and all $M\beta < m_0$.

Now let

$$\widetilde{G}_0(\delta) = \Phi(\delta, m_0), \qquad \widetilde{G}_1(\delta) = G_1(\delta) + 2m_0 G_2(\delta). \tag{8.4.19}$$

In order that $\Theta_{M\beta}$ given in (8.4.12) be positive, it is necessary that

$$\Phi(\delta, M\beta) = \widetilde{G}_0 + (M\beta - m_0)\widetilde{G}_1(\delta) + (M\beta - m_0)^2 G_2(\delta)$$

be positive definite. For this to be true, we obtain, using the same arguments as above, that $\Theta_{M\beta} > 0$ is true for all $M\beta$ such that

$$M\beta < m_0 + \inf_{\delta\in(0,\mathsf{T}]} -\frac{\lambda_m(\widetilde{G}_0(\delta))}{\lambda_m(\widetilde{G}_1(\delta))}. \tag{8.4.20}$$

We repeat the above computation, replacing in (8.4.19) m_0 by the right-hand side of (8.4.20) until the increment of m_0 is negligible. Set $(M\beta)^*_{\text{opt}}$ equal to the final value of m_0. □

Remark 8.4.3 To determine $\overline{(M\beta)}_{\text{opt}}$, we compute $(M\beta)^*_{\text{opt}}$ for different state representations, $\widetilde{A} = SAS^{-1}, \widetilde{B} = SB, \widetilde{C} = CS^{-1}$, where S is a nonsingular matrix. In doing so, we choose a set of nonsingular matrices S, say Ω, using a random generator (e.g., the rand command in MATLAB). An estimate of $(M\beta)_{\text{opt}}$, $\overline{(M\beta)}_{\text{opt}}$, can be determined by setting $\overline{(M\beta)}_{\text{opt}} = \max_{S\in\Omega}(M\beta)^*_{\text{opt}}$. The above procedure is repeated, increasing the size of Ω, until no further improvements are realized. □

Remark 8.4.4 If M is allowed to assume *negative values* (corresponding to positive feedback in Figure 8.4.1), then similarly as above, we can obtain a *lower bound* for $M\beta$ given by

$$M\beta > \begin{cases} \sup\limits_{\delta\in(0,\mathsf{T}]} -\frac{1}{\lambda_M(G_1(\delta))}, & \text{if } \lambda_M(G_1(\delta)) > 0 \\ -\infty, & \text{otherwise} \end{cases}$$

where $G_1(\delta)$ is given in Theorem 8.4.2. □

Before giving two specific examples to demonstrate the applicability of the preceding results, we point out that results for the boundedness of solutions and the asymptotic stability of the trivial solution for type II PWM systems with linear plants that have one pole at the origin have also been established [23].

C. Examples

To demonstrate the applicability of the results established in Subsection B, and to illustrate how to compute estimates of upper bounds $\overline{(M\beta)}_{\text{opt}}$, we consider in the present subsection two examples. In order to be able to make comparisons with existing results, we choose one *identical* example that was considered by Balestrino *et al.* [5], and Gelig and Churilov [16]. However, before doing so, we outline in the following a procedure for computing an estimate for the optimal stability bound for $M\beta$, based on Theorem 8.4.2 and Remarks 8.4.2 and 8.4.3.

Stability Bound Procedure: An upper bound of $M\beta$ that satisfies (8.4.12) can be computed and optimized in the following manner.

(1) Determine P from $(e^{A\mathsf{T}})^T P e^{A\mathsf{T}} - P = -(e^{A\mathsf{T}})^T e^{A\mathsf{T}}$.

(2) Choose a precision level $\delta > 0$ and a correspondingly dense partition of $[0, \mathsf{T}]$, say the set $\{t_0 = 0, t_1, \ldots, t_N = \mathsf{T}\}$, where $0 < t_{j+1} - t_j < \delta$, $j = 0, 1, \ldots, N-1$.

(3) For each j, $j = 0, 1, \ldots, N$, compute $W(t_j)$, $G_1(t_j)$, and $G_2(t_j)$.

(4) Initialize m_0 by setting $m_0 = \min_{0\leq j\leq N} -\frac{1}{\lambda_m(G_1(t_j))}$.

(5) Let (see (8.4.19))

$$\widetilde{G}_0(t_j) = I + m_0 G_1(t_j) - m_0^2 G_2(t_j), \qquad \widetilde{G}_1(t_j) = G_1(t_j) - 2m_0 G_2(t_j),$$

$\widetilde{m}_0 = m_0$. Then let m_0 be (see (8.4.20))

$$m_0 = \widetilde{m}_0 + \min_{0 \le j \le N} - \frac{\lambda_m(\widetilde{G}_0(t_j))}{\lambda_m(\widetilde{G}_1(t_j))}.$$

(6) Repeat Step (5) until the increment of m_0 is negligible, say, $m_0 - \widetilde{m}_0 < \varepsilon$, where $\varepsilon > 0$ is a chosen precision level. Set $\overline{(M\beta)}^*_{\text{opt}} = m_0$, where $\overline{(M\beta)}^*_{\text{opt}}$ is an estimate of $(M\beta)^*_{\text{opt}}$.

(7) Repeat Steps (1)–(6), using finer partitions of the interval $[0, \mathsf{T}]$ (i.e, smaller δ), until there is no further significant improvement for $\overline{(M\beta)}^*_{\text{opt}}$.

(8) Repeat Steps (1)–(7) for different but equivalent matrices $\widetilde{A}$, $\widetilde{B}$, and $\widetilde{C}$. This can be done, for example, by generating a set Ω of random (nonsingular) matrices, and for each $S \in \Omega$ letting $\widetilde{A} = SAS^{-1}$, $\widetilde{B} = SB$, and $\widetilde{C} = CS^{-1}$. Determine an optimal upper bound for $M\beta$ by setting $\overline{(M\beta)}_{\text{opt}} = \max_{S \in \Omega} \overline{(M\beta)}^*_{\text{opt}}$. In general, the larger the size of Ω, the closer the computed value $\overline{(M\beta)}_{\text{opt}}$ to the *actual* upper bound of $M\beta$.

We are now in a position to consider two examples.

Example 8.4.1 (*First-order system*) In the present case, the plant is characterized by a transfer function of the form

$$G(s) = \frac{c}{s + a}, \qquad a > 0,$$

or by the state–space representation (8.4.3) with $A = -a, B = 1, C = c$. The upper bound of $M\beta$ that satisfies (8.4.12) is

$$\begin{cases} \inf\limits_{\delta_k \in (0,\mathsf{T}]} \dfrac{-G_1 - \sqrt{G_1^2 - 4G_2}}{2G_2} =\cdot \dfrac{1 - e^{-a\mathsf{T}}}{|c|}, & \text{if } c < 0, \\ \infty, & \text{if } c \ge 0. \end{cases}$$

The bound above is identical to the result reported in [5].

Using a method that employs averaging of the pulse-width modulator output, and assuming $M = 1$ and $c > 0$, the stability condition

$$\frac{1}{\beta} > \frac{2}{\pi} c + \frac{2}{\pi\sqrt{3}} ac\mathsf{T}$$

is obtained in [16]. For this particular example, the present result is clearly less conservative than that obtained in [16].

Note that the optimal bound obtained for $M\beta$ above is the exact value, because in the present case it was not necessary to invoke approximations to apply Theorem 8.4.2. □

Example 8.4.2 (*Second-order system*) In this case the plant is characterized by the transfer function

$$G(s) = \frac{K}{(s+1)(s+2)}.$$

The state–space representation is given by

$$A = S\begin{bmatrix} -1 & 0 \\ 0 & -2 \end{bmatrix} S^{-1}, \qquad B = S\begin{bmatrix} K \\ K \end{bmatrix}, \qquad C = [1 - 1]S^{-1}$$

where S is a nonsingular matrix. In applying the Stability Bound Procedure we let $\delta = 0.001$ and $\varepsilon = 0.0001$ (the improvements of the computed results were negligible for smaller δ and ε), and we generated 200 random matrices S to form the set Ω. In particular, when

$$S = \begin{bmatrix} -3.1887 & 4.8612 \\ 2.5351 & -2.1877 \end{bmatrix}$$

the upper and lower bounds for MK are computed to be 6.3004 and -0.9670, respectively; when

$$S = \begin{bmatrix} 1.6130 & -0.2781 \\ -1.1766 & 1.7069 \end{bmatrix},$$

the upper and lower bounds are computed to be 2.8447 and -1.9370, respectively.

It follows from Theorem 8.4.2 that the trivial solution of (8.4.3) is uniformly asymptotically stable in the large if $-1.9370 < MK < 6.3004$.

To determine the quality of the estimates of the bounds for MK obtained above, we note that if $MK = -2$, then $x(t) = (1, 0.5)^T$ is an equilibrium of system (8.4.3) with $\mathsf{T}_k = \mathsf{T} = 1$ for all k. Also, when $MK = 6.8$, system (8.4.3) has a limit cycle as shown in Figure 8.4.4. Therefore, the trivial solution of the PWM feedback system

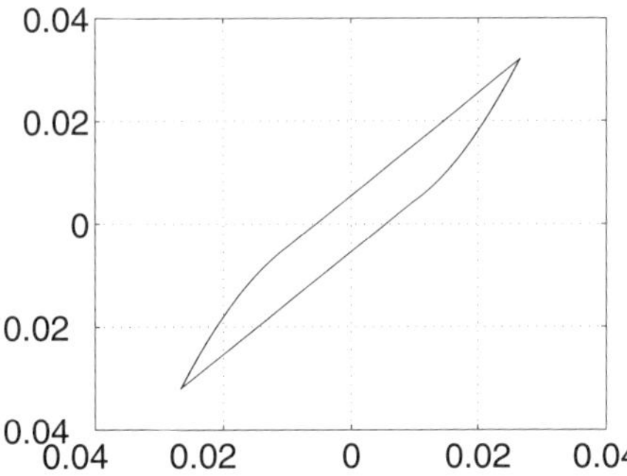

Figure 8.4.4: A limit cycle of the PWM feedback system when $\beta = 1$ and $MK = 6.8$.

(8.4.3) cannot be uniformly asymptotically stable in the large for the above two cases. This shows that our result, $-1.9370 < MK < 6.3004$, obtained by Theorem 8.4.2, is very close to the *actual* lower and upper bounds for MK that ensure stability. We would like to point out that the above result is very close to the result we obtained in [25] for a PWM feedback system having the above plant but with type I modulation ($-1.9789 < MK < 6.3278$).

Although the stability bounds for PWM feedback systems with type I and type II modulation are close, the states generally approach the trivial solution faster when using type II modulation. This can be seen in Figure 8.4.5. □

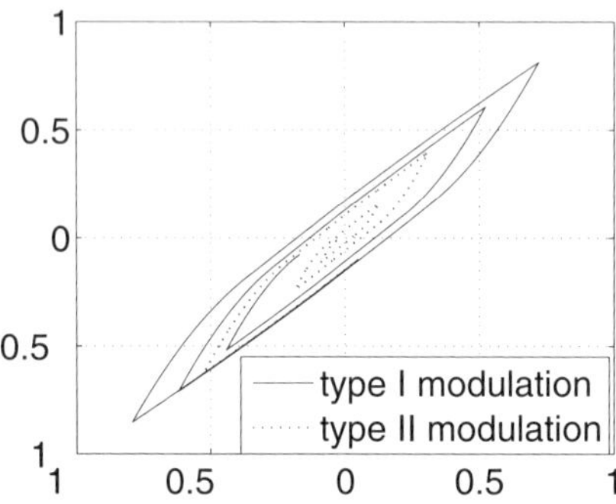

Figure 8.4.5: Example solutions of system (8.4.3) with type I and type II modulations ($MK = 6$, $B = 1$), respectively.

8.5 Digital Filters

In the final part of this chapter, we investigate stability properties of discrete-time systems described by equations of the form

$$x(k+1) = \text{sat}[Ax(k)], \qquad k \in \mathbb{N} \tag{8.5.1}$$

where $x(k) \in D^n = \{x \in \mathbb{R}^n : -1 \leq x_i \leq 1\}$, $A \in \mathbb{R}^{n\times n}$,

$$\text{sat}(x) = [\text{sat}(x_1), \text{sat}(x_2), \ldots, \text{sat}(x_n)]^T,$$

and

$$\text{sat}(x_i) = \begin{cases} 1, & x_i > 1 \\ x_i, & -1 \leq x_i \leq 1 \\ -1, & x_i < -1 \end{cases}.$$

We say that *system* (8.5.1) *is stable* if $x_e = 0$ is the only equilibrium of system (8.5.1) and $x_e = 0$ is globally asymptotically stable. Because we have saturation nonlinearities in (8.5.1), it is clear that for any $x(0) \notin D^n$, $x(k) \in D^n$, $k \geq 1$, will always be true. Thus, without loss of generality, we assume that $x(0) \in D^n$.

Equation (8.5.1) describes a class of discrete-time dynamical systems with symmetrically saturating states after normalization. Examples of such systems include control systems having saturation-type nonlinearities on the state (see, e.g., [15], [43], and [56]), neural networks defined on hypercubes (see, e.g., [42] and [53]), and digital filters using saturation overflow arithmetic (see, e.g., [43], [57], and [63]).

System (8.5.1) may be used to represent *control systems* with saturating states with no external inputs. In the analysis and design of such systems, the first fundamental question addresses stability: under what conditions is $x_e = 0$ an equilibrium and when is this equilibrium globally asymptotically stable?

The condition that A is a stable matrix, that is, every eigenvalue λ_i of A satisfies $|\lambda_i| < 1$, is not sufficient for system (8.5.1) to be stable. (It is easy to give examples for which A is a stable matrix, but system (8.5.1) is not stable.)

In many important applications, equation (8.5.1) may be used to represent *digital processing systems,* including *digital filters* and digital control systems (see, e.g., [15], [43], [56], [57], [63], and [72]) with finite wordlength arithmetic under zero external inputs. In such systems, saturation arithmetic is used to cope with the overflow. The absence of limit cycles in such systems is of great interest and can be guaranteed by the global asymptotic stability of the equilibrium $x_e = 0$ for (8.5.1). The Lyapunov theory has been found to be an appropriate method for solving such problems (see, e.g., [43], [57], and [72]).

A. A general result for discrete-time systems with state saturation nonlinearities

In establishing our results, we make use of Lyapunov functions for the linear systems corresponding to system (8.5.1), given by

$$w(k+1) = Aw(k), \qquad k \in \mathbb{N} \tag{8.5.2}$$

where $A \in \mathbb{R}^{n \times n}$ is defined in (8.5.1).

We recall that for a general autonomous system

$$x(k+1) = f(x(k)), \qquad k \in \mathbb{N}, \tag{8.5.3}$$

with $x(k) \in \mathbb{R}^n$ and $f\colon \mathbb{R}^n \to \mathbb{R}^n$, x_e is an equilibrium for (8.5.3) if and only if

$$x_e = f(x_e).$$

We assume, without loss of generality that $x_e = 0$ (refer to Subsection 6.1B). Recall also that the equilibrium $x_e = 0$ for system (8.5.3) is globally asymptotically stable, or asymptotically stable in the large, if there exists a continuous function $v\colon \mathbb{R}^n \to \mathbb{R}$ which is positive definite, radially unbounded, and along solutions of (8.5.3) satisfies the condition that

$$Dv_{(8.5.3)}(x(k)) \triangleq v(x(k+1)) - v(x(k)) = v(f(x(k))) - v(x(k)) \tag{8.5.4}$$

is negative definite for all $x(k) \in \mathbb{R}^n$ (refer to Theorem 6.3.2(a)).

In the stability analysis of the equilibrium $x_e = 0$ of system (8.5.1), we find it useful to employ Lyapunov functions v whose value for a given state vector $w \notin D^n$ is greater than the value for the corresponding saturated state vector sat(w). Specifically, we make the following assumption.

Assumption 8.5.1 Assume that for system (8.5.2), there exists a continuous function $v\colon \mathbb{R}^n \to \mathbb{R}$ with the following properties:

(i) v is positive definite, radially unbounded, and

$$Dv_{(8.5.2)}(w(k)) \triangleq v(w(k+1)) - v(w(k)) = v(Aw(k)) - v(w(k))$$

is negative definite for all $w(k) \in \mathbb{R}^n$ (and thus, the eigenvalues of A are within the unit circle).

(ii) For all $w \in \mathbb{R}^n$ such that $w \notin D^n$, it is true that

$$v(\text{sat}(w)) < v(w) \tag{8.5.5}$$

where D^n and sat$(\cdot)$ are defined in (8.5.1). □

An example of a function $v_1 : \mathbb{R}^2 \to \mathbb{R}$ that satisfies (8.5.5) is given by $v_1(w) = d_1 w_1^2 + d_2 w_2^2$, $d_1, d_2 > 0$. On the other hand, the function $v_2 \colon \mathbb{R}^2 \to \mathbb{R}$ given by $v_2(w) = w_1^2 + (2w_1 + w_2)^2$ does not satisfy (8.5.5). To see this, consider the point $w = [-0.99, 1.05]^T \notin D^2$ and note that $v_2(\text{sat}(w)) = 1.9405$ and $v_2(w) = 1.845$.

We are now in a position to prove the following result.

Theorem 8.5.1 If Assumption 8.5.1 holds, then the equilibrium $x_e = 0$ of system (8.5.1) is globally asymptotically stable.

Proof. Because Assumption 8.5.1 is true, there exists a positive definite and radially unbounded function v for system (8.5.2) such that (8.5.5) is true, which in turn implies that $v(\text{sat}(Aw)) \le v(Aw)$ for all $w \in \mathbb{R}^n$. Also, by Assumption 8.5.1, $v(Aw(k)) - v(w(k)) < 0$ for all $w(k) \neq 0$. Thus, along the solutions of system (8.5.1), we have

$$\begin{aligned} Dv_{(8.5.1)}(x(k)) &= v(x(k+1)) - v(x(k)) \\ &= v(\text{sat}[Ax(k)]) - v(x(k)) \\ &\le v(Ax(k)) - v(x(k)) \\ &< 0 \end{aligned}$$

for all $x(k) \neq 0$ and $Dv_{(8.5.1)}(x(k)) = 0$ if and only if $x(k) = 0$. Therefore, $v(x)$ is positive definite and radially unbounded, and $Dv_{(8.5.1)}(x)$ is negative definite for all x. Hence, in view of Theorem 6.3.2(a), the equilibrium $x_e = 0$ of system (8.5.1) is globally asymptotically stable. □

Remark 8.5.1 In particular, for fixed p, $1 \le p \le \infty$, let us choose

$$v(w) = |w|_p = \left(\sum_{i=1}^{n} |w_i|^p \right)^{1/p}$$

for system (8.5.2) and assume that $\|A\|_p < 1$, where $\|A\|_p$ denotes the matrix norm induced by the vector norm $|w|_p$. Under these conditions, Assumption 8.5.1 is true. To see this, note that v is positive definite and radially unbounded, that

$$v(Aw) = |Aw|_p \le \|A\|_p |w|_p < |w|_p = v(w),$$

and that $|\text{sat}(w)|_p < |w|_p$, for all $w \in \mathbb{R}^n$ such that $w \notin D^n$.

Therefore, the equilibrium $x_e = 0$ of system (8.5.1) is globally asymptotically stable if

$$\|A\|_p < 1 \tag{8.5.6}$$

for some p, $1 \leq p \leq \infty$.

In the case of digital filters, the above argument holds for *any* type of overflow nonlinearity $\varphi\colon \mathbb{R} \to [-1, 1]$. To see this, let $f(w) = [\varphi(w_1), \ldots, \varphi(w_n)]^T$ and note that in this case $|f(w)|_p < |w|_p$ for all $w \in \mathbb{R}^n$ such that $w \notin D^n$. □

B. Results involving quadratic Lyapunov functions

In order to generate quadratic form Lyapunov functions that satisfy Assumption 8.5.1 for systems described by equation (8.5.1), we find it convenient to utilize the next assumption.

Assumption 8.5.2 Let

$$x_s = \text{sat}(x) = [\text{sat}(x_1), \ldots, \text{sat}(x_n)]^T$$

for $x \in \mathbb{R}^n$ and let $H \in \mathbb{R}^{n\times n}$ denote a positive definite matrix. Assume that

$$x_s^T H x_s < x^T H x, \tag{8.5.7}$$

whenever $x \notin D^n$, $x \in \mathbb{R}^n$. □

An example of a matrix that satisfies Assumption 8.5.2 is any diagonal matrix with positive diagonal elements. On the other hand, the positive definite matrix H given by

$$H = \begin{bmatrix} 5 & 2 \\ 2 & 1 \end{bmatrix},$$

does not satisfy Assumption 8.5.2. (To see this, refer to the example following Assumption 8.5.1 by noting that $v_2(x) = x^T Hx$.)

The next result gives a *necessary and sufficient* condition for matrices to satisfy Assumption 8.5.2. This result is very useful in applications.

Lemma 8.5.1 An $n\times n$ positive definite matrix $H = [h_{ij}]$ satisfies Assumption 8.5.2 *if and only if*

$$h_{ii} \geq \sum_{j=1, j\neq i}^{n} |h_{ij}|, \qquad i = 1, \ldots, n. \tag{8.5.8}$$

Proof. This lemma is a special case of Lemma 8.5.2 (when $L = 1$). For the statement and proof of Lemma 8.5.2, refer to Subsection C of the present section. □

The following result is a direct consequence of Theorem 8.5.1.

Corollary 8.5.1 The equilibrium $x_e = 0$ of system (8.5.1) is *globally asymptotically stable* if there exists a matrix H that satisfies Assumption 8.5.2 such that

$$Q \triangleq H - A^T HA$$

is positive definite.

Proof. By choosing $v(x) = x^T Hx$, the proof follows from Theorem 8.5.1. □

In the next result, Theorem 8.5.2, we show that Corollary 8.5.1 is actually true when Q is only positive semidefinite, still assuming that A is stable.

Theorem 8.5.2 The equilibrium $x_e = 0$ of system (8.5.1) is *globally asymptotically stable* if A is stable and if there exists a matrix H that satisfies Assumption 8.5.2 such that

$$Q \triangleq H - A^T HA$$

is positive semidefinite.

Proof. Let us choose $v(x(k)) = x^T(k)Hx(k)$ for the system (8.5.1). The function v is clearly positive definite and radially unbounded. Also, because

$$\begin{aligned} Dv_{(8.5.1)}(x(k)) &= v(x(k+1)) - v(x(k)) \\ &= [\text{sat}(Ax(k))]^T H[\text{sat}(Ax(k))] - x^T(k)Hx(k) \\ &\leq x^T(k)(A^T HA - H)x(k), \end{aligned}$$

and because $H - A^T HA$ is positive semidefinite, $Dv_{(8.5.1)}(x(k))$ is negative semidefinite for all $x(k)$. Therefore, the equilibrium $x_e = 0$ is stable (refer to Theorem 6.3.1(a)). To show that it is asymptotically stable, we must show that $x(k) \to 0$ as $k \to \infty$ (refer to Definition 6.1.1(h)).

Let us consider an n consecutive step iteration for system (8.5.1), from $n_0 \geq 0$ to $n + n_0$. Without loss of generality, assume that system (8.5.1) saturates at $k = l$, $l \in [n_0, n + n_0)$. In view of Assumption 8.5.2, it follows that

$$\begin{aligned} v(x(l+1)) &= x^T(l+1)Hx(l+1) \\ &= [\text{sat}(Ax(l))]^T H[\text{sat}(Ax(l))] \\ &< [Ax(l)]^T HAx(l) \\ &\leq x^T(l)Hx(l) \\ &= v(x(l)). \end{aligned}$$

On the other hand, if no saturation occurs during this period, then, using the fact that if $H - A^T HA$ is positive semidefinite, then $H - (A^T)^n HA^n$ is positive definite for any $n > 1$ when A is stable (see, e.g., [72]), we have

$$\begin{aligned} v(x(n+n_0)) &= x^T(n+n_0)Hx(n+n_0) \\ &= [A^n x(n_0)]^T H A^n x(n_0) \\ &= x^T(n_0)(A^T)^n H A^n x(n_0) \\ &< x^T(n_0)Hx(n_0) \\ &= v(x(n_0)). \end{aligned}$$

Therefore, we can conclude that for the sequence $\{k\colon k = 1, 2, \dots\}$, there always exists an infinite subsequence $\{k_j\colon j = 1, 2, \dots\}$, such that $Dv_{(8.5.1)}(x(k_j))$ is negative for $x(k_j) \neq 0$ and that $v(x(k)) \leq v(x(k_j))$ for all $k \geq k_j$. Because v is a positive definite quadratic form, it follows that $v(x(k_j)) \to 0$ as $j \to \infty$, and therefore $v(x(k)) \to 0$ as $k \to \infty$. This in turn implies that $x(k) \to 0$ as $k \to \infty$. Thus, the equilibrium $x_e = 0$ of (8.5.1) is globally asymptotically stable. □

C. Stability of digital filters with generalized overflow nonlinearities

Because no limit cycles can exist in a digital filter if its trivial solution is globally asymptotically stable, we can use the results of this section to establish the following results for n-th order digital filters with saturation arithmetic.

Corollary 8.5.2 (i) A digital filter described by (8.5.1) is free of limit cycles if Assumption 8.5.1 is satisfied. (ii) A digital filter described by (8.5.1) is free of limit cycles if A is stable and if there exists a matrix H that satisfies Assumption 8.5.2, such that

$$Q \triangleq H - A^T H A$$

is positive semidefinite. □

We now consider nth-order digital filters described by equations of the form

$$x(k+1) = f[Ax(k)], \qquad k \in \mathbb{N} \tag{8.5.9}$$

where $x(k) \in \mathbb{R}^n$, $A \in \mathbb{R}^{n\times n}$,

$$f(x) = [\varphi(x_1), \varphi(x_2), \dots, \varphi(x_n)]^T, \tag{8.5.10}$$

and $\varphi\colon \mathbb{R} \to [-1, 1]$ is piecewise continuous. We call system (8.5.9) a *fixed-point digital filter using overflow arithmetic*. For such filters, we make the following assumption.

Assumption 8.5.3 Let f be defined as in (8.5.10). Assume that $H \in \mathbb{R}^{n\times n}$ is a positive definite matrix and that

$$f(x)^T H f(x) < x^T H x, \tag{8.5.11}$$

for all $x \in \mathbb{R}^n$, $x \notin D^n$. □

In what follows, we let the function φ in (8.5.10) be defined as (see Figure 8.5.1)

$$\varphi(x_i) = \begin{cases} L, & x_i > 1, \\ x_i, & -1 \le x_i \le 1, \\ -L, & x_i < -1, \end{cases} \tag{8.5.12}$$

or (see Figure 8.5.2)

$$\begin{cases} L \le \varphi(x_i) \le 1, & x_i > 1, \\ \varphi(x_i) = x_i, & -1 \le x_i \le 1, \\ -1 \le \varphi(x_i) \le -L, & x_i < -1, \end{cases} \tag{8.5.13}$$

where $-1 \le L \le 1$. We call the function φ defined in (8.5.12) and (8.5.13) a *generalized overflow characteristic*. Note that when defined in this way, the function φ includes as special cases the usual types of overflow arithmetic employed in practice, such as zeroing, two's complement, triangular, and saturation overflow characteristics.

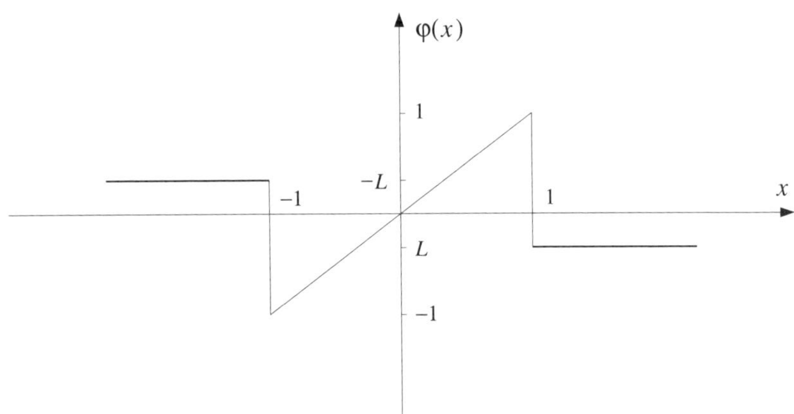

Figure 8.5.1: The generalized overflow nonlinearity described by (8.5.12).

To establish our next result, Theorem 8.5.3, we require the following preliminary result, Lemma 8.5.2.

Lemma 8.5.2 [43] Assume that f is defined in (8.5.10) and φ is given in (8.5.12) or in (8.5.13) with $-1 < L \le 1$. An $n \times n$ positive definite matrix $H = H^T = [h_{ij}]$ satisfies Assumption 8.5.3 if and only if

$$(1 + L)h_{ii} \ge 2 \sum_{j=1, j\ne i}^{n} |h_{ij}|, \qquad i = 1, \dots, n. \tag{8.5.14}$$

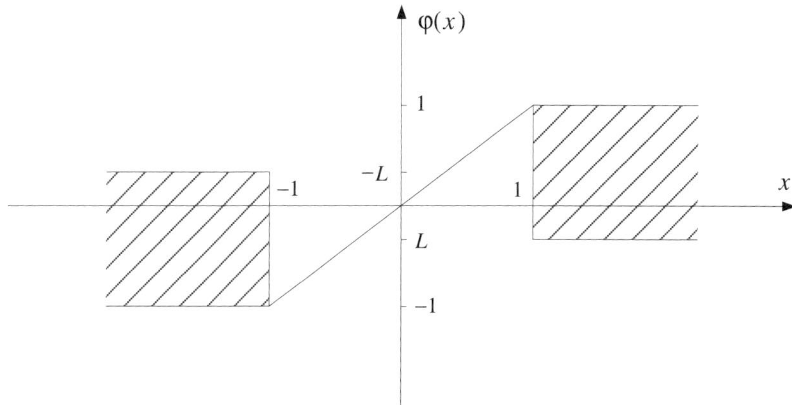

Figure 8.5.2: The generalized overflow nonlinearity described by (8.5.13).

Proof. We first prove this lemma for the overflow arithmetic given in Equation (8.5.12).

We introduce the following notation. For φ defined in (8.5.12), let us denote

$$f(x) = [\varphi(x_1), \dots, \varphi(x_n)]^T = Ex$$

where $E = \text{diag}[e_1, e_2, \dots, e_n]$, $e_i = 1$ if $|x_i| \leq 1$, and $e_i = L/|x_i|$ if $|x_i| > 1$. Then, we have

$$x^T Hx - f(x)^T Hf(x) = x^T(H - EHE)x.$$

Sufficiency: Suppose $x = [x_1, x_2, \dots, x_n]^T$, $|x_k| > 1$ and $|x_i| \leq 1$ for $i \neq k$ ($x \notin D^n$). We have $-1 < e_k < 1$ and $e_i = 1$ for $i \neq k$, and therefore,

$$H - EHE$$

$$= \begin{bmatrix} 0 & \cdots & 0 & h_{1k}(1-e_k) & 0 & \cdots & 0 \\ \vdots & \cdots & \vdots & \vdots & \vdots & \cdots & \vdots \\ 0 & \cdots & 0 & h_{k-1,k}(1-e_k) & 0 & \cdots & 0 \\ h_{k1}(1-e_k) & \cdots & h_{k,k-1}(1-e_k) & h_{kk}(1-e_k^2) & h_{k,k+1}(1-e_k) & \cdots & h_{kn}(1-e_k) \\ 0 & \cdots & 0 & h_{k+1,k}(1-e_k) & 0 & \cdots & 0 \\ \vdots & \cdots & \vdots & \vdots & \vdots & \cdots & \vdots \\ 0 & \cdots & 0 & h_{nk}(1-e_k) & 0 & \cdots & 0 \end{bmatrix}$$

and

$$x^T(H - EHE)x = (1 - e_k)\left(h_{kk}(1 + e_k)x_k^2 + 2\sum_{i=1, i\neq k}^{n} h_{ik}x_i x_k\right). \quad (8.5.15)$$

Note that in the above equation we have used the fact that $h_{ij} = h_{ji}$.

From $|x_i| \le 1$ for $i \neq k$, $|x_k| > 1$, $e_k|x_k| = L$ and $L > -1$, we have

$$(1+L)|x_i x_k| \le (1+L)|x_k| < (|x_k| + L)|x_k| = (1+e_k)x_k^2.$$

Hence, from (8.5.15), we have

$$\begin{aligned} x^T(H - EHE)x &\ge (1-e_k)\left(h_{kk}(1+e_k)x_k^2 - 2\sum_{i=1,i\neq k}^{n} |h_{ik}x_i x_k|\right) \\ &> (1-e_k^2)x_k^2\left(h_{kk} - \frac{2}{1+L}\sum_{i=1,i\neq k}^{n} |h_{ik}|\right) \\ &\ge 0; \end{aligned}$$

that is, $x^T Hx > x^T EHEx = f(x)^T Hf(x)$.

Denote $M = \{1, 2, \ldots, m\}$ for *any* m, $0 < m \le n$, and

$$N = \{k_i \colon 0 < k_i \le n,\ k_i \neq k_j,\ \text{when } i \neq j,\ i, j \in M\}.$$

Now suppose that $x = [x_1, x_2, \ldots, x_n]^T$, $|x_k| > 1$ for $k \in N$ and $|x_i| \le 1$ for $i \notin N$ $(x \notin D^n)$. Following the same procedure as above, we have

$$\begin{aligned} x^T(H - EHE)x &= \sum_{k\in N}(1-e_k)\left(h_{kk}(1+e_k)x_k^2 + 2\sum_{i=1,i\notin N}^{n} h_{ik}x_i x_k\right) \\ &\quad + \sum_{k\in N}\sum_{l\in N,l\neq k} h_{kl}x_k x_l(1-e_k e_l) \\ &\ge \sum_{k\in N}(1-e_k)\left(h_{kk}(1+e_k)x_k^2 - 2\sum_{i=1,i\notin N}^{n} |h_{ik}x_i x_k|\right) \\ &\quad + \sum_{k\in N}\sum_{l\in N,l\neq k} h_{kl}x_k x_l(1-e_k e_l) \\ &> \sum_{k\in N}(1-e_k^2)x_k^2\left(h_{kk} - \frac{2}{1+L}\sum_{i=1,i\notin N}^{n} |h_{ik}|\right) \\ &\quad + \sum_{k\in N}\sum_{l\in N,l\neq k} h_{kl}x_k x_l(1-e_k e_l) \\ &= \sum_{k\in N}(1-e_k^2)x_k^2\left(h_{kk} - \frac{2}{1+L}\sum_{i=1,i\neq k}^{n} |h_{ik}|\right) \\ &\quad + \frac{2}{1+L}\sum_{k\in N}(1-e_k^2)x_k^2 \sum_{i\in N,i\neq k} |h_{ik}| \\ &\quad + \sum_{k\in N}\sum_{l\in N,l\neq k} h_{kl}x_k x_l(1-e_k e_l). \end{aligned} \tag{8.5.16}$$

The first summation of the right-hand side in (8.5.16) is nonnegative, by assumption. Considering the last two terms in (8.5.16), by noting that $-1 < e_k < 1$ and $e_k|x_k| = L$ for $k \in N$, and $-1 < L \leq 1$, we have

$$\begin{aligned}
&\frac{2}{1+L}\sum_{k\in N}(1-e_k^2)x_k^2 \sum_{i\in N, i\neq k}|h_{ik}| + \sum_{k\in N}\sum_{l\in N, l\neq k} h_{kl}x_k x_l(1-e_k e_l)\\
&\geq \sum_{k\in N}\sum_{l\in N, l\neq k}(1-e_k^2)x_k^2|h_{kl}| - \sum_{k\in N}\sum_{l\in N, l\neq k}|h_{kl}x_k x_l|(1-e_k e_l)\\
&= \sum_{k\in N}\sum_{l\in N, l\neq k}|h_{kl}x_k|\big(|x_k| - e_k L - |x_l| + e_k L\big)\\
&= \sum_{k\in N}\sum_{l\in N, l\neq k}|h_{kl}|x_k^2 - \sum_{k\in N}\sum_{l\in N, l\neq k}|h_{kl}x_k x_l|\\
&= \sum_{k\in N}\sum_{l\in N, l> k}|h_{kl}|(x_k^2 + x_l^2) - 2\sum_{k\in N}\sum_{l\in N, l> k}|h_{kl}x_k x_l|\\
&= \sum_{k\in N}\sum_{l\in N, l> k}|h_{kl}|\big(|x_k| - |x_l|\big)^2 \geq 0.
\end{aligned}$$

Therefore,

$$x^T Hx - f(x)^T Hf(x) = x^T(H - EHE)x > 0,$$

for any $x \in R^n$ such that $x \notin D^n$.

This proves the sufficiency.

Necessity: It suffices to show that if (8.5.14) does not hold, there always exist some points $x \notin D^n$, such that

$$x^T Hx \leq f(x)^T Hf(x).$$

Suppose that (8.5.14) does not hold for $i = k$; that is,

$$\delta \stackrel{\triangle}{=} 2\sum_{j=1, j\neq k}^{n}|h_{kj}| - (1+L)h_{kk} > 0.$$

Let us choose $|x_k| = 1 + \xi, \xi > 0$, and $x_i = -\text{sign}(h_{ik}x_k)$, $i \neq k$, where

$$\text{sign}(y) = \begin{cases} 1, & y > 0,\\ 0, & y = 0,\\ -1, & y < 0. \end{cases}$$

Then, $x = [x_1, \ldots, x_n]^T \notin D^n$ and (8.5.15) becomes

$$\begin{aligned}
x^T(H - EHE)x &= (1 - e_k)\Bigg(h_{kk}(1+e_k)x_k^2 - 2\sum_{i=1, i\neq k}^{n}|h_{ik}x_k|\Bigg)\\
&= (1-e_k)|x_k|\Bigg(h_{kk}\xi + (1+L)h_{kk} - 2\sum_{i=1, i\neq k}^{n}|h_{ki}|\Bigg)\\
&= (1-e_k)|x_k|(h_{kk}\xi - \delta).
\end{aligned}$$

Clearly, when we choose $0 < \xi \leq \delta/h_{kk}$, we have

$$x^T Hx - f(x)^T Hf(x) = x^T(H - EHE)x \leq 0.$$

Note here that $h_{kk} > 0$ because H is positive definite.

This proves the necessity.

For the overflow nonlinearity given in (8.5.13), the proof of sufficiency is similar to the proof given above. To prove necessity, we note that for a given L, when $|x_i| > 1$, $\varphi(x_i)$ in (8.5.13) may assume *any* value in the crosshatched regions in 8.5.2 including $\pm L$ (which is the case for the arithmetic given by (8.5.12)). □

We note that condition (8.5.14) is usually called a *diagonal dominance condition* in the literature [51].

We are now in a position to prove the following result.

Theorem 8.5.3 The nth-order digital filter described by (8.5.9), in which φ is given in (8.5.12) or (8.5.13) with $-1 < L \leq 1$, is free of limit cycles, if A is stable and if there exists a positive definite matrix H that satisfies (8.5.14), such that

$$Q \triangleq H - A^T HA$$

is positive semidefinite.

Proof. We can follow the same procedure as in the proof of Theorem 8.5.2 to prove that under the present conditions, the equilibrium $x_e = 0$ of system (8.5.9) is globally asymptotically stable. Thus the digital filter described by (8.5.9) is free of limit cycles. □

For the two's complement and triangular overflow characteristics, we have the following.

Lemma 8.5.3 An $n \times n$ positive definite matrix $H = [h_{ij}]$ satisfies Assumption 8.5.3 when f represents the two's complement or the triangular arithmetic, *if and only if* H is a diagonal matrix with positive diagonal elements.

Proof. The proof is similar to the proof of Lemma 8.5.2. □

D. Examples

To demonstrate the applicability of the results in the previous two subsections, we now consider two specific examples.

Example 8.5.1 For system (8.5.1) with

$$A = \begin{bmatrix} 1 & 2^{-3} \\ -0.1 & 0.9 \end{bmatrix}, \tag{8.5.17}$$

we have $\|A\|_p > 1$, $p = 1, 2$, or ∞. Therefore, condition (8.5.6) fails as a global asymptotic stability test for this example, as shown in the following.

Assumption 8.5.2 is satisfied for this example by choosing

$$H = \begin{bmatrix} 1 & 0.5 \\ 0.5 & 0.8 \end{bmatrix}. \tag{8.5.18}$$

Inasmuch as

$$Q = H - A^T HA = \begin{bmatrix} 0.092 & 0.00325 \\ 0.00325 & 0.023875 \end{bmatrix}$$

is positive definite, all conditions of Theorem 8.5.2 are satisfied and the equilibrium $x_e = 0$ of system (8.5.1) with A specified by (8.5.17) is globally asymptotically stable. □

Example 8.5.2 For system (8.5.1) with A given by

$$A = \begin{bmatrix} -1 & 0 & 0.1 & 0 \\ 0.2 & -0.6 & 0 & 0.8 \\ -0.1 & 0.1 & 0.8 & 0 \\ 0.1 & 0 & 0.1 & -0.5 \end{bmatrix}, \tag{8.5.19}$$

it can easily be verified that $\|A\|_p > 1, p = 1, 2,$ or ∞. Hence, condition (8.5.6) fails again as a global asymptotic stability test for the present example.

Assumption 8.5.2 is satisfied for this example by choosing

$$H = \begin{bmatrix} 1.4 & 0 & -0.2 & 0.4 \\ 0 & 1.6 & 0.2 & -0.4 \\ -0.2 & 0.2 & 3.4 & 0.5 \\ 0.4 & -0.4 & 0.5 & 3 \end{bmatrix}. \tag{8.5.20}$$

Because

$$Q = H - A^T HA = \begin{bmatrix} 0.026 & 0.161 & -0.003 & 0.077 \\ 0.161 & 1.014 & -0.003 & 0.497 \\ -0.003 & -0.003 & 1.124 & 0.774 \\ 0.077 & 0.497 & 0.774 & 0.906 \end{bmatrix}$$

is positive definite, all conditions of Theorem 8.5.2 are satisfied, and the equilibrium $x_e = 0$ of system (8.5.1) with such a coefficient matrix is globally asymptotically stable. □

8.6 Notes and References

For further details concerning Luré-type of results (Theorem 8.1.1), refer to [45]. The Yacubovich–Kalman Lemma (Lemma 8.1.1) was independently established by Yacubovich [73] and Kalman [32]. The proof of Popov's criterion (Theorem 8.1.2), using the Yacubovich–Kalman Lemma, follows along the proof given in Lefschetz [40]. In the original proof of results of this type, Popov relied on functional analysis techniques [62]. In addition to [40], there are several other monographs on the absolute stability of regulator systems, including Aizerman and Gantmacher [2] and Narendra and Taylor [60]. Our treatment in Section 8.1 of the absolute stability of regulator systems is in the spirit of the presentation on this subject given in [52] and [54].

For background material and further qualitative studies of the class of artificial neural networks considered in Section 8.2, refer, for example, to [9], [11], [20], [37], [41], and [47]–[50]. The particular method used in this section was popularized by Hopfield [20]. The first rigorous proof of Theorem 8.2.1 (in a more general setting) was given in [11] by Cohen and Grossberg. For further results concerning associate memories realized by means of feedback artificial neural networks, refer to [47], [49], and [50]. The idea of viewing neural networks as interconnected systems is motivated by the viewpoints adopted in [19], [51], and [64]. Our presentation in Section 8.2 is primarily based on [48], [49], and [50].

The analysis and synthesis of *linear digital feedback control systems* with *one uniform sampling rate* have been of interest for a long time ([1], [3], [14], [15], [30]) and in recent works, systems with nonuniform sampling rates have also been addressed (e.g., [28], and [29]). The implementation of the controllers of such systems by digital computers, resulting in digital feedback control systems, has brought about several investigations of the effects of the quantization nonlinearities (e.g., [7], [10], [13], [21], [26], [27], [55], [56], [65], and [69]). Additionally, analyses of digital control systems with nonlinear plants have also been conducted (e.g., [6], [21], and [26]). These works address the particular conditions under which a linearization of the plant is permissible (i.e., under which conditions the stability properties of the feedback control systems with nonlinear plants can be deduced from the corresponding feedback control systems with the nonlinear plants replaced by their linearization). We note here that although similar in spirit, the results presented in [26] apply to a substantially larger class of systems than those considered in [6]. Our presentation in Section 8.3 follows closely the development given in [21] and [26].

The results of Section 8.4 concerning pulse-width-modulated feedback control systems are based on [22]. There have only been a few results ([5], [16], [23], [36], [38]) concerning PWM feedback systems with *type II modulation*. (The majority stability results reported in the literature are for PWM feedback systems with *type I modulation*, e.g., [5], [25], [58], and [71].) The examples treated in this section are from [5] and [16]. Our comparisons, using these examples, indicate that the stability results reported in Section 8.4 are less conservative than the results reported in the above references. The reason for this appears twofold. First, the results in Section 8.4 are based on the general stability results for DDS which require that the quadratic Lyapunov functions employed in the analysis decrease along the solutions of the PWM systems *only at instants when the PWM controller is turned off* (and satisfy certain bounds at the remaining times) whereas the results reported in [5] require that the Lyapunov functions that are utilized (usually quadratic ones) decrease along the solutions of the PWM system *at all times* and results in [16] use an averaging method combined with the Popov criterion or the circle criterion. Additionally, the stability results for PWM feedback control systems reported in Section 8.4 incorporate optimization procedures to decrease conservatism, which is not the case in the stability results cited above.

For additional references on PWM feedback control systems (not necessarily dealing with stability), the reader may want to consult [4], [12], [17], [18], [31], [33]–[35], [39], [46], [61], [66]–[68], and [70], [71].

For a discussion of stability results of systems endowed with saturation nonlinearities and their applications to digital filters, refer to the monograph [44] and the references cited therein. The material in Section 8.5 is based on results presented in [24] and [43].

Bibliography

[1] J. Ackerman, *Sampled-Data Control Systems: Analysis and Synthesis, Robust Systems Design*, New York: Springer-Verlag, 1985.

[2] M. A. Aizerman and F. R. Gantmacher, *Absolute Stability of Regular Systems*, San Francisco: Holden-Day, 1963.

[3] P. J. Antsaklis and A. N. Michel, *Linear Systems*, Boston: Birkhäuser, 2006.

[4] G. S. Axelby, "Analysis of pulse-width modulation with a variable transport lag in a rendezvoue radar," *IEEE Trans. Autom. Control*, vol. 9, pp. 514–525, 1964.

[5] A. Balestrino, A. Eisenberg, and L. Sciavicco, "A generalized approach to the stability analysis of PWM feedback control systems," *J. Franklin Institute*, vol. 298, pp. 45–48, 1974.

[6] A. T. Barabanov and Ye. F. Starozhilov, "Investigation of the stability of continuous discrete systems by Lyapunov's second methods," *Soviet J. Autom. Inf. Sci.*, vol. 21, pp. 35–41, 1988.

[7] J. E. Bertram, "The effects of quantization in sampled feedback systems," *Trans. AIEE Appl. Ind.*, Part 2, vol. 77, pp. 177–181, 1958.

[8] E. Bolzer, "The limitation of impulses in cardiac muscle," *Amer. J. Physiol.*, vol. 138, pp. 273–282, 1942.

[9] G. A. Carpenter, M. Cohen, and S. Grossberg, "Computing with neural networks," *Science*, vol. 235, pp. 1226–1227, 1987.

[10] J.-H. Chou, S.-H. Chen, and I.-R. Horng, "Robust stability bound on linear time-varying uncertainties for linear digital control systems under finite wordlength effects," *JSME Int. J.*, Series C, vol. 39, pp. 767–771, 1996.

[11] M. Cohen and S. Grossberg, "Absolute stability of global pattern formation and parallel memory storage by competitive networks," *IEEE Trans. Syst., Man, Cyberne.*, vol. 13, pp. 815–826, 1983.

[12] F. R. Delfelt and G. J. Murphy, "Analysis of pulse-width-modulated control systems," *IEEE Trans. Autom. Control*, vol. 6, pp. 283–292, 1961.

[13] J. A. Farrell and A. N. Michel, "Estimates of asymptotic trajectory bounds in digital implementations of linear feedback control systems," *IEEE Trans. Autom. Control*, vol. 34, pp. 1319–1324, 1989.

[14] B. A. Francis and T. T. Georgiou, "Stability theory for linear time-invariant plants with periodic digital controllers," *IEEE Trans. Autom. Control*, vol. 33, pp. 820–832, 1988.

[15] G. F. Franklin and J. D. Powell, *Digital Control of Dynamic Systems*, Reading, MA: Addison-Wesley, 1980.

[16] A. Kh. Gelig and A. N. Churilov, *Stability and Oscillations of Nonlinear Pulse-Modulated Systems*, Boston: Birkhäuser, 1998.

[17] M. Gouy, "On a constant temperature oven," *J. Physique*, vol. 6, pp. 479–483, 1897.

[18] S. C. Gupta and E. I. Jury, "Statistical study of pulse-width-modulated control systems," *J. Franklin Institute*, vol. 273, pp. 292–321, 1962.

[19] L. T. Grujić, A. A. Martynyuk, and M. Ribbens-Pavella, *Large Scale Systems Under Structural and Singular Perturbation*, Berlin: Springer-Verlag, 1987.

[20] J. J. Hopfield, "Neurons with graded response have collective computational properties like those of two-state neurons," *Proc. Nat. Acad. Sci. USA*, vol. 81, pp. 3088–3092, 1984.

[21] L. Hou, *Qualitative Analysis of Discontinuous Deterministic and Stochastic Dynamical Systems*, Ph.D. Dissertation, University of Notre Dame, Notre Dame, IN, 2000.

[22] L. Hou, "Stability analysis of pulse-width-modulated feedback systems with type 2 modulation," *Proc. 44th IEEE Conference on Decision and Control*, Paradise Island, Bahamas, Dec. 2004, pp. 2972–2977.

[23] L. Hou, "Stability analysis of pulse-width-modulated feedback systems with type 2 modulation: The critical case," *Proc. 2005 Int. Symposium on Circuits and Systems*, Tokyo, Japan, May 2005, pp. 3187–3190.

[24] L. Hou and A. N. Michel, "Asymptotic stability of systems with saturation constraints," *IEEE Trans. Autom. Control*, vol. 43, pp. 1148–1154, 1998.

[25] L. Hou and A. N. Michel, "Stability analysis of pulse-width-modulated feedback systems," *Automatica*, vol. 37, pp. 1335–1349, 2001.

[26] L. Hou, A. N. Michel, and H. Ye, "Some qualitative properties of sampled-data control systems," *Proc. 35th IEEE Conference on Decision and Control*, Kobe, Japan, Dec. 1996, pp. 911–917.

[27] B. Hu, Z. Feng, and A. N. Michel, "Quantized sampled-data feedback stabilization for linear and nonlinear control systems," *Proc. 38th IEEE Conference on Decision and Control*, Phoenix, AZ, Dec. 1999, pp. 4392–4397.

[28] B. Hu and A. N. Michel, "Robust analysis of digital control systems with time-varying sampling periods," *Proc. 1999 American Control Conference*, San Diego, CA, June 1999, pp. 3484–3488.

[29] B. Hu and A. N. Michel, "Stability analysis of digital feedback control systems with time-varying sampling periods," *Automatica*, vol. 36, pp. 897–905, 2000.

[30] P. A. Iglesias, "On the stability of sampled-date linear time-varying feedback systems," *Proc. 33rd IEEE Conference on Decision and Control*, Orlando, FL, Dec. 1994, pp. 219–224.

[31] T. T. Kadota and H. C. Bourne, "Stability conditions of pulse-width-modulated systems through the second method of Lyapunov," *IEEE Trans. Autom. Control*, vol. 6, pp. 266–276, 1961.

[32] R. E. Kalman, "Lyapunov functions for the problem of Luré in automatic control," *Proc. Nat. Acad. Sci. USA*, vol. 49, pp. 201–205, 1963.

[33] B. Katz, "The nerve impulse," *Sci. Amer.*, vol. 187, pp. 55–64, 1962.

[34] A. Khayatian and D. G. Taylor, "Feedback control of linear systems by multi-rate pulse-width modulation," *IEEE Trans. Autom. Control*, vol. 39, pp. 1292–1297, 1994.

[35] A. Khayatian and D. G. Taylor, "Multi-rate modeling and control design for switched-mode power converters," *IEEE Trans. Autom. Control*, vol. 39, pp. 1848–1852, 1994.

[36] A. I. Korshunov, "Remarks on the article by V.M. Kuntsevich 'Global asymptotic stability of two classes of control systems with pulse-duration and pulse-frequency modulations'," *Autom. Remote Control*, vol. 34, no. 8, pt. 2, pp. 1356–1357, Aug. 1973.

[37] B. Kosko, *Neural Networks and Fuzzy Systems*, Englewood Cliffs, NJ: Prentice Hall, 1992.

[38] V. M. Kuntsevich, "Global asymptotic stability of two classes of control systems with pulse duration and pulse frequency modulations," *Autom. Remote Control*, vol. 33, no. 7, pp. 1124–32, July 1972.

[39] V. M. Kuntsevich and Yu. N. Akekhovol, "Fundamentals of nonlinear control systems with pulse-frequency and pulse width modulation," *Automatica*, vol. 7, pp. 83–81, 1971.

[40] S. Lefschetz, *Stability of Nonlinear Control Systems*, New York: Academic Press, 1965.

[41] J.-H. Li, A. N. Michel, and W. Porod, "Qualitative analysis and synthesis of a class of neural networks," *IEEE Trans. Circ. Syst.*, vol. 35, pp. 976–986, 1988.

[42] J.-H. Li, A. N. Michel, and W. Porod, "Analysis and synthesis of a class of neural networks: Linear systems operating on a closed hypercube," *IEEE Trans. Circ. Syst.*, vol. 36, pp. 1405–1422, Nov. 1989.

[43] D. Liu and A. N. Michel, "Asymptotic stability of discrete-time systems with saturation nonlinearities with applications to digital filters," *IEEE Trans. Circ. Syst. I: Fund. Theor. Appl.*, vol. 39, pp. 798–807, Oct. 1992.

[44] D. Liu and A. N. Michel, *Dynamical Systems with Saturation Nonlinearities: Analysis and Design*, New York: Springer-Verlag, 1994.

[45] A. I. Luré, *On Some Nonlinear Problems in the Theory of Automatic Control*, London: H. M. Stationary Office, 1951.

[46] B. H. C. Matthews, "The nervous system as an electrical instrument," *J. Inst. Elec. Eng.*, vol. 95, pp. 397–402, 1948.

[47] A. N. Michel and J. A. Farrell, "Associate memories via artificial neural networks," *IEEE Control Sys. Mag.*, vol. 10, pp. 6–17, 1990.

[48] A. N. Michel, J. A. Farrell, and W. Porod, "Qualitative analysis of neural networks," *IEEE Trans. Circ. Syst.*, vol. 36, pp. 229–243, 1989.

[49] A. N. Michel, J. A. Farrell, and H. F. Sun, "Analysis and synthesis techniques for Hopfield type synchronous discrete time neural networks with applications to content addressable memory," *IEEE Trans. Circ. Syst.*, vol. 37, pp. 1356–1366, 1990.

[50] A. N. Michel and D. Liu, *Qualitative Analysis and Synthesis of Recurrent Neural Networks*, New York: Marcel Dekker, 2002.

[51] A. N. Michel and R. K. Miller, *Qualitative Analysis of Large Scale Dynamical Systems*, New York: Academic Press, 1977.

[52] A. N. Michel, K. Wang, and B. Hu, *Qualitative Theory of Dynamical Systems-The Role of Stability Preserving Mappings*, 2nd Edition, New York: Marcel Dekker, 2001.

[53] A. N. Michel, J. Si, and G. Yen, "Analysis and synthesis of a class of discrete-time neural networks described on hypercubes," *IEEE Trans. Neural Netw.*, vol. 2, pp. 32–46, Jan. 1991.

[54] R. K. Miller and A. N. Michel, *Ordinary Differential Equations*, New York: Academic Press, 1982.

[55] R. K. Miller, A. N. Michel, and J. A. Farrell, "Quantizer effects on steady-state error specifications of digital feedback control systems," *IEEE Trans. Autom. Control*, vol. 34, pp. 651–654, 1989.

[56] R. K. Miller, M. S. Mousa, and A. N. Michel, "Quantization and overflow effects in digital implementations of linear dynamic controllers," *IEEE Trans. Autom. Control*, vol. 33, pp. 698–704, July 1988.

[57] W. L. Mills, C. T. Mullis, and R. A. Roberts, "Digital filter realizations without overflow oscillations," *IEEE Trans. Acoust., Speech, Signal Proc.*, vol. ASSP-26, pp. 334–338, Aug. 1978.

[58] B. J. Min, C. Slivinsky, and R. G. Hoft, "Absolute stability analysis of PWM systems," *IEEE Trans. Autom. Control*, vol. 22, pp. 447–452, 1977.

[59] G. J. Murphy and S. H. Wu, "A stability criterion for pulse-width-modulated feedback control systems," *IEEE Trans. Autom. Control*, vol. 9, pp. 434–441, 1964.

[60] K. S. Narendra and J. H. Taylor, *Frequency Domain Stability for Absolute Stability*, New York: Academic Press, 1973.

[61] E. Polak, "Stability and graphical analysis of first-order pulse-width-modulated sampled-data regulator systems," *IEEE Trans. Autom. Control*, vol. 6, pp. 276–282, 1961.

[62] V. M. Popov, "Absolute stability of nonlinear systems of automatic control," *Autom. Remote Control*, vol. 22, pp. 857–895, 1961.

[63] I. W. Sandberg, "A theorem concerning limit cycles in digital filters," *Proc. 7th Annual Allerton Conference on Circuit and System Theory*, University of Illinois at Urbana-Champaign, Urbana, IL, pp. 63–68, Oct. 1969.

[64] D. D. Siljak, *Large-Scale Dynamical Systems: Stability and Structure*, New York: North Holland, 1978.

[65] V. Singh, "Elimination of overflow oscillations in fixed-point state-space digital filters using saturation arithmetic," *IEEE Trans. Circ. Syst.*, vol. 37, pp. 814–818, 1990.

[66] H. Sira-Ramirez, "A geometric approach to pulse-width modulated control in nonlinear dynamical systems," *IEEE Trans. Autom. Control*, vol. 34, pp. 184–187, 1989.

[67] H. Sira-Ramirez and L. S. Orestes, "On the dynamical pulse-width-modulation control of robotic manipulator systems," *Int. J. Robust Control*, vol. 6, pp. 517–537, 1996.

[68] H. Sira-Ramirez and M. T. Prada-Rizzo, "Nonlinear feedback regulator design for the Cuk converter," *IEEE Trans. Autom. Control*, vol. 37, pp. 1173–1180, 1992.

[69] J. E. Slaughter, "Quantization errors in digital control systems," *IEEE Trans. Autom. Control*, vol. 9, pp. 70–74, 1964.

[70] D. G. Taylor, "Pulse-width modulated control of electromechanical systems," *IEEE Trans. Autom. Control*, vol. 37, pp. 524–528, 1992. p. 70–74, 1964.

[71] S. G. Tsefastas, "Pulse width and pulse frequency modulated control systems," in *Simulation of Control Systems*, I. Troch, Editor, New York: North-Holland, 1978, pp. 41–48.

[72] P. P. Vaidyanathan and V. Liu, "An improved sufficient condition for absence of limit cycles in digital filters," *IEEE Trans. Circ. Syst.*, vol. CAS-34, pp. 319–322, Mar. 1987

[73] V. A. Yacubovich, "Solution of certain matrix inequalities encountered in nonlinear control theory," *Soviet Math. Doklady*, vol. 5, pp. 652–666, 1964.

Chapter 9

Infinite-Dimensional Dynamical Systems

In this chapter we address the Lyapunov stability and the boundedness of motions of infinite-dimensional dynamical systems determined by differential equations defined on Banach spaces and semigroups. As in Chapters 6, 7, and 8, we concentrate on the qualitative properties of equilibria and we consider continuous as well as discontinuous dynamical systems (DDS).

This chapter consists of eleven parts. In the *first section* we establish some of the notation used throughout this chapter. In the *second section* we present the principal Lyapunov stability and instability results of an equilibrium for dynamical systems determined by general differential equations in Banach spaces, as well as results for the boundedness of motions (Lagrange stability) for such systems. Most of these results are direct consequences of the corresponding results established in Chapter 3 for dynamical systems defined on metric spaces. We demonstrate the applicability of these results in the analysis of several specific classes of differential equations defined on different Banach spaces. In the *third section* we present converse theorems (necessary conditions) for most of the results (sufficient conditions) established in the second section. Most of these results are also direct consequences of corresponding results given in Chapter 3 for dynamical systems defined on metric spaces. In the *fourth section* we present the invariance theory for autonomous differential equations defined on Banach spaces and we apply these results in the analysis of specific classes of systems. In the *fifth section* we develop a comparison theory for general differential equations defined on Banach spaces and we apply these results in a subsequent section. In the *sixth section* we present stability results for composite systems defined on Banach spaces described by a mixture of different differential equations and we apply these results in the analysis of a specific class of systems. In the *seventh section* we apply the results developed in the fifth section in the stability analysis of a point kinetics model of a multicore nuclear reactor (described by Volterra integrodifferential equations). In the *eighth section* we present general stability, instability, and boundedness results for dynamical systems determined by retarded functional differential equations (RFDEs) (a special important class of differential equations in Banach spaces). In addition to these general results, we present results concerning the invariance theory for RFDEs and Razumikhin-type theorems for such systems. In the *ninth section* we apply the results of the eighth section in the analysis of a class of artificial neural networks with time delays. In the *tenth section* we address stability and boundedness results for discontinuous dynamical systems determined by differential equations in Banach spaces. We address local and global stability and instability results of an equilibrium and results for the boundedness of motions, as well as converse theorems. We apply these results in the analysis of several specific classes of systems. Finally, in the *eleventh section* we present stability results for discontinuous dynamical systems determined by linear and nonlinear semigroups and we apply these results in the analysis of several specific classes of infinite-dimensional DDS.

9.1 Preliminaries

Let X be a Banach space with norm $\|\cdot\|$, let C be a subset of X, let $0 \in C$ and let $F\colon \mathbb{R} \times C \to X$. Recall from Subsection 2.6A that the equation

$$\dot{x} = F(t, x) \tag{GE}$$

is called a *differential equation in Banach space*, where $\dot{x} = dx/dt$. A function $\varphi\colon [t_0, t_0 + c) \to C$, $c > 0$, is called a *solution of* (GE) if $\varphi \in C\,[[t_0, t_0 + c), C]$, if φ is differentiable with respect to t for $t \in [t_0, t_0 + c)$ and if φ satisfies the equation $(d\varphi/dt)(t) = F(t, \varphi(t))$ for all $t \in [t_0, t_0 + c)$. Associated with (GE) we have the *initial value problem* given by

$$\dot{x} = F(t, x), \qquad x(t_0) = x_0. \tag{I_{GE}}$$

Throughout this chapter we assume that for each $(t_0, x_0) \in \mathbb{R}^+ \times C$ there exists at least one solution of (I_{GE}), $\varphi(t, t_0, x_0)$, that satisfies the initial condition $x(t_0) = x_0 = \varphi(t_0, t_0, x_0)$. The reader should refer to Subsection 2.6A for a discussion concerning the existence and uniqueness of solutions of the initial value problem (I_{GE}).

As discussed throughout Chapter 2, special classes of differential equations on Banach spaces include *ordinary differential equations, functional differential equations, Volterra integrodifferential equations* and *partial differential equations*. All of these determine dynamical systems. We denote a dynamical system determined by (GE) by S_{GE}.

In a similar manner as in the case of finite-dimensional dynamical systems S_E determined by (E) (refer to Chapter 6), we use the phrase "*M is an invariant set of (GE)*" in place of the phrase "*M is an invariant set with respect to S_{GE}*", the phrase "*the invariant set M of (GE) is stable*" in place of "*the invariant set M is stable with respect to S_{GE}*", and similar phrases for all other remaining stability, instability, and boundedness types that we encounter.

As in Chapters 6 through 8, we continue to focus on *equilibria* for (GE); that is, $M = \{x_e\}$. We use the phrase "*the equilibrium x_e of (GE) is stable*" in place of "*the invariant set $\{x_e\}$ of (GE) is stable*", "*the equilibrium x_e of (GE) is asymptotically stable*" in place of "*the invariant set $\{x_e\}$ of (GE) is asymptotically stable*", and so forth.

We recall that in the case of finite-dimensional dynamical systems, all norms are topologically equivalent and therefore, when addressing convergence properties, (e.g., the stability of an equilibrium), the particular choice of norm plays no important role. This is in general not the case for dynamical systems defined on infinite-dimensional spaces. Specifically, for dynamical systems determined by (GE), the various stability and boundedness properties depend on the choice of the norm, because on infinite-dimensional normed linear spaces, different norms are in general not topologically equivalent. Accordingly, it is usually necessary to specify *explicitly* which norm is being used in a given result.

Similarly as in the case of finite-dimensional dynamical systems, we may assume without loss of generality that $x_e = 0$ is an equilibrium for (GE).

As in Chapters 6 and 7, we employ lowercase letters to denote scalar-valued Lyapunov functions (e.g., v) and uppercase letters to denote vector-valued Lyapunov functions (e.g., V). Also, we denote scalar Lyapunov functions that are the components of vector Lyapunov functions by lowercase letters (e.g., v_i).

We can characterize a scalar Lyapunov function $v \in C[B(r) \times \mathbb{R}^+, \mathbb{R}]$ (resp., $v \in C[X \times \mathbb{R}^+, \mathbb{R}]$) as being *positive definite* (*negative definite*), *positive semidefinite (negative semidefinite), decrescent*, or *radially unbounded*, by modifying Definitions 6.1.8–6.1.13 (resp., Theorems 6.1.1–6.1.3) in an appropriate way, replacing $\mathbb{R}^n$ by X, $\Omega \subset \mathbb{R}^n$ by $C \subset X$, and $B(r) \subset \Omega$ by $B(r) \subset C$. As in the case of the various stability and boundedness concepts, the above characterizations are tied to the choice of the particular Banach space being used. Thus, we may have to use phrases such as "v is positive definite with respect to the space X", and similar other phrases.

Let $\varphi(\cdot, t_0, x_0)$ denote a solution of (GE). For every function $v \in C[B(r) \times \mathbb{R}^+, \mathbb{R}]$, where $B(r) = \{x \in X : \|x\| < r\}$ with $r > 0$ and $B(r) \subset C$, we define the *upper-right derivative of v with respect to t along the solutions of (GE)* by

$$v'_{(GE)}(x,t) = \overline{\lim_{\Delta t \to 0^+}} \sup_{\varphi(t,t,x)=x} \frac{1}{\Delta t}\big[v(\varphi(t+\Delta t, t, x), t+\Delta t) - v(x,t)\big]. \quad (9.1.1)$$

When (GE) has a *unique* solution for every $x(t_0) = x_0 = \varphi(t_0, t_0, x_0)$ with $(t_0, x_0) \in \mathbb{R}^+ \times B(r)$, then (9.1.1) reduces to

$$v'_{(GE)}(x,t) = \overline{\lim_{\Delta t \to 0^+}} \frac{1}{\Delta t}\big[v(\varphi(t+\Delta t, t, x), t+\Delta t) - v(x,t)\big]. \quad (9.1.2)$$

If in addition to the above assumptions, v satisfies a *local Lipschitz condition in x*, that is, if for every $T > 0$ there exists an $L > 0$ such that

$$\big|v(x,t) - v(y,t)\big| \leq L\|x-y\| \quad (9.1.3)$$

for all $x, y \in B(r)$ and for all $t \in [0, T]$, then (9.1.2) can equivalently be expressed as

$$v'_{(GE)}(x,t) = \overline{\lim_{\Delta t \to 0^+}} \frac{1}{\Delta t}\big[v(x + \Delta t F(t,x), t + \Delta t) - v(x,t)\big]. \tag{9.1.4}$$

We ask the reader to verify relation (9.1.4) in the problem section.

9.2 The Principal Lyapunov Stability and Boundedness Results for Differential Equations in Banach Spaces

In the present section we address stability and boundedness properties of infinite-dimensional dynamical systems determined by differential equations in Banach spaces (GE).

A. Local results

We assume that $C \subset X$, $x_e = 0 \in C$, $x_e = 0$ is an equilibrium for (GE) and we let $\|\cdot\|$ denote the norm for the Banach space X.

Theorem 9.2.1 Assume that for system (GE) there exists a positive definite function $v \in C[B(r) \times \mathbb{R}^+, \mathbb{R}]$ such that $v'_{(GE)}$ is negative semidefinite where $B(r) \subset C$ for some $r > 0$. Then the following are true.

(a) The equilibrium $x_e = 0$ of (GE) is *stable*.

(b) If in addition to the above, v is decrescent, then $x_e = 0$ of (GE) is *uniformly stable*.

(c) If in (b), $v'_{(GE)}$ is negative definite, then $x_e = 0$ of (GE) is *uniformly asymptotically stable*.

(d) If in (c), v satisfies

$$c_1\|x\|^b \le v(x,t) \le c_2\|x\|^b$$

and

$$v'_{(GE)}(x,t) \le -c_3\|x\|^b$$

for all $(x,t) \in B(r) \times \mathbb{R}^+$, where c_1, c_2, c_3 and b are positive constants, then $x_e = 0$ of (GE) is *exponentially stable*.

Proof. The proof of this theorem follows from the proofs of Theorems 3.3.1–3.3.3 and follows along the lines of the proof of Theorems 6.2.1, 6.2.2, and 6.2.4. We omit the details in the interest of brevity. □

In order to apply Theorem 9.2.1 (and the other Lyapunov theorems that we consider) in the stability analysis of initial value and boundary value problems determined by partial differential equations, we need to show that such problems may be viewed as systems of differential equations (GE). To fix some of the ideas involved, we consider in the following an important specific class of initial value and boundary value problems determined by partial differential equations.

Let Ω be a bounded domain in $\mathbb{R}^n$ with smooth boundary $\partial\Omega$ (i.e., $\partial\Omega$ is of class C^∞ (refer to Section 2.10)), let ∂_x^α denote the operator $\partial^{|\alpha|}/(\partial_{x_1}^{\alpha_1}\cdots\partial_{x_n}^{\alpha_n})$ for $x=(x_1,\dots,x_n)$ and $\alpha=(\alpha_1,\dots,\alpha_n)$ with $|\alpha|=\alpha_1+\cdots+\alpha_n$, and let f denote a real n vector-valued C^∞ function of t, x, u, and $\partial_x^\alpha u$ for all α such that $|\alpha|\le 2m$, where $u=(u_1,\dots,u_l)^T\colon \mathbb{R}^+\times\Omega\to\mathbb{R}^l$ and $\partial_x^\alpha u=(\partial_{x_1}^{\alpha_1}u_1\cdots\partial_{x_l}^{\alpha_l}u_l)^T$. We now consider the class of partial differential equations with *initial conditions* and *boundary conditions* given by

$$\begin{cases} \dfrac{\partial u}{\partial t}(t,x)=f(t,x,u,\partial_x u,\dots,\partial_x^\alpha u,\dots), & & (t,x)\in[t_0,\infty)\times\Omega \\ \partial_x^\alpha u(t,x)=0, & |\alpha|<m, & (t,x)\in[t_0,\infty)\times\partial\Omega \\ u(t_0,x)=u_0(x), & & x\in\Omega \end{cases} \tag{9.2.1}$$

where $\partial u/\partial t=(\partial u_1/\partial t,\dots,\partial u_l/\partial t)^T$, $t_0\in\mathbb{R}^+$, and $u_0\in C^\infty[\overline{\Omega},\mathbb{R}^l]$ satisfies $\lim_{x\to\partial\Omega}\partial_x^\alpha u_0(x)=0$ for all $|\alpha|<m$.

Assume that for every $t_0\in\mathbb{R}^+$ and $u_0\in C^\infty[\overline{\Omega},\mathbb{R}^l]$, there exists at least one solution $u\in C^\infty[[t_0,\infty)\times\overline{\Omega},\mathbb{R}^l]$ that satisfies (9.2.1). Conditions that ensure this for *linear parabolic equations* are given in Theorem 2.10.2. For results that ensure the existence of certain classes of *nonlinear parabolic equations*, refer to [17] and [19].

Now let $C=C^\infty[\overline{\Omega},\mathbb{R}^l]\cap H_0^m(\Omega)$, let $U(t)=u(t,\cdot)\in C^\infty[\overline{\Omega},\mathbb{R}^l]$, and define $F\colon\mathbb{R}^+\times C\to C$ by

$$F(t,U(t))(x)=f(t,x,u,\partial_x u,\dots,\partial_x^\alpha u,\dots).$$

(Refer to Section 2.10 for the definition of $H_0^m(\Omega)$.) We can now rewrite (9.2.1) as

$$\frac{dU}{dt}=F(t,U),\qquad U(t_0)=u_0 \tag{9.2.2}$$

where $t\in[t_0,\infty)$, $(t_0,u_0)\in\mathbb{R}^+\times C$ and $F\colon\mathbb{R}^+\times C\to C$. Then (9.2.2) constitutes an *initial value problem* defined on the Banach space $X=H_0^m(\Omega)$ for the differential equation

$$\frac{dU}{dt}=F(t,U) \tag{9.2.3}$$

which is a special case of the differential equation (GE).

We conclude by noting that there are more general classes of initial value and boundary value problems than (9.2.1) which can be cast as special cases of (GE).

We now apply Theorem 9.2.1 in the stability analysis of a specific example of (9.2.1).

Example 9.2.1 A special case of (9.2.1) is the initial value and boundary value problem for the *fourth-order linear scalar parabolic partial differential equation* given by

$$\begin{cases} \dfrac{\partial u}{\partial t} = -\dfrac{\partial^4 u}{\partial x^4}(t,x), & (t,x) \in \mathbb{R}^+ \times (0,1) \\ u(0,x) = \psi(x), & x \in (0,1) \\ u(t,0) = u(t,1) = \dfrac{\partial u}{\partial x}(t,0) = \dfrac{\partial u}{\partial x}(t,1) = 0, & t \in \mathbb{R}^+. \end{cases} \tag{9.2.4}$$

In view of Theorem 2.10.2, for each $\psi \in X \triangleq H^4[(0,1),\mathbb{R}] \cap H_0^2[(0,1),\mathbb{R}]$ there exists a unique solution $u = u(t,x)$ of (9.2.4) such that $u(t,\cdot) \in X$ for each fixed $t \in \mathbb{R}^+$, and U, defined by $U(t) = u(t,\cdot)$, is a continuously differentiable function from $\mathbb{R}^+$ to X with respect to the H^2-norm (refer, e.g., to [33, p. 210]).

In view of our discussion of the class of systems (9.2.1), we may view (9.2.4) as an initial value problem for a differential equation of the form (GE) in the Banach space X with the H^2-norm. It is easily shown that $\varphi_e \equiv 0 \in X$ is an equilibrium for (9.2.4).

In applying Theorem 9.2.1, we choose the Lyapunov function

$$v(\psi) = \|\psi\|_{H^2}^2 = \int_0^1 \left[\left(\frac{\partial^2 \psi}{\partial x^2}\right)^2 + \left(\frac{\partial \psi}{\partial x}\right)^2 + \psi^2\right] dx \tag{9.2.5}$$

and we denote a solution of (9.2.4) by $u(t,x)$.

Along the solutions of (9.2.4) we have

$$\begin{aligned} \frac{dv}{dt}(u) &= 2\int_0^1 \left[\left(\frac{\partial^2 u}{\partial x^2}\right)\frac{\partial}{\partial t}\left(\frac{\partial^2 u}{\partial x^2}\right) + \left(\frac{\partial u}{\partial x}\right)\frac{\partial}{\partial t}\left(\frac{\partial u}{\partial x}\right) + u\frac{\partial u}{\partial t}\right] dx \\ &= -2\int_0^1 \left[\left(\frac{\partial^2 u}{\partial x^2}\right)\frac{\partial^6 u}{\partial x^6} + \left(\frac{\partial u}{\partial x}\right)\frac{\partial^5 u}{\partial x^5} + u\frac{\partial^4 u}{\partial x^4}\right] dx. \end{aligned} \tag{9.2.6}$$

In order to ascertain the definiteness of $(dv/dt)(u)$, we rewrite the three terms in brackets in (9.2.6). We first consider the second term. Recalling the expression for integration by parts,

$$\int_0^1 pdq = pq\Big|_0^1 - \int_0^1 qdp$$

and letting

$$\begin{aligned} p &= \frac{\partial \psi}{\partial x}, & \frac{\partial p}{\partial x} &= \frac{\partial^2 \psi}{\partial x^2}, & dp &= \frac{\partial^2 \psi}{\partial x^2}dx \\ q &= \frac{\partial^4 \psi}{\partial x^4}, & dq &= \left(\frac{\partial^5 \psi}{\partial x^5}\right)dx = d\left(\frac{\partial^4 \psi}{\partial x^4}\right) \end{aligned}$$

and applying the initial conditions, we obtain

$$\int_0^1 \left(\frac{\partial \psi}{\partial x}\right)\frac{\partial^5 \psi}{\partial x^5}dx = -\left[-\int_0^1 \left(\frac{\partial^3 \psi}{\partial x^3}\right)\left(\frac{\partial^3 \psi}{\partial x^3}\right)dx\right] = \int_0^1 \left(\frac{\partial^3 \psi}{\partial x^3}\right)^2 dx.$$

In a similar manner, we obtain for the first term in brackets in (9.2.6)

$$\int_0^1 \left(\frac{\partial^2 \psi}{\partial x^2}\right) \frac{\partial^6 \psi}{\partial x^6} dx = \int_0^1 \left(\frac{\partial^4 \psi}{\partial x^4}\right)^2 dx$$

and for the third term,

$$\int_0^1 \psi \frac{\partial^4 \psi}{\partial x^4} dx = \int_0^1 \left(\frac{\partial^2 \psi}{\partial x^2}\right)^2 dx.$$

Hence, we have

$$\frac{dv}{dt}(u) = -2 \int_0^1 \left[\left(\frac{\partial^4 u}{\partial x^4}\right)^2 + \left(\frac{\partial^3 u}{\partial x^3}\right)^2 + \left(\frac{\partial^2 u}{\partial x^2}\right)^2 \right] = -2 \left\| \frac{\partial^2 u}{\partial x^2} \right\|_{H^2}^2 .$$

Therefore, along the solutions of (9.2.4) we have

$$\frac{dv}{dt}(u) = -2 \left\| \frac{\partial^2 u}{\partial x^2} \right\|_{H^2}^2 \leq 0 \tag{9.2.7}$$

for all $u \in X$.

It now follows from (9.2.5), (9.2.7) and Theorem 9.2.1 that the equilibrium $\varphi_e = 0$ of (9.2.4) is *uniformly stable with respect to the H^2-norm.*

In Section 9.4 (see Example 9.4.2) we show, utilizing a result from the invariance theory for infinite-dimensional systems, that the equilibrium $\varphi_e = 0$ of system (9.2.4) is actually uniformly asymptotically stable in the large with respect to the H^1-norm. □

B. Global results

In the case of global results we let $C = X$.

Theorem 9.2.2 Assume that there exists a function $v \in C[X \times \mathbb{R}^+, \mathbb{R}]$, two functions $\psi_1, \psi_2 \in \mathcal{K}_\infty$, and a function $\psi_3 \in \mathcal{K}$ such that

$$\psi_1(\|x\|) \leq v(x,t) \leq \psi_2(\|x\|)$$

and

$$v'_{(GE)}(x,t) \leq -\psi_3(\|x\|)$$

for all $(x,t) \in X \times \mathbb{R}^+$. Then the equilibrium $x_e = 0$ of (GE) is *uniformly asymptotically stable in the large*. Furthermore, if there exist four positive constants c_1, c_2, c_3, and b such that

$$c_1\|x\|^b \leq v(x,t) \leq c_2\|x\|^b$$

and

$$v'_{(GE)}(x,t) \leq -c_3\|x\|^b$$

for all $(x,t) \in X \times \mathbb{R}^+$, then the equilibrium $x_e = 0$ of (GE) is *exponentially stable in the large.*

Proof. The proof of this theorem is a direct consequence of Theorems 3.3.6 and 3.3.7. □

We apply the above results in the analysis of the *heat equation*.

Example 9.2.2 A class of *initial and boundary value problems* determined by the *heat equation* is given by

$$\begin{cases} \dfrac{\partial u}{\partial t} = a^2 \Delta u, & (t,x) \in \mathbb{R}^+ \times \Omega \\ u(0,x) = \psi(x), & x \in \Omega \\ u(t,x) = 0, & (t,x) \in \mathbb{R}^+ \times \partial\Omega \end{cases} \tag{9.2.8}$$

where $\Omega \subset \mathbb{R}^n$ is a bounded domain with smooth boundary $\partial\Omega$, $a^2 > 0$ is a constant, and $\Delta = \sum_{i=1}^n \partial^2/\partial x_i^2$ denotes the *Laplacian*. For each $\psi \in X = H^2[\Omega,\mathbb{R}] \cap H_0^1[\Omega,\mathbb{R}]$ there exists a unique solution $u = u(t,x)$ of (9.2.8) such that $u(t,\cdot) \in X$ for each fixed $t \in \mathbb{R}^+$ and U, defined by $U(t) = u(t,\cdot)$, is a continuously differentiable function from $\mathbb{R}^+$ to X with respect to the H^1-norm (refer, e.g., to [33, p. 210]). Then (9.2.8) can be written as an *abstract Cauchy problem* in the space X with respect to the H^1-norm (refer to the discussion of (9.2.2)),

$$\dot{U}(t) = AU(t), \qquad t \geq 0$$

with initial condition $U(0) = \psi \in X$, where the operator A is linear and is defined by $A = \sum_{i=1}^n a^2 d^2/dx_i^2$.

In Chapter 3 we showed that $0 \in X$ is an equilibrium for (9.2.8). We now show, using Theorem 9.2.2, that the equilibrium $x_e = 0$ is *exponentially stable with respect to the H^1-norm*. In doing so, we make use of *Gauss' Divergence Theorem* [8], which we recall here in the context of the problem on hand: the volume integral of the divergence of any continuously differentiable vector Q is equal to the closed surface integral of the outward normal component of Q; that is,

$$\int_\Omega \nabla Q(x)dx = \int_{\partial\Omega} Q\hat{n}dx$$

where $Q = (Q_1,\ldots,Q_n)^T$, $\nabla Q = (\partial Q_1/\partial x_1,\ldots,\partial Q_n/\partial x_n)^T$ and $Q\hat{n}$ is the directional derivative of Q in the outward normal direction.

Now let $Q = u^T \nabla u$, where u is a solution of (9.2.8). Then for any $u(t,\cdot) \in H_0^1[\Omega,\mathbb{R}]$ we have

$$\int_\Omega \Big[(\nabla u)^T \nabla u + u\Delta u\Big] dx = \int_{\partial\Omega} u\frac{\partial u}{\partial x}dx = 0$$

which implies that

$$\int_\Omega u\Delta u dx = -\int_\Omega \left|\nabla u\right|^2 dx.$$

In applying Theorem 9.2.2, we choose as a Lyapunov function

$$v(\psi) = \|\psi\|_{H^1}^2 = \int_\Omega \left(|\nabla\psi|^2 + |\psi|^2\right)dx. \tag{9.2.9}$$

Let $u(t, x)$ denote a solution of (9.2.8). Then

$$\begin{aligned}\frac{dv}{dt}(u) &= \int_\Omega \frac{\partial}{\partial t}\Big[\sum_{i=1}^n \Big(\frac{\partial u}{\partial x_i}\Big)^2 + u^2\Big]dx \\ &= \int_\Omega \Big[\sum_{i=1}^n 2\Big(\frac{\partial u}{\partial x_i}\Big)\frac{\partial^2 u}{\partial x_i \partial t} + 2u\frac{\partial u}{\partial t}\Big]dx \\ &= -\sum_{i=1}^n 2\int_\Omega \frac{\partial^2 u}{\partial x_i^2}\frac{\partial u}{\partial t}dx + 2a^2\int_\Omega u\Delta u dx \\ &= -2a^2\int_\Omega (\Delta u)^2 dx - 2a^2\int_\Omega |\nabla u|^2 dx \\ &\le -2a^2\int_\Omega |\nabla u|^2 dx.\end{aligned}$$

By *Poincaré's inequality* [34], we have that

$$\int_\Omega |u|^2 dx \le \gamma^2 \int_\Omega |\nabla u|^2 dx$$

where γ can be chosen as $\delta/\sqrt{n}$ and Ω can be put into a cube of length δ. Hence, we have

$$\frac{dv}{dt}(u) \le -a^2\Big(\int_\Omega |\nabla u|^2 dx + \frac{1}{\gamma^2}\int_\Omega |u|^2 dx\Big) \le -c\|u\|_{H^1}^2$$

where $c = a^2 \min\{1, 1/\gamma^2\}$. Therefore,

$$v'_{(9.2.8)}(\psi) \le -c\|\psi\|_{H^1}^2 \tag{9.2.10}$$

for all $\psi \in X$.

It now follows from (9.2.9), (9.2.10), and Theorem 9.2.2 that the equilibrium $0 \in X$ of (9.2.8) is *exponentially stable in the large with respect to the H^1-norm.* □

Example 9.2.3 *Scalar linear Volterra integrodifferential equations* are of the form

$$\begin{cases} \dot{x}(t) = -ax(t) + \int_{-\infty}^t k(t-s)x(s)ds, & t \ge 0 \\ x(t) = \varphi(t), & t \le 0 \end{cases} \tag{9.2.11}$$

where $a > 0$ is a constant. As shown in Subsection 2.9D (refer to Example 2.9.3), such systems may be viewed as retarded functional differential equations, replacing the delay $[-r, 0]$ by $(-\infty, 0]$. System (9.2.11) can be rewritten as

$$\begin{cases} \dot{x}(t) = -ax_t(0) + \int_{-\infty}^0 k(-s)x_t(s)ds, & t \ge 0 \\ x(t) = \varphi(t), & t \le 0. \end{cases} \tag{9.2.12}$$

For (9.2.12) we define the fading memory space X as consisting of all measurable functions $\varphi\colon (-\infty, 0) \to \mathbb{R}$ with norm

$$\|\varphi\|_m^2 = |\varphi(0)|^2 + K\int_{-\infty}^0 |\varphi(s)|^2 e^{Ls} ds < \infty \tag{9.2.13}$$

with $K > 0$ to be determined as needed and $L > 0$ a fixed constant. We assume that $k(\cdot) \in X$ and we define $C_L = (\int_0^\infty |k(s)|^2 e^{Ls} ds)^{1/2}$.

If we define $F(t, \varphi) \equiv F(\varphi) = -\varphi(0) + \int_{-\infty}^0 k(-s)\varphi(s)ds$ for all $\varphi \in X$, then (9.2.12) is a special case of the differential equation in Banach space (GE) with the Banach space as specified above.

To obtain an estimate of solution bounds for (9.2.12) we choose for any $\varphi \in X$,

$$v(\varphi) = \|\varphi\|_m^2 \tag{9.2.14}$$

and we let $y(t) = v(x_t)$. Along the solutions of (9.2.12) we have

$$\dot{y}(t) = (K - 2a)|x(t)|^2 + 2C_L x(t)z(t) - KL|z(t)|^2 \tag{9.2.15}$$

where $z(t) = (\int_{-\infty}^0 |x_t(s)|^2 e^{Ls} ds)^{1/2}$. The right side of (9.2.15) is negative definite if and only if the matrix

$$P = \begin{bmatrix} 2a - K & -C_L \\ -C_L & KL \end{bmatrix} \tag{9.2.16}$$

is positive definite which is the case when (i) $0 \leq K < 2a$, (ii) $KL > 0$ (which is always true), and (iii) $C_L/\sqrt{L} \leq a$. Therefore, when $C_L/\sqrt{L} \leq a$, there exists a $K > 0$ such that the right-hand side of (9.2.15) is negative definite.

We want to show that there exists an $\alpha < 0$ such that

$$\dot{y}(t) \leq \alpha y(t). \tag{9.2.17}$$

Letting

$$Q = \begin{bmatrix} 1 & 0 \\ 0 & K \end{bmatrix}, \tag{9.2.18}$$

this is equivalent to finding an α such that $P \geq -\alpha Q$. It is easy to see that this is the case when

$$\alpha = -\frac{\lambda_m(P)}{\max\{1, K\}} < 0 \tag{9.2.19}$$

where $\lambda_m(P)$ denotes the smallest eigenvalue of P.

We conclude that if $C_L/\sqrt{L} \leq a$ and K is chosen appropriately, then there exists an $\alpha < 0$ such that $\dot{y}(t) \leq \alpha y(t)$. Therefore,

$$\|x_t\|_m \leq e^{(\alpha/2t)} \|x_0\|_m, \qquad \alpha < 0 \tag{9.2.20}$$

and we conclude that the equilibrium $\varphi_e = 0$ of system (9.2.12) is *exponentially stable in the large* with respect to the norm $\|\cdot\|_m$. □

Theorem 9.2.3 Assume that there exists a function $v \in C[X \times \mathbb{R}^+, \mathbb{R}]$ that satisfies the following conditions.

(i) There exist two continuous, real-valued and strictly increasing functions ψ_1, ψ_2 that are defined on $\mathbb{R}^+$ with $\lim_{r\to\infty} \psi_i(r) = +\infty$, $i = 1, 2$, and a constant $M > 0$ such that

$$\psi_1(\|x\|) \leq v(x, t) \leq \psi_2(\|x\|)$$

for all $\|x\| \geq M$ and $t \in \mathbb{R}^+$.

(ii) For all $\|x\| \geq M$ and $t \in \mathbb{R}^+$,

$$v'_{(GE)}(x,t) \leq 0.$$

Then the solutions of (GE) are *uniformly bounded.*

If in addition to hypotheses (i) and (ii) there exists a function $\psi_3 \in \mathcal{K}$ such that

$$v'_{(GE)}(x,t) \leq -\psi_3(\|x\|)$$

for all $\|x\| \geq M$ and $t \in \mathbb{R}^+$, then the solutions of (GE) are *uniformly ultimately bounded.*

Proof. The proof of this theorem is a direct consequence of Theorems 3.3.4 and 3.3.5. □

We conclude this subsection with two examples.

Example 9.2.4 We consider the dynamical system determined by (9.2.4) in Example 9.2.1. Because the function v defined in (9.2.5) is positive definite, decrescent, and radially unbounded with respect to the H^2-norm, it follows from (9.2.7) and Theorem 9.2.3 that the solutions of (9.2.4) are *uniformly bounded with respect to the H^2-norm.* □

Example 9.2.5 We consider the dynamical system determined by (9.2.8) in Example 9.2.2. Because the function v defined in (9.2.9) is positive definite, decrescent, and radially unbounded with respect to the H^1-norm, it follows from (9.2.10) and Theorem 9.2.3 that the solutions of (9.2.8) are *uniformly ultimately bounded with respect to the H^1-norm.* □

C. Instability

As in Subsection A, we assume in the following that $C \subset X$, $x_e = 0 \in C$, $x_e = 0$ is an equilibrium for (GE) and we let $\|\cdot\|$ denote the norm for the Banach space X.

Theorem 9.2.4 (*Lyapunov's First Instability Theorem for differential equations in Banach spaces*) The equilibrium $x_e = 0$ of (GE) is *unstable* if there exist a $t_0 \in \mathbb{R}^+$ and a decrescent function $v \in C[B(r) \times \mathbb{R}^+, \mathbb{R}]$ for some $r > 0$, $B(r) \subset C$, such that $v'_{(GE)}$ is positive definite (negative definite) and if in every neighborhood of the origin there is a point x such that $v(x, t_0) > 0$ ($v(x, t_0) < 0$). Moreover, if v is positive definite (negative definite), then the equilibrium $x_e = 0$ is *completely unstable.*

Proof. The proof of this result is a direct consequence of Theorem 3.3.8. □

We apply the above result in the stability analysis of the *backward heat equation.*

Example 9.2.6 Consider the *initial and boundary value problem* given by

$$\begin{cases} \dfrac{\partial u}{\partial t} = -a^2 \Delta u, & (t,x) \in \mathbb{R}^+ \times \Omega \\ u(0,x) = \psi(x), & x \in \Omega \\ u(t,x) = 0, & (t,x) \in \mathbb{R}^+ \times \partial\Omega \end{cases} \tag{9.2.21}$$

where $a^2 > 0$ is a constant, $\Omega \subset \mathbb{R}^n$ is a bounded domain with smooth boundary $\partial\Omega$, Δ denotes the Laplacian, and $\psi \in C_0[\Omega, \mathbb{R}]$. We may view (9.2.21) as a differential equation in the Banach space $X = H_0[\Omega, \mathbb{R}]$. (Refer to Section 2.10 for the definitions of $C_0[\Omega, \mathbb{R}]$ and $H_0[\Omega, \mathbb{R}]$.) It is easy to show that $0 \in X$ is an equilibrium of (9.2.21).

In applying Theorem 9.2.4 in the stability analysis of (9.2.21), we choose as a Lyapunov function

$$v(\varphi) = \int_\Omega |\varphi(x)|^2 dx = \|\varphi\|_{L_2}^2 \tag{9.2.22}$$

for all $\varphi \in X$. This function is clearly positive definite with respect to the L_2-norm. Along the solutions of (9.2.21) we have

$$v'_{(9.2.21)}(\varphi) = 2\int_\Omega \varphi \frac{d\varphi}{dt} dx = -2\int_\Omega \varphi(\Delta\varphi) dx.$$

By Gauss' Divergence Theorem and Poincaré's Inequality (refer to Example 9.2.2) we have

$$-2\int_\Omega \varphi(\Delta\varphi)dx = 2\int_\Omega |\nabla\varphi|^2 dx \geq \frac{2}{\gamma^2}\int_\Omega |\varphi|^2 dx = \frac{2}{\gamma^2}\|\varphi\|_{L_2}$$

for all $\varphi \in X$, where γ is a positive constant that depends on Ω (refer to Example 9.2.2). Therefore,

$$v'_{(9.2.21)}(\varphi) \geq \frac{2}{\gamma^2}\|\varphi\|_{L_2} \tag{9.2.23}$$

which shows that $v'_{(9.2.21)}$ is positive definite. It now follows from (9.2.22), (9.2.23), and Theorem 9.2.4 that the equilibrium $x_e = 0$ of (9.2.21) is *completely unstable with respect to the* L_2*-norm.* □

Theorem 9.2.5 (*Lyapunov's Second Instability Theorem for differential equations in Banach spaces*) Assume that for system (GE) there exists a bounded function $v \in C[B(\varepsilon) \times [t_0, \infty), \mathbb{R}]$ where $\varepsilon > 0$ and $B(\varepsilon) \subset C$, having the following properties.

(i) For all $(x,t) \in B(\varepsilon) \times [t_0, \infty)$,

$$v'_{(GE)}(x,t) \geq \lambda v(x,t)$$

where $\lambda > 0$ is a constant.

(ii) In every neighborhood of $x_e = 0$, there exists an x such that $v(x, t_1) > 0$ for a fixed $t_1 \geq t_0$.

Then the equilibrium $x_e = 0$ of (GE) is *unstable*.

Proof. The proof of this result is a direct consequence of Theorem 3.3.10. □

We demonstrate the applicability of Theorem 9.2.5 in the analysis of a specific example.

Example 9.2.7 Consider the initial value and boundary value problem given by

$$\begin{cases} \dfrac{\partial u_1}{\partial t} = u_1 + u_2 + \displaystyle\sum_{j=1}^{n} a_j \dfrac{\partial u_1}{\partial x_j} & (t,x) \in \mathbb{R}^+ \times \Omega \\ \dfrac{\partial u_2}{\partial t} = u_1 + u_2 + \displaystyle\sum_{j=1}^{n} b_j \dfrac{\partial u_2}{\partial x_j} + \Delta u_2, & (t,x) \in \mathbb{R}^+ \times \Omega \\ u_i(0,x) = \psi_i(x), & x \in \Omega,\ i = 1,2 \\ u_i(t,x) = 0, & (t,x) \in \mathbb{R}^+ \times \partial\Omega,\ i = 1,2 \end{cases} \tag{9.2.24}$$

where $\Omega \subset \mathbb{R}^n$ is a bounded domain with smooth boundary, $\Delta = \sum_{j=1}^{n} \partial^2/\partial x_j^2$ denotes the Laplacian, a_j, b_j are real constants, $j = 1, \ldots, n$, and $\psi_i \in C_0[\Omega, \mathbb{R}]$, $i = 1, 2$. (Refer to Section 2.10 for the definition of $C_0[\Omega, \mathbb{R}]$.)

Equations (9.2.24) may be viewed as differential equations in the Banach space $X = H_0[\Omega, \mathbb{R}] \times H_0[\Omega, \mathbb{R}]$. It is easily verified that the origin of X is an equilibrium of (9.2.24).

In the following, we show that the equilibrium $0 \in X$ of system (9.2.24) is *unstable* with respect to the L_2-norm, using Theorem 9.2.5.

We choose as a Lyapunov function,

$$v(\varphi) = \int_{\Omega} \left(|\varphi_1|^2 - |\varphi_2|^2\right) dx \tag{9.2.25}$$

where $\varphi = (\varphi_1, \varphi_2) \in X$. Along the solutions of (9.2.24) we have

$$\begin{aligned} v'_{(9.2.24)}(\varphi) &= 2\int_{\Omega} \varphi_1 \left[\varphi_1 + \varphi_2 + \sum_{j=1}^{n} a_j \frac{\partial \varphi_1}{\partial x_j}\right] dx \\ &\quad - 2\int_{\Omega} \varphi_2 \left[\varphi_1 + \varphi_2 + \sum_{j=1}^{n} b_j \frac{\partial \varphi_2}{\partial x_j} + \Delta\varphi_2\right] dx \\ &= 2v(\varphi) - 2\int_{\Omega} \varphi_2(\Delta\varphi_2) dx \end{aligned}$$

where in the last step we have used the fact that

$$2\int_{\Omega} \varphi_i \frac{\partial \varphi_i}{\partial x_j} dx = \int_{\Omega} \frac{\partial [\varphi_i^2]}{\partial x_j} dx = 0$$

$j = 1, \ldots, n$, for $\varphi_i \in C_0[\Omega, \mathbb{R}]$, $i = 1, 2$.

Invoking Gauss' Divergence Theorem, we have

$$2v(\varphi) - 2\int_{\Omega} \varphi_2(\Delta\varphi_2) dx = 2v(\varphi) + 2\int_{\Omega} |\nabla\varphi_2|^2 dx \geq 2v(\varphi)$$

for all $\varphi \in X$. Therefore,

$$v'_{(9.2.24)}(\varphi) \geq 2v(\varphi) \tag{9.2.26}$$

for all $\varphi \in X$.

Clearly, $v(\varphi)$ is bounded on

$$B(1) = \{\varphi = (\varphi_1, \varphi_2) \in X \colon \|\varphi\|^2 = \|\varphi_1\|_{L_2}^2 + \|\varphi_1\|_{L_2}^2 \leq 1\}$$

and $v(\varphi) > 0$ if $\varphi = (\varphi_1, 0)$ and $\|\varphi_1\|_{L_2} \neq 0$. Therefore, condition (ii) in Theorem 9.2.5 is satisfied. It follows from (9.2.26) and Theorem 9.2.5 that the equilibrium $\varphi_e = 0 \in X$ of (9.2.24) is *unstable with respect to the L_2-norm.* □

We now state the last result of the present section.

Theorem 9.2.6 (*Chetaev's Instability Theorem for differential equations in Banach spaces*) Assume that for system (GE) there exist a function $v \in C[B(r) \times \mathbb{R}^+, \mathbb{R}]$ for some $r > 0$, where $B(r) \subset C \subset X$, and a $t_0 \in \mathbb{R}^+$ such that the following conditions are satisfied.

(i) There exists a component D of the set $\{(x,t) \in B(r) \times \mathbb{R}^+ \colon v(x,t) < 0\}$ such that for every neighborhood of the origin there exists an x in this neighborhood such that $(x, t_0) \in D$.

(ii) v is bounded from below on D.

(iii) $v'_{(GE)}(x,t) \leq -\psi(|v(x,t)|)$ for all $(x,t) \in D$ where $\psi \in \mathcal{K}$.

Then the equilibrium $x_e = 0$ of (GE) is *unstable.* □

We ask the reader to prove Theorem 9.2.6.

9.3 Converse Theorems for Differential Equations in Banach Spaces

In the present section we establish converse results for some of the principal Lyapunov stability and boundedness results for differential equations in Banach spaces presented in Section 9.2. We recall the differential equation in Banach space given by

$$\dot{x} = F(t, x) \tag{GE}$$

where $F \colon \mathbb{R} \times C \to X$. We assume that $x_e = 0$ is an equilibrium for the dynamical system determined by (GE) and that the set C contains a neighborhood of the origin $x_e = 0$. Also, we assume that for every $(t_0, x_0) \in \mathbb{R}^+ \times C$, there exists a unique noncontinuable solution of (GE) with initial condition $x(t_0) = x_0$ that depends continuously on (t_0, x_0).

We present only local results. Our first result concerns uniform stability.

Theorem 9.3.1 If the equilibrium $x_e = 0$ of (GE) is *uniformly stable*, then there exists a positive definite and decrescent function $v \in C[B(r) \times \mathbb{R}^+, \mathbb{R}^+]$ for some $r > 0$, where $B(r) \subset C$, such that $v'_{(GE)}$ is negative semidefinite.

Proof. The proof of this result is a direct consequence of Theorem 3.6.1 and follows along the lines of the proof of Theorem 6.5.1. The continuity of v is a consequence of the assumed continuity of the solution $\varphi(t, t_0, x_0)$ of (GE) with respect to the initial data. □

The next result concerns uniform asymptotic stability.

Theorem 9.3.2 If the equilibrium $x_e = 0$ of (GE) is *uniformly asymptotically stable*, then there exists a positive definite and decrescent function $v \in C[B(r) \times \mathbb{R}^+, \mathbb{R}^+]$ for some $r > 0$, where $B(r) \subset C$, such that $v'_{(GE)}$ is negative definite.

Proof. The proof of this result is a consequence of Theorem 3.6.2, the continuity of the solutions of (GE) with respect to initial conditions and the continuity results given in Subsection 3.6B. □

As in the case for finite-dimensional systems (see Theorem 6.5.3), the next result, which addresses the exponential stability of the equilibrium $x_e = 0$ for (GE), is not symmetric to the exponential stability result given in Theorem 9.2.1d. Nevertheless, this result does provide a set of necessary conditions for exponential stability.

Theorem 9.3.3 If the equilibrium $x_e = 0$ of (GE) is *exponentially stable*, then there exists a positive definite and decrescent function $v \in C[B(r) \times \mathbb{R}^+, \mathbb{R}^+]$ for some $r > 0$, where $B(r) \subset C$, such that

$$v'_{(GE)}(x, t) \leq -cv(x, t)$$

for all $(x, t) \in B(r) \times \mathbb{R}^+$, where $c > 0$ is a constant.

Proof. The proof of this result is a direct consequence of Theorem 3.6.3 and the continuity of the solutions of (GE). □

We conclude by noting that converse theorems for system (GE) for *uniform boundedness*, *uniform ultimate boundedness*, *uniform asymptotic stability in the large*, *exponential stability in the large*, and *instability* can also be established, using the methodology employed in the preceding results.

9.4 Invariance Theory for Differential Equations in Banach Spaces

In the present section we consider infinite-dimensional dynamical systems determined by a class of autonomous differential equations in Banach space given by

$$\dot{x} = F(x) \tag{GA}$$

where $F: C \to X$, $C \subset X$, and X is a Banach space. We assume that $x_e = 0$ is an equilibrium for the dynamical system determined by (GA) and that C contains a neighborhood of the origin $x_e = 0$. Furthermore, we assume that for each $x_0 \in C$, there exists one and only one solution of (GA) for the initial condition $x(0) = x_0$. Under these conditions, the solutions of (GA) determine a semigroup and the invariance theory for dynamical systems developed in Section 4.2 is applicable to dynamical systems determined by (GA). Our aim in the present section is to improve some of the stability results presented in Section 9.2.

Theorem 9.4.1 Assume that for system (GA) there exists a function $v \in C[C, \mathbb{R}]$ such that $v'_{(GA)}(x) \leq 0$ for all $x \in C$. Let M be the largest invariant set with respect to the dynamical system determined by (GA) in the set

$$Z = \{x \in C \colon v'_{(GA)}(x) = 0\}. \tag{9.4.1}$$

Then for every solution φ of (GA) such that the closure of the trajectory of φ is compact, $\varphi(t)$ approaches M as $t \to \infty$.

Proof. The proof of this theorem is a direct consequence of Theorem 4.2.1, where X_1 is generated by taking the closure of every solution φ of (GA) having the property that the closure of φ is compact. □

We apply the above result in the analysis of a scalar *Volterra integrodifferential equation*.

Example 9.4.1 Consider the equation (refer to [9])

$$\dot{x}(t) = -\int_{-\infty}^{t} a(t-u)g(x(u))du \tag{9.4.2}$$

where $t \in \mathbb{R}^+$, $a \in C^2[\mathbb{R}^+, \mathbb{R}]$, $a(t) > 0$, $\dot{a}(t) < 0$, and $\ddot{a}(t) \geq 0$ for all $t \in \mathbb{R}^+$, and $\lim_{t\to\infty} t^2\dot{a}(t) = 0$ and $\int_0^\infty t^2\dot{a}(t)dt < \infty$. The *fading memory space* X for (9.4.2) is the Banach space consisting of all functions $\varphi \colon (-\infty, 0] \to \mathbb{R}$ such that

$$\|\varphi\|^2 = |\varphi(0)| + \int_{-\infty}^{0} k(\theta)|\varphi(\theta)|d\theta$$

is finite, where $k(\theta) > 0$ for $-\infty < \theta \leq 0$, $\int_{-\infty}^0 k(\theta)d\theta < \infty$, and $\dot{k}(\theta) \geq 0$. Assume that $g(x)$ has a finite number of zeros and that $g \in C^1[\mathbb{R}, \mathbb{R}]$, and that $\int_0^x g(s)ds \to \infty$ as $|x| \to \infty$.

In the following, we apply Theorem 9.4.1 to prove that every solution of (9.4.2) with initial condition $\varphi \in X$ satisfying $\sup_{-\infty<\theta\leq 0} |\varphi(\theta)| < \infty$ approaches an equilibrium of (9.4.2).

Choose

$$v(\varphi) = \int_0^{\varphi(0)} g(s)ds - \frac{1}{2}\int_{-\infty}^{0} \dot{a}(-\theta)\left(\int_\theta^0 g(\varphi(s))ds\right)^2 d\theta$$

for $\varphi \in X$. The second term in v is defined for all $\varphi \in X$ satisfying $\sup_{-\infty<\theta\leq 0}|\varphi(\theta)| = L < \infty$ because

$$\begin{aligned} 0 &\leq -\frac{1}{2}\int_{-\infty}^{0}\dot{a}(-\theta)\left(\int_{\theta}^{0}g(\varphi(s))ds\right)^2 d\theta \\ &\leq -\Big(\sup_{-L\leq s\leq L}|g(s)|\Big)^2\frac{1}{2}\int_{-\infty}^{0}\dot{a}(-\theta)\theta^2 d\theta \\ &< \infty. \end{aligned} \tag{9.4.3}$$

Therefore v is defined for all $\varphi \in X$ satisfying $\sup_{-\infty<\theta\leq 0}|\varphi(\theta)| < \infty$.

Suppose that $x(t, t_0, \varphi_0)$ is a solution of (9.4.2) with the initial condition $\varphi_0 \in X$ satisfying $\sup_{-\infty<\theta\leq t_0}|\varphi_0(\theta)| < \infty$. Then by the continuity of $x(t, t_0, \varphi_0)$ with respect to t, $\sup_{-\infty<\theta\leq t}|x(t, t_0, \varphi_0)| < \infty$ and hence, $v(x_t)$ is defined for all $t \geq t_0$.

Along the solutions of (9.4.2), we have

$$v'_{(9.4.2)}(x_t) = -\frac{1}{2}\int_{-\infty}^{0}\ddot{a}(-\theta)\left(\int_{\theta}^{0}g(x(t+s))ds\right)^2 d\theta \leq 0,$$

which implies that $v(x_t)$ is nonincreasing and hence, $v(x_t) \leq v(\varphi_0)$ for all $t > t_0$. By hypothesis, $\int_0^{x(t)} g(s)ds \to \infty$ if $|x(t)| \to \infty$. Therefore there exists an $L > 0$ such that $|x(t)| < L$ for all $-\infty < t < \infty$. We now have an estimate for the norm of x_t,

$$\|x_t\|^2 = |x(t)| + \int_{-\infty}^{0}k(\theta)|x(t+\theta)|d\theta \leq L\left(1 + \int_{-\infty}^{0}k(\theta)d\theta\right),$$

which implies that the trajectory $\gamma^+(x) = \{x_t(t_0, \varphi_0) : t \geq t_0\}$ is a bounded set in X.

Next we show that the closure of the trajectory $\gamma^+(x)$ is compact. Because in Banach spaces compactness is equivalent to sequential compactness and x_t is continuous with respect to t in X, we only need to show that there is a convergent subsequence for any sequence $\{x_{t_k}\}_{k\in\mathbb{N}}$ with $t_k \to \infty$ monotonically as $k \to \infty$. For any $A > 0$, $\{x_{t_k}\}$ with $t_k > A$ is equicontinuous on $[-A, 0]$ and uniformly bounded on $(-\infty, 0]$ ($|x(t)| < L$ was shown earlier). By the Ascoli–Arzela Lemma (refer to Problem 2.14.7), there exists a subsequence that converges uniformly to a function $\varphi_A \in C[[-A, 0], \mathbb{R}]$. Choosing $A = 1, 2, \ldots$, there exist subsequences $\{t_{Ak}\}$ such that $\{t_k\} \supset \{t_{1k}\} \supset \{t_{2k}\} \supset \cdots$ and the subsequences $\{x_{t_{Ak}}\}$ converge uniformly to a function $\varphi_A \in C[[-A, 0], \mathbb{R}]$ for all $A = 1, 2, \ldots$. Then the subsequence $\{x_{t_{kk}} : k = 1, 2, \ldots\}$ converges uniformly on all compact subsets of $(-\infty, 0]$ to a function $\varphi \in C[(-\infty, 0], \mathbb{R}]$. φ is bounded by the same bound L and thus, $\varphi \in X$. What is left to be shown is that $x_{t_{kk}} \to \varphi$ as $k \to \infty$ using the norm $\|\cdot\|$ in X. Let $\varepsilon > 0$ be arbitrary. There exists an $A > 0$ such that $\int_{-\infty}^{-A} k(\theta)d\theta < \varepsilon/(2L)$. We have

$$\begin{aligned} \|x_{t_{kk}} - \varphi\|^2 &= |x_{t_{kk}}(0) - \varphi(0)| + \int_{-\infty}^{0}k(\theta)|x(t_{kk}+\theta) - \varphi(\theta)|d\theta \\ &\leq |x(t_{kk}) - \varphi(0)| + \varepsilon + \int_{-A}^{0}k(\theta)|x(t_{kk}+\theta) - \varphi(\theta)|d\theta. \end{aligned}$$

On the compact set $[-A, 0]$, there exists an $m > 0$ such that the first and last term in the above expression are less than ε for all $k > m$. Hence $x_{t_{kk}} \to \varphi$ as $k \to \infty$.

Finally, if $\psi \in Z = \{x \in X : v'_{(9.4.2)}(x) = 0\}$, then

$$v'_{(9.4.2)}(\psi_t) = \int_{-\infty}^{0} \ddot{a}(-\theta)\left(\int_{\theta}^{0} g(\psi(t+s))ds\right)^2 d\theta = 0.$$

Because $\ddot{a}(t) \geq 0$, it must be true that

$$\int_{\theta}^{0} g(\psi(t+s))ds = 0 \qquad \text{for all } -\infty < t < \infty \tag{9.4.4}$$

whenever $\ddot{a}(-\theta) > 0$. From the assumptions that $t^2\dot{a}(t) \to 0$ as $t \to \infty$ and $\ddot{a}(t) \geq 0$, there must exist a t_0 and a δ with $t_0 > \delta > 0$ such that $\ddot{a}(s) > 0$ for all $s \in [t_0 - \delta, t_0 + \delta]$. In view of (9.4.4), we have

$$\int_{-s_2}^{-s_1} g(\psi(t+s))ds = 0 \qquad \text{for all } -\infty < t < \infty,$$

for all $s_1, s_2 \in [t_0 - \delta, t_0 + \delta]$. This is true if and only if $g(\psi(t)) \equiv 0$ for all t. Under the assumption that $g(\cdot)$ has a finite number of zeros, the set Z is comprised of these zeros, which are also the equilibria of (9.4.2). It now follows from Theorem 9.4.1 that x_t approaches an equilibrium of (9.4.2). □

In our next result we require the following concept.

Definition 9.4.1 Let X and $\widetilde{X}$ be two Banach spaces and assume that $X \subset \widetilde{X}$. The *embedding* $X \subset \widetilde{X}$ is said to be *compact* if every closed and bounded subset of X is compact in $\widetilde{X}$ with respect to the norm $\|\cdot\|_{\widetilde{X}}$ of $\widetilde{X}$. □

Now let $\Omega \subset \mathbb{R}^n$ be a bounded domain with smooth boundary $\partial\Omega$. One of *Sobolev's Embedding Theorems* (see, e.g., [6] and [33]) ensures that the embedding $H^m[\Omega, \mathbb{R}] \subset H^l[\Omega, \mathbb{R}]$ is compact and continuous for all $m > l$. Accordingly, the embedding $H_0^m[\Omega, \mathbb{R}] \subset H_0^l[\Omega, \mathbb{R}]$ is also compact and continuous for $m > l$.

The differential equation (GA) may be defined on different Banach spaces. This will always be true for the Banach spaces X and $\widetilde{X}$ when $X \subset \widetilde{X}$. Therefore, the Sobolev Embedding Theorem cited above suggests the next result, where $\|\cdot\|_{\widetilde{X}}$ denotes the norm for $\widetilde{X}$ and where we assume that $C = X$.

Theorem 9.4.2 Assume that for the dynamical system determined by (GA) there exists a Banach space $\widetilde{X} \supset X$ such that the embedding $X \subset \widetilde{X}$ is compact. Assume that there exists a function $v\colon X \to \mathbb{R}$ that is continuous with respect to the norm $\|\cdot\|_{\widetilde{X}}$ that satisfies the following conditions.

(i) $v'_{(GA)}(x) \leq 0$ for all $x \in X$.

(ii) $v(x) > v(0) = 0$ for all $x \in X - \{0\}$.

(iii) $v(x) \to \infty$ as $\|x\|_{\widetilde{X}} \to \infty$.

(iv) $\{0\}$ is the only invariant subset in

$$Z = \{x \in X : v'_{(GA)}(x) = 0\}.$$

Then the equilibrium $x_e = 0$ of (GA) is *uniformly asymptotically stable in the large* in X with respect to the norm $\|\cdot\|_{\widetilde{X}}$.

Proof. The proof of this theorem is a direct consequence of Corollary 4.2.2, where we let X be equipped with the norm $\|\cdot\|_{\widetilde{X}}$. We note that X is locally compact with respect to the norm $\|\cdot\|_{\widetilde{X}}$ and that in the proof of Corollary 4.2.2, the local compactness of X is used, rather than the compactness of X. □

Example 9.4.2 We now revisit system (9.2.4) given in Example 9.2.1 where $X = H^4[(0,1),\mathbb{R}] \cap H_0^2[(0,1),\mathbb{R}]$ with the H^2-norm. We choose $\widetilde{X} = H_0^1[(0,1),\mathbb{R}]$. In view of Sobolev's Embedding Theorem cited above, the embedding $X \subset \widetilde{X}$ is compact. In applying Theorem 9.4.2, we choose as a Lyapunov function

$$v(\psi) = \|\psi\|_{H^1}^2 = \|\psi\|_{\widetilde{X}}^2 = \int_0^1 \left[\left(\frac{\partial \psi}{\partial x}\right)^2 + \psi^2 \right] dx.$$

The function $v\colon X \to \mathbb{R}$ is clearly continuous with respect to the norm $\|\cdot\|_{H^1}$. In a similar manner as was done in Example 9.2.1, we compute

$$v'_{(9.2.4)}(\psi) = -2\int_0^1 \left[\left(\frac{\partial^3 \psi}{\partial x^3}\right)^2 + \left(\frac{\partial^2 \psi}{\partial x^2}\right)^2 \right] dx = -2 \left\| \frac{d^2\psi}{dx^2} \right\|_{H^1} \le 0$$

for all $x \in X$.

We have $v(0) = 0$ and $v(\psi) > 0$ for all $\psi \in \widetilde{X} - \{0\}$, and in particular, for all $\psi \in X - \{0\}$. Moreover, $v(\psi) \to \infty$ as $\|\psi\|_{H^1} \to \infty$ also holds. Finally,

$$Z = \{\psi \in X : v'_{(9.2.4)}(\psi) = 0\} = \{0\}$$

because $d^2\psi/dx^2 \equiv 0$ and $\psi(0) = \psi(1) = 0$ imply that $\psi \equiv 0$ for $\psi \in C^2[(0,1),\mathbb{R}]$ and because $C^2[(0,1),\mathbb{R}] \cap X$ is dense in X.

All hypotheses of Theorem 9.4.2 are satisfied and therefore, the equilibrium $\varphi_e = 0 \in X$ of system (9.2.4) is *uniformly asymptotically stable in the large with respect to the H^1-norm.* □

9.5 Comparison Theory for Differential Equations in Banach Spaces

In the present section we specialize the results of Section 4.3 to develop a comparison theory in the stability analysis of infinite-dimensional dynamical systems determined by differential equations in Banach spaces.

We consider once more a class of differential equations defined on a Banach space X of the form

$$\dot{x} = F(t, x) \qquad (GE)$$

where $F\colon \mathbb{R}^+ \times C \to X$, $C \subset X$. We assume that $x_e = 0 \in C$ and that $x_e = 0$ is an equilibrium of (GE),

For the l-dimensional Euclidean vector space, $(\mathbb{R}^l, |\cdot|)$, we let in the following $\mathbb{R}^l \supset B_E(r) = \{x \in \mathbb{R}^l\colon |x| < r\}$, and as usual, we let $B(r) = \{x \in X\colon \|x\| < r\}$. Also, for a vector-valued function $V\colon B(r) \times \mathbb{R}^+ \to \mathbb{R}^l$, where $B(r) \subset C \subset X$ for some $r > 0$, we use the notation

$$V(x, t) = [v_1(x, t), \ldots, v_l(x, t)]^T$$

and

$$V'_{(GE)}(x, t) = [v'_{1(GE)}(x, t), \ldots, v'_{l(GE)}(x, t)]^T.$$

For system (GE) we employ as a *comparison system* ordinary differential equations of the form

$$\dot{y} = g(t, y) \qquad (C)$$

where $t \in \mathbb{R}^+$, $y \in \mathbb{R}^l$, $g \in C[\mathbb{R}^+ \times B_E(r_1), \mathbb{R}^l]$ for some $r_1 > 0$, $B_E(r_1) \subset (\mathbb{R}^+)^l$, and $g(t, 0) = 0$ for all $t \in \mathbb{R}^+$. By making use of (C), we first establish *local results.*

Theorem 9.5.1 Assume that there exists a function $V \in C[B(r) \times \mathbb{R}^+, (\mathbb{R}^+)^l]$, where $B(r) \subset C \subset X$ for some $r > 0$, such that $|V(x, t)|$ is positive definite and decrescent, and there exists a quasi-monotone nondecreasing function $g \in C[\mathbb{R}^+ \times B_E(r_1), \mathbb{R}^l]$, where $B_E(r_1) \subset (\mathbb{R}^+)^l$ for some $r_1 > 0$, which satisfies the conditions that $g(t, 0) = 0$ for all $t \in \mathbb{R}^+$ and

$$V'_{(GE)}(x, t) \leq g(t, V(x, t))$$

for all $(x, t) \in B(r) \times \mathbb{R}^+$. In the above, inequality is to be interpreted componentwise. Then the following statements are true.

(a) The *stability, asymptotic stability, uniform stability* and *uniform asymptotic stability* of the equilibrium $y_e = 0$ of (C) implies the same corresponding types of stability of the equilibrium $x_e = 0$ of (GE).

(b) If in addition to the above assumptions, $|V(x, t)| \geq a|x|^b$ for all $(x, t) \in B(r) \times \mathbb{R}^+$, where $a > 0$ and $b > 0$, then the *exponential stability* of the equilibrium $y_e = 0$ of (C) implies the exponential stability of the equilibrium $x_e = 0$ of (GE).

Proof. The proofs of these results follow directly from Theorem 4.3.2(b) and (c). □

In the next theorem, where we address *global results*, we assume that $C = X$.

Theorem 9.5.2 Assume that for (GE) there exists a function $V \in C[X \times \mathbb{R}^+, (\mathbb{R}^+)^l]$ such that $|V(x, t)|$ is positive definite, decrescent, and radially unbounded, and that there exists a quasi-monotone nondecreasing function $g \in C[\mathbb{R}^+ \times (\mathbb{R}^+)^l, \mathbb{R}^l]$ such that $g(t, 0) = 0$ for all $t \in \mathbb{R}^+$ and such that the inequality

$$V'_{(GE)}(x,t) \leq g(t, V(x,t))$$

holds componentwise for all $(x, t) \in X \times \mathbb{R}^+$. Then the following statements are true.

(a) The *uniform asymptotic stability in the large* of the equilibrium $y_e = 0$ of (C) implies the uniform asymptotic stability in the large of the equilibrium $x_e = 0$ of (GE).

(b) If in addition to the above assumptions, $a_1|x|^b \leq |V(x,t)| \leq a_2|x|^b$ for all $(x, t) \in X \times \mathbb{R}^+$, where $a_2 \geq a_1 > 0$ and $b > 0$, then the *exponential stability in the large* of the equilibrium $y_e = 0$ of (C) implies the exponential stability in the large of the equilibrium $x_e = 0$ of (GE).

(c) The *uniform boundedness* and *uniform ultimate boundedness* of the solutions of (C) imply the same corresponding types of boundedness properties of the solutions of (GE). □

We leave the proofs of the above results as an exercise.

We note that when $l = 1$, the quasi-monotonicity condition in Theorems 9.5.1 and 9.5.2 are automatically satisfied because this is always true for scalar-valued functions (refer to Definition 3.8.1).

In applications, the comparison system given by

$$\dot{y} = Py + m(t, y) \tag{9.5.1}$$

is especially useful, where $P = [p_{ij}] \in \mathbb{R}^{l \times l}$ satisfies the condition that $p_{ij} \geq 0$ for $i, j = 1, \ldots, l$ and $i \neq j$, and where $m\colon \mathbb{R}^+ \times B_E(r_1) \to \mathbb{R}^l$ for some $r_1 > 0$ and $B_E(r_1) \subset (\mathbb{R}^+)^l$, is assumed to satisfy the condition

$$\lim_{|y| \to 0} \frac{|m(t,y)|}{|y|} = 0 \qquad \text{uniformly for } t \in \mathbb{R}^+.$$

Applying *Lyapunov's First Method* (Theorem 7.6.2), it follows that the equilibrium $y_e = 0$ of (9.5.1) is uniformly asymptotically stable if $-P$ is an M-matrix (refer to Definition 7.7.1 for the definition of M-matrix and the discussion following that definition for several equivalent characterizations of M-matrices).

We conclude by noting that in Section 9.7 we apply some of the results of the present section in the analysis of the model of a multicore nuclear reactor described by a class of Volterra integrodifferential equations.

9.6 Composite Systems

In Section 2.11 we introduced composite dynamical systems on metric spaces. We now address the stability analysis of such systems in the context of composite dynamical systems defined on normed linear spaces.

We let $(X, \|\cdot\|)$, $(X_i, \|\cdot\|_i)$, $i = 1, \ldots, l$, be normed linear spaces; we assume that $X = X_1 \times \cdots X_l$ and that there are constants $c_1 > 0$ and $c_2 > 0$ such that

$$c_1\|x\| \leq \sum_{i=1}^{l} \|x_i\|_i \leq c_2\|x\|$$

for all $x \in X$, where $x = (x_1, \ldots, x_l)^T$, $x_i \in X_i$, $i = 1, \ldots, l$. We can define the norm $\|\cdot\|$ on X in a variety of ways, including $\|x\| = \sum_{i=1}^{l} \|x_i\|_i$. We define a *composite dynamical system*, $\{\mathbb{R}^+, X, A, S\}$, defined on a normed linear space $(X, \|\cdot\|)$ by modifying Definition 2.11.1 in the obvious way, choosing for metric spaces (X, d), (X_i, d_i), $i = 1, \ldots, l$, normed linear spaces $(X, \|\cdot\|)$, $(X_i, \|\cdot\|_i)$, $i = 1, \ldots, l$, respectively.

In our first result, we define for $M_i \subset X_i$,

$$d_i(x_i, M_i) = \inf_{x \in M_i} \|x_i - x\|_i.$$

Theorem 9.6.1 Let $\{\mathbb{R}^+, X, A, S\}$ be a dynamical system where $X = X_1 \times \cdots \times X_l$ and $X_i, i = 1, \ldots, l$, are normed linear spaces with norms $\|\cdot\|_i, i = 1, \ldots, l$, respectively. Assume that $M = M_1 \times \cdots \times M_l$ is an invariant set (i.e., (S, M) is invariant), where $M_i \subset X_i$, $i = 1, \ldots, l$, and assume that the following hypotheses are satisfied.

(i) There exist $v_i \in C[X_i \times \mathbb{R}^+, \mathbb{R}]$ and $\psi_{i1}, \psi_{i2} \in \mathcal{K}$, $i = 1, \ldots, l$, such that

$$\psi_{i1}(d_i(x_i, M_i)) \leq v_i(x_i, t) \leq \psi_{i2}(d_i(x_i, M_i))$$

for all $x_i \in X_i$ and $t \in \mathbb{R}^+$.

(ii) Given v_i in hypothesis (i), there exist constants $a_{ij} \in \mathbb{R}$ and functions $\psi_{i4} \in \mathcal{K}$, $i, j = 1, \ldots, l$, such that

$$Dv_i(p_i(t, a, t_0), t) \leq \psi_{i4}(d_i(p_i, M_i)) \sum_{j=1}^{l} a_{ij}\psi_{j4}(d_j(p_j, M_j))$$

for all $p(\cdot, a, t_0) = [p_1(\cdot, a, t_0), \ldots, p_l(\cdot, a, t_0)]^T = [p_1, \ldots, p_l]^T \in S$ and $t \geq t_0$, where D denotes a fixed Dini derivative with respect to $t \in \mathbb{R}^+$.

(iii) There exists an l-vector $\alpha^T = (\alpha_1, \ldots, \alpha_l)$, $\alpha_i > 0$, $i = 1, \ldots, l$, such that the *test matrix* $B = [b_{ij}] \in \mathbb{R}^{l \times l}$ specified by

$$b_{ij} = \frac{1}{2}(\alpha_i a_{ij} + \alpha_j a_{ji}), \qquad i, j = 1, \ldots, l,$$

is either negative semidefinite or negative definite.

Then the following statements are true.

(a) If B is negative semidefinite, then (S, M) is *uniformly stable*.

(b) If B is negative definite, then (S, M) is *uniformly asymptotically stable*.

(c) If B is negative semidefinite, if M is bounded, and if $\psi_{i1}, \psi_{i2} \in \mathcal{K}_\infty$ for all $i = 1, \ldots, l$, then S is *uniformly bounded.*

(d) If in (c), B is negative definite, then S is *uniformly ultimately bounded* and furthermore, (S, M) is *uniformly asymptotically stable in the large.*

(e) If B is negative definite and if there exist positive numbers a_1, a_2, b, and c such that

$$a_1 r^b \leq \psi_{i1}(r) \leq \psi_{i2}(r) \leq a_2 r^b$$

and

$$\psi_{i4}^2(r) \geq c r^b$$

for all $r \in \mathbb{R}^+$, $i = 1, \ldots, l$, then (S, M) is *exponentially stable in the large.*

Proof. The proof of this theorem is a consequence of Theorems 3.3.1–3.3.7. We present here only the proofs of parts (a) and (b). The proofs of the remaining parts follow along similar lines.

We choose

$$v(x, t) = \sum_{i=1}^{l} \alpha_i v_i(x_i, t)$$

where $x = [x_1, \ldots, x_l]^T$ and $\alpha = [\alpha_1, \ldots, \alpha_l]^T$ is given in hypothesis (iii). By hypothesis (i), we have

$$\sum_{i=1}^{l} \alpha_i \psi_{i1}(d_i(x_i, M_i)) \leq v(x, t) \leq \sum_{i=1}^{l} \alpha_i \psi_{i2}(d_i(x_i, M_i))$$

for all $x = [x_1, \ldots, x_l]^T \in X_1 \times \cdots \times X_l = X$. Let $r = d(x, M)$ and $r_i = d(x_i, M_i)$, $1 \leq i \leq l$. We may assume without loss of generality that $r = \sum_{i=1}^{l} r_i$. We have that

$$\sum_{i=1}^{l} \alpha_i \psi_{i2}(r_i) \leq \sum_{i=1}^{l} \alpha_i \psi_{i2}(r) \triangleq \psi_2(r).$$

Also, we let $\widetilde{\psi}_1(r) = \min_{1 \leq i \leq l}\{\psi_{i1}(r)\}$ and $a = \min_{1 \leq i \leq l}\{\alpha_i\}$. Then

$$\begin{aligned}
\sum_{i=1}^{l} \alpha_i \psi_{i1}(r_i) &\geq \max_{1 \leq i \leq l}\{\alpha_i \psi_{i1}(r_i)\} \\
&\geq \max_{1 \leq i \leq l}\{\alpha_i \widetilde{\psi}_1(r_i)\} \\
&\geq a \max_{1 \leq i \leq l}\{\widetilde{\psi}_1(r_i)\} \\
&= a \widetilde{\psi}_1\big(\max_{1 \leq i \leq l}\{r_i\}\big) \\
&\geq a \widetilde{\psi}_1(r/l) \\
&\triangleq \psi_1(r).
\end{aligned}$$

Hence,

$$\psi_1(r) \leq v(x,t) \leq \psi_2(r) \tag{9.6.1}$$

for all $x \in X$ and $t \in T$, where $\psi_1, \psi_2 \in \mathcal{K}$. Let $r_i = d_i(p_i, M_i)$, $i = 1, \ldots, l$. Furthermore, in view of hypothesis (ii),

$$\begin{aligned} Dv(p(t,a,t_0),t) &\leq \sum_{i=1}^{l} \alpha_i \left(\psi_{i4}(r_i) \sum_{i=1}^{l} a_{ij}\psi_{j4}(r_j) \right) \\ &= w^T R w \\ &= w^T \left(\frac{R + R^T}{2} \right) w \\ &= w^T B w \\ &\leq \lambda_M(B)|w|^2 \end{aligned}$$

for all $t \geq t_0$ and $p(\cdot, a, t_0) \in S$, where $w = [\psi_{14}(r_1), \ldots, \psi_{l4}(r_l)]^T$, $R = [\alpha_i a_{ij}] \in \mathbb{R}^{l \times l}$, and $\lambda_M(B)$ denotes the largest eigenvalue of B.

Let $\widetilde{\psi_3}(r) = \min_{1 \leq i \leq l}\{\psi_{i4}^2(r)\}$ and $d(p, M) = r = \sum_{i=1}^{l} r_i$. Then

$$|w|^2 \geq \max_{1 \leq i \leq l}\{\psi_{i4}^2(r_i)\} \geq \max_{1 \leq i \leq l}\{\widetilde{\psi_3}(r_i)\} \geq \widetilde{\psi_3}(r/l) \triangleq \psi_3(r).$$

Thus,

$$Dv(p(t,a,t_0),t) \leq \lambda_M(B)\psi_3(d(p,M)). \tag{9.6.2}$$

In view of (9.6.1) and (9.6.2), it follows from Theorem 3.3.1 that (S, M) is uniformly stable if B is negative semidefinite and from Theorem 3.3.2 that (S, M) is uniformly asymptotically stable if B is negative definite. □

Our next result is easier to apply than Theorem 9.6.1; however, because it requires some restrictions on the hypotheses of that theorem, it is more conservative than Theorem 9.6.1.

Corollary 9.6.1 Assume that the hypotheses (i) and (ii) of Theorem 9.6.1 are true and that $-A = [-a_{ij}] \in \mathbb{R}^{l \times l}$ is an M-matrix where the a_{ij} are given in Theorem 9.6.1. Then the following statements are true.

(a) (S, M) is *uniformly asymptotically stable*.

(b) If in hypothesis (i) of Theorem 9.6.1, $\psi_{i1}, \psi_{i2} \in \mathcal{K}_\infty$, $i = 1, \ldots, l$, and if M is bounded then S is *uniformly bounded*, and *uniformly ultimately bounded*. Furthermore, under these conditions, (S, M) is *uniformly asymptotically stable in the large*.

(c) If in hypotheses (i) and (ii) of Theorem 9.6.1, $\psi_{ik} = a_{ik}r^{b_k}$ with $a_{ik} > 0, b_k > 0$ and $b_1 = b_2 = 2b_4$, $i = 1, \ldots, l$, $k = 1, 2, 4$, and $r \in \mathbb{R}^+$, then (S, M) is *exponentially stable in the large*.

Proof. The proofs of all parts are consequences of Theorem 9.6.1 by invoking the following property of M-matrices [25]: if $D \in \mathbb{R}^{l\times l}$ is an M-matrix, then there exists a matrix $\Lambda = \text{diag}[\alpha_1, \dots, \alpha_l]$, $\alpha_i > 0$, $i = 1, \dots, l$, such that the matrix $\Lambda D + D^T \Lambda$ is positive definite. Choosing $D = -A$, we see that hypothesis (iii) of Theorem 9.6.1 is satisfied by choosing $B = [b_{ij}]_{l\times l} = (\Lambda A + A^T \Lambda)/2$ (i.e., $b_{ij} = (\alpha_i a_{ij} + \alpha_j a_{ji})/2, 1 \leq i, j \leq j$), where B is negative definite. □

We now reconsider Example 2.11.1, which may be viewed as a control problem consisting of an infinite-dimensional plant (the heat equation) and a finite-dimensional controller (an ordinary differential equation), utilizing distributed control (in contrast to boundary control). As in Example 2.11.1, the state variables for the controller and the plant are denoted by $z_1(t)$ and $z_2(t, x)$, respectively.

Example 9.6.1 [25], [36] We consider the *composite system* given by

$$\begin{cases} \dot{z}_1(t) = Az_1(t) + b\displaystyle\int_\Omega f(x)z_2(t,x)dx, & t \in \mathbb{R}^+ \\ \dfrac{\partial z_2}{\partial t}(t,x) = \alpha\Delta z_2(t,x) + g(x)c^T z_1(t), & (t,x) \in \mathbb{R}^+ \times \Omega \\ z_2(t,x) = 0 & (t,x) \in \mathbb{R}^+ \times \partial\Omega \end{cases} \tag{9.6.3}$$

where $z_1 \in \mathbb{R}^m$, $z_2 \in \mathbb{R}$, $A \in \mathbb{R}^{m\times m}$, $b, c \in \mathbb{R}^m$, f and $g \in L_2[\Omega, \mathbb{R}]$, $\alpha > 0$, Ω is a bounded domain in $\mathbb{R}^n$ with smooth boundary $\partial\Omega$, and $\Delta = \sum_{i=1}^n \partial^2/\partial x_i^2$ denotes the Laplacian. System (9.6.3) may be viewed as a differential equation in the *product Banach space* $X = \mathbb{R}^m \times H_0[\Omega, \mathbb{R}]$ where $H_0[\Omega, \mathbb{R}]$ denotes the completion of $C_0[\Omega, \mathbb{R}]$ with respect to the L_2-norm and where $H_0[\Omega, \mathbb{R}] \subset L_2[\Omega, \mathbb{R}]$ (refer to Section 2.10). In [26] it is shown that this initial value and boundary value problem is well posed: for every initial condition $z_0 = (z_{10}, z_{20})^T \in \mathbb{R}^m \times H_0[\Omega, \mathbb{R}]$ there exists a unique solution $z(t, z_0)$ that depends continuously on z_0. It is easily shown that the set $\{(z_1, z_2)^T\} = \{(0, 0)^T\} \subset X$ is an invariant set for the dynamical system determined by (9.6.3).

Now assume that all the eigenvalues of A have negative real parts. Then there exists a positive definite matrix $P = P^T$ such that

$$PA + A^T P = C \tag{9.6.4}$$

where C is negative definite (refer to Section 7.4).

Next, we choose the Lyapunov functions

$$v_1(z_1) = z_1^T P z_1 \tag{9.6.5}$$

and

$$v_2(z_2) = \frac{1}{2}\|z_2\|_{L_2}^2 = \frac{1}{2}\int_\Omega |z_2(x)|^2 dx \tag{9.6.6}$$

where P is given in (9.6.4). Then

$$\lambda_m(P)|z_1|^2 \leq v_1(z_1) \leq \lambda_M(P)|z_1|^2 \tag{9.6.7}$$

for all $z_1 \in \mathbb{R}^m$, where $\lambda_M(P) > 0$ and $\lambda_m(P) > 0$ denote the largest and smallest eigenvalues of P, respectively. In the notation of Theorem 9.6.1, we now have, in view of (9.6.4)–(9.6.7), that $\psi_{11}(r) = \lambda_m(P)r^2$, $\psi_{12}(r) = \lambda_M(P)r^2$, and $\psi_{21}(r) = \psi_{22}(r) = r^2$ for all $r \in \mathbb{R}^+$.

Along the solutions of (9.6.3) we now have

$$\begin{aligned} v'_{1(9.6.3)}(z_1) &= z_1^T C z_1 + 2 z_1^T P b \int_\Omega f(x) z_2(x) dx \\ &\leq -\gamma |z_1|^2 + 2|b| \|P\|_2 \|f\|_{L_2} |z_1| \|z_2\|_{L_2} \end{aligned}$$

where $-\gamma < 0$ denotes the largest eigenvalue of C and $\|\cdot\|_2$ denotes the matrix norm induced by the Euclidean vector norm $|\cdot|$ defined on $\mathbb{R}^m$. Also, along the solutions of (9.6.3) we have, invoking Gauss' Divergence Theorem and Poincaré's Inequality (refer to Example 9.2.2),

$$\begin{aligned} v'_{2(9.6.3)}(z_2) &= \alpha \int_\Omega z_2(\Delta z_2) dx + \int_\Omega z_2 g c^T z_1 dx \\ &= -\alpha \int_\Omega |\nabla z_2|^2 dx + c^T z_1 \int_\Omega z_2 g dx \\ &\leq -\alpha\Gamma \|z_2\|_{L_2}^2 + |c| \|g\|_{L_2} |z_1| \|z_2\|_{L_2} \end{aligned} \tag{9.6.8}$$

for all $z = (z_1, z_2)^T \in X$, where $\Gamma \geq n^2/\delta^2$ and where Ω can be put into a hypercube of length δ.

It now follows that hypotheses (i) and (ii) of Theorem 9.6.1 are satisfied with $A \triangleq [a_{ij}]$ given by

$$A = \begin{bmatrix} -\gamma & 2|b| \ \|P\|_2 \|f\|_{L_2} \\ |c| \ \|g\|_{L_2} & -\alpha\Gamma \end{bmatrix}.$$

If $-A$ is an M-matrix, then the hypotheses of Corollary 9.6.1 are satisfied as well. It follows that the equilibrium $z_e = 0$ of system (9.6.3) is exponentially stable in the large if

$$\gamma\alpha\Gamma > 2|b| \ |c| \ \|P\|_2 \|f\|_{L_2} \|g\|_{L_2}.$$

□

9.7 Analysis of a Point Kinetics Model of a Multicore Nuclear Reactor

We now return to the point kinetics model of a multicore nuclear reactor presented in Example 2.8.1 described by the Volterra integrodifferential equations [25], [35]

$$\begin{aligned} \dot{p}_i(t) = &- K_i p_i(t) + \int_{-\infty}^t F_i(t-s) p_i(s) ds + p_i(t) \int_{-\infty}^t n_i(t-s) p_i(s) ds \\ &+ \sum_{j=1, i\neq j}^{l} \int_{-\infty}^t G_{ij}(t-s) p_j(s) ds, \qquad i = 1, \ldots, l, \end{aligned} \tag{9.7.1}$$

for $t \geq 0$. For the meaning of all the symbols given in (9.7.1), as well as background material concerning the above model, the reader should refer to Example 2.8.1. As pointed out in that example, the initial conditions for (9.7.1) given by $p_i(t) = \varphi_i(t)$ for $t \in (-\infty, 0]$ where $\varphi_i \in Z_i$, the fading memory space of all absolutely continuous functions ψ_i defined on $(-\infty, 0]$ such that

$$\|\psi_i\|^2 = |\psi_i(0)|^2 + \int_{-\infty}^{0} |\psi_i(s)|^2 e^{L_i s} ds < \infty, \tag{9.7.2}$$

where $L_i > 0$, $i = 1, \ldots, l$, are constants that are specified later. The set of all solutions of system (9.7.1), generated by varying φ_i over Z_i, $i = 1, \ldots, l$, determines a dynamical system in the Banach space X which is the completion of $Z_1 \times \cdots \times Z_l$ with respect to the norm defined by (9.7.2).

For $\varphi = [\varphi_1, \ldots, \varphi_l]^T \in X$ we now choose the vector Lyapunov function

$$V(\varphi) = [v_1(\varphi_1), \ldots, v_l(\varphi_l)]^T,$$

where

$$v_i(\varphi_i) = \left(\varphi_i(0)^2 + K_i \int_{-\infty}^{0} \varphi_i(u)^2 e^{L_i u} du\right)^{1/2}.$$

Now define $p_{it}(\theta) = p_i(t + \theta), \theta \in (-\infty, 0]$. For $p_{it} \in Z_i$, we have along the solutions of (9.7.1),

$$\begin{aligned}
v'_{i(9.7.1)}(p_{it}) &= \frac{1}{2v_i(p_{it})}\left\{2p_i(t)\dot{p}_i(t) + K_i \int_{-\infty}^{0} \left(\frac{d}{du}[p_i(t+u)]^2\right) e^{L_i s} du\right\} \\
&= \frac{1}{2v_i(p_{it})}\Bigg\{2p_{it}(0)\Bigg[-K_i p_{it}(0) + \int_{-\infty}^{0} F_i(-s)p_{it}(s)ds \\
&\quad + p_{it}(0)\int_{-\infty}^{0} n_i(-s)p_{it}(s)ds + \sum_{j=1, i\neq i}^{l} \int_{-\infty}^{0} G_{ij}(-s)p_{it}(s)ds\Bigg] \\
&\quad + K_i\left[[p_{it}(0)]^2 - L_i \int_{-\infty}^{0} [p_{it}(s)]^2 e^{L_i s} ds\right]\Bigg\}.
\end{aligned}$$

Now let

$$b_i \triangleq \left(K_i \int_{-\infty}^{0} [p_{it}(s)]^2 e^{L_i s} ds\right)^{1/2}$$

and assume that $L_i > 0$, $i = 1, \ldots, l$, are such that

$$c_i \triangleq \left(\int_{0}^{\infty} [F_i(s)]^2 e^{L_i s} ds\right)^{1/2} < \infty,$$

$$d_i \triangleq \left(\int_{0}^{\infty} [n_i(s)]^2 e^{L_i s} ds\right)^{1/2} < \infty,$$

and

$$c_{ij} \triangleq \left(\int_0^\infty [G_{ij}(s)]^2 e^{L_i s} ds \right)^{1/2} < \infty.$$

Then

$$\begin{aligned} v'_{i(9.7.1)}(p_{it}) &\le \frac{1}{2v_i(p_{it})} \left\{ -K_i p_{it}(0)^2 + \frac{2p_{it}(0)c_i b_i}{\sqrt{K_i}} + 2p_{it}(0)^2 \frac{d_i b_i}{\sqrt{K_i}} \right. \\ &\quad \left. + 2p_{it}(0) \sum_{j=1, i\neq j}^{l} c_{ij} \frac{b_j}{\sqrt{K_j}} + K_i p_{it}(0)^2 - L_i b_i^2 \right\} \\ &\le \frac{1}{2v_i(p_{it})} \left\{ -K_i p_{it}(0)^2 + \frac{2c_i}{\sqrt{K_i}} p_{it}(0) b_i - L_i b_i^2 \right. \\ &\quad \left. + \frac{2d_i}{\sqrt{K_i}} p_{it}(0)^2 b_i \right\} + \sum_{j=1, i\neq j}^{l} \frac{c_{ij}}{\sqrt{K_j}} v_j(p_{jt}). \end{aligned} \tag{9.7.3}$$

Now if

$$K_i \sqrt{L_i} > c_i, \tag{9.7.4}$$

then the symmetric matrix given by

$$D_i = \begin{bmatrix} K_i & -c_i/\sqrt{K_i} \\ -c_i/\sqrt{K_i} & L_i \end{bmatrix} \tag{9.7.5}$$

is positive definite. Let $\sigma_i > 0$ denote the smallest eigenvalue of D_i, $i = 1, \ldots, l$. Then for any $\varepsilon > 0$ with $\varepsilon < \sigma_i$, $i = 1, \ldots, l$, there exists an $r = r(\varepsilon) > 0$ such that if $\|\psi\| < r$, where $\|\cdot\|$ denotes the norm defined by (9.7.2), then

$$\frac{2d_i}{\sqrt{K_i}} \psi_i(0)^2 b_i < \varepsilon \big(\psi_i(0)^2 + b_i^2 \big)$$

where b_i is defined as above. From (9.7.3) we now obtain

$$\begin{aligned} v'_{i(9.7.1)}(\psi_i) &\le -\frac{\sigma_i - \varepsilon}{2v_i(\psi_i)} \big(\psi_i(0)^2 + b_i^2 \big) + \sum_{j=1, i\neq j}^{l} \frac{c_{ij}}{\sqrt{K_j}} v_j(\psi_j) \\ &= \frac{\sigma_i - \varepsilon}{2} v_i(\psi_i) + \sum_{j=1, i\neq j}^{l} \frac{c_{ij}}{\sqrt{K_j}} v_j(\psi_j), \end{aligned} \tag{9.7.6}$$

$i = 1, \ldots, l$. Letting

$$A = [a_{ij}], \qquad a_{ii} = \frac{\sigma_i}{2}, \qquad \text{and} \qquad a_{ij} = \frac{c_{ij}}{\sqrt{K_j}}, \qquad i \neq j, \tag{9.7.7}$$

for all $i, j = 1, \ldots, l$, we can rewrite (9.7.6) more compactly in matrix form as

$$V'_{(9.7.1)}(\psi) \le -(A - \varepsilon I) V(\psi) \tag{9.7.8}$$

for all $\|\psi\| < r$. In (9.7.8), inequality is interpreted componentwise and I denotes the $l \times l$ identity matrix.

Now assume that A in (9.7.8) is an M-matrix. Then there exists an $\varepsilon > 0$, sufficiently small, so that $A - \varepsilon I$ is also an M-matrix. It follows from Theorem 9.5.1 (refer also to the discussion concerning equation (9.5.1)) that the exponential stability of the equilibrium $y_e = 0$ of the comparison system

$$\dot{y} = -(A - \varepsilon I)y \tag{9.7.9}$$

implies the exponential stability of the equilibrium $\psi_e = 0$ of the dynamical system determined by (9.7.1).

We have proved (using the comparison theory of Section 9.5) that the equilibrium $\psi_e = 0$ of (9.7.1) is *exponentially stable* if (9.7.4) is true for all $i = 1, \dots, l$ and if the matrix A given in (9.7.7) is an M-matrix. (Refer to Definition 7.7.1 for the definition of M-matrix and the discussion following that definition for several equivalent characterizations of M-matrices.)

9.8 Results for Retarded Functional Differential Equations

Referring to Section 2.7, we recall dynamical systems determined by retarded functional differential equations given by

$$\dot{x}(t) = F(t, x_t) \tag{F}$$

where $F \in C[\mathbb{R}^+ \times C, \mathbb{R}^n]$, C is an open connected subset of $X = C_r = C[[-r, 0], \mathbb{R}^n]$, and $x_t \in C_r$ is determined by $x_t(s) = x(t + s)$, $s \in [-r, 0]$. On C_r we define the norm

$$\|\varphi\| = \max_{-r \le s \le 0} \{|\varphi(s)|\} \tag{9.8.1}$$

where $|\cdot|$ denotes a norm on $\mathbb{R}^n$. Then $(X, \|\cdot\|)$ is a Banach space.

From Section 2.7 we recall that a function $p \in C[[t_0 - r, r + c), \mathbb{R}^n]$, $c > 0$, is a solution of (F) if $(t, x_t) \in \mathbb{R} \times C$ for all $t \in [t_0 - r, r + c)$ and $\dot{p} = F(t, p_t)$ for $t \in [t_0, t_0 + c)$. The reader should refer to Section 2.7 for results that ensure the existence, uniqueness, and continuation of solutions of (F). We assume that $0 \in C$ and that $F(t, 0) = 0$ for all $t \in \mathbb{R}^+$, so that $\varphi_e = 0$ is an equilibrium for (F).

We employ functions $v \in C[C \times \mathbb{R}^+, \mathbb{R}]$ (resp., $v \in C[C_r \times \mathbb{R}^+, \mathbb{R}]$). Along the solutions of (F), the upper-right derivative of v with respect to t is given by

$$v'_{(F)}(\varphi, t) = \overline{\lim_{h \to 0^+}} \frac{1}{h} [v(x_{t+h}(t, \varphi), t + h) - v(\varphi, t)] \tag{9.8.2}$$

where $x_{t+h}(t, \varphi)$ denotes the solution of (F) through (t, φ).

Finally, we let $B(h) = \{\varphi \in C_r \colon \|\varphi\| < h\}$ for some $h > 0$. Throughout this section, all results constitute statements relative to the norm $\|\cdot\|$ given in (9.8.1).

The stability and boundedness results that we presented in the preceding sections for differential equations defined on Banach spaces are of course applicable in particular to dynamical systems determined by retarded functional differential equations. (Recall from Section 2.7 that system (F) can be transformed into an *equivalent system* $(\widetilde{F})$ which is a special case of the general differential equation (GE) defined on Banach spaces.) However, because a solution of (F) is a function of t with range in $\mathbb{R}^n$, it is reasonable to expect that one should be able to improve some of the results presented thus far for systems determined by (F). This is indeed the case.

A. Stability and boundedness results

We first present local results.

Theorem 9.8.1 (i) Assume that for every bounded set G in C_r, the range of F on the set $\mathbb{R}^+ \times G$ is bounded in $\mathbb{R}^n$. Assume that there exist a function $v \in C[B(h) \times \mathbb{R}^+, \mathbb{R}]$, with $h > 0$ and $B(h) \subset C$, and two functions $\psi_1, \psi_2 \in \mathcal{K}$, such that

$$\psi_1(|\varphi(0)|) \leq v(\varphi, t) \leq \psi_2(\|\varphi\|) \tag{9.8.3}$$

where $|\cdot|$ denotes a norm on $\mathbb{R}^n$, and

$$v'_{(F)}(\varphi, t) \leq 0 \tag{9.8.4}$$

for all $\varphi \in B(h)$ and $t \in \mathbb{R}^+$. Then the equilibrium $\varphi_e = 0$ of (F) is *uniformly stable*.

(ii) If in addition to the above conditions there exists a function $\psi_3 \in \mathcal{K}$ such that

$$v'_{(F)}(\varphi, t) \leq -\psi_3(|\varphi(0)|) \tag{9.8.5}$$

for all $\varphi \in B(h)$ and $t \in \mathbb{R}^+$, then the equilibrium $\varphi_e = 0$ of (F) is *uniformly asymptotically stable*.

Proof. (i) For given $\varepsilon > 0$ and $t_0 > 0$, we can assume without loss of generality that $\varepsilon < h$, and we choose $\delta = \min\{\varepsilon, \psi_2^{-1} \circ \psi_1(\varepsilon)\}$. Then $\psi_2(\delta) \leq \psi_1(\varepsilon)$. Because (9.8.4) is true by assumption, we know that for $\varphi \in B(h)$, $v(x_t(\cdot, t_0, \varphi), t)$ is nonincreasing for $t \geq t_0$, where $x_t(\cdot, t_0, \varphi)$ denotes a solution of (F) with initial condition $x_{t_0} = \varphi$. Therefore,

$$\begin{aligned}
\psi_1(|x_t(0, t_0, \varphi)|) &\leq v(x_t(\cdot, t_0, \varphi), t) \\
&\leq v(x_{t_0}(\cdot, t_0, \varphi), t_0) \\
&= v(\varphi, t_0) \\
&\leq \psi_2(\|\varphi\|) \\
&< \psi_2(\delta) \\
&\leq \psi_2(\varepsilon)
\end{aligned}$$

for all $t \geq t_0$ whenever $\varphi \in B(\delta)$. Thus, $|x_t(0, t_0, \varphi)| \leq \varepsilon$ for all $t \geq t_0$ whenever $\varphi \in B(\delta)$. Furthermore, because for all $t \geq t_0$,

$$\|x_t(\cdot, t_0, \varphi)\| = \max_{s \in [-r, 0]} |x_{t-s}(0, t_0, \varphi)|$$

and for $t_0 > t - s \geq t_0 - r$, $|x_{t-s}(0, t_0, \varphi)| = |\varphi(t - s - t_0)| \leq \|\varphi\| < \delta \leq \varepsilon$ if $\varphi \in B(\delta)$, then $\|x_t(\cdot, t_0, \varphi)\| < \varepsilon$ for all $t \geq t_0$ whenever $\varphi \in B(\delta)$. We have proved that the equilibrium $\varphi_e = 0$ is uniformly stable.

(ii) Because the uniform stability of the equilibrium $\varphi_e = 0$ of (F) has been proved above, we only need to prove the uniform attractivity of $\varphi_e = 0$ of (F). Specifically, we need to prove that there exists an $\eta > 0$ (independent of ε and t_0), and for every $\varepsilon > 0$ and for every $t_0 \in \mathbb{R}^+$, there exists a $T = T(\varepsilon)$ (independent of t_0) such that $\|x_t(\cdot, t_0, \varphi)\| < \varepsilon$ for all $t \geq t_0 + T$ whenever $\|\varphi\| < \eta$. By the uniform stability of $\varphi_e = 0$, there exists an $\eta > 0$ such that $\|x_t(\cdot, t_0, \varphi)\| < (r/2)$ for all $t \geq t_0$ whenever $\varphi \in B(\eta)$. We now claim that the equilibrium $\varphi_e = 0$ is uniformly attractive in $B(\eta)$. For if this were not true, then there would exist an $\varepsilon_0 > 0$, a $t_0 \in \mathbb{R}^+$, and a $\varphi_0 \in B(\eta)$ such that $\|x_{t_k}(\cdot, t_0, \varphi_0)\| \geq \varepsilon_0$ for a sequence of $t_k \geq t_0$ with $\lim_{k\to\infty} t_k = \infty$. Now $\|x_{t_k}(\cdot, t_0, \varphi_0)\| \geq \varepsilon_0$ implies that there exists a sequence $s_k \in [-r, 0]$ such that $|x(t_k + s_k, t_0, \varphi_0)| = |x_{t_k}(s_k, t_0, \varphi_0)| \geq \varepsilon_0$. Let $t'_k = t_k + s_k$ and assume without loss of generality that $t'_{k+1} \geq t'_k + 1$ (if this is not the case, then choose a subsequence of t'_k). By our assumption on F for the bounded set $B(r/2)$, there exists a constant L such that $|\dot{x}(t, t_0, \varphi_0)| = |F(t, x_t)| \leq L$ for all $t \geq t_0$, and such that $L \geq 2\delta$. We now have

$$|x(t, t_0, \varphi_0)| \geq \frac{\varepsilon_0}{2} \qquad \text{for } t \in \left[t'_k - \frac{\delta}{2L}, t'_k + \frac{\delta}{2L}\right].$$

Therefore, by (9.8.5), we have for $t \in [t'_k - (\delta/2L), t'_k + (\delta/2L)]$ that

$$v'_{(F)}(x_t(\cdot, t_0, \varphi_0), t) \leq -\psi_3(|x(t, t_0, \varphi_0)|) \leq -\psi_3\left(\frac{\varepsilon_0}{2}\right)$$

and if we let $h_k = t'_k + \delta/(2L)$ and $l_k = t'_k - \delta/(2L)$, then $v(x_{h_k}, h_k) - v(x_{l_k}, l_k) \leq -\psi_3(\varepsilon_0/2)\delta/L$, $k = 1, 2, \ldots$. Because $l_k \geq h_{k-1}$, it follows that $v(x_{l_k}, l_k) \leq v(x_{h_{k-1}}, h_{k-1})$. Thus, $v(x_{h_k}, h_k) - v(x_{l_1}, l_1) \leq -\psi_3(\varepsilon_0/2)\delta k/L$. For

$$k > \frac{v(x_{l_1}, l_1)L}{\psi_3(\varepsilon_0/2)\delta}$$

we now have $v(x_{h_k}, h_k) < 0$, which is in contradiction with (9.8.3). We have proved that the equilibrium $\varphi_e = 0$ of (F) is uniformly asymptotically stable. □

Next, we consider some global results.

Theorem 9.8.2 Assume that $C = C_r$ and that for every bounded set G in C_r, the range of F on the set $\mathbb{R}^+ \times G$ is a bounded set in $\mathbb{R}^n$. Assume that there exist a function $v \in C[C_r \times \mathbb{R}^+, \mathbb{R}]$ and two functions $\psi_1, \psi_2 \in \mathcal{K}_\infty$ such that

$$\psi_1(|\varphi(0)|) \leq v(\varphi, t) \leq \psi_2(\|\varphi\|) \tag{9.8.6}$$

and such that

$$v'_{(F)}(\varphi, t) \leq 0 \tag{9.8.7}$$

for all $\|\varphi\| \geq M$ for some $M > 0$ and for all $t \in \mathbb{R}^+$. Then S_F, the set of all solutions of (F), is *uniformly bounded.*

If in addition to the conditions given above, there exists a function $\psi_3 \in \mathcal{K}$ such that

$$v'_{(F)}(\varphi, t) \leq -\psi_3(|\varphi(0)|) \tag{9.8.8}$$

for all $\|\varphi\| \geq M$ and all $t \in \mathbb{R}^+$, then S_F is *uniformly ultimately bounded.*

Proof. (i) For every $\alpha > 0$, we choose $\beta = \max\{(\psi_1^{-1} \circ \psi_2)(\alpha), \alpha, 2M\}$. If $\varphi \in B(\alpha)$, then for all $t \geq t_0$ such that $\|x_t(0, t_0, \varphi)\| \geq M$, we have

$$\begin{aligned}\psi_1(|x_t(0, t_0, \varphi)|) &\leq v(x_t(\cdot, t_0, \varphi), t)\\ &\leq v(x_{t_0}(\cdot, t_0, \varphi), t_0)\\ &= v(\varphi, t_0)\\ &\leq \psi_2(\|\varphi\|)\\ &< \psi_2(\alpha)\\ &\leq \psi_1(\beta),\end{aligned}$$

and therefore, $|x_t(0, t_0, \varphi)| < \beta$. It follows that $\|x_t(\cdot, t_0, \varphi)\| < \beta$ for all $t \geq t_0$ whenever $\varphi \in B(\alpha)$. We have proved that S_F is uniformly bounded.

(ii) The proof of uniform ultimate boundedness can be accomplished in a similar manner as the proof of part (ii) of Theorem 9.8.1 and is left as an exercise. □

Theorem 9.8.3 Assume that $C = C_r$ and that for every bounded set G in C_r, the range of F on the set $\mathbb{R}^+ \times G$ is a bounded set in $\mathbb{R}^n$. Assume that there exist a function $v \in C[C_r \times \mathbb{R}^+, \mathbb{R}]$, two functions $\psi_1, \psi_2 \in \mathcal{K}_\infty$, and a function $\psi_3 \in \mathcal{K}$ such that

$$\psi_1(|\varphi(0)|) \leq v(\varphi, t) \leq \psi_2(\|\varphi\|) \tag{9.8.9}$$

and

$$v'_{(F)}(\varphi, t) \leq -\psi_3(|\varphi(0)|) \tag{9.8.10}$$

for all $\varphi \in C_r$ and $t \in \mathbb{R}^+$. Then the equilibrium $\varphi_e = 0$ of (F) is *uniformly asymptotically stable in the large.*

Proof. It follows from Theorems 9.8.1 and 9.8.2 that the equilibrium $\varphi_e = 0$ of (F) is uniformly stable and that S_F is uniformly bounded. To prove part (iii) of Definition 3.1.16, we repeat the proof given in part (ii) of Theorem 9.8.1 with η replaced by α and $r/2$ replaced by β, where α and β are the constants used above in the proof of Theorem 9.8.2. We omit the details. □

Before presenting instability results, we consider some specific examples.

Example 9.8.1 Let us consider the retarded functional differential equation

$$\dot{x}(t) = Ax(t) + Bx(t - r), \qquad r > 0 \tag{9.8.11}$$

where $x \in \mathbb{R}^n$ and $A, B \in \mathbb{R}^{n \times n}$. We assume that A is a stable matrix (i.e., all the eigenvalues of A have negative real parts).

If we let $B = 0$, then (9.8.11) reduces to the linear ordinary differential equation

$$\dot{x}(t) = Ax(t). \tag{9.8.12}$$

For (9.8.12) we can construct a Lyapunov function

$$w(x) = x^T P x \tag{9.8.13}$$

where $P = P^T \in \mathbb{R}^{n \times n}$ is a positive definite matrix (i.e., $P > 0$) such that

$$w'_{(9.8.12)}(x) = -x^T C x \tag{9.8.14}$$

where $C = C^T \in \mathbb{R}^{n \times n}$ is a positive definite matrix (i.e., $C > 0$). The validity of (9.8.13) and (9.8.14) follows from the fact that for *every* positive definite matrix C there exists a *unique* positive definite matrix P such that

$$A^T P + PA = -C. \tag{9.8.15}$$

When $B \neq 0$, we cannot use (9.8.13) as a Lyapunov function for (9.8.11), because this function would not capture the effects of the delayed term $Bx(t-r)$. To accomplish this, we append to (9.8.13) a term, resulting in the Lyapunov function

$$v(\varphi) = \varphi^T(0) P \varphi(0) + \int_{-r}^{0} \varphi^T(\theta) E \varphi(\theta) d\theta \tag{9.8.16}$$

where $E = E^T \in \mathbb{R}^{n \times n}$ is a positive definite matrix (i.e., $E > 0$).

Let δ_1 and δ_2 denote the smallest and largest eigenvalues of P, respectively, and let μ_1 and μ_2 denote the smallest and largest eigenvalues of E, respectively. Then

$$\begin{aligned} v(\varphi) &\leq \delta_2 |\varphi(0)|^2 + \int_{-r}^{0} \mu_2 |\varphi(\theta)|^2 d\theta \\ &\leq \delta_2 \|\varphi\|^2 + \mu_2 r \|\varphi\|^2 \\ &= (\delta_2 + \mu_2 r) \|\varphi\|^2. \end{aligned}$$

Also, $v(\varphi) \geq \delta_1 |\varphi(0)|^2$. Thus, there are constants $c_2 > c_1 > 0$ such that

$$c_1 |\varphi(0)|^2 \leq v(\varphi) \leq c_2 \|\varphi\|^2 \tag{9.8.17}$$

for all $\varphi \in C_r$.

Along the solutions of (9.8.11) we have

$$\begin{aligned} v'_{(9.8.11)}(\varphi) &= -\varphi^T(0) C \varphi(0) + 2\varphi^T(0) P B \varphi(-r) \\ &\quad + \varphi^T(0) E \varphi(0) - \varphi^T(-r) E \varphi(-r) \\ &= -\begin{bmatrix} \varphi^T(0) & \varphi^T(-r) \end{bmatrix} \begin{bmatrix} (C-E) & -PB \\ -(PB)^T & E \end{bmatrix} \begin{bmatrix} \varphi(0) \\ \varphi(-r) \end{bmatrix}. \end{aligned} \tag{9.8.18}$$

If the matrix $C - E$ is positive definite (i.e., $C - E > 0$ or $C > E$), and if λ_1 denotes the smallest eigenvalue of $C - E$, then

$$x^T(C - E)x \geq \lambda_1 |x|^2 \quad \text{and} \quad x^T E x \geq \mu_1 |x|^2 \tag{9.8.19}$$

for all $x \in \mathbb{R}^n$. Then

$$\begin{aligned} v'_{(9.8.11)}(\varphi) &\leq -\lambda_1|\varphi(0)|^2 + 2\|PB\|\,|\varphi(0)||\varphi(-r)| - \mu_1|\varphi(-r)|^2 \\ &= -\begin{bmatrix}|\varphi(0)| & |\varphi(-r)|\end{bmatrix}\begin{bmatrix}\lambda_1 & -\|PB\| \\ -\|PB\| & \mu_1\end{bmatrix}\begin{bmatrix}|\varphi(0)| \\ |\varphi(-r)|\end{bmatrix}. \end{aligned} \tag{9.8.20}$$

Thus, $v'_{(9.8.11)}$ is negative definite if $\lambda_1 > 0$ (which is given) and $\lambda_1\mu_1 - \|PB\|^2 > 0$. In this case we obtain

$$v'_{(9.8.11)}(\varphi) \leq -c_3\big(|\varphi(0)|^2 + |\varphi(-r)|^2\big) \leq -c_3|\varphi(0)|^2 \tag{9.8.21}$$

for all $\varphi \in C_r$, where $c_3 > 0$ denotes the smallest eigenvalue of the symmetric matrix given in (9.8.20).

It now follows from (9.8.17), (9.8.21), and Theorem 9.8.3 that under the above assumptions, the equilibrium $\varphi_e = 0$ of system (9.8.11) is *uniformly asymptotically stable in the large.*

We have shown that *if the matrix A in* (9.8.11) *is stable and if the matrix B is sufficiently small in norm, then the equilibrium $\varphi_e = 0$ of system* (9.8.11) *is uniformly asymptotically stable in the large for all $r > 0$.* □

Example 9.8.2 We now consider the system described by the scalar equation

$$\dot{x}(t) = -a(t)x(t) - b(t)x(t - r) \tag{9.8.22}$$

where $t \in \mathbb{R}^+, r > 0, a(\cdot)$ and $b(\cdot)$ are real-valued, bounded, and continuous functions and $a(t) > 0$ for all $t \in \mathbb{R}^+$. We choose as a Lyapunov function

$$v(\varphi) = \frac{1}{2}\varphi(0)^2 + \mu\int_{-r}^{0}\varphi(\theta)^2 d\theta \tag{9.8.23}$$

where μ is a positive constant. Then

$$c_1|\varphi(0)|^2 = \frac{1}{2}|\varphi(0)|^2 \leq v(\varphi) \leq \frac{1}{2}\|\varphi\|^2 + \mu r\|\varphi\|^2 = c_2\|\varphi\|^2 \tag{9.8.24}$$

for all $\varphi \in C_r$.

Along the solutions of (9.8.22) we have

$$v'_{(9.8.22)}(\varphi) = -\begin{bmatrix}\varphi(0) & \varphi(-r)\end{bmatrix}\begin{bmatrix}a(t) - \mu & b(t)/2 \\ b(t)/2 & \mu\end{bmatrix}\begin{bmatrix}\varphi(0) \\ \varphi(-r)\end{bmatrix} \tag{9.8.25}$$

for all $\varphi \in C_r$. Now assume that for some $\delta > 0$, $(a(t) - \mu)\mu - b(t)^2/4 \geq \delta > 0$ for all $t \in \mathbb{R}^+$, or equivalently, that

$$b(t)^2 - 4(a(t) - \mu)\mu \leq -\delta < 0 \tag{9.8.26}$$

for all $t \in \mathbb{R}^+$. Note that inequality (9.8.26) also implies that $a(t) > \mu > 0$ (and $a(t) \leq M$ for some $M > 0$, by assumption). Therefore, under these assumptions, the symmetric matrix in (9.8.25) is positive definite. From the characteristic equation

$$\lambda^2 - a(t)\lambda + (a(t) - \mu)\mu - b(t)^2/4 = 0,$$

the smallest eigenvalue of the above-mentioned matrix can be estimated as

$$\lambda_m = \frac{a(t) - \sqrt{a(t)^2 - (4(a(t) - \mu)\mu - b(t)^2)}}{2} \geq \frac{a(t) - \sqrt{a(t)^2 - \delta^2}}{2} \geq \frac{\delta}{4M}.$$

Therefore,

$$v'_{(9.8.22)}(\varphi) \leq -c_3\big(\varphi(0)^2 + \varphi(-r)^2\big) \leq -c_3\varphi(0)^2 \tag{9.8.27}$$

for all $\varphi \in C_r$, where $c_3 = \delta/4M$. Inequality (9.8.24) and (9.8.27) along with Theorem 9.8.3 imply that the equilibrium $\varphi_e = 0$ of system (9.8.22) is *uniformly asymptotically stable in the large* if there exist $\mu > 0$ and $\delta > 0$ such that inequality (9.8.26) is satisfied. In particular, these conditions are satisfied for $b(t) \equiv b$ and $a(t) \equiv a$ if $|b| < a$. In this case we choose $\mu = a/2$ and $\delta = a^2 - b^2$. □

Example 9.8.3 In this example we demonstrate the advantage of the stability results of the present section over those of Section 9.2, when applied to functional differential equations. To this end, we reconsider system (9.8.22), using the same Lyapunov function as before, restated here as

$$v(\varphi) = \frac{1}{2}\varphi(0)^2 + \mu\int_{-r}^{0}\varphi(\theta)^2 d\theta.$$

We have in the present case

$$c_1\|\varphi\|^2 \leq v(\varphi) \leq c_2\|\varphi\|^2 \tag{9.8.28}$$

where $c_1 = \min\{1/2, \mu\}$ and $c_2 = \max\{1/2, \mu\}$ and where the norm is given by (9.8.1).

As in Example 9.8.2, we have along the solutions of system (9.8.22) the estimate

$$v'_{(9.8.22)}(\varphi) \leq -c_3\big(\varphi(0)^2 + \varphi(-r)^2\big)$$

for all $\varphi \in C_r$. Therefore, $v'_{(9.8.22)}$ is negative semidefinite with respect to the norm $\|\cdot\|$ and we can conclude from Theorem 9.2.1(b) that the equilibrium $\varphi_e = 0$ of system (9.8.22) is *uniformly stable*. However, because we cannot show in the present case that $v'_{(9.8.22)}$ is negative definite with respect to the norm $\|\cdot\|$, we cannot apply Theorem 9.2.1(c) (resp., Theorem 9.2.2) to conclude that the equilibrium $\varphi_e = 0$ of system (9.8.22) is uniformly asymptotically stable (in the large). □

B. Instability results

We now present instability results for retarded functional differential equations (F) which in general will yield less conservative results than the corresponding instability results given in Subsection 9.2C.

Theorem 9.8.4 (*Lyapunov's First Instability Theorem for retarded functional differential equations*) Assume that there exist a function $v \in C[B(h) \times \mathbb{R}^+, \mathbb{R}]$ for some $h > 0$, where $B(h) \subset C$, and a $t_0 \in \mathbb{R}^+$, such that the following conditions are satisfied.

(i) There exists a function $\psi_1 \in \mathcal{K}$ defined on $[0, h]$ such that

$$v(\varphi, t) \leq \psi_1(|\varphi(0)|) \tag{9.8.29}$$

for all $\varphi \in B(h)$ and $t \in \mathbb{R}^+$.

(ii) There exists a function $\psi_2 \in \mathcal{K}$ defined on $[0, h]$ such that

$$v'_{(F)}(\varphi, t) \geq \psi_2(|\varphi(0)|) \tag{9.8.30}$$

for all $\varphi \in B(h)$ and $t \in \mathbb{R}^+$.

(iii) In every neighborhood of the origin $\varphi_e = 0 \in C_r$, there are points φ such that $v(\varphi, t_0) > 0$.

Then the equilibrium $\varphi_e = 0$ of (F) is *unstable*.

Proof. For a given $\varepsilon \in (0, h)$, let $\{\varphi_m\}_{m \in \mathbb{N}}$ be a sequence with $\varphi_m \in B(\varepsilon)$, such that $\|\varphi_m\| \to 0$ as $m \to \infty$, and $v(\varphi_m, t_0) > 0$, where we have used condition (iii). Let $x_t(\cdot, t_0, \varphi_m)$ be a solution of (F) with the initial condition $x_{t_0} = \varphi_m$, and let $v_m(t) \triangleq v(x_t(\cdot, t_0, \varphi_m), t)$. It suffices to prove that for every $m \in \mathbb{N}$, $x_t(\cdot, t_0, \varphi_m)$ must reach the boundary of $B(\varepsilon)$ in finite time. For otherwise, we would have $\|x_t(\cdot, t_0, \varphi_m)\| < \varepsilon$ for all $t \geq t_0$, and in particular, that $|x_t(0, t_0, \varphi_m)| < \varepsilon$ for all $t \geq t_0$. It follows from hypothesis (ii) that $v_m(t)$ is nondecreasing for $t \geq t_0$, and from hypothesis (i) it follows that

$$\psi_1(|x_t(0, t_0, \varphi_m)|) \geq v_m(t) \geq v_m(t_0) = v(\varphi_m, t_0) > 0$$

or

$$|x_t(0, t_0, \varphi_m)| \geq \psi_1^{-1}(v_m(t_0)) \triangleq \alpha_m > 0$$

for all $t \geq t_0$. Using hypothesis (iii), we now have

$$\begin{aligned}
\psi_1(\varepsilon) > \psi_1(|x_t(0, t_0, \varphi_m)|) &\geq v_m(t) \\
&\geq v_m(t_0) + \int_{t_0}^{t} \psi_2(\alpha_m) ds \\
&= v_m(t_0) + \psi_2(\alpha_m)(t - t_0)
\end{aligned}$$

for all $t \geq t_0$. But this is impossible. Therefore, the equilibrium $\varphi_e = 0$ of (F) is unstable. □

We demonstrate the applicability of Theorem 9.8.4 by means of a specific example.

Example 9.8.4 We consider the scalar retarded functional differential equation

$$\dot{x}(t) = -ax(t) - bx(t-r) \tag{9.8.31}$$

where $t \in \mathbb{R}^+, r > 0$ is a constant, and $a, b \in \mathbb{R}$. We choose as a Lyapunov function

$$v(\varphi) = \frac{1}{2}\varphi(0)^2 - \mu \int_{-r}^{0} \varphi(\theta)^2 d\theta \tag{9.8.32}$$

for all $\varphi \in C_r$, where $\mu > 0$ is a constant. Clearly,

$$v(\varphi) \leq \frac{1}{2}\varphi(0)^2 \tag{9.8.33}$$

for all $\varphi \in C_r$. Along the solutions of (9.8.31) we have

$$\begin{aligned} v'_{(9.8.31)}(\varphi) &= \begin{bmatrix}\varphi(0) & \varphi(-r)\end{bmatrix} \begin{bmatrix} -(a+\mu) & -b/2 \\ -b/2 & \mu \end{bmatrix} \begin{bmatrix} \varphi(0) \\ \varphi(-r) \end{bmatrix} \\ &\geq \lambda\big(\varphi(0)^2 + \varphi(-r)^2\big) \\ &\geq \lambda\varphi(0)^2 \end{aligned} \tag{9.8.34}$$

where λ denotes the smallest eigenvalue of the symmetric matrix given in (9.8.34). Now $\lambda > 0$ if and only if

$$a + \mu < 0 \quad \text{and} \quad -4(a+\mu)\mu > b^2. \tag{9.8.35}$$

The second inequality in (9.8.35) is equivalent to the inequality

$$(a + 2\mu)^2 + (b^2 - a^2) < 0.$$

Thus, the conditions in (9.8.35) hold for some $\mu > 0$ if and only if $-a > |b|$.

Hypothesis (iii) in Theorem 9.8.4 is clearly satisfied for the choice of $v(\varphi)$ given in (9.8.32). It now follows from (9.8.33), (9.8.34), and Theorem 9.8.4 that the equilibrium $\varphi_e = 0$ of system (9.8.31) is *unstable* if $-a > |b|$. □

Before addressing the next instability result, we note that *Lyapunov's Second Instability Theorem for functional differential equations* is identical in form to Theorem 9.2.5 (Lyapunov's Second Instability Theorem for differential equations in Banach space) and is not restated here.

Theorem 9.8.5 (*Chetaev's Instability Theorem for retarded functional differential equations*) Assume that there exist a function $v \in C[B(h) \times \mathbb{R}^+, \mathbb{R}]$ for some $h > 0$, where $B(h) \subset C$, a $t_0 \in \mathbb{R}^+$ and an $h_0 > 0$, such that the following conditions are satisfied.

(i) There exists a component G of the set

$$D = \big\{(\varphi, t) \in B(h) \times \mathbb{R}^+ \colon v(\varphi, t) < 0 \text{ and } \|\varphi\| < h_0\big\}$$

such that in every neighborhood of the origin $\varphi_e = 0 \in C_r$, there exists a φ with $(\varphi, t_0) \in G$.

(ii) $v(\varphi, t) \geq -\psi_1(\|\varphi\|)$ for all $(\varphi, t) \in G$, where $\psi_1 \in \mathcal{K}$.

(iii) $v'_{(F)}(\varphi, t) \leq -\psi_2(|\varphi(0)|)$ for all $(\varphi, t) \in G$, where $\psi_2 \in \mathcal{K}$.

Then the equilibrium $\varphi_e = 0$ of (F) is *unstable*.

Proof. Let $\{\varphi_m\}_{m\in\mathbb{N}}$ be a sequence in C_r such that $(\varphi_m, t_0) \in D$ and such that $\|\varphi_m\| \to 0$ as $m \to \infty$. The existence of such a sequence $\{\varphi_m\}$ is guaranteed by hypothesis (i). Let $x_t(\cdot, t_0, \varphi_m)$ be a solution of (F) with initial condition $x_{t_0} = \varphi_m$. It suffices to prove that for every $m \in \mathbb{N}$, $x_t(\cdot, t_0, \varphi_m)$ must reach the boundary of $B(h_0)$ in finite time. For otherwise, we would have $\|x_t(\cdot, t_0, \varphi_m)\| < h_0$ for all $t \geq t_0$. Hypothesis (iii) implies now that $v(x_t(\cdot, t_0, \varphi_m), t)$ is nonincreasing for all $t \geq t_0$ and that

$$\begin{aligned} v(x_t(\cdot, t_0, \varphi_m), t) &\leq v(\varphi_m, t_0) - \int_{t_0}^{t} (\psi_2 \circ \psi_1^{-1})(h_0) ds \\ &= v(\varphi_m, t_0) - (\psi_2 \circ \psi_1^{-1})(h_0)(t - t_0) \end{aligned}$$

where we have assumed without loss of generality that h_0 is in the range of ψ_1. (Should this not be the case, then we can always replace h_0 by a smaller number.) Therefore, $v(x_t(\cdot, t_0, \varphi_m), t) \to -\infty$ as $t \to \infty$. But this contradicts hypothesis (ii) which implies that v is bounded from below on D. This proves the theorem. □

We apply Theorem 9.8.5 in the stability analysis of a specific example.

Example 9.8.5 We consider a scalar retarded functional differential equation given by

$$\dot{x}(t) = -ax(t)^3 - bx(t - r)^3 \tag{9.8.36}$$

where $t \in \mathbb{R}^+$, $r > 0$, and $a, b \in \mathbb{R}$. We choose as a Lyapunov function

$$v(\varphi) = -\frac{\varphi(0)^4}{4} + \mu \int_{-r}^{0} \varphi(\theta)^6 d\theta \tag{9.8.37}$$

for all $\varphi \in C_r$, where $\mu > 0$ is a constant. Clearly,

$$v(\varphi) \geq -\frac{\|\varphi\|^4}{4} \tag{9.8.38}$$

for all $\varphi \in C_r$. Along the solutions of (9.8.36) we have

$$\begin{aligned} v'_{(9.8.36)}(\varphi) &= -\begin{bmatrix} \varphi(0)^3 & \varphi(-r)^3 \end{bmatrix} \begin{bmatrix} -(a+\mu) & -b/2 \\ -b/2 & \mu \end{bmatrix} \begin{bmatrix} \varphi(0)^3 \\ \varphi(-r)^3 \end{bmatrix} \\ &\leq -\lambda\big(\varphi(0)^6 + \varphi(-r)^6\big) \\ &\leq -\lambda\varphi(0)^6 \end{aligned} \tag{9.8.39}$$

for all $\varphi \in C_r$. Now $\lambda > 0$ (where λ denotes the smallest eigenvalue of the symmetric matrix given in (9.8.39) if and only if

$$a + \mu < 0 \quad \text{and} \quad -4(a + \mu)\mu > b^2. \tag{9.8.40}$$

In an identical manner as in Example 9.8.4, we can show that the conditions in (9.8.40) are satisfied if and only if $-a > |b|$.

For any $h_0 > 0$, let G_1 be any component of the set $D = \{\varphi \in B(h_0) : v(\varphi) < 0\}$ such that $\varphi_e = 0 \in \partial G_1$, and let $G = G_1 \times \mathbb{R}^+$. It now follows from Theorem 9.8.5 that the equilibrium $\varphi_e = 0$ of system (9.8.36) is *unstable* if $-a > |b|$. □

C. Invariance theory

We next address the stability analysis of dynamical systems determined by autonomous retarded functional differential equations given by

$$\dot{x}(t) = F(x_t) \tag{FA}$$

where $F \in C[C, \mathbb{R}^n]$ and C is an open connected subset of C_r with norm $\|\cdot\|$ defined in (9.8.1). In the present subsection we assume that F is *completely continuous*; that is, for any bounded closed set $B \subset C$, the closure of $F(B) = \{F(x) : x \in B\}$ is compact.

Theorem 9.8.6 Assume that F in (FA) is completely continuous and that there exists a function $v \in C[C, \mathbb{R}]$ such that $v'_{(FA)}(\varphi) \leq 0$ for all $\varphi \in C$. Let M be the largest invariant set with respect to (FA) in the set

$$Z = \{\varphi \in C : v'_{(FA)}(\varphi) = 0\}. \tag{9.8.41}$$

Then every bounded solution of (FA) approaches M as $t \to \infty$.

Proof. By Theorem 9.4.1 it suffices to prove that for every bounded solution $\varphi_t(\cdot)$ of (FA), the closure of the trajectory of $\{\varphi_t(\cdot)\}$ is compact in C_r. Given a bounded solution $\varphi_t(\cdot)$ of (FA), because F is completely continuous, there exists a constant $L > 0$ such that $|\dot{\varphi}(t)| \leq L$ for all $t \in \mathbb{R}^+$, where $\varphi(t) = \varphi_t(0)$. Therefore, by using the Ascoli–Arzela lemma (refer to Problem 2.14.7), we can prove that for every sequence $\varphi_{t_m}(\cdot)$, $t_m \in \mathbb{R}^+$, there exists a subsequence $\varphi_{t_{m_k}}(\cdot)$ that converges in C_r. This proves that the closure of the trajectory of $\{\varphi_t(\cdot)\}$ is compact in C_r. This completes the proof of the theorem. □

In the next result we assume that $\varphi_e = 0$ is an equilibrium of system (FA).

Theorem 9.8.7 Assume that $C = C_r$ and that F in (FA) is completely continuous. Assume that there exists a function $v \in C[C_r, \mathbb{R}]$ that satisfies the following conditions.

(i) $v'_{(FA)}(\varphi) \leq 0$ for all $\varphi \in C_r$.

(ii) There exists a function $\psi \in \mathcal{K}_\infty$ such that

$$\psi(|\varphi(0)|) \leq v(\varphi) \tag{9.8.42}$$

for all $\varphi \in C_r$.

(iii) $\{0\} \subset C_r$ is the only invariant subset in

$$Z = \{\varphi \in C_r : v'_{(FA)}(\varphi) = 0\}. \tag{9.8.43}$$

Then the equilibrium $\varphi_e = 0$ of system (FA) is *uniformly asymptotically stable in the large*.

Proof. Let $U_\eta = \{\varphi \in C_r \colon v(\varphi) < \eta\}$. If $\varphi \in U_\eta$, because $v'_{(FA)}(\varphi) \leq 0$, $x_t(\varphi, t_0) \in U_\eta$ for all $t \geq t_0$. It follows from (9.8.43) that $|x_t(\varphi, t_0)| \leq \psi_1^{-1}(\eta)$ for all $t \geq t_0$ if $\varphi \in U_\eta$, which implies that $x_t(\varphi, t_0)$ is uniformly bounded.

The uniform stability follows from Theorem 9.8.1. In applying Theorem 9.8.1 we note that if v is independent of t, then the condition $v(\varphi) \leq \psi_2(\|\varphi\|)$ can be deleted for uniform stability, inasmuch as in this case the continuity of v can be utilized instead in the proof of Theorem 9.8.1.

It now follows from Theorem 9.8.6 that the equilibrium $\varphi_e = 0$ of system (FA) is *uniformly asymptotically stable in the large*. □

In Example 9.8.2 we showed that for the system described by

$$\dot{x}(t) = -ax(t) - bx(t - r), \tag{9.8.44}$$

the equilibrium $\varphi_e = 0$ is uniformly asymptotically stable in the large if $|b| < a$ and in Example 9.8.4 we showed that the equilibrium $\varphi_e = 0$ of this system is unstable if $-a > |b|$. In the next example we address some of the critical cases for this system, using the results of the present subsection. Specifically, we show that if $a = b > 0$, then the equilibrium $\varphi_e = 0$ of the above system is uniformly asymptotically stable in the large and if $a = -b > 0$, then the solutions φ of this system must approach a constant as $t \to \infty$.

Example 9.8.6 [29] For system (9.8.44) we assume that $t \in \mathbb{R}^+$, $r > 0$, and $a, b \in \mathbb{R}$. We choose as a Lyapunov function

$$v(\varphi) = \frac{1}{2}\varphi(0)^2 + \frac{a}{2}\int_{-r}^{0} \varphi(\theta)^2 d\theta. \tag{9.8.45}$$

For $a > 0$, we have

$$v(\varphi) \geq \frac{1}{2}\,\varphi(0)^2 \tag{9.8.46}$$

and

$$v'_{(9.8.44)}(\varphi) = -\begin{bmatrix}\varphi(0) & \varphi(-r)\end{bmatrix}\begin{bmatrix} a & b \\ b & a\end{bmatrix}\begin{bmatrix}\varphi(0) \\ \varphi(-r)\end{bmatrix}. \tag{9.8.47}$$

The symmetric matrix in (9.8.47) is positive semidefinite if and only if $a^2 - b^2 \geq 0$ and therefore, $v'_{(9.8.44)}(\varphi) \leq 0$ for all $\varphi \in C_r$ if and only if $|b| \leq a$. In the following we address the critical case $a = |b|$. We accomplish this by considering the cases $a = b > 0$ and $a = -b > 0$.

(a) When $a = b > 0$, then

$$Z = \{\varphi \in C_r \colon v'_{(9.8.44)}(\varphi) = 0\} = \{\varphi \in C_r \colon \varphi(0) = -\varphi(-r)\}.$$

If M is the largest invariant subset in Z, then $x_t(\cdot) \in M$ implies that $x(t) = -x(t-r)$, and therefore, by invoking (9.8.44) it follows that $\dot{x}(t) = 0$. Therefore, $x(t) = c$, a constant, and in fact $c = 0$. It follows that $M = \{0\}$. All conditions of Theorem 9.8.7

are satisfied and we conclude that the equilibrium $\varphi_e = 0$ of system (9.8.44) is uniformly asymptotically stable in the large.

(b) When $a = -b > 0$, then

$$Z = \{\varphi \in C_r : v'_{(9.8.44)}(\varphi) = 0\} = \{\varphi \in C_r : \varphi(0) = \varphi(-r)\}.$$

Similarly as in part (a), we can show that the largest invariant subset in Z is given by $M = \{\varphi \in C_r : \varphi \equiv k\}$; that is, φ is a constant function. It follows from Theorem 9.8.2 that the solutions of system (9.8.44) are uniformly bounded. From the proof of Theorem 9.8.6 it follows that the trajectory of every solution of (9.8.44) must have a compact closure and from the proof of Lemma 4.2.2, that $v(x_t) \to c$, a constant, as $t \to \infty$. Therefore, the ω-limit set of x_t, $\omega(x_t)$, must be a subset of the set $v^{-1}(c) \cap M$. Now in the case when φ is a constant function, we have that $v(\varphi)$ is a quadratic polynomial in φ and $v^{-1}(c) \cap M$ consists of at most two constant functions. Therefore, x_t approaches a constant as $t \to \infty$. □

In the next section, we apply Theorem 9.8.7 further in the analysis of a class of artificial neural networks with time delays.

D. Razumikhin-type theorems

The stability analysis of dynamical systems determined by retarded functional differential equations (F) by the results presented thus far is in general more complicated than the analysis of dynamical systems determined by ordinary differential equations because the former involve hypotheses in the setting of the space C_r whereas the hypotheses of the latter involve assumptions defined on $\mathbb{R}^n$ (which is much simpler than the space C_r). Stability results of the Razumikhin-type circumvent such difficulties by requiring hypotheses that are defined exclusively on $\mathbb{R}^n$.

In the present subsection we return to dynamical systems determined by retarded functional differential equations (F), as described at the beginning of the present section. In the following results, we let $B_E(h) = \{x \in \mathbb{R}^n : |x| < h\} \subset \mathbb{R}^n$ for some $h > 0$, and as before $B(h) = \{\varphi \in C_r : \|\varphi\| < h\} \subset C \subset C_r$.

Theorem 9.8.8 Assume that for every bounded set G in C_r the range of F on the set $\mathbb{R}^+ \times G$ is a bounded set in $\mathbb{R}^n$. Assume that for (F) there exist a function $v \in C[B_E(h) \times [-r, \infty), \mathbb{R}]$ and two functions $\psi_1, \psi_2 \in \mathcal{K}$ and a nondecreasing function $\psi_3 \in C[\mathbb{R}^+, \mathbb{R}^+]$ such that

$$\psi_1(|x|) \le v(x,t) \le \psi_2(|x|) \tag{9.8.48}$$

and for all $t \ge -r$ and all $\varphi \in B(h)$, and

$$v'_{(F)}(\varphi(0), t) \le -\psi_3(|\varphi(0)|) \text{ if } v(\varphi(\theta), t+\theta) \le v(\varphi(0), t) \text{ for all } \theta \in [-r, 0]. \tag{9.8.49}$$

Then the following statements are true.

(i) The equilibrium $\varphi_e = 0$ of (F) is *uniformly stable*.

(ii) If $\psi_3 \in \mathcal{K}$ and there exists a nondecreasing function $f \in C[\mathbb{R}^+, \mathbb{R}^+]$ such that $f(s) > s$ for $s \in (0, h]$ such that for all $t \geq -r$ and all $\varphi \in B(h)$,

$$v'_{(F)}(\varphi(0), t) \leq -\psi_3(|\varphi(0)|) \quad \text{if } v(\varphi(\theta), t+\theta) \leq f(v(\varphi(0), t)) \text{ for all } \theta \in [-r, 0], \tag{9.8.50}$$

then the equilibrium $\varphi_e = 0$ of (F) is *uniformly asymptotically stable.*

Proof. (a) Let

$$\widetilde{v}(\varphi, t) \triangleq \sup_{\theta \in [-r,0]} v(\varphi(\theta), t+\theta)$$

for all $(\varphi, t) \in B(h) \times \mathbb{R}^+$.

If $\widetilde{v}(x_t(t_0, \varphi_0), t) = v(x(t, t_0, \varphi_0), t)$, that is,

$$v(x(t+\theta, t_0, \varphi_0), t+\theta) \leq v(x(t, t_0, \varphi_0), t) < f(v(x(t, t_0, \varphi_0), t)),$$

then $\widetilde{v}'_{(F)}(x_t(t_0, \varphi_0), t) \leq 0$ by hypothesis (ii). If $\widetilde{v}(x_t(t_0, \varphi_0), t) < v(x(t, t_0, \varphi_0), t)$, then for $\tau > 0$ sufficiently small

$$\widetilde{v}(x_{t+\tau}(t_0, \varphi_0), t+\tau) = \widetilde{v}(x_t(t_0, \varphi_0), t).$$

Hence $\widetilde{v}'_{(F)}(x_t(t_0, \varphi_0), t) = 0$. Therefore, $\widetilde{v}'(x_t(t_0, \varphi_0), t) \leq 0$ under the present assumptions.

It follows from (9.8.48) that $\psi_1(|\varphi(0)|) \leq \widetilde{v}(\varphi, t) \leq \psi_2(\|\varphi\|)$. From Theorem 9.8.1 we conclude that the equilibrium $\varphi_e = 0$ of (F) is *uniformly stable.*

(b) We first note that from part (a) and by (9.8.48) it follows that for a given $\varepsilon_0 > 0$ there exists a $\delta_0 > 0$ such that $\sup_{\theta \in [-r,0]} |x(t+\theta)| < \varepsilon_0$ for all $t \geq t_0 - r$ and for any solution $x(t)$ of (F) whenever $\sup_{\theta \in [-r,0]} |x(t_0+\theta)| < \delta_0$. To prove the uniform attractivity of the equilibrium $\varphi_e = 0$ of (F) we need to show that for every $\eta > 0$ there exists a $T = T(\eta, \delta_0) > 0$ such that $|x(t)| \leq \eta$ for all $t \geq t_0 + T$ whenever $\sup_{\theta \in [-r,0]} |x(t_0+\theta)| < \delta_0$.

Without loss of generality, assume that η is sufficiently small so that $\psi_1(\eta) < \psi_2(\delta_0)$. Then there exists an $a > 0$ such that $f(s) - s > a$ for all $s \in [\psi_1(\eta), \psi_2(\delta_0)]$. Also, there exists a positive integer N such that $\psi_1(\eta) + Na \geq \psi_2(\delta_0)$. For every fixed $\eta > 0$ and every fixed solution $x(t)$ of (F), define

$$F_1 = \{t \in [t_0, \infty) \colon v(x(t), t) > \psi_1(\eta) + (N-1)a\}.$$

Then for every $t \in F_1$, it follows that

$$\begin{aligned} f(v(x(t), t)) &> v(x(t), t) + a \\ &> \psi_1(\eta) + Na \\ &\geq \psi_2(\delta) \\ &\geq \psi_2(|x(t+\theta)|) \\ &\geq v(x(t+\theta), t+\theta). \end{aligned}$$

It follows from (9.8.50) that

$$v'_{(F)}(x(t),t) \leq -\psi_3(|x(t)|) < 0 \tag{9.8.51}$$

for all $t \in F_1$.

We next show that F_1 is bounded. Suppose that $F_1 \neq \emptyset$. Let $t_m = \inf\{t \in F_1\}$. It must be true that $v(x(t_m), t_m) \geq \psi_1(\eta) + (N-1)a$. If $t_m > t_0$, then (9.8.51) holds for $t = t_m$, which implies that $v(x(t_m - \Delta t), t_m - \Delta t) > v(x(t_m), t_m)$ for $\Delta t > 0$ sufficiently small. Therefore, $t_m - \Delta t \in F_1$. This contradicts the definition of t_m. Therefore, $t_m = t_0$ and furthermore, $t_0 < t_1 \in F_1$ implies that $[t_0, t_1] \subset F_1$. For any $t \in F_1$, we have

$$|x(t)| \geq \big(\psi_2^{-1} \circ v\big)(x(t),t) \geq \psi_2^{-1}\big(\psi_1(\eta) + (N-1)a\big).$$

Hence, for any $t_0 < t_1 \in F_1$,

$$v(x(t_1),t_1) \leq v(x(t_0),t_0) - \big(\psi_3 \circ \psi_2^{-1}\big)\big(\psi_1(\eta) + (N-1)a\big)(t_1 - t_0).$$

From this we conclude that F_1 is bounded, for otherwise for sufficiently large t_1, $v(x(t_1), t_1)$ will become negative, which contradicts the fact that v is positive definite.

For F_1 bounded there exists a $T_1 \geq t_0$ such that $v(x(t),t) \leq \psi_1(\eta) + (N-1)a$ for all $t \geq T_1$.

If $N > 1$, let

$$F_2 = \{t \in [T_1, \infty)\colon v(x(t),t) > \psi_1(\eta) + (N-2)a\}.$$

In a similar manner as for F_1, we can show that F_2 is bounded. Inductively, define F_3 if $N > 2, \dots, F_N$. Then F_N is bounded. Therefore, there exists a $T_N \geq \cdots \geq T_1 \geq t_0$ such that $\psi_1(|x(t)|) \leq v(x(t),t) \leq \psi_1(\eta)$ (i.e., $|x(t)| \leq \eta$ for all $t \geq T_N$). The proof is completed. □

We conclude the present section with a specific example that demonstrates the applicability of Theorem 9.8.8.

Example 9.8.7 We consider the scalar retarded functional differential equation

$$\dot{x}(t) = -a(t)x(t) - \sum_{j=1}^{n} b_j(t)x(t - r_j(t)) \tag{9.8.52}$$

where a, b_j, r_j, $j = 1, \dots, n$, are continuous functions on $\mathbb{R}^+$ that satisfy $a(t) \geq \delta$ for some $\delta > 0$ and $\sum_{j=1}^{n} |b_j(t)| < k\delta$, $0 < k < 1$, and $0 \leq r_j(t) \leq r$, $j = 1, \dots, n$, for all $t \in \mathbb{R}^+$.

We choose as a Lyapunov function

$$v(x) = \frac{1}{2}x^2.$$

Along the solutions of (9.8.52) we have

$$v'_{(9.8.52)}(x(t)) = -a(t)x(t)^2 - \sum_{j=1}^{n} b_j(t)x(t)x(t-r_j(t)).$$

Assume that $x(\theta)^2 < qx(t)^2$, $t - r \le \theta \le t$ and choose $f(s) = qs$, $q = 1/k > 1$. Then

$$v'_{(9.8.52)}(x(t)) \le -a(t)x(t)^2 + \sum_{j=1}^{n} |b_j(t)|qx(t)^2 \le \big(-\delta + k\delta\big)x(t)^2.$$

Therefore, the equilibrium $x(t) \equiv 0$ of system (9.8.52) is uniformly asymptotically stable in the large. □

9.9 Applications to a Class of Artificial Neural Networks with Time Delays

An important class of artificial recurrent neural networks, *Cohen–Grossberg neural networks*, is described by the set of ordinary differential equations,

$$\dot{x}_i(t) = -a_i(x_i(t))\Big[b_i(x_i(t)) - \sum_{j=1}^{n} t_{ij}s_j(x_j(t))\Big], \tag{9.9.1}$$

$i = 1, \dots, n$, where x_i denotes the state variable associated with the ith neuron, the function $a_i(\cdot)$ represents an amplification function, and $b_i(\cdot)$ is an arbitrary function; however, we require that $b_i(\cdot)$ be sufficiently well behaved to keep the solutions of (9.9.1) bounded. The matrix $T = [t_{ij}] \in \mathbb{R}^{n\times n}$ represents the *neuron interconnections* and the real function $s_i(\cdot)$ is a sigmoidal nonlinearity (specified later), representing the ith *neuron*. Letting $x^T = (x_1, \dots, x_n)$, $A(x) = \text{diag}[a_1(x), \dots, a_n(x)]$, $B(x) = [b_1(x_1), \dots, b_n(x_n)]^T$, and $S(x) = [s_1(x_1), \dots, s_n(x_n)]^T$, (9.9.1) can be rewritten as

$$\dot{x}(t) = -A(x(t))[B(x(t)) - TS(x(t))]. \tag{9.9.2}$$

If $T = T^T$, then (9.9.2) constitutes the *Cohen–Grossberg neural network model.*

Frequently, multiple time delays are incurred in such networks, either intentionally or unavoidably. Such networks are described by differential-difference equations of the form

$$\dot{x}_i(t) = -a_i(x_i(t))\Big[b_i(x_i(t)) - \sum_{j=1}^{n} t_{ij}^{(0)}s_j(x_j(t)) - \sum_{k=1}^{K}\sum_{j=1}^{n} t_{ij}^{(k)}s_j(x_j(t-\tau_k))\Big], \tag{9.9.3}$$

$i = 1, \dots, n$, where $t_{ij}^{(k)}$, $i, j = 1, \dots, n$, denote the interconnections that are associated with time delay τ_k, $k = 0, 1, \dots, K$. We assume without loss of generality that

$0 = \tau_0 < \tau_1 < \cdots < \tau_K$. The symbols x_i, $a_i(\cdot)$, $b_i(\cdot)$, and $s_i(\cdot)$, are the same as in (9.9.1). System (9.9.3) can now be expressed as

$$\dot{x}(t) = -A(x(t))\left[B(x(t)) - T_0 S(x(t)) - \sum_{k=1}^{K} T_k S(x(t-\tau_k))\right] \tag{9.9.4}$$

where x, $A(\cdot)$, $B(\cdot)$, and $S(\cdot)$ are defined similarly as in (9.9.2) and where T_k makes up the interconnections associated with delay τ_k, $k = 0, 1, \ldots, K$, so that $T = T_0 + T_1 + \cdots + T_K$.

Throughout this section we assume that the Cohen–Grossberg neural networks without delay, given by (9.9.1), and with delays, given by (9.9.3), satisfy the following assumptions.

Assumption 9.9.1

(i) The function $a_i(\cdot)$ is continuous, positive, and bounded.

(ii) The function $b_i(\cdot)$ is continuous.

(iii) $T = [t_{ij}]$ is symmetric; that is, $T = T^T$.

(iv) $s_j \in C^1[\mathbb{R}, \mathbb{R}]$ is a *sigmoidal function*; that is, $s_j(0) = 0$,

$$s_j'(x_j) \triangleq \frac{ds_j}{dx_j}(x_j) > 0,$$

$\lim_{x_j \to \infty} s_j(x_j) = 1$, $\lim_{x_j \to -\infty} s_j(x_j) = -1$, and $\lim_{|x_j| \to \infty} s_j'(x_j) = 0$.

(v) $\lim_{x_i \to \infty} b_i(x_i) = \infty$ and $\lim_{x_i \to -\infty} b_i(x_i) = -\infty$. □

Lemma 9.9.1 If Assumption 9.9.1 is satisfied, then the solutions of systems (9.9.1) and (9.9.3) are bounded.

Proof. Because system (9.9.1) may be viewed as a special case of system (9.9.3), we consider in our proof only system (9.9.3).

We know from Assumption 9.9.1 that the terms $s_j(x_j(t))$ and $s_j(x_j(t-\tau_k))$ are bounded for all $j = 1, \ldots, n$. Furthermore, because $\lim_{x_i \to \infty} b_i(x_i) = \infty$ and $\lim_{x_i \to -\infty} b_i(x_i) = -\infty$, there must exist an $M > 0$ such that

$$b_i(x_i(t)) - \sum_{j=1}^{n} t_{ij}^{(0)} s_j(x_j(t)) - \sum_{k=1}^{K}\sum_{j=1}^{n} t_{ij}^{(k)} s_j(x_j(t-\tau_k)) > 0$$

whenever $x_i(t) \geq M$ and

$$b_i(x_i(t)) - \sum_{j=1}^{n} t_{ij}^{(0)} s_j(x_j(t)) - \sum_{k=1}^{K}\sum_{j=1}^{n} t_{ij}^{(k)} s_j(x_j(t-\tau_k)) < 0$$

whenever $x_i(t) \leq -M$ for all $i = 1, \ldots, n$. Because $a_i(x_i(t))$ is positive by Assumption 9.9.1, we can conclude that for any solution $x(t)$ of (9.9.3), $\dot{x}_i(t) < 0$ whenever $x_i(t) \geq M$ and $\dot{x}_i(t) > 0$ whenever $x_i(t) \leq -M$ for all $i = 1, \ldots, n$. We may assume that for the initial condition $x_{t_0} \in C_{\tau_K}$, $\|x_{t_0}\| < M$. If this is not the case, we just pick a larger M. Therefore, we can conclude that $|x_i(t)| < M$ for all $t \geq 0$ and all $i = 1, \ldots, n$. □

If every nonequilibrium solution of (9.9.1) (and of (9.9.3)) converges to an equilibrium, then system (9.9.1) (and system (9.9.3)) is said to be *globally stable*. In order to ensure that the Cohen–Grossberg neural networks (9.9.1) and (9.9.3) are globally stable, we require that the sets of equilibria for these systems are discrete sets. It turns out that the next assumption ensures this automatically.

Assumption 9.9.2 For any equilibrium x_e of system (9.9.2), the matrix $J(x_e)$ is nonsingular, where

$$J(x) = -T + \text{diag}\left[\frac{b_1'(x_1)}{s_1'(x_1)}, \ldots, \frac{b_n'(x_n)}{s_n'(x_n)}\right]$$

and $b_i'(x_i) = (db_i/dx_i)(x_i)$, $i = 1, \ldots, n$. □

Using Sard's Theorem [1], it can be shown that for almost all $T \in \mathbb{R}^{n\times n}$ (except a set with Lebesgue measure zero), system (9.9.2) satisfies Assumption 9.9.2. Furthermore, by making use of the implicit function theorem (refer to Subsection 7.6A), it can be shown that the set of all equilibria of system (9.9.2) is a discrete set. Because the set of equilibria of system (9.9.2), $\{x_e(0)\} \subset \mathbb{R}^n$, and the set of vectors $\{\varphi_e(0)\} \subset \mathbb{R}^n$, determined by the set of equilibria $\{\varphi_k\} \subset C_{\tau_K}$ of system (9.9.4) are identical, we have the following result.

Lemma 9.9.2 If system (9.9.4) satisfies Assumption 9.9.2, then the set of equilibria of system (9.9.4) is a discrete set (i.e., with $T = T_0 + \sum_{k=1}^{K} T_k$, the set of points x_e such that $B(x_e) - TS(x_e) = 0$ is discrete, where $T = T^T$). Furthermore, system (9.9.4) satisfies Assumption 9.9.2 for all $T = T^T \in \mathbb{R}^{n\times n}$ except on a set of Lebesgue measure zero. □

For a proof of Lemma 9.9.2, the reader should refer to [21].

A. A global result

We are now in a position to prove the following result.

Theorem 9.9.1 [42] Suppose that for system (9.9.3) Assumptions 9.9.1 and 9.9.2 are satisfied and that

$$\sum_{k=1}^{K} \left(\tau_k \beta \|T_k\|\right) < 1 \tag{9.9.5}$$

where $\beta = \max_{x\in\mathbb{R}^n} \|A(x)S'(x)\|$ where $S'(x) = \text{diag}[s_1'(x_1), \ldots, s_n'(x_n)]$. Then system (9.9.3) is *globally stable*.

Proof. Because inequality (9.9.5) is satisfied, there must exist a sequence of positive numbers $(\alpha_1, \ldots, \alpha_K)$, such that

$$\sum_{k=1}^{K} \alpha_k = 1, \qquad \tau_k\beta\|T_k\| < \alpha_k \quad \text{for } k = 1, \ldots, K. \tag{9.9.6}$$

To prove the present result, we define for any $x_t \in C[[-\tau_K, 0], \mathbb{R}^n]$ an "*energy functional*" $E(x_t)$ associated with (9.9.3) by

$$E(x_t) = -S^T(x_t(0))TS(x_t(0)) + 2\sum_{i=1}^{n}\int_0^{[x_t(0)]_i} b_i(\sigma)s_i'(\sigma)d\sigma$$
$$+\sum_{k=1}^{K}\frac{1}{\alpha_k}\int_{-\tau_k}^{0}[S(x_t(\theta))-S(x_t(0))]^T T_k^T f_k(\theta)T_k[S(x_t(\theta))-S(x_t(0))]d\theta \tag{9.9.7}$$

where $(\alpha_1, \ldots, \alpha_K)$ is a sequence of positive numbers such that condition (9.9.6) is satisfied and $f_k(\theta) \in C^1[[-\tau_k, 0], \mathbb{R}^n]$, $k = 1, \ldots, K$, is specified later. After changing integration variables, (9.9.7) can be written as

$$E(x_t) = -S^T(x(t))TS(x(t)) + 2\sum_{i=1}^{n}\int_0^{x_i(t)} b_i(\sigma)s_i'(\sigma)d\sigma$$
$$+\sum_{k=1}^{K}\frac{1}{\alpha_k}\int_{t-\tau_k}^{t}[S(x(w))-S(x(t))]^T T_k^T f_k(w-t)T_k[S(x(w))-S(x(t))]dw. \tag{9.9.8}$$

The derivative of $E(x_t)$ with respect to t along any solution of (9.9.3) is computed as

$$E'_{(9.9.3)}(x_t)$$
$$= -2S^T(x(t))TS'(x(t))A(x(t))\Big[-B(x(t))+T_0S(x(t))+\sum_{k=1}^{K}T_kS(x(t-\tau_k))\Big]$$
$$+ 2x^T(t)B(x(t))S'(x(t))A(x(t))\Big[-B(x(t))+T_0S(x(t))+\sum_{k=1}^{K}T_kS(x(t-\tau_k))\Big]$$
$$-\sum_{k=1}^{K}\frac{1}{\alpha_k}\Big\{[S(x(t-\tau_k)) - S(x(t))]^T T_k^T f_k(-\tau_k)T_k[S(x(t-\tau_k)) - S(x(t))]$$
$$+\int_{t-\tau_k}^{t}[S(x(w)) - S(x(t))]^T T_k^T f_k'(w-t)T_k[S(x(w)) - S(x(t))]dw$$
$$+\int_{t-\tau_k}^{t}\Big[-B(x(t)) + T_0S(x(t)) + \sum_{k=1}^{K}T_kS(x(t-\tau_k))\Big]^T$$
$$\times A(x(t))S'(x(t))T_k^T f_k(w-t)T_k[S(x(w)) - S(x(t))]dw$$
$$+\int_{t-\tau_k}^{t}[S(x(w)) - S(x(t))]^T T_k^T f_k(w-t)T_kS'(x(t))A(x(t))$$
$$\times\Big[-B(x(t)) + T_0S(x(t)) + \sum_{k=1}^{K}T_kS(x(t-\tau_k))\Big]dw\Big\} \tag{9.9.9}$$

where $f'(\theta) = (df/d\theta)(\theta)$. If we adopt the notation

$$H_0 = -B(x(t)) + T_0 S(x(t)) + \sum_{k=1}^{K} T_k S(x(t-\tau_k)), \tag{9.9.10}$$

$$H_k = T_k[S(x(t-\tau_k)) - S(x(t))], \qquad k = 1, \dots, K, \tag{9.9.11}$$

$$G_k = T_k[S(x(w)) - S(x(t))], \qquad k = 1, \dots, K, \tag{9.9.12}$$

$$Q = A(x(t))S'(x(t)) = S'(x(t))A(x(t)), \tag{9.9.13}$$

(9.9.9) can be rewritten as

$$\begin{aligned}
&E'_{(9.9.3)}(x_t)\\
&= -2S^T(x(t))TQH_0 + 2x(t)^T B(x(t))QH_0 - \sum_{k=1}^{K} \frac{1}{\alpha_k}\Big\{H_k^T f_k(-\tau_k)H_k\\
&\quad + \int_{t-\tau_k}^{t} [G_k^T f_k'(w-t)G_k + H_0^T Q T_k^T f_k(w-t)G_k + G_k^T f_k(w-t)T_k Q H_0]dw\Big\}\\
&= -2H_0^T Q H_0 + 2\sum_{k=1}^{K} H_k^T Q H_0 - \sum_{k=1}^{K} \frac{1}{\alpha_k}\Big\{H_k^T f_k(-\tau_k)H_k\\
&\quad + \int_{t-\tau_k}^{t} [G_k^T f_k'(w-t)G_k + H_0^T Q T_k^T f_k(w-t)G_k + G_k^T f_k(w-t)T_k Q H_0]dw\Big\}
\end{aligned} \tag{9.9.14}$$

$$\begin{aligned}
&= \sum_{k=1}^{K} \Big[2H_k^T Q H_0 - \frac{1}{\alpha_k}\Big\{2H_0^T Q H_0 + H_k^T f_k(-\tau_k)H_k\\
&\quad + \int_{t-\tau_k}^{t} [G_k^T f_k'(w-t)G_k + H_0^T Q T_k^T f_k(w-t)G_k + G_k^T f_k(w-t)T_k Q H_0]dw\Big\}\Big]\\
&= -\sum_{k=1}^{K} \int_{-\tau_k}^{0} [\eta_k(x_t,\theta)]^T M_k(x_t,\theta)\eta_k(x_t,\theta)d\theta
\end{aligned}$$

where $[\eta_k(x_t,\theta)]^T = [H_0^T, H_k^T, \widetilde{G}_k^T]^T$ with H_0 and H_k given by (9.9.10) and (9.9.11),

$$\widetilde{G}_k = T_k[S(x(t+\theta)) - S(x(t))], \qquad k = 1, \dots, K, \tag{9.9.15}$$

$$M_k(x_t,\theta) = \begin{bmatrix} 2\alpha_k Q/\tau_k & -Q/\tau_k & Q T_k^T f_k(\theta)/\alpha_k \\ -Q/\tau_k & f_k(-\tau_k)I/(\tau_k\alpha_k) & 0 \\ f_k(\theta)T_k Q/\alpha_k & 0 & f_k'(\theta)I/\alpha_k \end{bmatrix} \tag{9.9.16}$$

and I denotes the $n \times n$ identity matrix. To obtain the last expression of (9.9.14), we changed the integration variables from w to θ.

We now show that if the hypotheses of Theorem 9.9.1 are satisfied, then $M_k(x_t, \theta)$ is positive definite for all $\theta \in [-\tau_k, 0]$ and all x_t that satisfy (9.9.3), for $k = 1, \ldots, K$. In doing so, we let $U = U_3 U_2 U_1$, where

$$U_1 = \begin{bmatrix} I/\sqrt{\alpha_k} & 0 & 0 \\ I/(2\sqrt{\alpha_k}) & \sqrt{\alpha_k} I & 0 \\ 0 & 0 & \sqrt{\alpha_k} I \end{bmatrix}$$

$$U_2 = \begin{bmatrix} I & 0 & 0 \\ 0 & I & 0 \\ -\tau_k f_k(\theta) T_k/(2\alpha_k) & 0 & I \end{bmatrix}$$

and

$$U_3 = \begin{bmatrix} I & 0 & 0 \\ 0 & I & 0 \\ 0 & f_k(\theta) T_k Q U_4/\alpha_4 & I \end{bmatrix}$$

where

$$U_4 = -\frac{1}{2}\left[\frac{f_k(-\tau_k)}{\tau_k} I - \frac{Q}{2\tau_k}\right]^{-1}.$$

It is not difficult to verify that $\widetilde{M}_k = U M_k(x_t, \theta) U^T$ is a diagonal matrix. In fact

$$\widetilde{M}_k = \operatorname{diag}[M_{k,1},\ M_{k,2},\ M_{k,3}] \tag{9.9.17}$$

where

$$M_{k,1} = \frac{2Q}{\tau_k} \tag{9.9.18}$$

$$M_{k,2} = \frac{f_k(-\tau_k)}{\tau_k} I - \frac{Q}{2\tau_k} \tag{9.9.19}$$

and

$$M_{k,3} = f_k'(\theta) I - \frac{f_k(\theta) T_k Q}{2\alpha_k}\left[\left(\frac{f_k(-\tau_k)}{\tau_k} I - \frac{Q}{2\tau_k}\right)^{-1} + 2\tau_k Q^{-1}\right]\frac{Q T_k^T f_k(\theta)}{2\alpha_k}. \tag{9.9.20}$$

It follows that $M_k(x_t, \theta)$ is positive definite if and only if $\widetilde{M}_k$ is positive definite and if and only if $M_{k,1}$, $M_{k,2}$, and $M_{k,3}$ are all positive definite.

We now show that if the condition $\tau_k \beta \|T_k\| < \alpha_k$ is satisfied, where

$$\beta = \max_{x \in \mathbb{R}} \|A(x) S'(x)\| = \max_{x \in \mathbb{R}} \|Q\|$$

then we can always find a suitable $f_k(\theta) \in C^1[[-\tau_k, 0], \mathbb{R}^+]$ such that $M_{k,1}$, $M_{k,2}$, and $M_{k,3}$ are positive definite for all x_t that satisfy (9.9.3) and for all $\theta \in [-\tau_k, 0]$. From this it follows that $M_k(x_t, \theta)$ is positive definite for all $k = 1, \ldots, K$ and therefore $E'_{(9.9.3)}(x_t) \leq 0$ along any solution x_t of (9.9.3).

By the assumption that $s_i'(x_i) > 0$ and $a_i(x_i) > 0$ for all $x_i \in \mathbb{R}$, the matrix $M_{k,1}$ is automatically positive definite. The matrix $M_{k,2}$ is always positive definite if the condition

$$2f_k(-\tau_k) - \beta > 0 \tag{9.9.21}$$

is satisfied. For $M_{k,3}$, it is easily shown that if

$$f_k'(\theta) > \frac{1}{4} f_k(\theta)^2 \frac{\|T_k\|^2}{\alpha_k^2} \left\| Q \left[\left(\frac{f_k(-\tau_k)}{\tau_k} I - \frac{Q}{2\tau_k} \right)^{-1} + 2\tau_k Q^{-1} \right] Q \right\| \tag{9.9.22}$$

is true, then $M_{k,3}$ is also positive definite. Notice that the matrix

$$D \triangleq Q \left[\left(\frac{f_k(-\tau_k)}{\tau_k} I - \frac{Q}{2\tau_k} \right)^{-1} + 2\tau_k Q^{-1} \right] Q$$

is a diagonal matrix; that is, $D = \text{diag}[d_1, \ldots, d_n]$. If we denote $Q = \text{diag}[q_1, \ldots, q_n]$, then it is easy to show that

$$d_i = \frac{4f_k(-\tau_k)q_i\tau_k}{2f_k(-\tau_k) - q_i} \qquad \text{for } i = 1, \ldots, n.$$

Because $q_i < \beta$ by the definitions of β and Q, we have, in view of (9.9.21), that

$$d_i < \frac{4f_k(-\tau_k)\beta\tau_k}{2f_k(-\tau_k) - \beta}.$$

Therefore, we obtain

$$\|D\| \leq \frac{4f_k(-\tau_k)\beta\tau_k}{2f_k(-\tau_k) - \beta}$$

and, furthermore, condition (9.9.22) is satisfied if (9.9.21) is satisfied and

$$f_k'(\theta) > \frac{1}{4} f_k(\theta)^2 \frac{\|T_k\|^2}{\alpha_k^2} \frac{4f_k(-\tau_k)\beta\tau_k}{2f_k(-\tau_k) - \beta} \tag{9.9.23}$$

is satisfied.

Next, we need to show that there is an $f_k \in C^1[[-\tau_k, 0], \mathbb{R}]$ such that conditions (9.9.21) and (9.9.23) are satisfied. We choose

$$f_k(-\tau_k) = \left[\beta\tau_k^2 \frac{\|T_k\|^2}{\alpha_k^2} \right]^{-1}. \tag{9.9.24}$$

Condition (9.9.21) is satisfied by the choice (9.9.24). Furthermore,

$$\left[f_k(-\tau_k) \frac{\|T_k\|}{\alpha_k} - \frac{\alpha_k}{\beta\tau_k\|T_k\|} \right]^2 + 1 - \frac{\alpha_k^2}{\beta^2\tau_k^2\|T_k\|^2} = 1 - \frac{\alpha_k^2}{\beta^2\tau_k^2\|T_k\|^2} < 0 \tag{9.9.25}$$

is true because $\beta\tau_k\|T_k\| < \alpha_k$. It follows from (9.9.25) that

$$\delta f_k(-\tau_k)\tau_k < 1 \tag{9.9.26}$$

where

$$\delta = \frac{\|T_k\|^2 f_k(-\tau_k)\beta\tau_k}{\alpha_k^2[2f_k(-\tau_k) - \beta]}. \tag{9.9.27}$$

Because $\delta f_k(-\tau_k)\tau_k < 1$, we can always find an l such that $0 < l < 1$, and $\delta f_k(-\tau_k)\tau_k < l$. Therefore, we always have $\gamma > 0$ where γ is given by

$$\gamma = \frac{l}{\delta f_k(-\tau_k)} - \tau_k. \tag{9.9.28}$$

We now choose $f_k(\theta)$ on $[-\tau_k, 0]$ as

$$f_k(\theta) = \frac{l}{\delta(\gamma - \theta)}. \tag{9.9.29}$$

It is easily verified that this choice is consistent with condition (9.9.24). Clearly, $f_k \in C^1[[-\tau_k, 0], \mathbb{R}^+]$ because $\gamma > 0$. The derivative of $f_k(\theta)$ is given by

$$f_k'(\theta) = \frac{l}{\delta(\gamma - \theta)^2} = \frac{\delta}{l} f_k(\theta)^2 > \delta f_k(\theta)^2 \tag{9.9.30}$$

because $l < 1$. Combining (9.9.27) and (9.9.30), we can verify that $f_k(\theta)$ satisfies condition (9.9.23).

Therefore, we have shown that if $\beta\tau_k\|T_k\| < \alpha_k$, then there exists an $f_k(\theta)$ (given by (9.9.29), where $f_k(-\tau_k)$, δ, and γ are given by (9.9.24), (9.9.27), and (9.9.28), respectively) such that conditions (9.9.21) and (9.9.23) are satisfied. Thus $M_k(x_t, \theta)$ is positive definite for all x_t satisfying (9.9.3) and all $\theta \in [-\tau_k, 0]$ for $k = 1, \ldots, K$. We have shown that

$$E'_{(9.9.3)}(x_t) \leq 0 \tag{9.9.31}$$

along any solution x_t of (9.9.3), where $E(x_t)$ is the "energy functional" given by (9.9.7).

We know from (9.9.14) that if $E'_{(9.9.3)}(x_t) = 0$, then $H_0 = 0$, $H_k = 0$, and $\widetilde{G}_k = 0$ for $k = 1, \ldots, K$, where H_0, H_k, and $\widetilde{G}_k$ are given by (9.9.10), (9.9.11), and (9.9.15), respectively. For any $\varphi \in C[[-\tau_k, 0], \mathbb{R}^n]$, we denote $\dot{E}_\varphi = 0$ if

$$-B(\varphi(0)) + T_0 S(\varphi(0)) + \sum_{k=1}^{K} T_k S(\varphi(-\tau_k)) = 0 \tag{9.9.32}$$

$$T_k[S(\varphi(-\tau_k)) - S(\varphi(0))] = 0, \qquad k = 1, \ldots, K \tag{9.9.33}$$

$$T_k[S(\varphi(-\theta)) - S(\varphi(0))] = 0 \quad \text{for all } \theta \in [-\tau_K, 0], \qquad k = 1, \ldots, K. \tag{9.9.34}$$

It is obvious that for any solution x_t of (9.9.3), $E'_{(9.9.3)}(x_t) = 0$ if and only if $\dot{E}_{x_t} = 0$.

Because for any x_t satisfying (9.9.3), x_t is bounded (Lemma 9.9.1) and because

$$E'_{(9.9.3)}(x_t) \leq 0,$$

it follows from the invariance theory (see Theorem 9.8.6) that the limit set of x_t as $t \to \infty$ is the invariant subset of the set $\Lambda = \{\varphi \in C[[-\tau_K, 0], \mathbb{R}^n]\colon\ \dot{E}_\varphi = 0\}$.

Therefore, we have $|x_t - \varphi| \to 0$ as $t \to \infty$ for some $\varphi \in \Lambda$. In particular, we have $x_t(0) \to \varphi(0)$ and $x_t(-\tau_k) \to \varphi(-\tau_k)$ as $t \to \infty$, $k = 1, \ldots, K$. Combining this with (9.9.32) and (9.9.33), we conclude that

$$-B(x_t(0)) + T_0 S(x_t(0)) + \sum_{k=1}^{K} T_k S(x_t(-\tau_k)) \to 0$$

and

$$T_k[S(x_t(\varphi(-\tau_k))) - S(x_t(0))] \to 0, \qquad k = 1, \ldots, K$$

as $t \to \infty$. It follows that

$$-B(x_t(0)) + TS(x_t(0)) \to 0,$$

or

$$-B(x(t)) + TS(x(t) \to 0,$$

as t approaches ∞. Now because x_t is bounded (Lemma 9.9.1), we conclude that any point in the limit set of $x(t)$ as $t \to \infty$ is an equilibrium of system (9.9.3) (or, equivalently, an equilibrium of system (9.9.1)). Furthermore, inasmuch as the set of equilibria of system (9.9.3) is a discrete set (Lemma 9.9.2), it follows that $x(t)$ approaches some equilibrium of system (9.9.3) as t tends to ∞. □

If $\tau_k = 0$ for $k = 1, \ldots, K$, then Theorem 9.9.1 reduces to a global stability result for Cohen–Grossberg neural networks without time delays: if for system (9.9.1) Assumptions 9.9.1 and 9.9.2 are satisfied, then system (9.9.1) is *globally stable*.

When the results given above apply, one can partition the state space, using the domains of attraction of the asymptotically stable equilibria of system (9.9.2) or (9.9.4). These partitions in turn determine equivalence relations that can be used as the basis for a variety of applications (e.g., in applications of associative memories to pattern recognition problems, classification of data, sorting problems, and the like). Algorithms have been established that provide estimates for the total number of equilibria and the total number of asymptotically stable equilibria (called stable memories). Also, algorithms have been developed that make it possible to place equilibria at desired locations and to minimize the number of undesired asymptotically stable equilibria (called spurious states). For additional material on these topics, the reader may wish to consult [24].

B. Local results

Good criteria that ensure the asymptotic stability of an equilibrium of system (9.9.3) are of great interest. We address this issue in the present subsection. By necessity, these results are local in nature.

We make use of the "*energy functional*" given in (9.9.7) which was used in the proof of Theorem 9.9.1. In the following, we require the following concept.

Definition 9.9.1 Let $\tau = \tau_K$. An element $\varphi \in C[[-\tau, 0], \mathbb{R}^n] = C_\tau$ is called a *local minimum* of the "energy functional" defined in (9.9.7) if there exists a $\delta > 0$ such that for any $\widetilde{\varphi} \in C_\tau$, $E(\varphi) \leq E(\widetilde{\varphi})$ whenever $\|\varphi - \widetilde{\varphi}\| < \delta$. □

Before stating and proving our next result, we recall that $x_e \in \mathbb{R}^n$ is an equilibrium of system (9.9.2) if

$$B(x_e) - TS(x_e) = 0.$$

Consistent with Lemma 9.9.2, $\varphi_{x_e} \in C_\tau$ is an equilibrium of system (9.9.4) if $\varphi_{x_e}(\theta) = x_e$, $-\tau \leq \theta \leq 0$, and

$$B(x_e) - TS(x_e) = 0,$$

where $T = T_0 + \sum_{k=1}^{K} T_k$.

Theorem 9.9.2 Suppose that the conditions of Theorem 9.9.1 are satisfied. If φ_{x_e} is an equilibrium of (9.9.4), then the following statements are equivalent.

(a) φ_{x_e} is a stable equilibrium of (9.9.4).

(b) φ_{x_e} is an asymptotically stable equilibrium of (9.9.4).

(c) φ_{x_e} is a local minimum of the "energy functional" E given by (9.9.7), where, as defined above, $\varphi_{x_e} \in C_\tau$ such that $\varphi_{x_e}(\theta) = x_e$, $-\tau \leq \theta \leq 0$.

(d) $J(x_e)$ is positive definite, where $J(x)$ is defined in Assumption 9.9.2.

Proof. (a) $\Longrightarrow$ (b). Because Assumption 9.9.2 is satisfied, the set of equilibria of system (9.9.4) is a discrete set by Lemma 9.9.2. Therefore, when $\varepsilon > 0$ is sufficiently small, there is no other equilibrium $\varphi_{x_e'}$ of (9.9.4) such that

$$x_e' \in U(x_e, \varepsilon) \triangleq \{x \in \mathbb{R}^n : |x - x_e| < \varepsilon\}. \tag{9.9.35}$$

Because φ_{x_e} is a stable equilibrium of (9.9.4), there exists an $\eta > 0$ such that for any $\varphi \in C_\tau$ satisfying $\|\varphi - \varphi_{x_e}\| < \eta$, $\|x_t - \varphi_{x_e}\| < \varepsilon$ for all $t > 0$, where x_t is the solution of (9.9.4) with initial condition φ. Thus $x_t \in C[[-\tau, 0], U(x_e, \varepsilon)]$ for all t. In view of Theorem 9.9.1 x_t will converge to some equilibrium of system (9.9.4). Because φ_{x_e} is the only equilibrium of (9.9.4) with $x_t \in C[[-\tau, 0], U(x_e, \varepsilon)]$, it follows that x_t converges to φ_{x_e}. Thus we have shown that φ_{x_e} is an attractive equilibrium of system (9.9.4). Therefore the stable equilibrium φ_{x_e} of (9.9.4) is an asymptotically stable equilibrium of system (9.9.4).

(b) $\Longrightarrow$ (c). Because φ_{x_e} is an asymptotically stable equilibrium of system (9.9.4), there exists an $\eta > 0$ such that for any $\varphi \in C_\tau$ satisfying $\|\varphi - \varphi_{x_e}\| < \eta$, x_t converges to φ_{x_e}, where x_t is the solution of (9.9.4) with initial condition φ. Therefore $E(\varphi_{x_e}) \leq E(x_t) \leq E(\varphi)$ for any $\varphi \in C_\tau$ satisfying $\|\varphi - \varphi_{x_e}\| < \eta$. Therefore, φ_{x_e} is a local minimum of the energy functional E.

(c) $\Longrightarrow$ (d). Let $\widetilde{E}$ be a function from $\mathbb{R}^n$ to $\mathbb{R}$ defined by

$$\widetilde{E}(x) \triangleq -S(x)^T TS(x) + 2\sum_{i=1}^{n} \int_0^{x_i} b_i(\sigma) s_i'(\sigma) d\sigma. \tag{9.9.36}$$

Comparing E with $\widetilde{E}$, we note that $\widetilde{E}$ is a function defined on $\mathbb{R}^n$, and E is a functional defined on C_τ. Because φ_{x_e} is a local minimum of E, x_e must be a local minimum of $\widetilde{E}$. Otherwise there would exist a sequence $\{x_n\} \subset \mathbb{R}^n$ such that $x_n \to x_e$ as $n \to \infty$ and $\widetilde{E}(x_n) < \widetilde{E}(x_e)$. Let φ_{x_n} denote the constant function $\varphi_{x_n} \equiv x_n$ in C_τ. Then $|\varphi_{x_n} - \varphi_{x_e}| \to 0$ as $n \to \infty$ and

$$E(\varphi_{x_n}) = \widetilde{E}(x_n) < \widetilde{E}(x_e) = E(\varphi_{x_e}).$$

This contradicts the fact that φ_{x_e} is a local minimum of E. Therefore, x_e is a local minimum of $\widetilde{E}$. Hence $\widetilde{J}(x_e)$ is positive semidefinite (see [21]), where $\widetilde{J}(x)$ is the Hessian matrix of $\widetilde{E}$ given by

$$\widetilde{J}(x) = \left[\frac{\partial^2 \widetilde{E}}{\partial x_i \partial x_j} \right]. \tag{9.9.37}$$

It can be shown that

$$\widetilde{J}(x) = 2S'(x)J(x)S'(x) \tag{9.9.38}$$

where

$$S'(x) = \text{diag}[s_1'(x_1), \dots, s_n'(x_n)] \tag{9.9.39}$$

and $J(x)$ is given in Assumption 9.9.2. Therefore, $J(x_e)$ is also positive semidefinite. By Assumption 9.9.2, $J(x_e)$ is a nonsingular matrix. Thus we have shown that $J(x_e)$ is positive definite.

(d) $\Longrightarrow$ (a). We need to prove that φ_{x_e} is a stable equilibrium of system (9.9.4); that is, for any $\varepsilon > 0$, there exists a $\delta > 0$ such that for any $\varphi \in C_\tau$, if $\|\varphi - \varphi_{x_e}\| < \delta$, then $\|x_t - \varphi_{x_e}\| < \varepsilon$, where x_t is the solution of (9.9.4) with initial condition φ.

Because $J(x_e)$ is positive definite, then $\widetilde{J}(x_e)$ must also be positive definite where $\widetilde{J}(x)$ is the Hessian matrix of $\widetilde{E}$ given by (9.9.38). Furthermore,

$$\nabla_x \widetilde{E}(x) = 2[-TS(x) + B(x)]^T S'(x)$$

where $S'(x)$ is given in (9.9.39). Therefore, $\nabla_x \widetilde{E}(x_e) = 0$ because φ_{x_e} is an equilibrium of (9.9.4). It follows (see [21]) that x_e is a local minimum of $\widetilde{E}$; that is, there exists a $\delta_1 > 0$, $\delta_1 < \varepsilon$, such that whenever $0 < |x - x_e| \le \delta_1$, $\widetilde{E}(x_e) < \widetilde{E}(x)$. Let $r = \min\{\widetilde{E}(x) \colon |x - x_e| = \delta_1\}$. Then it is true that $r > \widetilde{E}(x_e)$. Because $E(\varphi_{x_e}) = \widetilde{E}(x_e)$, it follows that $r > E(\varphi_{x_e})$. Note that E is a continuous functional. Therefore, there exists a $\delta \in (0, \delta_1)$ such that whenever $\|\varphi - \varphi_{x_e}\| < \delta$, where $\varphi \in C_\tau$, we have $E(\varphi) < r$. Suppose x_t is any solution of (9.9.4) with the initial condition φ such that $\|\varphi - \varphi_{x_e}\| < \delta$. We show that $\|x_t - \varphi_{x_e}\| < \delta_1 < \varepsilon$. Otherwise there would exist a $t_0 > 0$ such that $|x_{t_0}(0) - x_e| = \delta_1$ (i.e., $|x(t_0) - x_e| = \delta_1$). By the definition of E and $\widetilde{E}$, we have $E(x_{t_0}) \ge \widetilde{E}(x(t_0)) \ge r$. Therefore, we obtain $E(x_{t_0}) > E(\varphi)$, which contradicts the fact that E is monotonically decreasing along any solution of (9.9.4). Thus we have shown that φ_{x_e} is an asymptotically stable equilibrium of system (9.9.4). □

We note that statement (d) in Theorem 9.9.2 is independent of the delays τ_k, $k = 1, \dots, K$. Therefore, if system (9.9.4) satisfies Assumptions 9.9.1 and 9.9.2 and if the condition $\sum_{k=1}^{K} \tau_k \beta \|T_k\| < 1$ is satisfied, then the locations of the (asymptotically) stable equilibria of system (9.9.4) will not depend on the delays τ_k for $k = 1, \dots, K$. This is true if, in particular, $\tau_k = 0$, $k = 1, \dots, K$. Therefore, if $\sum_{k=1}^{K} \tau_k \beta \|T_k\| < 1$, then systems (9.9.4) and (9.9.2) (obtained by letting $\tau_k = 0$ for $k = 1, \dots, K$ in (9.9.4)) will have identical (asymptotically) stable equilibria. We state this in the form of a corollary.

Corollary 9.9.1 Under the conditions of Theorem 9.9.1, φ_{x_e} is an asymptotically stable equilibrium of system (9.9.4) if and only if x_e is an asymptotically stable equilibrium of system (9.9.2). This is true if and only if $J(x_e)$ is positive definite, where $J(x)$ is given in Assumption 9.9.2. □

Corollary 9.9.1 provides an effective criterion for testing the (asymptotic) stability of any equilibrium of Cohen–Grossberg neural networks with multiple delays described by (9.9.4). This criterion constitutes necessary and sufficient conditions, as long as

$$\sum_{k=1}^{K} \tau_k \beta \|T_k\| < 1.$$

9.10 Discontinuous Dynamical Systems Determined by Differential Equations in Banach Spaces

In the present section we address infinite-dimensional discontinuous dynamical systems (infinite-dimensional DDS), $\{T, X, A, S\}$, where $T = \mathbb{R}^+$, X is a Banach space with norm $\|\cdot\|$, $X \supset A$, and the motions S are determined by the solutions $x(\cdot, t_0, x_0)$ of discontinuous differential equations defined on Banach spaces, specified later. As in Chapter 3, we assume that the set of times at which discontinuities may occur is unbounded and discrete and is of the form

$$E_x = \{\tau_1^x, \tau_2^x, \dots : \tau_1^x < \tau_2^x < \cdots\}.$$

The notation E_x signifies that different motions may possess different sets of times at which discontinuities may occur. Usually, the particular set E_x in question is clear from context and accordingly, we are able to suppress the x-notation and simply write

$$E = \{\tau_1, \tau_2, \dots : \tau_1 < \tau_2 < \cdots\}.$$

As in Subsection 2.12C, we sometimes find it useful to express the motions (solutions) of infinite-dimensional DDS by

$$x(t, \tau_0, x_0) = x^{(k)}(t, \tau_k, x_k), \qquad \tau_k \leq t < \tau_{k+1}, \tag{9.10.1}$$

$k \in \mathbb{N}$, where τ_0 and x_0 are given initial conditions.

The most general class of infinite-dimensional DDS that we consider in the present section is generated by differential equations of the form (refer to Subsection 2.12C),

$$\begin{cases} \dot{x} = F_k(t, x(t)), & \tau_k \le t < \tau_{k+1} \\ x(\tau_{k+1}) = g_k(x(\tau_{k+1}^-)), & k \in \mathbb{N} \end{cases} \tag{SG}$$

where for each $k \in \mathbb{N}$, $F_k \colon \mathbb{R}^+ \times X \to X$, $\dot{x} = dx/dt$, $g_k \colon X \to X$, and $x(t^-) = \lim_{t' \to t, t' < t} x(t')$. As in Subsection 2.12C, associated with (SG), we consider the family of *initial value problems*, given by

$$\begin{cases} \dot{x}(t) = F_k(t, x(t)), & t \ge \tau_k \\ x(\tau_k) = x_k, \end{cases} \tag{SG_k}$$

$k \in \mathbb{N}$. For each $k \in \mathbb{N}$, we assume that for every $(\tau_k, x_k) \in \mathbb{R}^+ \times X$, ($SG_k$) possesses a unique solution $x^{(k)}(t, \tau_k, x_k)$ that exists for all $t \in [\tau_k, \infty)$. We express this by saying that (SG_k) is *well posed.*

Under the above assumptions, it is clear that for every $(\tau_0, x_0) \in \mathbb{R}^+ \times X$, ($SG$) has a unique solution $x(t, \tau_0, x_0)$ that exists for all $t \in [\tau_0, \infty)$. This solution is made up of a sequence of solution segments $x^{(k)}(t, \tau_k, x_k)$ defined over the intervals $[\tau_k, \tau_{k+1})$, $k \in \mathbb{N}$, where $x_k = x(\tau_k)$, $k = 1, 2, \dots$ and where (τ_0, x_0) are given. At points $\{\tau_{k+1}\}$, $k \in \mathbb{N}$, the solutions of (SG) *may* have discontinuities (determined by $g_k(\cdot)$).

In addition to the above, we assume that for every $k \in \mathbb{N}$, $F_k(t, 0) = 0$ for all $t \in \mathbb{R}^+$ and $g_k(0) = 0$. This ensures the existence of the zero solution $x^{(k)}(t, \tau_k, x_k) = 0$, $t \ge \tau_k$, with $x_k = 0$, which means that $x_e = 0$ is an equilibrium of (SG_k), $k \in \mathbb{N}$. Furthermore, $x_e = 0$ is also an equilibrium for (SG).

A. Local stability results

We first address local results.

Theorem 9.10.1 Assume that there exist a function $v \colon X \times \mathbb{R}^+ \to \mathbb{R}^+$ and functions $\psi_1, \psi_2 \in \mathcal{K}$ defined on $\mathbb{R}^+$ such that

$$\psi_1(\|x\|) \le v(x, t) \le \psi_2(\|x\|) \tag{9.10.2}$$

for all $x \in X$ and $t \in \mathbb{R}^+$.

(a) Assume that for every solution $x(\cdot, \tau_0, x_0)$ of (SG), $v(x(t, \tau_0, x_0), t)$ is continuous everywhere on $\mathbb{R}^+_{\tau_0} = \{t \in \mathbb{R}^+ \colon t \ge \tau_0\}$ except on an unbounded and discrete subset $E = \{\tau_1, \tau_2, \dots \colon \tau_1 < \tau_2 < \cdots\}$ of $\mathbb{R}^+_{\tau_0}$. Also, assume that there exists a neighborhood $U \subset X$ of the origin $0 \in X$ such that $v(x(\tau_k, \tau_0, x_0), \tau_k)$ is nonincreasing for all $x_0 \in U$ and all $k \in \mathbb{N}$, and assume that there exists a function $f \in C[\mathbb{R}^+, \mathbb{R}^+]$, independent of $x(\cdot, \tau_0, x_0)$, such that $f(0) = 0$ and that

$$v(x(t, \tau_0, x_0), t) \le f(v(x(\tau_k, \tau_0, x_0), \tau_k)), \qquad t \in (\tau_k, \tau_{k+1}), \tag{9.10.3}$$

$k \in \mathbb{N}$.

Then the equilibrium $x_e = 0$ of (SG) is *uniformly stable.*

(b) If in addition to the above assumptions, there exists a function $\psi_3 \in \mathcal{K}$ defined on $\mathbb{R}^+$ such that

$$Dv(x(\tau_k, \tau_0, x_0), \tau_k) \leq -\psi_3(\|x(\tau_k, \tau_0, x_0)\|) \tag{9.10.4}$$

for all $x_0 \in U$, $k \in \mathbb{N}$, where

$$Dv(x(\tau_k, \tau_0, x_0), \tau_k) = \frac{1}{\tau_{k+1} - \tau_k}\big[v(x(\tau_{k+1}, \tau_0, x_0), \tau_{k+1}) - v(x(\tau_k, \tau_0, x_0), \tau_k)\big], \tag{9.10.5}$$

then the equilibrium $x_e = 0$ of (SG) is *uniformly asymptotically stable.*

Proof. Parts (a) and (b) of this theorem are a direct consequence of Theorems 3.2.1 and 3.2.2, respectively. □

Theorem 9.10.2 Assume that there exist a function $v\colon X \times \mathbb{R}^+ \to \mathbb{R}^+$ and four positive constants c_1, c_2, c_3, and b such that

$$c_1\|x\|^b \leq v(x,t) \leq c_2\|x\|^b \tag{9.10.6}$$

for all $x \in X$ and $t \in \mathbb{R}^+$.

Assume that there exists a neighborhood U of the origin $x_e = 0$ such that for all solutions $x(\cdot, \tau_0, x_0)$ of (SG) with $x_0 \in U$, $v(x(t, \tau_0, x_0), t)$ is continuous everywhere on $\mathbb{R}^+_{\tau_0}$ except on an unbounded and discrete subset $E = \{\tau_1, \tau_2, \ldots : \tau_1 < \tau_2 < \cdots\}$ of $\mathbb{R}^+_{\tau_0}$. Furthermore, assume that there exists a function $f \in C[\mathbb{R}^+, \mathbb{R}^+]$, independent of $x(\cdot, \tau_0, x_0)$, such that $f(0) = 0$ and that

$$v(x(t, \tau_0, x_0), t) \leq f(v(x(\tau_k, \tau_0, x_0), \tau_k)), \qquad t \in (\tau_k, \tau_{k+1}), \tag{9.10.7}$$

for all $k \in \mathbb{N}$, and that for some positive q, $f(\cdot)$ satisfies

$$f(r) = o(r^q) \qquad \text{as } r \to 0^+ \tag{9.10.8}$$

(i.e., $\lim_{r\to 0^+}[f(r)/r^q] = 0$). Also, assume that for all $k \in \mathbb{N}$,

$$Dv(x(\tau_k, \tau_0, x_0), \tau_k) \leq -c_3\|x(\tau_k, \tau_0, x_0)\|^b \tag{9.10.9}$$

for all solutions $x(\cdot, \tau_0, x_0)$ of (SG) with $x_0 \in U$, where Dv is defined in (9.10.5). Then the equilibrium $x_e = 0$ of (SG) is *exponentially stable.*

Proof. This result is a direct consequence of Theorem 3.2.3. □

B. Global results

Next, we address global stability and boundedness results.

Theorem 9.10.3 (a) Assume that in Theorem 9.10.1, $\psi_1, \psi_2 \in \mathcal{K}_\infty$ and that $U = X$. Then the equilibrium $x_e = 0$ of (SG) is *uniformly asymptotically stable in the large*.

(b) Assume that in Theorem 9.10.2, $U = X$. Then the equilibrium $x_e = 0$ of (SG) is *exponentially stable in the large*.

Proof. Parts (a) and (b) of this theorem are a direct consequence of Theorems 3.2.6 and 3.2.7, respectively. □

Theorem 9.10.4 Assume that there exist a function $v\colon X \times \mathbb{R}^+ \to \mathbb{R}^+$ and two strictly increasing functions $\psi_1, \psi_2 \in C[\mathbb{R}^+, \mathbb{R}^+]$ with $\lim_{r\to\infty} \psi_i(r) = \infty$, $i=1,2$, such that

$$\psi_1(\|x\|) \le v(x,t) \le \psi_2(\|x\|) \tag{9.10.10}$$

for all $x \in X$ and all $t \in \mathbb{R}^+$ whenever $\|x\| \ge \Omega$, where Ω is a positive constant.

Assume that for all solutions $x(\cdot, \tau_0, x_0)$ of (SG), $v(x(t,\tau_0,x_0),t)$ is continuous everywhere on $\mathbb{R}^+_{\tau_0}$ except on an unbounded subset $E = \{\tau_1, \tau_2, \ldots : \tau_1 < \tau_2 < \cdots\}$ of $\mathbb{R}^+_{\tau_0}$. Also, assume that for every solution $x(\cdot, \tau_0, x_0)$ of (SG),

$$v(x(\tau_{k+1},\tau_0,x_0),\tau_{k+1}) \le v(x(\tau_k,\tau_0,x_0),\tau_k) \tag{9.10.11}$$

for all τ_k, whenever $\|x(\tau_k,\tau_0,x_0)\| \ge \Omega$.

Furthermore, assume that there exists a function $f \in C[\mathbb{R}^+, \mathbb{R}^+]$, independent of $x(\cdot,\tau_0,x_0)$, such that for all $k \in \mathbb{N}$ and all $x(\cdot,\tau_0,x_0)$

$$v(x(t,\tau_0,x_0),t) \le f(v(x(\tau_k,\tau_0,x_0),\tau_k)), \qquad t \in (\tau_k, \tau_{k+1}), \tag{9.10.12}$$

whenever $\|x(t,\tau_0,x_0)\| \ge \Omega$.

Moreover, assume that there exists a positive constant Γ such that

$$\|x(\tau_{k+1},\tau_0,x_0)\| \le \Gamma$$

whenever $\|x(\tau_k,\tau_0,x_0)\| \le \Omega$ for all solutions $x(\cdot,\tau_0,x_0)$ of (SG).

Then the solutions of (SG) are *uniformly bounded*.

Proof. This result is a direct consequence of Theorem 3.2.4. □

Theorem 9.10.5 If in addition to the assumptions of Theorem 9.10.4 there exists a function $\psi_3 \in \mathcal{K}$ defined on $\mathbb{R}^+$ such that for all solutions $x(\cdot,\tau_0,x_0)$ of (SG),

$$Dv(x(\tau_k,\tau_0,x_0),\tau_k) \le -\psi_3(\|x(\tau_k,\tau_0,x_0)\|) \tag{9.10.13}$$

for all τ_k, whenever $\|x(\tau_k,\tau_0,x_0)\| \ge \Omega$, where Dv is defined in (9.10.5), then the solutions of (SG) are *uniformly ultimately bounded*.

Proof. This result is a direct consequence of Theorem 3.2.5. □

C. Instability results

Next, we address instability results.

Theorem 9.10.6 Assume that for (SG) there exist a function $v\colon X \times \mathbb{R}^+ \to \mathbb{R}$ and a $\tau_0 \in \mathbb{R}^+$ that satisfy the following conditions.

(i) There exists a function $\psi_2 \in \mathcal{K}$ defined on $\mathbb{R}^+$ such that

$$v(x,t) \leq \psi_2(\|x\|) \tag{9.10.14}$$

for all $x \in X$ and $t \in \mathbb{R}^+$.

(ii) In every neighborhood of the origin $x_e = 0$ there is a point x such that $v(x, \tau_0) > 0$.

(iii) For any $x_0 \in X$ such that $v(x_0, \tau_0) > 0$ and any solution $x(\cdot, \tau_0, x_0)$ of (SG), $v(x(t, \tau_0, x_0), t)$ is continuous everywhere on $\mathbb{R}^+_{\tau_0}$ except on an unbounded and discrete subset $E = \{\tau_1, \tau_2, \dots : \tau_1 < \tau_2 < \cdots\}$ of $\mathbb{R}^+_{\tau_0}$. Assume that there exists a function $\psi_1 \in \mathcal{K}$ defined on $\mathbb{R}^+$ such that

$$Dv(x(\tau_k, \tau_0, x_0), \tau_k) \geq \psi_1(|v(x(\tau_k, \tau_0, x_0), \tau_k)|), \tag{9.10.15}$$

for all $k \in \mathbb{N}$, where Dv is defined in (9.10.5).

Then the equilibrium $x_e = 0$ of (SG) is *unstable*.

Proof. This result is a direct consequence of Theorem 3.2.8. □

Theorem 9.10.7 If in addition to the assumptions in Theorem 9.10.6, $v(x, \tau_0) > 0$ for all $x \neq 0$, then the equilibrium $x_e = 0$ of (SG) is *completely unstable*.

Proof. This result is a direct consequence of Theorem 3.2.9. □

D. Converse theorems

We now establish necessary stability and boundedness results for infinite-dimensional dynamical systems determined by differential equations in Banach spaces. Recall that we assume that (SG) possesses unique solutions $x(\cdot, \tau_0, x_0)$ for the initial conditions (τ_0, x_0).

Theorem 9.10.8 Assume that the equilibrium $x_e = 0$ of system (SG) is *uniformly stable* and that Assumption 3.5.1 holds. Then there exist neighborhoods A_1 and X_1 of $0 \in X$ such that $A_1 \subset X_1$ and a mapping $v\colon X_1 \times \mathbb{R}^+ \to \mathbb{R}^+$ that satisfies the following conditions.

(i) There exist $\psi_1, \psi_2 \in \mathcal{K}$ such that

$$\psi_1(\|x\|) \leq v(x,t) \leq \psi_2(\|x\|)$$

for all $t \in \mathbb{R}^+$ and $x \in X_1$.

(ii) For every solution $x(\cdot, \tau_0, x_0)$ of (SG) with $x_0 \in A_1$, $v(x(t, \tau_0, x_0), t)$ is nonincreasing for all $t \geq \tau_0$. □

The proof of Theorem 9.10.8 is identical to the proof of Theorem 9.3.1 for uniform stability of continuous dynamical systems determined by differential equations in Banach spaces. In the next result, we address uniform asymptotic stability.

Theorem 9.10.9 Assume that Assumptions 3.5.1 and 3.5.2 hold for system (SG). If the equilibrium $x_e = 0$ of system (SG) is *uniformly asymptotically stable*, then there exist neighborhoods A_1 and X_1 of $0 \in X$ such that $A_1 \subset X_1$ and a mapping $v\colon X_1 \times \mathbb{R}^+ \to \mathbb{R}^+$ that satisfies the following conditions.

(i) There exist $\psi_1, \psi_2 \in \mathcal{K}$ such that

$$\psi_1(\|x\|) \le v(x,t) \le \psi_2(\|x\|)$$

for all $t \in \mathbb{R}^+$ and $x \in X_1$.

(ii) There exists $\psi_3 \in \mathcal{K}$ such that for all solutions $x(\cdot, \tau_0, x_0)$ we have

$$Dv(x(\tau_k, \tau_0, x_0), \tau_k) \le -\psi_3(\|x(\tau_k, \tau_0, x_0)\|)$$

for all $k \in \mathbb{N}$, where $x_0 \in A_1$, and Dv is defined in (9.10.5).

(iii) There exists a function $f \in C[\mathbb{R}^+, \mathbb{R}^+]$ such that $f(0) = 0$ and

$$v(x(t, \tau_0, x_0), t) \le f(v(x(\tau_k, \tau_0, x_0), \tau_k))$$

for all $x(\cdot, \tau_0, x_0)$, $t \in (\tau_k, \tau_{k+1})$, $k \in \mathbb{N}$, $x_0 \in A_1$, and $\tau_0 \in \mathbb{R}^+$.

Proof. This result is a direct consequence of Theorem 3.5.2. □

The next result, where we address a converse result for the exponential stability of the equilibrium $x_e = 0$ of (SG), is not symmetric with the conditions given in Theorem 9.10.2 for exponential stability. Nevertheless, this result does provide us with a set of necessary conditions for exponential stability.

Theorem 9.10.10 Assume that Assumptions 3.5.1 and 3.5.2 hold for system (SG). If the equilibrium $x_e = 0$ of system (SG) is *exponentially stable*, then there exist neighborhoods A_1 and X_1 of $0 \in X$ such that $A_1 \subset X_1$ and a mapping $v\colon X_1 \times \mathbb{R}^+ \to \mathbb{R}^+$ that satisfies the following conditions.

(i) There exist $\psi_1, \psi_2 \in \mathcal{K}$ such that

$$\psi_1(\|x\|) \le v(x,t) \le \psi_2(\|x\|)$$

for all $t \in \mathbb{R}^+$ and $x \in X_1$.

(ii) There exists a constant $c > 0$ such that for all solutions $x(\cdot, \tau_0, x_0)$, we have

$$Dv(x(\tau_k, \tau_0, x_0), \tau_k) \le -cv(x(\tau_k, \tau_0, x_0), \tau_k)$$

for all $k \in \mathbb{N}$, where $x_0 \in A_1$ and Dv is defined in (9.10.5).

(iii) There exists a function $f \in C[\mathbb{R}^+, \mathbb{R}^+]$ such that

$$v(x(t, \tau_0, x_0), t) \le f(v(x(\tau_k, \tau_0, x_0), \tau_k))$$

for all $x(\cdot, \tau_0, x_0)$, $t \in (\tau_k, \tau_{k+1})$, $k \in \mathbb{N}$, $x_0 \in A_1$, and $\tau_0 \in \mathbb{R}^+$, and such that for some positive q, $f(\cdot)$ satisfies

$$f(r) = o(r^q) \quad \text{as } r \to 0^+.$$

Proof. This result is a direct consequence of Theorem 3.5.3. □

There are also converse results for *uniform asymptotic stability in the large*, *exponential stability in the large*, *instability*, and *complete instability* of the equilibrium $x_e = 0$ of system (SG), as well as for the *uniform boundedness* and *uniform ultimate boundedness* of solutions of (SG). We do not address these.

The converse theorems presented above involve Lyapunov functions that need not necessarily be continuous. In the next result, we show that under some additional very mild assumptions, the Lyapunov functions for the converse theorems are continuous with respect to initial conditions. (We consider only the case for Theorem 9.10.9.)

Theorem 9.10.11 If in addition to the assumptions given in Theorem 9.10.9, the motions in S are continuous with respect to initial conditions (in the sense of Definition 3.5.2), then there exists a *continuous* Lyapunov function that satisfies the conditions of Theorem 9.10.9.

Proof. The proof of this theorem is a direct consequence of Theorem 3.5.5. □

E. Examples

In the present subsection we apply the results of the preceding subsections in the analysis of several specific classes of infinite-dimensional discontinuous dynamical systems described by differential equations in Banach spaces.

Example 9.10.1 (*DDS system* (SG)) For system (SG_k) we assume that F_k satisfies the *Lipschitz condition*

$$\|F_k(t, x) - F_k(t, y)\| \le K_k \|x - y\| \tag{9.10.16}$$

for all $x, y \in X$ and $t \in \mathbb{R}^+$. Recalling our assumption that $F_k(t, 0) = 0$ for all $t \in \mathbb{R}^+$, and making use of the Gronwall inequality in a similar manner as was done in Example 6.4.5, we obtain the estimate

$$\|x^{(k)}(t, \tau_k, x_k)\| \le e^{K_k(t-\tau_k)} \|x_k\| \tag{9.10.17}$$

for all $t \ge \tau_k$ and all $x_k \in X$. We assume that

$$\sup_{k \in \mathbb{N}} K_k = K < \infty. \tag{9.10.18}$$

Recall that for system (SG) we assume that for all $k \in \mathbb{N}$, $g_k(0) = 0$. In addition, we assume that

$$\|g_k(x)\| \le \gamma_k \|x\| \tag{9.10.19}$$

for all $x \in X$, that

$$\sup_{k \in \mathbb{N}} \gamma_k = \Gamma < \infty, \tag{9.10.20}$$

and letting $\tau_{k+1} - \tau_k = \lambda_k$, that

$$\sup_{k \in \mathbb{N}} \lambda_k = \Lambda < \infty. \tag{9.10.21}$$

Proposition 9.10.1 Let K_k, γ_k, λ_k, K, Γ, and Λ be the parameters for system (SG) given in (9.10.16)–(9.10.21).

(a) If for all $k \in \mathbb{N}$, $\gamma_k e^{K_k \lambda_k} \le 1$, then the equilibrium $x_e = 0$ of (SG) is *uniformly stable.*

(b) If for all $k \in \mathbb{N}$, $\gamma_k e^{K_k \lambda_k} \le \alpha < 1$, where $\alpha > 0$ is a constant, then the equilibrium $x_e = 0$ of (SG) is *uniformly asymptotically stable in the large*, in fact, *exponentially stable in the large.*

Proof. We choose for system (SG) the Lyapunov function $v(x,t) \equiv v(x) = \|x\|$, $x \in X$, which when evaluated along the solutions of (SG) assumes the form

$$v(x(t,\tau_0,x_0)) = v(x^{(k)}(t,\tau_k,x_k)) = \|x^{(k)}(t,\tau_k,x_k)\|, \qquad \tau_k \le t < \tau_{k+1}, \tag{9.10.22}$$

$k \in \mathbb{N}$, where $x^{(k)}(\cdot,\tau_k,x_k)$ denotes the solution segment of the solution $x(\cdot,\tau_0,x_0)$ of (SG) over the interval $[\tau_k,\tau_{k+1})$. Clearly,

$$\psi_1(\|x\|) \le v(x) \le \psi_2(\|x\|) \tag{9.10.23}$$

for all $x \in X$, where $\psi_1(s) = \psi_2(s) = s \ge 0$; that is, $\psi_1, \psi_2 \in \mathcal{K}_\infty$.

Along the solutions of (SG) we have, in view of (9.10.17), that

$$\|x^{(k)}(t,\tau_k,x_k)\| \le e^{K_k(t-\tau_k)}\|x_k\| = e^{K_k(t-\tau_k)}\|x^{(k)}(\tau_k,\tau_k,x_k)\| \tag{9.10.24}$$

for $t \in [\tau_k,\tau_{k+1})$. At $t = \tau_{k+1}$ we have, in view of (9.10.19), that

$$\|x^{(k+1)}(\tau_{k+1},\tau_{k+1},x_{k+1})\| = \|g(x^{(k)}(\tau_{k+1}^-,\tau_k,x_k))\| \le \gamma_k\|x^{(k)}(\tau_{k+1}^-,\tau_k,x_k)\|. \tag{9.10.25}$$

Combining (9.10.24) and (9.10.25), we have

$$\|x^{(k+1)}(\tau_{k+1},\tau_{k+1},x_{k+1})\| \le \gamma_k e^{K_k\lambda_k}\|x^{(k)}(\tau_k,\tau_k,x_k)\|, \tag{9.10.26}$$

and because by assumption $\gamma_k e^{K_k\lambda_k} \le 1$, we have

$$\begin{aligned} v(x^{(k+1)}(\tau_{k+1},\tau_{k+1},x_{k+1})) &= \|x^{(k+1)}(\tau_{k+1},\tau_{k+1},x_{k+1})\| \\ &\le \|x^{(k)}(\tau_k,\tau_k,x_k)\| \\ &= v(x^{(k)}(\tau_k,\tau_k,x_k)). \end{aligned} \tag{9.10.27}$$

Because (9.10.27) holds for arbitrary $k \in \mathbb{N}$, it follows that $v(x(\tau_k, \tau_0, x_0))$ is nonincreasing.

Next, from (9.10.24) we have, recalling that $\sup_{k\in\mathbb{N}} K_k = K$ and $\sup_{k\in\mathbb{N}} \lambda_k = \Lambda$, that

$$\begin{aligned} v(x^{(k)}(t, \tau_k, x_k)) &= \|x^{(k)}(t, \tau_k, x_k)\| \\ &\leq e^{K\Lambda} v(x^{(k)}(\tau_k, \tau_k, x_k)) \\ &= f(v(x^{(k)}(\tau_k, \tau_k, x_k))), \end{aligned} \tag{9.10.28}$$

$t \in [\tau_k, \tau_{k+1})$, $k \in \mathbb{N}$, where $f(s) = e^{K\Lambda}s$. Therefore, all conditions of Theorem 9.10.1(a) are satisfied and we conclude that the equilibrium $x_e = 0$ of system (SG) is *uniformly stable*.

If in (9.10.26) we assume that $\gamma_k e^{K_k \lambda_k} \leq \alpha < 1$, we have

$$v(x^{(k+1)}(\tau_{k+1}, \tau_{k+1}, x_{k+1})) < \alpha v(x^{(k)}(\tau_k, \tau_k, x_k)) \tag{9.10.29}$$

and

$$\begin{aligned} &\big[v(x^{(k+1)}(\tau_{k+1}, \tau_{k+1}, x_{k+1})) - v(x^{(k)}(\tau_k, \tau_k, x_k))\big] / (\tau_{k+1} - \tau_k) \\ &\quad\leq [(\alpha - 1)/\Lambda] v(x^{(k)}(\tau_k, \tau_k, x_k)) \\ &\quad\triangleq -\psi_3(\|x^{(k)}(\tau_k, \tau_k, x_k)\|) \\ &\quad= -\psi_3(\|x(\tau_k, \tau_0, x_0)\|) \end{aligned} \tag{9.10.30}$$

for all $k \in \mathbb{N}$ and $(\tau_0, x_0) \in \mathbb{R}^+ \times X$. In (9.10.30) we have $\psi_3(s) = [(1 - \alpha)/\Lambda]s$, $s \geq 0$ (i.e., $\psi_3 \in \mathcal{K}_\infty$). Therefore, all conditions of Theorem 9.10.1(b) and Theorem 9.10.3(a) are satisfied and we conclude that the equilibrium $x_e = 0$ of system (SG) is *uniformly asymptotically stable in the large*.

Finally, from (9.10.23), it is clear that in relation (9.10.6) in Theorem 9.10.2 we have $c_1 = c_2 = b = 1$ and from (9.10.30), it is clear that in relation (9.10.9) in Theorem 9.10.2 we have $c_3 = (1 - \alpha)/\Lambda$. We have already shown that (9.10.7) of Theorem 9.10.2 is true, and clearly, for $f(s) = e^{K\Lambda}s$, we have $f(s) = \mathcal{O}(s^q)$ as $s \to 0$ for any $q \in (0, 1)$. Therefore, all the conditions of Theorems 9.10.2 and 9.10.3(b) are satisfied and we can conclude that the equilibrium $x_e = 0$ of (SG) is *exponentially stable in the large*. □

Example 9.10.2 (*Time-invariant linear functional differential equations*) If we let $X = C_r$ and $F_k(t, x) = L_k x_t$ where C_r, x_t, and L_k are defined as in Example 2.7.1, then (SG_k) assumes the form

$$\begin{cases} \dot{x}(t) = L_k x_t, \\ x_{\tau_k} = \varphi_k, \end{cases} \tag{9.10.31}$$

$k \in \mathbb{N}$, $t \in [\tau_k, \infty)$. If in (SG) we let $g_k(\eta) = G_k \eta$ where $G_k \colon C_r \to C_r$ is a linear operator, then (SG) assumes the form

$$\begin{cases} \dot{x}(t) = L_k x_t, & \tau_k \leq t < \tau_{k+1} \\ x_{\tau_{k+1}} = G_k x_{\tau_{k+1}^-}, & k \in \mathbb{N}. \end{cases} \tag{9.10.32}$$

For each $k \in \mathbb{N}$, L_k is defined, as in (2.7.2), by

$$L_k(\varphi) = \int_{-r}^{0} [dB_k(s)]\varphi(s), \tag{9.10.33}$$

where $B(s) = [b_{ij}(s)]$ is an $n \times n$ matrix whose entries are functions of bounded variation on $[-r, 0]$. Then L_k is Lipschitz continuous on C_r with Lipschitz constant K_k less than or equal to the variation of B_k, and as such, condition (9.10.17) still holds for (9.10.31). As in (2.9.14), the spectrum of L_k consists of all solutions of the equation

$$\det\left(\int_{-r}^{0} e^{\lambda_k s} dB_k(s) - \lambda_k I\right) = 0. \tag{9.10.34}$$

In accordance with (2.9.15), when all the solutions of (9.10.34) satisfy the relation $\text{Re}\lambda_k \leq -\alpha_0$, then for any positive $\alpha_k < \alpha_0$, there is a constant $M_k(\alpha_k) > 0$ such that the solutions of (9.10.31) allow the estimate

$$\|x_t^{(k)}(\tau_k, \varphi_k)\| \leq M_k(\alpha_k)e^{-\alpha_k(t-\tau_k)}\|\varphi_k\| \tag{9.10.35}$$

for all $t \geq \tau_k \geq 0$ and $\varphi_k \in C_r$. When the above assumption is not true, then in accordance with (9.10.17), the solutions of (9.10.31) still allow the estimate

$$\|x_t^{(k)}(\tau_k, \varphi_k)\| \leq e^{K_k(t-\tau_k)}\|\varphi_k\| \tag{9.10.36}$$

for all $t \geq \tau_k$ and $\varphi_k \in C_r$. Thus, in all cases we have

$$\|x_t^{(k)}(\tau_k, \varphi_k)\| \leq Q_k e^{w_k(t-\tau_k)}\|\varphi_k\| \tag{9.10.37}$$

for all $t \geq \tau_k \geq 0$ and $\varphi_k \in C_r$, where $Q_k = 1$ and $w_k = K_k$ when (9.10.36) applies and $Q_k = M_k(\alpha_k)$ and $w_k = -\alpha_k$, $\alpha_k > 0$, when (9.10.35) applies.

Finally, for each $k \in \mathbb{N}$, we have

$$\|G_k\eta\| \leq \|G_k\| \; \|\eta\| \tag{9.10.38}$$

for all $\eta \in C_r$, where $\|G_k\|$ is the norm of the linear operator G_k induced by the norm $\|\cdot\|$ defined on C_r.

In the following, we still assume that (9.10.18) and (9.10.21) hold and we assume that

$$\sup_{k\in\mathbb{N}} M_k(\alpha_k) = M < \infty. \tag{9.10.39}$$

Proposition 9.10.2 Let w_k, $\|G_k\|$, Q_k, λ_k, Λ, M, and K be the parameters for system (9.10.32) defined above.

(a) If for all $k \in \mathbb{N}$, $\|G_k\|Q_k e^{w_k\lambda_k} \leq 1$, then the equilibrium $x_e = 0$ of system (9.10.32) is *uniformly stable*.

(b) If for all $k \in \mathbb{N}$, $\|G_k\|Q_k e^{w_k\lambda_k} \leq \alpha < 1$, where $\alpha > 0$ is a constant, then the equilibrium $x_e = 0$ of system (9.10.32) is *uniformly asymptotically stable in the large*, and in fact, *exponentially stable in the large*.

Proof. Choosing $v(\varphi, t) \equiv v(\varphi) = \|\varphi\|$, $\varphi \in C_r$, we obtain the relations

$$v(x_t(\tau_0, x_0)) = v(x_t^{(k)}(\tau_k, \varphi_k)) = \|x_t^{(k)}(\tau_k, \varphi_k)\|, \qquad \tau_k \le t < \tau_{k+1} \quad (9.10.40)$$

$k \in \mathbb{N}$, and

$$\psi_1(\|x\|) \le v(x) \le \psi_2(\|x\|) \quad (9.10.41)$$

for all $x \in X$, where $\psi_1(s) = \psi_2(s) = s$, $s \ge 0$; that is, $\psi_1, \psi_2 \in \mathcal{K}_\infty$.

Along the solutions of (9.10.32) we have, in view of (9.10.37),

$$\|x_t^{(k)}(\tau_k, \varphi_k)\| \le Q_k e^{w_k(t-\tau_k)}\|\varphi_k\| = Q_k e^{w_k(t-\tau_k)}\|x_{\tau_k}^{(k)}(\tau_k, \varphi_k)\| \quad (9.10.42)$$

for $t \in [\tau_k, \tau_{k+1})$. At $t = \tau_{k+1}$ we have, when (9.10.36) applies

$$\|x_{\tau_{k+1}}^{(k)}(\tau_{k+1}, \varphi_{k+1})\| \le \|G_k\| \, \|x_{\tau_{k+1}^-}^{(k)}(\tau_k, \varphi_k)\| \le \|G_k\| e^{K_k\lambda_k}\|\varphi_k\| \quad (9.10.43)$$

and when (9.10.35) applies,

$$\|x_{\tau_{k+1}}^{(k+1)}(\tau_{k+1}, \varphi_{k+1})\| \le \|G_k\| M_k(\alpha_k) e^{-\alpha_k\lambda_k}\|\varphi_k\|. \quad (9.10.44)$$

Thus, in either case we have

$$\|x_{\tau_{k+1}}^{(k+1)}(\tau_{k+1}, \varphi_{k+1})\| \le \|G_k\| Q_k e^{w_k\lambda_k}\|x_{\tau_k}^{(k)}(\tau_k, \varphi_k)\|. \quad (9.10.45)$$

When $\|G_k\| Q_k e^{w_k\lambda_k} \le 1$, we obtain

$$\begin{aligned} v(x_{\tau_{k+1}}^{(k+1)}(\tau_{k+1}, \varphi_{k+1})) &= \|x_{\tau_{k+1}}^{(k+1)}(\tau_{k+1}, \varphi_{k+1})\| \\ &\le \|x_{\tau_k}^{(k)}(\tau_k, \varphi_k)\| \\ &= v(x_{\tau_k}^{(k)}(\tau_k, \varphi_k)), \quad k \in \mathbb{N}. \end{aligned} \quad (9.10.46)$$

Therefore, $v(x_{\tau_k}(\tau_0, \varphi))$, $k \in \mathbb{N}$, is nonincreasing.

Next, from (9.10.42), we have

$$v(x_t^{(k)}(\tau_k, \varphi_k)) \le Q_k e^{w_k(t-\tau_k)} v(x_{\tau_k}^{(k)}(\tau_k, \varphi_k)), \quad (9.10.47)$$

$t \in [\tau_k, \tau_{k+1})$, $k \in \mathbb{N}$. When (9.10.36) applies, $Q_k = 1$ and $w_k = K_k$ and when (9.10.35) applies, $Q_k = M_k(\alpha_k)$ and $w_k = -\alpha_k < 0$. Recall that $\sup_{k\in\mathbb{N}} K_k = K$, $\sup_{k\in\mathbb{N}} \lambda_k = \Lambda$, and $\sup_{k\in\mathbb{N}} M_k(\alpha_k) = M$. Let $P = \max\{e^{\Lambda K}, M\}$ and choose $f(s) = Ps$, $s \ge 0$. From (9.10.47) we now obtain

$$v(x_t^{(k)}(\tau_k, \varphi_k)) \le f(v(x_{\tau_k}^{(k)}(\tau_k, \varphi_k))), \quad (9.10.48)$$

$t \in [\tau_k, \tau_{k+1})$.

All conditions of Theorem 9.10.1(a) are satisfied and therefore the equilibrium $x_e = 0$ of system (9.10.32) is *uniformly stable*.

When $\|G_k\| Q_k e^{w_k\lambda_k} \le \alpha < 1$, we have

$$v\big(x_{\tau_{k+1}}^{(k+1)}(\tau_{k+1}, \varphi_{k+1})\big) < \alpha v\big(x_{\tau_k}^{(k)}(\tau_k, \varphi_k)\big)$$

and

$$\begin{aligned}
\big[v(x_{\tau_{k+1}}^{(k+1)}(\tau_{k+1},\varphi_{k+1})) - v(x_{\tau_k}^{(k)}(\tau_k,\varphi_k))\big]/(\tau_{k+1}-\tau_k)& \\
\leq [(\alpha-1)/\Lambda]v(x_{\tau_k}^{(k)}(\tau_k,\varphi_k))& \\
\triangleq -\psi_3(\|x_{\tau_k}^{(k)}(\tau_k,\varphi_k)\|)& \\
= -\psi_3(\|x_{\tau_k}(\tau_0,\varphi)\|)& \qquad (9.10.49)
\end{aligned}$$

for all $k \in \mathbb{N}$ and $\varphi \in C_r$. In (9.10.49), we have $\psi_3(s) = [(1-\alpha)/\Lambda]s$; that is, $\psi_3 \in \mathcal{K}_\infty$. Therefore, all conditions of Theorem 9.10.1(b) and Theorem 9.10.3(a) are satisfied and the equilibrium $x_e = 0$ of system (9.10.32) is *uniformly asymptotically stable in the large.*

In the notation of Theorems 9.10.2 and 9.10.3(b), we have $c_1 = c_2 = b = 1$ and $c_3 = (1-\alpha)/\Lambda$. Also, $f(s) = Ps$, so that $f(s) = o(s^q)$ as $s \to 0$ where $q \in (0,1)$. Therefore, all the conditions of Theorems 9.10.2 and 9.10.3(b) are satisfied and we conclude that the equilibrium $x_e = 0$ of system (9.10.32) is *exponentially stable in the large.* □

Example 9.10.3 (*Heat equation*) We consider a family of initial value and boundary value problems determined by the heat equation

$$\begin{cases} \dfrac{\partial u}{\partial t} = a_k^2 \Delta u, & (t,x) \in [\tau_k,\infty)\times\Omega \\ u(\tau_k,x) = \psi_k(x), & x \in \Omega \\ u(t,x) = 0, & (t,x) \in [\tau_k,\infty)\times\partial\Omega, \end{cases} \qquad (9.10.50)$$

$k \in \mathbb{N}$, where $\Omega \subset \mathbb{R}^n$ is a bounded domain with smooth boundary $\partial\Omega$, $a_k^2 > 0$ are constants, and $\Delta = \sum_{i=1}^n \partial^2/\partial x_i^2$ denotes the Laplacian. For a discussion and stability analysis of system (9.10.50), refer to Example 9.2.2.

Next, we consider a discontinuous dynamical system determined by

$$\begin{cases} \dfrac{\partial u}{\partial t} = a_k^2 \Delta u, & (t,x) \in [\tau_k,\tau_{k+1})\times\Omega \\ u(\tau_{k+1},\cdot) = g_k(u(\tau_{k+1}^-,\cdot)) & \\ u(t,x) = 0, & (t,x) \in \mathbb{R}^+\times\partial\Omega \end{cases} \qquad (9.10.51)$$

where all terms are defined similarly as in (9.10.50), $g_k\colon X \to X$,

$$X = H^2[\Omega,\mathbb{R}] \cap H_0^1[\Omega,\mathbb{R}]$$

with the H^1-norm (refer to Example 9.2.2), $k \in \mathbb{N}$. We assume that $g_k(0) = 0$ and that for each $k \in \mathbb{N}$, there exists a $\gamma_k > 0$ such that $\|g_k(\psi)\|_{H^1} \leq \gamma_k \|\psi\|_{H^1}$ for all $\psi \in X$.

If, similarly as in Example 9.2.2, we define $U^{(k)}(t) = u^{(k)}(t,\cdot)$ ($u^{(k)}(t,\psi)$ denotes the unique solution of system (9.10.50); see Example 9.2.2), we obtain the estimate

$$\|U^{(k)}(t)\|_{H^1} \leq e^{-c_k(t-\tau_k)/2}\|U^{(k)}(\tau_k)\|_{H^1} \qquad (9.10.52)$$

for $t \geq \tau_k$, where $c_k = \min\{a_k^2, a_k^2/\gamma\}$, where γ can be chosen as $\delta/\sqrt{n}$ and Ω can be put into a cube of length δ (refer to Example 9.2.2, in particular, (9.2.10)).

Each solution $u(t, x, \psi, \tau_0)$ of (9.10.51) is made up of a sequence of solution segments $u^{(k)}(t, x, \psi, \tau_0)$, defined on $[\tau_k, \tau_{k+1})$ for $k \in \mathbb{N}$, that are determined by (9.10.50) with $\varphi_k = u(\tau_k, \cdot)$.

Proposition 9.10.3 For system (9.10.51), let $w_k = -c_k/2$ and $\lambda_k = \tau_{k+1} - \tau_k$, $k \in \mathbb{N}$. Assume that $\sup_{k \in \mathbb{N}} \lambda_k = \Lambda < \infty$ and $\sup_{k \in \mathbb{N}} w_k = w < \infty$.

(a) If for all $k \in \mathbb{N}$, $\gamma_k e^{w_k \lambda_k} \leq 1$, then the equilibrium $\psi_e = 0 \in X$ for system (9.10.51) is *uniformly stable* (with respect to the H^1-norm).

(b) If for all $k \in \mathbb{N}$, $\gamma_k e^{w_k \lambda_k} \leq \alpha < 1$, where $\alpha > 0$ is a constant, then the equilibrium $\psi_e = 0$ for system (9.10.51) is *uniformly asymptotically stable in the large*, in fact, *exponentially stable in the large*.

Proof. We choose the Lyapunov function $v(\psi, t) \equiv v(\psi) = \|\psi\|_{H^1}$, $\psi \in X$, which when evaluated along the solutions $u(t, x, \psi, \tau_0)$ of (9.10.51) assumes the form

$$v(U(t, \tau_0, \psi)) = v(U^{(k)}(t, \tau_k, \psi_k)) = \|U^{(k)}(t, \tau_k, \psi_k)\|_{H^1}, \qquad \tau_k \leq t < \tau_{k+1} \tag{9.10.53}$$

$k \in \mathbb{N}$. Clearly,

$$\psi_1(\|\psi\|_{H^1}) \leq v(\psi) \leq \psi_2(\|\psi\|_{H^1}) \tag{9.10.54}$$

for all $\psi \in X$, where $\psi_1(s) = \psi_2(s) = s$, $s \geq 0$; that is, $\psi_1, \psi_2 \in \mathcal{K}_\infty$.

Along the solutions of (9.10.51) we have, in view of (9.10.52), that

$$\|U^{(k)}(t, \tau_k, \psi_k)\|_{H^1} \leq e^{-c_k(t-\tau_k)/2}\|\psi_k\|_{H^1} = e^{-c_k(t-\tau_k)/2}\|U^{(k)}(\tau_k, \tau_k, \psi_k)\|_{H^1} \tag{9.10.55}$$

for $t \in [\tau_k, \tau_{k+1})$. At $t = \tau_{k+1}$, we have, in view of (9.10.51), that

$$\begin{aligned} \|U^{(k+1)}(\tau_{k+1}, \tau_{k+1}, \psi_{k+1})\|_{H^1} &= \|g(U^{(k)}(\tau_{k+1}^-, \tau_k, \psi_k))\|_{H^1} \\ &\leq \gamma_k \|U^{(k)}(\tau_{k+1}^-, \tau_k, \psi_k)\|_{H^1}. \end{aligned} \tag{9.10.56}$$

Combining (9.10.55) and (9.10.56), we have

$$\|U^{(k+1)}(\tau_{k+1}, \tau_{k+1}, \psi_{k+1})\|_{H^1} \leq \gamma_k e^{-c_k \lambda_k/2}\|U^{(k)}(\tau_k, \tau_k, \psi_k)\|_{H^1} \tag{9.10.57}$$

where $-c_k/2 = w_k$ and because by assumption $\gamma_k e^{w_k \lambda_k} \leq 1$, we have

$$\begin{aligned} v(U^{(k+1)}(\tau_{k+1}, \tau_{k+1}, \psi_{k+1})) &= \|U^{(k+1)}(\tau_{k+1}, \tau_{k+1}, \psi_{k+1})\|_{H^1} \\ &\leq \|U^{(k)}(\tau_k, \tau_k, \psi_k)\|_{H^1} \\ &= v(U^{(k)}(\tau_k, \tau_k, \psi_k)). \end{aligned}$$

Because this holds for arbitrary $k \in \mathbb{N}$, it follows that $v(U(\tau_k, \tau_0, \psi))$, $k \in \mathbb{N}$, is nonincreasing.

Next, from (9.10.55) we have, recalling that $\sup_{k\in\mathbb{N}} w_k = w < \infty$ and

$$\sup_{k\in\mathbb{N}} \lambda_k = \Lambda < \infty,$$

that

$$\begin{aligned} v(U^{(k)}(t,\tau_k,\psi_k)) &= \|U^{(k)}(t,\tau_k,\psi_k)\|_{H^1} \\ &\le e^{w\Lambda} v(U^{(k)}(\tau_k,\tau_k,\psi_k)) \\ &\triangleq f(v(U^{(k)}(\tau_k,\tau_k,\psi_k))), \end{aligned}$$

$t \in [\tau_k, \tau_{k+1})$, $k \in \mathbb{N}$, where $f(s) = e^{w\Lambda}s$. Therefore, all conditions of Theorem 9.10.1(a) are satisfied and we conclude that the equilibrium $\psi_e = 0 \in X$ of system (9.10.51) is *uniformly stable*.

If in (9.10.57) we assume that $\gamma_k e^{w_k \lambda_k} \le \alpha < 1$, we have

$$v(U^{(k+1)}(\tau_{k+1},\tau_{k+1},\psi_{k+1})) < \alpha v(U^{(k)}(\tau_k,\tau_k,\psi_k))$$

and

$$\begin{aligned} \big[v(U^{(k+1)}(\tau_{k+1},\tau_{k+1},\psi_{k+1})) - v(U^{(k)}(\tau_k,\tau_k,\psi_k))\big]/(\tau_{k+1}-\tau_k) \\ \le [(\alpha-1)/\Lambda] v(U^{(k)}(\tau_k,\tau_k,\psi_k)) \\ \triangleq -\psi_3(\|U^{(k)}(\tau_k,\tau_k,\psi_k)\|) \\ = -\psi_3(\|U(\tau_k,\tau_0,\psi)\|) \end{aligned} \tag{9.10.58}$$

for all $k \in \mathbb{N}$ and $(\tau_0,\psi) \in \mathbb{R}^+ \times X$. In (9.10.58) we have $\psi_3(s) = [(1-\alpha)/\Lambda]s$, $s \ge 0$; that is, $\psi_3 \in \mathcal{K}_\infty$. Therefore, all conditions of Theorems 9.10.1(b) and 9.10.3(a) are satisfied and the equilibrium $\psi_e = 0 \in X$ of system (9.10.51) is *uniformly asymptotically stable in the large*.

In the notation of Theorems 9.10.2 and 9.10.3(b), we have $c_1 = c_2 = b = 1$ and $c_3 = (1-\alpha)/\Lambda$. We have already shown that $f(s) = e^{w\Lambda}s$ and thus, $f(s) = o(s^q)$ as $s \to 0$ for any $q \in (0,1)$. Therefore, all conditions of Theorems 9.10.2 and 9.10.3(b) are satisfied and we conclude that the equilibrium $\psi_e = 0$ of system (9.10.32) is *exponentially stable in the large*. □

Example 9.10.4 (*Linear scalar Volterra integrodifferential equation*) We consider a family of scalar linear Volterra integrodifferential equations given by

$$\begin{cases} \dot{x}(t) = -a_n x(t) + \int_{-\infty}^{t} k_n(t-s)x(s)ds, & t \ge \tau_n \\ x(\tau_n) = \varphi_n, \end{cases} \tag{9.10.59}$$

$n \in \mathbb{N}$, which can equivalently be expressed as

$$\begin{cases} \dot{x}(t) = -a_n x_t(0) + \int_{-\infty}^{0} k_n(-s)x_t(s)ds, & t \ge \tau_n \\ x(\tau_n) = \varphi_n. \end{cases} \tag{9.10.60}$$

In (9.10.60), all terms are defined similarly as in Example 9.2.3. In particular, the state space is the fading memory space X consisting of all measurable functions $\varphi\colon(-\infty,0)\to\mathbb{R}$ with norm given by

$$\|\varphi\|_m^2=|\varphi(0)|^2+K_n\int_{-\infty}^{0}|\varphi(s)|^2e^{Ls}ds<\infty, \tag{9.10.61}$$

where K_n is determined later. Let $C_{L,n}=\left(\int_0^\infty|k_n(s)|^2e^{Ls}ds\right)^{1/2}$ for some fixed $L>0$. From Example 9.2.3, when $C_{L,n}/\sqrt{L}\le a_n$, with appropriate K_n ($0<K_n<2a_n$), we can obtain $\alpha_n<0$ such that

$$\|x_t^{(n)}\|_m\le e^{w_n(t-\tau_n)}\|x_{\tau_n}^{(n)}\|_m \tag{9.10.62}$$

where $w_n=\alpha_n/2$, $\alpha_n=-\lambda_m(P_n)/\max\{1,K_n\}$, where

$$P_n=\begin{bmatrix}2a_n-K_n & -C_{L,n}\\ -C_{L,n} & K_nL\end{bmatrix} \tag{9.10.63}$$

is positive definite.

We now consider discontinuous dynamical systems described by

$$\begin{cases}\dot{x}(t)=-a_nx_t(0)+\displaystyle\int_{-\infty}^{0}k_n(-s)x_t(s)ds, & t\in[\tau_n,\tau_{n+1})\\ x_{\tau_{n+1}}=G_nx_{\tau_{n+1}}^{-}\end{cases} \tag{9.10.64}$$

$n\in\mathbb{N}$, where the G_n are bounded linear mappings. For (9.10.64), we assume that $\inf_{n\in\mathbb{N}}a_n=a>0$, and we assume that $C_{L,n}/\sqrt{L}\le a_n$ for all $n\in\mathbb{N}$. Then we can find an appropriate K ($0<K<2a$) such that (9.10.62) is true with $K_n=K$ for any $n\in\mathbb{N}$.

Proposition 9.10.4 For system (9.10.64), let $w_n=\alpha_n/2$ and $\lambda_n=\tau_{n+1}-\tau_n$, $n\in\mathbb{N}$. Assume that $\sup_{n\in\mathbb{N}}\lambda_n=\Lambda<\infty$ and $\sup_{n\in\mathbb{N}}w_k=w<\infty$.

(a) If for all $n\in\mathbb{N}$, $\|G_n\|e^{w_n\lambda_n}\le 1$, then the equilibrium $0\in X$ of system (9.10.64) is *uniformly stable*.

(b) If for all $n\in\mathbb{N}$, $\|G_n\|e^{w_n\lambda_n}\le\delta<1$, where $\delta>0$ is a constant, then the equilibrium $0\in X$ of system (9.10.64) is *uniformly asymptotically stable in the large*, and in fact, *exponentially stable in the large*. □

The proof is similar to the proof of Proposition 9.10.3 and is left as an exercise.

9.11 Discontinuous Dynamical Systems Determined by Semigroups

In this section we establish stability results for discontinuous dynamical systems determined by linear semigroups (C_0-semigroups) and nonlinear semigroups (refer to Section 2.9). In contrast to the preceding stability results for DDS, in the results of the present section we do not make use of Lyapunov functions, but instead, we bring to bear results known for linear and nonlinear semigroups.

A. DDS determined by semigroups

In the following, we require a given collection of C_0-semigroups $\mathcal{T} = \{T_i(t)\}$ (each $T_i(t)$ is defined on a Banach space X), a given collection of bounded linear operators $\mathcal{H} = \{H_j\}$ ($H_j\colon X \to X$), and a given unbounded and discrete set $E = \{\tau_0, \tau_1, \tau_2, \dots : \tau_0 < \tau_1 < \tau_2 < \cdots\} \subset \mathbb{R}^+$. The number of elements in $\mathcal{T}$ and $\mathcal{H}$ may be *finite* or *infinite*.

We now consider dynamical systems whose *motions* $y(\cdot, y_0, t_0)$ with *initial time* $t_0 = \tau_0 \in \mathbb{R}^+$ and *initial state* $y(t_0) = y_0 \in X$ are given by

$$\begin{cases} y(t, y_0, t_0) = T_k(t - \tau_k)y(\tau_k), & \tau_k \le t < \tau_{k+1} \\ y(t) = H_k y(t^-), & t = \tau_{k+1}, \quad k \in \mathbb{N} \end{cases} \tag{9.11.1}$$

where for each $k \in \mathbb{N}$, $T_k(t) \in \mathcal{T}$, $H_k \in \mathcal{H}$, and $\tau_k \in E$. We define the *discontinuous dynamical system determined by linear semigroups*, S_{DC_0}, as

$$\begin{aligned} S_{DC_0} = \big\{y = y(\cdot, x, t_0) \colon & y(t, x, t_0) = T_k(t - \tau_k)y(\tau_k), \ \tau_k \le t < \tau_{k+1}, \\ & y(t) = H_k y(t^-), \qquad t = \tau_{k+1}, \ k \in \mathbb{N}, \\ & t_0 = \tau_0 \in \mathbb{R}^+, \qquad y(\tau_0) = x \in X\big\}. \end{aligned}$$

Note that every motion $y(\cdot, x, t_0)$ is unique, with $y(t_0, x, t_0) = x$, exists for all $t \ge t_0$, and is continuous with respect to t on $[t_0, \infty) - \{\tau_1, \tau_2, \dots\}$, and that at $t = \tau_k$, $k = 1, 2, \dots$, $y(\cdot, x, t_0)$ *may* be discontinuous. We call the set $E_1 \triangleq \{\tau_1, \tau_2, \dots\}$ the *set of discontinuities* for the motion $y(\cdot, x, t_0)$. Because H_k and $T_k(t), t \in \mathbb{R}^+$, are linear, it follows that in particular $y(t, 0, t_0) = 0$ for all $t \ge t_0$. We call $x_e = 0$ the *equilibrium for the dynamical system* S_{DC_0} and $y(t, 0, t_0) = 0$, $t \ge t_0$, the *trivial motion*.

In the following, we require a given collection of nonlinear semigroups $\mathcal{T} = \{T_i(t)\}$ (each $T_i(t)$ is defined on $C \subset X$), a given collection of bounded continuous mappings $\mathcal{H} = \{H_j\}$ ($H_j\colon C \to C$) and a given unbounded and discrete set $E = \{t_0 = \tau_0, \tau_1, \tau_2, \dots : \tau_0 < \tau_1 < \tau_2 < \cdots\}$. As before, the number of elements in $\mathcal{T}$ and $\mathcal{H}$ may be *finite* or *infinite*.

We now consider dynamical systems whose motions $y(\cdot, y_0, t_0)$ with initial time $t = \tau_0 \in \mathbb{R}^+$ and initial state $y(t_0) = y_0 \in C \subset X$ are given by

$$\begin{cases} y(t, y_0, t_0) = T_k(t - \tau_k)(y(\tau_k)), & \tau_k \le t < \tau_{k+1} \\ y(t) = H_k(y(t^-)), & t = \tau_{k+1}, \qquad k \in \mathbb{N} \end{cases} \tag{9.11.2}$$

where for each $k \in \mathbb{N}$, $T_k(t) \in \mathcal{T}$, $H_k \in \mathcal{H}$, and $\tau_k \in E$. We define the *discontinuous dynamical system determined by nonlinear semigroups*, S_{DN}, similarly, by

$$S_{DN} = \{y = y(\cdot, x, t_0) \colon y(t, x, t_0) = T_k(t - \tau_k)(y(\tau_k)),\ \ \tau_k \le t < \tau_{k+1},\\ y(t) = H_k(y(t^-)),\ \ t = \tau_{k+1},\ \ k \in \mathbb{N},\\ t_0 = \tau_0 \in \mathbb{R}^+,\ \ y(\tau_0) = x \in C \subset X\}. \qquad (9.11.3)$$

We *assume* that the origin $y = 0$ is in the interior of C. Note that every motion $y(\cdot, x, t_0)$ is unique, with $y(t_0, x, t_0) = x$, exists for all $t \ge t_0$, is continuous with respect to t on $[t_0, \infty) - \{\tau_1, \tau_2, \dots\}$, and that at $t = \tau_k$, $k = 1, 2, \dots$, $y(\cdot, x, t_0)$ *may* be discontinuous. *Throughout*, we assume that $T_k(t)(x) = 0$ for all $t \ge 0$ if $x = 0$ and that $H_k(x) = 0$ if $x = 0$ for all $k \in \mathbb{N}$. From this it follows that $y(t, x, t_0) = 0$ for all $t \ge t_0$ if $x = 0 \in C$. We call $x_e = 0$ an *equilibrium* and $y(t, 0, t_0) = 0$, $t \ge t_0$, a *trivial motion* for the dynamical system S_{DN}.

Remark 9.11.1 For different initial conditions (x, t_0), resulting in different motions $y(\cdot, x, t_0)$, we allow the set of discontinuities $E_1 = \{\tau_1, \tau_2, \dots\}$, the set of semigroups $\{T_k\} \subset \mathcal{T}$, and the set of functions $\{H_k\} \subset \mathcal{H}$ to differ, and accordingly, the notation $E_1^{x,t_0} = \{\tau_1^{x,t_0}, \tau_2^{x,t_0}, \dots\}$, $\{T_k^{x,t_0}\}$, and $\{H_k^{x,t_0}\}$ might be more appropriate. However, because in all cases all meaning is clear from context, we do not use such superscripts. □

Remark 9.11.2 The DDS models considered herein (S_{DC_0} and S_{DN}) are very general and include large classes of finite-dimensional dynamical systems determined by ordinary differential equations and by large classes of infinite-dimensional dynamical systems determined by differential-difference equations, functional differential equations, Volterra integrodifferential equations, certain classes of partial differential equations, and more generally, differential equations and inclusions defined on Banach spaces. This generality allows analysis of distributed parameter systems, systems with delays, systems endowed with hysteresis effects, and the like. □

Remark 9.11.3 The dynamical system models S_{DC_0} and S_{DN} are very flexible, and include as special cases many of the DDS considered in the literature, as well as general autonomous continuous dynamical systems: (a) if $T_k(t) = T(t)$ for all k ($\mathcal{T}$ has only one element) and if $H_k = I$ for all k, where I denotes the identity transformation, then S_{DC_0} reduces to an autonomous, linear, continuous dynamical system and S_{DN} to an autonomous nonlinear, continuous dynamical system; (b) in the case of dynamical systems subjected to impulsive effects (considered in the literature for finite-dimensional systems; see, e.g., [2]), one would choose $T_k(t) = T(t)$ for all k whereas the impulse effects are captured by an infinite family of functions $\mathcal{H} = \{H_k\}$; (c) in the case of *switched systems*, frequently only a *finite number* of systems that are being switched are required, and so in this case one would choose a finite family of semigroups $\mathcal{T} = \{T_i(t)\}$ (see, e.g., [5], [22], [23], and [40]); and so forth. □

Remark 9.11.4 Perhaps it needs pointing out that even though systems S_{DN} and S_{DC_0} are *determined* by families of semigroups (and nonlinearities), by themselves they are *not* semigroups, because in general, they are time-varying and do not satisfy the hypotheses (i)–(iii) given in Definitions 2.9.1 and 2.9.5. However, each individual semigroup $T_k(t)$, used in describing S_{DN} or S_{DC_0}, does possess the semigroup properties, albeit, only over a finite interval (τ_k, τ_{k+1}), $k \in \mathbb{N}$. □

B. Qualitative characterizations of DDS

Recall that the DDS S_{DC_0} determined by linear semigroups, is defined on a Banach space X whereas the DDS given by S_{DN} is defined on $C \subset X$. Recall also that the origin 0 is assumed to be in the interior of C and that $y_e = 0$ is an equilibrium for both S_{DC_0} and S_{DN}. Because the following definitions pertain to both S_{DN} and S_{DC_0}, we refer to either of them simply as S.

Definition 9.11.1 The equilibrium $y_e = 0$ of S is *stable* if for every $\varepsilon > 0$ and every $t_0 \geq 0$, there exists a $\delta = \delta(\varepsilon, t_0) > 0$ such that for all $y(\cdot, y_0, t_0) \in S$, $\|y(t, y_0, t_0)\| < \varepsilon$ for all $t \geq t_0$, whenever $\|y_0\| < \delta$ (and $y_0 \in C$). The equilibrium $y_e = 0$ is *uniformly stable* if δ is independent of t_0; that is, $\delta = \delta(\varepsilon)$. The equilibrium $y_e = 0$ of S is *unstable* if it is not stable. □

Definition 9.11.2 The equilibrium $y_e = 0$ of S is *attractive* if there exists an $\eta = \eta(t_0) > 0$ such that

$$\lim_{t \to \infty} \|y(t, y_0, t_0)\| = 0 \tag{9.11.4}$$

for all $y(\cdot, y, t_0) \in S$ whenever $\|y_0\| < \eta$ (and $y_0 \in C$). □

We call the set of all $y_0 \in C$ such that (9.11.4) holds the *domain of attraction* of $y_e = 0$.

Definition 9.11.3 The equilibrium $y_e = 0$ of S is *asymptotically stable* if it is stable and attractive. □

Definition 9.11.4 The equilibrium $y_e = 0$ of S is *uniformly attractive* if for every $\varepsilon > 0$ and every $t_0 \geq 0$, there exist a $\delta > 0$, independent of t_0 and ε, and a $\mu = \mu(\varepsilon) > 0$, independent of t_0, such that $\|y(t, y_0, t_0)\| < \varepsilon$ for all $t \geq t_0 + \mu$ and for all $y(\cdot, y_0, t_0) \in S$, whenever $\|y_0\| < \delta$ (and $y_0 \in C$). □

Definition 9.11.5 The equilibrium $y_e = 0$ of S is *uniformly asymptotically stable* if it is uniformly stable and uniformly attractive. □

Definition 9.11.6 The equilibrium $y_e = 0$ of S is *exponentially stable* if there exists an $\alpha > 0$, and for every $\varepsilon > 0$ and every $t_0 \geq 0$, there exists a $\delta = \delta(\varepsilon) > 0$ such that $\|y(t, y_0, t_0)\| < \varepsilon e^{-\alpha(t-t_0)}$ for all $t \geq t_0$ and for all $y(\cdot, y_0, t_0) \in S$ whenever $\|y_0\| < \delta$ (and $y_0 \in C$). □

The preceding definitions concern *local characterizations* of an equilibrium. In the following, we address *global characterizations*. In this case we find it convenient to let $C = X$.

Definition 9.11.7 The equilibrium $y_e = 0$ of S is *asymptotically stable in the large* if

(i) it is stable; and

(ii) for every $y(\cdot, y_0, t_0) \in S$ and for all $(t_0, y_0) \in \mathbb{R}^+ \times X$, (9.11.4) holds. □

In this case, the domain of attraction of $y_e = 0$ is all of X.

Definition 9.11.8 The equilibrium $y_e = 0$ of S is *uniformly asymptotically stable in the large* if

(i) it is uniformly stable;

(ii) it is *uniformly bounded*; that is, for any $\alpha > 0$ and every $t_0 \in \mathbb{R}^+$, there exists a $\beta = \beta(\alpha) > 0$ (independent of t_0) such that if $\|y_0\| < \alpha$, then $\|y(t, y_0, t_0)\| < \beta$ for all $t \geq t_0$ for all $y(\cdot, y_0, t_0) \in S$; and

(iii) it is *uniformly attractive in the large*; that is, for every $\alpha > 0$ and every $\varepsilon > 0$, and for every $t_0 \geq 0$, there exists a $\mu = \mu(\varepsilon, \alpha) > 0$ (independent of t_0), such that if $\|y_0\| < \alpha$, then for all $y(\cdot, y_0, t_0) \in S$, $\|y(t, y_0, t_0)\| < \varepsilon$ for all $t \geq t_0 + \mu$. □

Definition 9.11.9 The equilibrium $y_e = 0$ of S is *exponentially stable in the large* if there exist an $\alpha > 0$ and a $\gamma > 0$, and for every $\beta > 0$, there exists a $k(\beta) > 0$ such that

$$\|y(t, y_0, t_0)\| \leq k(\beta)\|y_0\|^{\gamma} e^{-\alpha(t-t_0)} \tag{9.11.5}$$

for all $y(\cdot, y_0, t_0) \in S$, for all $t \geq t_0$, whenever $\|y_0\| < \beta$. □

C. The principal stability results for DDS determined by semigroups

In our first result we establish sufficient conditions for various stability properties for system S_{DN}. We assume in these results that for each nonlinear semigroup $T_k(t)$ there exist constants $M_k \geq 1$ and $\omega_k \in \mathbb{R}$, $k \in \mathbb{N}$, such that

$$\|T_k(t)(y)\| \leq M_k e^{\omega_k t}\|y\| \tag{9.11.6}$$

for all $y \in C$, $t \geq 0$. We recall from Subsection 2.9C (see (2.9.1)) that in particular, (9.11.6) is always satisfied for a *quasi-contractive semigroup* $T_k(t)$ for some *computable parameters* (M_k, ω_k), $M_k \geq 1$ and $\omega_k \in \mathbb{R}$, whereas for a *contractive semigroup* $T_k(t)$, inequality (9.11.6) is satisfied with $M_k \geq 1$ and $\omega_k \leq 0$.

Also, in our first results we let

$$\lambda_k = \tau_{k+1} - \tau_k, \qquad k \in \mathbb{N} \tag{9.11.7}$$

and we assume that each mapping $H_k \colon C \to C$ satisfies the condition

$$\|H_k(y)\| \leq c_k\|y\| \tag{9.11.8}$$

for all $y \in C$, $k \in \mathbb{N}$, where $c_k > 0$ is a constant.

We require some additional notation. For any given $l_0 \in \mathbb{N}$ and $l_k \in \mathbb{N}^+_{l_0+1} \triangleq \{l_0+1, l_0+2, \dots\}$, we let $\pi_{l_0,l_0} = 1$, and we let π_{l_k,l_0} and a_{l_k,l_0} denote the *finite products*

$$\begin{cases} \pi_{l_k,l_0} = \prod_{i=0}^{k-1} (c_{l_i} M_{l_i} e^{\omega_{l_i}\lambda_{l_i}}) \\ a_{l_k,l_0} = M_{l_k} e^{((\omega_{l_k}+|\omega_{l_k}|)/2)\lambda_{l_k}} \pi_{l_k,l_0}, \qquad k \in \mathbb{N}^+_1 = \{1, 2, \dots\}. \end{cases} \tag{9.11.9}$$

Theorem 9.11.1 (a) For system S_{DN}, under the conditions (9.11.6) and (9.11.8), assume that for any $l_0 \in \mathbb{N}$ there exists a constant $\nu(l_0) > 0$ such that

$$a_{l_k,l_0} \le \nu(l_0) \tag{9.11.10}$$

for all $k \in \mathbb{N}$, where a_{l_k,l_0} is defined in (9.11.9). Then the equilibrium $y_e = 0$ of S_{DN} is *stable*.

(b) If in part (a), $\nu(l_0) = \nu$ (i.e., $\nu(l_0)$ in (9.11.10)) can be chosen independent of $l_0 \in \mathbb{N}$, then the equilibrium $y_e = 0$ of S_{DN} is *uniformly stable*.

(c) If in part (a), (9.11.10) is replaced by

$$\lim_{k\to\infty} a_{l_k,l_0} = 0 \tag{9.11.11}$$

for all $l_0 \in \mathbb{N}$, then the equilibrium $y_e = 0$ of S_{DN} is *asymptotically stable*.

(d) If the conditions of part (b) are satisfied and if in part (c) relation (9.11.11) is satisfied uniformly with respect to $l_0 \in \mathbb{N}$ (i.e., for every $\varepsilon > 0$ and every $l_0 \in \mathbb{N}$ there exists a $K(\varepsilon) \in \mathbb{N}$, independent of $l_0 \in \mathbb{N}$, such that $a_{l_k,l_0} < \varepsilon$ for all $k \ge K(\varepsilon)$), then the equilibrium $y_e = 0$ of S_{DN} is *uniformly asymptotically stable*.

(e) Assume that in part (a), (9.11.10) is replaced by

$$a_{l_k,l_0} < a\rho^{l_k-l_0}, \qquad l_0 \in \mathbb{N}, \qquad k \in \mathbb{N} \tag{9.11.12}$$

where $a > 0$ and $0 < \rho < 1$. Assume also that

$$\lambda_k = \tau_{k+1} - \tau_k \le \theta, \qquad k \in \mathbb{N} \tag{9.11.13}$$

where $\theta > 0$ is a constant. Then the equilibrium $y_e = 0$ of S_{DN} is *exponentially stable*.

(f) If in parts (c), (d), and (e), respectively, conditions (9.11.6) and (9.11.8) hold for *all* $y \in X$, then the equilibrium $y_e = 0$ of S_{DN} is *asymptotically stable in the large*, *uniformly asymptotically stable in the large*, and *exponentially stable in the large*, respectively.

Proof. (a) For system S_{DN}, with $E = \{\tau_0, \tau_1, \tau_2, \dots\}$, we associate each interval $[\tau_k, \tau_{k+1})$ with the index $k \in \mathbb{N}$. We find it convenient to employ a *relabeling* of indices. To this end, let $l_0 = [t_0] = [\tau_0]$, where $[x]$ denotes the integer part of $x \in \mathbb{R}$, and let $l_{k+1} = l_k + 1$, $k \in \mathbb{N}$. Then we can relabel E as $\{\tau_{l_0}, \tau_{l_1}, \dots\}$ and $[\tau_k, \tau_{k+1})$ as $[\tau_{l_k}, \tau_{l_k+1})$.

If $y(t_0) = y(\tau_{l_0}) = y_0$ and $y_0 \in C$, we have

$$y(t) \leq M_{l_0} e^{\omega_{l_0}(t-\tau_{l_0})} \|y(\tau_{l_0})\| \leq M_{l_0} e^{((\omega_{l_0}+|\omega_{l_0}|)/2)\lambda_{l_0}} \|y_0\|$$

for $t \in [\tau_{l_0}, \tau_{l_0+1})$. Therefore, in view of (9.11.9),

$$\|y(t)\| \leq a_{l_0,l_0} \|y_0\|, \qquad t \in [\tau_{l_0}, \tau_{l_0+1}) \tag{9.11.14}$$

is true. It is clear that

$$\|y(\tau_{l_0+1})\| \leq c_{l_0} M_{l_0} e^{\omega_{l_0}\lambda_{l_0}} \|y_0\|.$$

Similarly, for $t \in [\tau_{l_k}, \tau_{l_k+1})$, $k \in \mathbb{N}_1^+ = \{1, 2, \dots\}$, if $y(\tau_{l_k}) \in C$, then

$$\|y(t)\| \leq M_{l_k} e^{\omega_{l_k}(t-\tau_{l_k})} \|y(\tau_{l_k})\| \leq M_{l_k} e^{((\omega_{l_k}+|\omega_{l_k}|)/2)\lambda_{l_k}} \|y(\tau_{l_k})\|$$

is true for $t \in [\tau_{l_k}, \tau_{l_k+1})$, and

$$\|y(\tau_{l_k+1})\| \leq c_{l_k} M_{l_k} e^{\omega_{l_k}\lambda_{l_k}} \|y_k\|.$$

Therefore, by (9.11.9) and (9.11.14), we have

$$\|y(t)\| \leq a_{l_k,l_0} \|y_0\|, \qquad t \in [\tau_{l_k}, \tau_{l_k+1}), \qquad k \in \mathbb{N}. \tag{9.11.15}$$

For any $\varepsilon > 0$ and $l_0 \in \mathbb{N}$, let $\delta'(\varepsilon, l_0) = \varepsilon/\nu(l_0)$. From (9.11.10) and (9.11.15), it now follows that $\|y(t)\| < \varepsilon$, $t \in [\tau_{l_k}, \tau_{l_k+1})$, $k \in \mathbb{N}$, whenever $\|y_0\| < \delta$ and $y_0 \in C$. Because $l_0 = [t_0]$ and because for *all* $l_0 \in \mathbb{N}$ and *all* $k \in \mathbb{N}$ we can equate $\delta(\varepsilon, t_0) = \delta'(\varepsilon, [t_0])$, $t_0 \geq 0$, it follows that the equilibrium $y_e = 0$ of S_{DN} is stable.

(b) In proving part (b), note that $\delta'(\varepsilon, l_0) = \delta'(\varepsilon) = \varepsilon/\nu$ can be chosen independent of $l_0 \in \mathbb{N}$, and consequently, $\delta(\varepsilon, t_0) = \delta'(\varepsilon, [t_0]) = \delta'(\varepsilon) = \delta(\varepsilon)$ can also be chosen independent of $t_0 \in \mathbb{R}^+$. Therefore, the equilibrium $y_e = 0$ of S_{DN} is uniformly stable.

(c) From the assumptions on $E = \{t_0 = \tau_0, \tau_1, \tau_2, \dots\}$ it follows that $\lim_{k\to\infty} \tau_k = \infty$. Hence, $\sum_{i=0}^{k-1} \lambda_i = \tau_k - \tau_0 \to \infty$ as $k \to \infty$. Because for any $t \in [\tau_k, \tau_{k+1})$ we have $t = t_0 + \sum_{i=0}^{k-1} \lambda_i + \xi_k = \tau_k + \xi_k$ for some $0 \leq \xi_k < \tau_{k+1} - \tau_k = \lambda_k$, then $t \to \infty$ when $k \to \infty$. Hence, it follows from (9.11.11) and (9.11.15) that (9.11.4) holds for all $y(\cdot, y_0, t_0) \in S_{DN}$ whenever $y_0 \in C$. Therefore, the equilibrium $y_e = 0$ of S_{DN} is attractive and its domain of attraction coincides with the entire set $C \subset X$. Because (9.11.10) follows from (9.11.11), then, as in part (a), $y_e = 0$ of S_{DN} is stable. Hence, the equilibrium $y_e = 0$ of S_{DN} is asymptotically stable.

(d) The conditions of part (b) are satisfied, and thus the equilibrium $y_e = 0$ of system S_{DN} is uniformly stable. Therefore, we only need to prove that $y_e = 0$ is uniformly attractive.

Choose $\delta > 0$ in such a way that $B_\delta \triangleq \{y_0 \colon \|y_0\| < \delta\} \subset C$. Because (9.11.11) is satisfied uniformly with respect to $l_0 \in \mathbb{N}$, then for every $\varepsilon^* > 0$ and every $l_0 \in \mathbb{N}$ there exists a $K^*(\varepsilon^*) \in \mathbb{N}$ (independent of $l_0 \in \mathbb{N}$) such that $a_{l_k,l_0} < \varepsilon^*$ for all $k > K^*(\varepsilon^*)$. Hence, from (9.11.15), we have $\|y(t)\| \leq a_{l_k,l_0}\|y_0\| < \varepsilon^*\delta$ for all

$t \in [\tau_{l_k}, \tau_{l_k+1})$ and for all $k \geq K^*(\varepsilon^*)$. Let $\varepsilon^* = \varepsilon/\delta$. Then $K^*(\varepsilon^*) = K^*(\varepsilon/\delta) = K(\varepsilon)$ and $\|y(t)\| < \varepsilon$ for all $t \geq \tau_{l_0+K(\varepsilon)}$. If we let $\mu(\varepsilon) = \tau_{l_0+K(\varepsilon)} - \tau_{l_0}$, then we have that $\|y(t, y_0, t_0)\| < \varepsilon$ for all $t \geq t_0 + \mu$ and for all $y(\cdot, y_0, t_0) \in S_{DN}$, whenever $\|y_0\| < \delta$. Hence, the equilibrium $y_e = 0$ of S_{DN} is uniformly attractive and uniformly asymptotically stable.

(e) To prove part (e), note that as was shown in the proofs of parts (a) and (c), for any $t_0 \in \mathbb{R}^+$ and any $t \geq t_0$, there exist an $l_0 \in \mathbb{N}$ and a $k \in \mathbb{N}$ such that $t \in [\tau_{l_k}, \tau_{l_k+1})$ and (9.11.15) holds. Because $t - t_0 < \tau_{l_k+1} - \tau_{l_0} = \sum_{i=l_0}^{l_k} \lambda_i$ and in view of (9.11.13), $\sum_{i=l_0}^{l_k} \lambda_i \leq (l_k - l_0 + 1)\theta$, and therefore, we have $l_k - l_0 > ((t - t_0)/\theta) - 1$. Hence, in view of (9.11.12), we have $\|y(t)\| < a\rho^{((t-t_0)/\theta)-1}\|y_0\|$. For any $\varepsilon > 0$, let $\delta = (\varepsilon\rho)/a$. Then for any $y_0 \in C$ with $\|y_0\| < \delta$, we have $\|y(t)\| < \varepsilon e^{-\alpha(t-t_0)}$, $t \geq t_0$, where $\alpha = (-\ln\rho)/\theta > 0$. Therefore, the equilibrium $y_e = 0$ of S_{DN} is exponentially stable.

(f) We note that if the estimates (9.11.6) and (9.11.8) hold for all $y \in X$, then inequality (9.11.15) is valid for *all* $y_0 \in X$.

(i) Repeating the reasoning in the proof of part (c) for any $y_0 \in X$ and any $t_0 \in \mathbb{R}^+$, we can conclude that in this case (9.11.4) holds for all $y(\cdot, y_0, t_0) \in S_{DN}$ whenever $y_0 \in X$ and $t_0 \in \mathbb{R}^+$. Therefore, the equilibrium $y_e = 0$ of S_{DN} is asymptotically stable in the large.

(ii) The equilibrium $y_e = 0$ is uniformly stable and (9.11.15) is valid for all $y_0 \in X$. Therefore, whenever $\|y_0\| < \delta$, then $\|y(t, y_0, t_0)\| < \varepsilon$ for all $t \geq t_0$, where $\varepsilon = \varepsilon(\delta)$. Therefore, for any $\alpha > 0$ and every $t_0 \in \mathbb{R}^+$, there exists a $\beta = \beta(\alpha) > 0$, independent of t_0, such that when $\|y_0\| < \delta$, then $\|y(t, y_0, t_0)\| < \beta$ for all $t \geq t_0$ and all $y(\cdot, y_0, t_0) \in S$. Therefore, the system S_{DN} is uniformly bounded.
Next, similarly as in the proof of part (d), for every $\alpha > 0$ and for every $\varepsilon > 0$ there exists a $K(\varepsilon, \alpha) \in \mathbb{N}$ (independent of $t_0 \geq 0$), such that $\|y(t)\| < \varepsilon$ for all $t \geq \tau_{l_0} + K(\varepsilon, \alpha)$. If we let $\mu(\varepsilon, \alpha) = \tau_{l_0+K(\varepsilon,\alpha)} - \tau_{l_o}$, then we have that $\|y(t, y_0, t_0)\| < \varepsilon$ for all $t \geq t_0 + \mu$ and for all $y(\cdot, y_0, t_0) \in S_{DN}$, whenever $\|y_0\| < \delta$. Hence, the equilibrium $y_e = 0$ of S_{DN} is uniformly asymptotically stable in the large.

(iii) For every $\beta > 0$ and for every $\|y_0\| < \beta$ we have similarly as in the proof of part (e) above that

$$\|y(t)\| < (\alpha/\rho)\rho^{(t-t_0)/\theta}\|y_0\| = (\alpha/\rho)\|y_0\|e^{-\alpha(t-t_0)}$$

for all $t \geq t_0 \geq 0$, where $\alpha = -\ln\rho/\theta > 0$. Let $k(\beta) = \alpha/\rho$. It now follows that the equilibrium $y_e = 0$ of S_{DN} is exponentially stable in the large. This completes the proof. □

Corollary 9.11.1 (a) For system S_{DN} assume that the following statements are true.

(i) Condition (9.11.6) holds (with parameters M_k, ω_k).

(ii) Condition (9.11.8) holds (with parameter c_k).

(iii) For all $k \in \mathbb{N}$, $\lambda_k = \tau_{k+1} - \tau_k \leq \theta < \infty$.

(iv) For all $k \in \mathbb{N}$, $M_k \leq M < \infty$ and $\omega_k \leq \omega < \infty$ where $M \geq 1$ and $\omega \in \mathbb{R}$ are constants.

(v) For all $k \in \mathbb{N}$,

$$c_k M_k e^{\omega_k \lambda_k} \leq 1. \qquad (9.11.16)$$

Then the equilibrium $y_e = 0$ of S_{DN} is *stable* and *uniformly stable*.

(b) If in part (a), hypothesis (v) is replaced by

$$c_k M_k e^{\omega_k \lambda_k} \leq \delta < 1 \qquad (9.11.17)$$

for all $k \in \mathbb{N}$, where $\delta > 0$, then the equilibrium $y_e = 0$ of S_{DN} is *asymptotically stable*, *uniformly asymptotically stable*, and *exponentially stable*.

(c) If in part (a) it is assumed that inequalities (9.11.6) and (9.11.8) hold for all $y \in X$ and inequality (9.11.16) is replaced by (9.11.17), then the equilibrium $y_e = 0$ of S_{DN} is *asymptotically stable in the large*, *uniformly asymptotically stable in the large*, and *exponentially stable in the large*.

Proof. (a) It is easily shown that in part (a) the estimate (9.11.10) is satisfied with $\nu(l_0) = \nu = Me^{((\omega+|\omega|)/2)\theta}$, independent of $l_0 \in \mathbb{N}$. Therefore, the conditions in parts (a) and (b) of Theorem 9.11.1 are satisfied. This proves part (a) of the corollary.

(b) In view of inequality (9.11.17) the estimate (9.11.12) is true with

$$a = (M + 1)e^{((\omega+|\omega|)/2)\theta}$$

and $\rho = \delta$. Therefore the limit relation (9.11.11) is satisfied uniformly with respect to $l_0 \in \mathbb{N}$. This proves part (b) of the corollary.

(c) The conclusions of part (c) of this corollary follow directly from part (f) of Theorem 9.11.1. □

From Theorem 2.9.1, we recall that for *any* C_0-semigroup $T_k(t)$, there will exist $\omega_k \geq 0$ and $\mu_k \geq 1$ such that

$$\|T_k(t)\| \leq \mu_k e^{\omega_k t}, \qquad t \geq 0. \qquad (9.11.18)$$

Furthermore, in accordance with Theorem 2.9.5, if $T_k(t)$ is a C_0-semigroup that is differentiable for $t > r$, if A_k is its infinitesimal generator, and if $\text{Re}\lambda_k \leq -\alpha_{k_0}$ for all $\lambda_k \in \sigma(A_k)$, then given any positive $\alpha_k < \alpha_{k_0}$, there is a constant $K(\alpha_k) > 0$ such that

$$\|T_k(t)\| \leq K(\alpha_k)e^{-\alpha_k t}, \qquad t > r. \qquad (9.11.19)$$

These facts simplify considerably the estimates of the analogous parts of Theorem 9.11.1 and Corollary 9.11.1, valid for C_0-semigroups. We state these results in the following. Their proofs are very similar to the corresponding proofs given in Theorem 9.11.1 and Corollary 9.11.1 and are omitted.

Similarly as in Theorem 9.11.1, we utilize in our next result the relation

$$\|T_k(t)\| \leq M_k e^{\omega_k t}, \qquad t \geq 0 \qquad (9.11.20)$$

where, depending on the situation on hand, the constants $M_k \geq 1$ and $\omega_k \in \mathbb{R}$ are obtained from either (9.11.18) or (9.11.19).

Similarly as in (9.11.9), we define in the case of DDS S_{DC_0} the *finite products*

$$\begin{cases} \pi_{l_k,l_0} = \prod_{i=0}^{k-1}(\|H_{l_i}\|M_{l_i}e^{\omega_{l_i}\lambda_{l_i}}) \\ a_{l_k,l_0} = M_{l_k}e^{((\omega_{l_k}+|\omega_{l_k}|)/2)\lambda_k}\pi_{l_k,l_0}, \end{cases} \tag{9.11.21}$$

$k \in \mathbb{N}_1^+ = \{1, 2, \dots\}$, where $\|H_k\|$, $k \in \mathbb{N}$, denotes the norm of the bounded linear operator H_k used in defining the DDS S_{DC_0} in (9.11.1).

Theorem 9.11.2 (a) For system S_{DC_0} assume that (9.11.20) is true and that for any $l_0 \in \mathbb{N}$ there exists a constant $\nu(l_0) > 0$ such that

$$a_{l_k,l_0} \leq \nu(l_0) \tag{9.11.22}$$

holds for all $k \in \mathbb{N}$, where a_{l_k,l_0} is defined in (9.11.21). Then the equilibrium $y_e = 0$ of S_{DC_0} is *stable*.

(b) If in part (a), $\nu(l_0) = \nu > 0$ can be chosen independent of $l_0 \in \mathbb{N}$, then the equilibrium $y_e = 0$ of S_{DC_0} is *uniformly stable*.

(c) If in part (a), hypothesis (9.11.22) is replaced by

$$\lim_{k\to\infty} a_{l_k,l_0} = 0 \tag{9.11.23}$$

for all $l_0 \in \mathbb{N}$, then the equilibrium $y_e = 0$ of S_{DC_0} is *asymptotically stable in the large*.

(d) If the conditions of part (b) are satisfied and in part (c), the limit relation (9.11.23) is satisfied uniformly with respect to $l_0 \in \mathbb{N}$, then the equilibrium $y_e = 0$ of S_{DC_0} is *uniformly asymptotically stable in the large*.

(e) If in part (d) relations (9.11.12) and (9.11.13) hold, then the equilibrium $y_e = 0$ of S_{DC_0} is *exponentially stable in the large*. □

Corollary 9.11.2 For system S_{DC_0} assume that

(i) For all $k \in \mathbb{N}$, $\lambda_k = \tau_{k+1} - \tau_k \leq \theta < \infty$.

(ii) For all $k \in \mathbb{N}$, $M_k \leq M < \infty$ and $\omega_k \leq \omega < \infty$ where $M \geq 1$ and $\omega \in \mathbb{R}$ are constants (M_k and ω_k are given in (9.11.20)).

(a) Assume that

$$\|H_k\|M_ke^{\omega_k\lambda_k} \leq 1$$

for all $k \in \mathbb{N}$. Then the equilibrium $y_e = 0$ of S_{DC_0} is *stable* and *uniformly stable*.

(b) Assume that

$$\|H_k\|M_ke^{\omega_k\lambda_k} \leq \delta < 1$$

for all $k \in \mathbb{N}$. Then the equilibrium $y_e = 0$ of S_{DC_0} is *asymptotically stable*, *uniformly asymptotically stable*, *uniformly asymptotically stable in the large*, *exponentially stable*, and *exponentially stable in the large*. □

Remark 9.11.5 Corollaries 9.11.1 and 9.11.2 are more conservative than Theorems 9.11.1 and 9.11.2 because in the case of the latter we put restrictions on partial products (see, e.g., (9.11.10)) whereas in the case of the former, we put corresponding restrictions on the individual members of the partial products (see, e.g., (9.11.16)). However, the corollaries are easier to apply than the theorems. □

Remark 9.11.6 In contrast to the stability results for DDS given in the preceding section, the results of the present section do not require determination of appropriate Lyapunov functions, which is not necessarily an easy task. Instead, in the application of Theorems 9.11.1 and 9.11.2 and Corollaries 9.11.1 and 9.11.2, we bring to bear the qualitative theory of semigroups in determining appropriate estimates of bounds of the norms of semigroups. It must be pointed out, however, that the determination of such estimates is not necessarily an easy task either. Moreover, the ambiguity involved in the search of Lyapunov functions in the application of the results for DDS involving such functions offers flexibility in efforts of reducing conservatism of results. □

D. Applications

We now apply the results of the present section in the stability analysis of three classes of discontinuous dynamical systems.

Example 9.11.1 (*Autonomous first-order retarded functional differential equations*)

(1) *Dynamical systems determined by nonlinear semigroups*

Consider initial value problems described by a system of *autonomous first-order retarded functional differential equations* (with *delay* r) given by

$$\begin{cases} \dot{x}(t) = f(x_t), & t > 0 \\ x(t) = \varphi(t), & -r \le t \le 0 \end{cases} \tag{9.11.24}$$

where $f\colon C \to \mathbb{R}$, $C \subset C_r$, $C_r = C[[-r,0],\mathbb{R}^n]$ is a Banach space with norm defined by

$$\|\varphi\| = \max\{|\varphi(t)|\colon -r \le t \le 0\} \tag{9.11.25}$$

and $x_t \in C$ is the function determined by $x_t(s) = x(t+s)$ for $-r \le s \le 0$. We assume that C is a neighborhood of the origin.

Assume that f satisfies a Lipschitz condition

$$|f(\xi) - f(\eta)| \le K\|\xi - \eta\| \tag{9.11.26}$$

for all $\xi, \eta \in C$. Under these conditions, the initial value problem (9.11.24) has a unique solution for every initial condition $\varphi \in C$, denoted by $\psi(t,\varphi)$ that exists for all $t \in \mathbb{R}^+$ (see Example 2.9.2). In this case, the mapping $T(t)\colon C \to C$ given by $T(t)(\varphi) = \psi_t(\cdot,\varphi)$, or equivalently, $(T(t)\varphi)(s) = \psi(t+s,\varphi)$, defines a nonlinear semigroup on $C \subset C_r$. In fact, $T(t)$ is a quasi-contractive semigroup, and

$$\|T(t)(\xi) - T(t)(\eta)\| \le e^{Kt}\|\xi - \eta\| \tag{9.11.27}$$

for all $t \in \mathbb{R}^+$ and $\xi, \eta \in C$ (see Example 2.9.2).

If we define $A: D(A) \to C$ by $A\varphi = \dot{\varphi}$, $D(A) = \{\varphi \in C: \dot{\varphi} \in C \text{ and } \dot{\varphi}(0) = f(\varphi)\}$, then $D(A)$ is dense in C, A is the generator and also the infinitesimal generator of $T(t)$, and $T(t)$ is differentiable for $t > r$ (see Example 2.9.2).

(2) *Discontinuous dynamical systems determined by nonlinear semigroups*

Now consider the system of *discontinuous retarded functional differential equations* given by

$$\begin{cases} \dot{x}(t) = F_k(x_t), & \tau_k \le t < \tau_{k+1} \\ x_t = H_k(x_{t^-}), & t = \tau_{k+1}, \quad k \in \mathbb{N} \end{cases} \tag{9.11.28}$$

where $\{F_k\}$ and $\{H_k\}$ are given collections of mappings $F_k: C \to \mathbb{R}^n$ and $H_k: C \to C$ and $E = \{t_0 = \tau_0, \tau_1, \tau_2, \dots : \tau_0 < \tau_1 < \tau_2 < \cdots\}$ is a given unbounded set. We assume that for all $k \in \mathbb{N}$, $H_k \in C[C, C]$, $H_k(0) = 0$, and

$$\|H_k(\xi)\| \le C_k \|\xi\| \tag{9.11.29}$$

for all $\xi \in C$, where $C_k > 0$ is a finite constant. Also, we assume that $F_k(0) = 0$ and that F_k satisfies the Lipschitz condition

$$|F_k(\xi) - F_k(\eta)| < K_k \|\xi - \eta\| \tag{9.11.30}$$

for all $\xi, \eta \in C$.

For every $k \in \mathbb{N}$, the initial value problem

$$\begin{cases} \dot{x}(t) = F_k(x_t), & t > \tau_k \\ x_t = \varphi^{(k)}, & t = \tau_k \end{cases} \tag{9.11.31}$$

possesses a unique solution $\psi_t^{(k)}(\cdot, \varphi^{(k)}, \tau_k)$ for every initial condition $\varphi^{(k)} \in C$ that exists for all $t \ge \tau_k$ with $\psi_{\tau_k}^{(k)}(\cdot, \varphi^{(k)}, \tau_k) = \varphi^{(k)}$. Therefore, it follows that for every $\varphi^{(0)}$, (9.11.28) possesses a unique solution that exists for all $t \ge t_0 = \tau_0 \ge 0$, given by

$$\psi_t(\cdot, \varphi^{(0)}, \tau_0) = \begin{cases} \psi_t^{(k)}(\cdot, \varphi^{(k)}, \tau_k), & \tau_k \le t < \tau_{k+1} \\ H_k(\psi_{t^-}^{(k)}(\cdot, \varphi^{(k)}, \tau_k)) = \varphi^{k+1}, & t = \tau_{k+1}, \ k \in \mathbb{N}. \end{cases} \tag{9.11.32}$$

Note that $\varphi^{(k)} = H_{k-1}(\psi_{\tau_k^-}^{(k-1)}(\cdot, \varphi^{(k-1)}, \tau_{k-1}))$, $k = 1, 2, \dots$. Also, note that $\psi_t(\cdot, \varphi^{(0)}, t_0)$ is continuous with respect to t on $[t_0, \infty) - \{\tau_1, \tau_2, \dots\}$ and that at $t = \tau_k$, $k = 1, 2, \dots$, $\psi_t(\cdot, \varphi^{(0)}, t_0)$ *may* be discontinuous. Furthermore, note that $\xi = 0$ is an equilibrium of (9.11.28) and that $\psi_t(\cdot, 0, t_0) = 0$ for all $t \ge t_0$.

Next, for the initial value problem (9.11.31) we define

$$\psi_t^{(k)}(\cdot, \varphi^{(k)}, \tau_k) = T_k(t - \tau_k)(\varphi^{(k)}),$$

$T_k(t - \tau_k): C \to C$, $t \ge \tau_k$. It follows that $T_k(s)$, $s \in \mathbb{R}^+$, is a *quasi-contractive semigroup*. This allows us to characterize system (9.11.28) as

$$\begin{cases} y(t, \varphi^{(0)}, t_0) = T_k(t - \tau_k)\varphi^{(k)}), & \tau_k \le t < \tau_{k+1} \\ y_t = H_k(y_{t^-}), & t = \tau_{k+1}, \qquad k \in \mathbb{N}. \end{cases} \tag{9.11.33}$$

Finally, it is clear that (9.11.28) (resp., (9.11.33)) determines a discontinuous dynamical system that is a special case of the DDS S_{DN}.

Proposition 9.11.1 (a) For system (9.11.28) (resp., (9.11.33)) assume the following.

(i) For each $k \in \mathbb{N}$, the function F_k satisfies the Lipschitz condition (9.11.30) with Lipschitz constant K_k for all $\xi, \eta \in C \subset C_r$, where C is a neighborhood of the origin.

(ii) For each $k \in \mathbb{N}$, the function H_k satisfies condition (9.11.29) with constant C_k for all $\xi \in C$.

(iii) For each $k \in \mathbb{N}$, $(\tau_{k+1} - \tau_k) \stackrel{\triangle}{=} \lambda_k \leq \theta < \infty$, $C_k \leq \gamma < \infty$, and $K_k \leq K < \infty$.

(iv) For all $k \in \mathbb{N}$,

$$C_k e^{K_k \lambda_k} \leq 1. \tag{9.11.34}$$

Then the equilibrium $\xi = 0$ of system (9.11.28) (resp., (9.11.33)) is *uniformly stable*.

(b) In part (a) above, replace (iv) by the following hypothesis.

(v) for all $k \in \mathbb{N}$,

$$C_k e^{K_k \lambda_k} \leq \delta < 1. \tag{9.11.35}$$

Then the equilibrium $\xi = 0$ of system (9.11.28) (resp., (9.11.33)) is *uniformly asymptotically stable* and *exponentially stable*.

(c) In part (a) above, replace (iv) by hypothesis (v) and assume that conditions (9.11.29) and (9.11.30) hold for $C = C_r$. Then the equilibrium $\xi = 0$ of system (9.11.28) (resp., (9.11.33)) is *uniformly asymptotically stable in the large* and *exponentially stable in the large*.

Proof. In view of (9.11.27), we have, because $F_k(0) = 0$,

$$\|T_k(t)(\xi)\| \leq e^{K_k t} \|\xi\| \tag{9.11.36}$$

for all $t \geq 0$, $k \in \mathbb{N}$, and $\xi \in C$, resp., $\xi \in C_r$. Setting $M_k = 1$, $c_k = C_k$, and $\omega_k = K_k$, we can see that all hypotheses of Corollary 9.11.1 are satisfied. This completes the proof. □

(3) *Dynamical systems determined by linear semigroups*

Now assume $C = C_r$. If in (9.11.24), $f = L$ is a linear mapping from C_r to $\mathbb{R}^n$ defined by the Stieltjes integral

$$L(\varphi) = \int_{-r}^{0} [dB(s)] \varphi(s), \tag{9.11.37}$$

we obtain the initial value problem (see Example 2.9.2)

$$\begin{cases} \dot{x}(t) = L(x_t), & t > 0, \\ x(t) = \varphi(t), & -r \leq t \leq 0. \end{cases} \tag{9.11.38}$$

In (9.11.37), $B(s) = [b_{ij}(s)]$ is an $n \times n$ matrix whose entries are assumed to be functions of bounded variation on $[-r, 0]$. Then L is Lipschitz continuous on C_r with Lipschitz constant K less than or equal to the variation of B in (9.11.37). In this case, the semigroup $T(t)$ is a C_0-semigroup. The *spectrum* of its generator consists of all solutions of the equation

$$\det\left(\int_{-r}^{0} e^{\lambda s} dB(s) - \lambda I\right) = 0. \tag{9.11.39}$$

If in particular, all the solutions of (9.11.39) satisfy the relation $\mathrm{Re}\lambda < -\alpha_0$ for some $\alpha_0 > 0$, then it follows from Theorem 2.9.5 that for any positive $\alpha < \alpha_0$, there is a constant $P(\alpha) > 0$ such that

$$\|T(t)\| \leq P(\alpha)e^{-\alpha t}, \qquad t \geq 0. \tag{9.11.40}$$

When the above assumptions do not hold, then in view of Theorem 2.9.1 we still have the estimate

$$\|T(t)\| \leq Qe^{\mu t}, \qquad t \geq 0, \tag{9.11.41}$$

for some $\mu \geq 0$ and $Q \geq 1$.

Next, let $F_k(x_t) = L_k x_t$ where $L_k \colon C_r \to \mathbb{R}^n$ is defined similarly as in (9.11.37) by $L_k(\varphi) = \int_{-r}^{0} [dB_k(s)]\varphi(s)$ and let $H_k(x_t) = G_k x_t$ where $G_k \in C[C_r, C_r]$ is assumed to be a bounded linear operator. Then system (9.11.28) assumes the form

$$\begin{cases} \dot{x}(t) = L_k x_t, & \tau_k \leq t < \tau_{k+1}, \\ x_t = G_k x_{t^-}, & t = \tau_{k+1}, \qquad k \in \mathbb{N}. \end{cases} \tag{9.11.42}$$

It is clear that (9.11.42) determines a *DDS determined by linear semigroups* that is a special case of S_{DC_0}.

In the following, when all the solutions of the characteristic equation

$$\det\left(\int_{-r}^{0} e^{\lambda_k s} d[B_k(s)] - \lambda_k I\right) = 0$$

satisfy the condition $\mathrm{Re}\lambda_k \leq -\alpha_{0k}$, then given any $0 < \alpha_k < \alpha_{0k}$, there is a constant $P_k(\alpha_k) > 0$ such that

$$\|T_k(t)\| \leq P_k(\alpha_k)e^{-\alpha_k t}, \qquad t \geq 0 \tag{9.11.43}$$

(see (9.11.40)). Otherwise, we still have the estimate

$$\|T_k(t)\varphi\| \leq Q_k e^{\mu_k t}, \qquad t \geq 0 \tag{9.11.44}$$

for some $Q_k \geq 1$, $\mu_k \geq 0$ (see (9.11.41)).

When (9.11.43) applies, we let in the following

$$M_k = P_k(\alpha_k), \qquad -\alpha_k = \omega_k \tag{9.11.45}$$

and when (9.11.44) applies, we let

$$M_k = Q_k, \qquad \mu_k = \omega_k. \tag{9.11.46}$$

Thus, in all cases we have the estimate

$$\|T_k(t)\| \leq M_k e^{\omega_k t}, \qquad t \geq 0. \tag{9.11.47}$$

Proposition 9.11.2 (a) For system (9.11.42) assume the following.

(i) For each $k \in \mathbb{N}$, $(\tau_{k+1} - \tau_k) \triangleq \lambda_k \leq \theta < \infty$, $M_k \leq M < \infty$, and $\omega_k \leq \omega < \infty$.

(ii) For each $k \in \mathbb{N}$,

$$\|G_k\| M_k e^{\omega_k \lambda_k} \leq 1 \tag{9.11.48}$$

where M_k and ω_k are given in (9.11.43)–(9.11.46).

Then the equilibrium $\xi = 0$ of system (9.11.42) is *uniformly stable*.

(b) In part (a) above, replace (9.11.48) by

$$\|G_k\| M_k e^{\omega_k \lambda_k} \leq \delta < 1. \tag{9.11.49}$$

Then the equilibrium $\xi = 0$ of system (9.11.42) is *uniformly asymptotically stable in the large* and *exponentially stable in the large*.

Proof. The proof follows directly from Corollary 9.11.2. □

Example 9.11.2 (*Heat equation*)

(1) *Dynamical systems determined by the heat equation*

We consider initial and boundary value problems described by equations of the form

$$\begin{cases} \dfrac{\partial u}{\partial t} = a^2 \Delta u, & (t, x) \in [t_0, \infty) \times \Omega \\ u(t_0, x) = \varphi(x), & x \in \Omega \\ u(t, x) = 0, & (t, x) \in [t_0, \infty) \times \partial\Omega \end{cases} \tag{9.11.50}$$

where $\Omega \subset \mathbb{R}^n$ is a bounded domain with smooth boundary $\partial\Omega$, $\Delta = \sum_{i=1}^{n} \partial^2/\partial x_i^2$ denotes the *Laplacian* and $a^2 > 0$ is a constant.

We assume that in (9.11.50), $\varphi \in X = H^2[\Omega, \mathbb{R}] \cap H_0^1[\Omega, \mathbb{R}]$ where $H_0^1[\Omega, \mathbb{R}]$ and $H^2[\Omega, \mathbb{R}]$ are *Sobolev spaces* (refer to Section 2.10). For any $\varphi \in X$, we define the H^1-norm by

$$\|\varphi\|_{H^1}^2 = \int_\Omega (|\nabla\varphi|^2 + |\varphi|^2) dx \tag{9.11.51}$$

where $\nabla\varphi^T = (\partial\varphi/\partial x_1, \ldots, \partial\varphi/\partial x_n)$.

It has been shown (see, e.g., [33]) that for each $\varphi \in X$ there exists a *unique solution* $u = u(t, x)$, $t \geq t_0$, $x \in \Omega$ for (9.11.50) such that $u(t, \cdot) \in X$ for each fixed $t \geq t_0$ and $u(t, \cdot) \in X$ is a continuously differentiable functions from $[t_0, \infty)$ to X with respect to the H^1-norm (9.11.51). In the present case, (9.11.50) can be cast as an initial value problem in the space X with respect to the H^1-norm, letting $u(t, \cdot) = U(t)$ and assuming, without loss of generality, that $t_0 = 0$,

$$\begin{cases} \dot{U}(t) = AU(t), & t \geq 0 \\ U(0) = \varphi \in X \end{cases} \tag{9.11.52}$$

where A is the linear operator determined by $A = \sum_{i=1}^{n} a^2 \partial^2 / \partial x_i^2$ with $U(t, \varphi)$, $t \geq 0$, denoting the solution of (9.11.52) with $U(0, \varphi) = \varphi$. Furthermore, it has been shown (e.g., [33]) that (9.11.52) determines a C_0-semigroup $T(t)\colon X \to X$, where for any $\varphi \in X$, $U(t, \varphi) = T(t)\varphi$. Because $U(t, 0) = 0$, $t \geq 0$, then $\varphi = 0 \in X$ is an equilibrium for (9.11.52) (resp., for (9.11.50)). Also (see (9.10.52) or [33]),

$$\|T(t)\|_{H^1} \leq e^{-(c/2)t}, \qquad t \geq 0 \tag{9.11.53}$$

where $c = \min\{a^2, a^2/\gamma^2\}$, $\gamma = \delta/\sqrt{n}$ and Ω can be put into a cube of length δ.

(2) *Discontinuous dynamical systems determined by the heat equation*

Now consider the DDS determined by the equations

$$\begin{cases} \dfrac{\partial u}{\partial t} = a_k^2 \Delta u, & (t, x) \in [\tau_k, \tau_{k+1}) \times \Omega \\ u(t, \cdot) = g_k(u(t^-, \cdot)) \stackrel{\triangle}{=} \varphi_{k+1}(\cdot), & t = \tau_{k+1} \\ u(t_0, x) = \varphi_0(x), & x \in \Omega \\ u(t, x) = 0, & (t, x) \in [t_0, \infty) \times \partial\Omega, \end{cases} \tag{9.11.54}$$

$k \in \mathbb{N}$, where all symbols are defined similarly as in (9.11.50), $a_k^2 > 0$, $k \in \mathbb{N}$, are constants, $\{g_k\}$ is a given family of mappings $g_k \in C[X, X]$, $k \in \mathbb{N}$, and

$$E = \{t_0 = \tau_0, \tau_1, \tau_2, \ldots : \tau_0 < \tau_1 < \tau_2 < \cdots\}$$

is a given unbounded set. We assume that $g_k(0) = 0$ and that there exists a constant $d_k > 0$ such that

$$\|g_k(\varphi)\|_{H^1} \leq d_k \|\varphi\|_{H^1} \tag{9.11.55}$$

for all $\varphi \in X$, $k \in \mathbb{N}$.

Associated with (9.11.54) we have a family of initial and boundary value problems determined by

$$\begin{cases} \dfrac{\partial u}{\partial t} = a_k^2 \Delta u, & (t, x) \in [\tau_k, \infty) \times \Omega \\ u(\tau_k, x) = \varphi_k(x), & x \in \Omega \\ u(t, x) = 0, & (t, x) \in [\tau_k, \infty) \times \partial\Omega \end{cases} \tag{9.11.56}$$

$k \in \mathbb{N}$. Because for every $k \in \mathbb{N}$ and every $(\tau_k, \varphi_k) \in \mathbb{R}^+ \times X$, the initial and boundary value problem (9.11.56) possesses a unique solution $u_k(t, \cdot)$ that exists for all $t \geq \tau_k$ with $u_k(\tau_k, x) = \varphi_k(x)$, it follows that for every $\varphi_0 \in X$, (9.11.54) possesses a unique solution $u(t, \cdot)$ that exists for all $t \geq \tau_0 \geq 0$, given by

$$u(t, \cdot) = \begin{cases} u_k(t, \cdot), & \tau_k \leq t < \tau_{k+1} \\ g_k(u_k(t^-, \cdot)) \stackrel{\triangle}{=} \varphi_{k+1}(\cdot), & t = \tau_{k+1}, \quad k \in \mathbb{N} \end{cases} \tag{9.11.57}$$

with $u(t_0, x) = \varphi_0(x)$. Notice that every solution $u(t, \cdot)$ is continuous with respect to t on $[t_0, \infty) - \{\tau_1, \tau_2, \dots\}$, and that at $t = \tau_k$, $k = 1, 2, \dots, u(t, \cdot)$ *may* be discontinuous. Furthermore, $\varphi_e = 0 \in X$ is an equilibrium for (9.11.54) and $u(t, \cdot)|_{\varphi=0} = 0$ for all $t \geq t_0 \geq 0$ is a trivial motion.

Next, as in the initial and boundary value problem (9.11.50), we can cast the initial and boundary value problems (9.11.56) as initial value problems (as in (9.11.52)) that determine C_0-semigroups $T_k(t - \tau_k)$, $t \geq \tau_k$, $k \in \mathbb{N}$, that admit the estimates

$$\|T_k(t - \tau_k)\|_{H^1} \leq e^{-(c_k/2)(t-\tau_k)} \tag{9.11.58}$$

where $c_k = \min\{a_k^2, a_k^2/\gamma^2\}$. Letting $u_k(t, \cdot) = T_k(t - \tau_k)u_k(\tau_k, \cdot)$ in (9.11.57), system (9.11.54) can be characterized as

$$\begin{cases} u(t, \cdot) = T_k(t - \tau_k)u_k(\tau_k, \cdot), & \tau_k \leq t < \tau_{k+1} \\ u(t, \cdot) = g_k(u_k(t^-, \cdot)), & t = \tau_{k+1}, \quad k \in \mathbb{N}. \end{cases} \tag{9.11.59}$$

Finally, it is clear that (9.11.54) (resp., (9.11.59)) determines a discontinuous dynamical system that is a special case of the DDS S_{DN}.

Proposition 9.11.3 For system (9.11.54) (resp., (9.11.59)) assume that

$$\lambda_k \triangleq \tau_{k+1} - \tau_k \leq \theta < \infty, \qquad \omega_k \triangleq -c_k/2 \leq \omega < \infty,$$

and $d_k \leq d < \infty$, $k \in \mathbb{N}$.

(a) If for each $k \in \mathbb{N}$,

$$d_k e^{\omega_k \lambda_k} \leq 1, \tag{9.11.60}$$

then the equilibrium $\varphi_e = 0$ of system (9.11.54) is *uniformly stable* with respect to the H^1-norm.

(b) If for all $k \in \mathbb{N}$,

$$d_k e^{\omega_k \lambda_k} \leq \delta < 1, \tag{9.11.61}$$

where $\delta > 0$ is a constant, then the equilibrium $\varphi_e = 0$ of system (9.11.54) is *uniformly asymptotically stable in the large* and *exponentially stable in the large*.

Proof. The proof follows directly from Corollary 9.11.1. □

9.12 Notes and References

The proofs of most of the results given in Sections 9.2–9.5, for dynamical systems determined by differential equations defined on Banach spaces, are direct consequences of corresponding results presented in Chapter 3, for dynamical systems defined on metric spaces. The results for composite systems presented in Section 9.6, which constitute generalizations of results established in [36], are motivated by results given in [29] for composite systems defined on metric spaces. The example given in

Section 9.6 is similar to an example considered in [36]. A good reference on point kinetics models of multicore nuclear reactors is [35]. Our presentation of the stability analysis of such models in Section 9.7 is based on the results established in [25]. Good references on retarded functional differential equations include [10], [15], and [43]. Razumikhin-type theorems (originally presented in [37] and [38]) are presented in [10] and [18]. The examples given in Section 9.8 are motivated by similar examples addressed in [10], [18], and [29]. Our analysis of the Cohen–Grossberg neural networks with delays in Section 9.9 is based on the results established in [42]. For additional results on this subject, the reader should consult [24] and [41]. Finally, Sections 9.10 and 9.11 are based on results established in [27] and [28], respectively. For related results concerning DDS determined by retarded functional differential equations, refer to [39]. Throughout this chapter we considered specific examples of infinite-dimensional dynamical systems determined by a variety of different types of equations. Material concerning these equations, along with many other specific classes of infinite-dimensional dynamical systems can be found in many references, including, for example, [3], [4], [6], [7], [9]–[18], [20], [25], [29], [33], and [43].

9.13 Problems

Problem 9.13.1 Similarly as in the case of finite-dimensional systems, show that if (GE) has an equilibrium, say $x_e \in X$, then we may assume without loss of generality that $x_e = 0$. □

Problem 9.13.2 Prove relation (9.1.4). □

Problem 9.13.3 Prove Theorem 9.2.6. □

Problem 9.13.4 Prove Theorem 9.5.2. □

Problem 9.13.5 Complete the proof of Theorem 9.6.1. □

Problem 9.13.6 In Theorem 9.6.1 let $M_i = \{0\} \subset X_i$, $i = 1, \dots, l$, let $M = \{0\} \subset X$, and replace hypothesis (i) by the following hypothesis.

(i′) Let $L = \{1, \dots, l\}$, $L = P \cup Q$, $P \cap Q = \emptyset$, and $Q \neq \emptyset$.

(a) For $i \in P$, assume there exists $v_i \in C[X_i \times \mathbb{R}^+, \mathbb{R}^+]$ and $\psi_{i1}, \psi_{i2} \in \mathcal{K}$ such that

$$\psi_{i1}(\|x_i\|_i) \leq v_i(x_i, t) \leq \psi_{i2}(\|x_i\|_i)$$

for all $x_i \in X_i$ and $t \in \mathbb{R}^+$.

(b) For $i \in Q$, assume there exist $v_i \in C[X_i \times \mathbb{R}^+, \mathbb{R}]$ and $\psi_{i2} \in \mathcal{K}$ such that $v_i(0, t) = 0$ for all $t \in \mathbb{R}^+$ and

$$\psi_{i2}(\|x_i\|_i) \leq -v_i(x_i, t)$$

for all $x_i \in X_i$ and $t \in \mathbb{R}^+$.

Assume that hypotheses (ii) and (iii) of Theorem 9.6.1 are true and that the matrix B is negative definite.

Prove that if the above assumptions are true, then the equilibrium $x_e = 0$ of the composite system $\{\mathbb{R}^+, X, A, S\}$ is *unstable* (i.e., $(S, \{0\})$ is unstable). If in addition, $Q = L$, then the equilibrium $x_e = 0$ is *completely unstable*. □

Problem 9.13.7 Assume that the hypotheses in Problem 9.13.6 are true and that $-A = [a_{ij}] \in \mathbb{R}^{l \times l}$ is an M-matrix, where the a_{ij} are given in Theorem 9.6.1 (and Corollary 9.6.1). Then the equilibrium $x_e = 0$ of the composite system $\{\mathbb{R}^+, X, A, S\}$ is *unstable*. □

Problem 9.13.8 Now let us reconsider the composite system (9.6.3) given in Example 9.6.1, except now assume that the matrix $A \in \mathbb{R}^{m \times m}$ has at least one eigenvalue with real part greater than zero and no eigenvalues with zero part. (We allow the possibility that A is completely unstable.) After an appropriate nonsingular transformation $w = Bz$, we obtain

$$BAB^{-1} = \begin{bmatrix} A_1 & 0 \\ 0 & A_2 \end{bmatrix} \tag{9.13.1}$$

where $-A_1$ is a stable $k \times k$ matrix and A_2 is a stable $j \times j$ matrix with $k + j = m$. Then system (9.6.3) can be rewritten as

$$\begin{cases} \dot{w}_1 = A_1 w_1 + b_1 \displaystyle\int_\Omega f(x) z_2(t,x)dx, & t \in \mathbb{R}^+ \\ \dot{w}_2 = A_2 w_2 + b_2 \displaystyle\int_\Omega f(x) z_2(t,x)dx, & t \in \mathbb{R}^+ \\ \dfrac{dz_2}{dt}(t,x) = \alpha \Delta z_2(t,x) + g(x)(c_1^T w_1 + c_2^T w_2), & (t,x) \in \mathbb{R}^+ \times \Omega \\ z_2(t,x) = 0 & (t,x) \in \mathbb{R}^+ \times \partial\Omega \end{cases} \tag{9.13.2}$$

where b_1, b_2, c_1, and c_2 are defined in the obvious way as consequences of the transformation given in (9.13.1) and all other symbols in (9.13.2) are defined in Example 9.6.1.

Because A_1 is completely unstable, there exists a matrix $P_1 = P_1^T > 0$ such that the matrix

$$(-A_1)^T P_1 + P_1(-A_1) = Q_1$$

is negative definite, and because A_2 is stable, there exists a matrix $P_2 = P_2^T > 0$ such that the matrix

$$A_2^T P_2 + P_2 A_2 = Q_2$$

is negative definite.

Let $\lambda_M(Q_1)$ and $\lambda_M(Q_2)$ denote the largest eigenvalues of Q_1 and Q_2, respectively, and let Γ be as defined in (9.6.8). Let

$$S = \begin{bmatrix} \lambda_M(Q_1) & 0 & 2\|P_1\|_2 |b_1| \, \|f\|_{L_2} \\ 0 & \lambda_M(Q_2) & 2\|P_2\|_2 |b_2| \, \|f\|_{L_2} \\ \|g\|_{L_2}|c_1| & \|g\|_{L_2}|c_2| & -\alpha\Gamma \end{bmatrix}.$$

Using the results given in Problems 9.13.6 and 9.13.7, prove that the equilibrium $(w_1^T, w_2^T, z_2) = 0$ of system (9.13.2) is *unstable* if the successive principal minors of the matrix $-S$ are positive.

Problem 9.13.9 Complete the proof of Theorem 9.8.2 for uniform ultimate boundedness.

Problem 9.13.10 Fill in the details for the proof of Theorem 9.8.3.

Problem 9.13.11 Consider the system

$$\dot{x}(t) = Ax(t) + Bx(t - r) \tag{9.13.3}$$

where $A, B \in \mathbb{R}^{n \times n}$. Assume that the matrix $W = (A + B) + (A + B)^T$ is negative definite. Choose as a Lyapunov function $v = x^T x$. Using Theorem 9.8.8, show that the equilibrium $\varphi_e = 0 \in C_\tau$ of system (9.13.3) is uniformly asymptotically stable if

$$2q\|B\| + \|B + B^T\| < \lambda$$

where $q > 1$ is a constant and $\lambda_M(W) = -\lambda$ ($\lambda_M(W)$ denotes the largest eigenvalue of W). □

Problem 9.13.12 We recall from Chapter 8 the model for Hopfield neural networks, given by

$$\dot{x} = -Bx + TS(x) + I \tag{H}$$

where $x = (x_1, \dots, x_n)^T \in \mathbb{R}^n$, $B = \text{diag}[b_1, \dots, b_n]$, $b_i > 0$, $i = 1, \dots, n$, $T = [T_{ij}] \in \mathbb{R}^{n \times n}$, $T^T = T$, $S(x) = [s_1(x_1), \dots, s_n(x_n)]^T$, and $I = [I, \dots, I_n]^T \in \mathbb{R}^n$. The x_i, $i = 1, \dots, n$ denotes the state variable associated with the ith neuron, the b_i, $i = 1, \dots, n$ represent self-feedback coefficients, the T_{ij} represent interconnection weights among the neurons, the I_i, $i = 1, \dots, n$ denote external inputs and bias terms, and the $s_i(\cdot)$, $i = 1, \dots, n$ are *sigmoidal functions* that represent the neurons. In the present case we have $s_i \in C[\mathbb{R}, (-1, 1)]$, $s_i(\cdot)$ is strictly increasing, $x_i s_i(x_i) > 0$ for all $x_i \neq 0$, and $s_i(0) = 0$.

Frequently, time delays are introduced intentionally or unavoidably into the interconnection structure of (H), resulting in neural networks described by equations of the form

$$\dot{x}(t) = -Bx(t) + TS(x(t - \tau)) + I, \tag{HD}$$

where $\tau \geq 0$ denotes a time delay and all other symbols are as defined in (H).

Theorem 9.13.1 For system (HD) assume the following:

(i) T is symmetric.

(ii) For $S(x) = [s_1(x_1), \dots, s_n(x_n)]^T$, $s_i \in C^1[\mathbb{R}, (-1, 1)]$, and $(ds_i/dx_i)(x_i) \triangleq s_i'(x_i) > 0$ for all $x_i \in \mathbb{R}$, $i = 1, \dots, n$.

(iii) $\tau\beta\|T\|_2 < 1$, where $\|\cdot\|_2$ denotes the matrix norm induced by the Euclidean norm on $\mathbb{R}^n$ and $\beta = \sup_{x\in\mathbb{R}^n} \|\widetilde{S}(x)\|_2$ where

$$\widetilde{S}(x) \triangleq \text{diag}[s_1'(x_1), \ldots, s_n'(x_n)].$$

(iv) System (HD) has a finite number of equilibria.

Prove that for every solution φ of (HD), there exists an equilibrium x_e such that $\lim_{t\to\infty} \varphi(t) = x_e$.

Hint: Let $y = S(x)$ and $y_t = S(x_t) \in C\left[[-\tau, 0], \mathbb{R}^n\right]$ and choose as a Lyapunov function

$$\begin{aligned} v(x_t) = & -y_t^T(0)Ty_t(0) + 2\sum_{i=1}^{n} \int_0^{(y_t(0))_i} b_i s_i^{-1}(\sigma)ds \\ & -2y_t^T(0)I + \int_{-\tau}^{0} [y_t(\theta) - y_t(0)]^T T^T f(\theta) T^T [y_t(\theta) - y_t(0)]d\theta \end{aligned}$$

where $f \in C^1\left[[-\tau, 0], \mathbb{R}^+\right]$ is to be determined in such a manner that $v'_{(HD)}(x_t) \leq 0$ along any solution of (HD). Then apply Theorem 9.8.6. □

Problem 9.13.13 Prove Theorem 9.10.8. □

Problem 9.13.14 Prove Proposition 9.10.4. □

Problem 9.13.15 Prove Theorem 9.11.2. □

Problem 9.13.16 Prove Proposition 9.11.3. □

Problem 9.13.17 [32] Consider dynamical systems determined by *countably infinite systems of ordinary differential equations* given by

$$\dot{z}_n = h_n(t, x), \qquad n = 1, 2, \ldots. \tag{9.13.4}$$

Here x is the infinite-dimensional vector $x = (z_1^T, z_2^T, \ldots, z_n^T, \ldots)^T \in \mathbb{R}^\omega$, $z_n \in \mathbb{R}^{m_n}$, and $h_n\colon \mathbb{R}^+ \times \mathbb{R}^\omega \to \mathbb{R}^{m_n}$. The infinite product $\mathbb{R}^\omega = \mathbb{R}^{m_1} \times \cdots \times \mathbb{R}^{m_n} \times \cdots$ is equipped with the usual product topology, which is equivalent to introducing the metric

$$d(x, \bar{x}) = \sum_{i=1}^{\infty} \left(\frac{1}{2^n}\right) \frac{|z_n - \bar{z}_n|}{(1 + |z_n - \bar{z}_n|)} \tag{9.13.5}$$

so that $\mathbb{R}^\omega$ is a metric space (a convex Fréchet space).

A *solution* of (9.13.4) is a function $x\colon [a, b] \to D \subset \mathbb{R}^\omega$, $b > a \geq 0$ such that $z_n \in C^1\left[[a, b], \mathbb{R}^{m_n}\right]$ and $\dot{z}_n(t) = h_n(t, x(t))$ for all $t \in [a, b]$ and for all $n = 1, 2, 3, \ldots$.

Frequently we view system (9.13.4) as an *interconnected system* of the form (see, e.g., [25])

$$\dot{z}_n = f_n(t, z_n) + g_n(t, x), \qquad n = 1, 2, \ldots \tag{Σ}$$

where in the notation of (9.13.4), $h_n(t,x) \triangleq f_n(t,z_n) + g_n(t,x)$. We view (Σ) as an interconnection of countably infinitely many *isolated* or *free subsystems* described by equations of the form

$$\dot{w}_n = f_n(t, w_n), \qquad (\Sigma_n)$$

$n = 1, 2, \ldots$, where $w_n \in \mathbb{R}^{m_n}$. The terms $g_n(t,x)$, $n = 1, 2, \ldots$ comprise the *interconnecting structure* of system (Σ).

In the following, we let for some $r_n > 0$,

$$D_k = \{x = (z_1^T, z_2^T, \ldots)^T \in \mathbb{R}^\omega : |z_n| \leq k r_n, \ \ n = 1, 2, \ldots\}, \qquad (9.13.6)$$

$k > 0$, and we assume that for every initial condition $x(t_0) = x_0$ with $(t_0, x_0) \in \mathbb{R}^+ \times D_1$, system (Σ) has at least one solution that exists over a finite or an infinite interval. For conditions that ensure this, refer to [32].

The system of equations (Σ) determines a dynamical system $\{T, X, A, S\}$ where $T = \mathbb{R}^+$, $X = \mathbb{R}^\omega$, $A = D_1$, and $S = S_\Sigma$, the set of motions determined by the solutions of (Σ). We note that because $\mathbb{R}^\omega$ is a product of *infinitely many* Banach spaces $\mathbb{R}^{m_n}$, $n = 1, 2, \ldots$, the results for composite systems established in Section 9.6 are not applicable, because these systems are defined on a *finite* product of Banach spaces.

In the next result, we say that the trivial solution $x \equiv 0$ of system (Σ) is *uniformly stable with respect to a set* D (resp., $(S_{\Sigma|D}, \{0\})$ is uniformly stable) if for any $\varepsilon > 0$ there is a $\delta(\varepsilon) > 0$ such that when $c \in D$ and $d(c, 0) < \delta$, then $d(x(t, c, t_0), 0) < \varepsilon$ for all $t \geq t_0 \geq 0$. The *uniform asymptotic stability* of the trivial solution $x \equiv 0$ of system (Σ) with respect to set D is defined similarly.

Theorem 9.13.2 [32] Assume that for system (Σ), the following hypotheses are true.

(i) For each isolated subsystem (Σ_n), $n = 1, 2, \ldots$, there exist a function $v_n \in C^1[B_{m_n}(r_n) \times \mathbb{R}^+, \mathbb{R}]$, where $B_{m_n}(r_n) = \{z_n \in \mathbb{R}^{m_n} : |z_n| < r_n\}$ for some $r_n > 0$, and three functions $\psi_{1n}, \psi_{2n}, \psi_{3n} \in \mathcal{K}$, and a constant $\sigma_n \in \mathbb{R}$, such that

$$\psi_{1n}(|z_n|) \leq v_n(z_n, t) \leq \psi_{2n}(|z_n|)$$

and

$$v'_{n(\Sigma_n)}(z_n, t) \leq \sigma_n \psi_{3n}(z_n)$$

for all $|z_n| < r_n$ and $t \in \mathbb{R}^+$.

(ii) Given ψ_{3n} in hypothesis (i), there are constants $a_{nj} \in \mathbb{R}$ such that

$$\nabla v_n(z_n, t)^T g_n(t, (z_1, \ldots, z_N, 0, 0, \ldots))$$
$$\leq \psi_{3n}(|z_n|)^{1/2} \sum_{j=1}^{N} a_{nj} \psi_{3j}(|z_n|)^{1/2}$$

for all $|z_n| < r_n$ and all $t > 0$, and all $N = 1, 2, \ldots$.

(iii) There exists a sequence of positive numbers $\{\lambda_i\}$, $i = 1, 2, \ldots$, such that $\sum_{n=1}^{\infty} \lambda_n \psi_{2n}(r_n) < \infty$ and such that for each $N = 1, 2, \ldots$, the $N \times N$ matrix $B_N = [b_{ij}]$ defined by

$$b_{ij} = \begin{cases} \lambda_i(\sigma_i + a_{ii}), & i = j \\ (\lambda_i a_{ij} + \lambda_j a_{ji})/2, & i \neq j \end{cases}$$

is negative semidefinite.

Prove that $(S_{\Sigma|D_1}, \{0\})$ is *invariant* and *uniformly stable* (D_1 is defined in (9.13.6) for $k = 1$).

Hint: Note that because D_1 is compact, then for every function $w \in C[D_1, \mathbb{R}]$ such that $w(x) > 0$ for $x \in D_1 - \{0\}$, there must exist $\psi_1, \psi_2 \in \mathcal{K}$ such that

$$\psi_2(d(x, 0)) \geq w(x) \geq \psi_1(d(x, 0))$$

for all $x \in D_1$. Now follow the proof of Theorem 9.6.1. □

Problem 9.13.18 [32] Consider the countably infinite system of scalar differential equations

$$\begin{cases} \dot{z}_1 = -z_1 \\ \dot{z}_2 = -z_n + z_{n-1}, & n \geq 2. \end{cases} \tag{9.13.7}$$

Prove that the trivial solution $x_e = 0$ of (9.13.7) is *stable* with respect to D_k for any $k > 0$.

Hint: Apply Theorem 9.13.2, choosing $r_n = r > 0$, $v_n(z_n) = z_n^2/2$, and $\lambda_n = 1/2^n$. □

Problem 9.13.19 [32] (*Invariance theorem for* (Σ)) Assume that the functions f_n and g_n in (Σ) are independent of t, $n = 1, 2, \ldots$, and assume that any solution $x(t) \in D_k$ for some $k > 1$ and all $t \geq t_0$ whenever $x(t_0) \in D_1$, where D_k is defined in (9.13.6). Assume that there exists a function $v \in C[D_k, \mathbb{R}]$ such that $v'_{(\Sigma)}(x) \leq 0$ for all $x \in \mathbb{R}^\omega$. Let M be the largest invariant set with respect to (Σ) in the set $Z = \{x \in D_k : v'_{(\Sigma)}(x) = 0\}$. Prove that $x(t)$ approaches M as $t \to \infty$ whenever $x(t_0) \in D_1$.

Hint: Noting that D_k is compact, apply Theorem 4.2.1. □

Problem 9.13.20 [32] Consider the countably infinite system of scalar differential equations

$$\begin{cases} \dot{z}_1 = -2z_1 + z_2 \\ \dot{z}_n = z_{n-1} - 2z_n + z_{n+1}, & n \geq 2. \end{cases} \tag{9.13.8}$$

Prove that there exists a sequence of positive numbers $\{r_n\}$, $n = 1, 2, \ldots$ such that $\{S_{(9.13.8)|D_k}, \{0\}\}$ is *invariant* and *uniformly asymptotically stable.*

Hint: Let $r_n = 1/2^n$, choose $v(x) = \sum_{i=1}^{\infty} |z_n|$ for $x = (z_1, z_2, \ldots) \in D_k$, and show that $v'_{(9.13.8)}(x) = -|z_1| \leq 0$ for all $x \in D_k$. Next, show that the origin $0 \in \mathbb{R}^\omega$ is the only invariant set in $Z = \{x \in D_k : z_1 = 0\}$, and then, apply the result given in Problem 9.13.19. □

Bibliography

[1] A. Avez, *Differential Calculus*, New York: Wiley, 1986.

[2] D. D. Bainov and P. S. Simeonov, *Systems with Impulse Effects: Stability Theory and Applications*, New York: Halsted Press, 1989.

[3] R. Bellman and K. L. Cooke, *Differential-Difference Equations*, New York: Academic Press, 1963.

[4] J. L. Daleckii and M. G. Krein, *Stability of Solutions of Differential Equations in Banach Space*, Translations of Mathematical Monographs, vol. 43, Providence, RI: American Mathematical Society, 1974.

[5] R. DeCarlo, M. Branicky, S. Pettersson, and B. Lennartson, "Perspectives on the stability and stabilizability of hybrid systems," *Proc. IEEE*, vol. 88, pp. 1069–1082, 2000.

[6] A. Friedman, *Partial Differential Equations of Parabolic Type*, Englewood Cliffs, NJ: Prentice Hall, 1964.

[7] P. R. Garabedian, *Partial Differential Equations*, New York: Chelsea, 1986.

[8] R. Haberman, *Elementary Applied Differential Equations with Fourier Series and Boundary Value Problems*, Englewood Cliffs, NJ: Prentice Hall, 1998.

[9] J. K. Hale, "Dynamical systems and stability," *J. Math. Anal. Appl.*, vol. 26, pp. 39–59, 1969.

[10] J. K. Hale, *Functional Differential Equations*, New York: Springer-Verlag, 1971.

[11] J. K. Hale, "Functional differential equations with infinite delays," *J. Math. Anal. Appl.*, vol. 48, pp. 276–283, 1974.

[12] E. Hille and R. S. Phillips, *Functional Analysis and Semigroups*, American Math. Soc. Colloquium Publ. 33, Providence, RI: American Mathematical Society, 1957.

[13] L. Hörmander, *Linear Partial Differential Equations*, New York: Springer-Verlag, 1963.

[14] L. Hörmander, *The Analysis of Linear Partial Differential Operators*, vol. I, II, III, IV, New York: Springer-Verlag, 1983–1985.

[15] N. N. Krasovskii, *Stability of Motion*, Stanford, CA: Stanford University Press, 1963.

[16] S. G. Krein, *Linear Differential Equations in Banach Spaces*, Translation of Mathematical Monographs, vol. 29, Providence, RI: American Mathematical Society, 1970.

[17] N. V. Krylov, *Nonlinear Elliptic and Parabolic Equations of the Second Order*, Boston: D. Reidel, 1987.

[18] Y. Kuang, *Delay Differential Equations with Applications in Population Dynamics*, New York: Academic Press, 1993.

[19] T. Kurtz, "Convergence of sequences of semigroups of nonlinear equations with applications to gas kinetics," *Trans. Amer. Math. Soc.*, vol. 186, pp. 259–272, 1973.

[20] V. Lakshmikantham and S. Leela, *Differential and Integral Inequalities*, vol. I, II, New York: Academic Press, 1969.

[21] J.-H. Li, A. N. Michel, and W. Porod, "Qualitative analysis and synthesis of a class of neural networks," *IEEE Trans. Circ. Syst.*, vol. 35, pp. 976–987, 1988.

[22] A. N. Michel, "Recent trends in the stability analysis of hybrid dynamical systems," *IEEE Trans. Circ. Syst. I: Fund. Theor. Appl.*, vol. 46, pp. 120–134, 1999.

[23] A. N. Michel and B. Hu, "Towards a stability theory of general hybrid dynamical systems," *Automatica*, vol. 35, pp. 371–384, 1999.

[24] A. N. Michel and D. Liu, *Qualitative Analysis and Synthesis of Recurrent Neural Networks*, New York: Marcel Dekker, 2002.

[25] A. N. Michel and R. K. Miller, *Qualitative Analysis of Large Scale Dynamical Systems*, New York: Academic Press, 1977.

[26] A. N. Michel and R. K. Miller, "Qualitative analysis of interconnected systems described on Banach spaces: Well posedness, instability, and Lagrange stability," *Zeitschrift für angewandte Mathematik und Mechanik*, vol. 58, pp. 289–300, 1978.

[27] A. N. Michel and Y. Sun, "Stability of discontinuous Cauchy problems in Banach space," *Nonlinear Anal.*, vol. 65, pp. 1805–1832, 2006.

[28] A. N. Michel, Y. Sun, and A. P. Molchanov, "Stability analysis of discontinuous dynamical systems determined by semigroups," *IEEE Trans. Autom. Control*, vol. 50, pp. 1277–1290, 2005.

[29] A. N. Michel, K. Wang, and B. Hu, *Qualitative Theory of Dynamical Systems: The Role of Stability Preserving Mappings*, Second Edition, New York: Marcel Dekker, 2001.

[30] A. N. Michel, K. Wang, D. Liu, and H. Ye, "Qualitative limitations incurred in implementations of recurrent neural networks," *IEEE Control Syst. Mag.*, vol. 15, pp. 56–65, 1995.

[31] R. K. Miller, *Nonlinear Volterra Integral Equations*, New York: Benjamin, 1971.

[32] R. K. Miller and A. N. Michel, "Stability theory for countably infinite systems of differential equations," *Tôhoku Math. J.* vol. 32, pp. 155–168, 1980.

[33] A. Pazy, *Semigroups of Linear Operators and Applications to Partial Differential Equations*, New York: Springer-Verlag, 1983.

[34] I. C. Petrovskij, *Lectures on Partial Differential Equations*, 3rd Ed., Moscow: Fizwatgiz, 1961.

[35] H. Plaza and W. H. Kohler, "Coupled-reactor kinetics equations," *Nuclear Sci. Eng.*, vol. 22, pp. 419–422, 1966.

[36] R. D. Rasmussen and A. N. Michel, "Stability of interconnected dynamical systems described on Banach spaces," *IEEE Trans. Autom. Control*, vol. 21, pp. 464–471, 1976.

[37] B. S. Razumikhin, "On the stability of systems with a delay," *Prikl. Mat. Mek.*, vol. 20, pp. 500–512, 1956.

[38] B. S. Razumikhin, "Application of Liapunov's method to problems in the stability of systems with delay," *Avtomat. i Telemeh.*, vol. 21, pp. 740–774, 1960.

[39] Y. Sun, A. N. Michel, and G. Zhai, "Stability of discontinuous retarded functional differential equations with applications," *IEEE Trans. Autom. Control*, vol. 50, pp. 1090–1105, 2005.

[40] H. Ye, A. N. Michel, and L. Hou, "Stability theory for hybrid dynamical systems," *IEEE Trans. Autom. Control*, vol. 43, pp. 461–474, 1998.

[41] H. Ye, A. N. Michel, and K. Wang, "Global stability and local stability of Hopfield neural networks with delays," *Phys. Rev. E*, vol. 50, pp. 4206–4213, 1994.

[42] H. Ye, A. N. Michel, and K. Wang, "Qualitative analysis of Cohen-Grossberg neural networks with multiple delays," *Phys. Rev. E*, vol. 51, pp. 2611–2618, 1995.

[43] T. Yoshizawa, *Stability Theory by Lyapunov's Second Method*, Tokyo: Math. Soc. Japan, 1966.

Index

Systems & Control: Foundations & Applications

Series Editor
Tamer Başar
Coordinated Science Laboratory
University of Illinois at Urbana-Champaign
1308 W. Main St.
Urbana, IL 61801-2307
U.S.A.

Systems & Control: Foundations & Applications

Aims and Scope

The aim of this series is to publish top quality state-of-the art books and research monographs at the graduate and post-graduate levels in systems, control, and related fields. Both foundations and applications will be covered, with the latter spanning the gamut of areas from information technology (particularly communication networks) to biotechnology (particularly mathematical biology) and economics.

Readership

The books in this series are intended primarily for mathematically oriented engineers, scientists, and economists at the graduate and post-graduate levels.

Types of Books

Advanced books, graduate-level textbooks, and research monographs on current and emerging topics in systems, control and related fields.

Preparation of manuscripts is preferable in LATEX. The publisher will supply a macro package and examples of implementation for all types of manuscripts.

Proposals should be sent directly to the editor or to: Birkhäuser Boston, 675 Massachusetts Avenue, Cambridge, MA 02139, U.S.A. or to Birkhäuser Publishers, 40-44 Viadukstrasse, CH-4051 Basel, Switzerland

A Partial Listing of Books Published in the Series

Identification and Stochastic Adaptive Control
Han-Fu Chen and Lei Guo

Viability Theory
Jean-Pierre Aubin

Representation and Control of Infinite Dimensional Systems, Vol. I
A. Bensoussan, G. Da Prato, M. C. Delfour, and S. K. Mitter

Representation and Control of Infinite Dimensional Systems, Vol. II
A. Bensoussan, G. Da Prato, M. C. Delfour, and S. K. Mitter

Mathematical Control Theory: An Introduction
Jerzy Zabczyk

H_∞-Control for Distributed Parameter Systems: A State-Space Approach
Bert van Keulen

Disease Dynamics
Alexander Asachenkov, Guri Marchuk, Ronald Mohler, and Serge Zuev

Theory of Chattering Control with Applications to Astronautics, Robotics, Economics, and Engineering
Michail I. Zelikin and Vladimir F. Borisov

Modeling, Analysis and Control of Dynamic Elastic Multi-Link Structures
J. E. Lagnese, Günter Leugering, and E. J. P. G. Schmidt

First-Order Representations of Linear Systems
Margreet Kuijper

Hierarchical Decision Making in Stochastic Manufacturing Systems
Suresh P. Sethi and Qing Zhang

Optimal Control Theory for Infinite Dimensional Systems
Xunjing Li and Jiongmin Yong

Generalized Solutions of First-Order PDEs: The Dynamical Optimization Perspective
Andreĭ I. Subbotin

Finite Horizon H_∞ and Related Control Problems
M. B. Subrahmanyam

Control Under Lack of Information
A. N. Krasovskii and N. N. Krasovskii

H^∞-Optimal Control and Related Minimax Design Problems: A Dynamic Game Approach
Tamer Başar and Pierre Bernhard

Control of Uncertain Sampled-Data Systems
Geir E. Dullerud

Robust Nonlinear Control Design: State-Space and Lyapunov Techniques
Randy A. Freeman and Petar V. Kokotović

Adaptive Systems: An Introduction
Iven Mareels and Jan Willem Polderman

Sampling in Digital Signal Processing and Control
Arie Feuer and Graham C. Goodwin

Ellipsoidal Calculus for Estimation and Control
Alexander Kurzhanski and István Vályi

Minimum Entropy Control for Time-Varying Systems
Marc A. Peters and Pablo A. Iglesias

Chain-Scattering Approach to H^∞-Control
Hidenori Kimura

Output Regulation of Uncertain Nonlinear Systems
Christopher I. Byrnes, Francesco Delli Priscoli, and Alberto Isidori

High Performance Control
Teng-Tiow Tay, Iven Mareels, and John B. Moore

Optimal Control and Viscosity Solutions of Hamilton–Jacobi–Bellman Equations
Martino Bardi and Italo Capuzzo-Dolcetta

Stochastic Analysis, Control, Optimization and Applications: A Volume in Honor of W.H. Fleming
William M. McEneaney, G. George Yin, and Qing Zhang, Editors

Mutational and Morphological Analysis: Tools for Shape Evolution and Morphogenesis
Jean-Pierre Aubin

Stabilization of Linear Systems
Vasile Dragan and Aristide Halanay

The Dynamics of Control
Fritz Colonius and Wolfgang Kliemann

Optimal Control
Richard Vinter

Advances in Mathematical Systems Theory:
A Volume in Honor of Diederich Hinrichsen
Fritz Colonius, Uwe Helmke, Dieter Prätzel-Wolters, and Fabian Wirth, Editors

Nonlinear and Robust Control of PDE Systems:
Methods and Applications to Transport-Reaction Processes
Panagiotis D. Christofides

Foundations of Deterministic and Stochastic Control
Jon H. Davis

Partially Observable Linear Systems Under Dependent Noises
Agamirza E. Bashirov

Switching in Systems and Control
Daniel Liberzon

Matrix Riccati Equations in Control and Systems Theory
Hisham Abou-Kandil, Gerhard Freiling, Vlad Ionescu, and Gerhard Jank

The Mathematics of Internet Congestion Control
Rayadurgam Srikant

H^∞ Engineering and Amplifier Optimization
Jeffery C. Allen

Advances in Control, Communication Networks, and Transportation Systems:
In Honor of Pravin Varaiya
Eyad H. Abed

Convex Functional Analysis
Andrew J. Kurdila and Michael Zabarankin

Max-Plus Methods for Nonlinear Control and Estimation
William M. McEneaney

Uniform Output Regulation of Nonlinear Systems:
A Convergent Dynamics Approach
Alexey Pavlov, Nathan van de Wouw, and Henk Nijmeijer

Supervisory Control of Concurrent Systems: A Petri Net Structural Approach
Marian V. Iordache and Panos J. Antsaklis

Filtering Theory: With Applications to Fault Detection, Isolation, and Estimation
Ali Saberi, Anton A. Stoorvogel, and Peddapullaiah Sannuti

Representation and Control of Infinite-Dimensional Systems, Second Edition
Alain Bensoussan, Giuseppe Da Prato, Michel C. Delfour, and Sanjoy K. Mitter

Set-Theoretic Methods in Control
Franco Blanchini and Stefano Miani

Stability of Dynamical Systems: Continuous, Discontinuous, and Discrete Systems
Anthony N. Michel, Ling Hou, and Derong Liu

Control of Turbulent and Magnetohydrodynamic Channel Flows: Boundary and Stabilization and State Estimation
Rafael Vazquez and Miroslav Krstic

Printed in the United States of America.